André Platzer

D0912615

Logical Foundations
of Cyber-Physical Systems

 Springer

André Platzer
Computer Science Department
Carnegie Mellon University
Pittsburgh, Pennsylvania, USA

The content of the book and the image used on the book cover are based upon work performed at Carnegie Mellon University and supported by the National Science Foundation under NSF CAREER Award CNS-1054246.

ISBN 978-3-030-09697-7   ISBN 978-3-319-63588-0   (eBook)
https://doi.org/10.1007/978-3-319-63588-0

This Springer imprint is published by the registered company Springer Nature Switzerland AG
The registered company address is: Gewerbestrasse 11, 6330 Cham, Switzerland

Logical Foundations of Cyber-Physical Systems

Endorsements

This excellent textbook marries design and analysis of cyber-physical systems with a logical and computational way of thinking. The presentation is exemplary for finding the right balance between rigorous mathematical formalization and illustrative case studies rooted in practical problems in system design.

Rajeev Alur, University of Pennsylvania

This book provides a wonderful introduction to cyber-physical systems, covering fundamental concepts from computer science and control theory from the perspective of formal logic. The theory is brought to life through many didactic examples, illustrations, and exercises. A wealth of background material is provided in the text and in an appendix for each chapter, which makes the book self-contained and accessible to university students of all levels.

Goran Frehse, Université Grenoble Alpes

[The author] has developed major important tools for the design and control of those cyber-physical systems that increasingly shape our lives. This book is a 'must' for computer scientists, engineers, and mathematicians designing cyber-physical systems.

Anil Nerode, Cornell University

As computing interfaces increasingly with our physical world, resulting in so-called cyber-physical systems, our foundations of computing need to be enriched with suitable physical models. This book strikes a wonderful balance between rigorous foundations for this next era of computing with illustrative examples and applications that drive the developed methods and tools. A must read book for anyone interested in the development of a modern and computational system science for cyber-physical systems.

George J. Pappas, University of Pennsylvania

This definitive textbook on cyber-physical systems lays the formal foundations of their behavior in terms of a single logical framework. Platzer's logic stands out among all other approaches because it provides a uniform treatment of both the discrete and continuous nature of cyber-physical systems, and does not shy away from their complex behavior due to stochasticity, uncertainty, and adversarial agents in the environment. His computational thinking approach makes this work accessible to practicing engineers who need to specify and verify that cyber-physical systems are safe.

Jeannette M. Wing, Columbia University

Foreword

I first met André when he was just finishing his PhD and gave a job talk at CMU (he got the job). I was a visiting researcher and got to take the young faculty candidate out for lunch. André talked about verifying cyber-physical systems (CPS) using "differential dynamic logic" and theorem proving. I was skeptical, for one because related approaches had only seen modest success, and also because my money was on a different horse. A few years before, I had developed a model checker (PHAVer), and was working on a second one, called SpaceEx. At the time, these were the only verification tools that, on the push of a button, could verify certain benchmarks from CPS and other domains involving continuous variables that change with time. I was quite proud of them and, for me, algorithmic verification was the way to go. But André was determined to make theorem proving work in practice, and indeed, he advanced the field to an extent that I did not think possible. André and his team first developed the logical framework, then built a very capable theorem prover for CPS (KeYmaera), successfully applied it to industrial case studies like airplane collision avoidance, and, finally, addressed important application issues such as validating the model at runtime.

The book in front of you provides a comprehensive introduction on how to reason about cyber-physical systems using the language of logic and deduction. Along the way, you will become familiar with many fundamental concepts from computer science, applied mathematics, and control theory, all of which are essential for CPS. The book can be read without much prior knowledge, since all necessary background material is provided in the text and in appendices for many chapters. The book is structured in the following four parts. In the first part, you will learn how to model CPS with continuous variables and programming constructs, how to specify requirements and how to check whether the model satisfies the requirements using proof rules. The second part adds differential equations for modeling the physical world. The third part introduces the concept of an adversary, who can take actions that the system can not influence directly. In a control system, the adversary can be the environment, which influences the system behavior through noise and other disturbances. Making decisions in the presence of an adversary means trying to be prepared for the worst case. The fourth part adds further elements for reasoning soundly and efficiently about systems in applications, such as using real arithmetic and – my favorite – monitor conditions. Monitor conditions are checked while the system is in operation. As long as they hold, one can be sure that not only the model but also the actual CPS implementation satisfy the safety requirements.

By now André and his group have handled an impressive number of case studies that are beyond the capabilities of any model checker I know. Fortunately for me and my horse, the converse is also still true, since some problems can in practice only be solved numerically using algorithmic approaches. If your goal is to obtain a rock-solid foundation for CPS from the beautiful and elegant perspective of logics, then this is the book for you.

Goran Frehse, Associate Professor, Université Grenoble Alpes, Grenoble, 2017

Acknowledgements

This textbook is based on the lecture notes for the *Foundations of Cyber-Physical Systems* undergraduate course I taught in the Computer Science Department at Carnegie Mellon University. The textbook would have been impossible without the feedback from the students and helpful discussions with the teaching assistants João Martins, Annika Peterson, Nathan Fulton, Anastassia Kornilova, Brandon Bohrer, and especially Sarah Loos who TAed for the first instance in Fall 2013 and co-instructed for the intensive courses at ENS Lyon, France, in Spring 2014 and at MAP-i, Braga, Portugal in Summer 2014. Based on the experience with earlier Ph.D.-level courses, this course was originally designed as an undergraduate course but then extended to master's students and eventually Ph.D. students.

   I appreciate the feedback of all my students on this textbook, but also by my postdocs Stefan Mitsch, Jean-Baptiste Jeannin, Khalil Ghorbal, and Jan-David Quesel. I am especially thankful to Sarah Loos's formative comments on the earliest draft and Yong Kiam Tan's careful extensive feedback for the final version. I am also grateful to Jessica Packer's exhaustive consistency checking on the textbook structuring and to Julia Platzer for crucial advice on illustrations. I am most indebted to the developers Stefan Mitsch and Nathan Fulton of the KeYmaera X prover for verifying cyber-physical systems, and very much appreciate also the KeYmaera X contributions by Brandon Bohrer, Yong Kiam Tan, Jan-David Quesel, and Marcus Völp. For help with the book process, I am grateful to the copyeditor and Ronan Nugent from Springer. Especially, however, I thank my family, without whose patience and support this book would not exist.

   This textbook captures findings from the NSF CAREER Award on *Logical Foundations of Cyber-Physical Systems*, which I am very grateful to have received. I also benefitted from Helen Gill's advice as a program manager when this project started.

Funding

This material is based upon work supported by the National Science Foundation under NSF CAREER Award CNS-1054246.

   Any opinions, findings, and conclusions or recommendations expressed in this publication are those of the author(s) and do not necessarily reflect the views of the National Science Foundation.

Pittsburgh, December 2017   *André Platzer*

Disclaimer

This book is presented solely for educational purposes. While best efforts have been used in preparing this book, the author and publisher make no representations or warranties of any kind and assume no liabilities of any kind with respect to the accuracy or completeness of the contents and specifically disclaim any implied warranties of merchantability or fitness of use for a particular purpose. Neither the author nor the publisher shall be held liable or responsible to any person or entity with respect to any loss or incidental or consequential damages caused, or alleged to have been caused, directly or indirectly, by the information contained herein. No warranty may be created or extended by sales representatives or written sales materials.

Contents

List of Figures

List of Tables

List of Expeditions

List of Theorems

Chapter 1
Cyber-Physical Systems: Overview

Synopsis Cyber-physical systems combine cyber capabilities with physical capabilities to solve problems that neither part could solve alone. This chapter provides an informal introduction to cyber-physical systems, setting the stage for this textbook. The primary purpose is a lightweight overview of the technical and nontechnical characteristics of cyber-physical systems, an overview of some of their application domains, and a discussion of their prospects and challenges. The chapter also informally outlines and explains the approach taken in this book to address crucial safety challenges in cyber-physical systems.

1.1 Introduction

This chapter provides a lightweight introduction to *cyber-physical systems* (*CPS*), which combine cyber capabilities (computation and/or communication as well as control) with physical capabilities (motion or other physical processes).

> **Note 1 (CPS)** *Cyber-physical systems* combine cyber capabilities with physical capabilities to solve problems that neither part could solve alone.

Cars, aircraft, and robots are prime examples, because they move physically in space in a way that is determined by discrete computerized control algorithms that adjust the actuators (e.g., brakes) based on sensor readings of the physical state. Designing these algorithms to control CPSs is challenging due to their tight coupling with physical behavior. At the same time, it is vital that these algorithms be correct, since we rely on CPSs for safety-critical tasks such as keeping aircraft from colliding.

> How can we provide people with cyber-physical systems they can bet their lives on?
> – Jeannette Wing

© Springer International Publishing AG, part of Springer Nature 2018
A. Platzer, *Logical Foundations of Cyber-Physical Systems*,
https://doi.org/10.1007/978-3-319-63588-0_1

Since cyber-physical systems combine cyber and physical capabilities, we need to understand both to understand CPS. It is not enough to understand both capabilities only in isolation, though, because we also need to understand how the cyber and the physics elements work together, i.e., what happens when they interface and interact, because this is what CPSs are all about.

1.1.1 Cyber-Physical Systems Analysis by Example

Airplanes provide a rich source of canonical examples for cyber-physical systems analysis challenges. While they are certainly not the only source of examples, airplanes quickly convey both a spatial intuition for their motion and an appreciation for the resulting challenges of finding out where and how to fly an airplane.

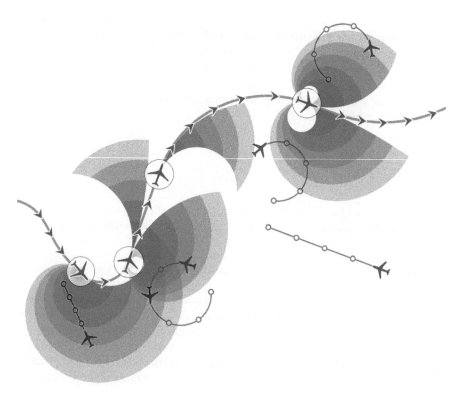

Fig. 1.1 Airplane example: Which control decisions are safe for aircraft collision avoidance?

If a pilot has gotten into a situation where her airplane is too close to other aircraft, see Fig. 1.1, then it would be immensely helpful to give the pilot good advice about how best to maneuver to resolve the situation. Of course, such advice needs to be given quickly and safely. There is not enough time to carefully plan out every possible trajectory of the the pilot's *ownship* and of all other *intruder* aircraft, but a quick response is needed right away, which is what computers are good at. But the advice also has to be safe such that it reliably separates the aircraft always under all relevant scenarios of when and how exactly the pilots will respond to the advice. For the ownship (following the thick blue trajectory), Fig. 1.1 gives a schematic illustration of unsafe zones (in shades of red, darker with more imminent danger) resulting from the given intruder aircraft (gray).

More generally, this begs the question of which control decisions are safe for aircraft collision avoidance. How can one predict right away whether given control decisions for the airplane and intruder aircraft are guaranteed to be safe in the future or whether they could possibly lead to a collision? How can a computer control program be designed that reaches safe decisions and gives good advice to pilots sufficiently quickly? What would constitute a safety argument for such a pilot decision support system that justifies why the system always gives safe collision avoidance advice that the pilot can follow confidently?

1.1.2 Application Domains

Cyber-physical systems provide prospects of improved safety and efficiency in numerous application domains [2, 29, 30, 60]. Examples include both fully autonomous self-driving cars and improved driver assistance technology for cars such as lane-keeping assistants or distance-keeping assistants [1, 11, 31, 34], where computer control technology helps people drive cars more safely and more efficiently. Both pilot decision support systems [22, 23, 57, 66] and full autopilots for unmanned drone aircraft fall under this paradigm. In the former, the computer focuses on an advisory rôle where it gives decision support to pilots, who are in charge. But autopilots also automate the flight during certain well-defined phases of the flight, such as in normal cruise flight or during landing. The case of drones provides more comprehensive automation where the computer is in primary control of the drone for extended periods of time and remote pilots limit themselves to only providing certain control decisions every once in a while. Other applications include train protection systems [35, 58], power plants [13], medical devices [25, 30], mobile robots that operate in the vicinity of humans [36, 41], and robotic surgery systems [7, 26]. Autonomous underwater vehicles (AUVs) also need computer control for sustained operation since their operating conditions only provide infrequent opportunities for human intervention. Many other application domains are of relevance, though, because the principle of using computer control to help physical systems is quite general.

1.1.3 Significance

Cyber-physical systems can help in many ways. Computers support the human driver in a car by taking control of the car either in full or partially for a certain period of time. For example, the computer can help prevent accidents by keeping the car in the lane in case the human driver is inattentive and it decelerates when the driver fails to notice that the car in front is braking. Of course, the tricky bit is that the computer needs to be able to reliably detect the circumstances where a correction of the car's trajectory is in order. In addition to the nontrivial challenges of reliably sensing other cars and lanes, the computer needs to distinguish user-intended lane changing from accidental lane departures, for example based on whether the driver signaled a lane change by a turn signal, and apply steering corrections appropriately.

In aerospace, computers can not only support pilots during figurative and literal fair-weather phases of the flight such as cruise flight, but can also help by providing pilots with quick collision avoidance advice whenever two or more aircraft get too close together. Since that is a very stressful situation for the pilots, good advice on how to get out of it again and avoid possible collisions is absolutely crucial. Likewise remote pilots cannot necessarily monitor all flight paths of drones closely all the time, such that computer assistance would help prevent collisions with commercial aircraft or other drones. Besides detection, the primary challenges are the uncertainties of when and how exactly the respective aircraft follow their trajectories and, of course, the need to prevent follow-on conflicts with other aircraft. While already quite challenging for two aircraft, this problem gets even more complicated in the presence of multiple aircraft, possibly with different flight characteristics.

For railway applications, technical safety controllers are also crucial, because the braking distances of trains[1] exceed the field of vision so that the brakes need to be applied long before another train is in sight. One challenge is to identify a safe braking distance that works reliably for the train and track conditions without reducing the expected overall performance by braking too early, which would limit operational suitability. Unlike a maximum use of the conventional service brake, full emergency brakes on a train may also damage the rails or wheels and are, thus, only used in the event of a true emergency.

1.1.4 The Importance of Safety

Wouldn't it be great if we could use computers to leverage the advances in safety and efficiency in the CPS application domains? Of course, the prerequisite is that the cyber-physical systems themselves need to be safe, otherwise the cure might be worse than the disease. Safety is paramount to ensure that the cyber-physical systems that are meant to improve safety and efficiency actually help. So, the key question is:

---

[1] Heavy freight trains as well as high-speed trains need a braking distance of 2.5 km to 3.3 km.

> How do we make sure cyber-physical systems make the world a better place?

Because the world is a difficult place, this is rather a difficult question to answer. An answer needs enough understanding of the world (in a model of the relevant part of the world), the control principles (what control actions are available and what their effect is on the physical world) and their implementation in a computer controller, as well as the requisite safety objectives (what precisely discriminates safe from potentially unsafe behavior). This leads to the following rephrasing [53]:

> How can we ensure that cyber-physical systems are guaranteed to meet their design goals?

Whether we can trust a computer to control physical processes depends on how it has been programmed and on what will happen if it malfunctions. When a lot is at stake, computers need to be guaranteed to interact correctly with the physical world.

The rationale pursued in this textbook argues that [53]:

1. Computers would perfectly earn our trust to control physical processes if only they came with suitable guarantees.
2. Safety guarantees require appropriate analytical foundations.
3. A foundational core that is common to *all* application domains is more useful than different mathematics for each area, e.g., a special mathematics for trains.
4. Foundations have already revolutionized the digital parts of computer science and, indirectly, the way our whole society works.
5. But we need even stronger foundations when software reaches out into our physical world, because they directly affect our physical environment.

These considerations lead to the following conclusion:

> Because of the impact that they can have on the real world, cyber-physical systems deserve proofs as safety evidence.

As has already been argued on numerous other occasions [2–6, 9, 10, 12, 18, 19, 27, 28, 32, 33, 37–40, 42–45, 60, 63–65, 68], the correctness of such systems needs to be verified, because testing may miss bugs. This problem is confounded, though, because the behavior of a CPS under one circumstance can radically differ from the behavior under another, especially when complex computer decisions for different objectives interact. Of course, due to their involvement in models of reality, the safety evidence should not be limited to proofs alone either, but needs to include appropriate testing as well. But without the generality resulting from mathematical proofs, it is ultimately impossible to obtain strong safety evidence beyond the isolated experience with the particular situations covered by the test data [49, 56, 69]. Even statistical demonstrations of safety by test driving are nearly impossible [24].

1.2 Hybrid Systems Versus Cyber-Physical Systems

While the defining criterion that cyber-physical systems combine cyber capabilities with physical capabilities makes it easy to recognize them in practice, this is hardly a precise mathematical criterion. For the characteristic behavior of a system, it should be mostly irrelevant whether it happens to be built literally by combining an actual computer with a physical system, or whether it is built in another way, e.g., by combining the physical system with a small embedded controller achieving the same performance, or maybe by exploiting a biochemical reaction to control a process.

Indeed, cyber-physical systems share mathematical characteristics, too, which are in many ways more important for our endeavor than the fact that they happen to be built from cyber components and from physical components. While a full understanding of the mathematical characteristics of cyber-physical systems will keep us busy for the better part of this book, it is reasonably straightforward to arrive at what is at the core of all mathematical models of cyber-physical systems. From a mathematical perspective, cyber-physical systems are hybrid systems (or extensions thereof):

> **Note 2 (Hybrid systems)** *Hybrid systems* are a mathematical model of dynamical systems that combine discrete dynamics with continuous dynamics. Their behavior includes both aspects that change discretely one step at a time and aspects that change continuously as continuous functions over time.

For example, the aircraft in Fig. 1.1 fly continuously along their trajectories as a continuous function of continuous time, because real aircraft do not jump around in space with discrete jumps. Every once in a while, though, the pilot and/or autopilot reaches a decision about turning in a different direction to avoid a possible collision with intruder aircraft. These discrete decisions are best understood as a discrete dynamics in discrete time, because they happen one step after another. The system reaches one discrete decision for a collision avoidance course, follows it continuously for a certain period of time, and then re-evaluates the resulting situation later to see whether a better decision becomes possible.

Similarly, a car controller decides to accelerate or brake, which is best understood as a discrete dynamics, because there is a discrete instant of time where that decision is reached and scheduled to take effect. The car's continuous motion down the road, instead, is best understood as a continuous dynamics, because it changes the position as a continuous function of time.

In the most naïve interpretation, the cyber components of cyber-physical systems directly correspond to the discrete dynamics of hybrid systems while the physical components of cyber-physical systems directly correspond to the continuous dynamics of hybrid systems. While possibly a good mental model initially, this view will ultimately turn out to be too simplistic. For example, there are events in physical models that are best described by a discrete dynamics even if they come from the physics. For instance, the touchdown of an airplane on the ground can be considered as causing a discrete state change by a discrete dynamics from flying to driving even if the runway that the aircraft touches down on is quite physical and not a cyber construct at all. Conversely, for some purposes, some of the computations happen so frequently and so quickly that we best understand them as if they were running continuously even if that is not entirely true. For instance, a digital PID controller[2] for an inner-loop flight controller that quickly adjusts the ailerons, rudders, and elevators of an aircraft can sometimes be considered as having a continuous effect even if it is actually implemented as a digital device with a fast clock cycle.

In fact, this is one of the most liberating effects of understanding the world from a hybrid systems perspective [53]. Since the mathematical principles of hybrid systems accept both discrete and continuous dynamics, we do not have to either coerce all aspects of a system model into the discrete to understand it with discrete mathematics or force all system aspects into a continuous understanding to analyze it with continuous techniques. Instead, hybrid systems make it perfectly acceptable to have some aspects discrete (such as the decision steps of a digital controller) and others continuous (such as continuous-time motion), while allowing modeling decisions about ambivalent aspects. For some purposes, it might be better to model the touchdown of an aircraft as a discrete state change from in-the-air to on-the-ground. For other purposes, such as developing an autopilot for landing, it is important to take a more fine-grained view, because the aircraft will simply take off again if it is still going too fast even if its state changes to on-the-ground. Hybrid systems enable such tradeoffs.

Overall, hybrid systems are *not* the same as cyber-physical systems. Hybrid systems are mathematical models of complex (often physical) systems, while cyber-physical systems are defined by their technical characteristics. Nevertheless, exhibiting a hybrid systems dynamics is such a common feature of cyber-physical systems that we will take the liberty of using the notions cyber-physical system and hybrid system quite interchangeably in Parts I and II of this book.

Despite this linguistic simplification, you should note that hybrid systems can be nontechnical. For example, certain biological mechanisms can be captured well with hybrid system models [65] or genetic networks [17] even if they have nothing to do with cyber-physical systems. Conversely, a number of cyber-physical systems feature additional aspects beyond hybrid systems, such as adversarial dynamics (studied in Part III), distributed dynamics [47], or stochastic dynamics [46].

1.3 Multi-dynamical Systems

Owing to the fact that cyber-physical systems can have more dynamical aspects than just those of hybrid systems, this book follows the more general multi-dynamical systems principle [48, 53] of understanding cyber-physical systems as a combination of multiple elementary dynamical aspects.

---

[2] Proportional-integral-derivative or PID controllers control a system by a linear combination of the error, the integral of the error over time, and the derivative of the error.

> **Note 3 (Multi-dynamical system)** *Multi-dynamical systems* [48] are mathe-matical models of dynamical systems characterized by multiple facets of dynamical systems, schematically summarized in Fig. 1.2.

Fig. 1.2 Multi-dynamical systems aspects of CPS

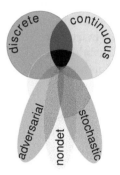

CPSs involve computer control decisions and are, thus, *discrete*. CPSs are also *continuous*, because they evolve along differential equations of motion or of other physical processes. CPSs are often *uncertain*, because their behavior is subject to choices coming from either environmental variability or from intentional uncertainties that simplify their model. This uncertainty can manifest in different ways. Uncertainties make CPSs *stochastic* when good information about the distribution of choices is available [46]. Uncertainties make CPSs *nondeterministic* when no commitment about the resolution of choices is made. Uncertainties make CPSs *adversarial* when they involve multiple agents with potentially conflicting goals or even active competition in a game [52, 55]. Verifying that CPSs work correctly requires dealing with many of these dynamical features at the same time. Sometimes, CPSs require even more dynamical features, such as *distributed* dynamics [47].

Hybrid systems are the special case of multi-dynamical systems that combine discrete and continuous dynamics, and will be considered in Parts I and II. *Hybrid games* are multi-dynamical systems that combine discrete, continuous, and adversarial dynamics that will be studied in Part III. *Stochastic hybrid systems* are multi-dynamical systems that combine discrete, continuous, and stochastic dynamics, but are beyond the scope of this book [8, 46]. *Distributed hybrid systems* are multi-dynamical systems combining discrete, continuous, and distributed dynamics [47].

Multi-dynamical systems study complex CPSs as a combination of multiple elementary dynamical aspects. Throughout this textbook, we will come to appreciate how this approach helps to tame the complexity of CPSs by understanding that their complexity just comes from combining lots of simple dynamical effects with one another. The overall system is quite complex, but each of its pieces is better behaved, since it only has one dynamics as opposed to all of them at once. What miracle translates this *descriptive simplification* of a CPS described as a combination of multiple dynamical aspects into an *analytic simplification* of multiple dynamical systems that can be considered side by side during analysis? The descriptive simplification is a helpful modeling advantage to disentangle different dynamical aspects of the system into separate aspects of a model. But the biggest impact of multi-dynamical systems is in how they enable an analytic simplification of studying and analyzing the individual dynamical aspects separately while still yielding results about the combined multi-dynamical system. How does the descriptive advantage of a multi-dynamical systems composition carry over to an analytic advantage?

The key to this mystery is to integrate the CPS dynamics all within a single, compositional logic [48, 53]. Compositionality means that the meaning of a construct is a simple function of the meaning of the pieces [61]. For example, the meaning of the logical conjunction operator $\wedge$ (read as "and") is a simple function of the meaning of its pieces. The formula $A \wedge B$ (read as "A and B") is true exactly if A is true and B is true, too. Another way to say this is that the set of states of a system in which formula $A \wedge B$ is true is exactly the intersection of the set of states in which A is true with the set of states in which B is true, because it is this intersection of states in which both A and B are true. This (simple) insight will already enable us to analyze a system separately for the questions of whether A is true and whether B is true in order to find out whether the conjunction $A \wedge B$ is true. Achieving compositionality for other CPS operators is more demanding but equally impactful.

Since compositionality is an intrinsic feature starting from the very semantics of logic [14, 16, 20, 21, 59, 62], logics naturally reason compositionally, too. For example, a proof of the formula $A \wedge B$ consists of a combination of a proof of A together with a proof of B, because the two of those proofs together justify that A and B are both true, which means that $A \wedge B$ is true. This makes it possible to take advantage of the compositionality in the formulas of a logic also when reasoning about formulas in the logic. A proof of $A \wedge B$ decomposes into a proof of the simpler subformula A together with a proof of the simpler subformula B.

With suitable generalizations of logics to embrace multi-dynamical systems [42, 44, 46, 47, 49, 52, 54, 55], this compositionality generalizes to CPS. We "just" need to make compositionality work for the CPS operators, which are, of course, more complicated than a mere logical $\wedge$ operator. Verification works by constructing a proof in such a multi-dynamical systems logic. The whole proof verifies a complex CPS. Yet, each proof step only reasons separately about one dynamical aspect at a time, for example, an isolated discrete assignment or the separate local dynamics of a differential equation, each captured in a separate, modular reasoning principle.

Multi-dynamical systems also impact and simplify the presentation of the Logical Foundations of Cyber-Physical Systems. The compositionality principles of logic and multi-dynamical systems considerably tame the conceptual complexity of CPS by making it possible to focus on one aspect at a time, one chapter after another, without losing the ability to combine the understanding attained for each aspect. This gradual approach effectively conveys the principles for a successful separation of concerns for CPS.

1.4 How to Learn About Cyber-Physical Systems

There are two primary ways of learning about cyber-physical systems.

Onion Model

The *Onion Model* follows the natural dependencies of the layers of mathematics going outside in, peeling off one layer at a time, and progressing to the next layer when all prerequisites have been covered. This would require the CPS student to first study all relevant parts of computer science, mathematics, and engineering, and then return to CPS in the big finale. That would require the first part of this book to cover real analysis, the second part differential equations, the third part conventional discrete programming, the fourth part classical discrete logic, the fifth part theorem proving, and finally the last part cyber-physical systems. In addition to the significant learning perseverance that the Onion Model requires, a downside is that it misses out on the integrative effects of cyber-physical systems that can bring different areas of science and engineering together, and which provide a unifying motivation for studying them in the first place.

Scenic Tour Model

This book follows the *Scenic Tour Model*, which starts at the heart of the matter, namely cyber-physical systems, going on scenic expeditions in various directions to explore the world around as we find the need to understand the respective subject matter. The textbook directly targets CPS right away, beginning with simpler layers that the reader can understand in full before moving on to the next challenge.

For example, the first layer comprises CPSs without feedback control, which allow simple finite open-loop controls to be designed, analyzed, and verified without the technical challenges considered in later layers of CPS. Likewise, the treatment of CPS is first limited to cases where the dynamics can be solved in closed form, such as straight-line accelerated motion of Newtonian dynamics, before generalizing to systems with more challenging differential equations that can no longer be solved explicitly. This gradual development where each level is mastered and understood and practiced in full before moving to the next level is helpful to tame complexity. The Scenic Tour Model has the advantage that we stay on cyber-physical systems the whole time, and leverage CPS as the guiding motivation for understanding more and more about the connected areas. It has the disadvantage that the resulting gradual development of CPS does not necessarily always present matters in the same way that an after-the-fact compendium would treat it. This textbook compensates by providing appropriate technical summaries and by highlighting important results for later reference in boxes, with a list of theorems and lemmas in the table of contents. A gradual development can also be more effective at conveying the ideas, reasons, and rationales behind the development compared to a final compendium.

Besides the substantial organizational impact that this "CPS first" approach has throughout the presentation of this book, the Scenic Tour Model is most easily noticeable in the Expedition boxes that this textbook provides. Every part of this textbook is written in a simple style bringing mathematical results in as needed, and with an emphasis on intuition. The Expedition boxes invite the reader to additionally connect to other areas of science that are of no crucial relevance for the immediate study of CPS nor the remainder of the textbook, but still provide a link to another area, in case the reader happens to be familiar with it or takes this link as an inspiration to explore that other area of science further.

Prerequisites

Even if deliberately light on prerequisites, this textbook cannot start from zero either. Its primary assumptions are some prior exposure to basic programming and elementary mathematics. Specifically, the textbook assumes that the reader has had some prior experience with computer programming (such as what is covered in a first semester undergraduate course taught in any programming language covering if-then-else conditionals and loops).

While Chap. 2 starts out with an intuitive and a rigorous treatment of differential equations and provides a few conceptually important meta-results in its appendix, this book is no replacement for a differential equations course. But it also does not have to be. The concepts required for CPS from differential equations will be picked up and expanded upon at a light pace throughout this textbook. The textbook does, however, assume that the reader is comfortable with simple derivative and differential equation notation. For example, Chap. 2 will discuss how $x' = v, v' = a$ is a differential equation, in which the time-derivative x' of position x equals velocity v, whose time-derivative v' in turn equals the acceleration a. This differential equation characterizes accelerated motion of a point x with velocity v and acceleration a along a straight line.

While a good deal of the interest in this textbook comes from its general applicability, it is also structured to minimize dependency on prerequisites. In particular, Part I of this book can already be understood if the reader is familiar with the differential equation $x' = v, v' = a$ for accelerated motion of point x along a straight line. While Part II provides analytic tools for studying systems with significantly more general differential equations, it is enough to have an intuition for the differential equation $x' = y, y' = -x$ characterizing rotation of the point (x, y) around the origin. Of course, this textbook studies other differential equations in some illustrative examples as well, but those are not on the critical path to understanding the rest of this book.

Most crucially, however, the textbook assumes that the reader has been exposed to some form of mathematical reasoning before (such as *either* in a calculus or analysis course *or* in a matrix or linear algebra course *or* a mathematics course for computer scientists or engineers). The particular contents covered in such a prior course are not at all as important as the mathematical experience itself with mathematical concept developments and proofs. This textbook develops a fair amount of logic on its own as part of the way of understanding cyber-physical systems. A prior understanding of logic is, thus, not necessary for the study of this book. And, in fact, the *Foundations of Cyber-Physical Systems* undergraduate course that the author teaches at Carnegie Mellon University and on which this textbook is based counts as fulfilling a Logics/Languages elective or Programming Languages requirement without prior background in either.

1.5 Computational Thinking for Cyber-Physical Systems

The approach that this book follows takes advantage of Computational Thinking [67] for cyber-physical systems [50]. Due to their subtleties and the intricate interactions of complex control software with the physical world, cyber-physical systems are notoriously challenging. Logical scrutiny, formalization, and thorough safety and correctness arguments are, thus, critical for cyber-physical systems. Because cyber-physical system designs are so easy to get wrong, these logical aspects are an integral part of CPS design and critical to understanding their complexities.

The primary attention of this book, thus, is on the foundations and core principles of cyber-physical systems. The book tames some of the complexities of cyber-physical systems by focusing on a simple core programming language for CPS. The elements of the programming language are introduced hand in hand with their reasoning principles, which makes it possible to combine CPS program design with their safety arguments. This is important, not just because abstraction is a key factor for success in CPS, but also because retrofitting safety is not possible in CPS.

To simplify matters, the chapters in this textbook are also organized to carefully reveal the complexities of cyber-physical systems in layers. Each layer will be covered in full, including its programmatic, semantic, and logical treatment, before proceeding to the next level of complexity. For example, the book first studies single-shot control before considering control loops, and only then proceeds to systems with differential equations that cannot be solved in closed form. Adversarial aspects are covered subsequently.

1.6 Learning Objectives

The respective learning objectives are identified at the beginning of each chapter, both textually and with a schematic diagram. They are organized along the three dimensions *modeling and control*, *computational thinking*, and *CPS skills*. The most important overall learning objectives throughout this textbook are the following.

Modeling and Control: In the area of *Modeling and Control* (MC), the most important goals are to

- *understand the core principles behind CPS.* The core principles are important for effectively recognizing how the integration of cyber and physical aspects can solve problems that no part could solve alone.
- *develop models and controls.* In order to understand, design, and analyze CPSs, it is important to be able to develop models of the various relevant aspects of a CPS design and to design controllers for the intended functionalities based on appropriate specifications.
- *identify the relevant dynamical aspects.* It is important to be able to identify which types of phenomena of a CPS have a relevant influence for the purpose of understanding a particular property of a particular system. These allow us to judge, for example, when it is important to manage adversarial effects, and when a nondeterministic model is sufficient.

Computational Thinking: In the area of *Computational Thinking* (CT), the most important goals are to

- *identify safety specifications and critical properties.* In order to develop correct CPS designs, it is important to identify what "correctness" means, how a design may fail to be correct, and how to make it correct if it is not correct yet.
- *understand abstraction in system designs.* The power of abstraction is essential for the modular organization of CPSs, and for the ability to reason about separate parts of a system independently. Because of the overwhelming practical challenges and numerous levels of detail, abstraction is even more critical than it already is in conventional software design.
- *express pre- and postconditions and invariants for CPS models.* Pre- and postconditions allow us to capture under which circumstance it is safe to run a CPS or a part of its design, and what safety entails. They allow us to achieve what abstraction and hierarchies achieve at the system level: decompose correctness of a full CPS into correctness of smaller pieces. The fundamental notion of invariants achieves a similar decomposition by establishing which relations of variables remain true no matter how long and how often the CPS runs.
- *use design-by-invariant.* In order to develop correct CPS designs, invariants are an important structuring principle guiding what the control has to maintain in order to preserve the invariant and, thereby, safety. This guidance simplifies the design process, because it applies locally at the level of individual localized control decisions that preserve invariants without explicitly having to take system-level closed-loop properties into account.
- *reason rigorously about CPS models.* Reasoning is required to ensure correctness and find flaws in a CPS design. Both informal reasoning and formal reasoning in a logic are important objectives for being able to establish correctness.
- *verify CPS models of appropriate scale.* This textbook covers the science of how to prove CPSs. You can gain practical experience through its exercises and appropriately scoped projects in the theorem prover KeYmaera X. This experience will help you learn how best to select the most interesting questions in formal verification and validation. Formal verification is not only critical but, given the right abstractions, quite feasible in high-level CPS control designs.

CPS Skills: In the area of *CPS skills*, the most important goals are to

- *understand the semantics of a CPS model.* What may be easy in a classical isolated program becomes very demanding when that program interfaces with effects in the physical world. A precise understanding of the nuanced meaning of a CPS model is fundamental to reasoning, along with an understanding of how it will execute. A deep understanding of the semantics of CPS models is also obtained by carefully relating their semantics to their reasoning principles and aligning them in perfect unison.
- *develop an intuition for operational effects.* Intuition for the joint operational effect of a CPS is crucial. For example, it is crucial to understand what the effect of a particular discrete computer control algorithm will be on a continuous plant.
- *identify control constraints.* An operational intuition guides our understanding of the operational effects and, along with their precise logical rendition, their impact on finding correct control constraints that make a CPS controller safe.
- *understand opportunities and challenges in CPS and verification.* While the beneficial prospects of CPS for society are substantial, it is crucial to also develop an understanding of their inherent challenges and of approaches to minimize the impact of potential safety hazards. Likewise, it is important to understand the ways in which formal verification can best help improve the safety of system designs.

This textbook will give the reader the required skills to formally analyze the cyber-physical systems that are all around us – from power plants to pacemakers and everything in between – so that when you contribute to the design of a CPS, you are able to understand important safety-critical aspects and feel confident designing and analyzing system models. Other beneficial by-products include that cyber-physical systems provide a well-motivated exposure to numerous other areas of mathematics and science in action.

identify safety specifications for CPS
rigorous reasoning about CPS
understand abstraction & architectures
programming languages for CPS
verify CPS models at scale

cyber+physics models & controls   semantics of CPS models
core principles of CPS   operational effects
relate discrete+continuous dynamics   identify control constraints
                                                                   opportunities and challenges

1.7 Structure of This Textbook

This textbook consists of four main parts, which develop different levels of the logical foundations of cyber-physical systems. You are now reading the introduction.

Elementary Cyber-Physical Systems

Part I studies elementary cyber-physical systems characterized by a hybrid system dynamics whose continuous dynamics can still be solved in closed form. Differential equations are studied as models of continuous dynamics, while control programs are considered for the discrete dynamics. Part I investigates differential dynamic logic for specifying properties and axioms for reasoning about CPS. It further investigates appropriate structuring principles for proofs and the handling of control loops via loop invariants, and discusses both event-triggered and time-triggered control. This part provides an extensive introduction to the wonders and challenges of cyber-physical systems, but still isolates most of the reasoning challenges in the search for discrete loop invariants since their differential equations can still be solved explicitly. While enabling interesting and challenging considerations about CPSs, Part I limits the level of interaction and subtlety in their safety arguments. The insights from Part I already enable, for example, a comprehensive study of controllers for safe acceleration and braking of a car along a straight lane.

Differential Equations Analysis

Part II considers advanced cyber-physical systems whose dynamics cannot be solved in explicit closed form. Most crucially, this necessitates indirect forms for analyzing the safety of a CPS, because solutions are no longer available. Based on the understanding of discrete induction for control loops from Part I, Part II develops induction techniques for differential equations. In addition to developing differential invariants as induction techniques for differential equations, this part studies differential cuts that make it possible to prove and then use lemmas about differential equations. It also considers so-called differential ghosts, which can simplify safety arguments by adding extra variables (ghost variables or auxiliary variables) with additional differential equations into the dynamics to balance out generalized energy invariants. Part II is required for handling safety arguments for CPS with nonsolvable dynamics such as robots racing on a circular race track or driving along curves in the plane or for aircraft flying along three-dimensional curves.

Adversarial Cyber-Physical Systems

Part III advances the understanding of cyber-physical systems to cover hybrid games mixing discrete dynamics, continuous dynamics, and adversarial dynamics. Based on the understanding of hybrid systems models for CPSs from Part I and invariants for differential equations from Part II, Part III shifts the focus to an exploration of hybrid games, in which the interaction of different players with different objectives is a dominant aspect. Unlike hybrid systems, in which all choices are nondeterministic, hybrid games give different choices to different players at different times. Part III is required for handling safety arguments for CPSs in which multiple agents interact with possibly conflicting goals, or with the same goals but possibly conflicting actions resulting from different perceptions of the world.

Comprehensive CPS Correctness

Part IV complements the CPS foundations from the previous parts with an account of what it takes to round out a comprehensive correctness argument for a cyber-physical system. Part IV condenses the logical reasoning principles of CPS from Parts I and II into a completely axiomatic style that makes it easy to implement logical reasoning with an extremely parsimonious logical framework based solely on uniform substitutions. Since the nuances of cyber-physical systems provide ample opportunity for subtle discrepancies, Part IV also investigates a logical way to tame the subtle relationship of CPS models to CPS implementations. The logical foundations of model safety transfer can synthesize provably correct monitor conditions that, if checked to hold at runtime, are guaranteed to imply that offline safety verification results about CPS models apply to the present run of the actual CPS implementation. Finally, this part considers logical elements of reasoning techniques for the real arithmetic that is used in CPS verification.

Online Material

The theory exercises provided at the end of the chapters are designed to actively check the understanding of the material and provide routes for further developments. In addition, the reader is invited to advance his or her understanding of the material by practicing CPS proving in the KeYmaera X verification tool [15], which is an aXiomatic Tactical Theorem Prover for Hybrid Systems that implements differential dynamic logic [48, 49, 51, 54]. For technical reasons, the concrete syntax in KeYmaera X has a slightly different ASCII notation, but, other than that, KeYmaera X closely follows the theory of differential dynamic logic as presented in this book. For educational purposes, this textbook also focuses on a series of instructive simpler examples instead of the technical complexities of full-blown applications that are reported elsewhere [22, 26, 31, 35, 36, 57, 58].

The Web page for this book is at the following URL:

```
http://www.lfcps.org/lfcps/
```

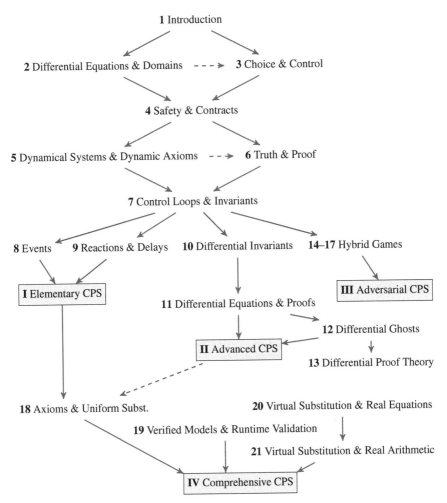

Fig. 1.3 Dependencies and suggested reading sequences of the chapters

Suggested Reading Sequence

Even if the basic suggested reading order in this book is linear, this textbook can be read in many different ways. Except for most of the foundation developed in Part I, the other parts of this book are independent and can be read in any order. The dependencies among the topics in the chapters are shown in Fig. 1.3. Weak dependencies on some small number of concepts are indicated as dashed lines, as these topics might be presented in a different order. The core of the textbook is the chapters that lead to Elementary CPS (Part I) in Fig. 1.3, including either Chaps. 8 or 9 or both. An integral part for Advanced CPS is Chaps. 10 and 11, along with an optional study of the topic of differential ghosts for advanced differential equations in Chap. 12.

Different reading sequences are possible for this textbook. The minimal core for an understanding of elementary cyber-physical systems consists of Chaps. 1–7 from Part I. A minimal course emphasizing experience with system modeling covers the Chaps. 1–9 that lead to Part I on Elementary CPS in Fig. 1.3. For a minimal course emphasizing CPS reasoning Chaps. 1–7 would be followed by Chaps. 10–11 from Part II, possibly including Chap. 12 for advanced reasoning techniques. The other chapter sequences are independent. After Chaps. 1–9, any sequence of the other topics following the reader's interest is possible since the hybrid game chapters Chaps. 14–17 in Part III are independent of the subsequent topics in Part IV, which are, in turn, mostly independent of one another.

This textbook features an active development leading the reader through a critical and self-propelled development of the core aspects of cyber-physical systems. Especially at places marked as follows ...

> Before you read on, see if you can find the answer for yourself.

... the reader is advised to work toward an answer before comparing it with the development pursued in the textbook. Of course, when comparing answers, the reader should keep in mind that there is more than one correct way of developing the material. The reader may have found a perfectly correct answer that just was not anticipated in the writing of the textbook. That represents a great opportunity to investigate advantages and downsides of the respective approaches.

1.8 Summary

This chapter gave an informal overview of application domains for cyber-physical systems, which combine cyber capabilities such as communication, computation, and control with physical capabilities such as motion or chemical process control. It motivated the need for careful designs and comprehensive safety analyses, which will be developed in this book. Closely related is the mathematical notion of hybrid systems, which are dynamical systems that combine discrete dynamics with continuous dynamics. Despite the fact that they are different notions (cyber-physical systems are based on the technical characteristics, while hybrid systems are a mathematical model), this textbook simplifies matters by using the two notions interchangeably in Parts I and II. More advanced models of cyber-physical systems will be deferred to Part III after the hybrid systems model has been understood well in Part I and Part II.

This chapter set the stage for the multi-dynamical systems approach that this textbook follows. Multi-dynamical systems are characterized by multiple facets of dynamical systems whose compositionality in a logic of dynamical systems enables a separation of concerns for CPS. The multi-dynamical systems view directly benefits the presentation in this book as well, by making it possible to focus on one aspect at a time without losing the ability to combine the understanding attained for each aspect.

References

[1] Matthias Althoff and John M. Dolan. Online verification of automated road vehicles using reachability analysis. *IEEE Trans. on Robotics* **30**(4) (2014), 903–918. DOI: `10.1109/TRO.2014.2312453`.

[2] Rajeev Alur. Formal verification of hybrid systems. In: *EMSOFT*. Ed. by Samarjit Chakraborty, Ahmed Jerraya, Sanjoy K. Baruah, and Sebastian Fischmeister. New York: ACM, 2011, 273–278. DOI: `10.1145/2038642.2 038685`.

[3] Rajeev Alur. *Principles of Cyber-Physical Systems*. Cambridge: MIT Press, 2015.

[4] Rajeev Alur, Costas Courcoubetis, Nicolas Halbwachs, Thomas A. Henzinger, Pei-Hsin Ho, Xavier Nicollin, Alfredo Olivero, Joseph Sifakis, and Sergio Yovine. The algorithmic analysis of hybrid systems. *Theor. Comput. Sci.* **138**(1) (1995), 3–34. DOI: `10.1016/0304-3975(94)00202-T`.

[5] Rajeev Alur, Thomas Henzinger, Gerardo Lafferriere, and George J. Pappas. Discrete abstractions of hybrid systems. *Proc. IEEE* **88**(7) (2000), 971–984.

[6] Michael S. Branicky. General hybrid dynamical systems: modeling, analysis, and control. In: *Hybrid Systems*. Ed. by Rajeev Alur, Thomas A. Henzinger, and Eduardo D. Sontag. Vol. 1066. LNCS. Berlin: Springer, 1995, 186–200. DOI: `10.1007/BFb0020945`.

[7] Davide Bresolin, Luca Geretti, Riccardo Muradore, Paolo Fiorini, and Tiziano Villa. Formal verification applied to robotic surgery. In: *Coordination Control of Distributed Systems*. Ed. by Jan H. van Schuppen and Tiziano Villa. Vol. 456. Lecture Notes in Control and Information Sciences. Berlin: Springer, 2015, 347–355. DOI: `10.1007/978-3-319-10407-2_40`.

[8] Luminita Manuela Bujorianu. *Stochastic Reachability Analysis of Hybrid Systems*. Berlin: Springer, 2012. DOI: `10.1007/978-1-4471-2795-6`.

[9] Edmund M. Clarke, E. Allen Emerson, and Joseph Sifakis. Model checking: algorithmic verification and debugging. *Commun. ACM* **52**(11) (2009), 74–84. DOI: `10.1145/1592761.1592781`.

[10] Jennifer M. Davoren and Anil Nerode. Logics for hybrid systems. *IEEE* **88**(7) (2000), 985–1010. DOI: `10.1109/5.871305`.

[11] Akash Deshpande, Aleks Göllü, and Pravin Varaiya. SHIFT: a formalism and a programming language for dynamic networks of hybrid automata. In: *Hybrid Systems*. Ed. by Panos J. Antsaklis, Wolf Kohn, Anil Nerode, and Shankar Sastry. Vol. 1273. LNCS. Springer, 1996, 113–133. DOI: `10.1007 /BFb0031558`.

[12] Laurent Doyen, Goran Frehse, George J. Pappas, and André Platzer. Verification of hybrid systems. In: *Handbook of Model Checking*. Ed. by Edmund M.

Clarke, Thomas A. Henzinger, Helmut Veith, and Roderick Bloem. Springer, 2018. Chap. 30. DOI: 10.1007/978-3-319-10575-8_30.

[13] G. K. Fourlas, K. J. Kyriakopoulos, and C. D. Vournas. Hybrid systems modeling for power systems. *Circuits and Systems Magazine, IEEE* **4**(3) (2004), 16–23. DOI: 10.1109/MCAS.2004.1337806.

[14] Gottlob Frege. *Begriffsschrift, eine der arithmetischen nachgebildete Formelsprache des reinen Denkens.* Halle: Verlag von Louis Nebert, 1879.

[15] Nathan Fulton, Stefan Mitsch, Jan-David Quesel, Marcus Völp, and André Platzer. KeYmaera X: an axiomatic tactical theorem prover for hybrid systems. In: *CADE*. Ed. by Amy Felty and Aart Middeldorp. Vol. 9195. LNCS. Berlin: Springer, 2015, 527–538. DOI: 10.1007/978-3-319-21401-6_36.

[16] Gerhard Gentzen. Untersuchungen über das logische Schließen I. *Math. Zeit.* **39**(2) (1935), 176–210. DOI: 10.1007/BF01201353.

[17] Radu Grosu, Grégory Batt, Flavio H. Fenton, James Glimm, Colas Le Guernic, Scott A. Smolka, and Ezio Bartocci. From cardiac cells to genetic regulatory networks. In: *CAV*. Ed. by Ganesh Gopalakrishnan and Shaz Qadeer. Vol. 6806. LNCS. Berlin: Springer, 2011, 396–411. DOI: 10.1007/978-3-642-22110-1_31.

[18] Thomas A. Henzinger. The theory of hybrid automata. In: *LICS*. Los Alamitos: IEEE Computer Society, 1996, 278–292. DOI: 10.1109/LICS.1996.561342.

[19] Thomas A. Henzinger and Joseph Sifakis. The discipline of embedded systems design. *Computer* **40**(10) (Oct. 2007), 32–40. DOI: 10.1109/MC.2007.364.

[20] David Hilbert. Die Grundlagen der Mathematik. *Abhandlungen aus dem Seminar der Hamburgischen Universität* **6**(1) (1928), 65–85. DOI: 10.1007/BF02940602.

[21] Charles Antony Richard Hoare. An axiomatic basis for computer programming. *Commun. ACM* **12**(10) (1969), 576–580. DOI: 10.1145/363235.363259.

[22] Jean-Baptiste Jeannin, Khalil Ghorbal, Yanni Kouskoulas, Aurora Schmidt, Ryan Gardner, Stefan Mitsch, and André Platzer. A formally verified hybrid system for safe advisories in the next-generation airborne collision avoidance system. *STTT* **19**(6) (2017), 717–741. DOI: 10.1007/s10009-016-0434-1.

[23] Taylor T. Johnson and Sayan Mitra. Parametrized verification of distributed cyber-physical systems: an aircraft landing protocol case study. In: *ICCPS*. Los Alamitos: IEEE, 2012, 161–170. DOI: 10.1109/ICCPS.2012.24.

[24] Nidhi Kalra and Susan M. Paddock. *Driving to Safety – How Many Miles of Driving Would It Take to Demonstrate Autonomous Vehicle Reliability?* Tech. rep. RAND Corporation, 2016. DOI: 10.7249/RR1478.

[25] BaekGyu Kim, Anaheed Ayoub, Oleg Sokolsky, Insup Lee, Paul L. Jones, Yi Zhang, and Raoul Praful Jetley. Safety-assured development of the GPCA infusion pump software. In: *EMSOFT*. Ed. by Samarjit Chakraborty, Ahmed

Jerraya, Sanjoy K. Baruah, and Sebastian Fischmeister. New York: ACM, 2011, 155–164. DOI: 10.1145/2038642.2038667.

[26] Yanni Kouskoulas, David W. Renshaw, André Platzer, and Peter Kazanzides. Certifying the safe design of a virtual fixture control algorithm for a surgical robot. In: *HSCC*. Ed. by Calin Belta and Franjo Ivancic. ACM, 2013, 263–272. DOI: 10.1145/2461328.2461369.

[27] Kim Guldstrand Larsen. Verification and performance analysis for embedded systems. In: *TASE 2009, Third IEEE International Symposium on Theoretical Aspects of Software Engineering, 29-31 July 2009, Tianjin, China*. Ed. by Wei-Ngan Chin and Shengchao Qin. IEEE Computer Society, 2009, 3–4. DOI: 10.1109/TASE.2009.66.

[28] Edward Ashford Lee and Sanjit Arunjumar Seshia. *Introduction to Embedded Systems — A Cyber-Physical Systems Approach*. Lulu.com, 2013.

[29] Insup Lee and Oleg Sokolsky. Medical cyber physical systems. In: *DAC*. Ed. by Sachin S. Sapatnekar. New York: ACM, 2010, 743–748.

[30] Insup Lee, Oleg Sokolsky, Sanjian Chen, John Hatcliff, Eunkyoung Jee, BaekGyu Kim, Andrew L. King, Margaret Mullen-Fortino, Soojin Park, Alex Roederer, and Krishna K. Venkatasubramanian. Challenges and research directions in medical cyber-physical systems. *Proc. IEEE* **100**(1) (2012), 75–90. DOI: 10.1109/JPROC.2011.2165270.

[31] Sarah M. Loos, André Platzer, and Ligia Nistor. Adaptive cruise control: hybrid, distributed, and now formally verified. In: *FM*. Ed. by Michael Butler and Wolfram Schulte. Vol. 6664. LNCS. Berlin: Springer, 2011, 42–56. DOI: 10.1007/978-3-642-21437-0_6.

[32] Jan Lunze and Françoise Lamnabhi-Lagarrigue, eds. *Handbook of Hybrid Systems Control: Theory, Tools, Applications*. Cambridge: Cambridge Univ. Press, 2009. DOI: 10.1017/CBO9780511807930.

[33] Oded Maler. Control from computer science. *Annual Reviews in Control* **26**(2) (2002), 175–187. DOI: 10.1016/S1367-5788(02)00030-5.

[34] Sayan Mitra, Tichakorn Wongpiromsarn, and Richard M. Murray. Verifying cyber-physical interactions in safety-critical systems. *IEEE Security & Privacy* **11**(4) (2013), 28–37. DOI: 10.1109/MSP.2013.77.

[35] Stefan Mitsch, Marco Gario, Christof J. Budnik, Michael Golm, and André Platzer. Formal verification of train control with air pressure brakes. In: *Reliability, Safety, and Security of Railway Systems. Modelling, Analysis, Verification, and Certification - Second International Conference, RSSRail 2017, Pistoia, Italy, November 14-16, 2017, Proceedings*. Ed. by Alessandro Fantechi, Thierry Lecomte, and Alexander Romanovsky. Vol. 10598. LNCS. Springer, 2017, 173–191. DOI: 10.1007/978-3-319-68499-4_12.

[36] Stefan Mitsch, Khalil Ghorbal, David Vogelbacher, and André Platzer. Formal verification of obstacle avoidance and navigation of ground robots. *I. J. Robotics Res.* **36**(12) (2017), 1312–1340. DOI: 10.1177/02783649177 33549.

[37]   Anil Nerode. Logic and control. In: *CiE*. Ed. by S. Barry Cooper, Benedikt
       Löwe, and Andrea Sorbi. Vol. 4497. LNCS. Berlin: Springer, 2007, 585–597.
       DOI: 10.1007/978-3-540-73001-9_61.

[38]   Anil Nerode and Wolf Kohn. Models for hybrid systems: automata, topolo-
       gies, controllability, observability. In: *Hybrid Systems*. Ed. by Robert L.
       Grossman, Anil Nerode, Anders P. Ravn, and Hans Rischel. Vol. 736. LNCS.
       Berlin: Springer, 1992, 317–356.

[39]   NITRD CPS Senior Steering Group. *CPS vision statement*. NITRD. 2012.

[40]   George J. Pappas. Wireless control networks: modeling, synthesis, robust-
       ness, security. In: *Proceedings of the 14th ACM International Conference on
       Hybrid Systems: Computation and Control, HSCC 2011, Chicago, IL, USA,
       April 12-14, 2011*. Ed. by Marco Caccamo, Emilio Frazzoli, and Radu Grosu.
       New York: ACM, 2011, 1–2. DOI: 10.1145/1967701.1967703.

[41]   Erion Plaku, Lydia E. Kavraki, and Moshe Y. Vardi. Hybrid systems: from
       verification to falsification by combining motion planning and discrete search.
       Form. Methods Syst. Des. **34**(2) (2009), 157–182. DOI: 10.1007/s10703
       -008-0058-5.

[42]   André Platzer. Differential dynamic logic for hybrid systems. *J. Autom. Reas.*
       41(2) (2008), 143–189. DOI: 10.1007/s10817-008-9103-8.

[43]   André Platzer. Differential Dynamic Logics: Automated Theorem Proving
       for Hybrid Systems. PhD thesis. Department of Computing Science, Univer-
       sity of Oldenburg, 2008.

[44]   André Platzer. Differential-algebraic dynamic logic for differential-algebraic
       programs. *J. Log. Comput.* **20**(1) (2010), 309–352. DOI: 10.1093/logcom/
       exn070.

[45]   André Platzer. *Logical Analysis of Hybrid Systems: Proving Theorems for
       Complex Dynamics*. Heidelberg: Springer, 2010. DOI: 10.1007/978-3-
       642-14509-4.

[46]   André Platzer. Stochastic differential dynamic logic for stochastic hybrid pro-
       grams. In: *CADE*. Ed. by Nikolaj Bjørner and Viorica Sofronie-Stokkermans.
       Vol. 6803. LNCS. Berlin: Springer, 2011, 446–460. DOI: 10.1007/978-
       3-642-22438-6_34.

[47]   André Platzer. A complete axiomatization of quantified differential dynamic
       logic for distributed hybrid systems. *Log. Meth. Comput. Sci.* **8**(4:17) (2012).
       Special issue for selected papers from CSL'10, 1–44. DOI: 10.2168/
       LMCS-8(4:17)2012.

[48]   André Platzer. Logics of dynamical systems. In: *LICS*. Los Alamitos: IEEE,
       2012, 13–24. DOI: 10.1109/LICS.2012.13.

[49]   André Platzer. The complete proof theory of hybrid systems. In: *LICS*. Los
       Alamitos: IEEE, 2012, 541–550. DOI: 10.1109/LICS.2012.64.

[50]   André Platzer. Teaching CPS foundations with contracts. In: *CPS-Ed*. 2013,
       7–10.

[51]   André Platzer. A uniform substitution calculus for differential dynamic logic.
       In: *CADE*. Ed. by Amy Felty and Aart Middeldorp. Vol. 9195. LNCS. Berlin:
       Springer, 2015, 467–481. DOI: 10.1007/978-3-319-21401-6_32.

[52]   André Platzer. Differential game logic. *ACM Trans. Comput. Log.* **17**(1) (2015), 1:1–1:51. DOI: `10.1145/2817824`.

[53]   André Platzer. Logic & proofs for cyber-physical systems. In: *IJCAR*. Ed. by Nicola Olivetti and Ashish Tiwari. Vol. 9706. LNCS. Berlin: Springer, 2016, 15–21. DOI: `10.1007/978-3-319-40229-1_3`.

[54]   André Platzer. A complete uniform substitution calculus for differential dynamic logic. *J. Autom. Reas.* **59**(2) (2017), 219–265. DOI: `10.1007/s108 17-016-9385-1`.

[55]   André Platzer. Differential hybrid games. *ACM Trans. Comput. Log.* **18**(3) (2017), 19:1–19:44. DOI: `10.1145/3091123`.

[56]   André Platzer and Edmund M. Clarke. The image computation problem in hybrid systems model checking. In: *HSCC*. Ed. by Alberto Bemporad, Antonio Bicchi, and Giorgio C. Buttazzo. Vol. 4416. LNCS. Springer, 2007, 473–486. DOI: `10.1007/978-3-540-71493-4_37`.

[57]   André Platzer and Edmund M. Clarke. Formal verification of curved flight collision avoidance maneuvers: a case study. In: *FM*. Ed. by Ana Cavalcanti and Dennis Dams. Vol. 5850. LNCS. Berlin: Springer, 2009, 547–562. DOI: `10.1007/978-3-642-05089-3_35`.

[58]   André Platzer and Jan-David Quesel. European Train Control System: a case study in formal verification. In: *ICFEM*. Ed. by Karin Breitman and Ana Cavalcanti. Vol. 5885. LNCS. Berlin: Springer, 2009, 246–265. DOI: `10.10 07/978-3-642-10373-5_13`.

[59]   Vaughan R. Pratt. Semantical considerations on Floyd-Hoare logic. In: *17th Annual Symposium on Foundations of Computer Science, 25-27 October 1976, Houston, Texas, USA*. Los Alamitos: IEEE, 1976, 109–121. DOI: `10 .1109/SFCS.1976.27`.

[60]   President's Council of Advisors on Science and Technology. *Leadership under challenge: information technology R&D in a competitive world*. An Assessment of the Federal Networking and Information Technology R&D Program. Aug. 2007.

[61]   Dana Scott and Christopher Strachey. *Towards a mathematical semantics for computer languages*. Tech. rep. PRG-6. Oxford Programming Research Group, 1971.

[62]   Raymond M. Smullyan. *First-Order Logic*. Mineola: Dover, 1968. DOI: `10 .1007/978-3-642-86718-7`.

[63]   Paulo Tabuada. *Verification and Control of Hybrid Systems: A Symbolic Approach*. Berlin: Springer, 2009. DOI: `10.1007/978-1-4419-0224-5`.

[64]   Ashish Tiwari. Abstractions for hybrid systems. *Form. Methods Syst. Des.* **32**(1) (2008), 57–83. DOI: `10.1007/s10703-007-0044-3`.

[65]   Ashish Tiwari. Logic in software, dynamical and biological systems. In: *LICS*. IEEE Computer Society, 2011, 9–10. DOI: `10.1109/LICS.201 1.20`.

[66]   Claire Tomlin, George J. Pappas, and Shankar Sastry. Conflict resolution for air traffic management: a study in multi-agent hybrid systems. *IEEE T. Automat. Contr.* **43**(4) (1998), 509–521. DOI: `10.1109/9.664154`.

[67]   Jeannette M. Wing. Computational thinking. *Commun. ACM* **49**(3) (2006), 33–35. DOI: 10.1145/1118178.1118215.

[68]   Jeannette M. Wing. Five deep questions in computing. *Commun. ACM* **51**(1) (2008), 58–60. DOI: 10.1145/1327452.1327479.

[69]   Paolo Zuliani, André Platzer, and Edmund M. Clarke. Bayesian statistical model checking with application to Simulink/Stateflow verification. *Form. Methods Syst. Des.* **43**(2) (2013), 338–367. DOI: 10.1007/s10703-013-0195-3.

Part I
Elementary Cyber-Physical Systems

Overview of Part I on Elementary Cyber-Physical Systems

The first part of this book studies elementary *cyber-physical systems* (CPSs) in a gradual way by developing their models and reasoning principles one layer at a time. This part first considers CPSs without feedback control, which allow designs and analysis with finite-horizon open-loop controls, before considering the challenges of feedback control mechanisms in which a system repeatedly takes sensor information into account to decide how best to act. The focus in this first part is on CPSs limited to cases where the dynamics can still be solved in closed form, which simplifies analytic understanding considerably. An analytic treatment of CPSs with more complicated dynamics will be covered subsequently in Part II.

Part I lays the foundation for cyber-physical systems by presenting both CPS programming language models and their semantics. It introduces differential dynamic logic dL as the logic of dynamical systems, which provides the fundamental basis for understanding cyber-physical systems and serves as the language of choice for rigorously specifying and verifying CPSs. Proof-structuring principles will be discussed as well to organize our CPS reasoning and make sure we do not lose track of what CPS correctness arguments need to show. The important control paradigms of open-loop control, closed-loop control, time-triggered control, and event-triggered control are discussed along with lessons about their models and common analysis insights.

Chapter 2
Differential Equations & Domains

Synopsis The primary goal of this chapter is to obtain a solid working intuition for the continuous dynamics part of cyber-physical systems. It provides a brief introduction to differential equations with evolution domains as models of continuous physical processes. While focusing on an intuitive development, this chapter lays the foundation for an operational understanding of continuous processes. For reference, it discusses some of the elementary theory of differential equations. This chapter also introduces the first-order logic of real arithmetic as a language for describing the evolution domains to which continuous processes are restricted when forming hybrid systems.

2.1 Introduction

Cyber-physical systems combine cyber capabilities with physical capabilities. You already have experience with models of computation and algorithms for the cyber part of CPS if you have seen the use of programming languages for computer programming. In CPS, we do not program computers, though, but rather program CPSs instead. Hence, we program computers that interact with physics to achieve their goals. In this chapter, we study models of physics and the most elementary part of how they can interact with the cyber part. Physics by and large is obviously a deep subject. But for CPS, one of the most fundamental models of physics is sufficient at first, that of ordinary differential equations.

While this chapter covers the most important parts of differential equations, it is not to be misunderstood as a diligent coverage of the fascinating area of ordinary differential equations. What you need to get started with the book is an intuition about differential equations, as well as an understanding of their precise meaning. This will be developed in the present chapter. Subsequently, we will return to the topic of differential equations for a deeper understanding of differential equations and their proof principles a number of times in later chapters, especially Part II. The other important aspect that this chapter develops is *first-order logic of real arith-*

© Springer International Publishing AG, part of Springer Nature 2018
A. Platzer, *Logical Foundations of Cyber-Physical Systems*,
https://doi.org/10.1007/978-3-319-63588-0_2

metic for the purpose of representing domains and domain constraints of differential equations, which is of paramount significance in hybrid systems. More detailed treatments of differential equations can be found, e.g., in the seminal book by Walter [10] or elsewhere [2, 4, 8, 9].

The most important learning goals of this chapter are:

Modeling and Control: We develop an understanding of one core principle behind CPS: the case of continuous dynamics and differential equations with evolution domains as models of the physics part of CPS. We introduce first-order logic of real arithmetic as the modeling language for describing evolution domains of differential equations.

Computational Thinking: Both the significance of meaning and the descriptive power of differential equations will play key roles, foreshadowing many important aspects underlying the proper understanding of cyber-physical systems. We will also begin to learn to carefully distinguish between syntax (which is notation) and semantics (what carries meaning), a core principle for both logic and computer science that continues to be crucial for CPS.

CPS Skills: We develop an intuition for the continuous operational effects of CPS and devote significant attention to understanding the exact semantics of differential equations, which has some subtleties in store for us.

semantics of differential equations
descriptive power of differential equations
syntax versus semantics

continuous dynamics   continuous operational effects
differential equations
evolution domains
first-order logic

2.2 Differential Equations as Models of Continuous Physical Processes

Differential equations model processes in which the state variables of a system evolve continuously in time. A differential equation describes quite concisely how the system evolves over time. It describes how the variables change locally, so it, basically, indicates the direction in which the variables evolve at each point in space.

Fig. 2.1 shows the respective directions in which the system evolves by a vector at each point and illustrates one solution as a curve in two-dimensional space that follows those vectors everywhere. Of course, the figure would be rather cluttered if we were to try to indicate the vector at literally each and every point, of which there are uncountably infinitely many. But this is a shortcoming only of our illustration, not of the mathematical realities. Differential equations actually define such a vector for the direction of evolution at *every* point in space.

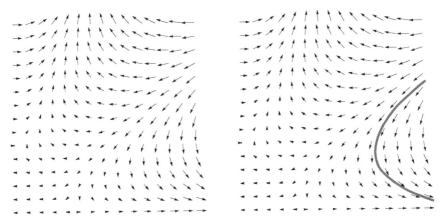

Fig. 2.1 Vector field (**left**) and vector field with one solution of a differential equation (**right**)

As an example, suppose we have a car whose position is denoted by x. Cars have a tendency to move, so the car's position x will change over time. How the value of variable x changes over time depends on how fast the car is driving. Let v denote the velocity of said car. Since v is the velocity of the car, its position x changes according to the velocity. So, the position x changes such that its derivative x' is v, which we denote by the differential equation $x' = v$. This differential equation means that the time-derivative of the position x equals the velocity v. So how x evolves depends on v. If the velocity is $v = 0$, then the position x does not change at all and the car might be parked or in a traffic jam. If $v > 0$, then the position x keeps on increasing over time. How fast x increases depends on the value of v, a bigger v give a quicker changes in x, because the time-derivative of x equals v in the differential equation $x' = v$.

Of course, the velocity v itself may also be subject to change over time. The car might accelerate, so let a denote its acceleration. Then the velocity v changes with time-derivative a, that is by the differential equation $v' = a$. Overall, the car then follows the differential equation (really a differential equation system):[1]

$$x' = v, v' = a$$

---

[1] Note that the value of x changes over time, so it is really a function of time. Hence, the notation $x'(t) = v(t), v'(t) = a$ is sometimes used. It is customary, however, to eloquently suppress the argument t for time and just write $x' = v, v' = a$ instead. In the physics literature, the notation $\dot{x}$ is often

That is, the position x of the car changes with time-derivative v, which, in turn, changes with time-derivative a.

What we mean by this differential equation, intuitively, is that the system always follows a direction (or vector field shown in Fig. 2.2) where, at all points (x, v), the direction vectors have their direction for positions point in a direction that equals the current v while their direction for velocities points in the same direction a. The system is always supposed to follow exactly in the direction of those direction vectors at every point. What does this mean exactly? How can we understand it doing that at all of the infinitely many points?

Fig. 2.2 Vector field with one solution of accelerated straight-line motion

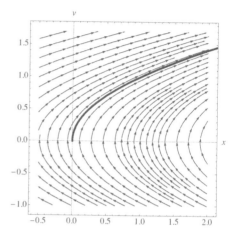

To sharpen our intuition for this aspect, consider a one-dimensional differential equation with a position x that changes over time t starting at initial state 1 at initial time 0:

$$\left(\begin{array}{l} x'(t) = \tfrac{1}{4}x(t) \\ x(0) = 1 \end{array} \right)$$

For a number of different time discretization steps $\Delta \in \{4, 2, 1, \tfrac{1}{2}\}$, Fig. 2.3 illustrates what an approximate pseudo-solution would look like that only respects the differential equation at the times that are integer multiples of Δ and is in blissful ignorance of the differential equation in between these grid points. Such a pseudo-solution corresponds to what is obtained by explicit Euler integration [3]. The true solution of the differential equation should, however, also respect the direction that the differential equation prescribes at all the other uncountably infinitely many time points in between. Because this differential equation is quite well behaved, the discretizations still approach the true continuous solution $x(t) = e^{\frac{t}{4}}$ as Δ gets smaller. But differential equations come with a lot of surprises when anyone attempts to understand them from a discrete perspective. No matter how small a discretization

---

used when referring to the time-derivative of x. We prefer the mathematical notation x', because dots are more easily overlooked, especially on longer names, and are hard to typeset in ASCII.

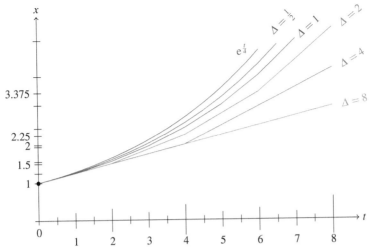

Fig. 2.3 Discretizations of differential equations with discretization time step Δ

$\Delta > 0$ we choose, that discretization will be arbitrarily far away from the true continuous solution for large t.

2.3 The Meaning of Differential Equations

We have obtained an intuitive understanding of how differential equations describe the direction of the evolution of a system as a vector field from Fig. 2.1. But some questions remain. What exactly is a vector field? What does it mean to describe directions of evolution at literally every point in space, of which there are uncountably infinitely many? Could these directions not *possibly contradict each other so that the description becomes ambiguous?* What is the exact meaning of a differential equation in the first place?

The only way to truly understand any system is to understand exactly what each of its pieces does. CPSs are demanding, and misunderstandings about their effect often have far-reaching consequences. So let us start by making the pieces of a CPS unambiguous. The first piece is differential equations.

> **Note 4 (Importance of meaning)** The physical impacts of CPSs do not leave much room for failure. We want to immediately get into the habit of always studying the behavior and exact meaning of all relevant aspects of a CPS.

An ordinary differential equation in explicit form is an equation $y'(t) = f(t, y)$ where $y'(t)$ is meant to be the derivative of y with respect to time t and f is a function of time t and the current state y. A solution is a differentiable function Y of time that satisfies this equation when substituted into the differential equation, i.e., when substituting $Y(t)$ for y and the time-derivative $Y'(t)$ of Y at t for $y'(t)$. That is, the time-derivative of the solution at each time is equal to the differential equation's right-hand side, as illustrated for time $t = -1$ in Fig. 2.4.

Fig. 2.4 Differential equation solution condition: time-derivative shown in red at $t = -1$ equals right-hand side of differential equation at all times

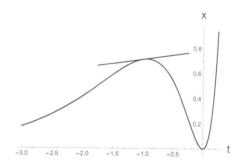

In the next chapter, we will study a more elegant definition of a solution of a differential equation that is well-attuned with the concepts in this book. But first, we consider the (equivalent) classical mathematical definition of a solution.

Definition 2.1 (Ordinary differential equation). Let $f : D \to \mathbb{R}^n$ be a function on a *domain* $D \subseteq \mathbb{R} \times \mathbb{R}^n$, i.e., an open and connected subset. The function $Y : J \to \mathbb{R}^n$ is a *solution* on the interval $J \subseteq \mathbb{R}$ of the *initial value problem*

$$\left(\begin{array}{l} y'(t) = f(t,y) \\ y(t_0) = y_0 \end{array} \right) \tag{2.1}$$

with *ordinary differential equation (ODE)* $y' = f(t,y)$, if, for all times $t \in J$

1. solution Y is in the domain $(t,Y(t)) \in D$,
2. time-derivative $Y'(t)$ exists and is $Y'(t) = f(t,Y(t))$, and
3. initial value $Y(t_0) = y_0$ is respected at the initial time, also $t_0 \in J$.

If $f : D \to \mathbb{R}^n$ is continuous, then $Y : J \to \mathbb{R}^n$ is continuously differentiable, because its derivative $Y'(t)$ is $f(t,Y(t))$, which is continuous as f is continuous and Y is differentiable so continuous. Similarly if f is k-times continuously differentiable then Y is $k+1$-times continuously differentiable. The definition is analogous for higher-order differential equations, i.e., those involving higher-order derivatives such as $y''(t)$ or $y^{(n)}(t)$ for $n > 1$.

Let us consider the intuition for this definition. A differential equation (system) can be thought of as a vector field such as the one in Fig. 2.1, where, at each point, the vector shows in which direction the solution evolves. At every point, the vector corresponds to the right-hand side of the differential equation. A solution of a differential equation adheres to this vector field at every point, i.e., the solution (e.g., the solid curve in Fig. 2.1) *locally* follows the direction indicated by the vector of the right-hand side of the differential equation. There are many solutions of the differential equation corresponding to the vector field illustrated in Fig. 2.1. For the initial value problem (2.1), however, solutions also have to start at the prescribed position y_0 at the initial time t_0 and then follow the differential equations or vector field from this point. In general, there can still be multiple solutions for the same initial value problem, but not for well-behaved differential equations (Sect. 2.9.2).

2.4 A Tiny Compendium of Differential Equation Examples

While cyber-physical systems do not necessitate a treatment and understanding of every differential equation you could ever think of, they do still benefit from a working intuition about differential equations and their relationships to their solutions. The following list of examples indicate by a * which differential equations play an important rôle in this book (Example 2.4, Example 2.5 and Example 2.7), compared to the ones that are merely listed to support a general intuition about the different possibilities that might happen when working with differential equations.

Example 2.1 (A constant differential equation). Some differential equations are easy to solve, especially those with constant right-hand sides. The initial value problem

$$\left(\begin{array}{l} x'(t) = \frac{1}{2} \\ x(0) = -1 \end{array} \right)$$

describes that x initially starts at -1 and always changes at the rate $1/2$. It has the solution $x(t) = \frac{1}{2}t - 1$ shown in Fig. 2.5. How can we verify that this is indeed a solution? This can be checked easily by inserting the solution into the differential equation and initial value equation and checking that they evaluate to the desired values according to the initial value problem:

$$\left(\begin{array}{l} (x(t))' = (\frac{1}{2}t - 1)' = \frac{1}{2} \\ x(0) = \frac{1}{2} \cdot 0 - 1 = -1 \end{array} \right)$$

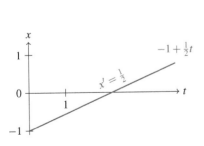

Fig. 2.5 Constant differential equation

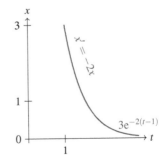

Fig. 2.6 Linear differential equation

Example 2.2 (A negative linear differential equation). Consider an initial value problem whose right-hand side is a linear function with a negative coefficient

$$\begin{pmatrix} x'(t) = -2x(t) \\ x(1) = 3 \end{pmatrix}$$

in which the rate of change of $x(t)$ depends on the current value of $x(t)$ and is in fact $-2x(t)$, so the rate of change gets smaller (more negative) as $x(t)$ gets bigger. This problem describes exponential decay and has the solution $x(t) = 3e^{-2(t-1)}$ shown in Fig. 2.6, which starts at the initial time $t = 1$. The test, again, is to insert the solution into the (differential) equations of the initial value problems and check:

$$\begin{pmatrix} (3e^{-2(t-1)})' = -6e^{-2(t-1)} = -2x(t) \\ x(1) = 3e^{-2(1-1)} = 3 \end{pmatrix}$$

Example 2.3 (A positive linear differential equation). The initial value problem

$$\begin{pmatrix} x'(t) = \frac{1}{4}x(t) \\ x(0) = 1 \end{pmatrix}$$

shown alongside different discretizations of it in Fig. 2.3 on p. 31 describes exponential growth and has the true continuous solution $x(t) = e^{\frac{t}{4}}$, which can be checked in the same way as for the previous example:

$$\begin{pmatrix} (e^{\frac{t}{4}})' = e^{\frac{t}{4}}(\frac{t}{4})' = e^{\frac{t}{4}}\frac{1}{4} = \frac{1}{4}x(t) \\ e^{\frac{0}{4}} = 1 \end{pmatrix}$$

Of course, none of the discretizations actually satisfies these equations, except at the discretization points (the multiples of the respective discretization step Δ). Since the discretizations only satisfy the equation $x'(t) = \frac{1}{4}x(t)$ at integer multiples of Δ and nowhere else, they do not agree with the actual differential equation solution $e^{\frac{t}{4}}$ anywhere other than at the initial time $t = 0$.

Example 2.4 (* Accelerated motion in a straight line). Consider the important differential equation system $x' = v, v' = a$ and the initial value problem

$$\begin{pmatrix} x'(t) = v(t) \\ v'(t) = a \\ x(0) = x_0 \\ v(0) = v_0 \end{pmatrix} \tag{2.2}$$

This differential equation states that the position $x(t)$ changes with a time-derivative equal to the respective current velocity $v(t)$, which, in turn, changes with a time-derivative equal to the acceleration a, which remains constant. The position and velocity start at the initial values x_0 and v_0. This initial value problem is a *symbolic initial value problem* with symbols x_0, v_0 as initial values (not specific numbers like 5 and 2.3). Moreover, the differential equation has a constant symbol a, and not a specific number like 0.6, in the differential equation. When concrete numbers $x_0 = 0, v_0 = 0, a = 0.5$ are chosen, the initial value problem (2.2) becomes numerical and has the vector field shown in Fig. 2.2. The initial value problem (2.2) corresponds to a vectorial differential equation with vector $y(t) := (x(t), v(t))$ of dimension $n = 2$:

$$\left(\begin{array}{l} y'(t) = \left(\begin{array}{c} x \\ v \end{array}\right)'(t) = \left(\begin{array}{c} v(t) \\ a \end{array}\right) \\ y(0) = \left(\begin{array}{c} x \\ v \end{array}\right)(0) = \left(\begin{array}{c} x_0 \\ v_0 \end{array}\right) \end{array} \right) \tag{2.3}$$

The solution of this initial value problem is

$$x(t) = \frac{a}{2}t^2 + v_0 t + x_0$$
$$v(t) = at + v_0$$

We can show that this is the solution by inserting the solution into the (differential) equations of the initial value problems and checking:

$$\left(\begin{array}{l} (\frac{a}{2}t^2 + v_0 t + x_0)' = 2\frac{a}{2}t + v_0 = v(t) \\ (at + v_0)' = a \\ x(0) = \frac{a}{2}0^2 + v_0 0 + x_0 = x_0 \\ v(0) = a0 + v_0 = v_0 \end{array} \right)$$

Example 2.5 (A two-dimensional linear differential equation for rotation).* In the important differential equation system $v' = w, w' = -v$ with initial value problem

$$\left(\begin{array}{l} v'(t) = w(t) \\ w'(t) = -v(t) \\ v(0) = 0 \\ w(0) = 1 \end{array} \right) \tag{2.4}$$

the rate of change of $v(t)$ gets bigger as $w(t)$ gets bigger but, simultaneously, the rate of change of $w(t)$ is $-v(t)$ so it gets smaller as $v(t)$ gets bigger and vice versa. This differential equation describes a rotational effect (Fig. 2.7) with solution

$$v(t) = \sin(t)$$
$$w(t) = \cos(t)$$

That this is the solution can also be checked by inserting the solution into the (differential) equations of the initial value problems and checking:

$$\left(\begin{array}{l} (\sin(t))' = \cos(t) = w(t) \\ (\cos(t))' = -\sin(t) = -v(t) \\ v(0) = \sin(0) = 0 \\ w(0) = \cos(0) = 1 \end{array} \right)$$

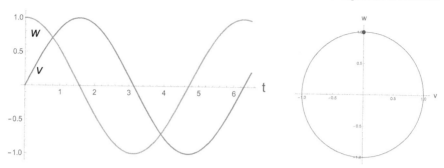

Fig. 2.7 A solution of the rotational differential equations v and w over time t **(left)** and in phase space with coordinates w over v **(right)**

Example 2.6 (A similar two dimensional linear differential equation). Consider the same differential equation system $v' = w, w' = -v$ from Example 2.5 but with different initial values than (2.4):

$$\begin{pmatrix} v'(t) = w(t) \\ w'(t) = -v(t) \\ v(0) = 1 \\ w(0) = 1 \end{pmatrix}$$

This differential equation still describes a rotational effect (Fig. 2.8), but the solution now is

$$v(t) = \cos(t) + \sin(t)$$
$$w(t) = \cos(t) - \sin(t)$$

Showing that this is the solution amounts to inserting the solution into the (differential) equations of the initial value problems and checking:

$$\begin{pmatrix} (\cos(t) + \sin(t))' = -\sin(t) + \cos(t) = w(t) \\ (\cos(t) - \sin(t))' = -\sin(t) - \cos(t) = -v(t) \\ v(0) = \cos(0) + \sin(0) = 1 \\ w(0) = \cos(0) - \sin(0) = 1 \end{pmatrix}$$

Example 2.7 (An adjustable linear differential equation for rotation).* In the important differential equation system $v' = \omega w, w' = -\omega v$ with initial value problem

$$\begin{pmatrix} v'(t) = \omega w(t) \\ w'(t) = -\omega v(t) \\ v(0) = 0 \\ w(0) = 1 \end{pmatrix} \qquad (2.5)$$

the rate of change of $v(t)$ gets bigger as $w(t)$ gets bigger but, simultaneously, the rate of change of $w(t)$ is $-v(t)$ so it gets smaller as $v(t)$ gets bigger and vice versa.

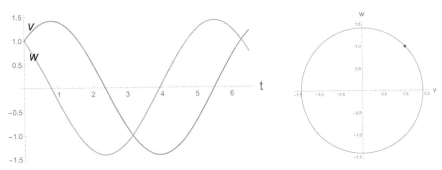

Fig. 2.8 Another solution of the rotational differential equations v and w over time t with initial values $v(0) = w(0) = 1$ (**left**) and in phase space with coordinates w over v (**right**)

But in all places, the rate of change is multiplied by a constant parameter ω, which represents the angular velocity. Bigger magnitudes of ω give faster rotations and positive ω gives clockwise rotations. This differential equation describes a rotational effect (Fig. 2.9) with solution

$$v(t) = \sin(\omega t)$$
$$w(t) = \cos(\omega t)$$

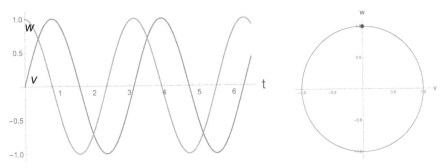

Fig. 2.9 A faster solution of the rotational differential equations v and w over time t with initial values $v(0) = 0, w(0) = 1$ and $\omega = 2$ (**left**) and in phase space with coordinates w over v (**right**)

Some differential equations mention the time variable t, which means that the required time-derivatives change over time.

Example 2.8 (Time square oscillator). Consider the following differential equation system $x'(t) = t^2 y, y'(t) = -t^2 x$, which explicitly mentions the time variable t, and the initial value problem

$$\begin{pmatrix} x'(t) = t^2 y \\ y'(t) = -t^2 x \\ x(0) = 0 \\ y(0) = 1 \end{pmatrix} \tag{2.6}$$

The solution shown in Fig. 2.10(left) illustrates that the system stays bounded but oscillates increasingly quickly. In this case, the solution is

$$\begin{pmatrix} x(t) = \sin\left(\frac{t^3}{3}\right) \\ y(t) = \cos\left(\frac{t^3}{3}\right) \end{pmatrix} \tag{2.7}$$

Note that there is no need to mention time variable t itself in the differential equa-

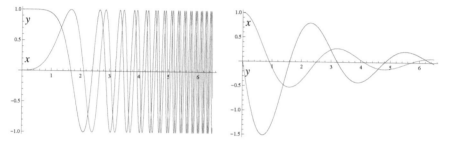

Fig. 2.10 A solution of the time square oscillator (**left**) and of the damped oscillator (**right**) up to time 6.5

tion. We could just as well have added an extra clock variable s with differential equation $s' = 1$ and initial value $s(0) = 0$ to serve as a proxy for time t. This leads to a system equivalent to (2.6) without explicit dependency on the time variable t, since the proxy s behaves in the same way as time t but is an ordinary state variable:

$$\begin{pmatrix} x'(t) = s^2 y \\ y'(t) = -s^2 x \\ s'(t) = 1 \\ x(0) = 0 \\ y(0) = 1 \\ s(0) = 0 \end{pmatrix}$$

This transformation to use s instead of t makes the differential equation *autonomous* because its right-hand side does not depend on the actual time variable t.

In this example, the solution oscillates increasingly quickly. The following example has no time-dependence and a constant frequency but the amplitude of the oscillation decreases over time.

Example 2.9 (Damped oscillator). Consider the linear differential equation $x' = y, y' = -4x - 0.8y$ and the initial value problem

$$\begin{pmatrix} x'(t) = y \\ y'(t) = -4x - 0.8y \\ x(0) = 1 \\ y(0) = 0 \end{pmatrix} \tag{2.8}$$

The solution shown in Fig. 2.10(right) illustrates that the dynamical system decays over time. In this case, the explicit global solution representing the dynamical system is more difficult to write down explicitly but a function that solves it still exists.

> **Note 5 (Descriptive power of differential equations)** As a very general phenomenon, observe that solutions of differential equations can be much more involved than the differential equations themselves, which is part of the representational and descriptive power of differential equations. Pretty simple differential equations can describe quite complicated physical processes.

2.5 Domains of Differential Equations

Now we understand precisely what a differential equation is and how it describes a continuous physical process. In CPS, however, physical processes are not running in isolation, but interact with cyber elements such as computers or embedded systems. When and how do physics and cyber elements interact?

The first thing we need to understand for that is how to describe when physics stops so that the cyber elements take control of what happens next. Obviously, physics does not literally stop evolving, but rather keeps on evolving all the time. Yet, the cyber parts only take effect every now and then, because they only provide input into physics by way of changing the actuators every once in a while. So, our intuition may imagine physics "pauses" for a period of duration 0 and lets the cyber take action to influence the inputs on which physics is based. In fact, cyber may interact with physics over a period of time or after computing for some time to reach a decision. But the phenomenon is still the same. At some point, cyber is done sensing and deliberating and deems it time to act (if cyber never acts, it's boring and will be discarded). At this moment of time, physics needs to "pause" for a conceptual period of time of imaginary duration 0 to give cyber a chance to act.

The cyber and the physics can interface in more than one way. Physics might evolve and the cyber elements interrupt physics periodically to make measurements of the current state of the system in order to decide what to do next (Chap. 9). Or the physics might trigger certain conditions or events that cause cyber elements to compute their respective responses to these events (Chap. 8). Another way to look at this is that a differential equation that a system follows forever without further intervention by anything would not describe a particularly well-controlled system. If physics on its own were already to drive cars safely, we would not need any cyber or any control in the first place. But since physics has not quite passed the driver's license test yet, proper control is rather on the crucial side.

All those ways have in common that our model of physics not only needs to explain how the state changes over time with a differential equation, but it also needs to specify when physics stops evolving to give cyber a chance to perform its task. This information is what is called an *evolution domain Q* of a differential equation, which describes a region that the system cannot leave while following that particular continuous mode of the system. If the system were ever about to leave this region, it would stop evolving right away (for the purpose of giving the cyber parts of the system a chance to act) before it leaves the evolution domain. Of course, the overall idea will be for physics to resume once cyber is done inspecting and acting, but that is a matter for Chap. 3.

Note 6 (Evolution domain constraint) A *differential equation $x' = f(x)$ with evolution domain Q* is denoted by

$$x' = f(x) \,\&\, Q$$

using a conjunctive notation (&) between the differential equation and its evolution domain. This notation $x' = f(x) \,\&\, Q$ signifies that the system obeys *both* the differential equation $x' = f(x)$ *and* the evolution domain Q. The system follows this differential equation for any duration while inside the region Q, but is never allowed to leave the region described by Q. So the system evolution has to stop while the state is still in Q.

If, e.g., t is a time variable with $t' = 1$, then $x' = v, v' = a, t' = 1 \,\&\, t \leq \varepsilon$ describes a system that follows the differential equation at most until time $t = \varepsilon$ and not any further, because the evolution domain $Q \stackrel{\text{def}}{\equiv} (t \leq \varepsilon)$ would be violated after time ε. That can be a useful model of the kind of physics that gives the cyber elements a chance to act at the latest at time ε, because physics is not allowed to continue beyond time $t = \varepsilon$. The evolution domain $Q \stackrel{\text{def}}{\equiv} (v \geq 0)$, instead, restricts the system $x' = v, v' = a \,\&\, v \geq 0$ to nonnegative velocities. Should the velocity ever be about to become negative while following the differential equation $x' = v, v' = a$, then the system stops evolving before the velocity becomes negative. In a similar way $x' = v, v' = a \,\&\, v \leq 10$ describes a physics that has its velocity limited by an upper bound of 10. But, honestly, keeping the velocity of a car below 10 will also end up requiring some effort on the part of the cyber controller, not just the physics, which is a phenomenon that we will come back to in Chaps. 8 and 9.

In the first two scenarios illustrated in Fig. 2.11, the system starts at time 0 inside the evolution domain Q, which is depicted as a shaded green region in Fig. 2.11. Then the system follows the differential equation $x' = f(x)$ for any period of time, but has to stop before it leaves Q. Here, it stops at different choices for the stopping time r.

In contrast, consider the scenario shown on the right of Fig. 2.11. The system stops at time r and is *not* allowed to evolve until time s, because—even if the system would be back in the evolution domain Q at that time—it has already left the evolution domain Q between time r and s (indicated by dotted lines), which is not

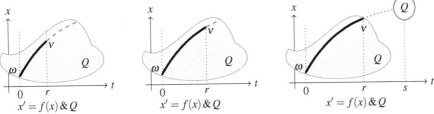

Fig. 2.11 System $x' = f(x) \& Q$ follows the differential equation $x' = f(x)$ for any duration r but cannot leave the (green) evolution domain Q

allowed. Physics $x' = f(x) \& Q$ cannot sneak out of its respective evolution domain Q hoping that we will not notice, not even temporarily. Consequently, the continuous evolution on the right of Fig. 2.11 will also stop at time r at the latest and cannot continue any further.

Now that we know what the evolution domain constraint Q of a differential equation is supposed to do, the question is how we can properly describe it in a CPS model? We will need some logic for that. For one thing, we should start getting precise about how to describe the evolution domain Q for a differential equation, just as we have become precise about the understanding of a differential equation itself. The most critical bit of an evolution domain is the question of which points satisfy Q and which ones doesn't, which is what logic is good at making precise.

2.6 Syntax of Continuous Programs

After these preparations for understanding differential equations and domains, we start developing a precise mathematical model. The differential equations with their evolution domains will ultimately need a way of interfacing with discrete computer control programs, because hybrid system models of cyber-physical systems combine discrete dynamics and continuous dynamics. The conceptually easiest and most compositional way to make that happen is to integrate continuous dynamics seamlessly into the computer control programs. This book develops the programming language of *hybrid programs*, which contain discrete features in addition to the differential equations. Hybrid programs and their analysis is developed in layers one after another. For now, we focus on the first layer of this programming language, which contains only the most crucial feature of continuous dynamics.

2.6.1 Continuous Programs

The first element of the syntax of hybrid programs is purely continuous programs.

Note 7 (Continuous programs) Layer 1 of *hybrid programs* (HPs) comprises *continuous programs*. If x is a variable, e any term possibly containing x, and Q a formula of first-order logic of real arithmetic, then continuous programs are of the form

$$\alpha ::= x' = e \,\&\, Q$$

Continuous programs consist of a single statement of the form $x' = e \,\&\, Q$. In later chapters, we will add more statements to form hybrid programs, but we just focus on differential equations for the continuous dynamics for now. The *continuous evolution* $x' = e \,\&\, Q$ expresses that, from the present value of variable x, the system follows the differential equation $x' = e$ for some amount of time within the *evolution domain constraint* Q. What form formula Q can take will be defined below. But it has to enable an unambiguous definition of the set of points that satisfy Q, because the continuous evolution is not allowed to ever leave that region. Further x is a variable, but is also allowed to be a vector of variables and, then, e is a vector of terms of the same dimension. This corresponds to the case of differential equation systems such as

$$x' = v, v' = a \,\&\, (v \geq 0 \wedge v \leq 10)$$

Differential equations are allowed without an evolution domain constraint Q as well:

$$x' = y, y' = x + y^2$$

which corresponds to choosing *true* for Q, since the formula *true* is true everywhere and, thus, actually imposes no condition on the state whatsoever, because every state easily satisfies the formula *true* with flying colors. Of course, we will have to be more precise about what terms e are allowed and what formulas Q are allowed, which is what we will pursue next.

2.6.2 Terms

A rigorous definition of the syntax of hybrid programs also depends on defining what a term e and what a formula Q of first-order logic of real arithmetic are. Terms e occur on the right-hand side of differential equations, and formulas Q are their evolution domains.

Definition 2.2 (Terms). A *term* e is a polynomial term defined by the grammar (where $e, \tilde{e}$ are terms, x is a variable, and c is a rational number constant)

$$e, \tilde{e} ::= x \mid c \mid e + \tilde{e} \mid e \cdot \tilde{e}$$

This grammar[2] means that a term e (or a term $\tilde{e}$) is either a variable x, or a rational number constant $c \in \mathbb{Q}$ such as 0 or 1 or $5/7$, or a sum of terms $e, \tilde{e}$, or it is

---

[2] From a formal languages perspective, it would be fine to use the equivalent grammar

a product of terms $e, \tilde{e}$, which are again built in this way recursively. The cases of variables x or constants c are called *atomic terms*. The other cases take two terms as input to produce more complex terms and are called *compound terms*. The addition term $e + \tilde{e}$, for example, consists of two terms e and $\tilde{e}$. So does the multiplication term $e \cdot \tilde{e}$. Subtraction $e - \tilde{e}$ is another useful case, but it turns out that it is already included, because the subtraction term $e - \tilde{e}$ is already definable by the term $e + (-1) \cdot \tilde{e}$. That is why we will not worry about subtraction in developing the theory, but use it in our examples regardless. Unary negation $-e$ is helpful, too, but also already included as $0 - e$. For example, $4 + x \cdot 2$ and $x \cdot 2 + y \cdot y$ are terms and $4 \cdot x - y \cdot y + 1$ will also be considered as a term even if it really should have been written as $((4 \cdot x) + (((-1) \cdot y) \cdot y)) + 1$. Definition 2.2 yields all polynomials. The polynomial $x^3 + 5x^2 - x + 4$ can, e.g., be represented by the term $x \cdot x \cdot x + 5 \cdot x \cdot x + (-1) \cdot x + 4$.

If you were to implement the syntax of terms in a computer program, you could implement the four cases of the syntax of terms in Definition 2.2 as constructors of a data type. An atomic term can either be constructed by providing a variable x or a computer representation of a rational number $c \in \mathbb{Q}$. Compound terms can be constructed with the sum constructor $e + \tilde{e}$ when providing two previously constructed terms $e, \tilde{e}$. Or they can be constructed with the product constructor $e \cdot \tilde{e}$. That way, every concrete term such as $x \cdot 2 + y \cdot y$ can be represented in the data structure by calling the respective constructors with the appropriate arguments.

2.6.3 First-Order Formulas

The formulas of first-order logic of real arithmetic are defined as usual in first-order logic, except that, being the logic for real arithmetic, it also uses the specific language of real arithmetic, for example $e \geq \tilde{e}$ for greater-or-equal. First-order logic supports the logical connectives not ($\neg$), and ($\wedge$), or ($\vee$), implies ($\rightarrow$), and bi-implication alias equivalence ($\leftrightarrow$), as well as quantifiers for all ($\forall$) and exists ($\exists$). In the first-order logic of *real* arithmetic, $\forall, \exists$ quantify over the reals $\mathbb{R}$.

Definition 2.3 (Formulas of first-order logic of real arithmetic). The *formulas* of *first-order logic of real arithmetic* are defined by the following grammar (where P, Q are formulas of first-order logic of real arithmetic, $e, \tilde{e}$ are terms, and x is a variable):

$$P, Q ::= e \geq \tilde{e} \mid e = \tilde{e} \mid \neg P \mid P \wedge Q \mid P \vee Q \mid P \rightarrow Q \mid P \leftrightarrow Q \mid \forall x P \mid \exists x P$$

$$e ::= x \mid c \mid e + e \mid e \cdot e$$

with a single nonterminal e. We use the slightly more verbose form with two redundant nonterminals $e, \tilde{e}$ just to emphasize directly that a term can be a sum $e + \tilde{e}$ of any arbitrary and possibly different terms $e, \tilde{e}$ and does not have to consist of sums $e + e$ of one and the same term e. The two presentations of the grammar are equivalent.

The usual abbreviations are allowed, such as $e \leq \tilde{e}$ for $\tilde{e} \geq e$ and $e < \tilde{e}$ for $\neg(e \geq \tilde{e})$ and will be used in examples even if the theory does not need to be bothered with them.

Example formulas we saw before as evolution domains are $t \leq \varepsilon$, which we mean as a bound on time t, as well as $v \geq 0$, which we used as a velocity bound. But these first-order logic formulas themselves do not tell us that t is a time and v a velocity, which is the rôle of the differential equations. However, the formula $v \geq 0$ quite precisely instructs us where it is true (which is what it semantics will define rigorously). Whenever the value of velocity v is greater-or-equal 0, the formula $v \geq 0$ will be *true*, otherwise *false*. The formula $t \leq \varepsilon \wedge v \geq 0$, which is the conjunction of the formulas $t \leq \varepsilon$ and $v \geq 0$, is used as a domain for systems that evolve at most till time ε and also cannot move backwards. An implication $t \geq 2 \rightarrow v \leq 5$ may be used as an evolution domain to say that if the system is beyond time 2, then the velocity is at most 5.

The quantifiers for all ($\forall$) and exists ($\exists$) over the reals are less relevant for evolution domains and become more important in later chapters of the book. They are already shown here because they are a characteristic part of first-order logic. For example, $\forall x \exists y\,(y > x)$ expresses that for all real numbers x there is a real number y that is larger than x. The formula $\exists x\,(x \cdot x = 2)$ expresses that there is a real number whose square is 2, which is true thanks to the real number $\sqrt{2}$. But the formula $\exists x\,(x \cdot x = -1)$ is not true over the reals, because the squares of all real numbers are nonnegative and the imaginary unit i, which satisfies $\mathrm{i}^2 = -1$, is not a real number. The formula $\exists y\,(x < y \wedge y < x+1)$ is also true, no matter what the value of x is, but would be false over the integers, which suffer from a clear lack of numbers between x and $x+1$.

Expedition 2.1 (Naming conventions)

In this book, we generally follow these naming conventions (programs, function, and predicate symbols will be introduced in later chapters):

Letters Convention

Letters	Convention
x, y, z	variables
$e, \tilde{e}$	terms
P, Q	formulas
α, β	programs
c	constant symbols
f, g, h	function symbols
p, q, r	predicate symbols

In any application context, it may be better to deviate from this convention and follow the naming conventions of the application. For example, x is often used for position, v for velocity, and a for acceleration, even if all are variables.

2.7 Semantics of Continuous Programs

This is the first of many occasions in this textbook where we observe a distinct dichotomy between syntax and semantics. The syntax defines the notation, i.e., what questions can be written down and how. But to the innocent bystander and any self-respecting logician, the syntax just provides a long list of funny, arbitrary symbols, until their intended meaning has been clarified. The syntax is only given meaning by the semantics, which defines what real or mathematical object each element of the syntax stands for. The symbols that the syntax uses are arbitrary and completely meaningless until the semantics defines what the symbols mean. Of course, the syntax is cleverly chosen to already remind us of what it is supposed to stand for.

> **Note 8 (Syntax versus semantics)** Syntax just defines arbitrary notation. Its meaning is defined by the semantics.

It is by way of this clear distinction of syntactic and semantic objects that we will ultimately develop a nuanced and accurate understanding of the relationships between mathematical meaning and computer-based insights. Without such a clear distinction, there can be no subsequent alignment and relation. Numerous mistakes in reasoning can be traced back to the lack of a clear separation of the syntactic object-level and semantic meta-level elements of a development. Object-level expressions are expressions in the language (for example first-order logic). Meta-level statements are statements about the language, phrased, for example, in mathematics or English.

2.7.1 Terms

The meaning of a continuous evolution $x' = e \,\&\, Q$ depends on understanding the meaning of term e. A term e is a syntactic expression. The meaning of a term e is given by the real number to which it evaluates. In a term, e, the symbol $+$, of course, means addition of real numbers, $\cdot$ means multiplication, and the meaning of the constant symbol $5/7$ is the rational number five sevenths.

But the overall value of term e also depends on how we interpret the variables appearing in the term e. What values those variables have changes depending on the current state of the CPS. A state needs to let us know what real values all variables of e have, before we are able to say what real value term e evaluates to in that state. In fact, a *state* ω *is* nothing but a mapping from variables to real numbers, such that it associates a real value $\omega(x)$ with each variable x. The *set of states* is denoted $\mathscr{S}$. In other words, if $\mathscr{V}$ is the set of all variables, then a state $\omega \in \mathscr{S}$ is a function $\omega : \mathscr{V} \to \mathbb{R}$ whose value $\omega(x) \in \mathbb{R}$ at $x \in \mathscr{V}$ indicates the real value that variable x has in state ω.

Since the value of a term depends on the state, we will use the notation $\omega[\![e]\!]$ for the real value that the term e evaluates to in state ω. This notation is reminiscent of function application $\omega(e)$, but when a state is a function $\omega : \mathcal{V} \to \mathbb{R}$, then $\omega(e)$ is only defined if e is a variable, not if it is a term $x+7$. The notation $\omega[\![e]\!]$, thus, lifts to terms e the value that the state ω assigns only to variables $x \in \mathcal{V}$.

Definition 2.4 (Semantics of terms). The *value of term* e in state $\omega \in \mathcal{S}$ is a real number denoted $\omega[\![e]\!]$ and is defined by induction on the structure of e:

$$\omega[\![x]\!] = \omega(x) \qquad\qquad \text{if } x \text{ is a variable}$$
$$\omega[\![c]\!] = c \qquad\qquad \text{if } c \in \mathbb{Q} \text{ is a rational constant}$$
$$\omega[\![e+\tilde{e}]\!] = \omega[\![e]\!] + \omega[\![\tilde{e}]\!]$$
$$\omega[\![e \cdot \tilde{e}]\!] = \omega[\![e]\!] \cdot \omega[\![\tilde{e}]\!]$$

That is, the value of a variable x in state ω is given directly by the state ω, which is a mapping from variables to real numbers. A rational constant c such as 0.5 evaluates to itself. The value of a sum term of the form $e + \tilde{e}$ in a state ω is the sum of the values of the subterms e and $\tilde{e}$ in ω, respectively. Those lines of Definition 2.4 already explain that $\omega[\![x+y]\!] = \omega(x) + \omega(y)$. Likewise, the value of a product term of the form $e \cdot \tilde{e}$ in a state ω is the product of the values of the subterms e and $\tilde{e}$ in ω, respectively. Each term has a value in every state, because each case of the syntactic form of terms (Definition 2.2) has been given a semantics in Definition 2.4. The semantics of every term, thus, is a mapping from states $\omega \in \mathcal{S}$ to the real value $\omega[\![e]\!]$ that the term e evaluates to in the respective state ω.

The value of a variable-free term like $4 + 5 \cdot 2$ does not depend on the state ω at all. In this case, the value is 14. The value of a term with variables, like $4 + x \cdot 2$, depends on what value the variable x has in state ω. Suppose $\omega(x) = 5$, then the value is also $\omega[\![4 + x \cdot 2]\!] = 4 + \omega(x) \cdot 2 = 14$. However, for $\nu(x) = 2$, it evaluates to $\nu[\![4 + x \cdot 2]\!] = 4 + \nu(x) \cdot 2 = 8$, instead. While, technically, the state ω is a mapping from all variables to real numbers, the values that ω gives to most variables are immaterial; only the values of the variables that actually occur in the term have any influence (Sect. 5.6.5). So while the value of $4 + x \cdot 2$ very much depends on the value of x, it does not depend on the value that variable y has since y does not even occur in the term $4 + x \cdot 2$. This is in contrast to the term $x \cdot 2 + y \cdot y$ whose value depends on the values of both x and y but not on z, so for $\omega(x) = 5$ and $\omega(y) = 4$, it evaluates to $\omega[\![x \cdot 2 + y \cdot y]\!] = \omega(x) \cdot 2 + \omega(y) \cdot \omega(y) = 26$.

The way that Definition 2.4 defines a semantics for terms directly corresponds to the definition of a recursive function in a functional programming language by distinguishing the respective constructors of the data types for terms. For each constructor, there is a corresponding case that defines its value in the argument state. And if that function makes a recursive call, as in the cases of $\omega[\![e + \tilde{e}]\!]$ and $\omega[\![e \cdot \tilde{e}]\!]$, it does so on terms that are smaller to make sure the function terminates and is well defined on all inputs.

Expedition 2.2 (Semantic brackets $[\![\cdot]\!] : \text{Trm} \rightarrow (\mathscr{S} \rightarrow \mathbb{R})$)

There are multiple equivalent ways of understanding Definition 2.4. The most elementary understanding is, as written, to understand it as defining the real value $\omega[\![e]\!]$ in a state ω for each term e by an inductive definition on the structure of e. If e is a variable, the first line is applicable, if e is a rational constant symbol the second line. If e is a sum term, then the third line, and the fourth line is applicable if e is a product term. Since every term fits to exactly one of those four shapes and the right-hand sides use the definition on smaller subterms of e that have already received a value by this definition, the definition is well defined.

More eloquently, Definition 2.4 can be read as defining an operator $[\![\cdot]\!]$ that defines the semantics of a term e as a mapping $[\![e]\!] : \mathscr{S} \rightarrow \mathbb{R}$ from states to real numbers such that the real value $\omega[\![e]\!]$ is computed according to the equations in Definition 2.4. That is, the function $[\![e]\!]$ is defined by induction on the structure of e:

$$[\![x]\!] : \mathscr{S} \rightarrow \mathbb{R}; \; \omega \mapsto \omega(x) \qquad \text{if } x \text{ is a variable}$$

$$[\![c]\!] : \mathscr{S} \rightarrow \mathbb{R}; \; \omega \mapsto c \qquad \text{if } c \in \mathbb{Q} \text{ is a rational constant}$$

$$[\![e + \tilde{e}]\!] : \mathscr{S} \rightarrow \mathbb{R}; \; \omega \mapsto \omega[\![e]\!] + \omega[\![\tilde{e}]\!]$$

$$[\![e \cdot \tilde{e}]\!] : \mathscr{S} \rightarrow \mathbb{R}; \; \omega \mapsto \omega[\![e]\!] \cdot \omega[\![\tilde{e}]\!]$$

The notation for evaluating $[\![e]\!]$ at state ω is still $\omega[\![e]\!]$. For example, the last line defines the function $[\![e \cdot \tilde{e}]\!]$ as the function $[\![e \cdot \tilde{e}]\!] : \mathscr{S} \rightarrow \mathbb{R}$ that maps state ω to the real value given by the product $\omega[\![e]\!] \cdot \omega[\![\tilde{e}]\!]$ of the values $\omega[\![e]\!]$ and $\omega[\![\tilde{e}]\!]$.

The two ways of understanding Definition 2.4 are equivalent. The former is more elementary. The latter generalizes more directly to defining a semantics for other syntactic objects when choosing a different semantic domain than $\mathscr{S} \rightarrow \mathbb{R}$. When Trm is the set of terms, the semantic brackets for terms define an operator $[\![\cdot]\!] : \text{Trm} \rightarrow (\mathscr{S} \rightarrow \mathbb{R})$ that defines the meaning $[\![e]\!]$ for each term $e \in \text{Trm}$, which, in turn, defines the real value $\omega[\![e]\!] \in \mathbb{R}$ for each state $\omega \in \mathscr{S}$. In a functional programming language, the difference between the two styles of definition of the semantics of terms is exactly currying, i.e., translating a function that takes multiple arguments into a sequence of functions each of which only takes a single argument.

2.7.2 First-Order Formulas

Unlike for terms, the value of a logical formula is not a real number but instead *true* or *false*. Whether a logical formula evaluates to *true* or *false* still depends on the interpretation of its symbols. In first-order logic of real arithmetic, the meaning of all symbols except the variables is fixed. The meaning of terms and of formulas of first-order logic of real arithmetic is as usual in first-order logic, except that $+$

really means addition, · means multiplication, $\geq$ means greater or equals, and the quantifiers $\forall x$ and $\exists x$ quantify over the reals. As always in first-order logic, the meaning of $\wedge$ is conjunction and that of $\vee$ is disjunction, etc. The meaning of the variables in the formula is again determined by the state ω of the CPS.

In direct analogy to the real-valued semantics $\omega[\![e]\!] \in \mathbb{R}$ of terms e, we might define a boolean-valued semantics $\omega[\![P]\!] \in \{true, false\}$ for formulas P that defines what truth-value (*true* or *false*) the formula P has in state ω (Exercise 2.10). However, our interest is ultimately in understanding which formulas are true, because the complement then also already tells us which formulas are false. That is why it simplifies matters if we define the semantics via the set of states $[\![P]\!]$ in which formula P is true. Then $\omega \in [\![P]\!]$ will say that formula P is true in state ω. The opposite $\omega \notin [\![P]\!]$ says that formula P is not true so false in state ω. For consistency with other books, this chapter uses the *satisfaction relation* notation $\omega \models P$ instead of $\omega \in [\![P]\!]$, but they mean the same thing.

> **Definition 2.5 (First-order logic semantics).** The first-order formula P is true in state ω, is written $\omega \models P$, and is defined inductively as follows:
>
> - $\omega \models e = \tilde{e}$ iff $\omega[\![e]\!] = \omega[\![\tilde{e}]\!]$
>   That is, an equation $e = \tilde{e}$ is true in a state ω iff the terms e and $\tilde{e}$ evaluate to the same number in ω according to Definition 2.4.
> - $\omega \models e \geq \tilde{e}$ iff $\omega[\![e]\!] \geq \omega[\![\tilde{e}]\!]$
>   That is, a greater-or-equals inequality is true in a state ω iff the term on the left evaluates to a number that is greater than or equal to the value of the right term.
> - $\omega \models \neg P$ iff $\omega \not\models P$, i.e., if it is not the case that $\omega \models P$
>   That is, a negated formula $\neg P$ is true in state ω iff the formula P itself is not true in ω.
> - $\omega \models P \wedge Q$ iff $\omega \models P$ and $\omega \models Q$
>   That is, a conjunction is true in a state iff both conjuncts are true in said state.
> - $\omega \models P \vee Q$ iff $\omega \models P$ or $\omega \models Q$
>   That is, a disjunction is true in a state iff either of its disjuncts is true in said state.
> - $\omega \models P \rightarrow Q$ iff $\omega \not\models P$ or $\omega \models Q$
>   That is, an implication is true in a state iff either its left-hand side is false or its right-hand side is true in said state.
> - $\omega \models P \leftrightarrow Q$ iff $(\omega \models P$ and $\omega \models Q)$ or $(\omega \not\models P$ and $\omega \not\models Q)$
>   That is, a bi-implication is true in a state iff both sides are true or both sides are false in said state.
> - $\omega \models \forall x P$ iff $\omega_x^d \models P$ for all $d \in \mathbb{R}$
>   That is, a universally quantified formula $\forall x P$ is true in a state iff its *kernel* P is true in all variations of the state, no matter what real number d the quantified variable x evaluates to in the variation ω_x^d defined below.

- $\omega \models \exists x P$ iff $\omega_x^d \models P$ for some $d \in \mathbb{R}$

That is, an existentially quantified formula $\exists x P$ is true in a state iff its kernel P is true in some variation of the state, for a suitable real number d that the quantified variable x evaluates to in the variation ω_x^d.

If $\omega \models P$, then we say that P is true at ω or that ω is a model of P. Otherwise, i.e., if $\omega \not\models Q$, we say that P is false at ω. A formula P is *valid*, written $\models P$, iff $\omega \models P$ for all states ω. Formula P is called *satisfiable* iff there is a state ω such that $\omega \models P$. Formula P is *unsatisfiable* iff there is no such ω. The set of states in which formula P is true is written $[\![P]\!] = \{\omega : \omega \models P\}$.

The definition of the semantics of quantifiers uses *state modifications*, i.e., ways of changing a given state ω around by changing the value of a variable x but leaving the values of all other variables alone. The notation $\omega_x^d \in \mathscr{S}$ denotes the state that agrees with state $\omega \in \mathscr{S}$ except for the interpretation of variable x, which is changed to the value $d \in \mathbb{R}$. That is, the state ω_x^d has the same values that the state ω has for all variables other than the variable x, and the value of the variable x in ω_x^d is d:

$$\omega_x^d(y) \stackrel{\text{def}}{=} \begin{cases} d & \text{if } y \text{ is the modified variable } x \\ \omega(y) & \text{if } y \text{ is another variable} \end{cases} \tag{2.9}$$

The formula $x > 0 \wedge x < 1$ is satisfiable, because all it takes for it to be true is a state ω in which, indeed, the value of x is a real number between zero and one, such as 0.592. The formula $x > 0 \wedge x < 0$ is unsatisfiable, because it is kind of hard (read: impossible) to find a state which satisfies both conjuncts at once. The formula $x > 0 \vee x < 1$ is valid, because there is no state in which it would not be true, because, surely, x will either be positive or smaller than one.

In the grand scheme of things, the most exciting formulas are the ones that are valid, i.e., $\models P$, because that means they are true no matter what state a system is in. Valid formulas, and how to find out whether a formula is valid, will keep us busy quite a while in this textbook. Consequences of a formula set Γ are also amazing, because, even if they may not be valid per se, they are true whenever Γ is. For this chapter, however, it is more important which formulas are true in a given state, because that is what we need to make sense of an evolution domain of a continuous evolution, which would be futile if it were true in all states.

For example, the formula $\exists y (y > x)$ is valid, so $\models \exists y (y > x)$, because it is true, i.e., $\omega \in [\![\exists y (y > x)]\!]$, in all states ω, as there always is a real number y that is a little larger than the value of x, whatever real value x might have in ω. The formula $t \leq \varepsilon$ is not valid. Its truth-value depends on the value of its variables t and ε. In a state ω with $\omega(t) = 0.5$ and $\omega(\varepsilon) = 1$, the formula is true, so $\omega \in [\![t \leq \varepsilon]\!]$. But in a state v with $v(t) = 0.5$ and $v(\varepsilon) = 0.1$, the formula is false, so $v \notin [\![t \leq \varepsilon]\!]$. In a state ω with $\omega(t) = 0.5$ and $\omega(\varepsilon) = 1$ and $\omega(v) = 5$, even the formula $t \leq \varepsilon \wedge v \geq 0$ is true, so $\omega \in [\![t \leq \varepsilon \wedge v \geq 0]\!]$, because both $\omega \in [\![t \leq \varepsilon]\!]$ and $\omega \in [\![v \geq 0]\!]$.

As will be elaborated in Sect. 5.6.5, the only relevant information from the state is the values of the free variables, i.e., those that occur outside the scope of quantifiers for that variables. For example the truth-value of $\exists y\, (y^2 \leq x)$ depends on the value that its free variable x has in the state, but does not depend on the value of variable y, because the existential quantifier $\exists y$ will give y a new value anyhow. For example, $\omega \in [\![\exists y\, (y^2 \leq x)]\!]$ for a state ω with $\omega(x) = 5$, regardless of what $\omega(y)$ is and regardless of $\omega(z)$, because z does not occur at all and y only occurs in the scope of a quantifier. But $\nu \notin [\![\exists y\, (y^2 \leq x)]\!]$ in state ν with $\nu(x) = -1$, because it is impossible to find a real number whose square is less than or equal to -1. Come to think of it, $x \geq 0$ would have been an easier way to state $\exists y\, (y^2 \leq x)$, because the two formulas are equivalent, i.e., have the same truth-value in every state. All states define real values for all variables, because states are (total) functions from the variables to the reals. But the only relevant values are the values of the free variables of the formula.

A formula P is valid iff the formula $\forall x P$ is valid, because validity means truth in all states, which includes all real values for all variables, in particular the variable x. Likewise, P is satisfiable iff the formula $\exists x P$ is satisfiable, because satisfiability means truth in some state, which provides a real value for all variables, even variable x. In a similar way, we could prefix universal quantifiers explicitly for all free variables when asking for validity, because that implicitly quantifies over all real values of all variables. We could also prefix existential quantifiers explicitly for the free variables when asking for satisfiability.

Of course, which quantifier we implicitly mean for a formula with free variables such as $\forall y\, (y^2 > x)$ depends. If we ask whether $\forall y\, (y^2 > x)$ is true in the state ω with $\omega(x) = -1$, there is no implicit quantifier, because we ask about that specific state, and the answer is yes, so $\omega \in [\![\forall y\, (y^2 > x)]\!]$. When asking whether $\forall y\, (y^2 > x)$ is true in the state ν with $\nu(x) = 0$, then the answer is no, since $\nu \notin [\![\forall y\, (y^2 > x)]\!]$. If we ask whether $\forall y\, (y^2 > x)$ is valid, we do not provide a specific state, because validity requires truth in all states, so implicitly quantifies all free variables universally. The answer is no, because $\forall y\, (y^2 > x)$ is not valid as witnessed by the above state ν. The variation $\forall y\, (y^2 \geq -x^2)$ is valid, written $\vDash \forall y\, (y^2 \geq -x^2)$. If we ask whether $\forall y\, (y^2 > x)$ is satisfiable, we do not provide a specific state either, because the question is whether there is a state ω with $\omega \in [\![\forall y\, (y^2 > x)]\!]$, so the free variables are implicitly quantified existentially. The answer is yes as witnessed by the above state ω.

With the semantics, we now know how to evaluate whether an evolution domain Q of a continuous evolution $x' = e\,\&\,Q$ is true in a particular state ω or not. If $\omega \in [\![Q]\!]$, then the evolution domain Q holds in that state. Otherwise (i.e., if $\omega \notin [\![Q]\!]$), Q does not hold in ω. Yet, in which states ω do we even need to check the evolution domain? We need to find some way of saying that the evolution domain constraint Q is checked for whether it is true (i.e., $\omega \in [\![Q]\!]$) in all states ω along the solution of the differential equation. That will be the next item on our agenda.

2.7.3 Continuous Programs

The semantics of continuous programs surely depends on the semantics of their pieces, which include terms and formulas. The latter have now both been defined, so the next step is giving continuous programs themselves a proper semantics.

In order to keep things simple, all we care about for now is the observation that running a continuous program $x' = e \,\&\, Q$ takes the system from an initial state ω to a new state ν. And, in fact, one crucial aspect to notice is that there is not only one state ν that $x' = e \,\&\, Q$ can reach from ω just as there is not only one solution of the differential equation $x' = e$. Even in cases where there is a unique solution of maximal duration, there are still many different solutions differing only in the duration of the solution. Thus, the continuous program $x' = e \,\&\, Q$ can lead from initial state ω to more than one possible state ν. Which states ν are reachable from an initial state ω along the continuous program $x' = e \,\&\, Q$ exactly? Well, these should be the states ν to which ω can be connected with a solution of the differential equation $x' = e$ that remains entirely within the set of states where the evolution domain constraint Q holds true, as illustrated in Fig. 2.12. Giving this a precise meaning requires going back and forth between syntax and semantics carefully.

Definition 2.6 (Semantics of continuous programs). The state ν is reachable from initial state ω by the continuous program $x'_1 = e_1, \ldots, x'_n = e_n \,\&\, Q$ iff there is a *solution* φ of some duration $r \geq 0$ along $x'_1 = e_1, \ldots, x'_n = e_n \,\&\, Q$ from state ω to state ν, i.e., a function $\varphi : [0, r] \to \mathscr{S}$ such that:

- initial and final states match: $\varphi(0) = \omega, \varphi(r) = \nu$;
- φ respects the differential equations: For each variable x_i, the value $\varphi(\zeta)[\![x_i]\!] = \varphi(\zeta)(x_i)$ of x_i at state $\varphi(\zeta)$ is continuous in ζ on $[0, r]$ and, if $r > 0$, has a time-derivative of value $\varphi(\zeta)[\![e_i]\!]$ at each time $\zeta \in [0, r]$, i.e.,

$$\frac{\mathrm{d}\varphi(t)(x_i)}{\mathrm{d}t}(\zeta) = \varphi(\zeta)[\![e_i]\!]$$

- the value of other variables $y \notin \{x_1, \ldots, x_n\}$ remains constant throughout the continuous evolution, that is $\varphi(\zeta)[\![y]\!] = \omega[\![y]\!]$ for all times $\zeta \in [0, r]$;
- and φ respects the evolution domain at all times: $\varphi(\zeta) \in [\![Q]\!]$ for each $\zeta \in [0, r]$.

Fig. 2.12 Illustration of the dynamics of continuous programs

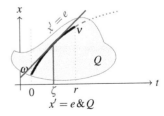

Observe that this definition is explicit about the fact that variables without differential equations do not change during a continuous program. The semantics is *explicit change*: nothing changes unless a program statement specifies how. Further observe the explicit passing from syntax to semantics[3] by the use of the semantics function $[\![\cdot]\!]$ in Definition 2.6. Finally note that for duration $r = 0$, no condition is imposed on the time-derivative, because there are no time-derivatives for a function that is only defined at 0. Consequently, the only conditions that Definition 2.6 imposes for duration 0 are that the initial state ω and final state ν agree and that the evolution domain constraint Q is respected at that state: $\omega \in [\![Q]\!]$. Later chapters will have a slightly refined understanding, but Definition 2.6 is sufficient for Part I.

2.8 Summary

This chapter gave a precise semantics to differential equations and presented first-order logic of real arithmetic, which we use for the evolution domain constraints within which differential equations are supposed to stay. The operators in first-order logic of real arithmetic and their informal meaning is summarized in Table 2.1.

While this chapter provided an important continuous foundation, all its elements will be revisited in more detail in subsequent chapters. The semantics of continuous programs will be revisited in their interaction with discrete programs in Chap. 3. First-order logic will be substantially generalized in more detail in Chap. 4. In fact, even the set of terms will be extended in Part II but will be with us throughout the book.

Table 2.1 Operators and meaning in first-order logic of real arithmetic (FOL)

FOL	Operator	Meaning
$e = \tilde{e}$	equals	true iff values of e and $\tilde{e}$ are equal
$e \geq \tilde{e}$	greater or equals	true iff value of e greater-or-equal to $\tilde{e}$
$\neg P$	negation / not	true iff P is false
$P \wedge Q$	conjunction / and	true iff both P and Q are true
$P \vee Q$	disjunction / or	true iff P is true or if Q is true
$P \rightarrow Q$	implication / implies	true iff P is false or Q is true
$P \leftrightarrow Q$	bi-implication / equivalent	true iff P and Q are both true or both false
$\forall x P$	universal quantifier / for all	true iff P is true for all values of real variable x
$\exists x P$	existential quantifier / exists	true iff P is true for some values of real variable x

---

[3] This important aspect is often overlooked. Informally, one might say that x obeys $x' = e$, but this cannot mean that the equation $x' = e$ holds true, because it is not even clear what the meaning of x' would be. A syntactic variable x has a meaning in a single state, but a time-derivative cannot. The semantical value of x along a function φ, instead, can have a well-defined derivative at time ζ. This requires passing back and forth between syntax and semantics.

2.9 Appendix

For your reference, this appendix already contains a short primer on some important results about differential equations from the literature [10]. While not crucial for the immediate technical development in subsequent chapters, this appendix is a helpful resource to come back to as desired for important meta-results for the general theory of differential equations. This appendix also lists useful counterexamples highlighting that the assumptions of the respective theorems are necessary.

The most crucial insights are that continuous differential equations have solutions (Sect. 2.9.1) and locally Lipschitz continuous differential equations (such as continuously differentiable differential equations) have unique solutions (Sect. 2.9.2).

2.9.1 Existence Theorems

Classical theorems guarantee existence and/or uniqueness of solutions of differential equations (not necessarily closed-form solutions with elementary functions, though). The existence theorem is due to Peano [6]. A proof can be found in the literature [10, Theorem 10.IX].

> **Theorem 2.1 (Peano's existence theorem).** *Let $f : D \to \mathbb{R}^n$ be a continuous function on an open, connected domain $D \subseteq \mathbb{R} \times \mathbb{R}^n$. Then, the initial value problem (2.1) with $(t_0, y_0) \in D$ has a solution. Every solution of (2.1) can be continued arbitrarily close to the boundary of D.*

Peano's theorem only proves that a solution exists, not for what duration it exists. Still, it shows that every solution can be *continued arbitrarily close to the boundary* of the domain. That is, the closure of the graph of the solution is not a compact subset of D, which means [10, §6.VII] that the solution exists either globally on $[0, \infty)$ or on a bounded interval $[0, r)$ with the solution approaching the boundary of D or an infinite norm at r. The *graph* graph(y) of a function $y : J \to D$ is the subset $\{(t, y(t)) : t \in J\}$ of $J \times D$.

Peano's theorem shows the existence of solutions of continuous differential equations on open, connected domains, but there can still be multiple solutions.

Example 2.10 (Nonunique solutions). The initial value problem with the following continuous differential equation

$$\begin{pmatrix} y' = \sqrt[3]{|y|} \\ y(0) = 0 \end{pmatrix}$$

has multiple solutions, for example

$$y(t) = 0$$

$$y(t) = \left(\frac{2}{3}t\right)^{\frac{3}{2}}$$

$$y(t) = \begin{cases} 0 & \text{for } t \leq s \\ \left(\frac{2}{3}(t-s)\right)^{\frac{3}{2}} & \text{for } t > s \end{cases}$$

where $s \geq 0$ can be any nonnegative real number.

2.9.2 Uniqueness Theorems

As usual in mathematics, $C^k(D, \mathbb{R}^n)$ denotes the space of k times continuously differentiable functions from domain D to $\mathbb{R}^n$. The Euclidean norm of a vector $v = (v_1, \ldots, v_n)$ is denoted by $\|v\| = \sqrt{\sum_{i=1}^{n} v_i^2}$.

If (the right-hand side of) the differential equation is continuously differentiable, then the Picard-Lindelöf theorem gives a stronger result than Peano's theorem. It shows that the solution is unique. For this, recall that a function $f : D \to \mathbb{R}^n$ with $D \subseteq \mathbb{R} \times \mathbb{R}^n$ is called *Lipschitz continuous* with respect to y iff there is an $L \in \mathbb{R}$ such that for all $(t, y), (t, \bar{y}) \in D$,

$$\|f(t, y) - f(t, \bar{y})\| \leq L\|y - \bar{y}\|.$$

If the partial derivative $\frac{\partial f(t,y)}{\partial y}$ exists and is bounded on D, then f is Lipschitz continuous with $L = \max_{(t,y) \in D} \|\frac{\partial f(t,y)}{\partial y}\|$ by the mean value theorem. Similarly, f is *locally Lipschitz continuous* iff for each $(t, y) \in D$, there is a neighborhood in which f is Lipschitz continuous. In particular, if f is continuously differentiable, i.e. $f \in C^1(D, \mathbb{R}^n)$, then f is locally Lipschitz continuous.

The Picard-Lindelöf theorem [5], which is also known as the Cauchy-Lipschitz theorem, guarantees existence and uniqueness of solutions (except, of course, that the restriction of any solution to a sub-interval is again a solution). A proof can be found in the literature [10, Theorem 10.VI]

> **Theorem 2.2 (Picard-Lindelöf uniqueness theorem).** *Let $f : D \to \mathbb{R}^n$ be a continuous function on an open, connected domain $D \subseteq \mathbb{R} \times \mathbb{R}^n$ that is locally Lipschitz continuous with respect to y (e.g., $f \in C^1(D, \mathbb{R}^n)$). Then the initial value problem (2.1) with $(t_0, y_0) \in D$ has a unique solution.*

The Picard-Lindelöf theorem does not indicate the duration of the solution. It only shows that the solution is unique on a nonempty open interval. Under the assumptions of the Picard-Lindelöf theorem, every solution can be extended to a solution of maximal duration arbitrarily close to the boundary of D by Peano's theorem.

Example 2.11 (Quadratic). The initial value problem

$$\begin{pmatrix} y' = y^2 \\ y(0) = 1 \end{pmatrix}$$

has the unique solution $y(t) = \frac{1}{1-t}$ of maximal duration on the domain $t < 1$. It cannot be extended to include its singularity at $t = 1$, though, but can get arbitrarily close. At the singularity it converges to the boundary $\pm\infty$ of the domain $\mathbb{R}$.

The following global uniqueness theorem shows a stronger property of a global solution on $[0, a]$ when the domain is a global stripe $[0, a] \times \mathbb{R}^n$. It is a corollary to Theorems 2.1 and 2.2, but used prominently in the proof of Theorem 2.2, and is of independent interest. A direct proof of the following global version of the Picard-Lindelöf theorem can be found in the literature [10, Proposition 10.VII].

> **Corollary 2.1 (Global uniqueness theorem of Picard-Lindelöf).** *Let* $f : [t_0, a] \times \mathbb{R}^n \to \mathbb{R}^n$ *be a continuous function that is Lipschitz continuous with respect to y. Then, there is a unique solution of the initial value problem (2.1) on* $[t_0, a]$.

As Example 2.11 illustrates, local Lipschitz continuity is not enough to guarantee the existence of a global solution that Corollary 2.1 concludes from global Lipschitz continuity.

2.9.3 Linear Differential Equations with Constant Coefficients

For the common class of linear differential equation systems with constant coefficients there is a well-established theory for obtaining closed-form solutions of initial value problems using classical techniques from linear algebra.

> **Proposition 2.1 (Linear differential equations with constant coefficients).** *For a constant matrix* $A \in \mathbb{R}^{n \times n}$, *the initial value problem*
>
> $$\begin{pmatrix} y'(t) = Ay(t) + b(t) \\ y(\tau) = \eta \end{pmatrix} \tag{2.10}$$
>
> *has the (unique) solution*
>
> $$y(t) = e^{A(t-\tau)}\eta + \int_\tau^t e^{A(t-s)} b(s)\, ds$$
>
> *where exponentiation of matrices is defined by the usual power series (generalized to matrices):*
>
> $$e^{At} = \sum_{n=0}^{\infty} \frac{1}{n!} A^n t^n$$

A proof, more details, and generalizations can be found in the literature [10, §18.VI]. If the matrix A is *nilpotent*, i.e., $A^n = 0$ for some $n \in \mathbb{N}$, and the terms $b(t)$ are polynomials in t, then the solution of the initial value problem is a polynomial function, because the exponential series stops at A^n and is then a finite polynomial in t

$$e^{At} = \sum_{k=0}^{\infty} \frac{1}{k!} A^k t^k = \sum_{k=0}^{n-1} \frac{1}{k!} A^k t^k$$

Since products and sums of polynomials are polynomials (polynomials form what is known as an algebra [1]) and polynomials in the variable t are closed under integration (meaning integrating a univariate polynomial in the variable t will yield another such polynomial), the solution identified in Proposition 2.1 is a polynomial. Furthermore, this solution is unique by Theorem 2.2. Such polynomials are especially useful for formal verification, because the resulting arithmetic is decidable. But the assumptions needed to guarantee such simple solutions are quite strong.

Example 2.12 (Accelerated motion in a straight line). In the initial value problem from Example 2.4 on p. 34, we guessed the solution of the differential equation system and then checked that it is the right solution by inserting it into the differential equations. But how do we compute the solution in the first place without having to guess? The differential equations $x' = v, v' = a$ from (2.2) are linear with constant coefficients. In vectorial notation where we denote $y(t) := (x(t), v(t))$, the vectorial equivalent (2.3) of (2.2) can be rewritten in explicit linear form (2.10) as follows:

$$\left(y'(t) = \begin{pmatrix} x \\ v \end{pmatrix}'(t) = \begin{pmatrix} 0 & 1 \\ 0 & 0 \end{pmatrix} \begin{pmatrix} x(t) \\ v(t) \end{pmatrix} + \begin{pmatrix} 0 \\ a \end{pmatrix} =: Ay(t) + b(t) \right)$$

$$\left(y(0) = \begin{pmatrix} x \\ v \end{pmatrix}(0) = \begin{pmatrix} x_0 \\ v_0 \end{pmatrix} =: \eta \right)$$

This linear differential equation system has the form of Proposition 2.1 with a constant coefficient matrix A. First, we compute the exponential series for the matrix A, which terminates quickly because $A^2 = 0$:

$$e^{At} = \sum_{n=0}^{\infty} \frac{1}{n!} A^n t^n = A^0 + At + \frac{1}{2!} \underbrace{A^2}_{0} t^2 + \underbrace{A^2}_{0} \sum_{n=3}^{\infty} \frac{1}{n!} A^{n-2} t^n$$

$$= \begin{pmatrix} 1 & 0 \\ 0 & 1 \end{pmatrix} + \begin{pmatrix} 0 & 1 \\ 0 & 0 \end{pmatrix} t = \begin{pmatrix} 1 & t \\ 0 & 1 \end{pmatrix}$$

Now Proposition 2.1 can be used to compute a solution of this differential equation:

$$y(t) = e^{At}\eta + \int_0^t e^{A(t-s)}b(s)\,ds$$

$$= \begin{pmatrix} 1 & t \\ 0 & 1 \end{pmatrix}\begin{pmatrix} x_0 \\ v_0 \end{pmatrix} + \int_0^t \begin{pmatrix} 1 & t-s \\ 0 & 1 \end{pmatrix}\begin{pmatrix} 0 \\ a \end{pmatrix}\,ds$$

$$= \begin{pmatrix} x_0 + v_0 t \\ v_0 \end{pmatrix} + \int_0^t \begin{pmatrix} at - as \\ a \end{pmatrix}\,ds$$

$$= \begin{pmatrix} x_0 + v_0 t \\ v_0 \end{pmatrix} + \begin{pmatrix} \int_0^t (at - as)\,ds \\ \int_0^t a\,ds \end{pmatrix}$$

$$= \begin{pmatrix} x_0 + v_0 t \\ v_0 \end{pmatrix} + \begin{pmatrix} ats - \frac{a}{2}s^2 \\ as \end{pmatrix}\Big|_{s=0}^{s=t}$$

$$= \begin{pmatrix} x_0 + v_0 t \\ v_0 \end{pmatrix} + \begin{pmatrix} at^2 - \frac{a}{2}t^2 \\ at \end{pmatrix} - \begin{pmatrix} a\cdot 0^2 - \frac{a}{2}\cdot 0^2 \\ a\cdot 0 \end{pmatrix}$$

$$= \begin{pmatrix} x_0 + v_0 t + \frac{a}{2}t^2 \\ v_0 + at \end{pmatrix}$$

The last equation is exactly the solution we guessed and checked in Example 2.4. Now we have computed it constructively. An alternative way of computing solutions of differential equations is by proof [7].

2.9.4 Continuation and Continuous Dependency

Occasionally it is helpful to know that solutions of differential equations can be continued by concatenation toward a maximal interval of existence. The following result is a componentwise generalization of a classical result [10, Proposition 6.VI] to vectorial differential equations and can be used to extend solutions.

> **Proposition 2.2 (Continuation of solutions).** *Let $f : D \to \mathbb{R}^n$ be a continuous function on the open, connected domain $D \subseteq \mathbb{R} \times \mathbb{R}^n$. If φ is a solution of differential equation $y' = f(t,y)$ on $[0,b)$ whose image $\varphi([0,b))$ lies within a compact set $A \subseteq D$, then φ can be continued to a solution on $[0,b]$. Furthermore, if φ_1 is a solution of differential equation $y' = f(t,y)$ on $[0,b]$ and φ_2 is a solution of $y' = f(t,y)$ on $[b,c]$ with $\varphi_1(b) = \varphi_2(b)$, then their concatenation*
>
> $$\varphi(t) := \begin{cases} \varphi_1(t) & \text{for } 0 \le t \le b \\ \varphi_2(t) & \text{for } b < t \le c \end{cases}$$
>
> *is a solution on $[0,c]$.*

The solution of a Lipschitz continuous initial value problem depends continuously on the initial values and permits error estimation from the Lipschitz constants. A proof and generalizations are elsewhere [10, Proposition 12.V].

> **Proposition 2.3 (Lipschitz estimation).** *Let $f : D \to \mathbb{R}^n$ be a continuous function that is Lipschitz continuous with Lipschitz constant L with respect to y on the open, connected domain $D \subseteq J \times \mathbb{R}^n$ with an interval J. Let y be a solution on J of the initial value problem $y'(t) = f(t, y(t)), y(0) = y_0$ and z an approximate solution in the sense that*
>
> $$\|z(0) - y(0)\| \leq \gamma, \quad \|z'(t) - f(t, z(t))\| \leq \delta$$
>
> *then for all $t \in J$:*
>
> $$\|y(t) - z(t)\| \leq \gamma e^{L|t|} + \frac{\delta}{L}(e^{L|t|} - 1)$$
>
> *Here J is any interval with $0 \in J$ and $graph(y), graph(z) \subseteq D$.*

Exercises

2.1. Suppose $\omega(x) = 7$, then explain why the value of $4 + x \cdot 2$ in ω is $\omega[\![4 + x \cdot 2]\!] = 18$. What is the value of the same term $4 + x \cdot 2$ in a state v with $v(x) = -4$? What is the value of the term $4 + x \cdot 2 + x \cdot x \cdot x$ in the same state v? How does its value change in a state where $v(x) = -4$ but also $v(y) = 7$? Suppose $\omega(x) = 7$ and $\omega(y) = -1$, explain why $x \cdot 2 + y \cdot y$ evaluates to $\omega[\![x \cdot 2 + y \cdot y]\!] = 15$. What is the value of $x \cdot 2 + y \cdot y$ in a state v with $v(x) = -4$ and $v(y) = 7$?

2.2. Subtraction $e - \tilde{e}$ is already implicitly available as a term, because it is definable via $e + (-1) \cdot \tilde{e}$. In practice, we can, thus, pretend $e - \tilde{e}$ is in the syntax, while theoretical investigations can ignore $e - \tilde{e}$ as it is not an official part of the syntax. What about negation $-e$? Is negation implicitly already available as well? What about division $e/\tilde{e}$ and powers $e^{\tilde{e}}$?

2.3 (Speeding). Consider a car that is driving on a road with a speed limit of either 35 mph or 50 km/h when a deer darts onto the road at a distance in front of the car that is just enough for the car to stop when driving at the speed limit. Suppose the car was speeding at 45 mph or 70 km/h, instead. How fast is the car still going when it meets the surprised deer? You may assume brakes of effective deceleration $a = -6m/s^2$, which is typical for some road conditions. How does the answer change when the driver needs a reaction time of 2 seconds?

2.4 (Changing accelerations). Some settings of idealized physics benefit from considering not just position x, its rate of change velocity v, and its rate of change acceleration a, but also the continuous rate of change of the acceleration a, which is called jolt or jerk j. Jolt may happen, for example, when changing gears abruptly or when not releasing the brakes of a car when it is about to come to a standstill. Solve the resulting differential equation of accelerated acceleration in a straight line:

$$x' = v, v' = a, a' = j$$

2.5 (Robot moving along a planar circular curve). This exercise develops differential equations for the continuous dynamics of a robot that is moving in the two-dimensional plane. Consider a robot at a point with coordinates (x, y) that is facing in direction (v, w). While the robot is moving along the dashed curve, this direction (v, w) is simultaneously rotating with angular velocity ω.

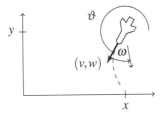

Develop a differential equation system describing how the position and direction of the robot change over time. Build your way up to that differential equation by first considering just the rotation of (v, w), then considering the motion of (x, y) in a fixed direction (v, w), and then putting both types of behavior together. Can you subsequently generalize the dynamics to also include an acceleration of the linear ground speed when the robot is speeding up?

2.6. A number of differential equations and some suggested solutions are listed in the following table. Are these correct solutions? Are these all solutions? Are there other solutions? In what ways are the solutions characteristically more complicated than their differential equations?

ODE	Solution
$x' = 1, x(0) = x_0$	$x(t) = x_0 + t$
$x' = 5, x(0) = x_0$	$x(t) = x_0 + 5t$
$x' = x, x(0) = x_0$	$x(t) = x_0 e^t$
$x' = x^2, x(0) = x_0$	$x(t) = \frac{x_0}{1 - t x_0}$
$x' = \frac{1}{x}, x(0) = 1$	$x(t) = \sqrt{1 + 2t}$
$y'(x) = -2xy, y(0) = 1$	$y(x) = e^{-x^2}$
$x'(t) = tx, x(0) = x_0$	$x(t) = x_0 e^{\frac{t^2}{2}}$
$x' = \sqrt{x}, x(0) = x_0$	$x(t) = \frac{t^2}{4} \pm t\sqrt{x_0} + x_0$
$x' = y, y' = -x, x(0) = 0, y(0) = 1$	$x(t) = \sin t, y(t) = \cos t$
$x' = 1 + x^2, x(0) = 0$	$x(t) = \tan t$
$x'(t) = \frac{2}{t^3} x(t)$	$x(t) = e^{-\frac{1}{t^2}}$ non-analytic
$x'(t) = e^{t^2}$	no elementary closed form

2.7 (**). Would the differential equation semantics defined in Definition 2.6 change when we only require the differential equation to be respected at all times $\zeta \in (0, r)$ in an open interval rather than at all times $\zeta \in [0, r]$ in a closed interval?

2.8. Review the theory of ordinary differential equations. Investigate which theorems from this chapter's appendix apply to the example differential equations given in this chapter.

2.9 (Valid quantified formulas). Using the semantics of quantifiers, show that the following first-order logic formulas are valid, i.e., true in all states:

$$(\forall x\, p(x)) \to (\exists x\, p(x))$$
$$(\forall x\, p(x)) \to p(e)$$
$$\forall x\, (p(x) \to q(x)) \to (\forall x\, p(x) \to \forall x\, q(x))$$

In the second formula, e is any term and $p(e)$ should be understood here as resulting from the formula $p(x)$ by replacing all (free) occurrences of variable x by term e. An occurrence of x in $p(x)$ within the scope of another quantifier for x is not free but bound, so will not be substituted. Some care is needed to avoid capture, i.e., that x does not occur in the scope of a quantifier binding a variable of e, because replacing x with e would then bind a free variable of e.

2.10 (* Two-valued semantics). The semantics of terms is defined by a real-valued mapping $[\![e]\!] : \mathscr{S} \to \mathbb{R}$ in Expedition 2.2. The semantics of formulas was given in Definition 2.5 by defining the set of states $[\![P]\!]$ in which formula P is true. Give an equivalent definition of the semantics of first-order formulas using a function $[\![P]\!]_{\mathbb{B}} : \mathscr{S} \to \{true, false\}$ that defines the truth-value $[\![P]\!]_{\mathbb{B}}(\omega)$ of formula P in each state ω. Prove equivalence of the semantics by showing that the new truth-value semantics evaluates to the truth-value *true* if and only if the formula P is true at ω:

$$[\![P]\!]_{\mathbb{B}}(\omega) = true \quad \text{iff} \quad \omega \in [\![P]\!]$$

2.11 (Term interpreter). In a programming language of your choosing, fix a recursive data structure for terms from Definition 2.2 and fix some finite representation for states where all variables have rational values instead of reals. Write a term interpreter as a computer program that, given a state ω and a term e, computes the value of $\omega[\![e]\!]$ by implementing Definition 2.4 as a recursive function. Write a similar interpreter for first-order formulas from Definition 2.3 that, given a state ω and a formula P, reports "yes" if and only if $\omega \in [\![P]\!]$. Which cases are problematic?

2.12 (** Set-valued semantics). There are at least two styles of giving a meaning to a logical formula. One way is to inductively define a satisfaction relation $\models$ that holds between a state ω and a dL formula P, written $\omega \models P$, whenever the formula P is true in the state ω. Its definition includes, among other cases, the following notational variant of Definition 2.5:

$$\omega \models e \geq \tilde{e} \text{ iff } \omega[\![e]\!] \geq \omega[\![\tilde{e}]\!]$$
$$\omega \models P \wedge Q \text{ iff } \omega \models P \text{ and } \omega \models Q$$

The other way is to inductively define, for each dL formula P, the set of states, written $[\![P]\!]$, in which P is true. Its definition will include, among other cases, the

following:
$$[\![e \ge \tilde{e}]\!] = \{\omega : \omega[\![e]\!] \ge \omega[\![\tilde{e}]\!]\}$$
$$[\![P \wedge Q]\!] = [\![P]\!] \cap [\![Q]\!]$$

Complete both styles of defining the semantics and prove that they are equivalent. That is, $\omega \models P$ iff $\omega \in [\![P]\!]$ for all states ω and all first-order formulas P.

Such a proof can be conducted by induction on the structure of P. That is, consider each case, say $P \wedge Q$, and prove $\omega \models P \wedge Q$ iff $\omega \in [\![P \wedge Q]\!]$ from the inductive hypothesis that the conjecture already holds for the smaller subformulas:

$$\omega \models P \text{ iff } \omega \in [\![P]\!]$$
$$\omega \models Q \text{ iff } \omega \in [\![Q]\!]$$

2.13. Explain which formulas you expect to be particularly common as evolution domain constraints in cyber-physical systems.

References

[1] Nicolas Bourbaki. *Algebra I: Chapters 1–3*. Elements of mathematics. Berlin: Springer, 1989.

[2] Kenneth Eriksson, Donald Estep, Peter Hansbo, and Claes Johnson. *Computational Differential Equations*. Cambridge: Cambridge University Press, 1996.

[3] Leonhard Euler. *Institutionum calculi integralis*. St Petersburg: Petropoli, 1768.

[4] Philip Hartman. *Ordinary Differential Equations*. Hoboken: John Wiley, 1964.

[5] M. Ernst Lindelöf. Sur l'application de la méthode des approximations successives aux équations différentielles ordinaires du premier ordre. *Comptes rendus hebdomadaires des séances de l'Académie des sciences* **114** (1894), 454–457.

[6] Giuseppe Peano. Demonstration de l'intégrabilité des équations différentielles ordinaires. *Mathematische Annalen* **37**(2) (1890), 182–228. DOI: 10.1007/BF01200235.

[7] André Platzer. A complete uniform substitution calculus for differential dynamic logic. *J. Autom. Reas.* **59**(2) (2017), 219–265. DOI: 10.1007/s10817-016-9385-1.

[8] William T. Reid. *Ordinary Differential Equations*. Hoboken: John Wiley, 1971.

[9] Gerald Teschl. *Ordinary Differential Equations and Dynamical Systems*. Providence: AMS, 2012.

[10] Wolfgang Walter. *Ordinary Differential Equations*. Berlin: Springer, 1998. DOI: 10.1007/978-1-4612-0601-9.

Chapter 3
Choice & Control

Synopsis This chapter develops the central dynamical systems model for describing the behavior of cyber-physical systems with a programming language. It complements the previous understanding of continuous dynamics with an understanding of the discrete dynamics caused by choices and controls in cyber-physical systems. The chapter interfaces the continuous dynamics of differential equations with the discrete dynamics of conventional computer programs by directly integrating differential equations with discrete programming languages. This leverages well-established programming language constructs around elementary discrete and continuous statements to obtain *hybrid programs* as a core programming language for cyber-physical systems. In addition to embracing differential equations, semantical generalizations to mathematical reals as well as operators for nondeterminism are important to make hybrid programs appropriate for cyber-physical systems.

3.1 Introduction

Chapter 2 saw the beginning of cyber-physical systems, yet emphasized only their continuous part in the form of differential equations $x' = f(x)$. The sole interface between continuous physical capabilities and cyber capabilities was by way of their evolution domain. The evolution domain Q in a continuous program $x' = f(x) \,\&\, Q$ imposes restrictions on how far or how long the system can evolve along that differential equation. Suppose a continuous evolution has succeeded, and the system stops following its differential equation, e.g., because the state would otherwise leave the evolution domain Q if it kept going. Then what happens now? How does the cyber part take control? How do we describe what the cyber elements compute afterwards? What descriptions explain how cyber interacts with physics?

An overall understanding of a CPS ultimately requires an understanding of the joint model with both its discrete dynamics and its continuous dynamics. It takes both to understand, for example, what effect a discrete car controller has, via its engine and steering actuators, on the continuous physical motion of a car down the

© Springer International Publishing AG, part of Springer Nature 2018
A. Platzer, *Logical Foundations of Cyber-Physical Systems*,
https://doi.org/10.1007/978-3-319-63588-0_3

road. Continuous programs are powerful for modeling continuous processes, such as continuous motion. They cannot—on their own—model discrete changes of variables, however.[1] Such discrete state change is a good model for the impact of computer decisions on cyber-physical systems, in which computation decides to, say, stop speeding up and apply the brakes instead. During the evolution along a differential equation, such as $x' = v, v' = a$ for accelerated motion along a straight line, all variables change continuously over time, because the solution of a differential equation is (sufficiently) smooth. Discontinuous change of variables, such as a change of acceleration from $a = 2$ to $a = -6$ by applying the brakes instead arises from a discrete change of state resulting from how computers compute decisions one step at a time. Time passes while differential equations evolve, but no time passes during an immediate discrete change (it is easy to model computations that take time by mixing both). What could be a model for describing such discrete changes in a system?

Discrete change can be described by different models. The most prominent ones are conventional programming languages, in which everything takes effect one discrete step at a time, just like computer processors operate one clock cycle at a time.

CPSs combine cyber and physics, though. In CPS, we do not program computers, but program cyber-physical systems instead. We program the computers that control the physics, which requires programming languages for CPSs to involve physics, and integrate differential equations seamlessly with discrete computer operations. The basic idea is that discrete statements are executed by a computer processor, while continuous statements are handled by the physical elements, such as wheels, engines, or brakes. CPS programs need a mix of both to accurately describe the combined discrete and continuous dynamics.

Does it matter which discrete programming language we choose to enrich with the continuous statements from Chap. 2? It might be argued that the hybrid aspects are more important for CPS than the discrete language. After all, there are many conventional programming languages that are Turing-equivalent, i.e., that compute the same functions [3, 10, 26]. Yet there are numerous significant differences even among discrete programming languages that make some more desirable than others [7]. For the particular purposes of CPS, we will identify additional desired features. We will develop what we need as we go, culminating in the programming language of *hybrid programs* [16–21], which plays a fundamental rôle in this book.

Other areas such as automata theory and the theory of formal languages [10] or Petri nets [15] also provide models of discrete change. There are ways of augmenting these models with differential equations as well [1, 4, 13, 14]. But programming languages are uniquely positioned to extend their virtues of built-in compositionality. Just as the meaning and effect of a conventional program is a function of its pieces, the meaning and operation of a hybrid program is also a function of its parts.

The most important learning goals of this chapter are:

---

[1] There is a much deeper sense [20] in which continuous dynamics and discrete dynamics have surprising similarities regardless. But even so, these similarities rest on the foundations of hybrid systems, which we need to understand first.

Modeling and Control: This chapter plays a pivotal rôle in understanding and designing models of CPSs. We develop an understanding of the core principles behind CPS by studying how discrete and continuous dynamics are combined and interact to model cyber and physics, respectively. We see the first example of how to develop models and controls for a simple CPS. Even if subsequent chapters will blur the overly simplistic categorization of cyber=discrete versus physics=continuous, it is useful to equate them for now, because cyber, computation, and decisions quickly lead to discrete dynamics, while physics naturally gives rise to continuous dynamics. Later chapters will show that some physical phenomena are better modeled with discrete dynamics, while some controller aspects also have a manifestation in the continuous dynamics.

Computational Thinking: We introduce and study the important phenomenon of nondeterminism, which is crucial for developing faithful models of a CPS's environment and helpful for developing effective models of the CPS itself. We emphasize the importance of abstraction, which is an essential modular organization principle in CPS as well as all other parts of computer science. We capture the core aspects of CPS in the programming language of hybrid programs.

CPS Skills: We develop an intuition for the operational effects of CPS. And we will develop an understanding for the semantics of the programming language of hybrid programs, which is the CPS model on which this textbook is based.

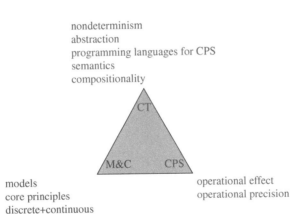

3.2 A Gradual Introduction to Hybrid Programs

This section gradually introduces the operations that hybrid programs provide, one step at a time. Its emphasis is on their motivation and an intuitive development before subsequent sections provide a comprehensive view. The motivating examples we consider now are naïve but still provide a good introduction to the world of CPS programming. With more understanding, we will later be able to augment their designs.

3.2.1 Discrete Change in Hybrid Programs

Discrete change happens immediately in computer programs when a new value is assigned to a variable. The statement $x := e$ assigns the value of term e to variable x by evaluating the term e and assigning the result to the variable x. It leads to a discrete, discontinuous change, because the value of x does not vary smoothly over time but radically when the value of e is suddenly assigned to x.

Fig. 3.1 An illustration of the behavior of an instantaneous discrete change at time $= 0$

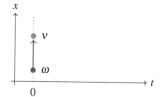

This gives us a discrete model of change, $x := e$, in addition to the continuous model of change, $x' = f(x) \,\&\, Q$, from Chap. 2. We can now model systems that are *either* discrete *or* continuous. Yet, how can we possibly model proper CPSs that combine cyber and physics with one another and that, thus, simultaneously combine discrete and continuous dynamics? We need such hybrid behavior every time a system has both continuous dynamics (such as the continuous motion of a car down the street) in addition to discrete dynamics (such as shifting gears).

3.2.2 Compositions of Hybrid Programs

One way cyber and physics can interact is if a computer provides input to physics. Physics may mention variables such as a for acceleration, and the computer program sets its value depending on whether the computer program wants to accelerate or brake. That is, cyber may set the values of actuators that affect physics.

In this case, cyber and physics interact in such a way that first the cyber part does something and physics then follows. Such a behavior corresponds to a sequential composition $(\alpha; \beta)$ in which first the HP α on the left of the sequential composition operator (;) runs and, when it's done, the HP β on the right of operator ; runs. The following HP[2]

$$a := a + 1; \{x' = v, v' = a\} \qquad (3.1)$$

will first let cyber perform a discrete change of setting acceleration variable a to $a + 1$ and then let physics follow the differential equation[3] $x'' = a$, which describes accelerated motion of the point x along a straight line. The overall effect is that cyber instantly increases the value of the acceleration variable a and physics then lets x evolve continuously with that acceleration a (increasing velocity v continuously with derivative a). HP (3.1) models a situation where the desired acceleration is commanded once to increase and the robot then moves with that fixed acceleration; see Fig. 3.2. The curve for position looks almost linear in Fig. 3.2, because the velocity difference is so small, which is a great example of how misleading visual representations can be compared to rigorous analysis methods. The sequential composition operator (;) has the same effect that it has in programming languages such as Java. It separates statements that are to be executed sequentially one after the other. If you look closely, you will find a minor subtle difference, because programming languages such as Java or C expect ; at the end of every statement, not just between sequentially composed statements. This syntactic difference is inconsequential, and a common trait of mathematical programming languages.

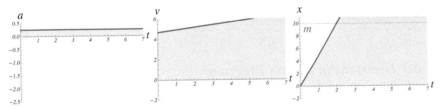

Fig. 3.2 Fixed acceleration a (**left**), velocity v (**middle**), and position x (**right**) change over time t

The HP in (3.1) executes control (it sets the acceleration for physics), but it has very little choice, or rather no choice at all. So only if the CPS is very lucky will an increase in acceleration be the right action to remain safe forever. Quite likely, the robot will have to change its mind ultimately, which is what we investigate next.

But first observe that the constructs we saw so far, assignments, sequential compositions, and differential equations, already suffice to exhibit typical hybrid systems dynamics. The behavior shown in Fig. 3.3 could be exhibited by this hybrid program:

---

[2] Note that the parentheses around the differential equation are redundant and will often be left out in the textbook or in scientific papers. HP (3.1) would be written $a := a + 1; x' = v, v' = a$. Round parentheses are often used in theoretical developments, while braces are useful for programs to disambiguate grouping in bigger CPS applications.

[3] We frequently use $x'' = a$ as an abbreviation for $x' = v, v' = a$, even if x'' is not officially permitted in the KeYmaera X theorem prover for hybrid systems.

$$a := -2; \quad \{x' = v, v' = a\};$$
$$a := 0.25; \quad \{x' = v, v' = a\};$$
$$a := -2; \quad \{x' = v, v' = a\};$$
$$a := 0.25; \quad \{x' = v, v' = a\};$$
$$a := -2; \quad \{x' = v, v' = a\};$$
$$a := 0; \quad \{x' = v, v' = a\}$$

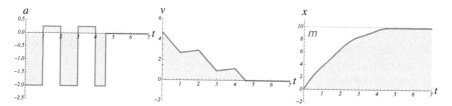

Fig. 3.3 Acceleration a (**left**), velocity v (**middle**), and position x (**right**) change over time t, with a piecewise constant acceleration changing discretely at instants of time while velocity and position change continuously over time

Can you already spot a question that comes up about how exactly we run this program? We will postpone the formulation of and answer to this question to Sect. 3.2.6.

3.2.3 Decisions in Hybrid Programs

In general, a CPS will have to check conditions on the state to see which action to take. Otherwise the CPS could not possibly be safe and, quite likely, will also not take the correct actions to get to its goal. One way of programming these conditions is the use of an if-then-else statement, as in classical discrete programs:

$$\text{if}(v < 4)\, a := a + 1 \,\text{else}\, a := -b;$$
$$\{x' = v, v' = a\} \tag{3.2}$$

This HP will check the condition $v < 4$ to see whether the current velocity is still less than 4. If it is, then a will be increased by 1. Otherwise, a will be set to $-b$ for some braking deceleration constant $b > 0$. Afterwards, i.e., when the if-then-else statement in the first line has run to completion, the HP will again evolve x continuously with acceleration a along the differential equation in the second line.

The HP (3.2) takes only the current velocity into account to reach a decision on whether to accelerate or brake. That is usually not enough information to guarantee safety, because a robot doing that would be so fixated on achieving its desired speed that it would happily speed into any walls or other obstacles along the way. Consequently, programs that control robots also take other state information into account, for example sufficient distance $x - m$ to an obstacle m from the robot's position x:

$$\text{if}(x - m > s)\, a := a + 1 \,\text{else}\, a := -b;$$
$$\{x' = v, v' = a\} \tag{3.3}$$

Whether that is safe depends on the choice of the required safety distance s. Controllers could also take both distance *and* velocity into account for the decision:

$$\text{if}(x - m > s \wedge v < 4)\, a := a + 1 \,\text{else}\, a := -b;$$
$$\{x' = v, v' = a\} \tag{3.4}$$

> **Note 9 (Iterative design)** To design serious controllers, you will usually develop a series of increasingly more intelligent controllers for systems that face increasingly challenging environments. Designing controllers for robots or other CPSs is a serious challenge. You will want to start with simple controllers for simple circumstances and only move on to more advanced challenges when you have fully understood and mastered the previous controllers, what behavior they guarantee and what functionality they are still missing. If a controller is not even safe under simple circumstances (for example when it only knows how to brake), it will not be safe in more complex cases either.

3.2.4 Choices in Hybrid Programs

A common feature of CPS models is that they often include only some but not all detail about the system. This is for good reasons, because full detail about everything can be overwhelming and is often a distraction from the really important aspects of a system. A (somewhat) more complete model of (3.4) might have the following shape, with some further formula S as an extra condition for checking whether to actually accelerate based on battery efficiency or secondary considerations which are not safety-critical:

$$\text{if}(x - m > s \wedge v < 4 \wedge S)\, a := a + 1 \,\text{else}\, a := -b;$$
$$\{x' = v, v' = a\} \tag{3.5}$$

Consequently, (3.4) is not actually a faithful model of (3.5), because (3.4) insists that the acceleration would always be increased just because $x - m > s \wedge v < 4$ holds, unlike (3.5), which also checks the additional condition S. Likewise, (3.3) certainly is no faithful model of (3.5). But it looks simpler.

How can we describe a model that is simpler than (3.5) because it ignores the details of S yet that is still faithful to the original system? What we want this model to do is characterize that the controller may either increase acceleration by 1 or brake. All acceleration should certainly only happen when certain safety-critical conditions are met. But the model should make less commitment than (3.3) about the precise circumstances under which braking is chosen. After all, braking may sometimes just be the right thing to do, for example when arriving at the goal. So we want a model that allows braking under more circumstances than (3.3) without having to model precisely under what circumstances that is. If a system with more behavior is safe, then the actual implementation will be safe as well, because it will only ever exercise some of the verified behavior [12]. The extra behavior in the system might, in fact, occur in reality whenever there are minor lags or discrepancies. So it is good to have the extra assurance that some flexibility in the execution of the system will not break its safety guarantees.

> **Note 10 (Abstraction)** Successful CPS models often include only the relevant aspects of the system and elide irrelevant detail. The benefit of doing so is that the model and its analysis become simpler, enabling us to focus on the critical parts without being bogged down in tangentials. This is the *power of abstraction*, probably the primary secret weapon of computer science. It does take considerable skill, however, to find the best level of abstraction for a system, a skill that you will continue to sharpen throughout your entire career.

Let us take the development of this model step by step. The first feature that the controller in the model has is a choice. The controller can choose to increase acceleration or to brake, instead. Such a choice between two actions is denoted by the choice operator $\cup$:

$$(a := a + 1 \cup a := -b);$$
$$\{x' = v, v' = a\} \tag{3.6}$$

When running this hybrid program, the first thing that happens is that the first statement (before the ;) runs, which is a choice ($\cup$) between whether to run $a := a + 1$ or whether to run $a := -b$. That is, the choice is whether to increase acceleration a by 1 or whether to reset a to $-b$ for braking. After this choice (i.e., after the ; sequential composition operator), the system follows the usual differential equation $x'' = a$ describing accelerated motion along a line.

Now, wait. There was a choice. Who chooses? How is the choice resolved?

> **Note 11 (Nondeterministic $\cup$)** The choice ($\cup$) is *nondeterministic*. That is, every time a choice $\alpha \cup \beta$ runs, exactly one of the two choices, α or β, is chosen to run. The choice is *nondeterministic*, i.e., there is no prior way of telling which of the two choices is going to be chosen. Both outcomes are perfectly possible and a safe system design needs to be prepared to handle either outcome.

The HP (3.6) is a *faithful abstraction* [12] of (3.5), because every way (3.5) can run can be mimicked by (3.6) so that the outcome of (3.6) corresponds to that of (3.5). Whenever (3.5) runs $a := a + 1$, which happens exactly if $x - m > s \wedge v < 4 \wedge S$ is *true*, (3.6) only needs to choose to run the left choice $a := a + 1$. Whenever (3.5) runs $a := -b$, which happens exactly if $x - m > s \wedge v < 4 \wedge S$ is *false*, (3.6) needs to choose to run the right choice $a := -b$. So all runs of (3.5) are possible runs of (3.6). Furthermore, (3.6) is much simpler than (3.5), because it contains less detail. It does not mention the complicated extra condition S. However, (3.6) is a little too permissive, because it suddenly allows the controller to choose $a := a+1$ even when it is already too fast or even at a small distance from the obstacle. That way, even if (3.5) was a safe controller, (3.6) is still unsafe, and, thus, not a very suitable abstraction.

3.2.5 Tests in Hybrid Programs

In order to build a faithful yet not overly permissive abstraction of (3.5), we need to restrict the permitted choices in (3.6) so that there is enough flexibility, but only so much that the acceleration choice $a := a+1$ can only be chosen when it is currently safe to do so. The way to do that is to use tests on the current state of the system.

A *test* $?Q$ is a statement that checks the truth-value of a first-order formula Q of real arithmetic in the current state. If Q holds in the current state, then the test passes, nothing happens, and the HP continues to run normally. If, instead, Q does not hold in the current state, then the test fails, and the system execution is aborted and discarded. That is, when ω is the current state, then $?Q$ runs successfully without changing the state when $\omega \in [\![Q]\!]$. Otherwise, i.e., if $\omega \notin [\![Q]\!]$, the run of $?Q$ is aborted and not considered any further, because it did not play by the rules of the system.

Of course, it can be difficult to figure out which control choice is safe under what circumstances, and the answer also depends on whether the safety goal is to limit speed or to remain at a safe distance from other obstacles. For the model in this chapter, we simply pretend that $v < 4$ is the appropriate safety condition and revisit the question of how to design and explain such conditions in later chapters.

The test statement $?(v < 4)$ alias $?v < 4$ can be used to change (3.6) so that it allows acceleration only when $v < 4$, while braking is still allowed always:

$$((?v < 4; a := a+1) \cup a := -b);$$
$$\{x' = v, v' = a\}$$
(3.7)

The first statement of (3.7) is a choice ($\cup$) between $(?v < 4; a := a+1)$ and $a := -b$. All choices in hybrid programs are nondeterministic, so either outcome is always possible. In (3.7), this means that the left choice can always be chosen, just as well as the right one. The first statement that happens in the left choice, however, is the test $?v < 4$, which the system run has to pass in order to be able to continue. In particular, if $v < 4$ is indeed *true* in the current state, then the system passes that test $?v < 4$ and the execution proceeds to after the sequential composition (;) to run $a := a+1$. If $v < 4$ is *false* in the current state, however, the system fails the test $?v < 4$ and that run is aborted and discarded. The right option to brake is always available, because it does not involve any tests to pass.

> **Note 12 (Discarding failed runs)** System runs that fail tests $?Q$ are discarded and not considered any further, because a failed run did not play by the rules of the system. It is as if those failed system execution attempts had never happened. Even if one execution attempt fails, other runs may still be successful. Operationally, you can imagine finding them by backtracking through all the possible choices in the system run and taking alternative choices instead.

In principle, there are always two choices when running (3.7). However, which ones actually run successfully depends on the current state. If the car is currently slow (so the test $?v < 4$ will succeed), then both options of accelerating and braking are possible and can execute successfully. Otherwise, only the braking choice executes, because trying the left choice will fail its test $?v < 4$ and be discarded. Both choices formally exist but only one will succeed in that case.

> **Note 13 (Successful runs)** Notice that only successfully executed runs of HPs will be considered, and all others will be discarded because they did not play by the rules. For example, $?v < 4; v := v + 1$ can only run in states where $v < 4$, otherwise there are no runs of this HP. Failed runs are discarded entirely, so the HP $v := v + 1; ?v < 4$ can also only run in states where $v < 3$. Operationally, you can imagine running the HP step by step and rolling all its changes back if any test ever fails. The velocity increases by $v := v + 1$, but this change is undone and the entire run discarded unless the subsequent test $?v < 4$ succeeds for the new value.

Comparing (3.7) with (3.5), we see that (3.7) is a faithful abstraction of the more complicated (3.5), because all runs of (3.5) can be mimicked by (3.7). Yet, unlike the intermediate guess (3.6), the improved HP (3.7) still retains the critical information that acceleration is only allowed by (3.5) when $v < 4$. Unlike (3.5), (3.7) does not restrict the cases where acceleration can be chosen to those that also satisfy $v < 4 \wedge S$. Hence, (3.7) is more permissive than (3.5). But (3.7) is also simpler and only contains crucial information about the controller. Hence, (3.7) is a more abstract faithful model of (3.5) that retains just the relevant detail. Studying the abstract (3.7) instead of the more concrete (3.5) has the advantage that only relevant details need to be understood while irrelevant aspects can be ignored. It also has the additional advantage that a safety analysis of the more abstract (3.7), which allows lots of behavior, will imply safety of the special concrete case (3.5) but also implies safety of other implementations of (3.7). For example, replacing S by a different condition in (3.5) still gives a special case of (3.7). So if all behaviors of (3.7) are safe, all behaviors of that different replacement will already be safe. With a single verification result about a more general, more abstract system, we can verify a whole class of systems rather than just one particular system. This important phenomenon [12] will be investigated in more detail in later parts of the book.

Of course, which details are relevant and which ones can be simplified depends on the analysis question at hand, a question that we will be better equipped to answer in a later chapter. For now, suffice it to say that (3.7) has the relevant level of abstraction for our purposes.

> **Note 14 (Broader significance of nondeterminism)** Nondeterminism comes up in the above cases for reasons of abstraction and to focus the system model on the most critical aspects of the system while suppressing irrelevant detail. This simplification is one important reason for introducing nondeterminism in system models, but there are other important reasons as well. Whenever a system includes models of its environment, nondeterministic models are crucial, because we often have only a partial understanding of what the environment will do. A car controller for example, will not always know for sure what other cars or pedestrians in its environment will do, exactly, so that nondeterministic models are the only faithful representations.

A pretty reasonable model of the controller of the acceleration c of another car in our environment is to nondeterministically either accelerate or brake, e.g., $c:=2 \cup c:=-b$, because we cannot perfectly predict which one is going to happen, anyhow.

Note the notational convention that sequential composition ; binds more strongly than nondeterministic choice $\cup$ so we can leave parentheses out without changing (3.7):

$$\left(?v < 4; a:=a+1 \cup a:=-b\right);$$
$$\{x' = v, v' = a\} \tag{3.7*}$$

3.2.6 Repetitions in Hybrid Programs

The hybrid programs above were interesting, but only allowed the controller to choose what action to take at most once. All controllers so far inspected the state in a test or in an if-then-else condition and then chose what to do once, only to let physics take control subsequently by following a differential equation. That makes for rather short-lived controllers. They have a job only once in their lives. And most decisions they reach may end up being bad ones at some point. Say, one of those controllers, e.g., (3.7), inspects the state and finds it still okay to accelerate. If it chooses $a:=a+1$ and then lets physics move with the differential equation $x'' = a$, there will probably come a time at which increased acceleration is no longer such a great idea. But the controller of (3.7) has no way to change its mind, because it has no more choices and cannot exercise any control anymore.

If the controller of (3.7) is supposed to be able to make a second control choice later after physics has followed the differential equation for a while, then (3.7) can simply be sequentially composed with itself:

$$
\begin{aligned}
&\big((?v < 4; a := a+1) \cup a := -b\big); \\
&\{x' = v, v' = a\}; \\
&\big((?v < 4; a := a+1) \cup a := -b\big); \\
&\{x' = v, v' = a\}
\end{aligned}
\tag{3.8}
$$

In (3.8), the cyber controller can first choose to accelerate or brake (depending on whether $v < 4$ is true in the present state), then physics evolves along differential equation $x'' = a$ for some while, then the controller can again choose whether to accelerate or brake (depending on whether $v < 4$ is true in the state reached then), and finally physics again evolves along $x'' = a$.

For a controller that is supposed to be allowed to have a third control choice, copy and paste replication would again help:

$$
\begin{aligned}
&\big((?v < 4; a := a+1) \cup a := -b\big); \\
&\{x' = v, v' = a\}; \\
&\big((?v < 4; a := a+1) \cup a := -b\big); \\
&\{x' = v, v' = a\}; \\
&\big((?v < 4; a := a+1) \cup a := -b\big); \\
&\{x' = v, v' = a\}
\end{aligned}
\tag{3.9}
$$

But this is neither a particularly concise nor a particularly useful modeling style. What if a controller might need 10 control decisions or 100? Or what if there is no way of telling ahead of time how many control decisions the cyber part will have to take to reach its goal? Think of how many control decisions you might need when driving in a car from Paris to Rome. Do you even know that ahead of time? Even if you do, do you want to model a system by explicitly replicating its controller that often?

> **Note 15 (Repetition)** As a more concise and more general way of describing repeated control choices, hybrid programs allow for the repetition operator *, which works like the Kleene star operator in regular expressions, except that it applies to a hybrid program α as in α^*. It repeats α any number $n \in \mathbb{N}$ of times, including 0, by a nondeterministic choice.

The programmatic way of summarizing (3.7), (3.8), (3.9) and the infinitely many more n-fold replications of (3.7), for any $n \in \mathbb{N}$, is by using a repetition operator:

$$
\left(\big((?v < 4; a := a+1) \cup a := -b\big); \atop \{x' = v, v' = a\}\right)^{*}
\tag{3.10}
$$

This HP can repeat (3.7) any number of times $(0,1,2,3,4,\dots)$. Of course, it would not be very meaningful to repeat a loop half a time or minus 5 times, so the repetition count $n \in \mathbb{N}$ still has to be some natural number.

But how often does a nondeterministic repetition like (3.10) repeat then? That choice is again nondeterministic.

> **Note 16 (Nondeterministic *)** Repetition (*) is *nondeterministic*. That is, program α^* can repeat α *any* number ($n \in \mathbb{N}$) of times. The choice how often to run α is *nondeterministic*, i.e., there is no prior way of telling how often α will be repeated.

However, hold on, every time the loop in (3.10) is run, how long does the continuous evolution along $\{x' = v, v' = a\}$ in that loop iteration take? Or, actually, even in the loop-free (3.8), how long does the first $x'' = a$ take before the controller has its second control choice? How long did the continuous evolution take in (3.7) even?

There is a choice even in following a single differential equation! However deterministic the solution of the differential equation itself may be. Even if the solution of the differential equation is unique (which it is in the sufficiently smooth cases that we consider, according to Chap. 2), it is still a matter of choice how long to follow that solution. The choice is, as always in hybrid programs, nondeterministic.

> **Note 17 (Nondeterministic $x' = f(x)$)** The duration of evolution of a differential equation $(x' = f(x) \& Q)$ is *nondeterministic* (except that the evolution can never be so long that the state leaves Q). That is, $x' = f(x) \& Q$ can follow the solution of $x' = f(x)$ for any amount of time ($0 \le r \in \mathbb{R}$) within the interval of existence of the solution within Q. The choice how long to follow $x' = f(x)$ is *nondeterministic*, i.e., there is no prior way of telling how long $x' = f(x)$ will evolve (except that it can never leave Q).

3.3 Hybrid Programs

Based on the above gradual motivation, this section formally defines the programming language of hybrid programs [18, 20], in which all of the operators motivated above are allowed.

3.3.1 Syntax of Hybrid Programs

Formal grammars have worked well to define the syntax of terms e and first-order logic formulas Q in Chap. 2, which is why we, of course, continue to use a grammar to define the syntax of hybrid programs.

Definition 3.1 (Hybrid program). *Hybrid programs* are defined by the following grammar (α, β are HPs, x is a variable, e is a term possibly containing x, e.g., a polynomial in x, and Q is a formula of first-order logic of real arithmetic):

$$\alpha, \beta ::= x := e \mid ?Q \mid x' = f(x) \& Q \mid \alpha \cup \beta \mid \alpha; \beta \mid \alpha^*$$

The first three cases are called *atomic HPs*, the last three *compound HPs*, because they are built out of smaller HPs. The *assignment* $x := e$ instantaneously changes the value of variable x to the value of term e with a discrete state change. The *differential equation* $x' = f(x) \& Q$ follows a continuous evolution from the present value of x along the differential equation $x' = f(x)$ for any amount of time but restricted to the domain of evolution Q, where x' denotes the time-derivative of x. It goes without saying that $x' = f(x) \& Q$ is an explicit differential equation, so no derivatives occur in $f(x)$ or Q. Recall that a differential equation $x' = f(x)$ without an *evolution domain constraint* is short for $x' = f(x) \& true$, since that imposes no restriction on the duration of the continuous evolution. The *test* action $?Q$ is used to define conditions. Its effect is that of a *no-op* if the formula Q is true in the current state; otherwise, like an *abort* statement would, it allows no transitions. That is, if the test succeeds because formula Q holds in the current state, then the state does not change (it was only a test), and the system execution continues normally. If the test fails because formula Q does not hold in the current state, however, then the system execution cannot continue, and is cut off, discarded, and not considered any further since it is a failed execution attempt that did not play by the rules of the HP.[4]

Nondeterministic choice $\alpha \cup \beta$, sequential composition $\alpha; \beta$, and nondeterministic repetition α^* of programs are as in regular expressions but generalized to the semantics of hybrid systems. *Nondeterministic choice* $\alpha \cup \beta$ expresses behavioral alternatives between the runs of α and β. That is, the HP $\alpha \cup \beta$ can choose nondeterministically to follow the runs of HP α, or, instead, to follow the runs of HP β. The *sequential composition* $\alpha; \beta$ models that the HP β starts running after HP α has finished (β never starts if α does not terminate successfully). In $\alpha; \beta$, the runs of α take effect first, until α terminates (if it does), and then β continues. Observe that, like repetitions, continuous evolutions within α can take more or less time, which causes uncountable nondeterminism. This nondeterminism occurs in hybrid systems because they can operate in so many different ways, which is reflected in HPs. *Nondeterministic repetition* α^* is used to express that the HP α repeats any number of times, including zero times. When following α^*, the runs of HP α can be repeated over and over again, any nondeterministic number of times (≥ 0).

---

[4] The effect of the test $?Q$ is the same as that of if(Q) skip else abort where skip has no effect and abort aborts and discards the system run. Indeed, skip is equivalent to the trivial test $?true$ and abort is equivalent to the impossible test $?false$. But then we would have to add if-then-else, skip and abort, which HPs already provide for free.

Expedition 3.1 (Operator precedence for hybrid programs)

In practice, it is useful to save parentheses by agreeing on notational *operator precedences*. Unary operators (including repetition *) bind more strongly than binary operators and ; binds more strongly than $\cup$, so $\alpha; \beta \cup \gamma \equiv (\alpha; \beta) \cup \gamma$ and $\alpha \cup \beta; \gamma \equiv \alpha \cup (\beta; \gamma)$. Especially, $\alpha; \beta^* \equiv \alpha; (\beta^*)$.

3.3.2 Semantics of Hybrid Programs

After having developed a syntax for CPS and an operational intuition for its effects, we seek operational precision in its effects. That is, we will pursue one important leg of computational thinking and give an unambiguous meaning to all operators of HPs. We will do this in pursuit of the realization that the only way to be precise about an analysis of CPS is to first be precise about the meaning of the models of CPS. Furthermore, we will leverage another important leg of computational thinking rooted in logic by exploiting that the right way of understanding something is to understand it compositionally as a function of its pieces [6]. So we will give meaning to hybrid programs by giving a meaning to each of their operators. Thereby, a meaning of a large HP is merely a function of the meaning of its pieces. This is the style of denotational semantics for programming languages due to Scott and Strachey [25].

There is more than one way to define the meaning of a program, including defining a denotational semantics [24], an operational semantics [24], a structural operational semantics [22], or an axiomatic semantics [9, 23]. For our purposes, what is most relevant is how a hybrid program changes the state of the system. Consequently, the semantics of hybrid programs considers what (final) state v is reachable by running an HP α from an (initial) state ω. Semantical models that expose more detail, e.g., about the internal states during the run of an HP, are possible [11] but can be ignored for most purposes in this book.

Recall that a *state* $\omega : \mathcal{V} \to \mathbb{R}$ is a mapping from variables to $\mathbb{R}$, which assigns a real value $\omega(x) \in \mathbb{R}$ to each variable $x \in \mathcal{V}$. The *set of states* is denoted $\mathcal{S}$. The meaning of an HP α is given by a reachability relation $[\![\alpha]\!] \subseteq \mathcal{S} \times \mathcal{S}$ on states. So $(\omega, v) \in [\![\alpha]\!]$ means that final state v is reachable from initial state ω by running HP α. From any initial state ω, there might be many states v that are reachable because the HP α may involve nondeterministic choices, repetitions, or differential equations, so there may be many different states v for which $(\omega, v) \in [\![\alpha]\!]$. From other initial states ω, there might be no reachable states v at all for which $(\omega, v) \in [\![\alpha]\!]$. So $[\![\alpha]\!]$ is a proper relation, not a function.

HPs have a compositional semantics [17–19]. Recall from Chap. 2 that the value of term e in state ω is denoted by $\omega[\![e]\!]$. Further, $\omega \in [\![Q]\!]$ denotes that first-order formula Q is true in state ω, where $[\![Q]\!] \subseteq \mathcal{S}$ is the set of all states in which formula Q is *true*. The semantics of an HP α is then defined by its reachability relation $[\![\alpha]\!] \subseteq \mathcal{S} \times \mathcal{S}$. The notation α^* for loops comes from the notation ρ^* for the re-

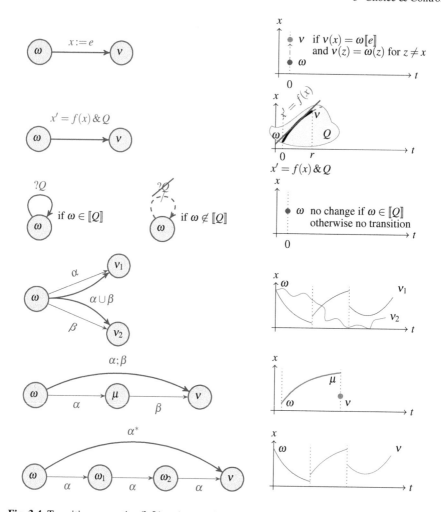

Fig. 3.4 Transition semantics (**left**) and example dynamics (**right**) of hybrid programs

flexive, transitive closure of a relation ρ. Graphical illustrations of the transition semantics of hybrid programs defined below and possible example dynamics are depicted in Fig. 3.4. The left of Fig. 3.4 illustrates the generic shape of the transition structure $[\![\alpha]\!]$ for transitions along various cases of hybrid programs α from state ω to state v. The right of Fig. 3.4 shows examples of how the value of a variable x may evolve over time t when following the dynamics of the respective hybrid program α.

Definition 3.2 (Transition semantics of HPs). Each HP α is interpreted semantically as a binary reachability relation $[\![\alpha]\!] \subseteq \mathscr{S} \times \mathscr{S}$ over states, defined inductively by:

1. $[\![x := e]\!] = \{(\omega, v) \; : \; v = \omega \text{ except that } v[\![x]\!] = \omega[\![e]\!]\}$
   That is, final state v differs from initial state ω only in its interpretation of the variable x, which v changes to the value that the right-hand side term e has in the initial state ω.

2. $[\![?Q]\!] = \{(\omega, \omega) \; : \; \omega \in [\![Q]\!]\}$
   That is, the final state ω is the same as the initial state ω (no change) but there is such a transition only if test formula Q holds in ω, otherwise no transition is possible at all and the system is stuck because of a failed test.

3. $[\![x' = f(x) \& Q]\!] = \{(\omega, v) : \varphi(0) = \omega \text{ except at } x' \text{ and } \varphi(r) = v \text{ for a solution } \varphi : [0, r] \to \mathscr{S} \text{ of any duration } r \text{ satisfying } \varphi \models x' = f(x) \wedge Q\}$
   That is, the final state $\varphi(r)$ is connected to the initial state $\varphi(0)$ by a continuous function of some duration $r \geq 0$ that solves the differential equation and satisfies Q at all times; see Definition 3.3.

4. $[\![\alpha \cup \beta]\!] = [\![\alpha]\!] \cup [\![\beta]\!]$
   That is, $\alpha \cup \beta$ can do exactly any of the transitions that α can do as well as any of the transitions that β is capable of. Every run of $\alpha \cup \beta$ has to choose whether it follows α or β, but cannot follow both at once.

5. $[\![\alpha; \beta]\!] = [\![\alpha]\!] \circ [\![\beta]\!] = \{(\omega, v) \; : \; (\omega, \mu) \in [\![\alpha]\!], (\mu, v) \in [\![\beta]\!]\}$
   That is, the meaning of $\alpha; \beta$ is the composition[a] $[\![\alpha]\!] \circ [\![\beta]\!]$ of relation $[\![\beta]\!]$ after $[\![\alpha]\!]$. Thus, $\alpha; \beta$ can do any transitions that go through any intermediate state μ to which α can make a transition from the initial state ω and from which β can make a transition to the final state v.

6. $[\![\alpha^*]\!] = [\![\alpha]\!]^* = \bigcup\limits_{n \in \mathbb{N}} [\![\alpha^n]\!]$ with $\alpha^{n+1} \equiv \alpha^n; \alpha$ and $\alpha^0 \equiv ?true$.
   That is, α^* can repeat α any number of times, i.e., for any $n \in \mathbb{N}$, α^* can act like the n-fold sequential composition $\alpha^n \equiv \underbrace{\alpha; \alpha; \alpha; \ldots; \alpha}_{n \text{ times}}$ would.

---

[a] The notational convention for composition of relations is flipped compared to the composition of functions. For functions f and g, the function $f \circ g$ is the composition f after g that maps x to $f(g(x))$. For relations R and T, the relation $R \circ T$ is the composition of T after R, so first follow relation R to an intermediate state and then follow relation T to the final state.

To keep things simple, this definition uses simplifying abbreviations for differential equations. Chapter 2 provides full detail, including the definition for differential equation systems. The semantics of loops can also be rephrased equivalently as:

$$[\![\alpha^*]\!] = \bigcup\limits_{n \in \mathbb{N}} \{(\omega_0, \omega_n) : \omega_0, \ldots, \omega_n \text{ are states such that } (\omega_i, \omega_{i+1}) \in [\![\alpha]\!] \text{ for all } i < n\}$$

For later reference, we repeat the definition of the semantics of differential equations separately:

Definition 3.3 (Transition semantics of ODEs).

$$[\![x' = f(x)\,\&\,Q]\!] = \{(\omega, v) : \varphi(0) = \omega \text{ except at } x' \text{ and } \varphi(r) = v \text{ for a solution}$$

$$\varphi{:}[0,r] \to \mathscr{S} \text{ of any duration } r \text{ satisfying } \varphi \models x' = f(x) \wedge Q\}$$

where $\varphi \models x' = f(x) \wedge Q$, iff for all times $0 \le z \le r$: $\varphi(z) \in [\![x' = f(x) \wedge Q]\!]$
with $\varphi(z)(x') \stackrel{\text{def}}{=} \frac{d\varphi(t)(x)}{dt}(z)$ and $\varphi(z) = \varphi(0)$ except at x, x'.

The condition that $\varphi(0) = \omega$ *except at* x' is explicit about the fact that the initial state ω and the first state $\varphi(0)$ of the continuous evolution have to be identical (except for the value of x', for which Definition 3.3 only provides a value along φ). Part I of this book does not track the values of x' except during continuous evolutions. But that will change in Part II, for which Definition 3.3 is already prepared appropriately.

Observe that $?Q$ cannot run from an initial state ω with $\omega \notin [\![Q]\!]$, in particular $[\![?false]\!] = \emptyset$. Likewise, $x' = f(x)\,\&\,Q$ cannot run from an initial state ω with $\omega \notin [\![Q]\!]$, because no solution of any duration, not even duration 0, starting in ω will always stay in the evolution domain Q if it already starts outside Q. A nondeterministic choice $\alpha \cup \beta$ cannot run from an initial state from which neither α nor β can run. Similarly, $\alpha;\beta$ cannot run from an initial state from which α cannot run, nor from an initial state from which all final states after α make it impossible for β to run. Assignments and repetitions can always run, e.g., by repeating 0 times.

Example 3.1. When α denotes the HP in (3.8) on p. 74, its semantics $[\![\alpha]\!]$ is a relation on states connecting the initial to the final state along the differential equation with two control decisions according to the nondeterministic choice, one at the beginning and one after following the first differential equation. How long is that, exactly? Well, that's nondeterministic, because the semantics of differential equations is such that any final state after any permitted duration is reachable from a given initial state. So the duration for the first differential equation in (3.8) could be one second or two or 424 or half a second or zero or π or any other nonnegative real number. This would be very different for an HP whose differential equation has an evolution domain constraint, because that limits how long a continuous evolution can take. The exact duration is still nondeterministic, but it cannot ever evolve outside its evolution domain.

By plugging one transition structure pattern into another, Fig. 3.4 illustrates the generic shape of transition structures for more complex HPs. For example, Fig. 3.5 illustrates the transition structure of $(\alpha;\beta)^*$ and Fig. 3.6 illustrates $(\alpha \cup \beta)^*$. This plugging in is directly analogous to how the semantics of bigger programs is defined by recursively following their semantics in Definition 3.2 based on their respective top-level operator.

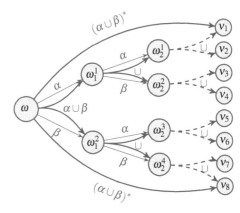

Fig. 3.5 Nested transition semantics pattern for $(\alpha;\beta)^*$

Fig. 3.6 Nested transition semantics pattern for $(\alpha \cup \beta)^*$

Expedition 3.2 (HP semantics $\llbracket \cdot \rrbracket : \mathrm{HP} \to \wp(\mathscr{S} \times \mathscr{S})$)

The semantics of an HP α from Definition 3.2 directly defines the transition relation $\llbracket \alpha \rrbracket \subseteq \mathscr{S} \times \mathscr{S}$ of initial and final states inductively, for each HP α:

$$\llbracket x := e \rrbracket = \{(\omega, \nu) \; : \; \nu = \omega \text{ except that } \nu\llbracket x \rrbracket = \omega\llbracket e \rrbracket\}$$
$$\llbracket ?Q \rrbracket = \{(\omega, \omega) \; : \; \omega \in \llbracket Q \rrbracket\}$$
$$\llbracket x' = f(x) \& Q \rrbracket = \{(\omega, \nu) : \varphi(0) = \omega \text{ except at } x' \text{ and } \varphi(r) = \nu \text{ for a solution}$$
$$\varphi:[0,r] \to \mathscr{S} \text{ of any duration } r \text{ satisfying } \varphi \models x' = f(x) \wedge Q\}$$
$$\llbracket \alpha \cup \beta \rrbracket = \llbracket \alpha \rrbracket \cup \llbracket \beta \rrbracket$$
$$\llbracket \alpha;\beta \rrbracket = \llbracket \alpha \rrbracket \circ \llbracket \beta \rrbracket = \{(\omega, \nu) : (\omega, \mu) \in \llbracket \alpha \rrbracket, (\mu, \nu) \in \llbracket \beta \rrbracket\}$$
$$\llbracket \alpha^* \rrbracket = \llbracket \alpha \rrbracket^* = \bigcup_{n \in \mathbb{N}} \llbracket \alpha^n \rrbracket \text{ with } \alpha^{n+1} \equiv \alpha^n; \alpha \text{ and } \alpha^0 \equiv ?true$$

When HP is the set of hybrid programs, the semantic brackets define an operator $\llbracket \cdot \rrbracket : \mathrm{HP} \to \wp(\mathscr{S} \times \mathscr{S})$ that defines the meaning $\llbracket P \rrbracket$ for each hybrid program $\alpha \in \mathrm{HP}$, which, in turn, defines the transition relation $\llbracket \alpha \rrbracket \subseteq \mathscr{S} \times \mathscr{S}$ where $(\omega, \nu) \in \llbracket \alpha \rrbracket$ indicates that final state ν is reachable from initial state ω when running HP α. The *powerset* $\wp(\mathscr{S} \times \mathscr{S})$ is the set of all subsets of the Cartesian product $\mathscr{S} \times \mathscr{S}$ of the set of states $\mathscr{S}$ with itself. The powerset $\wp(\mathscr{S} \times \mathscr{S})$, thus, is the set of binary relations on $\mathscr{S}$.

3.4 Hybrid Program Design

This section discusses some early lessons on good and bad modeling choices in hybrid systems. As our understanding of the subject matter advances throughout this textbook, we will find additional insights into tradeoffs and caveats. The aspects that can easily be understood on a pure modeling level will be discussed now.

3.4.1 To Brake, or Not to Brake, That Is the Question

As a canonical example for a system that has to make a choice, consider a ground robot at position x moving with velocity v and acceleration a along a straight line. This results in the differential equation $x' = v, v' = a$. When driving along a straight line, the ground robot has the control decision to either set the acceleration to a positive value $A > 0$ by the discrete assignment $a := A$ or set it to a negative value $-b < 0$ by the discrete assignment $a := -b$. The control question is when to brake and when not to brake (so accelerate since, for simplicity, this example does not allow coasting with a constant velocity). Let's call the condition under which acceleration can be chosen Q_A and the condition under which braking can be chosen Q_b:

$$((?Q_A; a := A \cup ?Q_b; a := -b); \{x' = v, v' = a\})^* \tag{3.11}$$

What concrete formulas to best use for the tests $?Q_A$ and $?Q_b$ depends on the control objectives and is often quite nontrivial to determine. If the system can stay in its continuous evolution for an unbounded amount of time, then it is virtually never safe to accelerate, so Q_A will need to be *false*. Consequently, we assume that ε is the reaction time, so the maximal amount of time that a continuous evolution can take before it stops and gives the discrete controller a chance to react to the new situation again. The HP, thus, includes a clock t measuring the progress of time along $t' = 1$, which is reset with $t := 0$ before the differential equation and bounded by the evolution domain $t \leq \varepsilon$. Finally, if the controller chooses a negative acceleration with $a := -b$, then the differential equation system $x' = v, v' = a$ will ultimately move backwards when $v < 0$. In order to model that braking will not make the robot move backwards, we add $v \geq 0$ to the evolution domain constraint.

Taking these thoughts about braking and acceleration into account, we refine HP (3.11) with a clock and a lower velocity bound 0:

$$((?Q_A; a := A \cup ?Q_b; a := -b); t := 0; \{x' = v, v' = a, t' = 1 \& v \geq 0 \wedge t \leq \varepsilon\})^* \tag{3.12}$$

The structure of the transition semantics for HP (3.12) according to the patterns in Fig. 3.4 is shown in Fig. 3.7. Any path through Fig. 3.7 that passes all tests along the way corresponds to an execution of HP (3.12), and vice versa. In particular, if a test fails, such as $?Q_A$, then only the other nondeterministic choices are available, which is why it is important that at least one choice is always allowed. If both tests $?Q_A$ and $?Q_b$ pass in the current state, then both nondeterministic choices are available.

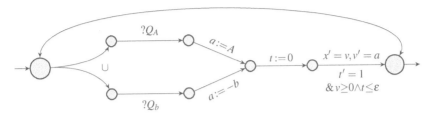

Fig. 3.7 Transition structure of the acceleration/braking example (3.12)

For verification purposes, it is often a good idea to design conditions Q_A and Q_b that overlap. Executing a controller may best be done in a deterministic way, which would argue for a disjoint design of Q_A and Q_b. But verification models benefit from nondeterminism, because if highly nondeterministic models have no unsafe behaviors then all more specific refinements have no unsafe behaviors either [12]. In particular, the controller is only guaranteed to remain responsive if there is always at least one option that can be chosen, which is obvious when choosing the trivial test $?true$ for $?Q_b$, which always allows the control to choose braking. Of course, when the control objective is to avoid collision with a moving obstacle in front of our robot, then it is always safe to brake, but only sometimes safe to accelerate. So, Q_A has to be some condition that ensures sufficient distance to the moving obstacle, which depends on how good the brakes are and several other parameters as well (Exercise 3.7).

3.4.2 A Matter of Choice

Let us change the HP from (3.8) and consider the following modification instead:

$$
\begin{aligned}
&?v < 4;\, a := a+1;\\
&\{x' = v, v' = a\};\\
&?v < 4;\, a := a+1;\\
&\{x' = v, v' = a\}
\end{aligned}
\tag{3.13}
$$

Then some behavior that was possible in (3.8) is no longer possible for (3.13). Let β denote the HP in (3.13), then the semantics $[\![\beta]\!]$ of β now only includes relations between initial and final states which can be reached by acceleration choices (because there are no braking choices in β). Note that the duration of the first differential equation in (3.13) is suddenly bounded, because if x keeps on accelerating for too long during the first differential equation, the intermediate state reached then will violate the test $?v < 4$, which, according to the semantics of tests, will fail and be discarded.

That is what makes (3.13) a bad model, because it truncates and discards behavior that the real system still possesses. Even if the controller in the third line of (3.13) is not prepared to handle the situation where the test $?v < 4$ fails, it might fail in reality. In that case, the controller in (3.13) simply runs out of choices. A more realistic and permissive controller, thus, also handles the case if that test fails, at which point we are back at (3.8).

Similarly, $Q_b \overset{\text{def}}{=} \mathit{false}$ is a bad controller design for (3.12), because it categorically disallows braking and, unrealistically, assumes $?Q_A$ to hold all the time.

Note 18 (Controllers cannot discard cases) While subsequent chapters discuss cases where hybrid programs use tests $?Q$ in crucial ways to discard non-permitted behaviors of the environment, great care needs to be exercised that controllers also handle the remaining cases. A bad controller

$$?(v < 4);\ \alpha$$

only handles the case where $v < 4$ and ignores all other circumstances, which renders the controller incapable of reacting and, thus, unsafe when $v \geq 4$. A better controller design always considers the case when a condition is not satisfied and handles it appropriately as well:

$$(?(v < 4); \alpha) \cup (?(v \geq 4); \ldots)$$

Liveness proofs can tell the two cases of controllers apart, but appropriate design principles of being prepared for both outcomes of each test go a long way in improving the controllers.
Similarly bad controller designs result from careless evolution domains:

$$a := -b;\ \{x' = v, v' = a \,\&\, v > 4\}$$

The differential equations in this controller silently assume the velocity will always stay above 4, which is clearly not always the case when braking. Accidental divisions by zero are another source of trouble in CPS controllers.

3.5 Summary

This chapter introduced hybrid programs as a model of cyber-physical systems, summarized in Table 3.1. Hybrid programs combine differential equations with conventional program constructs and discrete assignments. The programming language of hybrid programs embraces nondeterminism as a first-class citizen and features differential equations that can be combined to form hybrid systems using the compositional operators of hybrid programs.

Table 3.1 Statements and effects of hybrid programs (HPs)

HP Notation	Operation	Effect
$x := e$	discrete assignment	assigns current value of term e to variable x
$x' = f(x) \& Q$	continuous evolution	follow differential equation $x' = f(x)$ within evolution domain Q for any duration
$?Q$	state test / check	test first-order formula Q at current state
$\alpha; \beta$	seq. composition	HP β starts after HP α finishes
$\alpha \cup \beta$	nondet. choice	choice between alternatives HP α and HP β
α^*	nondet. repetition	repeats HP α any $n \in \mathbb{N}$ times

Hybrid programs take advantage of the structuring principles of programming languages and emphasize the composition operators ; and $\cup$ and *, which combine smaller hybrid programs to make bigger hybrid programs and have a simple compositional semantics. Hybrid programs resemble regular expressions, except that they start from discrete assignments, tests, and differential equations as a basis instead of individual letters of a formal language. Regular expressions allow a language-theoretic study of formal languages and have an automata-theoretic counterpart called finite automata [10]. Similarly, hybrid programs have *hybrid automata* [2, 8] as an automata-theoretic counterpart, which are explored in Exercise 3.18.

3.6 Appendix: Modeling the Motion of a Robot Around a Bend

This appendix develops a hybrid program model describing how a robot can drive along a sequence of lines and circular curve segments in the two-dimensional plane. This dynamics is called Dubins car dynamics [5], because it also describes the high-level motion of a car in the plane or of an aircraft remaining at the same altitude (Fig. 3.8).

Suppose there is a robot at a point with coordinates (x, y) that is facing in direction (v, w). The robot moves into the direction (v, w), whose norm $\sqrt{v^2 + w^2}$, thus, simultaneously determines the constant linear speed: how quickly the robot is moving on the ground. Suppose that direction (v, w) is simultaneously rotating with an angular velocity ω as in Example 2.7 (Fig. 3.9). The differential equations describing the motion of this robot are

$$x' = v, y' = w, v' = \omega w, w' = -\omega v$$

The time-derivative of the x coordinate is the component v of the direction and the time-derivative of the y coordinate is the component w of the direction. The angular velocity ω determines how fast the direction (v, w) rotates. Bigger magnitudes of

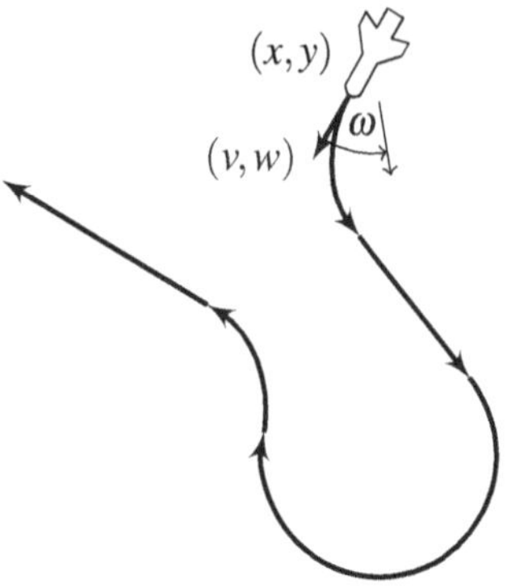

Fig. 3.8 Illustration of a
Dubins path consisting of a
sequence of lines and maximally curved circle segments

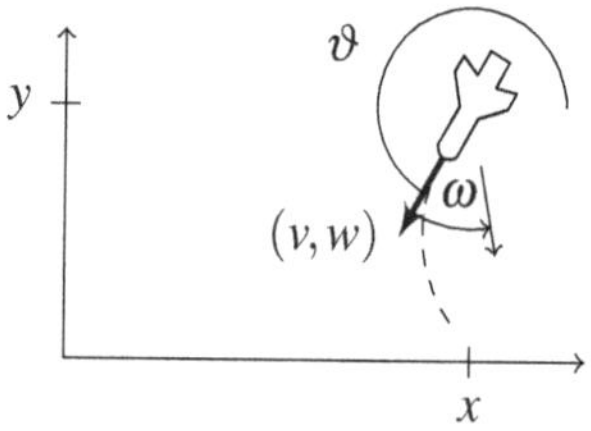

(x,y)

(v,w) ω

Fig. 3.9 Illustration of the
Dubins dynamics of a point
(x,y) moving in direction
(v,w) along a dashed curve
with angular velocity ω

ϑ

y

(v,w) ω

x

ω give faster rotations of (v,w) so tighter curves for the position (x,y). Positive ω
make the robot drive a right curve (when indeed plotting x as the x-axis and y as the
y-axis as in Fig. 3.9). The robot will follow a straight line when $\omega = 0$, because the
direction (v,w) then does not change.

Now if a robot can steer, its controller can change the angular velocity to make a
left curve ($\omega < 0$), a right curve ($\omega > 0$), or drive straight ahead ($\omega = 0$). One might
very well imagine a robot controller that additionally chooses sharp turns (ω with
big magnitude) or gentle turns (ω with small magnitude), but let's not consider this
yet. After all, Lester Dubins proved that the shortest curve connecting two points
with such a dynamics consists of a sequence of line segments and maximally sharp
turns [5].[5] If we simply assume that 1 and -1 are the extreme angular velocities, the
hybrid program for the ground robot steering and moving in the plane is

$$((\omega := -1 \cup \omega := 1 \cup \omega := 0); \{x' = v, y' = w, v' = \omega w, w' = -\omega v\})^* \quad (3.14)$$

Repeatedly, in a loop, this HP allows a choice of extreme left curves ($\omega := -1$),
extreme right curves ($\omega := 1$), or motion in a straight line ($\omega := 0$). After this discrete
controller, the robot follows the continuous motion described by the differential
equations for a nondeterministic period of time.

If the safety goal of the robot is never to collide with any obstacles, then HP
(3.14) cannot possibly be safe, because it allows arbitrary left and right curves and

---

[5] If you build a self-driving car following such a path with straight lines and maximal curvature,
don't be surprised if no passenger stays for a second ride. But robots are less fixated on comfort.

straight-line motion for any arbitrary amounts of time under any arbitrary conditions. HP (3.14) would even allow the controller to choose a left curve if that is the only direction that will immediately make the robot collide with an obstacle.

Consequently, each of the three control actions in HP (3.14) is only acceptable under certain conditions. After solving Exercise 3.9 you will have found logical formulas Q_{-1}, Q_1, Q_0 such that Q_ω indicates when it is safe to drive along the curve corresponding to angular velocity ω, which will transform the unsafe HP (3.14) into the following HP with a more constrained controller:

$$((?Q_{-1}; \omega := -1 \cup ?Q_1; \omega := 1 \cup ?Q_0; \omega := 0); \{x' = v, y' = w, v' = \omega w, w' = -\omega v\})^*$$

It is perfectly fine if multiple of the conditions Q_{-1}, Q_1, Q_0 are true in the same state, because that gives the controller a number of different control options to choose from. For example, there might be many states in which both a left curve and driving straight are safe. Of course, it would not be useful at all if there is a state in which all conditions Q_{-1}, Q_1, Q_0 are false, because the controller runs out of control choices and is then completely stuck, which is not very safe either. Grotesquely useless would be a controller that chose an impossible condition like $1 < 0$ for all three formulas Q_{-1}, Q_1, Q_0, because that robot can then never move anywhere, which is incredibly boring even for the most polite and patient robots. Yet, a robot that is initially stopped and never even begins to move at least does not bump into any walls. What is much worse is a controller that happily begins to drive and then fails to offer any acceptable control choices. That is why it is important that the disjunction $Q_{-1} \vee Q_1 \vee Q_0$ is true in every state, because the robot then always has at least one permitted choice.

Well, can this disjunction be true in every state? Since the conditions Q_ω are supposed to guarantee that the robot will never collide with an obstacle when following the trajectory with angular velocity ω, none of them can be true in a state where the robot has already collided to begin with. It is up to the robot controller to ensure that such a collision state is never reached. Hence, the disjunction $Q_{-1} \vee Q_1 \vee Q_0$ should be true in every collision-free state. How to design the Q_ω is your challenge for Exercise 3.9.

Exercises

3.1. The semantics of an HP α is its reachability relation $[\![\alpha]\!]$. For example,

$$[\![x := 2 \cdot x; x := x + 1]\!] = \{(\omega, v) : v(x) = 2 \cdot \omega(x) + 1 \text{ and } v(z) = \omega(z) \text{ for all } z \neq x\}$$

Describe the reachability relation of the following HPs in similarly explicit ways:

1. $x := x + 1; x := 2 \cdot x$
2. $x := 1 \cup x := -1$
3. $x := 1 \cup ?(x \leq 0)$

4. $x := 1; ?(x \leq 0)$
5. $?(x \leq 0)$
6. $x := 1 \cup x' = 1$
7. $x := 1; x' = 1$
8. $x := 1; \{x' = 1 \& x \leq 1\}$
9. $x := 1; \{x' = 1 \& x \leq 0\}$
10. $v := 1; x' = v$
11. $v := 1; \{x' = v\}^*$
12. $\{x' = v, v' = a \& x \geq 0\}$

3.2. The semantics of hybrid programs (Definition 3.2) requires evolution domain constraints Q to hold always throughout a continuous evolution. What exactly happens if the system starts in a state where Q does not hold to begin with?

3.3 (If-then-else). Sect. 3.2.3 considered if-then-else statements for hybrid programs. But they no longer showed up in the grammar of hybrid programs. Is this a mistake? Can you define if$(P) \alpha$ else β from the operators that HPs do provide?

3.4 (If-then-else). Suppose we add the if-then-else-statement if$(P) \alpha$ else β to the syntax of HPs. Define a semantics $[\![\text{if}(P) \alpha \text{ else } \beta]\!]$ for if-then-else statements and explain how it relates to Exercise 3.3.

3.5 (Switch-case). Define a switch statement that runs the statement α_i if formula P_i is true, and chooses *nondeterministically* if multiple conditions are true:

$$\text{switch } ($$
$$\text{case } P_1 : \alpha_1$$
$$\text{case } P_2 : \alpha_2$$
$$\vdots$$
$$\text{case } P_n : \alpha_n$$
$$)$$

What would need to be changed to make sure only the statement α_i with the first true condition P_i executes?

3.6 (While). Suppose we add the while loop while$(P) \alpha$ to the syntax of HPs. As usual, while$(P) \alpha$ is supposed to run α if P holds, and, whenever α finishes, repeat again if P holds. Define a semantics $[\![\text{while}(P) \alpha]\!]$ for while loops. Can you define a program that is equivalent to while$(P) \alpha$ from the original syntax of HPs?

3.7 (To brake, or not to brake, that is the question). Besides the positions of the robot and the position of the obstacle in front of it, what other parameters does the acceleration condition in (3.12) depend on if it is supposed to ensure that the robot does not collide with any obstacles on the straight line that it is driving along? Can you determine a corresponding formula Q_A that would guarantee safety?

3.8 (Two cars). Develop a model of the motion of two cars along a straight line, each of which has its own position, velocity, and acceleration. Develop a controller model that allows the leader car to accelerate or brake freely while limiting the choices of the follower car such that it will never collide with the car in front of it.

3.9 (Runaround robot). You are in control of a robot moving with constant ground speed in the two-dimensional plane, as in Sect. 3.6. It can follow a left curve ($\omega := -1$), a right curve ($\omega := 1$), or go straight ($\omega := 0$). Your job is to find logical formulas Q_{-1}, Q_1, Q_0 such that Q_ω indicates when it is safe to drive along the curve corresponding to angular velocity ω:

$$((?Q_{-1}; \omega := -1 \cup ?Q_1; \omega := 1 \cup ?Q_0; \omega := 0); \{x' = v, y' = w, v' = \omega w, w' = -\omega v\})^*$$

For the purpose of this exercise, fix one point (o_x, o_y) as an obstacle and consider this HP safe if it can never reach that obstacle. Does your HP always have at least one choice remaining or can it get stuck such that no choice is permitted? Having succeeded with these challenges, can you generalize your robot model and safety constraints to one where the robot can accelerate to speed up or brake to slow down?

3.10 (Other programming languages). Consider your favorite programming language and discuss in what ways it introduces discrete change and discrete dynamics. Can it model all behavior that hybrid programs can describe? Can your programming language model all behavior that hybrid programs without differential equations can describe? How about the other way around? And what would you need to add to your programming language to cover all of hybrid systems? How would you best do that?

3.11 (Choice vs. sequence). Can you find a discrete controller *ctrl* and a continuous program *plant* such that the following two HPs have different behaviors?

$$(ctrl; plant)^* \quad \text{versus} \quad (ctrl \cup plant)^*$$

3.12 (Nondeterministic assignments). Suppose we add a new statement $x := *$ for nondeterministic assignment to the syntax of HPs. The *nondeterministic assignment* $x := *$ assigns an *arbitrary* real number to the variable x. Define a semantics $[\![x := *]\!]$ for the $x := *$ statement.

3.13 (Nondeterministic choices from nondeterminism and if-then-else). Exercise 3.3 explored that if-then-else can be defined from nondeterministic choices. Once we add nondeterministic assignments, however, we could have defined it conversely. Using an auxiliary variable z, show that $\alpha \cup \beta$ has the same behavior as:

$$z := *; \text{ if}(z > 0) \, \alpha \, \text{else} \, \beta$$

3.14 (Nondeterministic repetitions from nondeterminism and while). The Exercise 3.6 explored that while loops can be defined from nondeterministic repetitions. Once we add nondeterministic assignments, however, we could have defined it conversely. Using an auxiliary variable z, show that α^* has the same behavior as:

$$z := *; \ \mathsf{while}(z > 0) \, (z := *; \ \alpha)$$

3.15 (Set-valued semantics). The semantics of hybrid programs (Definition 3.2) is defined as a transition relation $[\![\alpha]\!] \subseteq \mathscr{S} \times \mathscr{S}$ on states. Define an equivalent semantics using functions $R(\alpha) : \mathscr{S} \to 2^{\mathscr{S}}$ from the initial state to the *set* of all final states, where $2^{\mathscr{S}}$ denotes the powerset of $\mathscr{S}$, i.e., the set of all subsets of $\mathscr{S}$. Define this set-valued semantics $R(\alpha)$ without referring to the transition relation semantics $[\![\alpha]\!]$, and prove that it is equivalent, i.e.,

$$v \in R(\alpha)(\omega) \quad \text{iff} \quad (\omega, v) \in [\![\alpha]\!]$$

Likewise, define an equivalent semantics based on functions $\varsigma(\alpha) : 2^{\mathscr{S}} \to 2^{\mathscr{S}}$ from the set of possible final states to the set of initial states that can end in the given set of final states. Prove that it is equivalent, i.e., for all sets of states $X \subseteq \mathscr{S}$

$$\omega \in \varsigma(\alpha)(X) \quad \text{iff} \quad \text{there is a state } v \in X \text{ such that } (\omega, v) \in [\![\alpha]\!]$$

3.16 (Switched systems). Hybrid programs come in different classes; see Table 3.2. A continuous program is an HP that only consists of one continuous evolution of the form $x' = f(x) \,\&\, Q$. A discrete system corresponds to an HP that has no differential equations. A switched continuous system corresponds to an HP that has no assignments, because it does not have any instant changes of state variables but merely switches mode (possibly after some tests) from one continuous mode into another.

Table 3.2 Classification of hybrid programs and correspondence to dynamical systems

HP class	Dynamical systems class
only ODE	continuous dynamical systems
no ODE	discrete dynamical systems
no assignment	switched continuous dynamical systems
general HP	hybrid dynamical systems

Consider an HP in which the variables are partitioned into state variables (x, v), sensor variables (m), and controller variables (a):

$$\Big(\big((?x < m - 5; a := A) \cup a := -b \big); \\ \{x' = v, v' = a\} \Big)^{*}$$

Transform this HP into a switched program that has the same behavior on the observable state and sensor variables but is a switched system, so does not contain any assignments. The behavior of controller variables is considered irrelevant for the purpose of this transformation as long as the behavior of the other state variables x, v is unchanged.

3.17 (** Program interpreter). In a programming language of your choosing, fix a recursive data structure for hybrid programs from Definition 3.1 and fix some finite representation for states where all variables have rational values instead of reals. Write a program interpreter as a computer program that, given an initial state ω and a program α, successively enumerates possible final states v that can be reached by α from ω, that is $(\omega, v) \in [\![\alpha]\!]$, by implementing Definition 3.2. Resolve nondeterministic choices in the transition either by user input or by randomization. What makes the differential equation case particularly challenging?

3.18 (Hybrid automata).** The purpose of this exercise is to explore hybrid automata [2, 8], which are an automata-theoretic model for hybrid systems. Instead of the compositional language operators of hybrid programs, hybrid automata emphasize different continuous modes with discrete transitions between them. The system follows a continuous evolution while the automaton is in a node (called a location). Discrete jumps happen when following an edge of the automaton from one location to another. Hybrid automata augment finite automata with the specification of a differential equation and evolution domain in each location and with a description of the discrete transition (called a reset) for each edge in addition to a condition (called a guard) specifying when that edge can be taken.

Fig. 3.10 Hybrid automaton for a car that can accelerate or brake

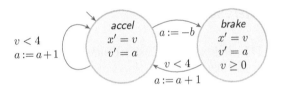

Figure 3.10 shows a hybrid automaton with two locations and three transitions between them, which starts in location *accel* as indicated by the initial arrow. While in location *accel*, the system follows the differential equation $x' = v, v' = a$. While in *brake* it follows $x' = v, v' = a \,\&\, v \geq 0$. When the automaton is in location *brake* and guard condition $v < 4$ is satisfied, then it can transition along the edge to location *accel*, which will change a by executing reset $a := a + 1$. When the automaton is in *accel*, then it can transition along the edge to *brake*, which will execute $a := -b$. This transition is always possible, because that edge has no guard. If $v < 4$, then the automaton can also transition from *accel* back to *accel*, which will execute $a := a + 1$.

1. Modify the hybrid automaton in Fig. 3.10 so that it directly corresponds to the hybrid program (3.10).
2. Draw a hybrid automaton for the hybrid program in Sect. 3.6.
3. Define the syntax of hybrid automata consisting of a (finite) set X of state variables and a (finite) set *Loc* of locations interconnected by a (finite) set *Edg* of edges where each location $\ell \in Loc$ has a differential equation $Flow(\ell)$ and evolution domain constraint $Inv(\ell)$ and where each edge $e \in Edg$ has a guard condition $Guard(e)$ and a reset $Reset(e)$ list of assignments. Also define the initial conditions by a formula $Init(\ell)$ per location $\ell \in Loc$ specifying in which region (if any) the hybrid automaton is allowed to start initially.

4. A state of a hybrid automaton is a pair (ℓ, ω) consisting of a location $\ell \in Loc$ and an assignment $\omega : X \to \mathbb{R}$ of real numbers to the variables X. Define the semantics of hybrid automata by defining which states (k, v) are reachable from initial state (ℓ, ω) by running the hybrid automaton.

5. Every finite automaton can be implemented in imperative programming languages with the help of a variable q storing the present location of the automaton that is updated to reflect the transition from one location to another. By analogy, show how every hybrid automaton can be implemented as a hybrid program with one additional variable q for the location. If we assume that the locations Loc are a set of distinct real numbers, the fact that the hybrid automaton is in location ℓ will correspond to the location variable q having value ℓ in the hybrid program.

6. Explain how the states (ℓ, ω) reachable by a hybrid automaton correspond to the states ω_q^ℓ reachable by the corresponding hybrid program in which location variable q has value ℓ.

References

[1] Rajeev Alur, Costas Courcoubetis, Nicolas Halbwachs, Thomas A. Henzinger, Pei-Hsin Ho, Xavier Nicollin, Alfredo Olivero, Joseph Sifakis, and Sergio Yovine. The algorithmic analysis of hybrid systems. *Theor. Comput. Sci.* **138**(1) (1995), 3–34. DOI: 10.1016/0304-3975(94)00202-T.

[2] Rajeev Alur, Costas Courcoubetis, Thomas A. Henzinger, and Pei-Hsin Ho. Hybrid automata: an algorithmic approach to the specification and verification of hybrid systems. In: *Hybrid Systems*. Ed. by Robert L. Grossman, Anil Nerode, Anders P. Ravn, and Hans Rischel. Vol. 736. LNCS. Berlin: Springer, 1992, 209–229. DOI: 10.1007/3-540-57318-6_30.

[3] Alonzo Church. A note on the Entscheidungsproblem. *J. Symb. Log.* **1**(1) (1936), 40–41.

[4] René David and Hassane Alla. On hybrid Petri nets. *Discrete Event Dynamic Systems* **11**(1-2) (2001), 9–40. DOI: 10.1023/A:1008330914786.

[5] Lester Eli Dubins. On curves of minimal length with a constraint on average curvature, and with prescribed initial and terminal positions and tangents. *American Journal of Mathematics* **79**(3) (1957), 497–516. DOI: 10.2307/2372560.

[6] Gottlob Frege. *Begriffsschrift, eine der arithmetischen nachgebildete Formelsprache des reinen Denkens*. Halle: Verlag von Louis Nebert, 1879.

[7] Robert Harper. *Practical Foundations for Programming Languages*. 2nd ed. Cambridge Univ. Press, 2016. DOI: 10.1017/CBO9781316576892.

[8] Thomas A. Henzinger. The theory of hybrid automata. In: *LICS*. Los Alamitos: IEEE Computer Society, 1996, 278–292. DOI: 10.1109/LICS.1996.561342.

[9]   Charles Antony Richard Hoare. An axiomatic basis for computer programming. *Commun. ACM* **12**(10) (1969), 576–580. DOI: 10.1145/363235.3 63259.

[10]   John E. Hopcroft, Rajeev Motwani, and Jeffrey D. Ullman. *Introduction to Automata Theory, Languages, and Computation.* 3rd ed. Pearson, Marlow, 2006.

[11]   Jean-Baptiste Jeannin and André Platzer. dTL^2: differential temporal dynamic logic with nested temporalities for hybrid systems. In: *IJCAR*. Ed. by Stéphane Demri, Deepak Kapur, and Christoph Weidenbach. Vol. 8562. LNCS. Berlin: Springer, 2014, 292–306. DOI: 10.1007/978-3-319-0 8587-6_22.

[12]   Sarah M. Loos and André Platzer. Differential refinement logic. In: *LICS*. Ed. by Martin Grohe, Eric Koskinen, and Natarajan Shankar. New York: ACM, 2016, 505–514. DOI: 10.1145/2933575.2934555.

[13]   Anil Nerode and Wolf Kohn. Models for hybrid systems: automata, topologies, controllability, observability. In: *Hybrid Systems*. Ed. by Robert L. Grossman, Anil Nerode, Anders P. Ravn, and Hans Rischel. Vol. 736. LNCS. Berlin: Springer, 1992, 317–356.

[14]   Xavier Nicollin, Alfredo Olivero, Joseph Sifakis, and Sergio Yovine. An approach to the description and analysis of hybrid systems. In: *Hybrid Systems*. Ed. by Robert L. Grossman, Anil Nerode, Anders P. Ravn, and Hans Rischel. Vol. 736. LNCS. Berlin: Springer, 1992, 149–178. DOI: 10.1007/3-540 -57318-6_28.

[15]   Ernst-Rüdiger Olderog. *Nets, Terms and Formulas: Three Views of Concurrent Processes and Their Relationship.* Cambridge: Cambridge University Press, 1991, 267.

[16]   André Platzer. Differential dynamic logic for verifying parametric hybrid systems. In: *TABLEAUX*. Ed. by Nicola Olivetti. Vol. 4548. LNCS. Berlin: Springer, 2007, 216–232. DOI: 10.1007/978-3-540-73099-6_17.

[17]   André Platzer. Differential dynamic logic for hybrid systems. *J. Autom. Reas.* **41**(2) (2008), 143–189. DOI: 10.1007/s10817-008-9103-8.

[18]   André Platzer. *Logical Analysis of Hybrid Systems: Proving Theorems for Complex Dynamics.* Heidelberg: Springer, 2010. DOI: 10.1007/978-3-642-14509-4.

[19]   André Platzer. Logics of dynamical systems. In: *LICS*. Los Alamitos: IEEE, 2012, 13–24. DOI: 10.1109/LICS.2012.13.

[20]   André Platzer. The complete proof theory of hybrid systems. In: *LICS*. Los Alamitos: IEEE, 2012, 541–550. DOI: 10.1109/LICS.2012.64.

[21]   André Platzer. A complete uniform substitution calculus for differential dynamic logic. *J. Autom. Reas.* **59**(2) (2017), 219–265. DOI: 10.1007/s108 17-016-9385-1.

[22]   Gordon D. Plotkin. *A structural approach to operational semantics.* Tech. rep. DAIMI FN-19. Denmark: Aarhus University, 1981.

[23]   Vaughan R. Pratt. Semantic considerations on Floyd-Hoare logic. In: *17th Annual Symposium on Foundations of Computer Science, 25-27 October*

1976, Houston, Texas, USA. Los Alamitos: IEEE, 1976, 109–121. DOI: 10 .1109/SFCS.1976.27.

[24] Dana S. Scott. *Outline of a Mathematical Theory of Computation.* Technical Monograph PRG–2. Oxford: Oxford University Computing Laboratory, Nov. 1970.

[25] Dana Scott and Christopher Strachey. *Towards a mathematical semantics for computer languages.* Tech. rep. PRG-6. Oxford Programming Research Group, 1971.

[26] Alan M. Turing. On computable numbers, with an application to the Entscheidungsproblem. *Proc. Lond. Math. Soc.* **42**(1) (1937), 230–265. DOI: 10.11 12/plms/s2-42.1.230.

Chapter 4
Safety & Contracts

Synopsis This chapter provides a lightweight introduction to safety specification techniques for cyber-physical systems. It discusses how program contracts generalize to CPS by declaring expectations on the initial states together with guarantees for all possible final states of a CPS model. Since assumptions and guarantees can be quite subtle for CPS applications, it is important to capture them early during a CPS design. This chapter introduces *differential dynamic logic*, a logic for specifying and verifying hybrid systems, which provides a formal underpinning for the precise meaning of CPS contracts. In subsequent chapters, differential dynamic logic plays a central rôle in rigorous verification of CPSs as well. This chapter also develops the running example of Quantum the bouncing ball, which is a hopelessly impoverished CPS but still features many of the important dynamical aspects of CPS in a perfectly intuitive setting.

4.1 Introduction

In the previous chapters, we have studied models of cyber-physical systems and the use of hybrid programs as their programming language [21, 22, 25, 30]. The distinguishing feature of hybrid programs are differential equations and nondeterminism alongside the usual classical control structures and discrete assignments. Together, these features provide powerful and flexible ways of modeling even very challenging systems and very complex control principles. This chapter will start the study of ways of making sure that the resulting behavior, however flexible and powerful it may be, also meets the required safety and correctness standards. Powerful and flexible CPSs would not be very useful if they failed to meet certain crucial safety demands.

In case you have already experienced contracts in conventional discrete programming languages, you will have observed how they make properties and input requirements of programs explicit. You will probably have seen how contracts can be checked dynamically at runtime, and, if they fail, this alerts you right away to flaws

A. Platzer, *Logical Foundations of Cyber-Physical Systems*,
https://doi.org/10.1007/978-3-319-63588-0_4

in the design of the program. In that case, you have experienced first hand that it is much easier to find and fix problems in programs starting from the first contract that failed in the middle of the program, rather than from the mere observation about the symptoms that ultimately surface when the final output of the program is not as expected. In particular, unless you check the output dynamically with a contract or manually every time, you may not even notice that something is wrong.

Another aspect of contracts that you may or may not have had the opportunity to observe is that they can be used in proofs, which show that *every* program run will satisfy the contracts, as opposed to just the ones you have tried. Every time the requirements hold for the input, the output will promise to meet its guarantees. Unlike in dynamic checking, the scope of correctness arguments with proofs extends far beyond the test cases that have been tried, however cleverly the tests may have been chosen. After all, testing can only show the presence of bugs, never quite their absence [32, 40]. Both uses of contracts, dynamic checking and rigorous proofs, are very helpful to check whether a system does what we intend it to, as has been argued repeatedly in the literature [5, 14, 17, 18, 34, 36, 39].

The principles of contracts help cyber-physical systems [3, 21, 22, 27] as well. Yet, their use in proving may, arguably, be more important than their use in dynamic checking. The reason has to do with the physical impact of CPS as well as the non-negotiability of the laws of physics. The reader is advised to imagine a situation where a self-driving car is propelling him or her down the street. Suppose the car's control software is covered with contracts all over, but all of them are exclusively for dynamic checking, none have been proved. If that self-driving car speeds up to 100 mph on a 55 mph highway and drives up very close to a car in front of it, then dynamically checking the contract "keep at least a distance of 1 meter to the car in front" no longer helps. If that contract fails, the car's software will know that it made a mistake, but it has become too late to do anything about it, because the brakes of the car cannot possibly slow the car down quickly enough. The car is "trapped in its own physics," in the sense that it has run out of all safe control options and can only brace itself for impact. While there are effective ways of making use of dynamic contract checking also in CPS [19], the design of those contracts requires proof to ensure that they respond early enough and that safety is always maintained.

For those reasons, this textbook will focus on the rôle of proofs as correctness arguments for CPS contracts much more than on their use in dynamical checking. Due to the physical consequences of malfunctions, correctness requirements on CPS are also more stringent. Subtle nuances in their behavior may require significantly more challenging arguments than, e.g., mere array-bounds checking for classical programs. For those reasons, we will approach CPS proofs with a fair amount of rigor. But such rigorous reasoning is already a story for a later chapter.

The focus of this chapter will first be to just understand CPS contracts themselves. As a useful modeling and specification exercise, we will develop a model of a bouncing ball and identify all requirements for it to be safe. Along the way, however, this chapter develops an intuitive understanding of the rôle of requirements and contracts in CPS as well as important ways of formalizing CPS properties and their analysis. The material in this chapter provides an intuitive gradual introduction to correctness specification techniques for CPS [20–22, 25, 30].

The most important learning goals of this chapter are:

Modeling and Control: We deepen our understanding of the core principles behind CPS by internalizing discrete and continuous aspects of CPS in logical formulas, which enable us to make precise statements about the expected behavior of a CPS. This forms a crucial stepping stone toward analytic reasoning principles.

Computational Thinking: We develop an example that is equally simple and instructive, to learn how to identify specifications and critical properties of CPS. Even if the example we look at, the bouncing ball, is a hopelessly impoverished CPS, it still conveys the formidable subtleties involved in hybrid systems models, which are crucial for understanding CPS. This chapter is devoted to contracts in the form of pre- and postconditions for CPS models. We will begin to rigorously specify our requirements and expectations for CPS models, which is critical to getting CPS right. In order to enable mathematically rigorous and unambiguous specifications, this chapter introduces *differential dynamic logic* dL [21, 22, 25, 30] as the specification and verification language for CPS that we will be using throughout this textbook.

CPS Skills: We will begin to deepen our understanding of the semantics of CPS models by tentatively relating it to their reasoning principles. This alignment will only be fully covered in the next chapter, though.

rigorous specification
contracts
preconditions
postconditions
differential dynamic logic

discrete+continuous
analytic specification

model semantics
reasoning principles

4.2 A Gradual Introduction to CPS Contracts

This section provides a gradual and informal introduction to contracts for cyber-physical systems. Its focus is on an intuitive development of the need for contracts, which provides the motivation for a subsequent rigorous development of a logic for CPS in which every aspect has an unambiguously well-defined meaning. This section also introduces the running example of Quantum the bouncing ball, which will be with us as a simple intuitive yet surprisingly representative example throughout the book.

A model of accelerated motion along a straight line with a choice of increasing acceleration or braking was considered in Sect. 3.2. That model did perform interesting control choices and we could continue to study it in this chapter. In order to sharpen our intuition about CPS, we will, however, prefer to study a very simple but also very intuitive system instead. Developing this example will be yet another welcome modeling exercise.

4.2.1 The Adventures of Quantum the Bouncing Ball

Once upon a time, there was a little bouncing ball called *Quantum*. Day in, day out, Quantum had nothing else to do but bounce up and down the street until he was tired of doing even that, which, in fact, rarely happened, because bouncing was such a joy for him (Fig. 4.1). Quantum the bouncing ball really was not much of a CPS, because bouncing balls do not actually have any interesting decisions to make. At least, Quantum was quite content without having to face any decisions, although, he will discover in Chap. 8 how empowering, subtle, and intriguing decisions can be. For one thing, Quantum did not even bring his computer, so he already lacked one simple indicator of qualifying as a cyber-physical system.

Fig. 4.1 Sample trajectory of a bouncing ball (plotted as height over time)

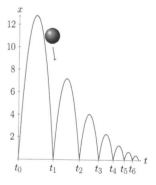

But Quantum nevertheless forms a perfectly reasonable hybrid system, because, after a closer look, the system turns out to involve both discrete and continuous dynamics. The continuous dynamics is caused by gravity, which is pulling the ball down and makes it fall down from the sky in the first place. The discrete dynamics comes from the singular discrete event of what happens when the ball hits the ground and bounces back up. There are a number of ways of modeling the ball and its impact on the ground with physics. They include a whole range of different, and either less or more realistic, physical effects such as gravity, aerodynamic resistance, the elastic deformation on the ground, and so on and so forth. But the little bouncing ball, Quantum, didn't study enough physics to know anything about those effects. And so Quantum had to go about understanding the world in easier terms. Quantum was a clever little bouncing ball, though, so he had also experienced the phenomenon of sudden change and was trying to use that to his advantage.

When looking for a very simple model of what the bouncing ball does, it is easier to describe it as a hybrid system. The ball at height x is falling subject to gravity g:

$$x'' = -g$$

That is, the height x changes with the second time-derivative $-g$ along the differential equation $x'' = -g$ with constant gravity factor g. When it hits the ground, which is assumed at height $x = 0$, the ball bounces back and jumps back up in the air. Yet, as every child knows, the ball tends to come back up a little less high than before. Given enough time to bounce around, it will ultimately even lie flat on the ground forever until it is picked up again and thrown high up in the air. Quantum was no stranger to this common experience on the physical effects of bouncing.

So Quantum went ahead to model the impact on the ground as a discrete phenomenon and sought ways of describing what happens to make the ball jump back up. One attempt to understand this could be to make the ball jump back up rather suddenly by increasing its height x by, say, 10 when the ball hits the ground $x = 0$:

$$x'' = -g;$$
$$\text{if}(x = 0)\, x := x + 10 \tag{4.1}$$

This HP first follows the differential equation in the first line continuously for some time and then, after the sequential composition (;), performs the discrete computation in the second line to increase x by 10 if it is presently on the ground $x = 0$. Such a model may be useful to describe other systems, but would be rather at odds with our physical experience with bouncing balls, because the ball in reality slowly climbs back up rather than suddenly starting out way up in the air again.

Quantum ponders about what happens when he hits the ground. Quantum does not suddenly get teleported to a new position above ground as HP (4.1) would suggest. Instead, the ball suddenly changes its direction but not position. A moment ago, Quantum used to fall down with a negative velocity (i.e., one that is pointing down toward the ground) and then, all of a sudden, he climbs back up with a positive velocity (pointing up into the sky). In order to be able to write such a model, the velocity v will be made explicit in the bouncing ball's differential equation system:

$$\{x' = v, v' = -g\};$$
$$\text{if}(x = 0)\, v := -v \tag{4.2}$$

Now the differential equation system $\{x' = v, v' = -g\}$ expresses that the time-derivative of height x is the vertical velocity v whose time-derivative is $-g$. Of course, something keeps happening after the bouncing ball reverses its direction because it hits the ground. Physics continues until it hits the ground again:

$$
\begin{aligned}
&\{x' = v, v' = -g\}; \\
&\text{if}(x = 0)\, v := -v; \\
&\{x' = v, v' = -g\}; \\
&\text{if}(x = 0)\, v := -v
\end{aligned}
\tag{4.3}
$$

Then, of course, physics moves on again, so the model actually involves a repetition:

$$
\begin{aligned}
\big(\{x' = v, v' = -g\}; \\
\text{if}(x = 0)\, v := -v\big)^{*}
\end{aligned}
\tag{4.4}
$$

It is good that repetitions (*) in HP are nondeterministic, because Quantum has no way of knowing ahead of time how many iterations of the control loop he will take. Yet, Quantum is now rather surprised. For if he follows HP (4.4), it seems as if he should always be able to climb back up to his initial height again. Excited about that possibility, Quantum tries and tries again, but he never really succeeds to bounce back up quite as high as he was before. So there must have been something wrong with the model in (4.4), Quantum concludes, and sets out to fix (4.4).

Having observed himself rather carefully when bouncing around for a while, Quantum concludes that he feels just a little bit slower when bouncing back up than he used to when falling down. Indeed, Quantum feels less energetic on his way up. So his velocity must not only flip direction from down to up at a bounce on the ground, but also seems to shrink in magnitude. Quantum swiftly calls the corresponding damping factor c and quickly comes up with a better model of himself:

$$
\begin{aligned}
\big(\{x' = v, v' = -g\}; \\
\text{if}(x = 0)\, v := -cv\big)^{*}
\end{aligned}
\tag{4.5}
$$

Now, if the ball is on the ground, its velocity will flip but its magnitude willshrink by damping factor c. Yet, running that model in clever ways, Quantum observes that model (4.5) could make him fall through the cracks in the ground. Terrified at that thought, Quantum quickly sets the physics right, lest he falls through the cracks in space before he has a chance to fix his very own model of physics. The issue with (4.5) is that its differential equation isn't told when to stop, so it could evolve for too long a time below ground and just fail the subsequent test $x = 0$ as shown in Fig. 4.2. Yet, Quantum luckily remembers from Chap. 2 that this purpose is exactly what evolution domains were meant for. Above ground is where he wants to remain, and so the evolution domain constraint $x \geq 0$ is what Quantum asks dear physics to kindly obey, since the floor underneath Quantum is of rather sturdy build. Unlike in poor Alice's case [2], the floor comes without rabbit holes to fall through:

$$
\begin{aligned}
\big(\{x' = v, v' = -g \,\&\, x \geq 0\}; \\
\text{if}(x = 0)\, v := -cv\big)^{*}
\end{aligned}
\tag{4.6}
$$

Fig. 4.2 Sample trajectory of a bouncing ball (plotted as height over time) with a crack in the floor

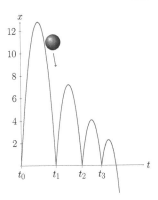

Now, indeed, physics will have to stop evolving before gravity has made our little bouncing ball Quantum fall through the ground. Yet, physics could still choose to stop evolving while the ball is high up in the sky. In that case, the ball will not yet be on the ground and line 2 of (4.6) would have no effect because $x \neq 0$ still. This is not exactly a catastrophe, however, because the loop in (4.6) could simply repeat, which would allow physics to continue to evolve along the same differential equation just a little bit further. Thankfully, the evolution domain constraint $\ldots \& x \geq 0$ prevents all ill-fated attempts to follow the differential equation below ground, because those would not satisfy $x \geq 0$ all the time.

Being quite happy with model (4.6), the bouncing ball Quantum goes on to explore whether the model does what he expects it to do. Of course, had Quantum read this book already, he would have marched right into a rigorous analysis of the model. Since Quantum is still a CPS rookie, he takes a detour along visualization road, and first shoots a couple of pictures of what happens when simulating model (4.6). Thanks to a really good simulator, these simulations all come out looking characteristically similar to Fig. 4.1.

4.2.2 How Quantum Discovered a Crack in the Fabric of Time

After a little while idly simulating his very own personal model, Quantum decides to take out his temporal magnifying glasses and zoom in really close to see what actually happens when his model (4.6) bounces on the ground ($x = 0$). At that point in time, the differential equation is forced to stop due to the evolution domain $x \geq 0$. So the continuous evolution stops and a discrete action happens that inspects the height and, if $x = 0$, discretely changes the velocity to $-cv$ instantly in no time.

At the continuous point in time of the first bounce—Quantum records the time t_1—the ball observes a succession of different states. First at continuous time t_1, Quantum has position $x = 0$ and velocity $v = -5$. But then, after the discrete assignment of (4.6) runs, yet still at real time t_1, it has position $x = 0$ and velocity $v = 4$. This temporal chaos cannot possibly go on like that, thought Quantum, and decided to give an extra natural number index $j \in \mathbb{N}$ to distinguish the two successive occurrences of continuous time t_1. So, for the sake of illustration, he called $(t_1, 0)$ the first point in time where Quantum was in state $x = 0, v = -5$, and then went on to call $(t_1, 1)$ the second point in time where he was in state $x = 0, v = 4$.

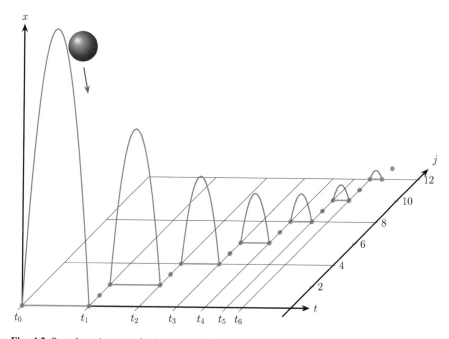

Fig. 4.3 Sample trajectory of a bouncing ball plotted as position x over its hybrid time domain with discrete time step j and continuous time t

In fact, Quantum's temporal magnifying glasses worked so well that he suddenly discovered he had accidentally invented an extra dimension for time: the discrete time step $i \in \mathbb{N}$ in addition to the continuous time coordinate $t \in \mathbb{R}$. Quantum plotted the continuous $\mathbb{R}$-valued time coordinate in the t axis of Fig. 4.3 while separating the $\mathbb{N}$-valued discrete step count of the hybrid time into the j axis and leaving the x axis for position. In Fig. 4.3, Quantum now observed the first simulation of model (4.6) with his temporal magnifiers activated to fully appreciate its hybrid nature in its full blossom. And, indeed, if Quantum looks at the hybrid time simulation from Fig. 4.3 and turns his temporal magnifiers off again, the extra dimension of discrete steps j vanishes again, leaving behind only the pale shadow of the execution in the x over t face, which agrees with the layman's simulation shown in Fig. 4.1. Even the projection of the hybrid time simulation from Fig. 4.3 to the j over t time face leads to a curious illustration, shown in Fig. 4.4, of what the temporal magnifying glasses revealed about how hybrid time has evolved in this particular simulation.

Armed with the additional intuition about the operations of HP (4.6) that these sample executions in Fig. 4.1 and its hybrid version Fig. 4.3 provide, Quantum now

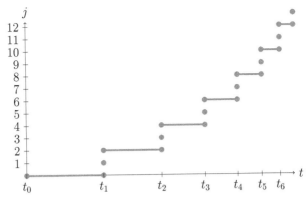

Fig. 4.4 Hybrid time domain for the sample trajectory of a bouncing ball with discrete time step j and continuous time t

feels prepared to ask the deeper questions. Will his model (4.6) *always* do the right thing? Or was he just lucky in the particular simulation shown in Fig. 4.1 and magnified in Fig. 4.3? What even is the right behavior for a proper bouncing ball? What are its most important properties? And for what purpose even? How could they be specified unambiguously? And, ultimately, how can Quantum possibly convince himself that these properties are true before he ever bounces around again idly and at potential risk of harm?

4.2.3 How Quantum Learned to Deflate

While half of Quantum's mind is already in hot pursuit of the pressing questions of correctness and personal safety he raised in the previous section, the other half is still wondering what went wrong with his model if the model does not even know how to lie still any more. Scrutinizing each and every simulation (Figs. 4.1 and 4.3) of the bouncing ball's model from every angle of time again, he learns that every which way balls bounce according to HP (4.6), they do not ever seem to stop bouncing. Bewildered about this distinct possibility, Quantum gave it a try and bounced around like he had never bounced before. After a number of unsuccessful attempts, he lay down to think, realizing that something must be rather counterfactual about the model if it seriously predicted he would always bounce, when, in reality, Quantum ultimately always runs out of steam and just lies flat on his back as in Fig. 4.5.

If Quantum wants to describe reality, the model (4.6) needs some fixing. Halfway through his sophisticated development of better models with increasingly high-fidelity mixes of elastic and plastic deformation of balls at a bounce, friction, and the rôle of energy loss, Quantum suddenly has a pretty clever idea. Of course, all these models of deformations and frictions and energies and whatnot would be needed for a model of highly precise physics. Yet, if Quantum is merely trying to describe the

Fig. 4.5 Sample trajectory
of a bouncing ball (plotted
as height over time) that
ultimately lies down flat

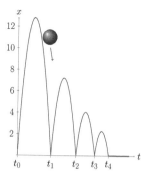

qualitative behavior of a bouncing ball, he might as well try to leverage the power
of abstraction that he read about in Chap. 3. On a big ideas level, when Quantum
bounces on the ground, he either bounces back up with reduced velocity $v := -cv$
or he just lies flat with no velocity at all $v := 0$. Figuring out exactly which case
happens when would put Quantum back into the mode of describing increasingly
precise physics and measuring all kinds of specific coefficients and parameters in
those models. But just describing the fact that one of the two options can happen
when on the ground is quite easy just with a nondeterministic choice ($\cup$):

$$\left(\{ x' = v, v' = -g \,\&\, x \geq 0 \}; \right.$$
$$\left. \text{if}(x = 0)\,(v := -cv \cup v := 0) \right)^{*} \tag{4.7}$$

This new and improved HP now allows with reasonable accuracy all behavior that
Quantum observes when trying to bounce around. The model also allows some extra
behavior that he never quite got actual physics to do for him, such as never stopping
bouncing around, so (4.7) is an overapproximation. Comparing notes with his high-
fidelity physics models, though, Quantum cannot help but appreciate the relative
simplicity of (4.7) and would much rather seek to analyze the simpler hybrid systems model (4.7) than try an analysis of a more precise but also significantly more
difficult physical model. Quantum now sees first hand how abstraction can create
simplicity and will be sure to keep that in mind in all future endeavors.

Speaking of simplicity: even if Quantum now appreciates HP (4.7) as the better
model compared to the simpler HP (4.6) on account of giving the physics an option
to have the ball lie flat after some bounce, Quantum still prefers to first investigate
the simpler model (4.6) for the pressing correctness questions that now have Quantum's full attention. It is also a distinct possibility that Quantum already took a peek
at Exercise 4.16, in which the more involved model (4.7) will be considered.

4.2.4 Postcondition Contracts for CPS

Having developed an elegant hybrid systems model of bouncing balls, Quantum subsequently proceeds to state his expectations about its behavior. Hybrid programs α are useful models for CPS. They describe with a program the behavior of a CPS, ultimately captured by their semantics $[\![\alpha]\!] \subseteq \mathscr{S} \times \mathscr{S}$, which is a reachability relation on states (Chap. 3). Yet, reliable development of CPS also needs a way of ensuring that this behavior will be as expected. For one thing, we want the behavior of a CPS to always satisfy certain crucial safety properties. A robot, for example, should never do anything unsafe such as running over a human being.

Expedition 4.1 (Three Laws of Robotics)

Safety of robots has been aptly defined by Isaac Asimov's Three Laws of Robotics [1]:
① A robot may not injure a human being or, through inaction, allow a human being to come to harm.
② A robot must obey the orders given to it by human beings, except where such orders would conflict with the First Law.
③ A robot must protect its own existence as long as such protection does not conflict with the First or Second Law.
Sadly, their exact rendition in logic or anything similarly precise still remains quite a challenge due to language ambiguities and similar minor nuisances that kept scientists busy for a good deal of a century since. The Three Laws of Robotics are not the answer. They are the inspiration!

Even if Quantum, the little bouncing ball, may be less safety-critical than a proper CPS, he is still quite interested in his own safety. Quantum wants to make sure that he couldn't ever fall through the cracks in the ground. And even though he would love to jump all the way up to the moon, Quantum turns out to be rather terrified of big heights. Come to think of it, he would never want to jump any higher than he was in the very beginning. So, when H denotes initial height, Quantum wants to know whether his height will always stay within $0 \leq x \leq H$ when following HP (4.6).

Scared of what otherwise might happen to him if $0 \leq x \leq H$ should ever be violated, Quantum decides to make his goals for HP (4.6) explicit. Fortunately, Quantum excelled in basic programming courses, where contracts have been used to make behavioral expectations for programs explicit. Even though Quantum clearly no longer deals with plain conventional programs, but rather a hybrid program, Quantum still decides to put an **ensures**(F) contract in front of HP (4.6) to express that all runs of that HP are expected to lead only to states in which logical formula F is true. Quantum even uses two postconditions, one for each of his expectations. For the time being, Quantum temporarily uses the following notation to indicate the two expected postconditions that the bouncing ball HP is supposed to ensure:

$$\textbf{ensures}(0 \leq x)$$
$$\textbf{ensures}(x \leq H)$$
$$(\{x' = v, v' = -g \,\&\, x \geq 0\}; \tag{4.8}$$
$$\text{if}(x = 0)\, v := -cv)^*$$

Subsequent sections will quickly abandon this **ensures**(F) notation in favor of a more elegant and more logical approach. But for now, Quantum is quite happy with documenting what his model is expected to achieve.

4.2.5 Precondition Contracts for CPS

Having read up a lot about conventional program contracts, Quantum immediately begins to wonder whether the **ensures**() contracts in HP (4.8) would, in fact, always be true after running that HP. After all, acrophobic Quantum would really like to rely on this contract never failing. In fact, he prefers to see that logical contract met before he even dares to try another careless bounce ever again.

Quantum's deliberations eventually lead him to conclude that whether **ensures**() contract in (4.8) works out will depend on the initial values that the bouncing ball starts out with. Quantum thinks of H as the initial height, but HP (4.8) cannot know that. Indeed, the contracts would be rather hard to fulfill if $H = -5$, because $0 \leq x$ and $x \leq H$ could not possibly both be true then.

So, Quantum decides he should demand a **requires**($x = H$) contract with the precondition $x = H$ to say that the height, x, of the bouncing ball is initially H. Since that still does not ensure that $0 \leq x$ has a chance of holding, Quantum requires $0 \leq H$ to hold initially as well, leading to

$$\textbf{requires}(x = H)$$
$$\textbf{requires}(0 \leq H)$$
$$\textbf{ensures}(0 \leq x)$$
$$\textbf{ensures}(x \leq H) \tag{4.9}$$
$$(\{x' = v, v' = -g \,\&\, x \geq 0\};$$
$$\text{if}(x = 0)\, v := -cv)^*$$

With this informal introduction to the need for and purpose of contracts in cyber-physical systems, it is about time for the development to become more general again. We will, thus, develop a systematic approach to CPS contracts before returning to a further study of whether the contracts expressed in Quantum's HP (4.9) really hold.

Expedition 4.2 (Invariant contracts for CPS)

In addition to preconditions and postconditions, loop invariants play a prominent rôle in contracts for conventional imperative programs, because they constitute the major logical mode for understanding loops. Preconditions state what is expected to hold before the program runs. Postconditions state what is guaranteed to hold after the program runs. And loop invariants indicate what is true every time a loop body executes, so before and after every run of the loop body. In C-style programs, for example, invariants are associated with loops:

```
i = 0;
while (i < 10)
    // loop_invariant(0 <= i && i <= 10)
    {
        i++;
    }
```

Dijkstra's algorithm for computing the greatest common divisor of a and b needs a loop invariant and a precondition, as gcd(5,0) would not terminate:

```
// requires(x!=0 && y!=0)
x=a; y=b; u=b; v=a;
while (x!=y)
    // loop_invariant(2*a*b == u*x + v*y)
    {
        if (x>y) {
            x=x-y; v=v+u;
        } else {
            y=y-x; u=u+v;
        }
    }
```

Such loop invariants will also play an equally important rôle in CPS (Chap. 7), but they first require additional developments to become meaningful.

4.3 Logical Formulas for Hybrid Programs

CPS contracts play a very useful rôle in the development of CPS programs or, in fact, any other CPS models. Using them as part of their design right from the very beginning is a good idea, probably much more crucial than it is when developing conventional programs, because CPSs have more stringent requirements on safety.

However, we do not only want to program CPSs, we also want to and will have to understand thoroughly what CPS programs and their contracts mean, and how we convince ourselves that the CPS contracts are respected by the CPS program. This is where mere contracts are at a disadvantage compared to the full features of logic.

> **Note 19 (Logic is for specification and reasoning)** Logic allows not only the specification of a whole CPS program, but also an analytic inspection of its parts as well as argumentative relations between contracts and program parts.

Logic was invented for precise statements, justifications, and ways of systematizing rational human thought and mathematical reasoning [6, 8–13, 16, 37, 38]. Logic saw influential generalizations to enable precise statements and reasoning about conventional discrete programs [5, 14, 34], and other aspects, including modes of truth such as necessity and possibility [15] or temporal relations of truth [33, 35].

What cyber-physical systems need, though, is, instead, a logic for precise statements and reasoning about their dynamical systems. So CPSs need logics of dynamical systems [25, 29], of which the most fundamental representative is *differential dynamic logic* (dL) [20–22, 25, 26, 30], the logic of hybrid systems. Differential dynamic logic allows direct logical statements about hybrid programs and, thus, serves as the logic of CPS programs in Parts I and II of this textbook for both specification and verification purposes, and still forms the basis for Parts III and IV. Additional multi-dynamical systems aspects beyond hybrid systems are discussed elsewhere [23–25, 28, 29, 31], some of which will be picked up in Part III of this textbook.

The most important feature of differential dynamic logic for our purposes is that it allows us to refer to hybrid systems. Chapter 2 introduced first-order logic of real arithmetic, which was used to describe evolution domain constraints of differential equations, and made it possible to refer to conjunctions or disjunctions of comparisons of (polynomial) terms with quantifiers over real-valued variables.

> **Note 20 (Limits of first-order logic for CPS)** First-order logic of real arithmetic is a crucial basis for describing what is true and false about CPSs, because it allows us to refer to real-valued quantities such as positions and velocities and their arithmetic relations. Yet, that is not quite enough, because first-order logic describes what is true in a single state of a system. It has no way of referring to what will be true in future states of a CPS, nor of describing the relationship of the initial state of the CPS to the final state of the CPS. Without such a capability, it is impossible to refer to what preconditions were true before the CPS started and how this relates to what postconditions are true afterwards.

Recall from Sect. 3.3.2 that the relation $[\![\alpha]\!] \subseteq \mathscr{S} \times \mathscr{S}$ is what ultimately constitutes the semantics of HP α. It defines which new state $v \in \mathscr{S}$ is reachable from which initial state $\omega \in \mathscr{S}$ in HP α, in which case we write $(\omega, v) \in [\![\alpha]\!]$.

> **Note 21 (Differential dynamic logic principle)** Differential dynamic logic, which is denoted dL, extends first-order logic of real arithmetic with operators that refer to the future states of a CPS, that is to the states that are reachable by running a given HP. The logic dL provides a modal operator $[\alpha]$, parametrized by HP α, which refers to all states reachable by this HP α according to the reachability relation $[\![\alpha]\!] \subseteq \mathscr{S} \times \mathscr{S}$ of its semantics. For any HP α, this modal

operator $[\alpha]$ can be placed in front of any dL formula P. The resulting dL formula

$$[\alpha]P$$

expresses that *all* states reachable by HP α satisfy formula P.
The logic dL also provides another modal operator $\langle\alpha\rangle$, parametrized by HP α, that can be placed in front of any dL formula P. The dL formula

$$\langle\alpha\rangle P$$

expresses that *there is at least one* state reachable by HP α for which P holds. The modalities $[\alpha]$ and $\langle\alpha\rangle$ can be used to express necessary or possible properties of the transition behavior of α, since they refer to all or some runs of α. The formula $[\alpha]P$ is pronounced "α box P" and $\langle\alpha\rangle P$ is "α diamond P."

With the help of dL's modalities, an **ensures**(E) postcondition for an HP α can be expressed directly as a logical formula in differential dynamic logic:

$$[\alpha]E$$

In particular, the first CPS postcondition **ensures**$(0 \le x)$ for the bouncing ball HP in (4.8) can be stated as a dL formula:

$$[(\{x' = v, v' = -g \,\&\, x \ge 0\}; \text{if}(x = 0)\,v := -cv)^*]\,0 \le x \qquad (4.10)$$

The second CPS postcondition **ensures**$(x \le H)$ for the bouncing ball HP in (4.8) can be stated as a dL formula as well:

$$[(\{x' = v, v' = -g \,\&\, x \ge 0\}; \text{if}(x = 0)\,v := -cv)^*]\,x \le H \qquad (4.11)$$

The logic dL allows all other logical operators from first-order logic, including conjunction ($\wedge$). So, the two dL formulas (4.10) and (4.11) can be stated together as a single dL formula consisting of the logical conjunction of (4.10) and (4.11):

$$
\begin{aligned}
&[(\{x' = v, v' = -g \,\&\, x \ge 0\}; \text{if}(x = 0)\,v := -cv)^*]\,0 \le x \\
\wedge\; &[(\{x' = v, v' = -g \,\&\, x \ge 0\}; \text{if}(x = 0)\,v := -cv)^*]\,x \le H
\end{aligned}
\qquad (4.12)
$$

Stepping back, we could have combined the two postconditions **ensures**$(0 \le x)$ and **ensures**$(x \le H)$ into a single postcondition **ensures**$(0 \le x \wedge x \le H)$ using a conjunction in the postcondition instead. The translation of that into dL would have gotten us an alternative way of combining both statements about the lower and upper bound on the height of the bouncing ball into a single dL formula:

$$[(\{x' = v, v' = -g \,\&\, x \ge 0\}; \text{if}(x = 0)\,v := -cv)^*]\,(0 \le x \wedge x \le H) \qquad (4.13)$$

Which way of representing what we expect bouncing balls to do is better? Like (4.12) or like (4.13)? Are they equivalent? Or do they express different things?

Before you read on, see if you can find the answer for yourself.

There is a very simple argument within the logic dL that shows that dL formulas (4.12) and (4.13) are equivalent. It even shows that the same equivalence holds not just for these particular formulas but for any dL formulas of the same form:

$$[\alpha]P \wedge [\alpha]Q \text{ is equivalent to } [\alpha](P \wedge Q) \qquad (4.14)$$

The equivalence of $[\alpha]P \wedge [\alpha]Q$ and $[\alpha](P \wedge Q)$ can in turn be expressed as a logical formula with the equivalence operator ($\leftrightarrow$) giving a formula that is true in all states:

$$[\alpha]P \wedge [\alpha]Q \leftrightarrow [\alpha](P \wedge Q)$$

This equivalence will be investigated in more detail in a later chapter, but it is useful to observe it now already in order to sharpen our intuition about dL and anticipate possible use cases for its flexibility.

Having said that, do we believe dL formula (4.12) should be valid so true in all states? Should (4.13) be valid? Well, they should certainly either agree to both be valid or agree to both not be valid since they are equivalent by (4.14). But is (4.12) valid now or is it not? Before we study this question in any further detail, the first question should be what it means for a modal formula $[\alpha]P$ to be true. What is its semantics? Better yet, what exactly is its syntax in the first place?

4.4 Differential Dynamic Logic

Based on the gradual motivation and informal introduction in this chapter, this section now defines *differential dynamic logic* [20–22, 25, 26, 30], which plays a central rôle as an unambiguous notation and our basis for rigorous reasoning techniques for CPS throughout this book. Differential dynamic logic uses the terms from Sect. 2.6.2 on p. 42 and the hybrid programs from Sect. 3.3.1 on p. 76.

4.4.1 Syntax of Differential Dynamic Logic

The formulas of differential dynamic logic are defined like the formulas of first-order logic of real arithmetic (Sect. 2.6.3) with the additional capability of using *modal operators* $[\alpha]$ and $\langle\alpha\rangle$ for any hybrid program α. The dL formula $[\alpha]P$ expresses that all states after all runs of HP α satisfy dL formula P. The dL formula $\langle\alpha\rangle P$ expresses that there is a run of HP α that leads to a state in which dL formula P is true. In the *modal formulas* $[\alpha]P$ and $\langle\alpha\rangle P$, the formula P is called the *postcondition*.

Definition 4.1 (dL formula). The *formulas of differential dynamic logic* (dL) are defined by the following grammar (where P, Q are dL formulas, $e, \tilde{e}$ are (polynomial) terms, x is a variable, and α is an HP):

$$P, Q ::= e = \tilde{e} \mid e \geq \tilde{e} \mid \neg P \mid P \wedge Q \mid P \vee Q \mid P \rightarrow Q \mid \forall x P \mid \exists x P \mid [\alpha]P \mid \langle \alpha \rangle P$$

Operators $>, \leq, <, \leftrightarrow$ are definable, e.g., $P \leftrightarrow Q \equiv (P \rightarrow Q) \wedge (Q \rightarrow P)$.

Of course, all first-order real-arithmetic formulas from Chap. 2 are also dL formulas and will mean exactly the same. Occasionally, we will use the backwards implication $P \leftarrow Q$, which is just alternative notation for the implication $Q \rightarrow P$.

The dL formula $[x := 5] x > 0$ expresses that x is always positive after assigning 5 to it, which is quite trivially true, because the new value 5 that x assumes is, indeed, positive. The formula $[x := x + 1] x > 0$ expresses that x is always positive after incrementing it by one in a discrete change, which is only *true* in some states but *false* in others whose value of x is too small. With an extra implication, the dL formula $x \geq 0 \rightarrow [x := x + 1] x > 0$ is valid, so *true* in all states, though, because x will certainly be positive after an increment if it was nonnegative initially. This implication expresses that in all states satisfying the left-hand side assumption $x \geq 0$, the right-hand side $[x := x + 1] x > 0$ is *true*, which, in turn, says that after incrementing x, it will have become positive. The implication is trivially true in all states falsifying the assumption $x \geq 0$, because implications are *true* if their left-hand side is *false*.

The program $x := x + 1$ can only be run in exactly one way, so quantifying over all runs of $x := x + 1$ in the box modality $[x := x + 1]$ is indistinguishable from quantifying over one run of $x := x + 1$ in the diamond modality $\langle x := x + 1 \rangle$. More interesting things happen for other HPs. The formula $[x := 0; (x := x + 1)^*] x \geq 0$ is valid, since incrementing x *any* number of times after assigning 0 to x will still yield a nonnegative number. The conjunction $[x := x + 1] x > 0 \wedge [x := x - 1] x < 0$ is *true* in an initial state where x is 0.5, but is not valid, because it is *false* when x starts out at -10.

Likewise, the dL formula $[x' = 2] x > 0$ is not valid, because it is *false* in initial states with negative x values. But $x > 0 \rightarrow [x' = 2] x > 0$ is valid, since x will always increase along the differential equation $x' = 2$. Similarly, $[x := 1; x' = 2] x > 0$ is valid, because x remains positive along ODE $x' = 2$ after first assigning 1 to x.

Quantifiers and modalities can be mixed as well. For example, the dL formula $x > 0 \rightarrow \exists d [x' = d] x > 0$ is valid, since if x starts positive, then there is a value for variable d, e.g., 2, which will always keep x positive along $x' = d$. The formula $\exists x \exists d [x' = d] x > 0$ is valid, since there is an initial value for x, e.g., 1 and a value for d, e.g., 2, such that x stays positive always after any continuous evolution along $x' = d$ starting from that initial value of x. Even a conjunction $\exists x \exists d [x' = d] x > 0 \wedge \exists x \exists d [x' = d] x < 0$ is valid, because there, indeed, is an initial value for x and slope d such that x always stays positive along $x' = d$ for those choices, but there also is an initial value for x, e.g., -1, and a d, e.g., -2, such that x always stays negative along $x' = d$. In this case, different values have to be chosen for x and d in the two conjuncts to make the whole formula evaluate to *true*, but that is perfectly allowed for different quantifiers in different places. Universal quantifiers and modalities combine as well. For example, $\forall x \, [x := x^2; x' = 2] \, x \geq 0$ is valid, because for all values of x, after assigning the (nonnegative) square of x to x, following differential equation $x' = 2$ for any amount of time keeps x nonnegative.

The box modality $[\alpha]$ in the dL formula $[\alpha]P$ expresses that its postcondition P is true after *all* runs of HP α. By contrast, the diamond modality $\langle \alpha \rangle$ in the dL formula $\langle \alpha \rangle P$ expresses that its postcondition P is true after *at least one* run of HP α. Even if the diamond modality is not yet that important in the earlier parts of this book, some examples are already discussed here.

For example, dL formula $\langle x' = 2 \rangle \, x > 0$ is valid, since whatever initial value x has, it will ultimately be positive at some point after just following the differential equation $x' = 2$ for long enough. The dL formula $\langle x' = d \rangle \, x > 0$ is not valid, but it is at least true in an initial state whose d value is 2 or otherwise positive. In fact, $\langle x' = d \rangle \, x > 0$ is equivalent to $x > 0 \vee d > 0$. In particular, $\exists d \, \langle x' = d \rangle \, x > 0$ is valid, because there is a choice of d for which x will eventually be positive after following the ODE $x' = d$ long enough. Even the following conjunction is valid: $\exists d \, \langle x' = d \rangle \, x > 0 \wedge \exists d \, \langle x' = d \rangle \, x < 0$, because the first conjunct is true by the previous argument and the second conjunct is true since a possibly different choice for the initial value of d will ultimately get x negative along ODE $x' = d$ with an appropriate (and different) choice of d such as -2.

Modalities can also be nested, because any dL formula can be used as a postcondition. For example, dL formula $x > 0 \rightarrow [x' = 2] \langle x' = -2 \rangle \, x < 0$ is valid, because if x starts positive, then no matter how long one follows the differential equation $x' = 2$, there is a way of subsequently following the differential equation $x' = -2$ for some amount of time such that x becomes negative. All it takes is sufficient patience. In fact, for the same reason, even the following dL formula is valid:

$$x > 0 \rightarrow [x' = 2](x > 0 \wedge \langle x' = -2 \rangle \, x = 0)$$

It expresses that, from any positive initial value of x, any duration of following $x' = 2$ will lead to a state in which x is still positive but for which it is also still possible to follow $x' = -2$ for some (larger) amount of time to make the final value of x zero.

Before you get yourself confused, beware that equality signs can occur in differential dynamic logic formulas and their hybrid programs in different rôles now:

Expression	Rôle
$x := e$	Discrete assignment in HP assigning new value of e to variable x
$x' = f(x)$	Differential equation in HP for continuous evolution of variable x
$x = e$	Equality comparison in dL formula that can be *true* or *false*
$?x = e$	Testing x for equality comparison in an HP, only continue if *true*

4.4.2 Semantics of Differential Dynamic Logic

For dL formulas that are also formulas of first-order real arithmetic (i.e., formulas without modalities), the semantics of dL formulas is the same as that of first-order

---

Expedition 4.3 (Operator precedence for differential dynamic logic)

To save parentheses, the notational conventions have unary operators (including[a] $\neg$, quantifiers $\forall x, \exists x$, modalities $[\alpha], \langle \alpha \rangle$ as well as HP operator *) bind more strongly than binary operators. We let $\wedge$ bind more strongly than $\vee$, which binds more strongly than $\rightarrow, \leftrightarrow$, and let ; bind more strongly than $\cup$. Arithmetic operators $+, -, \cdot$ associate to the left. All logical and program operators associate to the right.

These precedences imply that quantifiers and modal operators bind strongly, i.e., their scope only extends to the formula immediately after. So, $[\alpha]P \wedge Q \equiv ([\alpha]P) \wedge Q$ and $\forall x P \wedge Q \equiv (\forall x P) \wedge Q$ and $\forall x P \rightarrow Q \equiv (\forall x P) \rightarrow Q$. They imply $\alpha; \beta \cup \gamma \equiv (\alpha; \beta) \cup \gamma$ and $\alpha \cup \beta; \gamma \equiv \alpha \cup (\beta; \gamma)$ and $\alpha; \beta^* \equiv \alpha; (\beta^*)$ like in regular expressions. All logical and program operators associate to the right, most crucially $P \rightarrow Q \rightarrow R \equiv P \rightarrow (Q \rightarrow R)$. To avoid confusion, we do not adopt precedence conventions between $\rightarrow, \leftrightarrow$ but expect explicit parentheses. So $P \rightarrow Q \leftrightarrow R$ is illegal and explicit parentheses are required to distinguish $P \rightarrow (Q \leftrightarrow R)$ from $(P \rightarrow Q) \leftrightarrow R$. Likewise $P \leftrightarrow Q \rightarrow R$ is illegal and explicit parentheses are required to distinguish $P \leftrightarrow (Q \rightarrow R)$ from $(P \leftrightarrow Q) \rightarrow R$.

---

[a] It is debatable whether quantifiers are unary operators: $\forall x$ is a unary operator on formulas but $\forall$ is an operator with mixed arguments (one variable and one formula). In higher-order logic with λ-abstractions $\lambda x.P$ for the function that maps x to P, the operator $\forall$ can be understood by considering $\forall x P$ as an operator on functions: $\forall(\lambda x.P)$. Similar cautionary remarks apply to the understanding of modalities as unary operators. The primary reason for adopting this convention is that it mnemonically simplifies the precedence rules.

---

real arithmetic. The semantics in Chap. 2 inductively defined the satisfaction relation $\omega \models P$, which holds iff formula P is true in state ω, and then collected the set of states $[\![P]\!]$ in which P is true. Now, we define the semantics of dL formulas right away by simultaneously defining the set of states $[\![P]\!]$ in which formula P is true. The two styles of definition are equivalent (Exercise 4.13), but the latter is more convenient here.

The semantics of modalities $[\alpha]$ and $\langle \alpha \rangle$ quantifies over all (for the box modality $[\alpha]$) or over some (for the diamond modality $\langle \alpha \rangle$) of the (final) states reachable by following HP α, respectively.

Definition 4.2 (dL semantics). The *semantics* of a dL formula P is the set of states $[\![P]\!] \subseteq \mathscr{S}$ in which P is true, and is defined inductively as follows:

1. $[\![e = \tilde{e}]\!] = \{\omega : \omega[\![e]\!] = \omega[\![\tilde{e}]\!]\}$
   That is, an equation is true in the set of states ω in which the terms on the two sides evaluate to the same real number according to Definition 2.4.
2. $[\![e \geq \tilde{e}]\!] = \{\omega : \omega[\![e]\!] \geq \omega[\![\tilde{e}]\!]\}$
   That is, a greater-or-equals inequality is true in states ω where the term on

the left evaluates to a number that is greater than or equal to that on the right.

3. $[\![\neg P]\!] = ([\![P]\!])^{\complement} = \mathscr{S} \setminus [\![P]\!]$

    That is, a negated formula $\neg P$ is true in the complement of the set of states in which the formula P itself is true. So $\neg P$ is true iff P is false.

4. $[\![P \wedge Q]\!] = [\![P]\!] \cap [\![Q]\!]$

    That is, a conjunction is true in the intersection of the states where both conjuncts are true. So $P \wedge Q$ is true iff P and Q are true.

5. $[\![P \vee Q]\!] = [\![P]\!] \cup [\![Q]\!]$

    That is, a disjunction is true in the union of the set of states where either of its disjuncts is true. So $P \vee Q$ is true iff P or Q (or both) are true.

6. $[\![P \rightarrow Q]\!] = [\![P]\!]^{\complement} \cup [\![Q]\!]$

    That is, an implication is true in the states where its left-hand side is false or its right-hand side is true. So $P \rightarrow Q$ is true iff P is false or Q is true.

7. $[\![P \leftrightarrow Q]\!] = ([\![P]\!] \cap [\![Q]\!]) \cup ([\![P]\!]^{\complement} \cap [\![Q]\!]^{\complement})$

    That is, a bi-implication is true in the states where both sides are true or both sides are false. So $P \leftrightarrow Q$ is true iff P, Q are both false or both true.

8. $[\![\forall x P]\!] = \{\omega \; : \; \nu \in [\![P]\!]$ for all states ν that agree with ω except on $x\}$

    That is, a universally quantified formula $\forall x P$ is true in a state iff its kernel P is also true in all variations of the state that have other real values for x.

9. $[\![\exists x P]\!] = \{\omega \; : \; \nu \in [\![P]\!]$ for some state ν that agrees with ω except on $x\}$

    That is, an existentially quantified formula $\exists x P$ is true in a state iff its kernel P is true in some variation of the state that has a potentially different real value for x.

10. $[\![[\alpha]P]\!] = \{\omega \; : \; \nu \in [\![P]\!]$ for all states ν such that $(\omega, \nu) \in [\![\alpha]\!]\}$

    That is, a box modal formula $[\alpha]P$ is true in state ω iff its postcondition P is true in all states ν that are reachable by running α from ω.

11. $[\![\langle \alpha \rangle P]\!] = [\![\alpha]\!] \circ [\![P]\!] = \{\omega \; : \; \nu \in [\![P]\!]$ for some state ν with $(\omega, \nu) \in [\![\alpha]\!]\}$

    That is, diamond modal formula $\langle \alpha \rangle P$ is true in state ω iff its postcondition P is true in at least one state ν that is reachable by running α from ω.

If $\omega \in [\![P]\!]$, then we say that P is true in state ω. The literature sometimes also uses the *satisfaction relation* notation $\omega \models P$ synonymously for $\omega \in [\![P]\!]$. A formula P is *valid*, written $\models P$, iff it is true in all states, i.e., $[\![P]\!] = \mathscr{S}$, so $\omega \in [\![P]\!]$ for all states ω. A formula P is a *consequence* of a set of formulas Γ, written $\Gamma \models P$, iff, for each state ω: If ($\omega \in [\![Q]\!]$ for all $Q \in \Gamma$) then $\omega \in [\![P]\!]$.

The semantics of modal formulas $[\alpha]P$ and $\langle \alpha \rangle P$ in differential dynamic logic is illustrated in Fig. 4.6, showing how the truth of P at (all or some) states ν_i reachable by α from initial state ω relates to the truth of $[\alpha]P$ or $\langle \alpha \rangle P$ at state ω.

In a state where HP α cannot run, the formula $[\alpha]P$ is (vacuously) true, because all states reached after following α satisfy the formula P simply because there are no such states. For example, $[?x \geq 5]false$ is exactly true in all states in which the value of x is less than 5, because the HP $?x \geq 5$ cannot execute successfully at all then, because it fails the test. This is to be contrasted with the formula $\langle ?x \geq 5 \rangle false$,

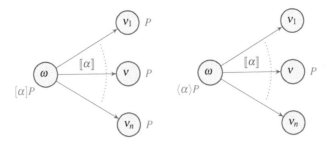

Fig. 4.6 Transition semantics of modalities in dL formulas

Expedition 4.4 (Set-valued dL semantics $[\![\cdot]\!] : \mathrm{Fml} \to \wp(\mathscr{S})$)

The semantics of a term e directly defines the real-valued function $[\![e]\!] : \mathscr{S} \to \mathbb{R}$ from states to the real value that the term evaluates to in that state (Expedition 2.2 on p. 47). Similarly, Definition 4.2 directly defines inductively, for each dL formula P, the set of states, written $[\![P]\!] \subseteq \mathscr{S}$, in which P is true:

$$[\![e \geq \tilde{e}]\!] = \{\omega \ : \ \omega[\![e]\!] \geq \omega[\![\tilde{e}]\!]\}$$
$$[\![P \wedge Q]\!] = [\![P]\!] \cap [\![Q]\!]$$
$$[\![P \vee Q]\!] = [\![P]\!] \cup [\![Q]\!]$$
$$[\![\neg P]\!] \ = [\![P]\!]^{\complement} = \mathscr{S} \setminus [\![P]\!]$$
$$[\![\langle \alpha \rangle P]\!] = [\![\alpha]\!] \circ [\![P]\!] = \{\omega \ : \ v \in [\![P]\!] \text{ for some state } v \text{ such that } (\omega, v) \in [\![\alpha]\!]\}$$
$$[\![[\alpha]P]\!] = [\![\neg \langle \alpha \rangle \neg P]\!] = \{\omega \ : \ v \in [\![P]\!] \text{ for all states } v \text{ such that } (\omega, v) \in [\![\alpha]\!]\}$$
$$[\![\exists x P]\!] \ = \{\omega \ : \ v \in [\![P]\!] \text{ for some state } v \text{ that agrees with } \omega \text{ except on } x\}$$
$$[\![\forall x P]\!] \ = \{\omega \ : \ v \in [\![P]\!] \text{ for all states } v \text{ that agree with } \omega \text{ except on } x\}$$

When Fml is the set of dL formulas, the semantic brackets for formulas define an operator $[\![\cdot]\!] : \mathrm{Fml} \to \wp(\mathscr{S})$ that defines the meaning $[\![P]\!]$ for each dL formula $P \in \mathrm{Fml}$, which, in turn, defines the set of states $[\![P]\!] \subseteq \mathscr{S}$ in which P is true. The *powerset* $\wp(\mathscr{S})$ is the set of all subsets of the set of states $\mathscr{S}$.

---

which is not true in any state, because it claims the existence of an execution of HP $?x \geq 5$ to a state where *false* holds true, which never holds. Contrast this with the formula $\langle ?x \geq 5 \rangle x < 7$, which is true exactly in all states satisfying $x \geq 5$ and $x < 7$.

4.5 CPS Contracts in Logic

Now that we know what truth and validity are in differential dynamic logic, let's go back to the previous question. Is dL formula (4.12) valid? Is (4.13) valid? Actually, let's first ask whether they are equivalent, i.e., whether the dL formula

(4.12) $\leftrightarrow$ (4.13) is valid. Expanding the abbreviations this is the question of whether the following dL formula is valid:

$$\Big([(\{x'=v,v'=-g\,\&\,x\geq 0\};\ \text{if}(x=0)\,v:=-cv)^*]\,0\leq x$$

$$\wedge\,[(\{x'=v,v'=-g\,\&\,x\geq 0\};\ \text{if}(x=0)\,v:=-cv)^*]\,x\leq H\Big) \tag{4.15}$$

$$\leftrightarrow\,[(\{x'=v,v'=-g\,\&\,x\geq 0\};\ \text{if}(x=0)\,v:=-cv)^*]\,(0\leq x\wedge x\leq H)$$

Exercise 4.1 gives you an opportunity to convince yourself that the equivalence (4.12) $\leftrightarrow$ (4.13) is indeed valid.[1] So if (4.12) is valid, then (4.13) is valid as well (Exercise 4.2). But is (4.12) valid?

> Before you read on, see if you can find the answer for yourself.

Certainly, (4.12) is not true in a state ω where $\omega(x) < 0$, because from that initial state, zero repetitions of the loop (which is allowed by nondeterministic repetition, Exercise 4.4) lead to the same state ω in which $0 \leq x$ is still false. The initial state is a possible final state for any HP of the form α^*, because it can repeat 0 times. Thus, (4.12) only has a chance of being true in initial states that satisfy further assumptions, including $0 \leq x$ and $x \leq H$. That is what the preconditions were meant for in Sect. 4.2.5. How can we express a precondition contract in a dL formula?

Preconditions serve a very different rôle than postconditions do. Postconditions of HP α are expected to be true after every run of α, which is difficult to express in first-order logic (to say the least), but straightforward using the modalities of dL. Do we also need any additional logical operator to express preconditions?

The meaning of precondition **requires**(A) of HP α is that A is assumed to hold before the HP starts. *If A holds when α starts, then* its postcondition **ensures**(B) holds after all runs of HP α. What if A does not hold when the HP starts?

If precondition A does not hold initially, then all bets are off, because the person who started the HP did not obey the requirements that need to be met before the HP can be started safely. The effects of ignoring precondition A are about as useful and predictable as what happens when ignoring the operating requirements "operate in a dry environment only" for a robot when it is submerged in the deep sea. If you are lucky, it will come out unharmed, but the chances are that its electronics will suffer considerably. The CPS contract **requires**(A) **ensures**(B) for an HP α promises that B will always hold after running α if A was true initially when α started. Thus, the meaning of a precondition can be expressed easily using an implication

$$A \rightarrow [\alpha]B \tag{4.16}$$

---

[1] This equivalence also foreshadows the fact that CPS provides ample opportunity for questions about how multiple system models relate to one another. The dL formula (4.15) relates three different properties of three occurrences of one and the same hybrid program, for example. The need to relate different properties of different CPSs will arise frequently throughout this book even if it may lie dormant for the moment. You are advised to already take notice that this is possible, because dL can form any arbitrary combination and nesting of all its logical operators.

because an implication is valid if, in every state in which the left-hand side is true, the right-hand side is also true. The implication (4.16) is valid ($\vDash A \rightarrow [\alpha]B$), if, indeed, for every state ω in which precondition A holds ($\omega \in [\![A]\!]$), it is the case that all runs of HP α lead to states v (with $(\omega, v) \in [\![\alpha]\!]$) in which postcondition B holds (so $v \in [\![B]\!]$). By the nature of implication, the dL formula (4.16) does not say what happens in states ω in which the precondition A does not hold (so $\omega \notin [\![A]\!]$).

How does formula (4.16) talk about the runs of an HP and postcondition B again? Recall that the dL formula $[\alpha]B$ is true in exactly those states in which all runs of HP α lead only to states in which postcondition B is true. The implication in (4.16), thus, ensures that this holds in all (initial) states that satisfy precondition A.

Note 22 (Contracts to dL Formulas) Consider HP α with a CPS contract using a single **requires**(A) precondition and single **ensures**(B) postcondition:

$$\textbf{requires}(A)$$
$$\textbf{ensures}(B)$$
$$\alpha$$

This CPS contract can be expressed directly as a logical formula in dL:

$$A \rightarrow [\alpha]B$$

dL formulas of this shape are very common and correspond to *Hoare triples* [14] but for hybrid systems instead of conventional programs. Hoare triples, in turn, are modeled after Aristotle's syllogisms.

CPS contracts with multiple preconditions and multiple postconditions can directly be expressed as dL formulas as well (Exercise 4.5).

Recall HP (4.9), which is shown here with both preconditions combined into one joint precondition and both postconditions combined into one postcondition:

$$\textbf{requires}(0 \leq x \wedge x = H)$$
$$\textbf{ensures}(0 \leq x \wedge x \leq H)$$
$$(\{x' = v, v' = -g \,\&\, x \geq 0\};$$
$$\text{if}(x = 0)\, v := -cv)^* \tag{4.17}$$

The dL formula expressing that the CPS contract for HP (4.17) holds is:

$$0 \leq x \wedge x = H \rightarrow \left[(\{x' = v, v' = -g \,\&\, x \geq 0\}; \text{ if}(x = 0)\, v := -cv)^* \right] (0 \leq x \wedge x \leq H) \tag{4.18}$$

So to find out whether (4.17) satisfies its CPS contract, we need to ask whether the corresponding dL formula (4.18) is valid. In other words, dL gives CPS contracts an unambiguous meaning.

We will need some operational way that allows us to tell whether such a dL formula is valid, i.e., true in all states, because mere inspection of the semantics alone is not a particularly scalable way of approaching validity questions. Such an operational way of determining the validity of dL formulas by proof will be pursued in the next chapter.

4.6 Identifying Requirements of a CPS

Before trying to prove any formulas to be valid, it is a pretty good idea to check whether all required assumptions have been found that are necessary for the formula to hold. Otherwise, the proof will fail and we will need to start over after having identified the missing requirements from the failed proof attempt. So let us scrutinize dL formula (4.18) and ponder whether there are any circumstances under which it is not true. Even though the bouncing ball is a rather impoverished CPS (it noticeably suffers from a lack of control), its immediate physical intuition still makes the ball a particularly insightful example for illustrating how critical it is to identify the right requirements. Besides, unlike for heavy-duty CPS, we trust you have had ample opportunities to become familiar with the behavior of bouncing balls before.

Maybe the first thing to notice is that the HP mentions g, which is meant to represent the standard gravitational constant, but the formula (4.18) never actually says that. Certainly, if gravity were negative ($g < 0$), bouncing balls would function rather differently in quite an astonishing way. They would suddenly become floating balls disappearing into the sky and would lose all the joy of bouncing around; see Fig. 4.7.

Fig. 4.7 Sample trajectory of a bouncing ball in an anti-gravity field with $g < 0$

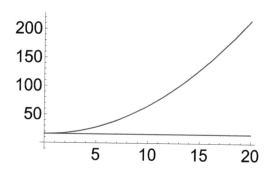

So let's modify (4.18) to assume $g = 9.81$:

$$0 \leq x \wedge x = H \wedge g = 9.81 \rightarrow$$
$$\left[(\{x' = v, v' = -g \,\&\, x \geq 0\}; \text{if}(x = 0)\, v := -cv)^* \right] (0 \leq x \wedge x \leq H) \quad (4.19)$$

Let's undo unnecessarily strict requirements right away, though. What would the bouncing ball do if it were set loose on the moon instead of on Earth? Would it still fall? Things are much lighter on the moon! Yet, they still fall down, which is what gravity is all about, just with a different constant (1.6 on the moon and 25.9 on Jupiter). Besides, none of those constants was particularly precise. Earth's gravity is more like 9.8067. The behavior of the bouncing ball depends on the value of that parameter g. But its qualitative behavior and whether it obeys (4.18) does not.

Note 23 (Parameters) A common feature of CPSs is that their behavior is subject to parameters, which can have quite a non-negligible impact. It is very hard to determine precise values for all parameters by measurements. When one particular concrete numeric value for a parameter has been assumed to prove a property of a CPS, it is not clear whether that property holds for the true system, which may in reality have a slightly different parameter value. Instead of concrete numerical values for a parameter, our analysis can proceed just fine by treating the parameter as a *symbolic parameter*, i.e., a variable such as g, which is not assumed to hold a specific numerical value like 9.81. Instead, we only assume certain constraints about the parameter, say $g > 0$ without choosing a specific value. If we then analyze the CPS with this symbolic parameter g, all analysis results will continue to hold for any concrete choice of g respecting its constraints (here $g > 0$). This results in a stronger statement about the system, which is less fragile, because it does not break down just because the true g is ≈ 9.8067 rather than the previously assumed $g = 9.81$. More general statements with symbolic parameters can even be easier to prove than statements about systems with specific magic numbers chosen for their parameters, because their assumptions are explicit.

In light of these thoughts, we might assume $9 < g < 10$ to be the gravitational constant for Earth. Yet, we can also just consider all bouncing balls on all planets in the solar system or elsewhere at once by assuming only $g > 0$ instead of $g = 9.81$ as in (4.19), since this is the only aspect of gravity that the usual behavior of a bouncing ball depends on:

$$0 \leq x \wedge x = H \wedge g > 0 \rightarrow$$
$$\left[\left(\{x' = v, v' = -g \,\&\, x \geq 0\}; \text{if}(x = 0)\, v := -cv \right)^* \right] (0 \leq x \wedge x \leq H) \quad (4.20)$$

Do we expect dL formula (4.20) to be valid, i.e., true in all states? What could possibly go wrong? The insight from modifying (4.18) to (4.19) and finally to (4.20) started with the observation that (4.18) did not include any assumptions about its parameter g. It is worth noting that (4.20) also does not assume anything about c. Bouncing balls clearly would not work as expected if $c > 1$, because such anti-damping would cause the bouncing ball to jump back up higher and higher and higher and ultimately as high up as the moon, clearly falsifying (4.20); see Fig. 4.8. Being a damping factor, we also expect $c \geq 0$ (despite Exercise 4.15). Yet, (4.20) really only has a chance of being true when we assume that c is not too big:

Fig. 4.8 Sample trajectory of a bouncing ball with anti-damping $c > 1$

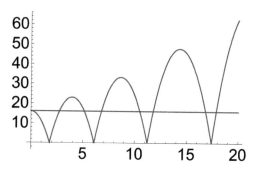

$$0 \leq x \wedge x = H \wedge g > 0 \wedge 1 \geq c \geq 0 \rightarrow$$

$$\left[\left(\{x' = v, v' = -g \,\&\, x \geq 0\};\ \text{if}(x = 0)\, v := -cv\right)^*\right] (0 \leq x \wedge x \leq H) \quad (4.21)$$

Is (4.21) valid now? Or does its truth still depend on assumptions that have not been identified yet? Is there some requirement we forgot about? Or did we find them all?

Before you read on, see if you can find the answer for yourself.

Now, all parameters (H, g, c) have some assumptions in (4.21), which is a good thing. But what about velocity variable v? Why is there no assumption about it yet? Should there be one? Unlike g and c, velocity v changes over time. What is its initial value allowed to be? What could go wrong?

Indeed, the initial velocity v of the bouncing ball could be positive ($v > 0$), which would make the bouncing ball climb initially, clearly exceeding its initial height H; see Fig. 4.9. This would correspond to the bouncing ball being thrown high up in the air in the beginning, so that its initial velocity v is upwards from its initial height $x = H$. Consequently, (4.21) has to be modified to assume $v \leq 0$ holds initially:

$$0 \leq x \wedge x = H \wedge v \leq 0 \wedge g > 0 \wedge 1 \geq c \geq 0 \rightarrow$$

$$\left[\left(\{x' = v, v' = -g \,\&\, x \geq 0\};\ \text{if}(x = 0)\, v := -cv\right)^*\right] (0 \leq x \wedge x \leq H) \quad (4.22)$$

Fig. 4.9 Sample trajectory of a bouncing ball climbing with upwards initial velocity $v > 0$

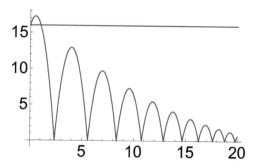

Now there finally are assumptions about all parameters and variables of (4.22). That does not mean that we found the right assumptions! But it is still a good sanity check. Before wasting cycles on trying to prove or otherwise justify (4.22), let's see once more whether we can find an initial state ω that satisfies all assumptions $v \leq 0 \wedge 0 \leq x \wedge x = H \wedge g > 0 \wedge 1 \geq c \geq 0$ on the left-hand side of the implication in (4.22) such that ω nevertheless does not satisfy the right-hand side of the implication in (4.22). Such an initial state ω falsifies (4.22) and would, thus, represent a *counterexample* to formula (4.22). Is there still a counterexample to (4.22)? Or have we successfully identified all assumptions so that it is now valid?

Before you read on, see if you can find the answer for yourself.

Formula (4.22) still has a problem. Even if the initial state satisfies all requirements in the antecedent of (4.22), the bouncing ball might still jump higher than it ought to, i.e., higher than its initial height H. That happens if the bouncing ball initially has a very large downwards velocity, so if v is a lot smaller than 0 (sometimes written $v \ll 0$). If v is a little smaller than 0, then the damping c will eat up enough of the ball's kinetic energy so that it cannot jump back up higher than it was initially (H). But if v is a lot smaller than 0, then it starts falling down with so much kinetic energy that the damping on the ground does not slow it down enough, so the ball will come bouncing back higher than it was originally, like when dribbling a basketball; see Fig. 4.10. Under which circumstance this happens depends on the relationship of the initial velocity and height to the damping coefficient.

Fig. 4.10 Sample trajectory of a bouncing ball dribbling with fast initial velocity $v < 0$

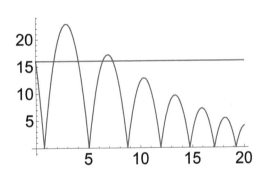

We could explore this relationship in more detail. But it is easier to infer it by conducting a proof. So we modify (4.22) to simply assume $v = 0$ initially:

$$0 \leq x \wedge x = H \wedge v = 0 \wedge g > 0 \wedge 1 \geq c \geq 0 \rightarrow$$
$$\left[(\{x' = v, v' = -g \,\&\, x \geq 0\}; \text{if}(x = 0) \, v := -cv)^* \right] (0 \leq x \wedge x \leq H) \quad (4.23)$$

Is dL formula (4.23) valid now? Or does it still have a counterexample?

Before you read on, see if you can find the answer for yourself.

It seems that all required assumptions have been identified to make the dL formula (4.23) valid so that the bouncing ball described in (4.23) satisfies the postcondition $0 \leq x \leq H$. But after so many failed starts and missing assumptions and requirements for the bouncing ball, it is a good idea to prove (4.23) once and for all beyond any doubt. It certainly is a good idea to prove dL formulas about more subtle CPS models, too.

In order to be able to prove dL formula (4.23), however, we need to investigate how proving works in CPS. How can dL formulas be proved? And, since first-order formulas are dL formulas as well, one part of the question will be: how can first-order formulas be proved? How can real arithmetic be proved? How can requirements for the safety of CPS be identified systematically? All these questions will be answered in this textbook, but not all of them already in this chapter.

In order to make sure we only need proof techniques for a minimal set of operators of dL, let's simplify (4.23) by getting rid of its if-then-else (Exercise 4.17):

$$0 \leq x \wedge x = H \wedge v = 0 \wedge g > 0 \wedge 1 \geq c \geq 0 \rightarrow$$
$$\left[\left(\{x' = v, v' = -g \,\&\, x \geq 0\}; \; (?x = 0; v := -cv \cup ?x \neq 0) \right)^* \right] (0 \leq x \wedge x \leq H)$$
$$\text{(4.24)}$$

Having added these crucial assumptions to the dL formula, Quantum quickly rephrases contract (4.17) by incorporating what we learned for dL formula (4.24):

$$\textbf{requires}(0 \leq x \wedge x = H \wedge v = 0)$$
$$\textbf{requires}(g > 0 \wedge 1 \geq c \geq 0)$$
$$\textbf{ensures}(0 \leq x \wedge x \leq H) \qquad\qquad\qquad \text{(4.25)}$$
$$\left(\{x' = v, v' = -g \,\&\, x \geq 0\}; \right.$$
$$\left. \text{if}(x = 0)\, v := -cv \right)^*$$

Observing the non-negligible difference between the original conjecture (4.19) and the revised and improved conjecture (4.24) leads us to adopt the principle of Cartesian Doubt from Expedition 4.5.

4.7 Summary

This chapter introduced differential dynamic logic (dL), whose operators and their informal meaning is summarized in Table 4.1. Their precise semantics was reported in Sect. 4.4.2 on p. 112. The most important aspect of differential dynamic logic is its ability to directly refer to possible or necessary reachability properties of hybrid programs. The fact that it is possible for hybrid program α to reach a state where formula P is true is directly expressed by the dL formula $\langle \alpha \rangle P$. That HP α necessarily only leads to (final) states where formula P is true is expressed by $[\alpha]P$. Differential dynamic logic is closed under all operators, so HP modalities, proposi-

> **Expedition 4.5 (Principle of Cartesian Doubt)**
>
> In 1641, René Descartes suggested an attitude of systematic doubt where he would be skeptical about the truth of all beliefs until he found a reason that the beliefs were justified [4]. This influential principle is now known as *Cartesian Doubt* or skepticism.
>
> We will have perfect justifications: proofs. But until we have found proof, it is helpful to adopt the principle of Cartesian Doubt in a weak and pragmatic form. Before setting out on the journey to prove a conjecture, we first scrutinize it to see whether we can find a counterexample that would make it false. Such a counterexample will not only save us a lot of misguided effort in trying to prove a false conjecture, but also helps us identify missing assumptions in conjectures and justifies our assumptions to be necessary. If, without making assumption A, a counterexample to a conjecture exists, then A is necessary.

tional connectives, and quantifiers can be nested arbitrarily often, leading to a pretty flexible specification language for CPS. Beyond its capability to capture the meaning of CPS contracts rigorously, subsequent chapters will develop the specification logic dL into a verification logic that can be used to prove dL specifications.

For future chapters, we should also keep the bouncing ball example and its surprising subtleties in mind.

Table 4.1 Operators and (informal) meaning in differential dynamic logic (dL)

dL	Operator	Meaning
$e = \tilde{e}$	equals	true iff values of e and $\tilde{e}$ are equal
$e \geq \tilde{e}$	greater equals	true iff value of e greater-or-equal to $\tilde{e}$
$\neg P$	negation / not	true iff P is false
$P \wedge Q$	conjunction / and	true iff both P and Q are true
$P \vee Q$	disjunction / or	true iff P is true or if Q is true
$P \rightarrow Q$	implication / implies	true iff P is false or Q is true
$P \leftrightarrow Q$	bi-implication / equivalent	true iff P and Q are both true or both false
$\forall x\, P$	universal quantifier / for all	true iff P is true for all values of variable x
$\exists x\, P$	existential quantifier / exists	true iff P is true for some values of variable x
$[\alpha]P$	[·] modality / box	true iff P is true after all runs of HP α
$\langle \alpha \rangle P$	$\langle \cdot \rangle$ modality / diamond	true iff P is true after at least one run of HP α

4.8 Appendix

This appendix already features some first reasoning aspects of CPS even if a fully systematic account of CPS reasoning will be pursued from scratch in more elegant ways in subsequent chapters, one operator at a time. Especially for readers who have seen the Floyd-Hoare calculus for conventional programs [5, 14] or who prefer to start with a concrete example, this appendix can be a useful stepping stone for reaching that level of generality. This appendix begins a semiformal study of the bouncing ball, which is an optional but useful preparation for the next chapter.

4.8.1 Intermediate Conditions for a Proof of Sequential Compositions

Before proceeding any further with ways of proving dL formulas, let's simplify (4.24) grotesquely by removing the loop:

$$0 \leq x \wedge x = H \wedge v = 0 \wedge g > 0 \wedge 1 \geq c \geq 0 \rightarrow$$
$$\left[\{x' = v, v' = -g \,\&\, x \geq 0\}; (?x = 0; v := -cv \cup ?x \neq 0) \right] (0 \leq x \wedge x \leq H) \quad (4.26)$$

Removing the loop clearly changes the behavior of the bouncing ball quite radically. It no longer bounces particularly well. All it can do now is fall and, if it reaches the floor, have its velocity reversed without actually ever climbing back up. So if we manage to prove (4.26), we certainly have not shown the actual dL formula (4.24). But it's a start, because the behavior modeled in (4.26) is a part of the behavior of (4.24). So it is useful (and easier) to understand the loop-free HP (4.26) first.

The dL formula (4.26) has a number of assumptions $0 \leq x \wedge x = H \wedge v = 0 \wedge g > 0 \wedge 1 \geq c \geq 0$ that can be used during the proof. It claims that the postcondition $0 \leq x \wedge x \leq H$ holds after all runs of the HP in the $[\cdot]$ modality. The top-level operator inside the modality of (4.26) is a sequential composition (;), for which we need to find a proof argument.

The HP in (4.26) first follows a differential equation and then a discrete program $(?x = 0; v := -cv \cup ?x \neq 0)$. This leads to different intermediate states after the differential equation and before the discrete program.

> **Note 24 (Intermediate states of sequential compositions)** The first HP α in a sequential composition $\alpha; \beta$ may reach a whole range of states, which represent intermediate states for the sequential composition $\alpha; \beta$, i.e., states that are final states for α and initial states for β. The intermediate states of $\alpha; \beta$ are the states μ in the semantics $[\![\alpha; \beta]\!]$ from Chap. 3:
>
> $$[\![\alpha; \beta]\!] = [\![\alpha]\!] \circ [\![\beta]\!] = \{(\omega, \nu) : (\omega, \mu) \in [\![\alpha]\!], (\mu, \nu) \in [\![\beta]\!] \text{ for some } \mu\}$$

Can we find a way of summarizing what all intermediate states between the differential equation and the discrete program of (4.26) have in common? They differ by how long the CPS has followed the differential equation.

If the system has followed the differential equation of (4.26) for time t, then the resulting velocity $v(t)$ at time t and height $x(t)$ at time t will be

$$v(t) = -gt, \quad x(t) = H - \frac{g}{2}t^2 \tag{4.27}$$

This answer can be found by integrating or solving the differential equations (Example 2.4 on p. 34). This knowledge (4.27) is useful but it is not (directly) clear how to use it to describe what all intermediate states have in common, because the time t in (4.27) is not available as a variable in the HP (4.26).[2] Can the intermediate states be described by a relation of the variables that (unlike t) are actually in the system? That is, an arithmetic formula relating variables x, v, g, H?

Before you read on, see if you can find the answer for yourself.

One way of producing a relation from (4.27) is to get the units aligned and get rid of time t. Time drops out of the "equation" when squaring the identity for velocity:

$$v(t)^2 = g^2 t^2, \quad x(t) = H - \frac{g}{2}t^2$$

and multiplying the identity for position by $2g$:

$$v(t)^2 = g^2 t^2, \quad 2gx(t) = 2gH - 2\frac{g^2}{2}t^2$$

Then substituting the first equation into the second yields

$$2gx(t) = 2gH - v(t)^2$$

This equation does not depend on time t, so we expect it to hold after all runs of the differential equation irrespective of t:

$$2gx = 2gH - v^2 \tag{4.28}$$

We conjecture the intermediate condition (4.28) to hold in the intermediate state of the sequential composition in (4.26). In order to prove (4.26) we can, thus, decompose our reasoning into two parts. The first part will prove that the intermediate condition (4.28) indeed holds after all runs of the first differential equation. The second part will assume (4.28) to hold and prove that all runs of the discrete program in (4.26) from any state satisfying (4.28) satisfy the postcondition $0 \le x \wedge x \le H$.

> **Note 25 (Intermediate conditions as contracts for sequential composition)**
> For an HP that is a sequential composition $\alpha; \beta$ an *intermediate condition* is a formula that characterizes the intermediate states in between HPs α and β. That is, for a dL formula
> $$A \to [\alpha; \beta]B$$

---

[2] Following these thoughts a bit further reveals how (4.27) can actually be used perfectly well to describe intermediate states when changing HP (4.26) a little bit. But working with solutions is still not the way that gets us to the goal the quickest, usually, because of their difficult arithmetic.

an intermediate condition is a formula E such that the following dL formulas are valid:

$$A \rightarrow [\alpha]E \quad \text{and} \quad E \rightarrow [\beta]B$$

The first dL formula expresses that intermediate condition E characterizes the intermediate states accurately, i.e., E actually holds after all runs of HP α from states satisfying A. The second dL formula says that the intermediate condition E characterizes intermediate states well enough, i.e., E is all we need to know about a state to conclude that all runs of β end up in B. That is, from all states satisfying E (including those that result by running α from a state satisfying A), B holds after all runs of β.

To prove (4.26), we conjecture that (4.28) is an intermediate condition, which requires us to prove the following two dL formulas:

$$0 \leq x \wedge x = H \wedge v = 0 \wedge g > 0 \wedge 1 > c \geq 0 \rightarrow [x' = v, v' = -g \,\&\, x \geq 0] 2gx = 2gH - v^2$$

$$2gx = 2gH - v^2 \rightarrow [?x = 0; v := -cv \cup ?x \neq 0] (0 \leq x \wedge x \leq H)$$
$$\tag{4.29}$$

Let's focus on the second formula in (4.29). Do we expect to be able to prove it? Do we expect it to be valid?

> Before you read on, see if you can find the answer for yourself.

The second formula of (4.29) claims that $0 \leq x$ holds after all runs of the hybrid program $?x = 0; v := -cv \cup ?x \neq 0$ from all states that satisfy $2gx = 2gH - v^2$. That is a bit much to hope for, however, because $0 \leq x$ is not even ensured in the precondition of this second formula. So the second formula of (4.29) is not valid. How can this problem be resolved? By adding $0 \leq x$ to the intermediate condition (4.28), thus, requiring us to prove these two formulas:

$$0 \leq x \wedge x = H \wedge v = 0 \wedge g > 0 \wedge 1 \geq c \geq 0 \rightarrow$$
$$[x' = v, v' = -g \,\&\, x \geq 0] (2gx = 2gH - v^2 \wedge x \geq 0) \quad (4.30)$$
$$2gx = 2gH - v^2 \wedge x \geq 0 \rightarrow [?x = 0; v := -cv \cup ?x \neq 0] (0 \leq x \wedge x \leq H)$$

Proving the first formula in (4.30) requires us to handle differential equations, which we will get to in Chap. 5. The second formula in (4.30) is the one whose proof is discussed first.

4.8.2 A Proof of Choice

The second formula in (4.30) has a nondeterministic choice ($\cup$) as the top-level operator in its $[\cdot]$ modality. How can we prove a formula of the form

$$A \rightarrow [\alpha \cup \beta]B \tag{4.31}$$

Recalling its semantics from Chap. 3,

$$[\![\alpha \cup \beta]\!] = [\![\alpha]\!] \cup [\![\beta]\!]$$

HP $\alpha \cup \beta$ has two possible behaviors. It can run as HP α does or as HP β does. And it is chosen nondeterministically which of the two behaviors happens. Since the behavior of $\alpha \cup \beta$ can be either α or β, proving (4.31) requires proving B to hold after α and after β. More precisely, (4.31) assumes A to hold initially, otherwise (4.31) is vacuously true. Thus, proving (4.31) allows us to assume A and requires us to prove that B holds after all runs of α (which is permitted behavior for $\alpha \cup \beta$) and to prove that, assuming A holds initially, B holds after all runs of β (which is also permitted behavior of $\alpha \cup \beta$).

> **Note 26 (Proving choices)** For an HP that is a nondeterministic choice $\alpha \cup \beta$, we can prove
>
> $$A \to [\alpha \cup \beta]B$$
>
> by proving the following dL formulas:
>
> $$A \to [\alpha]B \quad \text{and} \quad A \to [\beta]B$$

Using these thoughts on the second formula of (4.30), we can prove that formula if we manage to prove both of the following dL formulas:

$$2gx = 2gH - v^2 \wedge x \geq 0 \to [?x = 0; v := -cv]\,(0 \leq x \wedge x \leq H)$$
$$2gx = 2gH - v^2 \wedge x \geq 0 \to [?x \neq 0]\,(0 \leq x \wedge x \leq H) \tag{4.32}$$

4.8.3 A Proof of Tests

Consider the second formula of (4.32). Proving it requires us to understand how to handle a test $?Q$ in a modality $[?Q]$. The semantics of a test $?Q$ from Chap. 3

$$[\![?Q]\!] = \{(\omega, \omega) \,:\, \omega \in [\![Q]\!]\} \tag{4.33}$$

says that a test $?Q$ completes successfully without changing the state in any state ω in which Q holds (i.e., $\omega \in [\![Q]\!]$) and fails to run in all other states (i.e., where $\omega \notin [\![Q]\!]$). How can we prove a formula with a test

$$A \to [?Q]B \tag{4.34}$$

This formula expresses that, from all initial states that satisfy A, all runs of $?Q$ reach states satisfying B. When is there such a run of $?Q$ at all? There is a run of $?Q$ from state ω if and only if Q holds in ω. So the only cases to worry about are initial states that satisfy Q; otherwise, the HP in (4.34) cannot execute and fails miserably so that the run is discarded. Hence, we get to assume Q holds, as HP $?Q$ does not otherwise execute. In all states that the HP $?Q$ reaches from states satisfying A, (4.34) conjectures that B holds. By (4.33), the final states that $?Q$ reaches are the same as the initial state (as long as they satisfy Q so that HP $?Q$ can be executed at all). That is, postcondition B needs to hold in all states from which $?Q$ runs (i.e., that satisfy Q) and that satisfy the precondition A. So (4.34) can be proved by proving

$$A \wedge Q \to B$$

> **Note 27 (Proving tests)** For an HP that is a test $?Q$, we can prove
>
> $$A \to [?Q]B$$
>
> by proving the following dL formula:
>
> $$A \wedge Q \to B$$

Using this for the second formula of (4.32), Note 27 reduces proving the second formula of (4.32)

$$2gx = 2gH - v^2 \wedge x \geq 0 \to [?x \neq 0]\,(0 \leq x \wedge x \leq H)$$

to proving

$$2gx = 2gH - v^2 \wedge x \geq 0 \wedge x \neq 0 \to 0 \leq x \wedge x \leq H \qquad (4.35)$$

Now we are left with arithmetic that we need to prove. Proofs for arithmetic and propositional logical operators such as $\wedge$ and $\to$ will be considered in a later chapter. For now, we notice that the formula $0 \leq x$ in the right-hand side of $\to$ is justified by assumption $x \geq 0$ if we flip the inequality around. Yet, $x \leq H$ does not follow from the left-hand side of (4.35), because we lost our assumptions about H somewhere.

How could that happen? We used to know $x \leq H$ in (4.26). We still knew about it in the first formula of (4.30). But we somehow let it disappear from the second formula of (4.30), because we chose an intermediate condition that was too weak.

This is a common problem in trying to prove properties of CPSs or of any other mathematical statements. One of our intermediate steps might have been too weak, so that our attempt to prove the property fails and we need to revisit how we got there. For sequential compositions, this is actually a nonissue as soon as we move on (in the next chapter) to a proof technique that is more useful than the intermediate conditions from Note 25. But similar difficulties can arise in other parts of proof attempts.

In this case, the fact that we lost $x \leq H$ can be fixed by including it in the intermediate conditions, because it can be shown to hold after the differential equation. Other crucial assumptions have also suddenly disappeared in our reasoning. An extra assumption $1 \geq c \geq 0$, for example, is crucially needed to justify the first formula of (4.32). It is much easier to see why that particular assumption can be added to the intermediate contract without changing the argument much. The reason is that c never ever changes during the system run so if $1 \geq c$ was true initially, it is still true.

> **Note 28 (Changing assumptions in a proof)** It is very difficult to come up with bug-free code. Just thinking about your assumptions really hard does not ensure correctness. But we can gain confidence that our system does what we want it to by proving that certain properties are satisfied.
>
> Assumptions and arguments in a hybrid program frequently need to be changed during the search for a proof of safety. It is easy to make subtle mistakes in informal arguments such as "I need to know C here and I would know C if I had included it here or there, so now I hope the argument holds". This is one of many reasons why we are better off if our CPS proofs are rigorous, because we would rather not end up in trouble because of a subtle flaw in a correctness argument. The rigorous, formal proof calculus for differential dynamic logic (dL) that we develop in Chaps. 5 and 6 will help us avoid the pitfalls of informal arguments. The theorem prover KeYmaera X [7] implements a proof calculus for dL, which supports such mathematical rigor.
>
> A related observation from our informal arguments in this chapter is that we desperately need a way to keep an argument consistent as a single argument justifying one conjecture, quite the contrary to the informal loose threads of argumentation we have pursued in this chapter for the sake of developing intuition. Consequently, we will investigate what holds all arguments together and what constitutes an actual proof in Chap. 6, a proof in which the relationship of premises to conclusions via proof steps is rigorous.

Moreover, there are two loose ends in our arguments. For one, the differential equation in (4.30) is still waiting for an argument that can help us prove it. Also, the assignment in (4.32) still needs to be handled and its sequential composition needs an intermediate contract (Exercise 4.18). Both will be pursued in the next chapter, where we move to a much more systematic and rigorous reasoning style for CPS.

Exercises

4.1. Show that (4.15) is valid. It is okay to focus only on this example, even though the argument is more general, because the following dL formula is valid for any hybrid program α:

$$[\alpha]F \wedge [\alpha]G \leftrightarrow [\alpha](F \wedge G)$$

4.2 (Equivalence). Let A, B be dL formulas. Suppose $A \leftrightarrow B$ is valid and A is valid and show that B is then valid, too. Suppose $A \leftrightarrow B$ is valid and you replace an occurrence of A in another formula P with B to obtain formula Q. Are P and Q then equivalent, i.e., is $P \leftrightarrow Q$ valid? Why?

4.3. Let A, B be dL formulas. Suppose $A \leftrightarrow B$ is true in state ω and A is true in state ω. That is, $\omega \in [\![A \leftrightarrow B]\!]$ and $\omega \in [\![A]\!]$. Is B true in state ω? Prove or disprove. Is B valid? Prove or disprove.

4.4. Let α be an HP and let ω be a state. Prove or disprove each of the following cases:

1. If $\omega \notin [\![P]\!]$ then does $\omega \notin [\![[\alpha^*]P]\!]$ have to hold?
2. If $\omega \notin [\![P]\!]$ then does $\omega \notin [\![\langle\alpha^*\rangle P]\!]$ have to hold?
3. If $\omega \in [\![P]\!]$ then does $\omega \in [\![[\alpha^*]P]\!]$ have to hold?
4. If $\omega \in [\![P]\!]$ then does $\omega \in [\![\langle\alpha^*\rangle P]\!]$ have to hold?

4.5 (Multiple pre/postconditions). Suppose you have an HP α with a CPS contract using multiple preconditions $A_1, \ldots, A_n$ and multiple postconditions $B_1, \ldots, B_m$:

$$\textbf{requires}(A_1)$$
$$\textbf{requires}(A_2)$$
$$\vdots$$
$$\textbf{requires}(A_n)$$
$$\textbf{ensures}(B_1)$$
$$\textbf{ensures}(B_2)$$
$$\vdots$$
$$\textbf{ensures}(B_m)$$
$$\alpha$$

How can this CPS contract be expressed in a dL formula? If there are multiple alternative was to express it, discuss the advantages and disadvantages of each option.

4.6 (Late contracts). dL formula (4.18) represents the canonical way of turning the precondition of a contract into an implication and putting the postcondition after the modality. There are other ways of capturing contract (4.17) as a dL formula. The following formula initially only assumes $x = H$ but has an implication as a postcondition. Is it also a correct logical rendition of contract (4.17)?

$$x = H \rightarrow \left[\left(\{x' = v, v' = -g \,\&\, x \geq 0\}; \text{if}(x=0)\, v := -cv\right)^*\right] (0 \leq H \rightarrow 0 \leq x \wedge x \leq H) \tag{4.36}$$

4.7 (Systematically late contracts). This question compares canonical and late phrasings of precondition/postcondition contracts by shifting formulas from preconditions to postconditions. For simplicity assume that x is the only variable that α modifies. Are the following two dL formulas equivalent?

$$A(x) \rightarrow [\alpha]B(x)$$
$$x = x_0 \rightarrow [\alpha](A(x_0) \rightarrow B(x))$$

For example, is (4.36) logically equivalent to (4.18)?

4.8. For each of the following dL formulas, determine whether they are valid, satisfiable, *and/or* unsatisfiable:

1. $[?x \geq 0]x \geq 0$.
2. $[?x \geq 0]x \leq 0$.
3. $[?x \geq 0]x < 0$.
4. $[?true]true$.
5. $[?true]false$.
6. $[?false]true$.
7. $[?false]false$.
8. $[x' = 1 \& true]true$.
9. $[x' = 1 \& true]false$.
10. $[x' = 1 \& false]true$.
11. $[x' = 1 \& false]false$.
12. $[(x' = 1 \& true)^*]true$.
13. $[(x' = 1 \& true)^*]false$.
14. $[(x' = 1 \& false)^*]true$.
15. $[(x' = 1 \& false)^*]false$.
16. $x \geq 0 \rightarrow [x' = v, v' = a]x \geq 0$.
17. $x > 0 \rightarrow [x' = x^2]x > 0$.
18. $x > 0 \rightarrow [x' = y]x > 0$.
19. $x > 0 \rightarrow [x' = x]x > 0$.
20. $x > 0 \rightarrow [x' = -x]x > 0$.

4.9. For each of the following dL formulas, determine whether they are valid, satisfiable, *and/or* unsatisfiable:

1. $x > 0 \rightarrow [x' = 1]x > 0$
2. $x > 0 \rightarrow [x' = -1]x < 0$
3. $x > 0 \rightarrow [x' = -1]x \geq 0$
4. $x > 0 \rightarrow [(x := x + 1)^*]x > 0$
5. $x > 0 \rightarrow [(x := x + 1)^*]x > 1$
6. $[x := x^2 + 1; x' = 1]x > 0$.
7. $[(x := x^2 + 1; x' = 1)^*]x > 0$.
8. $[(x := x + 1; x' = -1)^*; ?x > 0; x' = 2]x > 0$
9. $x = 0 \rightarrow [x' = 1; x' = -2)]x < 0$.
10. $x \geq 0 \land v \geq 0 \rightarrow [x' = v, v' = 2]x \geq 0$.

4.10. For each of the following dL formulas, determine whether they are valid, satisfiable, *and/or* unsatisfiable:

1. $\langle x' = -1 \rangle x < 0$
2. $x > 0 \land \langle x' = 1 \rangle x < 0$
3. $x > 0 \land \langle x' = -1 \rangle x < 0$
4. $x > 0 \rightarrow \langle x' = 1 \rangle x > 0$
5. $[(x := x + 1)^*]\langle x' = -1 \rangle x < 0$.
6. $x > 0 \rightarrow [x' = 2](x > 0 \land [x' = 1]x > 0 \land \langle x' = -2 \rangle x = 0)$
7. $\langle x' = 2 \rangle [x' = -1]\langle x' = 5 \rangle x > 0$
8. $\forall x \langle x' = -1 \rangle x < 0$
9. $\forall x [x' = 1]x \geq 0$

10. $\exists x\, [x' = -1]\, x < 0$
11. $\forall x \exists d\, (x \ge 0 \rightarrow [x' = d]\, x \ge 0)$
12. $\forall x\, (x \ge 0 \rightarrow \exists d\, [x' = d]\, x \ge 0)$
13. $[x' = 1]\, (x \ge 0 \rightarrow [x' = 2]\, x \ge 0)$
14. $[x' = 1]\, x \ge 0 \rightarrow [x' = 2]\, x \ge 0$
15. $[x' = 2]\, x \ge 0 \rightarrow [x' = 1]\, x \ge 0$
16. $\langle x' = 2 \rangle\, x \ge 0 \rightarrow [x' = 1]\, x \ge 0$

4.11. For each $j, k \in \{\text{satisfiable}, \text{unsatisfiable}, \text{valid}\}$ answer whether there is a formula that is j but not k. Also answer for each such j, k whether there is a formula that is j but its negation is not k. Briefly justify each answer.

4.12. Replace α with a concrete HP that makes the following dL formulas valid or explain why such an HP does not exist. For an extra challenge do not use assignments in α.

$$[\alpha]\mathit{false}$$
$$[\alpha^*]\mathit{false}$$
$$[\alpha]x > 0 \leftrightarrow \langle \alpha \rangle x > 0$$
$$[\alpha]x > 0 \leftrightarrow [\alpha]x > 1$$
$$[\alpha]x > 0 \leftrightarrow \neg[\alpha \cup \alpha]x > 0$$
$$[\alpha]x = 1 \wedge [\alpha]x = 2$$

4.13 (Set-valued semantics). There are at least two styles of giving a meaning to a logical formula. One way is to inductively define a satisfaction relation $\models$ that holds between a state ω and a dL formula P, written $\omega \models P$, whenever the formula P is true in the state ω. Its definition includes, among other cases, the following:

$$\omega \models P \wedge Q \text{ iff } \omega \models P \text{ and } \omega \models Q$$
$$\omega \models \langle \alpha \rangle P \text{ iff } v \models P \text{ for some state } v \text{ such that } (\omega, v) \in [\![\alpha]\!]$$
$$\omega \models [\alpha]P \text{ iff } v \models P \text{ for all states } v \text{ such that } (\omega, v) \in [\![\alpha]\!]$$

The other way is to directly inductively define, for each dL formula P, the set of states, written $[\![P]\!]$, in which P is true. Its definition includes, among other cases, the following:

$$
\begin{aligned}
[\![e \ge \tilde{e}]\!] &= \{\omega : \omega[\![e]\!] \ge \omega[\![\tilde{e}]\!]\} \\
[\![P \wedge Q]\!] &= [\![P]\!] \cap [\![Q]\!] \\
[\![\neg P]\!] &= [\![P]\!]^{\complement} = \mathscr{S} \setminus [\![P]\!] \\
[\![\langle \alpha \rangle P]\!] &= [\![\alpha]\!] \circ [\![P]\!] = \{\omega : v \in [\![P]\!] \text{ for some state } v \text{ such that } (\omega, v) \in [\![\alpha]\!]\} \\
[\![[\alpha]P]\!] &= [\![\neg \langle \alpha \rangle \neg P]\!] = \{\omega : v \in [\![P]\!] \text{ for all states } v \text{ such that } (\omega, v) \in [\![\alpha]\!]\} \\
[\![\exists x P]\!] &= \{\omega : v \in [\![P]\!] \text{ for some state } v \text{ that agrees with } \omega \text{ except on } x\} \\
[\![\forall x P]\!] &= \{\omega : v \in [\![P]\!] \text{ for all states } v \text{ that agree with } \omega \text{ except on } x\}
\end{aligned}
$$

Prove that both styles of defining the semantics are equivalent. That is $\omega \models P$ iff $\omega \in [\![P]\!]$ for all states ω and all dL formulas P.

Such a proof can be conducted by induction on the structure of P. That is, you consider each case, say $P \wedge Q$, and prove $\omega \models P \wedge Q$ iff $\omega \in [\![P \wedge Q]\!]$ from the inductive hypothesis that the conjecture already holds for the smaller subformulas:

$$\omega \models P \text{ iff } \omega \in [\![P]\!]$$
$$\omega \models Q \text{ iff } \omega \in [\![Q]\!]$$

4.14 (Rediscover Quantum). Help Quantum the bouncing ball with a clean-slate approach. Pull up a blank sheet of paper and double check whether you can help Quantum identify all the requirements that imply the following formula:

$$\left[(\{x' = v, v' = -g \,\&\, x \geq 0\}; \right.$$
$$\left. \text{if}(x = 0) \, v := -cv)^* \right] (0 \leq x \leq H)$$

What are the requirements for the following formula to be true in which the order of the sequential composition is swapped so that the discrete step comes first?

$$\left[(\text{if}(x = 0) \, v := -cv; \right.$$
$$\left. \{x' = v, v' = -g \,\&\, x \geq 0\})^* \right] (0 \leq x \leq H)$$

4.15. What would happen with the bouncing ball if $c < 0$? Consider a variation of the arguments in Sect. 4.6 where instead of the assumption in (4.21), you assume $c < 0$. Is the formula valid? What happens with a bouncing ball of damping $c = 1$?

4.16 (Deflatable Quantum). Help Quantum the bouncing ball identify all requirements that imply the following formula, in which the bouncing ball might deflate and lie flat using the model from Sect. 4.2.3:

$$\left[(\{x' = v, v' = -g \,\&\, x \geq 0\}; \right.$$
$$\left. \text{if}(x = 0) \, (v := -cv \cup v := 0))^* \right] (0 \leq x \leq H)$$

4.17. We went from (4.23) to (4.24) by removing an if-then-else. Explain how this works and justify why it is okay to do this transformation. It is okay to focus only on this case, even though the argument is more general.

4.18 (*). Find an intermediate condition for proving the first formula in (4.32). The proof of the resulting formulas is complicated significantly by the fact that assignments have not yet been discussed in this chapter. Can you find a way of proving the resulting formulas before the next chapter develops how to handle assignments?

4.19 (**). Sect. 4.8.1 used a mix of systematic and ad hoc approaches to produce an intermediate condition that was based on solving and combining differential equations. Can you think of a more systematic rephrasing?

4.20 (**). Note 25 in Sect. 4.8.1 gave a way of showing a property of a sequential composition

$$A \to [\alpha; \beta]B$$

by identifying an intermediate condition E and showing

$$A \to [\alpha]E \qquad \text{and} \qquad E \to [\beta]B$$

Can you already see a way of exploiting the operators of differential dynamic logic to show the same formula without having to be creative by inventing a clever intermediate condition E?

4.21 (**). How could formula (4.30) be proved using its differential equation?

4.22 (Direct velocity control). Real cars have proper acceleration and braking. What would happen if cars had direct control of velocity with instantaneous effect? Your job is to fill in the blanks with a test condition that makes sure the car with position x and velocity v and reaction time ε cannot exceed the stoplight m.

$$x \le m \wedge V \ge 0 \to$$
$$\big[((?\underline{\qquad}; v := V \cup v := 0);$$
$$t := 0;$$
$$\{x' = v, t' = 1 \& t \le \varepsilon\}$$
$$)^* \big] x \le m$$

References

[1] Isaac Asimov. Runaround. *Astounding Science Fiction* (Mar. 1942).

[2] Lewis Carroll. *Alice's Adventures in Wonderland*. London: Macmillan, 1865.

[3] Patricia Derler, Edward A. Lee, Stavros Tripakis, and Martin Törngren. Cyber-physical system design contracts. In: *ICCPS*. Ed. by Chenyang Lu, P. R. Kumar, and Radu Stoleru. New York: ACM, 2013, 109–118. DOI: 10.1145/2502524.2502540.

[4] René Descartes. *Meditationes de prima philosophia, in qua Dei existentia et animae immortalitas demonstratur*. 1641.

[5] Robert W. Floyd. Assigning meanings to programs. In: *Mathematical Aspects of Computer Science, Proceedings of Symposia in Applied Mathematics*. Ed. by J. T. Schwartz. Vol. 19. Providence: AMS, 1967, 19–32. DOI: 10.1007/978-94-011-1793-7_4.

[6] Gottlob Frege. *Begriffsschrift, eine der arithmetischen nachgebildete Formelsprache des reinen Denkens*. Halle: Verlag von Louis Nebert, 1879.

[7] Nathan Fulton, Stefan Mitsch, Jan-David Quesel, Marcus Völp, and André Platzer. KeYmaera X: an axiomatic tactical theorem prover for hybrid systems. In: *CADE*. Ed. by Amy Felty and Aart Middeldorp. Vol. 9195. LNCS.

Berlin: Springer, 2015, 527–538. DOI: 10.1007/978-3-319-21401-6_36.

[8] Gerhard Gentzen. Untersuchungen über das logische Schließen I. *Math. Zeit.* **39**(2) (1935), 176–210. DOI: 10.1007/BF01201353.

[9] Gerhard Gentzen. Die Widerspruchsfreiheit der reinen Zahlentheorie. *Math. Ann.* **112** (1936), 493–565. DOI: 10.1007/BF01565428.

[10] Kurt Gödel. Über die Vollständigkeit des Logikkalküls. PhD thesis. Universität Wien, 1929.

[11] Kurt Gödel. Über formal unentscheidbare Sätze der Principia Mathematica und verwandter Systeme I. *Monatshefte Math. Phys.* **38**(1) (1931), 173–198. DOI: 10.1007/BF01700692.

[12] Kurt Gödel. Über eine bisher noch nicht benützte Erweiterung des finiten Standpunktes. *Dialectica* **12**(3-4) (1958), 280–287. DOI: 10.1111/j.1746-8361.1958.tb01464.x.

[13] David Hilbert and Wilhelm Ackermann. *Grundzüge der theoretischen Logik.* Berlin: Springer, 1928.

[14] Charles Antony Richard Hoare. An axiomatic basis for computer programming. *Commun. ACM* **12**(10) (1969), 576–580. DOI: 10.1145/363235.363259.

[15] Saul A. Kripke. Semantical considerations on modal logic. *Acta Philosophica Fennica* **16** (1963), 83–94.

[16] Gottfried Wilhelm Leibniz. *Generales inquisitiones de analysi notionum et veritatum.* 1686.

[17] Francesco Logozzo. Practical verification for the working programmer with codecontracts and abstract interpretation - (invited talk). In: *VMCAI.* Ed. by Ranjit Jhala and David A. Schmidt. Vol. 6538. LNCS. Berlin: Springer, 2011, 19–22. DOI: 10.1007/978-3-642-18275-4_3.

[18] Bertrand Meyer. Applying design by contract. *Computer* **25**(10) (Oct. 1992), 40–51. DOI: 10.1109/2.161279.

[19] Stefan Mitsch and André Platzer. ModelPlex: verified runtime validation of verified cyber-physical system models. *Form. Methods Syst. Des.* **49**(1-2) (2016). Special issue of selected papers from RV'14, 33–74. DOI: 10.1007/s10703-016-0241-z.

[20] André Platzer. Differential dynamic logic for verifying parametric hybrid systems. In: *TABLEAUX.* Ed. by Nicola Olivetti. Vol. 4548. LNCS. Berlin: Springer, 2007, 216–232. DOI: 10.1007/978-3-540-73099-6_17.

[21] André Platzer. Differential dynamic logic for hybrid systems. *J. Autom. Reas.* **41**(2) (2008), 143–189. DOI: 10.1007/s10817-008-9103-8.

[22] André Platzer. *Logical Analysis of Hybrid Systems: Proving Theorems for Complex Dynamics.* Heidelberg: Springer, 2010. DOI: 10.1007/978-3-642-14509-4.

[23] André Platzer. Stochastic differential dynamic logic for stochastic hybrid programs. In: *CADE.* Ed. by Nikolaj Bjørner and Viorica Sofronie-Stokkermans. Vol. 6803. LNCS. Berlin: Springer, 2011, 446–460. DOI: 10.1007/978-3-642-22438-6_34.

[24] André Platzer. A complete axiomatization of quantified differential dynamic logic for distributed hybrid systems. *Log. Meth. Comput. Sci.* **8**(4:17) (2012). Special issue for selected papers from CSL'10, 1–44. DOI: 10.2168/LMCS-8(4:17)2012.

[25] André Platzer. Logics of dynamical systems. In: *LICS*. Los Alamitos: IEEE, 2012, 13–24. DOI: 10.1109/LICS.2012.13.

[26] André Platzer. The complete proof theory of hybrid systems. In: *LICS*. Los Alamitos: IEEE, 2012, 541–550. DOI: 10.1109/LICS.2012.64.

[27] André Platzer. Teaching CPS foundations with contracts. In: *CPS-Ed*. 2013, 7–10.

[28] André Platzer. Differential game logic. *ACM Trans. Comput. Log.* **17**(1) (2015), 1:1–1:51. DOI: 10.1145/2817824.

[29] André Platzer. Logic & proofs for cyber-physical systems. In: *IJCAR*. Ed. by Nicola Olivetti and Ashish Tiwari. Vol. 9706. LNCS. Berlin: Springer, 2016, 15–21. DOI: 10.1007/978-3-319-40229-1_3.

[30] André Platzer. A complete uniform substitution calculus for differential dynamic logic. *J. Autom. Reas.* **59**(2) (2017), 219–265. DOI: 10.1007/s10817-016-9385-1.

[31] André Platzer. Differential hybrid games. *ACM Trans. Comput. Log.* **18**(3) (2017), 19:1–19:44. DOI: 10.1145/3091123.

[32] André Platzer and Edmund M. Clarke. The image computation problem in hybrid systems model checking. In: *HSCC*. Ed. by Alberto Bemporad, Antonio Bicchi, and Giorgio C. Buttazzo. Vol. 4416. LNCS. Springer, 2007, 473–486. DOI: 10.1007/978-3-540-71493-4_37.

[33] Amir Pnueli. The temporal logic of programs. In: *FOCS*. IEEE, 1977, 46–57.

[34] Vaughan R. Pratt. Semantical considerations on Floyd-Hoare logic. In: *17th Annual Symposium on Foundations of Computer Science, 25-27 October 1976, Houston, Texas, USA*. Los Alamitos: IEEE, 1976, 109–121. DOI: 10.1109/SFCS.1976.27.

[35] Arthur Prior. *Time and Modality*. Oxford: Clarendon Press, 1957.

[36] Dana S. Scott. Logic and programming languages. *Commun. ACM* **20**(9) (1977), 634–641. DOI: 10.1145/359810.359826.

[37] Thoralf Skolem. Logisch-kombinatorische Untersuchungen über die Erfüllbarkeit oder Beweisbarkeit mathematischer Sätze nebst einem Theorem über dichte Mengen. *Videnskapsselskapets skrifter, 1. Mat.-naturv. klasse* **4** (1920), 1–36.

[38] Alfred North Whitehead and Bertrand Russell. *Principia Mathematica*. Cambridge: Cambridge Univ. Press, 1910.

[39] Dana N. Xu, Simon L. Peyton Jones, and Koen Claessen. Static contract checking for Haskell. In: *POPL*. Ed. by Zhong Shao and Benjamin C. Pierce. New York: ACM, 2009, 41–52. DOI: 10.1145/1480881.1480889.

[40] Paolo Zuliani, André Platzer, and Edmund M. Clarke. Bayesian statistical model checking with application to Simulink/Stateflow verification. *Form. Methods Syst. Des.* **43**(2) (2013), 338–367. DOI: 10.1007/s10703-013-0195-3.

Chapter 5
Dynamical Systems & Dynamic Axioms

Synopsis This central chapter develops a logical characterization of the dynamics of hybrid programs in differential dynamic logic. It investigates fundamental compositional reasoning principles that capture how the truth of a property of a more complex hybrid program relates to the truth of corresponding properties of simpler program fragments. This leads to dynamic axioms for dynamical systems, with one axiom for each type of dynamics. These dynamic axioms enable rigorous reasoning about CPS models and begin an axiomatization of differential dynamic logic, which turns the specification logic dL into a verification logic for CPS. While more advanced aspects of loops and differential equations will be discussed in subsequent chapters, this chapter lays a pivotal foundation for all dynamical aspects of differential dynamic logic and its hybrid programs.

5.1 Introduction

Chap. 4 demonstrated how useful and crucial CPS contracts are for CPS. Their rôle and understanding goes beyond dynamic testing. In CPS, proven CPS contracts are infinitely more valuable than dynamically tested contracts, because, without sufficient care, dynamical tests of contracts at runtime of a CPS generally leave open very little flexibility for reacting to them in any safe way. After all, the failure of a contract indicates that some safety condition that was expected to hold is no longer true. Unless provably sufficient safety margins and fallback plans remain, the CPS is already in trouble then.[1]

Consequently, CPS contracts really shine in relation to how they are proved for CPS. Understanding how to prove CPS contracts requires us to understand the dynamical effects of hybrid programs in more detail. This deeper understanding of

---

[1] However, in combination with formal verification, the Simplex architecture can be understood as exploiting the relationship of dynamic contracts for safety purposes [14]. ModelPlex, which is based on differential dynamic logic, lifts this observation to a fully verified link from verified CPS models to verified CPS executions [4].

© Springer International Publishing AG, part of Springer Nature 2018
A. Platzer, *Logical Foundations of Cyber-Physical Systems*,
https://doi.org/10.1007/978-3-319-63588-0_5

the effects of hybrid program operators is not only useful for conducting proofs, but also for developing and sharpening our intuition about hybrid programs. This phenomenon illustrates the more general point that *proof and effect (and/or meaning) are intimately linked*. Truly understanding effect is ultimately the same as, as well as a prerequisite to, understanding how to prove properties of that effect [6, 8, 9, 11]. You may have seen this point demonstrated already in other treatises on programming languages, but it will shine in this chapter.

The route we choose to get to this level of understanding involves a closer look at the structure of the effect that hybrid programs have on states. This will enable us to devise authoritative proof principles for differential dynamic logic and hybrid programs [5, 6, 8, 9, 11]. This chapter will give us the essential reasoning tools for cyber-physical systems and is, thus, of crucial relevance.

The focus of this chapter is on a systematic development of the basic reasoning principles for cyber-physical systems. The goal is to cover all cyber-physical systems by identifying one fundamental reasoning principle for each of the operators of differential dynamic logic and, specifically, its hybrid programs. Once we have such a reasoning principle for each of the operators, the basic idea is that any arbitrary cyber-physical system can be analyzed by combining the various reasoning principles with one another, compositionally, by inspecting one operator at a time.

Note 29 (Logical guiding principle: Compositionality) Since every CPS is modeled by a hybrid program[a] and all hybrid programs are combinations of simpler hybrid programs using one of a handful of program operators (such as $\cup$ and ; and *), all CPSs can be analyzed if only we identify one suitable analysis technique for each of the operators.

---

[a] To faithfully represent complex CPSs, some models need an extension of hybrid programs, e.g., to hybrid games [10] or distributed hybrid programs [7], in which case suitable generalizations of the logical approach presented here work.

With enough understanding, this guiding principle ultimately succeeds [9–11]. It does, however, take more than one chapter to get there. This chapter settles for a systematic development of the reasoning principles for elementary operators in hybrid programs, leaving a detailed development of the others to later chapters.

This chapter is of central significance for the Foundations of Cyber-Physical Systems. It is the first of many chapters in this textbook where we observe a logical trichotomy between syntax, semantics, and axiomatics.

Note 30 (Logical trinity) The concepts developed in this chapter illustrate the more general relation of *syntax* (which is notation), *semantics* (which carries meaning), and *axiomatics* (which internalizes semantic relations into universal syntactic transformations). These concepts and their relations jointly form the significant *logical trinity* of syntax, semantics, and axiomatics.

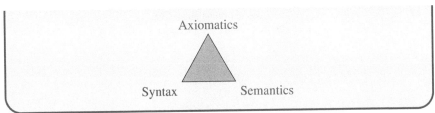

For example, the syntax for conjunction is $A \wedge B$. The semantics of $A \wedge B$ is that $A \wedge B$ is *true* iff A is *true* and B is *true*. Its axiomatics will tell us that a proof of $A \wedge B$ consists of a proof of A together with a proof of B, which is what will be explored in Chap. 6. Since the semantics is compositional (the meaning of $A \wedge B$ is that it is *true* whenever both A and B are *true*), the reasoning will be compositional, too, so that a proof of $A \wedge B$ can be decomposed into a proof of A and a separate proof of B. This chapter sets out to make the same kind of logical compositionality happen for all other operators of cyber-physical systems.

The most important learning goals of this chapter are:

Modeling and Control: We will understand the core principles behind CPS by understanding analytically and semantically how cyber and physical aspects are integrated and interact in CPS. This chapter will also begin to explicitly relate discrete and continuous systems, which ultimately leads to a fascinating view on understanding hybridness [9].

Computational Thinking: This chapter is devoted to the core aspects of reasoning rigorously about CPS models, which is critical to getting CPS right. CPS designs can be flawed for very subtle reasons. Without sufficient rigor in their analysis it can be impossible to spot the flaws, and it can be even more challenging to say for sure whether and why a design is no longer faulty. This chapter systematically develops one reasoning principle for each of the operators of hybrid programs. This chapter begins an *axiomatization* of differential dynamic logic dL [8, 9, 11] to lift dL from a specification language to a verification language for CPS.

CPS Skills: We will develop a deep understanding of the semantics of CPS models by carefully relating their semantics to their reasoning principles and aligning them in perfect unison. This understanding will also enable us to develop a better intuition for the operational effects involved in CPS.

5.2 Intermediate Conditions for CPS

Recall the bouncing ball from p. 122 in Chap. 4:

$$0 \le x \wedge x = H \wedge v = 0 \wedge g > 0 \wedge 1 \ge c \ge 0 \rightarrow$$
$$\left[\left(\{x' = v, v' = -g \, \& \, x \ge 0\}; \ (?x = 0; v := -cv \cup ?x \ne 0) \right)^* \right] (0 \le x \wedge x \le H) \tag{4.24*}$$

rigorous reasoning about CPS
dL as a verification language

cyber+physics interaction
relate discrete+continuous

align semantics+reasoning
operational CPS effects

To simplify the subsequent discussion, let's again drop the repetition (*) for now:

$$0 \leq x \wedge x = H \wedge v = 0 \wedge g > 0 \wedge 1 \geq c \geq 0 \rightarrow$$
$$\left[\{x' = v, v' = -g \,\&\, x \geq 0\}; (?x = 0; v := -cv \cup ?x \neq 0) \right] (0 \leq x \wedge x \leq H) \quad (5.1)$$

Of course, dropping the repetition grotesquely changes the behavior of the bouncing ball. It cannot even really bounce any longer now. It can merely fall and reverse its velocity vector when on the ground but is then stuck. The single-hop bouncing ball can only follow the first blue hop but not the gray remainder hops in Fig. 5.1. This degenerate model fragment is, nevertheless, an insightful stepping stone toward a proof of the full model. If we manage to prove (5.1), we certainly have not shown the full bouncing-ball formula (4.24) with its loop. But it's a start, because the behavior modeled in (5.1) is a part of the behavior of (4.24). So it is useful (and easier for us) to understand (5.1) first.

Fig. 5.1 Sample trajectory of a single-hop bouncing ball (plotted as height over time) that can follow the first blue hop but is incapable of following the remaining hops shown in gray

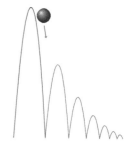

The dL formula (5.1) has assumptions $0 \leq x \wedge x = H \wedge v = 0 \wedge g > 0 \wedge 1 \geq c \geq 0$ that can be used during the proof. It claims that the postcondition $0 \leq x \wedge x \leq H$ holds after all runs of the HP in the [·] modality. The top-level operator in the modality of (5.1) is a sequential composition (;), for which we need to find a proof argument.[2]

---

[2] The way we proceed here to prove (5.1) is actually not the recommended way. We will develop an easier way. But it is instructive to understand the more verbose approach we take first. The first approach also prepares us for the challenges that lie ahead when proving properties of loops.

The HP in (5.1) follows a differential equation first and then, after the sequential composition (;), proceeds to run a discrete program $(?x = 0; v := -cv \cup ?x \neq 0)$. Depending on how long the HP follows its differential equation, the intermediate state after the differential equation and before the discrete program will be different.

> **Note 31 (Intermediate states of sequential compositions)** The first HP α in a sequential compositions $\alpha; \beta$ may reach a whole range of states, which represent intermediate states for the sequential composition $\alpha; \beta$, i.e., states that are final states for α and initial states for β. The intermediate states of $\alpha; \beta$ are the states μ in the semantics $[\![\alpha; \beta]\!]$ from Chap. 3:
>
> $$[\![\alpha; \beta]\!] = [\![\alpha]\!] \circ [\![\beta]\!] = \{(\omega, \nu) : (\omega, \mu) \in [\![\alpha]\!], (\mu, \nu) \in [\![\beta]\!] \text{ for some } \mu\}$$

One can summarize what all intermediate states between the differential equation and the discrete program of (5.1) have in common. They differ by how long the CPS has followed the differential equation. But the intermediate states still have in common that they satisfy a logical formula E. Which logical formula that is, is, in fact, instructive to find out, but of no immediate concern for the rest of this chapter. So we invite you to find out how to choose E for (5.1) before you compare your answer to the one we developed in Sect. 4.8.1 already.

For an HP that is a sequential composition $\alpha; \beta$ an *intermediate condition* is a formula that characterizes the intermediate states in between HP α and β. That is, for a dL formula

$$A \to [\alpha; \beta]B$$

an intermediate condition is a formula E such that the following dL formulas are valid:

$$A \to [\alpha]E \qquad \text{and} \qquad E \to [\beta]B$$

The first dL formula expresses that intermediate condition E characterizes the intermediate states accurately, i.e., E actually holds after all runs of HP α from states satisfying A. The second dL formula says that the intermediate condition E characterizes intermediate states well enough, i.e., E is all we need to know about a state to conclude that all runs of β end up in B. That is, from all states satisfying E (in particular from those that result by running α from a state satisfying A), B holds after all runs of β. Depending on the precision of the intermediate condition E, this argument may require showing that B holds after all runs of β from extra states that are not reachable from ω by running α but happen to satisfy E (unlabeled nodes in Fig. 5.2).

Intermediate condition contracts for sequential compositions are captured more concisely in the following proof rule by Tony Hoare [3]:

$$\text{H;} \quad \frac{A \to [\alpha]E \quad E \to [\beta]B}{A \to [\alpha; \beta]B}$$

The two dL formulas above the bar of the rule are called *premises*. The dL formula below the bar is called the *conclusion*. The above argument (informally) justifies the

Fig. 5.2 Intermediate conditions for sequential compositions

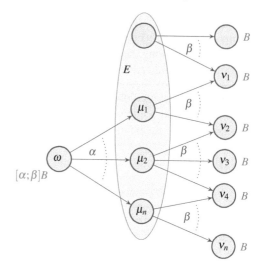

proof rule: if both premises are valid then the conclusion is valid, too. So, if we have a proof for each of the two premises, rule H; gave us a proof of the conclusion.

Since we will soon identify a better way of proving properties of sequential compositions, we do not pursue rule H; further now. Yet, there are some circumstances under which an intermediate condition as in H; actually simplifies your reasoning.

For now, we remark that, given an intermediate condition E, the rule H; splits a proof of (5.1) into a proof of the following two premises, which we saw in Chap. 4:

$$0 \leq x \wedge x=H \wedge v=0 \wedge g > 0 \wedge 1 \geq c \geq 0 \rightarrow \left[x' = v, v' = -g \& x \geq 0 \right] E \qquad (4.29^*)$$

$$E \rightarrow \left[?x = 0; v := -cv \cup ?x \neq 0 \right] (0 \leq x \wedge x \leq H) \qquad (4.30^*)$$

5.3 Dynamic Axioms for Dynamical Systems

This section develops axioms for decomposing dynamical systems, which are of central significance in this textbook. Each axiom describes the effect of one operator on hybrid programs in terms of simpler hybrid programs, thereby simultaneously serving as an explanation and as a basis for rigorous reasoning.

5.3.1 Dynamic Axioms for Nondeterministic Choices

By the logical guiding principle of compositionality (Note 29), the next operator that we need to understand in order to proceed with a proof of (5.1) is the nondeterministic choice $?x = 0; v := -cv \cup ?x \neq 0$, which is the top-level operator in the modality of (4.30). By the compositionality principle, we zero in on the nondeterministic choice operator $\cup$ and pretend we already know how to handle all other operators in the formula. If we succeed in reducing the property of the nondeterministic choice in formula (4.30) to properties of its subprograms, then we can subsequently develop axioms that actually handle the remaining operators.

Recall the semantics of nondeterministic choice from Sect. 3.3.2:

$$\llbracket \alpha \cup \beta \rrbracket = \llbracket \alpha \rrbracket \cup \llbracket \beta \rrbracket \tag{5.2}$$

Remember that $\llbracket \alpha \rrbracket$ is a reachability relation on states, where $(\omega, \nu) \in \llbracket \alpha \rrbracket$ iff HP α can run from state ω to state ν. Let us illustrate graphically what (5.2) means:

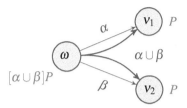

According to the reachability relation $\llbracket \alpha \rrbracket$, a number of states ν_i are reachable by running HP α from some initial state ω.[3] According to $\llbracket \beta \rrbracket$, a number of (possibly different) states ν_i are reachable by running HP β from the same initial state ω. By the semantic equation (5.2), running $\alpha \cup \beta$ from ω can give us exactly any of those possible outcomes that result from either running α or β. And there was nothing special about the initial state ω. The same principle holds for all other states as well.

> **Note 32** ($\cup$) The nondeterministic choice $\alpha \cup \beta$ can lead to exactly the states to which either α could take us or to which β could take us or to which both could lead. The dynamic effect of a nondeterministic choice $\alpha \cup \beta$ is that running it at any time results in a behavior either of α or of β, nondeterministically. So both the behaviors of α and β are possible when running $\alpha \cup \beta$.

If we want to understand whether and where dL formula $[\alpha \cup \beta]P$ is true, we need to understand which states the modality $[\alpha \cup \beta]$ refers to. In which states does P have to be true so that $[\alpha \cup \beta]P$ is true in state ω?

By definition of the semantics, P needs to be true in all states that $\alpha \cup \beta$ can reach from ω according to $[\alpha \cup \beta]$ for $[\alpha \cup \beta]P$ to be true in ω. Referring to the semantics (5.2) or looking at its illustration shows us that this includes all states that α can reach from ω according to $\llbracket \alpha \rrbracket$, hence $[\alpha]P$ has to be true in ω. It also includes all states that β can reach from ω, hence $[\beta]P$ has to be true in ω.

Consequently,

$$\omega \in \llbracket [\alpha]P \rrbracket \quad \text{and} \quad \omega \in \llbracket [\beta]P \rrbracket \tag{5.3}$$

are necessary conditions for

---

[3] The diagram only illustrates one such state ν_1 for visual conciseness. But ν_1 should be thought of as a generic representative for any such state that α can reach from the initial state ω.

$$\omega \in [\![[\alpha \cup \beta] P]\!] \tag{5.4}$$

That is, unless (5.3) holds, (5.4) cannot possibly hold. So (5.3) is necessary for (5.4). Are there any states missing? Are there any states that (5.4) would require to satisfy P, which (5.3) does not already ensure? No, because, by (5.2), $\alpha \cup \beta$ does not admit any behavior that neither α nor β can exhibit. Hence (5.3) is also sufficient for (5.4), i.e., (5.3) implies (5.4). So (5.3) and (5.4) are equivalent.

When adopting a more logical language again, this justifies

$$\omega \in [\![[\alpha \cup \beta] P \leftrightarrow [\alpha] P \wedge [\beta] P]\!]$$

This reasoning did not depend on the particular state ω but holds for all ω. Therefore, the formula $[\alpha \cup \beta] P \leftrightarrow [\alpha] P \wedge [\beta] P$ is valid, written:

$$\vDash [\alpha \cup \beta] P \leftrightarrow [\alpha] P \wedge [\beta] P$$

Exciting! We have just proved our first axiom to be sound (a proof is in Sect. 5.3.2).

Lemma 5.1 ([∪] axiom of nondeterministic choice). *The axiom of (nonde-terministic) choice is sound, i.e., all its instances are valid:*

$$[\cup] \quad [\alpha \cup \beta] P \leftrightarrow [\alpha] P \wedge [\beta] P$$

Nondeterministic choices split into their alternatives in axiom [∪]. From right to left: If all α runs lead to states satisfying P (i.e., $[\alpha] P$ is true) and all β runs lead to states satisfying P (i.e., $[\beta] P$ is true), then all runs of HP $\alpha \cup \beta$, which may choose between following α and following β, also lead to states satisfying P (i.e., $[\alpha \cup \beta] P$ is true). The converse implication from left to right holds, because $\alpha \cup \beta$ can run all runs of α and all runs of β, so all runs of α (and of β) lead to states satisfying P if $[\alpha \cup \beta] P$. We will mark the structurally complex formula in blue.

Armed with this axiom [∪] at our disposal, we can now easily make the following inference just by invoking the equivalence that [∪] justifies:

$$[\cup] \frac{A \to [\alpha] B \wedge [\beta] B}{A \to [\alpha \cup \beta] B}$$

Let's elaborate. If we want to prove the conclusion at the bottom

$$A \to [\alpha \cup \beta] B \tag{5.5}$$

then we can instead prove the premise at the top

$$A \to [\alpha] B \wedge [\beta] B \tag{5.6}$$

because by [∪], or rather an instance of [∪] formed by using B for P, we know

$$[\alpha \cup \beta] B \leftrightarrow [\alpha] B \wedge [\beta] B \tag{5.7}$$

Since (5.7) is a valid equivalence, its left-hand side and its right-hand side are equivalent. Wherever its left-hand side occurs, we can equivalently replace it with its right-hand side, since the two are equivalent.[4] Thus, replacing the left-hand side of (5.7) in the place where it occurs in (5.5) with the right-hand side of (5.7) gives us the formula (5.6), which is equivalent to (5.5). After all, according to the valid equivalence (5.7) justified by axiom [∪], (5.6) can be obtained from (5.5) just by replacing a formula with one that is equivalent (recall Exercise 4.2).

Actually, stepping back, the same argument can be made to go from (5.6) to (5.5) instead of from (5.5) to (5.6), because (5.7) is an equivalence. Both ways of using [∪] are perfectly correct, although the direction that gets rid of the ∪ operator is more useful, because it makes progress (getting rid of an HP operator).

Yet axiom [∪] can also be useful in many more situations. For example, axiom [∪] also justifies the inference

$$[\cup] \frac{[\alpha]A \wedge [\beta]A \to B}{[\alpha \cup \beta]A \to B}$$

which follows from the left-to-right implication of equivalence axiom [∪].

For the bouncing ball, axiom [∪] will decompose formula (4.30) and reduce it to

$$E \to \left[?x = 0; v := -cv\right](0 \leq x \wedge x \leq H) \wedge \left[?x \neq 0\right](0 \leq x \wedge x \leq H) \qquad (5.8)$$

A general design principle behind all dL axioms is most noticeable in axiom [∪]. All equivalence axioms of dL are primarily intended to be used by reducing the blue formula on the left to the (structurally simpler) formula on the right. Such a reduction symbolically decomposes a property of a more complicated system $\alpha \cup \beta$ into separate properties of smaller fragments α and β. While we might end up with more subproperties (like we do in the case of axiom [∪]), each of them is structurally simpler, because it involves fewer program operators. This decomposition of systems into their fragments makes the verification problem tractable and is good for scalability purposes, because it reduces the study of complex systems successively to a study of many but smaller subsystems, of which there are only finitely many. For these symbolic structural decompositions, it is very helpful that dL is a full logic that is closed under all logical operators [10, 11], including disjunction and conjunction, for then both sides in axiom [∪] are dL formulas again (unlike in Hoare logic [3]). This also turns out to be an advantage for computing invariants [2, 6, 12].

Axiom [∪] allows us to understand and handle $[\alpha \cup \beta]P$ properties. If we find appropriate axioms for all the other operators of hybrid programs, including $;, ^*, := , x'$, then we have a way of handling all hybrid programs, because we merely need to simplify the verification question by subsequently decomposing it with the respective axiom. Even if a full account of this principle is significantly more complicated, such recursive decomposition indeed ultimately succeeds [9, 11].

---

[4] This will be made formal in Chap. 6 following contextual equivalence reasoning [11].

5.3.2 Soundness of Axioms

The definition of soundness in Lemma 5.1 was not specific to axiom [∪], but applies to all dL axioms, so we discuss soundness once and for all.

> **Definition 5.1 (Soundness).** An axiom is *sound* iff all its instances are valid, i.e., true in all states.

From now on, every time we see a formula of the form $[\alpha \cup \beta]P$, we remember that axiom [∪] identifies the corresponding formula $[\alpha]P \wedge [\beta]P$ that is equivalent to it. Whenever we find a formula $[\gamma \cup \delta]Q$, we also remember that axiom [∪] says that formula $[\gamma]Q \wedge [\delta]Q$ is equivalent to it, just by instantiation [11] of axiom [∪]. The fact that axiom [∪] is sound ensures that we do not need to worry about whether such reasoning is correct every time we need it. Soundness of [∪] guarantees that every instance of [∪] is sound [11]. We can, thus, treat axiom [∪] syntactically and mechanically and apply it as needed, like a machine would.

But because soundness is such a big deal (a *conditio sine qua non* in logic, i.e., something without which logic could not be), we will prove the soundness of axiom [∪] carefully, even if we essentially already did that in our informal argument above.

Proof (of Lemma 5.1). The fact that axiom [∪] is sound can be proved as follows. Since $[\![\alpha \cup \beta]\!] = [\![\alpha]\!] \cup [\![\beta]\!]$, we have that $(\omega, v) \in [\![\alpha \cup \beta]\!]$ iff $(\omega, v) \in [\![\alpha]\!]$ or $(\omega, v) \in [\![\beta]\!]$. Thus, $\omega \in [\![[\alpha \cup \beta]P]\!]$ iff both $\omega \in [\![[\alpha]P]\!]$ and $\omega \in [\![[\beta]P]\!]$. □

Why is soundness so critical? Well, because, without it, we could accidentally declare a system safe that is not in fact safe, which would defeat the whole purpose of verification and possibly put human lives in jeopardy when they are entrusting their lives to an unsafe CPS. Unfortunately, soundness is actually not granted in all verification techniques for hybrid systems. But we will make it a point in this book to only ever use sound reasoning and to scrutinize all verifications for soundness right away. Soundness is something that is relatively easy to establish in a logic and proof approach, because it localizes into the separate study of soundness of each of its axioms.

5.3.3 Dynamic Axioms for Assignments

Part of the dL formula (5.8) that remains to be shown after using the [∪] axiom involves another sequential composition $?x = 0; v := -cv$. But even if Sect. 5.2 already discussed one possibility for reducing safety properties of sequential compositions to properties of the parts using appropriate intermediate conditions, we still need a way of handling the remaining assignment and test. Let's start with the assignment.

HPs may involve discrete assignments. Recall their semantics from Sect. 3.3.2:

$$[\![x := e]\!] = \{(\omega, v) \ : \ v = \omega \text{ except that } v[\![x]\!] = \omega[\![e]\!]\}$$

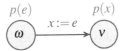

How can dL formula $[x:=e]p(x)$ be rephrased equivalently in simpler ways? It expresses that $p(x)$ holds always after assigning the value of term e to variable x. Well, in fact, there is exactly one way of assigning e to x. So, the formula $[x:=e]p(x)$ expresses that $p(x)$ holds after changing the value of x to that of e.

> **Lemma 5.2 ([:=] assignment axiom).** *The* assignment axiom *is sound:*
>
> $$[:=] \quad [x:=e]p(x) \leftrightarrow p(e)$$

The assignment axiom $[:=]$ expresses that $p(x)$ is true after the discrete assignment assigning term e to x iff $p(e)$ was already true before that change, since the assignment $x:=e$ will change the value of variable x to the value of e.

For example, axiom $[:=]$ immediately allows us to conclude that the dL formula $[x:=a\cdot x]x\cdot(x+1)\geq 0$ is equivalent to the first-order formula $(a\cdot x)\cdot(a\cdot x+1)\geq 0$, which is formed by replacing all free occurrences of x in postcondition $x\cdot(x+1)\geq 0$ with its new value $a\cdot x$.

If we succeed in splitting (5.8) into pieces, including its sequential composition $[?x=0;v:=-cv](0\leq x\wedge x\leq H)$ with yet another intermediate condition F according to Sect. 5.2, we will eventually have to prove a number of dL formulas including

$$F \to [v:=-cv](0\leq x\wedge x\leq H)$$

The assignment axiom $[:=]$ equivalently reduces this formula to $F \to 0\leq x\wedge x\leq H$, because the affected variable v does not occur in the postcondition. That is quite peculiar for the bouncing ball, because it will make the damping coefficient c disappear entirely from the proof. While we are always happy to see complexity reduced, this should make us pause our train of thought for a moment. Chapter 4 taught us that the safety of the bouncing ball depends on the damping coefficient being $c\leq 1$. How could the proof ever possibly succeed if we misplace c in the proof?

Before you read on, see if you can find the answer for yourself.

While the bouncing ball exhibits unsafe behavior if $c > 1$, that counterexample requires the ball to bounce back up from the ground, which is a capability that the simplified single-hop bouncing ball (5.1) lost compared to the full model with its repetition. That explains why our attempt to prove (5.1) can succeed when it removes c, but also indicates that we will need to make sure not to ignore the velocity v, whose change depends on c in a proof of the repetitive bouncing ball (Chap. 7).

> **Expedition 5.1 (Admissibility caveats for the $p(x)$ notation in axioms)**
>
> There is a simple elegant way of understanding the notation $p(x)$ and $p(e)$ in axiom [:=], which, unfortunately, needs more elaboration than this chapter provides. After that sophistication, uniform substitutions [11], which we investigate in Part IV, justify that both can be read as predicate symbol p applied to the variable x to form $p(x)$ and applied to the term e to form $p(e)$, respectively. Uniform substitutions replace predicate symbols (similarly for function symbols) as long as no $p(e)$ occurs in the scope of a quantifier or modality binding a variable of their replacements other than the occurrences in e itself.
>
> Till then, we need a simple, intuitive, but correct reading of $p(x)$ and $p(e)$ in axiom [:=] and elsewhere. The basic idea is that $p(e)$ stands for the same formula that $p(x)$ does, except that all free occurrences of x are replaced by e and that this substitution requires that x does not occur in $p(x)$ in a quantifier for or a modality with an assignment or with a differential equation for x or a variable of e. For example, $[x := x+y]x \le y^2 \leftrightarrow x+y \le y^2$ is an instance of [:=], but $[x := x+y]\forall y\,(x \le y^2) \leftrightarrow \forall y\,(x+y \le y^2)$ is not, because a variable of $x+y$ is bound in $p(x)$. Indeed, y would refer to different variables in the two sides so needs to be renamed first. Likewise, $[x := x+y][y := 5]x \ge 0 \leftrightarrow [y := 5]x+y \ge 0$ is no instance of [:=], because a variable of $x+y$ is bound by $y := 5$ in $p(x)$. Free and bound variables will be defined in Sect. 5.6.5.

5.3.4 Dynamic Axioms for Differential Equations

Some of the decompositions of the bouncing-ball safety property led to safety properties of differential equations, for example (4.29) discussed again at the end of Sect. 5.2.

HPs often involve differential equations. Recall their semantics from Sect. 3.3.2.

Definition 3.3 (Transition semantics of ODEs).

$$[\![x' = f(x)\,\&\,Q]\!] = \{(\omega, \nu) : \varphi(0) = \omega \text{ except at } x' \text{ and } \varphi(r) = \nu \text{ for a solution}$$
$$\varphi{:}[0,r] \to \mathscr{S} \text{ of any duration } r \text{ satisfying } \varphi \models x' = f(x) \wedge Q\}$$

where $\varphi \models x' = f(x) \wedge Q$, iff for all times $0 \le z \le r$: $\varphi(z) \in [\![x' = f(x) \wedge Q]\!]$ with $\varphi(z)(x') \overset{\text{def}}{=} \frac{\mathrm{d}\varphi(t)(x)}{\mathrm{d}t}(z)$ and $\varphi(z) = \varphi(0)$ except at x, x'.

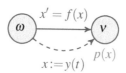

One possible approach to prove properties of differential equations is to work with a solution if one is available (and expressible in dL). Indeed, probably the first thing you learned about what to do with differential equations was to solve them.

Lemma 5.3 (['] solution axiom). *The* solution axiom schema *is sound:*

$$['] \ [x' = f(x)]p(x) \leftrightarrow \forall t{\geq}0\,[x{:=}y(t)]p(x) \quad (y'(t) = f(y))$$

where $y(\cdot)$ *solves the symbolic initial value problem* $y'(t) = f(y), y(0) = x$.

Solution $y(\cdot)$ is unique since $f(x)$ is smooth (Theorem 2.2). Given such a solution $y(\cdot)$, continuous evolution along differential equation $x' = f(x)$ can be replaced by a discrete assignment $x := y(t)$ with an additional quantifier for the evolution time t. It goes without saying that variables such as t are fresh in axiom ['] and other axioms. Conventional initial value problems (Definition 2.1) are numerical with concrete numbers as initial values, not symbolic variables x. This is not enough for our purpose, because we need to consider all states in which the ODE can start, which may be uncountably many. That is why axiom ['] solves one symbolic initial value problem, instead, since we can hardly solve uncountably many numerical initial value problems. Observe that axiom ['] refers to $y(t)$ at all times $t \geq 0$, which implies that the global solution y needs to exist for all times for axiom ['].

Note 33 (Discrete vs. continuous dynamics) Notice something quite intriguing and peculiar about axiom [']. It relates a property of a continuous system to a property of a discrete system. The HP on the left-hand side describes a smoothly changing continuous process, while the right-hand side describes an abruptly, instantaneously changing discrete process. Still, their respective properties coincide, thanks to the time quantifier. This is the beginning of an astonishingly intimate relationship of discrete and continuous dynamics [9].

What we have seen so far about the dynamics of differential equations does not yet help us prove properties of differential equations with evolution domain constraints $(x' = f(x) \& q(x))$. It also does not yet tell us what to do if we cannot solve the differential equation or if the solution is too complicated to be expressed as a term. We will get to those matters in more detail in Part II. But the evolution domain constraint $q(x)$ can be handled directly as well by adding a condition checking that the evolution domain was always true until the point in time of interest:

Lemma 5.4 (['] solution with domain axiom). *This axiom schema is sound:*

$$['] \ [x' = f(x) \& q(x)]p(x) \leftrightarrow \forall t{\geq}0\,\big((\forall 0{\leq}s{\leq}t\,q(y(s))) \rightarrow [x{:=}y(t)]p(x)\big)$$

where $y(\cdot)$ *solves the symbolic initial value problem* $y'(t) = f(y), y(0) = x$.

The effect of the additional constraint on $q(x)$ is to restrict the continuous evolution such that its solution $y(s)$ remains in the evolution domain $q(x)$ at all intermediate times $s \leq t$. This constraint simplifies to *true* if the evolution domain $q(x)$ is *true*, which makes sense, because there are no special constraints on the evolution (other than the differential equations) if the evolution domain is described by *true*; hence it is the full state space. In fact, because both axioms from Lemmas 5.3 and 5.4 give essentially the same result if $q(x)$ is *true*, we gave both the same name $[']$. For axiom schema $[']$, it is important, however, that $x' = f(x) \& q(x)$ is an explicit differential equation, so no x' occurs in $f(x)$ or $q(x)$ (or $p(x)$), because otherwise the notions of solutions become more complicated.

Axiom $[']$ explains the rôle of evolution domain constraints quite directly. The dL formula $[x' = f(x) \& q(x)]p(x)$ is true iff the postcondition $p(x)$ is true in all states that can be reached by following the solution of the differential equation $x' = f(x)$ *if* that solution has always been in the evolution domain $q(x)$ at all times.

In order to build our way up to using axiom $[']$ on the formula (4.29) that was repeated in Sect. 5.2, first consider a case with only one differential equation:

$$A \rightarrow [x' = v]E$$

Axiom $[']$ swiftly uses the unique solution $x(t) = x + vt$ to reduce this formula to:

$$A \rightarrow \forall t \geq 0 [x := x + vt]E$$

Let's add the second differential equation, the one for velocity v, back in:

$$A \rightarrow [x' = v, v' = -g]p(x) \tag{5.9}$$

Axiom $[']$ simplifies this formula, too, but things become a bit more complicated:

$$A \rightarrow \forall t \geq 0 [x := x + vt - \frac{g}{2}t^2]p(x) \tag{5.10}$$

First of all, the solution for x became more complex, because the velocity v now keeps changing over time in (5.9). That is perfectly in line with what we have learned about the solutions of differential equations in Chap. 2. Thinking carefully, we notice that (5.10) is the correct reduction of (5.9) by axiom schema $[']$ only if its postcondition $p(x)$ indeed only mentions the position x and not the velocity v. In fact, this is the reason why the postcondition was somewhat suggestively phrased as $p(x)$ in (5.9) to indicate that it is a condition on x. For a postcondition $p(x, v)$ on x and v or a general postcondition E that can mention any variable, axiom schema $[']$ for differential equation systems solves all such differential equations. Thus,

$$A \rightarrow [x' = v, v' = -g]E \tag{5.11}$$

simplifies by axiom schema $[']$ to:

$$A \rightarrow \forall t \geq 0 [x := x + vt - \frac{g}{2}t^2; v := v - gt]E \tag{5.12}$$

Now, the only remaining issue is that even this insight does not quite take care of handling the bouncing ball's gravity property (4.29), since that is restricted to the evolution domain constraint $x \geq 0$. But adapting these thoughts to the presence of an evolution domain constraint has now become as easy as switching from using axiom schema $[']$ from Lemma 5.3 without evolution domains to, instead, using axiom schema $[']$ from Lemma 5.4 with evolution domains. Then

$$A \rightarrow [x' = v, v' = -g \& x \geq 0]E$$

simplifies by the evolution domain solution axiom schema $[']$ from Lemma 5.4 to

$$A \rightarrow \forall t{\geq}0 \left(\left(\forall 0{\leq}s{\leq}t \left(x + vs - \frac{g}{2}s^2 \geq 0 \right) \right) \rightarrow [x := x + vt - \frac{g}{2}t^2; v := v - gt]E \right)$$

Significantly more advanced techniques for proving properties of more complicated differential equations, including differential equations without closed-form solutions, will be explored in Part II.

5.3.5 Dynamic Axioms for Tests

The bouncing-ball formula includes test statements, which we need to handle with an appropriate axiom as well. Recall their semantics from Sect. 3.3.2:

$$[\![?Q]\!] = \{(\omega, \omega) \ : \ \omega \in [\![Q]\!]\}$$

if $\omega \in [\![Q]\!]$

if $\omega \notin [\![Q]\!]$

How can we equivalently rephrase $[?Q]P$, which expresses that P always holds after running the test $?Q$ successfully? Tests do not change the state but impose conditions on the current state. Hence, P is always true after running $?Q$ only if P is already true initially, but there is only a way of running $?Q$ if Q is also true.

Lemma 5.5 ($[?]$ **test axiom**). *The* test axiom *is sound:*

$$[?] \ \ [?Q]P \leftrightarrow (Q \rightarrow P)$$

Tests in $[?Q]P$ are proven by assuming that the test succeeds with an implication in axiom $[?]$, because test $?Q$ can only make a transition when condition Q actually holds true. In states where test $?Q$ fails, no transition is possible and the failed attempt to run the system is discarded. If no transition exists for an HP α, there is nothing to show for $[\alpha]P$ formulas, because their semantics requires P to hold in all states reachable by running α, which is vacuously true if no states are reachable. From left to right, axiom $[?]$ for dL formula $[?Q]P$ assumes that formula Q holds true (otherwise there is no transition and thus nothing to show) and shows that P holds after the resulting no-op. The converse implication from right to left is by case distinction. Either Q is false, so $?Q$ cannot make a transition and there is nothing to show, or Q is true, but then also P is true according to the implication.

For example, the part $[?x \neq 0](0 \leq x \wedge x \leq H)$ of dL formula (5.8) can be replaced equivalently by the first-order formula $x \neq 0 \rightarrow 0 \leq x \wedge x \leq H$ using axiom $[?]$. After all, the two formulas are equivalent according to the equivalence in axiom $[?]$.

5.3.6 Dynamic Axioms for Sequential Compositions

For sequential compositions $\alpha; \beta$, Sect. 5.2 proposed the use of an intermediate condition E characterizing all intermediate states between α and β by way of Hoare's proof rule following the idea from Fig. 5.2:

$$\text{H;} \quad \frac{A \rightarrow [\alpha]E \quad E \rightarrow [\beta]B}{A \rightarrow [\alpha;\beta]B}$$

This proof rule can, indeed, sometimes be useful, but it comes with a significant cost compared to the simplicity and elegance of axiom $[\cup]$. When using rule H; from the desired conclusion to the premises, it does not say how to choose the intermediate condition E. Using rule H; successfully requires us to find the right intermediate condition E, for if we don't, the proof won't succeed, as we had seen in Sect. 4.8.1. If rule H; were all we have at our disposal for sequential compositions, then we would have to invent a good intermediate condition E for every single sequential composition in the system.

Fortunately, differential dynamic logic provides a better way that we also identify by investigating the dynamical system resulting from $\alpha; \beta$. Recall from Sect. 3.3.2:

$$[\![\alpha;\beta]\!] = [\![\alpha]\!] \circ [\![\beta]\!] \stackrel{\text{def}}{=} \{(\omega, v) : (\omega, \mu) \in [\![\alpha]\!], (\mu, v) \in [\![\beta]\!] \text{ for some } \mu\} \quad (5.13)$$

By its semantics, the dL formula $[\alpha;\beta]P$ is true in a state ω iff P is true in all states that $\alpha; \beta$ can reach according to $[\![\alpha;\beta]\!]$ from ω, i.e., all those states for which $(\omega, v) \in [\![\alpha;\beta]\!]$. Which states are those? And how do they relate to the states reachable by α or by β alone? They do not relate to those in a way that is as direct as for axiom $[\cup]$. But they still relate, and they do so by way of (5.13).

Postcondition P has to be true in all states reachable by $\alpha; \beta$ from ω for $[\alpha;\beta]P$ to be true at ω. By (5.13), those are exactly the states v to which we can get by running β from an intermediate state μ to which we have gotten from ω by running α. Thus, for $[\alpha;\beta]P$ to be true at ω, it is necessary that P holds in all states v to which we can get by running β from an intermediate state μ to which we get by running α from ω. Consequently, $[\alpha;\beta]P$ is only true at ω if $[\beta]P$ holds in all those intermediate states μ to which we can get from ω by running α. How do we characterize those states? How can we then express these thoughts in a single logical formula of dL?

Before you read on, see if you can find the answer for yourself.

If we want to express that $[\beta]P$ holds in all states μ to which we can get from ω by running α, then that is exactly what the truth of dL formula $[\alpha][\beta]P$ at ω means, because this is precisely the semantics of the modality $[\beta]$:

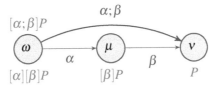

Consequently, if $[\alpha][\beta]P$ is true in ω, then so is $[\alpha;\beta]P$, as $\mu \in [\![[\beta]P]\!]$ at all such μ:

$$\omega \in [\![[\alpha][\beta]P \to [\alpha;\beta]P]\!]$$

Reexamining the argument backwards, we see the converse implication also holds,

$$\omega \in [\![[\alpha;\beta]P \to [\alpha][\beta]P]\!]$$

The same argument works for all states ω, so both implications are even valid.

Lemma 5.6 ([;] composition axiom). *The* composition axiom *is sound:*

$$[;] \quad [\alpha;\beta]P \leftrightarrow [\alpha][\beta]P$$

Proof. Since $[\![\alpha;\beta]\!] = [\![\alpha]\!] \circ [\![\beta]\!]$, we have that $(\omega, v) \in [\![\alpha;\beta]\!]$ iff $(\omega, \mu) \in [\![\alpha]\!]$ and $(\mu, v) \in [\![\beta]\!]$ for some intermediate state μ. Hence, $\omega \in [\![[\alpha;\beta]P]\!]$ iff $\mu \in [\![[\beta]P]\!]$ for all μ with $(\omega, \mu) \in [\![\alpha]\!]$. That is $\omega \in [\![[\alpha;\beta]P]\!]$ iff $\omega \in [\![[\alpha][\beta]P]\!]$. □

Sequential compositions are proven using nested modalities in axiom [;]. From right to left: If, after all α-runs, it is the case that all β-runs lead to states satisfying P (i.e., $[\alpha][\beta]P$ holds), then all runs of the sequential composition $\alpha;\beta$ lead to states satisfying P (i.e., $[\alpha;\beta]P$ holds), because $\alpha;\beta$ cannot do anything but follow α through some intermediate state to run β. The converse implication uses the fact that if after all α-runs all β-runs lead to P (i.e., $[\alpha][\beta]P$), then all runs of $\alpha;\beta$ lead to P (that is, $[\alpha;\beta]P$), because the runs of $\alpha;\beta$ are exactly those that first do any α-run, followed by any β-run. Again, it is crucial that dL is a full logic that considers reachability statements as modal operators, which can be nested, for then both sides in axiom [;] are dL formulas.

Axiom [;] directly explains sequential composition $\alpha;\beta$ in terms of a structurally simpler formula, one with nested modal operators but simpler hybrid programs. Using axiom [;] by reducing occurrences of its left-hand side to its right-hand side decomposes formulas into structurally simpler pieces, thereby making progress. One of the many ways of using axiom [;] is, therefore, captured in this proof rule:

$$[;]\mathrm{R} \quad \frac{A \to [\alpha][\beta]B}{A \to [\alpha;\beta]B}$$

Rule [;]R is easily justified from axiom [;] just by applying the equivalence [;]. Comparing rule [;]R to Hoare's rule H;, the new rule [;]R is easier to use, because it does not require us to first identify and provide an intermediate condition E like rule H; would. It does not branch into two premises, which helps to keep the proof lean. Is there a way of reuniting [;]R with H; by using the expressive power of dL?

Before you read on, see if you can find the answer for yourself.

Yes, indeed, there is a smart choice for the intermediate condition E that makes H; behave almost as the more efficient [;]R would. The clever choice $E \stackrel{\text{def}}{\equiv} [\beta]B$:

$$[;]R \frac{A \to [\alpha][\beta]B \quad [\beta]B \to [\beta]B}{A \to [\alpha;\beta]B}$$

which trivializes the right premise, because all formulas imply themselves, and makes the left premise identical to that of rule [;]R. Differential dynamic logic internalizes ways of expressing necessary and possible properties of hybrid programs and makes both first-class citizens in the logic. That cuts down on the amount of input that is needed when conducting proofs. Referring back to Fig. 5.2, rule [;]R corresponds to the case of rule H; when using $[\beta]B$ as the most precise intermediate condition E that implies that all runs of β satisfy B; see Fig. 5.3.

The sequential composition axiom [;] can be used to justify rule [;]R, which is why the latter rule will not be mentioned any more. The axiom can also be used directly to justify replacing any subformula of the form $[\alpha;\beta]P$ with the corresponding $[\alpha][\beta]P$ or vice versa. For example, axiom [;] can be used to simplify formula (5.8) to the following equivalent:

$$E \to [?x = 0][v := -cv](0 \le x \wedge x \le H) \wedge [?x \ne 0](0 \le x \wedge x \le H)$$

This formula can be simplified further by axioms [?] and [:=] to a first-order formula:

$$E \to (x = 0 \to 0 \le x \wedge x \le H) \wedge (x \ne 0 \to 0 \le x \wedge x \le H)$$

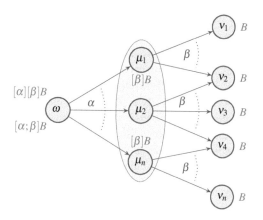

Fig. 5.3 Illustration of dynamic axiom for sequential composition

5.3.7 Dynamic Axioms for Loops

At this point, all HP operators have an axiom except for repetitions. Recall the semantics of loops from Sect. 3.3.2:

$$[\![\alpha^*]\!] = [\![\alpha]\!]^* = \bigcup_{n\in\mathbb{N}} [\![\alpha^n]\!] \quad \text{with} \quad \alpha^{n+1} \equiv \alpha^n; \alpha \text{ and } \alpha^0 \equiv ?true$$

How can we prove the property $[\alpha^*]P$ of a loop? Is there a way of reducing properties of loops to properties of simpler systems in similar ways to the other axioms of differential dynamic logic?

> Before you read on, see if you can find the answer for yourself.

It turns out that repetitions do not support such a straightforward decomposition into obviously simpler pieces as the other HP operators did. Why is that? Running a nondeterministic choice $\alpha \cup \beta$ amounts to running either HP α or β, both of which are smaller than the original $\alpha \cup \beta$. Running a sequential composition $\alpha; \beta$ amounts to first running HP α and then running β, both of which are smaller. But running a nondeterministic repetition α^* amounts to either not running anything at all or running α at least one time, so running α once and then subsequently running α^* to run any number of repetitions of α again. The latter is hardly a simplification compared to what HP α^* started out with. Nevertheless, there is a way of casting these thoughts into an axiom that reduces dL formula $[\alpha^*]P$ to an equivalent one at least in some sense of the word "reduction."

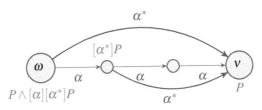

Lemma 5.7 ([*] iteration axiom). *The iteration axiom is sound:*

$$[*] \quad [\alpha^*]P \leftrightarrow P \wedge [\alpha][\alpha^*]P$$

Proof. Since loops repeat either zero times or at least one time, it is easy to see from their semantics that:

$$[\![\alpha^*]\!] = [\![\alpha^0]\!] \cup [\![\alpha; \alpha^*]\!]$$

Let $\omega \in [\![[\alpha^*]P]\!]$, then $\omega \in [\![P]\!]$ by choosing zero iterations and $\omega \in [\![[\alpha][\alpha^*]P]\!]$ by choosing at least one iteration. Conversely, let $\omega \in [\![P \wedge [\alpha][\alpha^*]P]\!]$. Then consider a run of α^* from ω to ν with $n \in \mathbb{N}$ iterations, i.e., $(\omega, \nu) \in [\![\alpha^n]\!]$. The proof shows $\nu \in [\![P]\!]$ by considering all cases of n:

0. Case $n = 0$: Then $v = \omega$ satisfies $v \in [\![P]\!]$ by the first conjunct.
1. Case $n \geq 1$: There is a state μ with $(\omega, \mu) \in [\![\alpha]\!]$ and $(\mu, v) \in [\![\alpha^{n-1}]\!]$. By the second conjunct, $\mu \in [\![[\alpha^*]P]\!]$. Hence, $v \in [\![P]\!]$, since $(\mu, v) \in [\![\alpha^{n-1}]\!] \subseteq [\![\alpha^*]\!]$.

$\square$

Axiom [*] is the iteration axiom, which partially unwinds loops. It uses the fact that P always holds after repeating α (i.e., $[\alpha^*]P$), if P holds at the beginning (so P holds after zero repetitions), and if, after one run of α, P holds after every number of repetitions of α, including zero repetitions (i.e., $[\alpha][\alpha^*]P$). So axiom [*] expresses that $[\alpha^*]P$ holds iff P holds immediately and after one or more repetitions of α.

The same axiom [*] can be used to unwind loops $N \in \mathbb{N}$ times, which corresponds to Bounded Model Checking [1], which Chap. 7 will investigate. If the formula is not valid, a bug has been found, otherwise N increases. An obvious issue with this simple approach is that we can never stop increasing how often we unroll the loop if the formula is actually valid, because we can never find a bug then. Chap. 7 will discuss proof techniques for repetitions based on loop invariants that are not subject to this issue. In particular, axiom [*] is characteristically different from the other axioms discussed in this chapter. Unlike the other axioms, axiom [*] does not exactly get rid of the formula on the left-hand side. It just puts it in a different syntactic place, which does not sound like much progress.[5]

5.3.8 Axioms for Diamonds

All previous axioms were for box modalities $[\alpha]$ with a specific shape of the hybrid program α. The diamond modalities $\langle \alpha \rangle$ also deserve axioms in order to equip them with rigorous reasoning principles. The most economical way of doing that is to understand, once and for all, for any arbitrary HP α, the relationship of the diamond modality $\langle \alpha \rangle$ to the box modality $[\alpha]$.

Recall the semantics of modalities from Definition 4.2:

$$[\![\langle \alpha \rangle P]\!] = \{\omega \ : \ v \in [\![P]\!] \text{ for some state } v \text{ such that } (\omega, v) \in [\![\alpha]\!]\}$$
$$[\![[\alpha]P]\!] = \{\omega \ : \ v \in [\![P]\!] \text{ for all states } v \text{ such that } (\omega, v) \in [\![\alpha]\!]\}$$

Both modalities are based on the reachability relation $[\![\alpha]\!] \subseteq \mathscr{S} \times \mathscr{S}$ corresponding to the HP α. The difference is that $[\alpha]P$ is true in a state ω iff P is true in *all* states v that HP α can reach from ω, while $\langle \alpha \rangle P$ is true in a state ω iff P is true in *at least one* state v that HP α can reach from ω. This makes the modalities $\langle \alpha \rangle$ and $[\alpha]$ duals. If $\langle \alpha \rangle P$ is true in a state, so there is a way of running α to a state where P is true, then $[\alpha]\neg P$ could not possibly be true in that state as well, because P is already true in one state that α reaches, so $\neg P$ cannot be true in the same state. Likewise, if $\langle \alpha \rangle P$ is false in a state, then $[\alpha]\neg P$ must be true, because, apparently, there is no way

---

[5] With a much more subtle and tricky analysis, it is possible to prove that [*] still makes sufficient progress by revealing the loop's recursion [10]. But this is far beyond the scope of this book.

of running α to a state where P is true, so $\neg P$ must be true in all states reached after running α (which could even be the empty set of states). These thoughts lead to the duality axiom, which is the direct counterpart of the equivalence $\exists x P \leftrightarrow \neg \forall x \neg P$, just for modalities.

> **Lemma 5.8 ($\langle \cdot \rangle$ duality axiom).** *The* diamond duality axiom *is sound:*
>
> $$\langle \cdot \rangle \quad \langle \alpha \rangle P \leftrightarrow \neg [\alpha] \neg P$$

5.4 A Proof of a Short Bouncing Ball

Now that we have understood so many axioms, let us use them to prove the (single-hop) bouncing ball that we have begun to consider:

$$0 \leq x \wedge x = H \wedge v = 0 \wedge g > 0 \wedge 1 \geq c \geq 0 \to$$
$$\left[\{x' = v, v' = -g \,\&\, x \geq 0\}; (?x = 0; v := -cv \cup ?x \neq 0) \right] (0 \leq x \wedge x \leq H) \quad (5.1^*)$$

Before proceeding, let's modify the hybrid program ever so subtly in two ways so that there's no longer an evolution domain, just so that we do not have to deal with evolution domains yet. We boldly drop the evolution domain constraint and make up for it by modifying the condition in the second test:

$$0 \leq x \wedge x = H \wedge v = 0 \wedge g > 0 \wedge 1 \geq c \geq 0 \to$$
$$\left[\{x' = v, v' = -g\}; (?x = 0; v := -cv \cup ?x \geq 0) \right] (0 \leq x \wedge x \leq H) \quad (5.14)$$

Hold on, why is that okay? Doesn't our previous investigation say that Quantum could suddenly fall through the cracks in the floor if physics evolves for hours before giving the poor bouncing-ball controller a chance to react? To make sure Quantum does not panic in light of this threat, solve Exercise 5.12 to investigate.

To fit things on the page more succinctly, we use some abbreviations:

$$A \stackrel{\text{def}}{\equiv} 0 \leq x \wedge x = H \wedge v = 0 \wedge g > 0 \wedge 1 \geq c \geq 0$$
$$B(x, v) \stackrel{\text{def}}{\equiv} 0 \leq x \wedge x \leq H$$
$$\{x'' = -g\} \stackrel{\text{def}}{\equiv} \{x' = v, v' = -g\}$$

With these abbreviations, (5.14) turns into

$$A \to [x'' = -g; (?x = 0; v := -cv \cup ?x \geq 0)] B(x, v) \quad (5.14^*)$$

Successively applying the axioms is all it takes to obtain a proof for bouncing balls:

$$\cfrac{A \to \forall t\geq 0\left((H-\frac{g}{2}t^2=0\to B(H-\frac{g}{2}t^2,-c(-gt)))\wedge(H-\frac{g}{2}t^2\geq 0\to B(H-\frac{g}{2}t^2,-gt))\right)}{\cfrac{A \to \forall t\geq 0[x:=H-\frac{g}{2}t^2]\left((x=0\to B(x,-c(-gt)))\wedge(x\geq 0\to B(x,-gt))\right)}{\cfrac{A \to \forall t\geq 0[x:=H-\frac{g}{2}t^2][v:=-gt]\left((x=0\to B(x,-cv))\wedge(x\geq 0\to B(x,v))\right)}{\cfrac{A \to \forall t\geq 0[x:=H-\frac{g}{2}t^2;v:=-gt]\left((x=0\to B(x,-cv))\wedge(x\geq 0\to B(x,v))\right)}{\cfrac{A \to [x''=-g]\left((x=0\to B(x,-cv))\wedge(x\geq 0\to B(x,v))\right)}{\cfrac{A \to [x''=-g]\left((x=0\to [v:=-cv]B(x,v))\wedge(x\geq 0\to B(x,v))\right)}{\cfrac{A \to [x''=-g]\left([?x=0][v:=-cv]B(x,v)\wedge[?x\geq 0]B(x,v)\right)}{\cfrac{A \to [x''=-g]\left([?x=0;v:=-cv]B(x,v)\wedge[?x\geq 0]B(x,v)\right)}{\cfrac{A \to [x''=-g][?x=0;v:=-cv\cup ?x\geq 0]B(x,v)}{A \to [x''=-g;(?x=0;v:=-cv\cup ?x\geq 0)]B(x,v)}}}}}}}}}$$

The dL axioms indicated on the left justify that the dL formulas in the two adjacent rows are equivalent. Since each step in this proof is justified by using a dL axiom, the conclusion at the very bottom of this derivation is proved if the premise at the very top can be proved, because truth is then inherited by the bottom from the top by soundness of the axioms. That premise at the top

$$A \to \forall t\geq 0\left((H-\frac{g}{2}t^2=0\to B(H-\frac{g}{2}t^2,cgt))\wedge(H-\frac{g}{2}t^2\geq 0\to B(H-\frac{g}{2}t^2,-gt))\right)$$

expands out to a real arithmetic formula when expanding the abbreviations:

$$0\leq x\wedge x=H\wedge v=0\wedge g>0\wedge 1\geq c\geq 0\to$$
$$\forall t\geq 0\left((H-\frac{g}{2}t^2=0\to 0\leq H-\frac{g}{2}t^2\wedge H-\frac{g}{2}t^2\leq H)\right.$$
$$\left.\wedge(H-\frac{g}{2}t^2\geq 0\to 0\leq H-\frac{g}{2}t^2\wedge H-\frac{g}{2}t^2\leq H)\right)$$

In this case, this remaining premise can easily be seen to be valid. The assumption $H-\frac{g}{2}t^2=0$ in the middle line directly implies the first conjunct of the middle line's right-hand side

$$0\leq H-\frac{g}{2}t^2\wedge H-\frac{g}{2}t^2\leq H$$

and reduces the remaining second conjunct to $0\leq H$, which the assumption in the first line assumed ($0\leq x=H$). Similarly, the assumption $H-\frac{g}{2}t^2\geq 0$ in the last line implies the first conjunct of its right-hand side

$$0\leq H-\frac{g}{2}t^2\wedge H-\frac{g}{2}t^2\leq H$$

and its second conjunct holds by assumption $g>0$ from the first line and the real-arithmetic fact that $t^2\geq 0$.

How exactly first-order logic and first-order real-arithmetic formulas such as this one can be proved in general, however, is an interesting topic for a later chapter. For now, we are happy to report that we have just formally verified our very first CPS. We have found a proof of (5.14). Exciting!

Okay, admittedly, the CPS we just verified was only a bouncing ball. And all we know about it now is that it won't fall through the cracks in the ground or jump high up to the moon. But big steps for mankind start with a small step by someone.

Yet, before we get too carried away in the excitement, we still need to remember that the formula (5.14) that we proved only expresses safety of a single-hop bouncing ball. So there's still an argument to be made about what happens if the bouncing ball repeats. And a rather crucial argument too, because bouncing balls let loose in the air tend not to jump any higher anyhow without hitting the ground first, which is where the model (5.14) stops prematurely, because it is missing a repetition. So let's put worrying about loops on the agenda for an upcoming chapter (Chap. 7).

Yet, there is one more pressing issue with the proof for the bouncing ball that we derived. It worked in a somewhat undisciplined chaotic way, by using dL axioms all over the place. This liberal proof style can be useful for manual proofs and creative shortcuts. Since the dL axioms are sound, even such a liberal proof is still a proof. And liberal proofs can even be very creative. But liberal proofs are also somewhat unfocused and non-systematic, which makes them unreasonable for automation purposes and can also get humans lost if the problems at hand are more complex than the single-hop bouncing ball. That is the reason why we will investigate more focused, more systematic, and more algorithmic proofs in Chap. 6.

The other thing to observe is that the above proof, however liberal it might have been, already had a lot more structure to it than we have made explicit so far. This structure will be uncovered in the next chapter as well.

5.5 Summary

The differential dynamic logic axioms that we have seen in this chapter are summarized in Fig. 5.4. These are dynamic axioms for dynamical systems, i.e., axioms in differential dynamic logic (dL) that characterize dynamical systems operators in terms of structurally simpler dL formulas. All it takes to understand the bigger system is to apply the appropriate axiom and investigate the smaller remainders. To summarize the individual soundness lemmas of this chapter, the axioms of dL that are listed in Fig. 5.4 are sound, i.e., valid and so are all their instances [9, 11].

> **Theorem 5.1 (Soundness).** *The dL axioms listed in Fig. 5.4 are sound. That is, all their instances are valid, i.e., true in all states.*

There are further axioms and proof rules of differential dynamic logic that later chapters will examine, which are also all proved sound in prior work [8, 9, 11]. The reasoning principles and axioms identified so far are fundamental and we will carry them with us throughout the whole textbook. Axiomatics crystallizes a semantic relationship in a syntactic expression (the axiom), which we justify to be sound from the semantics once and then use mechanically any number of times in our CPS proofs. In fact, this will turn out to work so well that differential dynamic logics also enjoy completeness guarantees about provability of all valid formulas with its axioms from elementary properties [5, 9, 11]. But we will not pursue completeness until we have reached the most appropriate level of understanding in a generalization in Chap. 16.

The equivalence axioms in Fig. 5.4 are primarily meant to be used by replacing the left-hand side (marked in blue) with the structurally simpler right-hand side. With the notable exception of iteration axiom [*], using these equivalences from left to right decomposes a property of a more complex HP into properties of obviously simpler subprograms.

$$[:=]\ [x:=e]p(x) \leftrightarrow p(e)$$

$$[?]\ [?Q]P \leftrightarrow (Q \rightarrow P)$$

$$[']\ [x'=f(x)]p(x) \leftrightarrow \forall t{\geq}0\,[x:=y(t)]p(x) \quad (y'(t)=f(y))$$

$$[\cup]\ [\alpha\cup\beta]P \leftrightarrow [\alpha]P \wedge [\beta]P$$

$$[;]\ [\alpha;\beta]P \leftrightarrow [\alpha][\beta]P$$

$$[*]\ [\alpha^*]P \leftrightarrow P \wedge [\alpha][\alpha^*]P$$

$$\langle\cdot\rangle\ \langle\alpha\rangle P \leftrightarrow \neg[\alpha]\neg P$$

Fig. 5.4 Summary of sound differential dynamic logic axioms from this chapter

5.6 Appendix

This appendix provides additional axioms for dL that provide conceptual insights and are sometimes useful for shortcuts in proofs. These additions are not on the critical path for most CPSs, but still advance our general structural understanding.

5.6.1 Modal Modus Ponens Has Implications on Boxes

Each of the axioms discussed in this chapter was specific to one operator of dL. But there are also axioms that are common to all hybrid programs α.

The next axiom provides a way of shoving implications through box modalities. If we would like to show $[\alpha]Q$ but only know $[\alpha]P$, then what do we need to know about the relationship of P and Q for $[\alpha]P$ to imply $[\alpha]Q$?

Before you read on, see if you can find the answer for yourself.

It would suffice if P were to imply Q unconditionally, because if all runs of α satisfy P, then all runs of α also satisfy Q when P implies Q. But, in fact, it even suffices if P only implies Q after all runs of α, because $[\alpha]Q$ only claims that Q holds after all runs of α, not in general.

Lemma 5.9 (K modal modus ponens axiom). *The* modal modus ponens axiom *is sound:*

$$\text{K } [\alpha](P \to Q) \to ([\alpha]P \to [\alpha]Q)$$

Proof. To show soundness, consider any state ω in which the assumed left-hand side $[\alpha](P \to Q)$ of the implication is true, so $\omega \in [\![[\alpha](P \to Q)]\!]$, and show that the right-hand side is true as well. In order to show that $[\alpha]P \to [\alpha]Q$ is true in ω, assume that its respective left-hand side is true, so $\omega \in [\![[\alpha]P]\!]$, and show that its right-hand side $[\alpha]Q$ is true in ω. In order to show the latter $\omega \in [\![[\alpha]Q]\!]$, we consider any state ν with $(\omega, \nu) \in [\![\alpha]\!]$ and need to show $\nu \in [\![Q]\!]$. Using the assumption $\omega \in [\![[\alpha]P]\!]$, we know that $\nu \in [\![P]\!]$ since $(\omega, \nu) \in [\![\alpha]\!]$. Using the assumption $\omega \in [\![[\alpha](P \to Q)]\!]$, we also know that $\nu \in [\![P \to Q]\!]$ since $(\omega, \nu) \in [\![\alpha]\!]$. Now $\nu \in [\![P]\!]$ and $\nu \in [\![P \to Q]\!]$ imply $\nu \in [\![Q]\!]$, which concludes the proof of $\omega \in [\![[\alpha]Q]\!]$, because ν was an arbitrary state with $(\omega, \nu) \in [\![\alpha]\!]$. $\qquad\square$

Axiom K distributes an implication over a box modality. It can be used to show postcondition Q of HP α by instead proving $[\alpha]P$ if the new postcondition P implies the original postcondition Q after all runs of the respective program α. For example, $x^2 > 0$ implies $x > 0$ after all runs of $x := x \cdot x$, but not in general, because that particular program happens to assign a nonnegative value to x, which is required for $x^2 > 0$ to imply $x > 0$. But showing $x^2 > 0$ after $x := x \cdot x$ would still suffice by K to also show postcondition $x > 0$ for this program. Admittedly, that is no easier in this particular case, but may help in others. Of course, if $P \to Q$ is valid, so true in all states, then it is also true after all runs of α (Gödel's generalization rule G explored in Sect. 5.6.3), such that formula $[\alpha]P$ will imply $[\alpha]Q$ by axiom K.

One way of using axiom K is from right to left in order to reduce a proof of $[\alpha]Q$ to a proof of $[\alpha]P$ together with a proof of $[\alpha](P \to Q)$. Such a use of axiom K can be understood as showing postcondition Q by, instead, proving another postcondition P and then merely showing that the already-established postcondition Q implies the originally desired postcondition P after all runs of HP α.

Implication is not the only operator that box modalities distribute over. They also distribute over conjunctions.

Lemma 5.10 ([]∧ boxes distribute over conjunctions). *This axiom is sound:*

$$[]\wedge \ [\alpha](P \wedge Q) \leftrightarrow [\alpha]P \wedge [\alpha]Q$$

The formula []∧ would be a sound axiom. The only reason why it is not officially adopted as an axiom is because it already follows from axiom K (Exercise 5.17).

5.6.2 Vacuous State Change if Nothing Relevant Ever Changes

The axioms discussed in the main part of this chapter were not just specific to one operator of dL, but also, for good reasons, are very precise about capturing the exact effect of the respective programs. That is not always required. Sometimes a coarse overapproximation of the effect of a program suffices for an argument. That happens if the variables that a postcondition p depends on are not even modified by program α. In that case, p is always true after running α if p was true before, because its truth-value does not change when running a program α that does not change the free variables of p. This reasoning is captured in the vacuous axiom V.

Lemma 5.11 (V vacuous axiom). *The* vacuous axiom *is sound:*

$$\text{V} \quad p \to [\alpha]p \quad (FV(p) \cap BV(\alpha) = \emptyset)$$

where no free variable of p is bound (written) in α.

Axiom V makes it possible to prove that the truth of a formula p is preserved when running an HP α that does not modify the free variables of p. For example, if $x > z$ holds initially, then it continues to hold always after running $y' = x$, because that program changes neither x nor z but merely y (and y'). Indeed, $x > z \to [y' = x]x > z$ by axiom V. Of course, because y is changed by the HP $y' = x$ and read by postcondition $x > y$, axiom V does not apply to $x > y \to [y' = x]x > y$, which is indeed false in an initial state where $x = 1$ and $y = 0$.

5.6.3 Gödel Generalizes Validities into Boxes

Axiom V expresses that the truth of a formula is preserved when running HP α if the formula has no free variables modified by α. There is also a way to show that a formula P always holds after running an HP α even if that HP modifies free variables of P. But that requires more than just the truth of the formula P in the initial state. After all, any arbitrary HP α might very well affect the truth-value of a formula if it modifies the free variables of P.

But if formula P is not just true initially but valid, so true in all states, then it will certainly still be true always after running any HP α no matter what variables that program modifies. So, if P is valid, then $[\alpha]P$ is valid as well. Obviously: If P is true in all states, then it is also true in all states that can be reached after running α.

Lemma 5.12 (G Gödel generalization rule). *The* Gödel rule *is sound:*

$$\text{G} \quad \frac{P}{[\alpha]P}$$

Proof. Soundness for proof rules means that validity of the premise implies validity of the conclusion. If premise P is valid, then it is true in all states. Then the conclusion $[\alpha]P$ is valid, because it is true in any state ω, since $\nu \in [\![P]\!]$ for all states ν including all states ν for which $(\omega, \nu) \in [\![\alpha]\!]$. $\qquad\square$

This is our first example of a proof rule. If we can prove its premise P above the bar, then it is valid, so the conclusion $[\alpha]P$ below the bar is proved by rule G. For example, $x^2 \geq 0$ is valid, so by G is also true after any HP, which implies validity of $[x' = x^3 + 1]x^2 \geq 0$. The truth of the postcondition $x^2 \geq 0$ simply does not depend on the HP $x' = x^3 + 1$ at all, because the postcondition is true in any state.

5.6.4 Monotonicity of Postconditions

Gödel's generalization rule G is a global version of the axiom V in the sense that rule G expresses that validity of P in all states is preserved after an $[\alpha]$ modality while axiom V expresses that the truth of p (in the initial state) is preserved after an $[\alpha]$ modality if α does not modify variables of p.

Similarly, the proof rule $M[\cdot]$ is a global version of axiom K in the sense that rule $M[\cdot]$ uses validity of an implication $P \to Q$, while axiom K merely requires truth of $P \to Q$ (albeit after all runs of α). Rule $M[\cdot]$ is the monotonicity rule expressing that if P implies Q, so Q is true in every state where P is true (premise), then $[\alpha]P$ also implies $[\alpha]Q$, so $[\alpha]Q$ is true in ever state where $[\alpha]P$ is true (conclusion). Obviously: if Q is true in every state in which P is true, then $[\alpha]Q$ is true in every state in which $[\alpha]P$ is true, because P being true after all runs of α implies that Q is then also true after all runs of α.

Lemma 5.13 ($M[\cdot]$ monotonicity rule). *The monotonicity rules are sound:*

$$M[\cdot] \ \frac{P \to Q}{[\alpha]P \to [\alpha]Q} \qquad M \ \frac{P \to Q}{\langle\alpha\rangle P \to \langle\alpha\rangle Q}$$

Proof. If the premise $P \to Q$ is valid, then Q is true in every state in which P is true in, because $[\![P]\!] \subseteq [\![Q]\!]$. Consequently, in every state in which $[\alpha]P$ is true, $[\alpha]Q$ is true, too, because if all runs of α lead to states satisfying P then they also all lead to states satisfying Q. In any state ω for which $\omega \in [\![[\alpha]P]\!]$, it is the case that $\nu \in [\![P]\!]$, hence also $\nu \in [\![Q]\!]$, for all states ν with $(\omega, \nu) \in [\![\alpha]\!]$. That implies $\omega \in [\![[\alpha]Q]\!]$. Soundness of rule M is similar. Alternatively, rule $M[\cdot]$ can also be proved sound by deriving it from Gödel's generalization rule G using axiom K (Exercise 5.19). $\qquad\square$

For example, $[x:=1; x' = x^2 + 2x^4]x^3 \geq x^2$ is difficult to prove directly. But proving $[x:=1; x' = x^2 + 2x^4]x \geq 1$ will turn out to be surprisingly easy in Part II based on the intuition that the right-hand side $x^2 + 2x^4$ of the ODE will only make x increase above 1, never fall below 1. And the postcondition $x \geq 1$ easily implies the original postcondition $x^3 \geq x^2$. This example shows how monotonicity rule M[·] can simplify proofs by reducing the proof of a difficult formula $[x := 1; x' = x^2 + 2x^4] x^3 \geq x^2$ to the proof of formula $[x := 1; x' = x^2 + 2x^4] x \geq 1$ with a different postcondition $x \geq 1$, which implies the original postcondition $x^3 \geq x^2$.

Observe how this proof with monotonicity rule M[·] is characteristically different from what Gödel's generalization rule G gives. Gödel's G gives us a proof for a postcondition that is true in all states and, thus, not a particularly informative postcondition specific to the particular HP. Monotonicity rule M gives us an insightful postcondition $x^3 \geq x^2$ from a proof of another postcondition $x \geq 1$ that is informative, because it is always true after HP $x := 1; x' = x^2 + 2x^4$ but not after HP $x' = -1$.

Unlike Gödel's generalization rule G, monotonicity rule M[·] has a direct counterpart, rule M, for diamond modalities. A diamond counterpart for rule G cannot exist, because even if P is valid, that does not mean that $\langle \alpha \rangle P$ is valid, because there may not be a way of running α to any final state at all. For example, $\langle ?false \rangle true$ is not valid, because there is no way of satisfying its test $?false$. That problem does not come up in the diamond montononicity rule M, because its conclusion already assumes $\langle \alpha \rangle P$ so there is a way of running HP α.

5.6.5 Of Free and Bound Variables

The vacuous axiom V from Sect. 5.6.2 and our understanding of the assignment axiom [:=] from Sect. 5.3.3 used a concept of free and bound variables. Free variables are all those that the value or semantics of an expression depends on. Bound variables are all those that can change their value during an HP.

For example, the free variables of term $x \cdot 2 + y \cdot y$ are $\{x, y\}$ but not z, because that term does not even mention z so its value depends only on the values of x, y but not on that of z. Likewise, the free variables of HP $a := -b; x' = v, v' = a$ are $\{b, x, v\}$ but not a, because its value is only read after a is written to. The bound variables of that HP are a, x, v (and also x', v' if you look closely because both change their value by virtue of what it means to solve a differential equation) but not b, because b is only read but never written to. The free variables of the formula $\forall x (x^2 \geq y + z)$ are just $\{y, z\}$, not x, which receives a new value by the quantifier.

> **Definition 5.2 (Static semantics).** The *static semantics* defines the *free variables*, which are all variables that the value of an expression depends on. It also defines the *bound variables*, which can change their value during the evaluation of an expression.

$$
\begin{aligned}
\mathrm{FV}(e) &= \big\{x \in \mathscr{V} : \text{there are } \omega = \tilde{\omega} \text{ on } \{x\}^{\complement} \text{ such that } \omega[\![e]\!] \neq \tilde{\omega}[\![e]\!]\big\} \\
\mathrm{FV}(P) &= \big\{x \in \mathscr{V} : \text{there are } \omega = \tilde{\omega} \text{ on } \{x\}^{\complement} \text{ such that } \omega \in [\![P]\!] \not\ni \tilde{\omega}\big\} \\
\mathrm{FV}(\alpha) &= \big\{x \in \mathscr{V} : \text{there are } \omega, \tilde{\omega}, v \text{ with } \omega = \tilde{\omega} \text{ on } \{x\}^{\complement} \text{ and } (\omega, v) \in [\![\alpha]\!] \\
&\qquad \text{but no } \tilde{v} \text{ with } v = \tilde{v} \text{ on } \{x\}^{\complement} \text{ such that } (\tilde{\omega}, \tilde{v}) \in [\![\alpha]\!]\big\} \\
\mathrm{BV}(\alpha) &= \big\{x \in \mathscr{V} : \text{there are } (\omega, v) \in [\![\alpha]\!] \text{ such that } \omega(x) \neq v(x)\big\}
\end{aligned}
$$

A variable $x \in \mathscr{V}$ is a free variable of a term e if the value of e depends on the value of x, i.e., there are two states ω and $\tilde{\omega}$ that only differ in the value of variable x (so they agree on the complement $\{x\}^{\complement}$ of the singleton set $\{x\}$) where e has different values in the two states ω and $\tilde{\omega}$. Likewise $x \in \mathscr{V}$ is a free variable of a formula P if there are two different states agreeing on $\{x\}^{\complement}$ such that P is true in one but false in the other. For an HP α, a variable $x \in \mathscr{V}$ is a free variable if, from two different states agreeing on $\{x\}^{\complement}$, there is a run of α in one of them but no corresponding run of α in the other (with merely a different value of x). A variable $x \in \mathscr{V}$ is a bound variable of HP α if there is a run of α that changes the value of x.

While every state defines a real value for every variable, the truth-value of a formula only depends on the values of its free variables. Likewise, the real value of a term only depends on the values of its free variables, and likewise for programs [11]. Only the bound variables of an HP α can change their value when running α.

The dynamic semantics gives a precise meaning to HPs (Chap. 3) and dL formulas (Chap. 4) but is inaccessible for effective reasoning purposes (unlike the syntactic axioms that this chapter provides). By contrast, the static semantics of dL and HPs defines only simple aspects of the dynamics concerning the variable usage that can also be read off more directly from the syntactic structure without running the programs or evaluating their dynamical effects, as we will see next. In fact, Definition 5.2 is somewhere in between. In a single definition, it captures quite concisely what the free and bound variables of a term, formula, or program are. But it does so using the dynamic semantics, which would have to be evaluated. The next section shows how (sound overapproximations of) the free and bound variables can be computed by distinguishing the cases of how a term, formula, or program was built.

5.6.6 Free and Bound Variable Analysis

Computing the set of free and bound variables according to Definition 5.2 precisely is impossible, because every nontrivial property of programs is undecidable [13]. For example, it takes at least a moment's thought to see that, despite first appearances, x is not actually read and y not actually written to in the HP

$$z := 0; (y := x + 1; z' = 1; ?z < 0 \cup z := z + 1)$$

Fortunately, any supersets of the sets of free and bound variables will work correctly, for example for the conditions for the vacuous axiom V in Sect. 5.6.2. Such supersets of the sets of free variables and bound variables are quite easily computable directly from the syntax. The easiest approach would be to simply take the set of all variables that are ever read anywhere, which is a simple superset of the free variables. The set of variables that are ever written to is an obvious superset of the bound variables. Those may yield too many variables, because $a \in \mathcal{V}$ is not actually free in the following HP $a := -b$; $x' = v, v' = a$ even if it is read somewhere, because it is written to first, so receives its value during the computation.

With more thought, more precise sound (super-)sets of free and bound variables can be computed equally easily by keeping track of variables that are written before they are read [11], but they will inevitably still be overapproximations. Bound variables x of a formula are those that are bound by $\forall x$ or $\exists x$, but also those that are bound by modalities such as $[x := 5y]$ or $\langle x' = 1 \rangle$ or $[x := 1 \cup x' = 1]$ or $[x := 1 \cup ?true]$ because of the assignment to x or differential equation for x they contain. The scope of the bound variable x is limited to the quantified formula or to the postcondition and remaining program of the modality.

Definition 5.3 (Bound variable). The set $\mathrm{BV}(P) \subseteq \mathcal{V}$ of (syntactically) *bound variables* of dL formula P is defined inductively as

$$\mathrm{BV}(e \geq \tilde{e}) = \emptyset \qquad\qquad \text{accordingly for } =, >, \leq, <$$
$$\mathrm{BV}(\neg P) = \mathrm{BV}(P)$$
$$\mathrm{BV}(P \wedge Q) = \mathrm{BV}(P) \cup \mathrm{BV}(Q) \qquad \text{accordingly for } \vee, \rightarrow, \leftrightarrow$$
$$\mathrm{BV}(\forall x P) = \mathrm{BV}(\exists x P) = \{x\} \cup \mathrm{BV}(P)$$
$$\mathrm{BV}([\alpha]P) = \mathrm{BV}(\langle \alpha \rangle P) = \mathrm{BV}(\alpha) \cup \mathrm{BV}(P)$$

The set $\mathrm{BV}(\alpha) \subseteq \mathcal{V}$ of (syntactically) *bound variables* of HP α, i.e., all those that may potentially be written to, is defined inductively as

$$\mathrm{BV}(x := e) = \{x\}$$
$$\mathrm{BV}(?Q) = \emptyset$$
$$\mathrm{BV}(x' = f(x) \,\&\, Q) = \{x, x'\}$$
$$\mathrm{BV}(\alpha \cup \beta) = \mathrm{BV}(\alpha; \beta) = \mathrm{BV}(\alpha) \cup \mathrm{BV}(\beta)$$
$$\mathrm{BV}(\alpha^*) = \mathrm{BV}(\alpha)$$

In a differential equation $x' = f(x)$ both x and x' are bound, both change their value.

The free variables of a quantified formula are defined by removing its bound variables, e.g., $\mathrm{FV}(\forall x P) = \mathrm{FV}(P) \setminus \{x\}$, since all occurrences of x in P are bound. The bound variables of a program in a modality act in a similar way, except that the program itself may read variables during the computation, so its free variables need to be taken into account. By analogy to the quantifier case, it is often suspected that $\mathrm{FV}([\alpha]P)$ might be defined as $\mathrm{FV}(\alpha) \cup (\mathrm{FV}(P) \setminus \mathrm{BV}(\alpha))$. But that would be unsound, because $[x := 1 \cup y := 2] x \geq 1$ would have no free variables then, contradicting the fact that its truth-value depends on the initial value of x. The reason is that x is a bound variable of that program, but only written to on some but not on all paths. So the initial value of x may be needed to evaluate the truth of the postcondition $x \geq 1$ on some execution paths. If a variable is *must-bound*, so written to on *all paths* of the program, however, it can safely be removed from the free variables of the postcondition. The static semantics, thus, first defines the subset of variables that are must-bound (MBV(α)), so must be written to on all execution paths of α.

Definition 5.4 (Must-bound variable). The set $\mathrm{MBV}(\alpha) \subseteq \mathrm{BV}(\alpha) \subseteq \mathscr{V}$ of (syntactically) *must-bound variables* of HP α, i.e., all those that must be written to on all paths of α, is defined inductively as

$$\mathrm{MBV}(x := e) = \{x\}$$
$$\mathrm{MBV}(?Q) = \emptyset$$
$$\mathrm{MBV}(x' = f(x) \& Q) = \{x, x'\}$$
$$\mathrm{MBV}(\alpha \cup \beta) = \mathrm{MBV}(\alpha) \cap \mathrm{MBV}(\beta)$$
$$\mathrm{MBV}(\alpha; \beta) = \mathrm{MBV}(\alpha) \cup \mathrm{MBV}(\beta)$$
$$\mathrm{MBV}(\alpha^*) = \emptyset$$

Finally, the static semantics defines which variables are free so may be read. The definition of free variables is simultaneously inductive for formulas (FV(P)) and programs (FV(α)) owing to their mutually recursive syntactic structure. The set $\mathrm{FV}(e) \subseteq \mathscr{V}$ of (syntactically) *free variables* of term e is the set of those that occur in e, at least till Chap. 10. The differential terms $(e)'$ that will be introduced in Chap. 10 have as free variables also all differential symbols that are primed versions of the free variables of e, so $\mathrm{FV}((e)') = \mathrm{FV}(e) \cup \mathrm{FV}(e)'$, e.g., $\mathrm{FV}((x+y)') = \{x, x', y, y'\}$.

Definition 5.5 (Free variable). The set FV(P) of (syntactically) *free variables* of dL formula P, i.e., all that occur in P outside the scope of quantifiers or modalities binding them, is defined inductively as

$$\mathrm{FV}(e \geq \tilde{e}) = \mathrm{FV}(e) \cup \mathrm{FV}(\tilde{e}) \qquad \text{accordingly for } =, \leq$$
$$\mathrm{FV}(\neg P) = \mathrm{FV}(P)$$
$$\mathrm{FV}(P \wedge Q) = \mathrm{FV}(P) \cup \mathrm{FV}(Q)$$
$$\mathrm{FV}(\forall x P) = \mathrm{FV}(\exists x P) = \mathrm{FV}(P) \setminus \{x\}$$
$$\mathrm{FV}([\alpha]P) = \mathrm{FV}(\langle\alpha\rangle P) = \mathrm{FV}(\alpha) \cup (\mathrm{FV}(P) \setminus \mathrm{MBV}(\alpha))$$

The set $\mathrm{FV}(\alpha) \subseteq \mathscr{V}$ of (syntactically) *free variables* of HP α, i.e., all those that may potentially be read, is defined inductively as

$$\mathrm{FV}(x := e) = \mathrm{FV}(e)$$
$$\mathrm{FV}(?Q) = \mathrm{FV}(Q)$$
$$\mathrm{FV}(x' = f(x) \& Q) = \{x\} \cup \mathrm{FV}(f(x)) \cup \mathrm{FV}(Q)$$
$$\mathrm{FV}(\alpha \cup \beta) = \mathrm{FV}(\alpha) \cup \mathrm{FV}(\beta)$$
$$\mathrm{FV}(\alpha; \beta) = \mathrm{FV}(\alpha) \cup (\mathrm{FV}(\beta) \setminus \mathrm{MBV}(\alpha))$$
$$\mathrm{FV}(\alpha^*) = \mathrm{FV}(\alpha)$$

The *variables* of dL formula P, whether free or bound, are $V(P) = FV(P) \cup BV(P)$. The *variables* of HP α, whether free or bound, are $V(\alpha) = FV(\alpha) \cup BV(\alpha)$.

Both x and x' are bound in $x' = f(x) \& Q$ since both change their value. Only x is added to the free variables, because the behavior of the differential equation depends on the initial value of x, not on that of x' (as a careful look at Definition 3.3 reveals).

   These syntactic definitions are correct [11], i.e., they compute supersets of the semantic definitions of the static semantics from Definition 5.2 (Exercise 5.23). Of course [13], syntactic computations may give bigger sets, e.g., $FV(x^2 - x^2) = \{x\}$ even if the value of this unsimplified term really depends on no variable, and $BV(x := x) = \{x\}$ even if this change is a no-op.

Exercises

5.1 (Necessity of assumptions). Identify which of the assumptions of (5.14) are actually required for the proof of (5.14). Which formulas could we have dropped from $0 \le x \wedge x = H \wedge v = 0 \wedge g > 0 \wedge 1 \ge c \ge 0$ and still been able to prove:

$$0 \le x \wedge x = H \wedge v = 0 \wedge g > 0 \wedge 1 \ge c \ge 0 \rightarrow$$
$$[x'' = -g; (?x = 0; v := -cv \cup ?x \ge 0)](0 \le x \wedge x \le H)$$

5.2. Show that the following axiom would be sound

$$Q \wedge P \rightarrow [?Q]P$$

5.3. Would the following be a sound axiom? Prove or disprove.

$$[x := e]P \leftrightarrow \langle x := e \rangle P$$

5.4. Use the axioms developed in this chapter to prove validity of this dL formula:

$$x \ge 0 \rightarrow [v := 1; (v := v + 1 \cup x' = v)]x \ge 0$$

5.5 (Solutions at the endpoint). Prove that the following axiom is sound when y is the unique global solution:

$$\forall t \ge 0 [x := y(t)](q(x) \rightarrow p(x)) \rightarrow [x' = f(x) \& q(x)]p(x) \qquad (y'(t) = f(y))$$

5.6. When misplacing the parentheses in axiom ['] to obtain the following formula, would it be a sound axiom, too? Prove or disprove.

$$[x' = f(x) \& q(x)]p(x) \leftrightarrow \forall t \ge 0 \forall 0 \le s \le t \left(q(y(s)) \rightarrow [x := y(t)]p(x) \right) \qquad (y'(t) = f(y))$$

As in axiom ['], you can assume y to be the unique global solution for the corresponding symbolic initial value problem.

5.7 (Solutions of systems). The dL formula (5.12) was suggested as a result of using axiom schema ['] on dL formula (5.11). Would the converse order of solutions have worked as well to reduce (5.11) to the following formula instead?

$$A \to \forall t \geq 0 \, [v := v - gt; x := x + vt - \frac{g}{2}t^2]E$$

5.8. The axioms of dynamic logic are also useful to prove the correctness of discrete programs. Find a way of proving the following formula which expresses that a triple of clever assignments swaps the values of two variables in place:

$$x = a \wedge y = b \to [x := x + y; y := x - y; x := x - y](x = b \wedge y = a)$$

5.9. Would either of the following axioms be a good replacement for the [*] axiom? Are they sound? Are they useful?

$$[\alpha^*]P \leftrightarrow P \wedge [\alpha^*]P$$
$$[\alpha^*]P \leftrightarrow [\alpha^*](P \wedge [\alpha][\alpha^*]P)$$

5.10 (Soundness of dynamic axioms). All axioms need to be proved to be sound. This textbook only gave a proper proof for some axioms, because proofs are published elsewhere [9]. Turn the informal arguments for the other axioms into proper soundness proofs using the semantics of dL formulas.

5.11 (Diamond axioms). This chapter identified axioms for all formulas of the form $[\alpha]P$ but none for formulas of the form $\langle \alpha \rangle P$. Identify and justify these missing axioms. Explain how they relate to the ones given in Fig. 5.4. Find out whether you made a mistake by proving them sound.

5.12 (Give bouncing ball back its evolution domain). Explain why the subtle transformation from (5.1) to (5.14) was okay *in this particular case.*

5.13 (Nondeterministic assignments). Continuing Exercise 3.12, suppose a new statement $x := *$ for nondeterministic assignment is added to HPs, which assigns an arbitrary real number to the variable x. The new syntactic construct of nondeterministic assignment needs a semantics to become meaningful:

7. $[\![x := *]\!] = \{(\omega, \nu) : \nu = \omega \text{ except for the value of } x, \text{ which can be any real}\}$

Develop an axiom for $[x := *]P$ and an axiom for $\langle x := * \rangle P$ that rephrase both in terms of simpler logical connectives, and prove soundness of these axioms.

5.14 (Differential assignments). Hybrid programs allow discrete assignments $x := e$ to any variable $x \in \mathcal{V}$. In Part II, we will discover the important rôle that differential symbols x' play. Part II will end up considering differential symbols x' as variables and will allow discrete assignments $x' := e$ to differential symbols x' that instantaneously change the value of x' to that of e. Develop a semantics for these *differential assignments* $x' := e$. Develop an axiom for $[x' := e]p(x)$ and an axiom for $\langle x' := e \rangle p(x)$, and prove soundness of these axioms.

5.15 (If-then-else). Exercise 3.4 defined a semantics for the if-then-else statement $\mathsf{if}(Q)\,\alpha\,\mathsf{else}\,\beta$ that was added into the syntax of HPs for this purpose. Develop an axiom for $[\mathsf{if}(Q)\,\alpha\,\mathsf{else}\,\beta]P$ and an axiom for $\langle\mathsf{if}(Q)\,\alpha\,\mathsf{else}\,\beta\rangle P$ that decompose the effect of the if-then-else statement in logic. Then prove soundness of these axioms.

5.16 ($K_{\langle\cdot\rangle}$ modal modus ponens for $\langle\cdot\rangle$). Develop an analogue of the modal modus ponens axiom K from Sect. 5.6.1 but for $\langle\alpha\rangle$ modalities instead of $[\alpha]$ modalities.

$$\text{K}\quad [\alpha](P\to Q)\to([\alpha]P\to[\alpha]Q)$$

5.17 (K knows that boxes distribute over conjunctions). Show that axiom $[]\wedge$, which distributes boxes over conjunctions

$$[]\wedge\quad [\alpha](P\wedge Q)\leftrightarrow[\alpha]P\wedge[\alpha]Q$$

can be derived from the modal modus ponens axiom K from Sect. 5.6.1.

5.18 (Distributivity and non-distributivity). Axioms K and $[]\wedge$ show how box modalities distribute over conjunctions and implications. Do they also distribute over other logical connectives? Which of the following formulas are valid?

$$[\alpha](P\vee Q)\to[\alpha]P\vee[\alpha]Q$$
$$[\alpha]P\vee[\alpha]Q\to[\alpha](P\vee Q)$$
$$[\alpha](P\leftrightarrow Q)\to([\alpha]P\leftrightarrow[\alpha]Q)$$
$$([\alpha]P\leftrightarrow[\alpha]Q)\to[\alpha](P\leftrightarrow Q)$$
$$[\alpha]\neg P\to\neg[\alpha]P$$
$$\neg[\alpha]P\to[\alpha]\neg P$$

How about diamond modalities? Which of the following are valid?

$$\langle\alpha\rangle(P\to Q)\to(\langle\alpha\rangle P\to\langle\alpha\rangle Q)$$
$$(\langle\alpha\rangle P\to\langle\alpha\rangle Q)\to\langle\alpha\rangle(P\to Q)$$
$$\langle\alpha\rangle(P\wedge Q)\to\langle\alpha\rangle P\wedge\langle\alpha\rangle Q$$
$$\langle\alpha\rangle P\wedge\langle\alpha\rangle Q\to\langle\alpha\rangle(P\wedge Q)$$
$$\langle\alpha\rangle(P\vee Q)\to\langle\alpha\rangle P\vee\langle\alpha\rangle Q$$
$$\langle\alpha\rangle P\vee\langle\alpha\rangle Q\to\langle\alpha\rangle(P\vee Q)$$
$$\langle\alpha\rangle(P\leftrightarrow Q)\to(\langle\alpha\rangle P\leftrightarrow\langle\alpha\rangle Q)$$
$$(\langle\alpha\rangle P\leftrightarrow\langle\alpha\rangle Q)\to\langle\alpha\rangle(P\leftrightarrow Q)$$
$$\langle\alpha\rangle\neg P\to\neg\langle\alpha\rangle P$$
$$\neg\langle\alpha\rangle P\to\langle\alpha\rangle\neg P$$

5.19 (Monotonicity rules). Prove that the monotonicity rule $M[\cdot]$ (Sect. 5.6.4) derives from Kripke's modal modus ponens axiom K (Sect. 5.6.1) and Gödel's generalization rule G (Sect. 5.6.3). That is, find a proof of the conclusion $[\alpha]P \to [\alpha]Q$ of rule M[·] assuming that you already have a proof of its premise $P \to Q$.

5.20 ($\langle \cdot \rangle$ monotonicity rule). Give a direct semantic soundness proof for the monotonicity rule M for the diamond modality in the same style, in which soundness was proved for the monotonicity rule M[·] for the box modality (Sect. 5.6.4). Then propose a diamond modality version of axiom K that would make it possible to derive rule M similarly to how rule M[·] was derived from axiom K and rule G in Exercise 5.19. Prove soundness of that newly proposed axiom.

5.21 (Sound and not-so-sound suggested axioms). Are any of the following suggested axioms sound? Are any of them useful?

Resignment axiom	$[x := e]P \leftrightarrow P$
Detest axiom	$[?Q]P \leftrightarrow [?P]Q$
Axiom of nondeterment choice	$[\alpha \cup \beta]P \leftrightarrow [\alpha]P$
Axiom of sequential confusion	$[\alpha;\beta]P \leftrightarrow [\beta]P$
Axiom of reiteration	$[\alpha^*]P \leftrightarrow [\alpha^*][\alpha^*]P$
Duality axiom	$\langle\alpha\rangle P \leftrightarrow \neg[\alpha]P$
Coconditional axiom	$[\text{if}(Q)\,\alpha]P \leftrightarrow (Q \to [\alpha]P)$
Unassignment axiom	$[x := e]p \leftrightarrow p$ (x is not free in p)
Reassignment axiom	$[x := e][x := e]p(x) \leftrightarrow p(e)$
Ж modal modus nonsens	$\langle\alpha\rangle(P \to Q) \to (\langle\alpha\rangle P \to \langle\alpha\rangle Q)$

In each case, prove soundness or construct a counterexample, i.e., an instance of the suggested axiom that is not a valid formula.

5.22 (Extra axioms and extra proof rules). Show that the additional axioms and proof rules listed in Fig. 5.5 are sound. If possible, try to derive them directly from other axioms, otherwise give a soundness proof using the semantics of dL.

M $\langle\alpha\rangle(P \vee Q) \leftrightarrow \langle\alpha\rangle P \vee \langle\alpha\rangle Q$

B $\exists x \langle\alpha\rangle P \leftrightarrow \langle\alpha\rangle \exists x P$ $(x \not\in \alpha)$

VK $p \to ([\alpha]\textit{true} \to [\alpha]p)$ $(\text{FV}(p) \cap \text{BV}(\alpha) = \emptyset)$

R $\dfrac{P_1 \wedge P_2 \to Q}{[\alpha]P_1 \wedge [\alpha]P_2 \to [\alpha]Q}$

Fig. 5.5 Additional axioms and proof rules for hybrid systems

5.23 (***). Show that the definitions of (syntactically) free and bound variables from Sect. 5.6.6 are correct, i.e., they are supersets of the semantic definitions of free and bound variables from Sect. 5.6.5. Under what circumstances are they proper supersets, i.e., contain additional variables? As a first step, simplify the definitions in Sect. 5.6.6 by producing simpler supersets if this simplifies your correctness proof.

References

[1]   Edmund M. Clarke, Armin Biere, Richard Raimi, and Yunshan Zhu. Bounded model checking using satisfiability solving. *Form. Methods Syst. Des.* **19**(1) (2001), 7–34. DOI: 10.1023/A:1011276507260.

[2]   Khalil Ghorbal and André Platzer. Characterizing algebraic invariants by differential radical invariants. In: *TACAS*. Ed. by Erika Ábrahám and Klaus Havelund. Vol. 8413. LNCS. Berlin: Springer, 2014, 279–294. DOI: 10.1007/978-3-642-54862-8_19.

[3]   Charles Antony Richard Hoare. An axiomatic basis for computer programming. *Commun. ACM* **12**(10) (1969), 576–580. DOI: 10.1145/363235.363259.

[4]   Stefan Mitsch and André Platzer. ModelPlex: verified runtime validation of verified cyber-physical system models. *Form. Methods Syst. Des.* **49**(1-2) (2016). Special issue of selected papers from RV'14, 33–74. DOI: 10.1007/s10703-016-0241-z.

[5]   André Platzer. Differential dynamic logic for hybrid systems. *J. Autom. Reas.* **41**(2) (2008), 143–189. DOI: 10.1007/s10817-008-9103-8.

[6]   André Platzer. *Logical Analysis of Hybrid Systems: Proving Theorems for Complex Dynamics*. Heidelberg: Springer, 2010. DOI: 10.1007/978-3-642-14509-4.

[7]   André Platzer. A complete axiomatization of quantified differential dynamic logic for distributed hybrid systems. *Log. Meth. Comput. Sci.* **8**(4:17) (2012). Special issue for selected papers from CSL'10, 1–44. DOI: 10.2168/LMCS-8(4:17)2012.

[8]   André Platzer. Logics of dynamical systems. In: *LICS*. Los Alamitos: IEEE, 2012, 13–24. DOI: 10.1109/LICS.2012.13.

[9]   André Platzer. The complete proof theory of hybrid systems. In: *LICS*. Los Alamitos: IEEE, 2012, 541–550. DOI: 10.1109/LICS.2012.64.

[10] André Platzer. Differential game logic. *ACM Trans. Comput. Log.* **17**(1) (2015), 1:1–1:51. DOI: 10.1145/2817824.

[11] André Platzer. A complete uniform substitution calculus for differential dynamic logic. *J. Autom. Reas.* **59**(2) (2017), 219–265. DOI: 10.1007/s10817-016-9385-1.

[12] André Platzer and Edmund M. Clarke. Computing differential invariants of hybrid systems as fixedpoints. *Form. Methods Syst. Des.* **35**(1) (2009). Special issue for selected papers from CAV'08, 98–120. DOI: 10.1007/s10703-009-0079-8.

[13] H. Gordon Rice. Classes of recursively enumerable sets and their decision problems. *Trans. AMS* **74**(2) (1953), 358–366. DOI: 10.2307/1990888.

[14] Danbing Seto, Bruce Krogh, Lui Sha, and Alongkrit Chutinan. The Simplex architecture for safe online control system upgrades. In: *American Control Conference*. Vol. 6. 1998, 3504–3508. DOI: 10.1109/ACC.1998.703255.

Chapter 6
Truth & Proof

Synopsis This chapter augments the dynamic axioms for dynamical systems from the previous chapter with the full mathematical rigor of a proof system. This proof system enables rigorous, systematic proofs for cyber-physical systems by providing systematic structuring mechanisms for their correctness arguments. The most important goals of such a proof system are that it guarantees to cover all cases of a correctness argument, so all possible behavior of a CPS, and that it provides guidance on which proof rules to apply. Its most important feature is the ability to use the dynamic axioms for dynamical systems that we already identified for rigorous reasoning about hybrid programs. A high-level interface of proofs with reasoning for real arithmetic as well as techniques for logically simplifying real-arithmetic questions are discussed as well.

6.1 Introduction[1]

Chap. 5 investigated dynamic axioms for dynamical systems, i.e., axioms in differential dynamic logic (dL) that characterize dynamical systems operators in terms of structurally simpler dL formulas. All it takes to understand the bigger system, thus, is to apply the appropriate axiom and investigate the smaller remainders. That chapter did not quite show all important axioms yet, but it still revealed enough to prove a property of a bouncing ball. While that chapter showed exactly how all the respective local properties of the system dynamics could be proved by invoking the corresponding dynamic axiom, it has not become clear yet how these individual inferences are best tied together to obtain a well-structured proof. That is what this chapter will identify.

---

[1] By both sheer coincidence and by higher reason, the title of this chapter turns out to be closely related to the subtitle of a well-known book on mathematical logic [1], which summarizes the philosophy pursued here in a way that is impossible to improve upon any further: *To truth through proof.*

© Springer International Publishing AG, part of Springer Nature 2018
A. Platzer, *Logical Foundations of Cyber-Physical Systems*,
https://doi.org/10.1007/978-3-319-63588-0_6

After all, there's more to proofs than just axioms. Proofs also have proof rules for combining fragments of arguments into a bigger proof by well-structured proof steps. Proofs, thus, are defined by the glue that holds axioms together into a single cohesive argument justifying its conclusion.

Granted, the working principle we followed with the axioms in Chap. 5 was quite intuitive. We repeatedly identified a subformula that we could simplify to an equivalent formula by applying any of the dL equivalence axioms from left to right. Since all dL axioms reduce more complex formulas on the left to structurally simpler formulas on the right, successively using them also simplified the conjecture correspondingly. That is quite systematic for such a simple mechanism.

Recall that our proof about the (single-hop) bouncing ball from the previous chapter still suffered from at least two issues, though. While it was a sound proof and an interesting proof, the way we came up with it was somewhat undisciplined. We just applied axioms seemingly at random at all kinds of places all over the logical formula. After we see such a proof, that is not a concern, because we can just follow its justifications and appreciate the simplicity and elegance of the steps it took to justify the conclusion.[2] But better structuring would certainly help us find proofs more constructively in the first place. The second issue was that the axioms for the dynamics that Chap. 5 showed us did not actually help in proving the propositional logic and arithmetic parts remaining at the end. So we were left with informal justifications of the resulting arithmetic, which leaves plenty of room for subtle mistakes in correctness arguments.

The present chapter addresses both issues by imposing more structure on proofs and, as part of that, handle the operators of first-order logic that differential dynamic logic inherits (propositional *connectives* such as $\wedge, \vee, \neg, \rightarrow$) and quantifiers ($\forall, \exists$). As part of the structuring, we will make ample and crucial use of the dynamic axioms from Chap. 5. Howevr, they will be used in a more structured way than so far, in a way that focuses their use on the top level of the formula and in the direction that actually simplifies the formula.

While Chap. 5 laid down the most fundamental cornerstones of the Foundations of Cyber-Physical Systems and their rigorous reasoning principles, the present chapter revisits these fundamental principles and shapes them into a systematic proof approach. The chapter is loosely based on previous work [14, Section 2.5.2]. The most important learning goals of this chapter are:

Modeling and Control: This chapter deepens our understanding from the previous chapter of how discrete and continuous systems relate to one another in the presence of evolution domain constraints, a topic that the previous chapter only touched upon briefly. It also makes precise how proofs can reason soundly when only assuming evolution domains to hold in the end compared to the fact that evolution domains have to hold always throughout a continuous evolution.

---

[2] Indeed, the proof in Chap. 5 was creative in that it used axioms quite carefully in an order that minimizes the notational complexity. But it is not easy to come up with such (nonsystematic) shortcut proofs even if the KeYmaera X prover makes this relatively straightforward with its proof-by-pointing feature [8].

Computational Thinking: Based on the core rigorous reasoning principles for CPS developed in Chap. 5, this chapter is devoted to reasoning not only rigorously but also *systematically* about CPS models. Systematic ways of reasoning rigorously about CPS are, of course, critical to getting more complex CPS right. The difference between the axiomatic way of reasoning rigorously about CPSs [15] as put forth in Chap. 5 and the systematic way [13, 14] developed here is not a big difference conceptually, but more a difference in pragmatics.

That does not make it less important, though, and the occasion to revisit it gives us a way of deepening our understanding of systematic CPS analysis principles. This chapter explains ways of developing CPS proofs systematically and is an important ingredient for verifying CPS models of appropriate scale. The chapter also adds a fourth leg to the logical trinity of syntax, semantics, and axiomatics considered in Chap. 5. It adds pragmatics, by which we mean the question of how to use axiomatics to justify the syntactic renditions of the semantical concepts of interest. That is, how best to go about conducting a proof to justify truth of a CPS conjecture. An understanding of such pragmatics follows from a more precise understanding of what a proof is and what arithmetic does.

CPS Skills: This chapter is mostly devoted to sharpening our analytic skills for CPS. We will also develop a slightly better intuition for the operational effects involved in CPS in that we understand in which order we should worry about operational effects and whether that has an impact on the overall understanding.

systematic reasoning for CPS
verifying CPS models at scale
pragmatics: how to use axiomatics to justify truth
structure of proofs and their arithmetic

discrete+continuous relation   analytic skills for CPS
with evolution domains

6.2 Truth and Proof

Truth is defined by the semantics of logical formulas. The semantics gives a mathematical meaning to formulas that, in theory, could be used to establish the truth of a logical formula by expanding all semantic definitions. In practice, this is quite infeasible, for one thing, because quantifiers of differential dynamic logic quantify over real numbers (after all their variables may represent real quantities such as velocity and position). Yet, there are (uncountably) infinitely many of those, so determining the truth value of a universally quantified logical formula directly by working with its semantics is impossibly challenging since that would require instantiating it with infinitely many real numbers, which would keep us busy for quite a while.

The semantics is even more challenging to deal with in the case of the hybrid systems in the modalities of differential dynamic logic formulas, because hybrid systems have so many possible behaviors and are highly nondeterministic. Literally following all possible behaviors to check all reachable states hardly sounds like a way that would ever enable us to stop and conclude the system is safe. Except, of course, if we happened to be lucky and found a bug during just one execution, because that would be enough to falsify the formula. Yet, in fact, even following just one particular execution of a hybrid system can be tricky, because that still involves the need to compute a solution of its differential equations and check their evolution domain constraints at all times.

We are, nevertheless, interested in establishing whether a logical formula is true, no matter how complicated that may be, because we would very much like to know whether the hybrid systems they refer to can be used safely. Or, come to think of it, we are interested in finding out whether the formula is valid, since truth of a logical formula depends on the state (see the definition of semantics $\omega \in [\![P]\!]$ in Definition 4.2) whereas validity of a logical formula is independent of the state (see the definition of validity $\vDash P$), because validity means truth in all states. Validity of formulas is what we ultimately care about, because we want our safety analysis to hold in all permitted initial states of the CPS, not just one particular initial state ω, because we may not even know the exact initial state of the CPS. In that sense, valid logical formulas are the most valuable ones. We should devote all of our efforts to finding out what is valid, because that will allow us to draw conclusions about all states, including the real-world state as well.

While exhaustive enumeration and simulation is hardly an option for systems as challenging as CPSs, the validity of logical formulas can be established by other means, namely by producing a proof of that formula. Like the formula itself, but unlike its semantics, a proof is a syntactical object that is amenable, e.g., to representation and manipulation in a computer. The finite syntactical argument represented in a proof witnesses the validity of the logical formula that it concludes. Proofs can be produced in a machine. They can be stored to be recalled as evidence for the validity of their conclusion. And they can be checked by humans or machines for correctness. Proofs can even be inspected for analytic insights about the reasons for the validity of a formula, which goes beyond the mere factual statement of validity. A proof justifies the judgment that a logical formula is valid, which, without such a proof as evidence, is no more than an empty claim. Empty claims are hardly useful foundations for building any cyber-physical systems on.

Truth and proof should be related intimately, however, because we only want to accept proofs that actually imply truth, i.e., proofs that imply their consequences to be valid if their premises are. That is, proof systems should be *sound* in order to allow us to draw reliable conclusions from the existence of a proof. This textbook will exercise great care to identify sound reasoning principles. The converse and equally intriguing question is that of *completeness*, i.e., whether all valid formulas can be proved, which turns out to be much more subtle [13, 16–18] and won't concern us until much later in this textbook.

6.2.1 Sequents

The proof built from axioms in Sect. 5.4 to justify a safety property of a bouncing ball was creative and insightful, but also somewhat spontaneous or maybe even disorganized. In fact, it has not even quite become particularly obvious what exactly a proof was, except that it is somehow supposed to glue axioms together into a single cohesive argument. But that is not a definition of a proof.[3]

In order to have a chance of conducting more complex proofs, we need a way of structuring the proofs and keeping track of all questions that come up while working on a proof as well as all assumptions that are available. But despite all the lamenting about the proof in Sect. 5.4, it has, secretly, been much more systematic than we were aware of at the time. Even if it went in a non-systematic order as far as the application order of the axioms was concerned, we still structured the proof quite well (unlike the ad hoc arguments in Sect. 4.8). So part of what this chapter needs to establish is to turn this lucky coincidence of a proper proof structure into an intentional principle. Rather than just coincidentally structuring the proof well, we want to structure all proofs well and make them all systematic by design.

Throughout this textbook, we will use *sequents*, which give us a structuring mechanism for conjectures and proofs. Sequent calculus was originally developed by Gerhard Gentzen [9, 10] for studying properties of natural deduction calculi, but sequent calculi have had tremendous success for numerous other purposes since.

In a nutshell, sequents are a standard form for logical formulas that is convenient for proving purposes, because it neatly aligns all available assumptions on the left of the sequent turnstile $\vdash$ and gathers what needs to be shown on the right.

> **Definition 6.1 (Sequent).** A *sequent* is of the form
>
> $$\Gamma \vdash \Delta$$
>
> where the *antecedent* Γ and *succedent* Δ are finite sets of dL formulas. The semantics of $\Gamma \vdash \Delta$ is that of the dL formula $\bigwedge_{P \in \Gamma} P \to \bigvee_{Q \in \Delta} Q$.

The antecedent Γ can be thought of as the list of formulas we assume to be true, whereas the succedent Δ can be understood as formulas for which we want to show that at least one of them is true, assuming all formulas of Γ are true. So for proving a sequent $\Gamma \vdash \Delta$, we assume all Γ and want to show that one of the Δ is true. For some simple sequents of the form $\Gamma, P \vdash P, \Delta$ where, among another set of formulas

---

[3] It would have been very easy to define, though, by inductively defining formulas to be provable if they are either instances of axioms or follow from provable formulas using modus ponens [15].

Γ in the antecedent and another set of formulas Δ in the succedent, literally the same formula P is in the antecedent and the succedent, we directly know that they are valid, because we can certainly show P if we already assume P. In fact, we will use this as a way of finishing a proof. For other sequents, it is more difficult to see whether they are valid (true under all circumstances) and it is the purpose of a proof calculus to provide a means to find out.

The basic idea in sequent calculus is to successively transform all formulas such that Γ forms a list of all assumptions and Δ the set of formulas that we would like to conclude from Γ (or, to be precise, the set Δ whose disjunction we would like to conclude from the conjunction of all formulas in Γ). For example, when a formula of the form $P \wedge Q$ is in the antecedent, we will identify a proof rule that simplifies $P \wedge Q$ in the sequent $\Gamma, P \wedge Q \vdash \Delta$ by replacing it with its two subformulas P and Q to lead to $\Gamma, P, Q \vdash \Delta$, because assuming the two formulas P and Q separately is the same as assuming the conjunction $P \wedge Q$, but involves smaller formulas.

Arguably the easiest way of understanding sequent calculus would be to interpret $\Gamma \vdash \Delta$ as the task of proving one of the formulas in the succedent Δ from all of the formulas in the antecedent Γ. But since dL is a classical logic, not an intuitionistic logic, we need to keep in mind that it is actually enough for proving a sequent $\Gamma \vdash \Delta$ to just prove the disjunction of all formulas in Δ from the conjunction of all formulas in Γ. For the proof rules of real arithmetic, we will later make use of this fact by considering the sequent $\Gamma \vdash \Delta$ as an abbreviation for the formula $\bigwedge_{P \in \Gamma} P \rightarrow \bigvee_{Q \in \Delta} Q$, because the two have the same semantics in dL. Indeed, a proof of the sequent $z = 0 \vdash x \geq z, x < z^2$ is only possible with this disjunctive interpretation of the succedent. We cannot say whether $x \geq z$ is true or whether $x < z^2$ is true, but if $z = 0$ is assumed, it is a classical triviality that their disjunction is true.

Empty conjunctions $\bigwedge_{P \in \emptyset} P$ are equivalent to *true*. Empty disjunctions $\bigvee_{P \in \emptyset} P$ are equivalent to *false*.[4] Hence, the sequent $\vdash A$ means the same as the formula A. The empty sequent $\vdash$ means the same as the formula *false*. The sequent $A \vdash$ means the same as formula $A \rightarrow false$. Starting off a proof question is easy, too, because if we would like to prove a dL formula P, we turn it into a sequent with no assumptions, since we do not initially have any, and set out to prove the sequent $\vdash P$.

> **Note 34 (Nonempty trouble with empty sequents)** If you ever reduce a conjecture about your CPS to proving the empty sequent $\vdash$, then you are in trouble, because the empty sequent $\vdash$ means the same as the formula *false* and it is impossible to prove *false*, since *false* isn't ever true. In that case, either you have taken a wrong turn in your proof, e.g., by discarding an assumption that was actually required for the conjecture to be true, or your CPS might take a wrong turn, because its controller can make a move that is actually unsafe.

---

[4] Note that *true* is the neutral element for the operation $\wedge$ and *false* the neutral element for the operation $\vee$. That is $A \wedge true$ is equivalent to A for any A and $A \vee false$ is equivalent to A. So *true* plays the same rôle for $\wedge$ that 1 plays for multiplication. And *false* plays the rôle for $\vee$ that 0 plays for addition. Another aspect of sequents $\Gamma \vdash \Delta$ that is worth mentioning is that other notations such as $\Gamma \Longrightarrow \Delta$ or $\Gamma \longrightarrow \Delta$ are also sometimes used in the literature.

In order to develop sequent calculus proof rules, we will again follow the logical guiding principle of compositionality from Chap. 5 by devising one suitable proof rule for each of the relevant operators. Only this time, we have two cases to worry about for each operator. We will need one proof rule for the case where the operator occurs in the antecedent so that it is available as an assumption. The corresponding rule for $\wedge$ will be called the $\wedge$L rule since it operates on the left of the $\vdash$ sequent turnstile. And we will need another proof rule for the case where that operator occurs in the succedent so that it is available as an option to prove. That rule for $\wedge$ will be called the $\wedge$R rule since it is for $\wedge$ and operates on the right of the $\vdash$ sequent turnstile. Fortunately, we will find a clever way of simultaneously handling all of the modality operators at once in sequent calculus by using the dL axioms from Chap. 5.

6.2.2 Proofs

Before developing any particular proof rules for sequent calculus, let us first understand what a proof is, what it means to prove a logical formula, and how we know whether a proof rule is sound so that it actually implies what it tries to prove.

> **Definition 6.2 (Global soundness).** A *sequent calculus proof rule* of the form
>
> $$\frac{\Gamma_1 \vdash \Delta_1 \quad \dots \quad \Gamma_n \vdash \Delta_n}{\Gamma \vdash \Delta}$$
>
> is *sound* iff validity of all *premises* (i.e., the sequents $\Gamma_i \vdash \Delta_i$ above the rule bar) implies validity of the *conclusion* (i.e., the sequent $\Gamma \vdash \Delta$ below the rule bar):
>
> $$\text{If } \vDash (\Gamma_1 \vdash \Delta_1) \text{ and } \dots \text{ and } \vDash (\Gamma_n \vdash \Delta_n) \text{ then } \vDash (\Gamma \vdash \Delta)$$

Recall that the meaning of a sequent $\Gamma \vdash \Delta$ is $\bigwedge_{P \in \Gamma} P \to \bigvee_{Q \in \Delta} Q$ by Definition 6.1, so that $\vDash (\Gamma \vdash \Delta)$ stands for $\vDash \left(\bigwedge_{P \in \Gamma} P \to \bigvee_{Q \in \Delta} Q \right)$ in Definition 6.2.

A formula P is provable or derivable (in the dL sequent calculus) if we can find a dL proof for it that concludes the sequent $\vdash P$ at the bottom *from no premises* and that has only used dL sequent proof rules to connect the premises to their conclusion. The rules id, $\top$R and $\bot$L we discuss below will prove particularly obvious sequents such as $\Gamma, P \vdash P, \Delta$ from no premises and, thereby, provide a way of finishing a proof. The shape of a dL *sequent calculus proof*, thus, is a tree with axioms at the top leaves and the formula that the proof proves at the bottom root.

When constructing proofs, however, we start with the desired goal $\vdash P$ at the bottom, since we want $\vdash P$ as the eventual conclusion of the proof. We work our way backwards to the subgoals until they can be proven to be valid. Once all subgoals have been proven to be valid, they entail their respective conclusions, which, recursively, entail the original goal $\vdash P$. This property of preserving truth or preserving validity is called soundness (Definition 6.2). When constructing proofs, we work bottom-up from the goal to the subgoals and apply all proof rules from the desired conclusion to the required premises. Once we have found a proof, we justify formulas conversely from the leaves top-down to the original goal at the bottom, because validity transfers from the premises to the conclusion with sound proof rules.

$$\text{construct proofs upwards} \quad \uparrow \quad \frac{\Gamma_1 \vdash \Delta_1 \quad \ldots \quad \Gamma_n \vdash \Delta_n}{\Gamma \vdash \Delta} \quad \downarrow \quad \text{validity transfers downwards}$$

We write $\vdash_{dL} P$ iff dL formula P can be *proved* with dL rules from dL axioms. That is, a dL formula is inductively defined to be *provable* in the dL sequent calculus iff it is the conclusion (below the rule bar) of an instance of one of the dL sequent proof rules, whose premises (above the rule bar) are all provable. In particular, since we will make sure that all dL proof rules are sound, the conclusion at the very bottom of the proof will be valid, because all its premises had a (shorter) proof and were, thus, valid by soundness of the respective proof rule that was used. A formula Q is *provable* from a set Φ of formulas, denoted by $\Phi \vdash_{dL} Q$, iff there is a finite subset $\Phi_0 \subseteq \Phi$ of formulas for which the sequent $\Phi_0 \vdash Q$ is provable.

6.2.3 Propositional Proof Rules

The first logical operators encountered during proofs are usually propositional logical connectives, because many dL formulas use shapes such as $A \rightarrow [\alpha]B$ to express that all behaviors of HP α lead to safe states satisfying B when starting the system in initial states satisfying A. For propositional logic, dL uses the standard propositional rules with the cut rule, which are listed in Fig. 6.1. These propositional rules decompose the propositional structure of formulas and neatly divide everything up into assumptions (which will ultimately be moved to the antecedent) and what needs to be shown (which will be moved to the succedent). The rules will be developed one at a time in the order that is most conducive to their intuitive understanding.

$$\neg R \frac{\Gamma, P \vdash \Delta}{\Gamma \vdash \neg P, \Delta} \qquad \wedge R \frac{\Gamma \vdash P, \Delta \quad \Gamma \vdash Q, \Delta}{\Gamma \vdash P \wedge Q, \Delta} \qquad \vee R \frac{\Gamma \vdash P, Q, \Delta}{\Gamma \vdash P \vee Q, \Delta}$$

$$\neg L \frac{\Gamma \vdash P, \Delta}{\Gamma, \neg P \vdash \Delta} \qquad \wedge L \frac{\Gamma, P, Q \vdash \Delta}{\Gamma, P \wedge Q \vdash \Delta} \qquad \vee L \frac{\Gamma, P \vdash \Delta \quad \Gamma, Q \vdash \Delta}{\Gamma, P \vee Q \vdash \Delta}$$

$$\rightarrow R \frac{\Gamma, P \vdash Q, \Delta}{\Gamma \vdash P \rightarrow Q, \Delta} \qquad \text{id} \frac{}{\Gamma, P \vdash P, \Delta} \qquad \text{TR} \frac{}{\Gamma \vdash true, \Delta}$$

$$\rightarrow L \frac{\Gamma \vdash P, \Delta \quad \Gamma, Q \vdash \Delta}{\Gamma, P \rightarrow Q \vdash \Delta} \qquad \text{cut} \frac{\Gamma \vdash C, \Delta \quad \Gamma, C \vdash \Delta}{\Gamma \vdash \Delta} \qquad \bot L \frac{}{\Gamma, false \vdash \Delta}$$

Fig. 6.1 Propositional proof rules of sequent calculus

Rules for Propositional Connectives

Proof rule ∧L is for handling conjunctions $(P \wedge Q)$ as one of the assumptions in the antecedent on the left of the sequent turnstile ($\vdash$). Assuming the conjunction $P \wedge Q$ is the same as assuming each conjunct P as well as Q separately.

$$\wedge\text{L} \; \frac{\Gamma, P, Q \vdash \Delta}{\Gamma, P \wedge Q \vdash \Delta}$$

Rule ∧L expresses that if a conjunction $P \wedge Q$ is among the list of available assumptions in the antecedent, then we might just as well assume both conjuncts (P and Q, respectively) separately. Assuming a conjunction $P \wedge Q$ is the same as assuming both conjuncts P and Q. So, if we set out to prove a sequent of the form in the conclusion ($\Gamma, P \wedge Q \vdash \Delta$), then we can justify this sequent by instead proving the sequent in the corresponding premise ($\Gamma, P, Q \vdash \Delta$), where the only difference is that the two assumptions P and Q are now assumed separately in the premise rather than jointly as a single conjunction, as in the conclusion.

If we keep on using proof rule ∧L often enough, then all conjunctions in the antecedent will ultimately have been split into their smaller pieces. Recall that the order of formulas in a sequent $\Gamma \vdash \Delta$ is irrelevant because Γ and Δ are sets, so we can always pretend that the formula to which we want to apply the rule ∧L is last in the antecedent. Rule ∧L takes care of all conjunctions that appear as top-level operators in antecedents even if its notation seems to indicate it would expect $P \wedge Q$ at the end of the antecedent. Of course, ∧L does not say how to prove $A \vee (B \wedge C) \vdash C$ or $A \vee \neg(B \wedge C) \vdash C$, because here the conjunction $B \wedge C$ is not a top-level formula in the antecedent but merely occurs somewhere deep inside as a subformula. But there are other logical operators to worry about as well, whose proof rules will decompose the formulas and ultimately reveal $B \wedge C$ at the top-level somewhere in the sequent.

Proof rule ∧R is for handling conjunction $P \wedge Q$ in the succedent by proving P and, in a separate premise, also proving Q:

$$\wedge\text{R} \; \frac{\Gamma \vdash P, \Delta \quad \Gamma \vdash Q, \Delta}{\Gamma \vdash P \wedge Q, \Delta}$$

Rule ∧R has to prove two premises, because if we are trying to prove a sequent $\Gamma \vdash P \wedge Q, \Delta$ with a conjunction $P \wedge Q$ in its succedent, it would not be enough at all to just prove $\Gamma \vdash P, Q, \Delta$, because the meaning of the succedent is a disjunction, so it would only enable us to conclude the weaker $\Gamma \vdash P \vee Q, \Delta$. Proving a conjunction in the succedent as in the conclusion of ∧R, thus, requires proving both conjuncts. It needs a proof of $\Gamma \vdash P, \Delta$ *and* a proof of $\Gamma \vdash Q, \Delta$. This is why rule ∧R splits the proof into two premises, one for proving $\Gamma \vdash P, \Delta$ and one for proving $\Gamma \vdash Q, \Delta$. If both premises of rule ∧R are valid then so is its conclusion. To see this, it is easier to first consider the case where Δ is empty. A proof of $\Gamma \vdash P$ together with a proof of $\Gamma \vdash Q$ implies that $\Gamma \vdash P \wedge Q$ is valid, because the conjunction $P \wedge Q$ follows from the assumptions Γ if both P and Q individually follow from Γ. Rule ∧R is justified by arguing by cases, once for the case where the disjunction corresponding to Δ is false (in which case the argument for $\Gamma \vdash P \wedge Q$ suffices) and once where it is true (in which case the conclusion is true without $P \wedge Q$). Overall, proof rule $\wedge$R captures that proving a conjunction $P \wedge Q$ amounts to proving both P and Q separately.

Proof rule $\vee$R is similar to rule $\wedge$L but for handling disjunctions in the succedent. If we set out to prove the sequent $\Gamma \vdash P \vee Q, \Delta$ in the conclusion with a disjunction $P \vee Q$ in the succedent, then we might as well split the disjunction into its two disjuncts and prove the premise $\Gamma \vdash P, Q, \Delta$ instead, since the succedent has a disjunctive meaning anyhow, so both sequents mean the same formula.

Proof rule $\vee$L handles a disjunction in the antecedent. When the assumptions listed in the antecedent of a sequent contain a disjunction $P \vee Q$, then there is no way of knowing which of the two disjuncts can be assumed, merely that at least one of them is assumed to be true. Rule $\vee$L, thus, splits the proof into cases. The left premise considers the case where the assumption $P \vee Q$ held because P was true. The right premise considers the case where assumption $P \vee Q$ held because Q was true. If both premises are valid (because we can find a proof for them), then, either way, the conclusion $\Gamma, P \vee Q \vdash \Delta$ will be valid no matter which of the two cases applies. Overall, rule $\vee$L captures that assuming a disjunction $P \vee Q$ requires two separate proofs that assume each disjunct instead.

Proof rule $\to$R handles implications in the succedent by using the implicational meaning of sequents. The way to understand it is to recall how we would prove an implication in mathematics. In order to prove an implication $P \to Q$, we would assume the left-hand side P (which rule $\to$R pushes into the assumptions listed in the antecedent) and try to prove its right-hand side Q (which $\to$R thus leaves in the succedent). This is how left-hand sides of implications ultimately end up as assumptions in the antecedent. Rule $\to$R, thus, captures that proving an implication $P \to Q$ amounts to assuming the left-hand P and proving the right-hand Q.

Proof rule $\to$L is more involved. It it used to handle assumptions that are implications $P \to Q$. When assuming an implication $P \to Q$, we can only assume its right-hand side Q (second premise) after we have shown its respective assumption P on its left-hand side (first premise). Another way to understand it is to recall that classical logic obeys the equivalence $(P \to Q) \equiv (\neg P \vee Q)$ and then to use the other propositional rules. Rule $\to$L captures that using an assumed implication $P \to Q$ allows us to assume its right-hand side Q if we can prove its left-hand side P.

Proof rule $\neg$R proves a negation $\neg P$ by, instead, assuming P. Again, the easiest way of understanding this rule is for an empty Δ in which case rule $\neg$R expresses that the way to prove a negation $\neg P$ in the succedent of the conclusion is to instead assume P in the antecedent in the premise and prove a contradiction, which is the formula *false* that an empty succedent means. When Δ is not empty, arguing by cases of whether the disjunction Δ is true or false will again do the trick. Alternatively, rule $\neg$R can be understood using the semantics of sequents from Definition 6.1, since a conjunct P on the left-hand side of an implication is semantically equivalent to a disjunct $\neg P$ on the right-hand side in classical logic. Overall, rule $\neg$R captures that to prove a negation $\neg P$, it is enough to assume P and prove a contradiction (or the remaining options Δ).

Proof rule $\neg$L handles a negation $\neg P$ among the assumptions in the antecedent of the conclusion by, instead, pushing P into the succedent of the premise. Indeed, for the case of empty Δ, if P were shown to hold from the remaining assumptions Γ, then Γ and $\neg P$ imply a contradiction in the form of the empty sequent, which is *false*. A semantic argument using the semantics of sequents also justifies ¬L directly since a conjunct $\neg P$ on the left-hand side of an implication is semantically equivalent to a disjunct P on the right-hand side in classical logic.

Identity and Cut Rules

All these propositional rules make progress by splitting operators. There is exactly one proof rule for each propositional logical connective on each side of the turnstile. All it takes is to look at the top-level operator of a formula and use the appropriate propositional sequent calculus rule from Fig. 6.1 to split the formula into its pieces. Such splitting will ultimately lead to *atomic formulas*, i.e., formulas without any logical operators. But there is no way to ever stop the proof yet. That is what the identity rule id from Fig. 6.1 is meant for. The identity rule id closes a goal (there are no further subgoals, which we sometimes mark by a ∗ instead of a sequent to indicate that we didn't just forget to finish the proof), because assumption P in the antecedent trivially implies P in the succedent (the sequent $\Gamma, P \vdash P, \Delta$ is a simple syntactic tautology). If, in our proving activities, we ever find a sequent of the form $\Gamma, P \vdash P, \Delta$, for any formula P, we can immediately use the identity rule id to close this part of the proof. The proof attempt succeeds if all premises are closed by id, or by other closing rules such as $\top$R (it is trivial to prove the valid formula *true*) or $\bot$L (assuming the unsatisfiable formula *false* means assuming the impossible).

Rule cut is Gentzen's *cut* rule [9, 10], which can be used for case distinctions or to prove a lemma and then use it. The right premise assumes any additional formula C in the antecedent that the left premise shows in the succedent. Semantically: regardless of whether C is actually true or false, both cases are covered by proof branches. Alternatively, and more intuitively, the cut rule is a fundamental lemma rule. The left premise proves an auxiliary lemma C in its succedent, which the right premise then assumes in its antecedent for the rest of the proof (again consider the case of empty Δ first to understand why this is sound). We only use cuts in an orderly fashion to derive simple rule dualities and to simplify meta-proofs. In practical applications, cuts are not needed in principle. In practice, complex CPS proofs still make use of cuts for efficiency reasons. Cuts can be used, for example, to substantially simplify arithmetic, or to first prove lemmas and then make ample use of them, in a number of places in the remaining proof.

Even though we write sequent rules as if the *principal formula* (which is the one that the sequent rule acts on such as $P \wedge Q$ in rules $\wedge$R and $\wedge$L) were at the end of the antecedent or at the beginning of the succedent, respectively, the sequent proof rules can be applied to other formulas in the antecedent or succedent, respectively, because we consider their order to be irrelevant. Antecedents and succedents are finite sets.

Sequent Proof Example

Even if the propositional sequent proof rules could hardly be the full story behind reasoning for cyber-physical systems, they still provide a solid basis and deserve to be explored with a simple example.

Example 6.1. A propositional proof of the exceedingly simple formula

$$v^2 \le 10 \wedge b > 0 \to b > 0 \wedge (\neg(v \ge 0) \vee v^2 \le 10) \tag{6.1}$$

is shown in Fig. 6.2. The proof starts with the desired goal as a sequent at the bottom:

$$\vdash v^2 \le 10 \wedge b > 0 \to b > 0 \wedge (\neg(v \ge 0) \vee v^2 \le 10).$$

and proceeds by applying suitable sequent proof rules upwards until we run out of subgoals and have finished the proof (the notation $*$ is used to indicate when there are no subgoals, which happens after rules id, $\top$R, $\bot$L).

$$
\cfrac{
  \cfrac{
    \cfrac{
      \cfrac{*}{\text{id}\ \ v^2 \le 10, b > 0 \vdash b > 0}
    }{\wedge\text{L}\ \ v^2 \le 10 \wedge b > 0 \vdash b > 0}
    \quad
    \cfrac{
      \cfrac{
        \cfrac{*}{\text{id}\ \ v^2 \le 10, b > 0 \vdash \neg(v \ge 0), v^2 \le 10}}
        {\wedge\text{L}\ \ v^2 \le 10 \wedge b > 0 \vdash \neg(v \ge 0), v^2 \le 10}
      }{\vee\text{R}\ \ v^2 \le 10 \wedge b > 0 \vdash \neg(v \ge 0) \vee v^2 \le 10}
  }{\wedge\text{R}\ \ v^2 \le 10 \wedge b > 0 \vdash b > 0 \wedge (\neg(v \ge 0) \vee v^2 \le 10)}
}{\to\text{R}\ \ \vdash v^2 \le 10 \wedge b > 0 \to b > 0 \wedge (\neg(v \ge 0) \vee v^2 \le 10)}
$$

Fig. 6.2 A simple propositional example proof in sequent calculus

The first (i.e., bottom-most) proof step applies proof rule →R to move the implication (→) to the sequent level by moving its left-hand side into the assumptions tracked in the antecedent. The next proof step applies rule ∧R to split the proof into the left branch for showing that conjunct $b > 0$ follows from the assumptions in the antecedent and the right branch for showing that conjunct $\neg(v \ge 0) \vee v^2 \le 10$ follows from the antecedent also. On the left branch, the proof closes with an axiom id after splitting the conjunction ∧ in the antecedent into its conjuncts with rule ∧L. We mark closed proof goals with $*$, to indicate that we did not just stop writing but that a subgoal is actually proved successfully. Of course, the left branch closes with rule id, because its assumption $b > 0$ in the antecedent trivially implies the formula $b > 0$ in the succedent, as the two formulas are identical. The right branch closes with rule id after splitting the disjunction (∨) in the succedent with rule ∨R and then splitting the conjunction (∧) in the antecedent with rule ∧L. On the right branch, the first assumption formula $v^2 \le 10$ in the antecedent trivially implies the last formula in the succedent, $v^2 \le 10$, because the two are identical, so rule id applies.

Now that all branches of the proof have closed (with id and marked by $*$), we know that all leaves at the top are valid. Since the premises are valid, each application of a proof rule ensures that their respective conclusions are valid also, by soundness. By recursively following this proof from the leaves at the top to the original root at the bottom, we conclude that the original goal at the bottom is valid and formula (6.1) is, indeed, true in all states. The conjecture that formula (6.1) is valid is exactly what the proof in Fig. 6.2 justifies.

While this proof does not prove any particularly exciting formula, it still shows how a proof can be built systematically in the dL calculus and gives an intuition as to how validity is inherited from the premises to the conclusions. The proof has been entirely systematic. All we did to come up with it was successively inspect the top-level operator in one of the logical formulas in the sequent and apply its corresponding propositional proof rule to find the resulting subgoals. All the while we were doing this, we carefully watched to see if the same formula shows up in the antecedent and succedent, for then the rule id closes that subgoal. There would be no point in proceeding with any other proof rule if the rule id closes a subgoal.

Most interesting formulas will not be provable with the sequent proof rules we have seen so far, because those were only for propositional connectives. So, next, set out to find proof rules for the other operators of differential dynamic logic.

6.2.4 Soundness of Proof Rules

Before proceeding with an investigation of additional sequent calculus proof rules, notice that the sequent proof rules for propositional logic are sound [9, 10, 14] according to the global soundness notion defined in Definition 6.2. We consider only one of the proof rules here to show how soundness works. Soundness is crucial, however, so you are invited to prove soundness for the other rules (Exercise 6.7).

Lemma 6.1 (∧R conjunction rule). *Proof rule ∧R is sound.*

Proof. Consider any instance for which both premises $\Gamma \vdash P, \Delta$ and $\Gamma \vdash Q, \Delta$ are valid and show that the conclusion $\Gamma \vdash P \wedge Q, \Delta$ is valid. To show the latter, consider any state ω. If there is a formula $G \in \Gamma$ in the antecedent that is not true in ω (i.e., $\omega \notin \llbracket G \rrbracket$) there is nothing to show, because $\omega \in \llbracket \Gamma \vdash P \wedge Q, \Delta \rrbracket$ then holds trivially, because not all assumptions in Γ are satisfied in ω. Likewise, if there is a formula $D \in \Delta$ in the succedent that is true in ω (i.e., $\omega \in \llbracket D \rrbracket$) there is nothing to show, because $\omega \in \llbracket \Gamma \vdash P \wedge Q, \Delta \rrbracket$ then holds trivially, because one of the formulas in the succedent is already satisfied in ω. Hence, the only interesting case to consider is the case where all formulas in $G \in \Gamma$ are true in ω and all formulas $D \in \Delta$ are false. In that case, since both premises were assumed to be valid, and Γ is true in ω but Δ false in ω, the left premise implies that $\omega \in \llbracket P \rrbracket$ and the right premise implies that $\omega \in \llbracket Q \rrbracket$. Consequently, $\omega \in \llbracket P \wedge Q \rrbracket$ by the semantics of $\wedge$. Thus, $\omega \in \llbracket \Gamma \vdash P \wedge Q, \Delta \rrbracket$. As the state ω was arbitrary, this implies $\vDash (\Gamma \vdash P \wedge Q, \Delta)$, i.e., the conclusion of the considered instance of ∧R is valid. □

In the rest of this chapter and, in fact, the whole textbook, we will scrutinize each proof rule to make sure it is sound according to Definition 6.2. We also make sure that all dL axioms are sound according to Definition 5.1. This implies that the dL proof calculus will only ever prove valid dL formulas, which is a *conditio sine qua non* in logic, that is, a condition without which logic could not be.

Recall from Sect. 6.2.2 that we write $\vdash_{dL} P$ iff dL formula P can be *proved* with dL rules from dL axioms. And recall from Chap. 4 that we write $\vDash P$ iff formula P is valid, i.e., true in all states.

> **Theorem 6.1 (Soundness).** *The* dL *sequent calculus is* sound. *That is, if a* dL *formula P has a proof in* dL*'s sequent calculus, i.e.,* $\vdash_{dL} P$, *then P is valid, i.e.,* $\vDash P$.

Proof. We only consider a schematic proof focusing on the structure of the soundness argument based on previous soundness considerations such as Lemma 5.1 and Lemma 6.1. Elaborate proofs of soundness for all cases are elsewhere [13, 16, 18]. A dL formula P is proved iff there is a proof of the sequent $\vdash P$ in dL's sequent calculus. Since more general shapes of sequents occur during the sequent proof, we show the stronger statement that every sequent $\Gamma \vdash \Delta$ that has a proof in dL's sequent calculus is valid (that is $\vDash (\Gamma \vdash \Delta)$ in the sense of Definition 6.1).

We show this by induction on the structure of the sequent calculus proof. That is, we prove that all base cases for small proofs with zero proof steps have valid conclusions. Then we consider all bigger proofs, and assume the induction hypothesis that all smaller proofs already have valid conclusions to show that one more proof step will still make the new conclusion valid.

0. The only proofs without proof steps are the ones that consist only of a dL axiom. Each dL axiom has been proved to be sound, for example in Chap. 5. All instances of sound axioms are valid dL formulas by Definition 5.1.
1. Consider a proof ending in a proof step with some number of premises $n \geq 0$:

$$\frac{\Gamma_1 \vdash \Delta_1 \quad \ldots \quad \Gamma_n \vdash \Delta_n}{\Gamma \vdash \Delta} \tag{6.2}$$

The respective subproofs for the premises $\Gamma_i \vdash \Delta_i$ are smaller, since they have one fewer proof step. So, by induction hypothesis, their respective conclusions $\Gamma_i \vdash \Delta_i$ are all valid (i.e., their corresponding dL formula from Definition 6.1 is valid):

$$\vDash (\Gamma_i \vdash \Delta_i) \quad \text{for all } i \in \{1, 2, \ldots, n\}$$

Because all of the finitely many dL proof rules that could be used in proof step (6.2) are sound (e.g., Lemma 6.1), the definition of soundness of proof rules (Definition 6.2) implies that the conclusion of the proof rule used in (6.2) is valid, which is $\vDash (\Gamma \vdash \Delta)$. □

Of course, the obligation is on us to ensure that we will, indeed, only ever add sound axioms (Definition 5.1) and sound proof rules (Definition 6.2) to the dL calculus, or else we will lose the crucial soundness theorem Theorem 6.1 and no longer have any faith in any proofs.

6.2.5 Proofs with Dynamics

Now that we have identified a left and a right proof rule for all propositional connectives we can literally continue the logical guiding principle of connectivity and proceed to also identify a left and a right proof rule for all top-level operator in all modalities.

Sequent Calculus Proof Rules for Dynamics

We could add a pair of sequent calculus proof rules for nondeterministic choices in box modalities, one in the antecedent (rule $[\cup]R$) and one in the succedent ($[\cup]L$):

$$[\cup]R \quad \frac{\Gamma \vdash [\alpha]P \wedge [\beta]P, \Delta}{\Gamma \vdash [\alpha \cup \beta]P, \Delta}$$

$$[\cup]L \quad \frac{\Gamma, [\alpha]P \wedge [\beta]P \vdash \Delta}{\Gamma, [\alpha \cup \beta]P \vdash \Delta}$$

These rules directly follow from the axioms from Chap. 5, though, and, thus, lead to quite a lot of unnecessary duplication of concepts.[5] Furthermore, such a list of sequent rules is less flexible than the axioms from Chap. 5. The sequent rules $[\cup]R, [\cup]L$ can only be applied when a nondeterministic choice is at the top-level position of a sequent, not when it occurs somewhere in a subformula, such as at the underlined position in the following sequent near the bottom of the proof of single-hop bouncing balls from Sect. 5.4:

$$A \vdash [x'' = -g]\underline{[?x = 0; v := -cv \cup ?x \geq 0]B(x, v)} \tag{6.3}$$

Substituting Equals for Equals

Thus, instead of writing down a pair of (rather redundant and quite inflexible) sequent rules for each dynamic axiom, we instead cover all axioms at once. The key observation was already foreshadowed in Chap. 5:

> **Note 35 (Substituting equals for equals)** If an equivalence $P \leftrightarrow Q$ is a valid formula, then any occurrence of its left-hand side P in any subformula can be replaced by its right-hand side Q (or vice versa), equivalently.

For example, using at the underlined position in the middle of dL formula (6.3) the equivalence

$$\underline{[?x = 0; v := -cv \cup ?x \geq 0]B(x, v)} \leftrightarrow [?x = 0; v := -cv]B(x, v) \wedge [?x \geq 0]B(x, v) \tag{6.4}$$

---

[5] One subsequent difference will be that applying rule $\wedge R$ to the premise of rule $[\cup]R$ will split the proof into two premises while subsequently applying $\wedge L$ to the premise of rule $[\cup]L$ will not.

which is a direct instance of axiom $[\cup]$ $[\alpha \cup \beta]P \leftrightarrow [\alpha]P \wedge [\beta]P$ from Chap. 5, the formula (6.3) is equivalent to

$$A \vdash [x'' = -g] \big([?x = 0; v := -cv]B(x,v) \wedge [?x \geq 0]B(x,v) \big) \qquad (6.5)$$

since (6.5) is constructed from (6.3) by replacing the left-hand side of equivalence (6.4) by its right-hand side in the middle of formula (6.3) at the indicated position.

Contextual Equivalence

The intuition of substituting equals for equals serves us well and is perfectly sufficient for all practical purposes. Logic is ultimately about precision, though, which is why we elaborate Note 35 as follows [18].

Lemma 6.2 (Contextual equivalence). *The contextual equivalence rewriting rules are sound:*

$$\text{CER} \;\; \frac{\Gamma \vdash C(Q), \Delta \quad \vdash P \leftrightarrow Q}{\Gamma \vdash C(P), \Delta} \qquad \text{CEL} \;\; \frac{\Gamma, C(Q) \vdash \Delta \quad \vdash P \leftrightarrow Q}{\Gamma, C(P) \vdash \Delta}$$

Proof. Rules CER and CEL derive with a cut from the contextual equivalence rule:

$$\text{CE} \;\; \frac{P \leftrightarrow Q}{C(P) \leftrightarrow C(Q)}$$

□

That is, if the equivalence $P \leftrightarrow Q$ in the second premise is proved, then P can be replaced by Q in any context $C(\_)$ anywhere in the succedent (rule CER) or in the antecedent (rule CEL) in the first premise. Here we read $C(\_)$ as the *context* in which the formula P occurs in the formula $C(P)$ and read $C(Q)$ as the result of replacing P in that context $C(\_)$ by Q. While a concise technical treatment and precise definitions of contexts and soundness proof for CER,CEL is surprisingly simple [18], this intuitive understanding is enough for our purposes here. If P and Q are equivalent (second premise of CER and of CEL), then we can replace P by Q no matter in what context $C(\_)$ they occur in the sequents in the succedent (CER) or antecedent (CEL), respectively. These contextual equivalence rules provide the perfect lifting device to use all equivalence axioms from Chap. 5 in any context in any proof.

Of course, it is crucial that P and Q are actually equivalent (second premise of CER and CEL) unconditionally without any assumptions from Γ, because those assumptions from Γ may no longer hold in the context $C(\_)$. For example, even if $x = 1$ and $x^2 = 1$ are equivalent when assuming $x \geq 0$, that assumption is no longer available in the context $[x := -1]\_$, so the following cannot be proved by CER:

$$\frac{x \geq 0 \vdash [x := -1]x^2 = 1 \quad x \geq 0 \vdash x = 1 \leftrightarrow x^2 = 1}{x \geq 0 \vdash [x := -1]x = 1}$$

This inference would, indeed, be unsound (written $\not\vdash$), because the premises are valid but the conclusion is not.

The flexible device of contextual equivalence rewriting by CER,CEL enables flexible and intuitive reasoning steps. Of course, we should still take care to use the axioms in the direction that actually simplifies the problem at hand. The dL axioms such as axiom $[\cup]$ are primarily meant to be used for replacing the left-hand side $[\alpha \cup \beta]P$ with the structurally simpler right-hand side $[\alpha]P \wedge [\beta]P$, because that direction of use assigns meaning to $[\alpha \cup \beta]P$ in logically simpler terms, i.e., as a structurally simpler logical formula. Furthermore, that direction reduces a dL formula to a formula with more formulas but smaller hybrid programs, which will terminate after finitely many such reductions, because every hybrid program only has finitely many subprograms.

Finally note that we will usually not explicitly mention the use of CEL and CER in proofs but only mention the axiom that they invoked. For example, the sequent proof step reducing conclusion (6.3) to premise (6.5) using axiom $[\cup]$ (and, of course, the implicit rule CER) is simply written as

$$[\cup]\frac{A \vdash [x''=-g]\left([?x=0;v:=-cv]B(x,v) \wedge [?x \geq 0]B(x,v)\right)}{A \vdash [x''=-g][?x=0;v:=-cv \cup ?x \geq 0]B(x,v)}$$

In fact the full proof in Sect. 5.4 can suddenly be understood as a sequent proof in this way by adding a sequent turnstile $\vdash$ and implicitly using CER in addition to the respective indicated dynamic axioms.

$$[:=]\frac{[:=]\frac{[:=]\frac{\vdash v^2 \leq 10 \wedge -(-b) > 0 \to b > 0 \wedge (\neg(v \geq 0) \vee v^2 \leq 10)}{\vdash [c:=10]\left(v^2 \leq 10 \wedge -(-b) > 0 \to b > 0 \wedge (\neg(v \geq 0) \vee v^2 \leq c)\right)}}{\vdash [a:=-b][c:=10]\left(v^2 \leq 10 \wedge -a > 0 \to b > 0 \wedge (\neg(v \geq 0) \vee v^2 \leq c)\right)}}{\vdash [a:=-b;c:=10]\left(v^2 \leq 10 \wedge -a > 0 \to b > 0 \wedge (\neg(v \geq 0) \vee v^2 \leq c)\right)}$$

Fig. 6.3 A simple example proof with dynamics in sequent calculus

Sequent Proof Example with Dynamics

See Fig. 6.3 for a simple example proof. This proof is not very interesting. Incidentally, though, the proof in Fig. 6.3 ends with a premise at the top that is identical to the (provable) conclusion at the bottom of Fig. 6.2. So gluing the two proofs together leads to a proof of the conclusion at the bottom of Fig. 6.3:

$$[a:=-b;c:=10]\left(v^2 \leq 10 \wedge -a > 0 \to b > 0 \wedge (\neg(v \geq 0) \vee v^2 \leq c)\right)$$

Since this completes the proof (no more premises) and dL proof rules and axioms are sound, this conclusion is valid, so true in all states. Most crucially, this dL formula now has the proof as a finite and entirely syntactic justification of why it is valid. That is certainly more practical than enumerating all of the infinitely many possible real values for the variables and checking whether the semantics evaluates to true.

A minor wrinkle foreshadowing further developments is that the proof in Fig. 6.3 ends in a formula mentioning $-(-b) > 0$ while the proof in Fig. 6.2 starts with a formula mentioning $b > 0$ in the same place. The two formulas are, of course, equivalent, but, in order to really glue both proofs togther, we still need to add some proof rule that justifies this arithmetic transformation. We could add the following special purpose proof rule for this purpose, but will ultimately decide on adding a much more powerful proof rule instead (Sect. 6.5):

$$\frac{\Gamma, \theta > 0 \vdash \Delta}{\Gamma, -(-\theta) > 0 \vdash \Delta}$$

6.2.6 Quantifier Proof Rules

When trying to make the proof for the bouncing ball from Sect. 5.4 systematic by turning it into a sequent calculus proof, the first propositional step succeeds with rule →R, then a couple of steps succeed in splitting the hybrid program with dynamic axioms from Chap. 5, but, ultimately, the differential equation solution axiom ['] produces a quantifier for time that still needs to be handled. Of course, even a mere inspection of the syntax of dL shows that there are logical operators that have no proof rules yet, namely the universal and existential quantifiers.

$$\vR \; \frac{\Gamma \vdash p(y), \Delta}{\Gamma \vdash \forall x\, p(x), \Delta} \; (y \notin \Gamma, \Delta, \forall x\, p(x)) \qquad \exists R \; \frac{\Gamma \vdash p(e), \Delta}{\Gamma \vdash \exists x\, p(x), \Delta} \; (\text{arbitrary term } e)$$

$$\forall L \; \frac{\Gamma, p(e) \vdash \Delta}{\Gamma, \forall x\, p(x) \vdash \Delta} \; (\text{arbitrary term } e) \qquad \exists L \; \frac{\Gamma, p(y) \vdash \Delta}{\Gamma, \exists x\, p(x) \vdash \Delta} \; (y \notin \Gamma, \Delta, \exists x\, p(x))$$

Fig. 6.4 Quantifier sequent calculus proof rules

The quantifier proof rules are listed in Fig. 6.4 and work much as in mathematics. In the proof rule ∀R, we want to show a universally quantified property. When a mathematician wants to show a universally quantified property $\forall x\, p(x)$ to hold, he could choose a fresh symbol[6] y and set out to prove that $p(y)$ holds. Once he has found a proof for $p(y)$, the mathematician remembers that y was arbitrary and his proof did not assume anything special about the value of y. So he concludes that $p(y)$ must indeed hold for all y since y was arbitrary, and that, hence, $\forall x\, p(x)$ holds true. For example, to show that the square of all numbers is nonnegative, a mathematician could start out by saying "let y be an arbitrary number," prove $y^2 \geq 0$ for

---

[6] In logic, these fresh symbols are known as *Skolem function symbols* [20] or Herbrand function symbols [11], except that here we can just use fresh variables for the same purpose.

that y, and then conclude $\forall x (x^2 \geq 0)$, since y was arbitrary. Proof rule $\forall R$ makes this reasoning formally rigorous. It chooses a *new* variable symbol y and replaces the universally quantified formula in the succedent by a formula for y. Notice, of course, how crucially important it is to actually choose a new symbol y that has not been used free anywhere else in the sequent before. Otherwise, we would assume special properties about y in Γ, Δ that we would not be justified to assume. In fact, it is enough if the variable y just is no free variable in the sequent $\Gamma \vdash \forall x\, p(x), \Delta$, in which case the variable x itself can be used for the fresh symbol y; see Sect. 5.6.5.

In proof rule $\exists R$, we want to show an existentially quantified property. When a mathematician proves $\exists x\, p(x)$, he can directly produce any specific term e as a witness for this existential property and prove that, indeed, $p(e)$, for then he would have shown $\exists x\, p(x)$ with this witness. For example, to show that there is a number whose cube is less than its square, a mathematician could start by saying "let me choose, say, $\frac{2-1}{2}$ and show the property for $\frac{2-1}{2}$." Then he can prove $(\frac{2-1}{2})^3 < (\frac{2-1}{2})^2$, because $0.125 < 0.25$, and conclude that there, thus, is such a number, i.e., $\exists x (x^3 < x^2)$, because $\frac{2-1}{2}$ was a perfectly good witness. This reasoning is exactly what proof rule $\exists R$ enables. It allows the choice of *any* term e for x and accepts a proof of $p(e)$ as a proof of $\exists x\, p(x)$. Unlike in rule $\forall R$, it is perfectly normal for the witness e to mention other variables. For example, a witness for $a > 0 \vdash \exists x (x > y^2 \wedge x < y^2 + a)$ is $y^2 + \frac{a}{2}$; any such witness depends on y and a.

However note that the claim "e is a witness" may turn out to be wrong, for example, the choice 2 for x would have been a pretty bad start for attempting to show $\exists x (x^3 < x^2)$. Consequently, proof rule $\exists R$ is sometimes discarded in favor of a rule that keeps both options $p(e)$ and $\exists x\, p(x)$ in the succedent. KeYmaera X instead allows proof steps to be undone if a proof attempt failed. If the proof with e is successful, the sequent is valid and the part of this proof can be closed successfully. If the proof with e later turns out to be unsuccessful, another proof attempt can be started.

This approach already hints at a practical problem. If we are very smart about our choice of the witness e, rule $\exists R$ leads to very short and elegant proofs. If not, we may end up going in circles without much progress in the proof. That is why KeYmaera X allows you to specify a witness if you can find one (and you should if you can, because that gives significantly faster proofs) but also allows you to keep going without a witness, e.g., by applying axioms to the formula $p(e)$ without touching the quantifier.

Rules $\forall L, \exists L$ are dual to $\exists R, \forall R$. In proof rule $\forall L$, we have a universally quantified formula in the assumptions (antecedent) that we can use, instead of in the succedent, which we want to show. In mathematics, when we know a universal fact, we can use this knowledge for any particular instance. If we know that all positive numbers have a square root, then we can also use the fact that 5 has a square root, because 5 is a positive number. Hence from assumption $\forall x (x > 0 \to \exists y (x = y^2))$ in the antecedent, we can also assume the particular instance $5 > 0 \to \exists y (5 = y^2)$ that uses 5 for x. Rule $\forall L$ can produce an instance $p(e)$ of the assumption $\forall x\, p(x)$ for an arbitrary term e. We sometimes need the universal fact $\forall x\, p(x)$ for multiple instantiations with e_1, e_2, e_3 during the proof. Fortunately, rule $\forall L$ is also sound when it keeps the assumption $\forall x\, p(x)$ in the antecedent so that it can be used repeatedly to obtain different instances.

In proof rule $\exists$L, we can use an existentially quantified formula from the antecedent. If we know an existential fact in mathematics, then we can give a name to the object that we then know does exist. If we know that there is a smallest integer less than 10 that is a square, we can call it y, but we cannot denote it by a different term like 5 or 4+2, because they may be (and in fact are) the wrong answer. Rule $\exists$L gives a fresh name y to the object that was assumed to exist. Since it does not make sense to give a different name to the same object later, $\exists x\, p(x)$ is removed from the antecedent when rule $\exists$L adds $p(y)$.

Note how the quantifier proof rules in Fig. 6.4 continue the trend of the propositional sequent calculus rules in Fig. 6.1: they decompose logical formulas into simpler subformulas. Admittedly, the instances e chosen in rules $\exists$R,$\forall$L can be rather large terms. But that is a matter of perspective. All it takes is for us to understand that concrete terms, no matter how large, are still structurally simpler than quantifiers.

6.3 Derived Proof Rules

The universal quantifier rule $\forall$L for the antecedent shown in Fig. 6.4 does not retain the universal assumption $\forall x\, p(x)$ in the antecedent even though it could. The following proof rule helps in cases where multiple instantiations of a universal assumption are needed, because it can be used repeatedly to produce $p(e)$ and $p(\tilde{e})$:

$$\forall\forall\text{L}\ \frac{\Gamma, \forall x\, p(x), p(e) \vdash \Delta}{\Gamma, \forall x\, p(x) \vdash \Delta}$$

But it is not very practical to adopt every possible proof rule. Instead, the newly suggested proof rule $\forall\forall$L is a *derived rule*, which means that it can be proved using the other proof rules already. Obviously, the only other proof rule that can produce an assumption $p(e)$ from the assumption $\forall x\, p(x)$ is rule $\forall$L from Fig. 6.4, and that rule swallows said assumption.

What we can do to derive $\forall\forall$L is to first copy the assumption $\forall x\, p(x)$ to obtain a duplicate, and then use $\forall$L to turn one copy into $p(e)$, leaving the other copy of $\forall x\, p(x)$ for later use. Only how do we copy assumptions?

Would it even be fine to duplicate assumptions in a sequent? Fortunately, sequents consist of a finite set of assumptions Γ and a finite set Δ, so that assuming the same formula twice does not change the meaning of the sequent (Sect. 6.5.4).

Operationally, assumptions can be duplicated by the cut rule to prove the formula $\forall x\, p(x)$ as a new lemma, which is trivial because it is among the assumptions, and we can then subsequently work with the extra assumption. This derives rule $\forall\forall$L by the following sequent calculus proof:

$$\text{cut}\ \frac{\text{id}\dfrac{*}{\Gamma, \forall x\, p(x) \vdash \forall x\, p(x), \Delta} \qquad \forall\text{L}\dfrac{\Gamma, \forall x\, p(x), p(e) \vdash \Delta}{\Gamma, \forall x\, p(x), \forall x\, p(x) \vdash \Delta}}{\Gamma, \forall x\, p(x) \vdash \Delta}$$

This sequent calculus proof starts with the conclusion of derived rule $\forall\forall$L at the bottom and ends with only the premises that rule $\forall\forall$L has at the top. What makes rule $\forall\forall$L a derived rule is that we can use it in any proof and expand it into the above more verbose proof using rules cut,id,$\forall$L instead. The big advantage of derived rules over new proof rules is that derived rules do not need a soundness proof from the semantics, because they merely combine other proof rules that were already established to be sound (Theorem 6.1).

6.4 A Sequent Proof for the Single-Hop Bouncing Ball

Recall the bouncing-ball abbreviations from Sect. 5.4:

$$A \overset{\text{def}}{\equiv} 0 \leq x \wedge x = H \wedge v = 0 \wedge g > 0 \wedge 1 \geq c \geq 0$$

$$B(x,v) \overset{\text{def}}{\equiv} 0 \leq x \wedge x \leq H$$

$$\{x'' = -g\} \overset{\text{def}}{\equiv} \{x' = v, v' = -g\}$$

And consider the single-hop bouncing-ball formula again:

$$A \rightarrow [x'' = -g; (?x = 0; v := -cv \cup ?x \geq 0)]B(x,v) \qquad (5.14^*)$$

Sect. 5.4 already had a proof of (5.14) with the dynamic axioms from Chap. 5. By simply adding a sequent turnstile $\vdash$ this happens to be a sequent calculus proof, too. Instead of repeating this proof in sequent calculus style, we consider a similar property where we now include the evolution domain but leave out the discrete part:

$$A \rightarrow [x'' = -g \,\&\, x \geq 0]B(x,v) \qquad (6.6)$$

To prove (6.6), we convert it to a sequent and conduct the sequent calculus proof shown in Fig. 6.5. This proof boldly states that the first premise closes, except that

$$
\cfrac{
\cfrac{
\cfrac{
\cfrac{
\cfrac{
\cfrac{
\cfrac{
\cfrac{*}{\mathbb{R}\ \ A, r \geq 0 \vdash 0 \leq r \leq r} \quad
\cfrac{A, r \geq 0, H - \frac{g}{2}r^2 \geq 0 \vdash B(H - \frac{g}{2}r^2, -gt)}{{}_{[:=]}\ A, r \geq 0, [x := H - \frac{g}{2}r^2]x \geq 0 \vdash [x := H - \frac{g}{2}r^2]B(x,v)}
}{\substack{\to L}\ A, r \geq 0, 0 \leq r \leq r \to [x := H - \frac{g}{2}r^2]x \geq 0 \vdash [x := H - \frac{g}{2}r^2]B(x,v)}
}{\substack{\forall L}\ A, r \geq 0, \forall 0 \leq s \leq r[x := H - \frac{g}{2}s^2]x \geq 0 \vdash [x := H - \frac{g}{2}r^2]B(x,v)}
}{\substack{\to R}\ A, r \geq 0 \vdash \forall 0 \leq s \leq r[x := H - \frac{g}{2}s^2]x \geq 0 \to [x := H - \frac{g}{2}r^2]B(x,v)}
}{\substack{\to R}\ A \vdash r \geq 0 \to (\forall 0 \leq s \leq r[x := H - \frac{g}{2}s^2]x \geq 0 \to [x := H - \frac{g}{2}r^2]B(x,v))}
}{\substack{\forall R}\ A \vdash \forall t \geq 0 (\forall 0 \leq s \leq t[x := H - \frac{g}{2}s^2]x \geq 0 \to [x := H - \frac{g}{2}t^2]B(x,v))}
}{{}_{['\,]}\ A \vdash [x'' = -g \,\&\, x \geq 0]B(x,v)}
}{\substack{\to R}\ \vdash A \rightarrow [x'' = -g \,\&\, x \geq 0]B(x,v)}
$$

Fig. 6.5 Sequent calculus proof for gravity above ground

$$A, r \geq 0 \vdash 0 \leq r \leq r$$

is not exactly an instance of the rule id. Even here we need simple arithmetic to conclude that $0 \leq r \leq r$ is equivalent to $r \geq 0$ by reflexivity and flipping sides, at which point the first premise turns into a formula that can be closed by the id rule:

$$\text{id} \frac{*}{A, r \geq 0 \vdash r \geq 0}$$

A full formal proof and a KeYmaera X proof, thus, need an extra proof step of arithmetic in the left premise (marked by rule ℝ). In paper proofs, we will frequently accept such minor steps as abbreviations but always take note of the reason. In the above example, we might, for example remark alongside ℝ the arithmetic reason "by reflexivity of $\leq$ and flipping $0 \leq r$ to $r \geq 0$."

The second remaining premise in the above proof is

$$A, r \geq 0, H - \frac{g}{2} r^2 \geq 0 \vdash B\left(H - \frac{g}{2} r^2, -gt\right)$$

which, when resolving abbreviations turns into

$$0 \leq x \wedge x = H \wedge v = 0 \wedge g > 0 \wedge 1 \geq c \geq 0, r \geq 0, H - \frac{g}{2} r^2 \geq 0 \vdash 0 \leq H - \frac{g}{2} r^2 \wedge H - \frac{g}{2} r^2 \leq H$$

This sequent is proved using rule ∧R plus simple arithmetic for its branch

$$0 \leq x \wedge x = H \wedge v = 0 \wedge g > 0 \wedge 1 \geq c \geq 0, r \geq 0, H - \frac{g}{2} r^2 \geq 0 \vdash 0 \leq H - \frac{g}{2} r^2$$

We again cite the arithmetic reason as "by flipping $0 \leq H - \frac{g}{2} r^2$ to $H - \frac{g}{2} r^2 \geq 0$." Some more arithmetic is needed on the respective right branch resulting from ∧R:

$$0 \leq x \wedge x = H \wedge v = 0 \wedge g > 0 \wedge 1 \geq c \geq 0, r \geq 0, H - \frac{g}{2} r^2 \geq 0 \vdash H - \frac{g}{2} r^2 \leq H$$

where we note the arithmetic reason "$g > 0$ and $r^2 \geq 0$." Finishing the above sequent proof as discussed for the second premise, thus, shows that dL formula (6.6) at the conclusion of the proof is provable. This time, we have a well-structured and entirely systematic sequent calculus proof in which proof rules and axioms are only used on the top level.

In order to make sure you do not forget why some arithmetic facts are true, you are strongly advised to write down such arithmetic reasons in your paper proofs to justify that the arithmetic is valid. KeYmaera X provides a number of ways for proving real arithmetic that will be discussed next.

6.5 Real Arithmetic

What, in general, can be done to prove real arithmetic? We managed to convince ourselves with ad-hoc arithmetic reasons that the simple arithmetic in the above proofs was fine. But that is neither a proper proof rule nor should we expect to get away with such simple arithmetic arguments for the full complexity of CPS.

Chapters 20 and 21 in Part IV will discuss the handling of real arithmetic in much more detail. For now, the focus is on the most crucial elements for proving CPSs. Differential dynamic logic and KeYmaera X make use of a fascinating miracle: the fact that first-order logic of real arithmetic, however challenging it might sound, is perfectly decidable according to a seminal result by Alfred Tarski [22]. *First-order logic of real arithmetic* (FOL$_\mathbb{R}$) is the fragment of dL consisting of quantifiers over reals and propositional connectives of polynomial (or rational) term arithmetic with (real-valued) variables and rational constant symbols such as $\frac{5}{7}$, but no modalities. The most immediate way of incorporating uses of real-arithmetic reasoning into our proofs is, thus, by the rule $\mathbb{R}$, which proves all sequents whose corresponding formulas in FOL$_\mathbb{R}$ are valid, which is decidable.

> **Lemma 6.3 ($\mathbb{R}$ real arithmetic).** *First-order logic of real arithmetic is decidable so that all valid facts of* FOL$_\mathbb{R}$ *are obtained by this proof rule:*
>
> $$\mathbb{R} \; \frac{}{\Gamma \vdash \Delta} \quad (\textit{if } \bigwedge_{P \in \Gamma} P \to \bigvee_{Q \in \Delta} Q \textit{ is valid in FOL}_\mathbb{R})$$

The proof rule $\mathbb{R}$ is remarkably different from all other proof rules we ever consider in this book. All other axioms and proof rules provide straightforward syntactic transformations that are easily implementable on a computer, for example in the theorem prover KeYmaera X [8]. The real arithmetic proof rule $\mathbb{R}$, however, has a side condition about a formula being valid in real arithmetic, which it is not at all obvious how to check, but fortunately is still decidable. It is the conceptually simplest interface between proof-theoretic logic on the one side and model-theoretic algebraic decision procedures for real arithmetic on the other side. The real arithmetic proof rule $\mathbb{R}$ proves exactly the sequents representing valid formulas in first-order real arithmetic. But the formula actually has to be in first-order real arithmetic, so cannot contain any modalities or differential equations, which are out of scope for Tarski's result.

Example 6.2. For example, proof rule $\mathbb{R}$ proves the following list of sequents because they represent valid first-order real arithmetic formulas:

$$\mathbb{R} \frac{*}{\vdash x^2 \geq 0} \qquad\qquad \mathbb{R} \frac{*}{x > 0 \vdash x^3 > 0}$$

$$\mathbb{R} \frac{*}{\vdash x > 0 \leftrightarrow \exists y\, x^5 y^2 > 0} \qquad \mathbb{R} \frac{*}{a > 0, b > 0 \vdash y \geq 0 \to a x^2 + b y \geq 0}$$

But rule $\mathbb{R}$ does not prove $x^2 > 0 \vdash x > 0$, because it is not valid. Rule $\mathbb{R}$ does not prove $x \geq 0, v > 0 \vdash [x' = v]x \geq 0$, either, because it is not in pure real arithmetic.

6.5.1 Real Quantifier Elimination

On the surface, proof rule $\mathbb{R}$ represents all we need to know at this stage about first-order real arithmetic. How does that miracle work, though? Without any doubt, the most complex features of first-order real arithmetic are its quantifiers. And even if a real-arithmetic formula has no quantifiers, we can pretend it does by prefixing it with universal quantifiers for all free variables (forming the *universal closure*). After all, if we want to show a formula is valid, then we need to show it is true for all values of all its variables, which semantically corresponds to having all universal quantifiers in front. That is why an understanding of first-order real arithmetic proceeds by understanding the rôle of quantifiers over the reals.

In a nutshell, the notation $QE(P)$ denotes the use of real-arithmetic reasoning on formula P to obtain a formula $QE(P)$ over the same free variables that is equivalent to P but simpler, because $QE(P)$ is quantifier-free. When starting with a first-order real-arithmetic formula P in which all variables are quantified, the quantifier-free equivalent $QE(P)$ has no variables, so directly evaluates to either *true* or *false*.

Example 6.3. Real quantifier elimination yields, e.g., the following equivalence:

$$QE(\exists x\,(ax+b=0)) \equiv (a \neq 0 \vee b = 0) \tag{6.7}$$

The two sides are easily seen to be equivalent, i.e.,

$$\vDash \exists x\,(ax+b=0) \leftrightarrow (a \neq 0 \vee b = 0) \tag{6.8}$$

because a linear equation with nonzero inhomogeneous part has a solution iff its linear part is nonzero as well. And a constant equation (with $a = 0$) only has a solution if $b = 0$. The left-hand side of the equivalence may be hard to evaluate, because it conjectures the existence of an x and it is not clear how we might get such a real number for x, since there are so many reals. The right-hand side, instead, is trivial to evaluate, because it is quantifier-free and directly says to compare the values of a and b to zero and that an x such that $ax+b=0$ will exist if and only if $a \neq 0$ or $b = 0$. This is easy to check at least if a, b are either concrete numbers or fixed parameters for your CPS. Then all you need to do is make sure your choices for those parameters satisfy these constraints. If a or b is a symbolic term (not mentioning x otherwise the equivalence (6.8) is false and QE gives a different result), then (6.8) still identifies the conditions for the existence of an x such that $ax+b=0$.

Example 6.4. Quantifier elimination also handles universal quantifiers:

$$QE(\forall x\,(ax+b \neq 0)) \equiv (a = 0 \wedge b \neq 0)$$

Expedition 6.1 (Quantifier elimination)

One of Alfred Tarski's many seminal results from the 1930s proves quantifier elimination and decidability for real arithmetic [22].

> **Definition 6.3 (Quantifier elimination).** A first-order logic theory (such as first-order logic $\text{FOL}_\mathbb{R}$ over the reals) admits *quantifier elimination* if, for each formula P, a quantifier-free formula $\text{QE}(P)$ can be effectively associated with P that is equivalent, i.e., $P \leftrightarrow \text{QE}(P)$ is valid.

> **Theorem 6.2 (Tarski's quantifier elimination).** *The first-order logic of real arithmetic admits quantifier elimination and is, thus, decidable.*

That is, there is an algorithm that accepts any first-order real-arithmetic formula P in $\text{FOL}_\mathbb{R}$ as input and computes a formula $\text{QE}(P)$ in $\text{FOL}_\mathbb{R}$ that is equivalent to P but quantifier-free (and does not mention new variables or function symbols either).

The operation QE can be assumed to evaluate ground formulas (i.e., without variables) such as $\frac{1+9}{4} < 2+1$, yielding a decision procedure for closed formulas of this theory (i.e., formulas without free variables, which one obtains when forming the universal closure by prefixing the formula with universal quantifiers for all free variables). For a closed formula P, all it takes is to compute its quantifier-free equivalent $\text{QE}(P)$ by quantifier elimination. The closed formula P is closed, so has no free variables or other free symbols, and neither will its quantifier-free equivalent $\text{QE}(P)$. Hence, P as well as its equivalent $\text{QE}(P)$ are equivalent to either *true* or *false*. Yet, $\text{QE}(P)$ is quantifier-free, so which one it is can be found out simply by evaluating the (variable-free) concrete arithmetic in $\text{QE}(P)$.

While a full account of the nuances of quantifier elimination [2, 3, 5–7, 12, 19, 21–23] is beyond the scope of this book, one useful procedure for quantifier elimination in real arithmetic will be investigated in Chaps. 20 and 21.

Again, both sides are easily seen to be equivalent, because all x ensure $ax + b \neq 0$ only if b is nonzero and no x can cancel b since $a = 0$. This proves the validity:

$$\vDash \forall x\,(ax + b \neq 0) \leftrightarrow (a = 0 \wedge b \neq 0)$$

Overall, if we have quantifiers, QE can remove them for us. But we first need such quantifiers. Rules $\forall$R,$\exists$R,$\forall$L,$\exists$R went through a lot of trouble to get rid of the quantifiers in the first place. Oh my! That makes it kind of hard to eliminate them equivalently later on. Certainly the proof rules in Fig. 6.4 have not been particularly careful about eliminating quantifiers equivalently. Just think of what might happen if we did try to use rule $\exists$R with the wrong witness. That is certainly cheaper than quantifier elimination, but hardly as precise and useful.

Yet, if we misplaced a quantifier using the ordinary quantifier rules from Fig. 6.4, then all we need to do is to dream it up again and we are back in business for eliminating quantifiers by QE. The key to understanding how that works is to recall that the fresh (Skolem) variable symbols introduced by rule $\forall$R were originally universal. And, in fact, whether they were or were not, we can always prove a property by proving it with an extra universal quantifier $\forall x$ in front.

Lemma 6.4 (i$\forall$ reintroducing universal quantifiers). *This rule is sound:*

$$\text{i}\forall \; \frac{\Gamma \vdash \forall x P, \Delta}{\Gamma \vdash P, \Delta}$$

With the rule i$\forall$, we can reintroduce a universal quantifier, which can then promptly be eliminated again by QE.

Example 6.5. Together with rule i$\forall$, quantifier elimination can decide whether FOL$_\mathbb{R}$ formula $\exists x(ax+b=0)$ is valid. The equivalence (6.7) already indicates that there are values of a and b that falsify $\exists x(ax+b=0)$, because there are values that falsify the equivalent formula $a \neq 0 \vee b = 0$. The direct way to decide this formula by quantifier elimination first uses i$\forall$ for the remaining free variables a, b and then handles the fully quantified universal closure by quantifier elimination to obtain a quantifier-free equivalent (with the same free variables, so none):

$$\text{QE}(\forall a \forall b \exists x(ax+b=0)) \equiv \textit{false}$$

So rule i$\forall$ can reintroduce a universal quantifier, which can then be eliminated again by QE. Wait, why did it make sense to first swallow a quantifier with the lightweight rule $\forall$R and then later reintroduce it with rule i$\forall$ and then eliminate it once again with the big steamroller in the form of QE?

Before you read on, see if you can find the answer for yourself.

It can be pretty useful to get quantifiers out of the way first using the quick rules $\forall$R,$\exists$R,$\forall$L,$\exists$L, because other sequent rules such as propositional rules only work at the top level, so quantifiers need to be moved out of the way before any other proof rules can be applied.[7] If the formula underneath the quantifier contains modalities with hybrid programs, then it is too much to ask QE to solve them for us as well. The key is to first get rid of quantifiers by using extra symbols, work out the proof arguments for the remaining hybrid program modalities and then reintroduce quantifiers using i$\forall$ to ask QE for the answer to the remaining real arithmetic.

Example 6.6. The following sequent proof illustrates how a quantifier is first handled by rule $\forall$R, then dynamic axioms handle the modalities and finally a universal quantifier is reintroduced using rule i$\forall$ before quantifier elimination proves the resulting arithmetic. In fact, the top most use of rule i$\forall$ also introduces a universal

---

[7] The exception are contextual equivalence rules CER,CEL, which, fortunately, can even proceed within the context of a quantifier. This can be particularly helpful for existential quantifiers.

quantifier for x, which was never quantified in the original goal. All free variables are implicitly universally quantified, which fits with the fact that we seek to prove validity, so truth in all states for all real values of all variables. Besides, rule i$\forall$ can always introduce a universal quantifier to prove a formula if that succeeds.

$$
\dfrac{
\dfrac{
\dfrac{
\dfrac{
\dfrac{
\dfrac{
\dfrac{*}{\vphantom{X}\ \ \vdash \forall x \forall d\,(d \geq -x \rightarrow 0 \geq 0 \wedge x+d \geq 0)}\ \mathbb{R}
}{\vdash \forall d\,(d \geq -x \rightarrow 0 \geq 0 \wedge x+d \geq 0)}\ \text{i}\forall
}{\vdash d \geq -x \rightarrow 0 \geq 0 \wedge x+d \geq 0}\ \text{i}\forall
}{\vdash d \geq -x \rightarrow 0 \geq 0 \wedge [x:=x+d]\,x \geq 0}\ [:=]
}{\vdash d \geq -x \rightarrow [x:=0]\,x \geq 0 \wedge [x:=x+d]\,x \geq 0}\ [:=]
}{\vdash d \geq -x \rightarrow [x:=0 \cup x:=x+d]\,x \geq 0}\ [\cup]
}{\vdash \forall d\,(d \geq -x \rightarrow [x:=0 \cup x:=x+d]\,x \geq 0)}\ \forall\mathbb{R}
$$

While this is a rather canonical proof structure, dynamic axioms can be applied anywhere. So, in this case, we could have skipped the rule $\forall$R and directly apply the dynamic axioms, bypassing also the need to reintroduce $\forall d$ using rule i$\forall$ later.

Example 6.7. Even if quantifier elimination handles existential quantifiers just as well as universal quantifiers, some care is needed with existential quantifiers. The additional complication is that when we turn an existential quantifier into a witness with rule $\exists$R, even with a variable as a witness, then there is no way of getting said existential quantifier back later, but only a stronger universal quantifier using rule i$\forall$. A formula with a genuine existential quantifier, though, usually cannot be proved by using the same formula with a universal quantifier instead, even if it would be sound to do so. That is why the following sequent proof uses dynamic axioms in the middle of the formula directly until the remaining formula is pure arithmetic such that rule $\mathbb{R}$ can handle it:

$$
\dfrac{
\dfrac{
\dfrac{
\dfrac{
\dfrac{*}{\vphantom{X}\ \ \vdash \forall x\,\big(x \geq 0 \rightarrow \exists d\,(d \geq 0 \wedge 0 \geq 0 \wedge x+d \geq 0)\big)}\ \mathbb{R}
}{\vdash x \geq 0 \rightarrow \exists d\,(d \geq 0 \wedge 0 \geq 0 \wedge x+d \geq 0)}\ \text{i}\forall
}{\vdash x \geq 0 \rightarrow \exists d\,(d \geq 0 \wedge 0 \geq 0 \wedge [x:=x+d]\,x \geq 0)}\ [:=]
}{\vdash x \geq 0 \rightarrow \exists d\,(d \geq 0 \wedge [x:=0]\,x \geq 0 \wedge [x:=x+d]\,x \geq 0)}\ [:=]
}{\vdash x \geq 0 \rightarrow \exists d\,(d \geq 0 \wedge [x:=0 \cup x:=x+d]\,x \geq 0)}\ [\cup]
$$

6.5.2 Instantiating Real-Arithmetic Quantifiers

Real arithmetic can be very challenging. That does not come as a surprise, because cyber-physical systems and the behavior of dynamical systems themselves is challenging. On the contrary, it is pretty amazing that differential dynamic logic reduces challenging questions about CPSs to just plain real arithmetic. Of course, that means that you may be left with challenging arithmetic, of quite noticeable computational complexity. This is one part where you can use your creativity to master challenging verification questions by helping the KeYmaera X prover figure them out. While there will soon be more tricks in your toolbox to overcome the challenges of arithmetic, we discuss some of them in this chapter.

Providing instantiations for quantifier rules ∃R,∀L can significantly speed up real-arithmetic decision procedures. The proof in Sect. 6.4 instantiated the universal quantifier ∀s for an evolution domain constraint by the endpoint r of the time interval using quantifier proof rule ∀L. This is a very common simplification that speeds up arithmetic significantly (Note 36). It does not always work, though, because the instance one guesses may not always be the right one. Even worse, there may not always be a single instance that is sufficient for the proof.

Note 36 (Extreme instantiation) The proof rules ∀L for universal quantifiers in the antecedent and ∃R for existential quantifiers in the succedent allow instantiation of the quantified variable x with any term e. Such an instantiation is very helpful if only a single instance e is important for the argument.

For quantifiers coming from the handling of evolution domains in axiom ['] from Lemma 5.4, most uses only require a single time instance, where an extremal value for time is often all it takes. The proof steps that often helps then is instantiation of the intermediate time s by the end time t:

$$
\dfrac{
  \dfrac{
    \dfrac{
      \dfrac{
        \dfrac{
          \dfrac{
            \dfrac{\quad * \quad}{^{\mathbb{R}}\Gamma,t{\ge}0 \vdash 0{\le}t{\le}t,\ldots} \qquad \overline{\Gamma,t{\ge}0,q(y(t)) \vdash [x:=y(t)]p(x)}
          }{{}^{{\to}\mathrm{L}}\,\overline{\Gamma,t{\ge}0,0{\le}t{\le}t \to q(y(t)) \vdash [x:=y(t)]p(x)}}
        }{{}^{\forall\mathrm{L}}\,\overline{\Gamma,t{\ge}0,\forall 0{\le}s{\le}t\,q(y(s)) \vdash [x:=y(t)]p(x)}}
      }{{}^{{\to}\mathrm{R}}\,\overline{\Gamma,t{\ge}0 \vdash (\forall 0{\le}s{\le}t\,q(y(s))) \to [x:=y(t)]p(x)}}
    }{{}^{{\to}\mathrm{R}}\,\overline{\Gamma \vdash t{\ge}0 \to \big((\forall 0{\le}s{\le}t\,q(y(s))) \to [x:=y(t)]p(x)\big)}}
  }{{}^{\forall\mathrm{R}}\,\overline{\Gamma \vdash \forall t{\ge}0\big((\forall 0{\le}s{\le}t\,q(y(s))) \to [x:=y(t)]p(x)\big)}}
}{{}^{[']}\,\overline{\Gamma \vdash [x'=f(x)\,\&\,q(x)]p(x)}}
$$

This happens so frequently that KeYmaera X defaults to just using this instantiation. Similar instantiations can simplify arithmetic in other cases as well.

6.5.3 Weakening Real Arithmetic by Removing Assumptions

It can be very useful to just drop arithmetic assumptions that are irrelevant for the proof to make sure they are no distraction for real-arithmetic decision procedures.

In the proof in Sect. 6.4, the left premise was

$$A, r{\ge}0 \vdash 0{\le}r{\le}r$$

The proof of this sequent did not make use of A at all. Here, the proof worked easily. But if A were a very complicated formula, then proving the same sequent might have been very difficult, because our proving attempts could have been distracted by the presence of A and all the lovely assumptions it provides. We might have applied lots of proof rules to A before finally realizing that the sequent is proved because of $r \geq 0 \vdash 0 \leq r \leq r$ alone.

While quantifier elimination is not based on applying propositional proof rules, unnecessary assumptions can still cause considerable distraction [14, Chapter 5]. Think of how much easier it is to see that $ax^2 + bx \geq 0$ is true if somebody only tells you the relevant assumptions $a \geq 0, bx \geq 0$ rather than listing a lot of other true but presently useless assumptions about the values of a, b, and x. Consequently, it often saves a lot of proof effort to simplify irrelevant assumptions away as soon as they have become unnecessary. Fortunately, sequent calculus already comes with a general-purpose proof rule for the job called weakening (WL,WR, which will be elaborated in Sect. 6.5.4), which we can use on our example from the left premise in the proof of Sect. 6.4 to remove the assumption A:

$$\text{WL} \frac{r \geq 0 \vdash 0 \leq r \leq r}{A, r \geq 0 \vdash 0 \leq r \leq r}$$

> **Note 37 (Occam's assumption razor)** Think how hard it would be to prove a theorem with all the facts in all books of mathematics as assumptions. Compare this to a proof from just the two facts that matter for that proof.

You are generally advised to get rid of assumptions that you no longer need. This will help you manage the relevant facts about your CPS, will make sure you stay on top of your CPS agenda, and will also help the arithmetic in KeYmaera X to succeed much more quickly. Just be careful not to discard an assumption that you still need. But if you accidentally do, then that alone can also be a valuable insight, because you just found out what the safety of your system critically depends on.

Finally, recall how the real-arithmetic proof of the first premise in Note 36 did not need the potentially long list of unnecessary assumptions in Γ. And, in fact, the proof also weakened away the modal formula $[x := y(t)]p(x)$ from the sequent with WR to make the sequent arithmetic and amenable to real-arithmetic rule $\mathbb{R}$.

6.5.4 Structural Proof Rules in Sequent Calculus

The antecedent and succedent of a sequent are considered as sets. That implies that the order of formulas is irrelevant, and we implicitly adopt what is called the *exchange rule* and do not distinguish between the following two sequents

$$\Gamma, A, B \vdash \Delta \quad \text{and} \quad \Gamma, B, A \vdash \Delta$$

ultimately since $A \wedge B$ and $B \wedge A$ are equivalent. Nor do we distinguish between

$$\Gamma \vdash C, D, \Delta \quad \text{and} \quad \Gamma \vdash D, C, \Delta$$

ultimately since $C \vee D$ and $D \vee C$ are equivalent. Antecedent and succedent are considered to be sets, not multisets, so we implicitly adopt what is called the *contraction rule* and do not distinguish between the two sequents

$$\Gamma, A, A \vdash \Delta \qquad \text{and} \qquad \Gamma, A \vdash \Delta$$

because $A \wedge A$ and A are equivalent. It does not matter whether we make an assumption A once or multiple times. Nor do we distinguish between

$$\Gamma \vdash C, C, \Delta \qquad \text{and} \qquad \Gamma \vdash C, \Delta$$

because $C \vee C$ and C are equivalent. We could adopt these exchange rules and contraction rules explicitly, but usually leave them implicit:

$$\text{PR} \ \frac{\Gamma \vdash Q, P, \Delta}{\Gamma \vdash P, Q, \Delta} \qquad \text{cR} \ \frac{\Gamma \vdash P, P, \Delta}{\Gamma \vdash P, \Delta}$$

$$\text{PL} \ \frac{\Gamma, Q, P \vdash \Delta}{\Gamma, P, Q \vdash \Delta} \qquad \text{cL} \ \frac{\Gamma, P, P \vdash \Delta}{\Gamma, P \vdash \Delta}$$

The only structural rule of sequent calculus that we will find reason to use explicitly in practice is the *weakening* proof rule (alias *hide* rule) that can be used to remove formulas from the antecedent (WL) or succedent (WR), respectively:

$$\text{WR} \ \frac{\Gamma \vdash \Delta}{\Gamma \vdash P, \Delta}$$

$$\text{WL} \ \frac{\Gamma \vdash \Delta}{\Gamma, P \vdash \Delta}$$

Weakening rules are sound, since it is always fine to prove a sequent with more formulas in the antecedent or succedent by a proof that uses only some of those formulas. Proof rule WL proves the conclusion $\Gamma, P \vdash \Delta$ from the premise $\Gamma \vdash \Delta$, which dropped the assumption P. Surely, if premise $\Gamma \vdash \Delta$ is valid, then conclusion $\Gamma, P \vdash \Delta$ is valid as well, because it even has one more (unused) assumption available, namely P. Proof rule WR proves the conclusion $\Gamma \vdash P, \Delta$ from the premise $\Gamma \vdash \Delta$, which is fine because $\Gamma \vdash \Delta$ just has one less (disjunctive) option in its succedent. To see why that is sound, recall the disjunctive meaning of succedents.

At first sight, weakening may sound like a stupid thing to do in any proof, because rule WL discards available assumptions (P in the antecedent) and rule WR discards available options (P in the succedent) for proving the statement. This seems to make it harder to prove the statement after using a weakening rule. But weakening is actually useful for managing computational and conceptual proof complexity by enabling us to throw away irrelevant assumptions. These assumptions may have been crucial for another part of the proof, but have just become irrelevant for the particular sequent at hand, which can, thus, be simplified to $\Gamma \vdash \Delta$. Weakening, thus, streamlines proofs, which also helps speed up arithmetic immensely (Sect. 6.5.3).

Of course, the opposite of the weakening rules would be terribly unsound. We cannot just invent extra assumptions out of thin air just because we feel like wanting to have them at our disposal. But once we have the assumptions, we are free to not use them. That is, the premise of WL implies the conclusion but *not* vice versa.

6.5.5 Substituting Equations into Formulas

If we have an equation $x = e$ among our assumptions (in the antecedent), it is often significantly more efficient to use that equation for substituting e for all other occurrences of x instead of waiting for a real-arithmetic decision procedure to figure this out. If we have $x = e$ among our assumptions, then any (free) occurrence of x can be replaced by e, both in the succedent as well as in the antecedent:

$$=R \ \frac{\Gamma, x = e \vdash p(e), \Delta}{\Gamma, x = e \vdash p(x), \Delta} \qquad =L \ \frac{\Gamma, x = e, p(e) \vdash \Delta}{\Gamma, x = e, p(x) \vdash \Delta}$$

It would be okay to use the equation in the other direction for replacing all occurrences of e by x, because the equation $e = x$ is equivalent to $x = e$ by symmetry. Both proof rules, =R and =L, apply an equation $x = e$ from the antecedent to an occurrence of x in the antecedent or succedent to substitute e for x. By using the proof rule sufficiently often, multiple occurrences of x in Γ and Δ can be substituted. Especially if x does not occur in e, then using the proof rules =R,=L exhaustively and weakening $x = e$ away by rule WL removes the variable x entirely, which is what quantifier elimination will otherwise have to achieve by a complex algorithm.

Quantifier elimination would have been able to prove the same fact, but with significantly more time and effort. So you are advised to exploit these proof shortcuts whenever you spot them. Of course, KeYmaera X is clever enough to spot certain uses of equality rewriting as well, but you may be a better judge of how you would like to structure your proof, because you are more familiar with your CPS of interest.

6.5.6 Abbreviating Terms to Reduce Complexity

The opposite of exhaustively substituting in equations by rules =L,=R can also be helpful sometimes. When there are complicated terms whose precise relation to the other variables is not important, then a new variable can be introduced as an abbreviation for the complicated term.

For example, the following sequent looks complicated but becomes easy when we abbreviate all occurrences of the complex term $\frac{a}{2}t^2 + vt + x$ by a new variable z:

$$a \geq 0, v \geq 0, t \geq 0, 0 \leq \underbrace{\frac{a}{2}t^2 + vt + x}_{z}, \underbrace{\frac{a}{2}t^2 + vt + x}_{z} \leq d, d \leq 10 \vdash \underbrace{\frac{a}{2}t^2 + vt + x}_{z} \leq 10$$

The sequent resulting from that abbreviation lost how exactly the value of the new variable z relates to the values of a, t, v, x but exposes the simple transitivity argument that easily proves the sequent by rule $\mathbb{R}$:

$$a \geq 0, v \geq 0, t \geq 0, 0 \leq z, z \leq d, d \leq 10 \vdash z \leq 10$$

This is especially obvious after another few weakening steps to discard the now obviously irrelevant assumptions $a \geq 0, v \geq 0, t \geq 0$.

A proof rule for introducing such abbreviations will be investigated in Chap. 12. In fact, the proof rule for introducing such abbreviations will turn out to be just the inverse of another useful proof rule for assignments, rule $[:=]_=$.

Using the assignment axiom $[:=]$ will substitute the right-hand side e of an assignment $x := e$ for the variable x (if that is admissible). An alternative is the equational assignment proof rule $[:=]_=$, which turns an assignment $x := e$ into an equation $y = e$ for a fresh variable y that has not been used in the sequent yet.

Lemma 6.5 ($[:=]_=$ equational assignment rule). *This is a derived rule:*

$$[:=]_= \quad \frac{\Gamma, y = e \vdash p(y), \Delta}{\Gamma \vdash [x := e]p(x), \Delta} \quad (y \text{ new})$$

Proof. The proof deriving rule $[:=]_=$ from the other axioms and proof rules, especially the assignment axiom $[:=]$, is shown in prior work [18, Theorem 40]. $\square$

Of course, it is important for soundness of rule $[:=]_=$ that the variable y is fresh and not used in Γ, Δ, or even e, because the following would, otherwise, incorrectly prove an invalid formula from a premise that is only valid because of the impossible assumption $y = y + 1$ that rule $[:=]_=$ introduces:

$$\frac{y = y + 1 \vdash y > 5}{\vdash [x := y + 1]x > 5}$$

6.5.7 Creatively Cutting Real Arithmetic to Transform Questions

Weakening is not the only propositional proof rule that can help accelerate arithmetic. The cut rule is not just a logical curiosity, but can actually be shockingly helpful in practice [4]. It can speed up real arithmetic a lot to use a cut to replace a difficult arithmetic formula with a simpler one that is sufficient for the proof.

For example, suppose $p(x)$ is a big and very complicated formula of first-order real arithmetic. Then proving the following formula

$$(x - y)^2 \leq 0 \wedge p(y) \rightarrow p(x)$$

by real arithmetic will turn out to be surprisingly difficult and can take ages (even if it ultimately terminates). Yet, upon closer inspection, $(x - y)^2 \leq 0$ implies that $y = x$, which makes the rest of the proof easy since $p(y)$ easily implies $p(x)$ if, indeed, $x = y$. How do we exhibit a proof based on these thoughts?

The critical idea for such a proof work is to use a creative cut with suitable arithmetic. Choosing $x = y$ as the cut formula C, we use the rule cut and proceed:

$$
\cfrac{
\cfrac{
\cfrac{
\cfrac{
\cfrac{\ast}{(x-y)^2 \leq 0 \vdash x = y}\;\mathbb{R}
}{(x-y)^2 \leq 0 \vdash x = y, p(x)}\;\text{WR}
}{(x-y)^2 \leq 0, p(y) \vdash x = y, p(x)}\;\text{WL}
\qquad
\cfrac{
\cfrac{
\cfrac{\ast}{p(y), x = y \vdash p(y)}\;\text{id}
}{p(y), x = y \vdash p(x)}\;\text{=R}
}{(x-y)^2 \leq 0, p(y), x = y \vdash p(x)}\;\text{WL}
}{
\cfrac{
\cfrac{(x-y)^2 \leq 0, p(y) \vdash p(x)}{(x-y)^2 \leq 0 \wedge p(y) \vdash p(x)}\;\wedge\text{L}
}{\vdash (x-y)^2 \leq 0 \wedge p(y) \to p(x)}\;\to\text{R}
}\;\text{cut}
$$

Indeed, the left premise is proved easily using real arithmetic. The right premise is proved trivially by the equality substitution proof rule =R to propagate that x is y, and then rule id. Observe that proofs like this one benefit substantially from weakening to get rid of superfluous assumptions, thereby simplifying the resulting arithmetic.

6.6 Summary

The sequent proof rules for differential dynamic logic that this chapter showed are summarized in Fig. 6.6. They are sound [13, 16, 18].

> **Theorem 6.1 (Soundness).** *The* dL *sequent calculus is sound. That is, if a* dL *formula P has a proof in* dL*'s sequent calculus, i.e.,* $\vdash_{dL} P$*, then P is valid, i.e.,* $\vDash P$*.*

The primary responsibility of the sequent calculus is to organize our thoughts and proofs to ensure that a proof is never finished before all cases in all premises are proved. Its most crucial aspect is its direct ability to use all the dynamic axioms from Chap. 5 and later parts of this book by replacing one side of the equivalences with the other. Formally, this is what the contextual equivalence rules CEL,CER allow, but we can also happily work with the mental model of substituting equals for equals. That is, any equivalence $P \leftrightarrow Q$ proved from the axioms allows us to replace P with Q.

There are further proof rules of differential dynamic logic that later chapters will examine [13, 15, 16, 18], but this chapter laid a rock-solid foundation for CPS verification. In addition to having seen the foundation and working principles of how systematic CPS proofs assemble arguments, this chapter discussed techniques to tame the complexity of real arithmetic.

$$\neg R \frac{\Gamma, P \vdash \Delta}{\Gamma \vdash \neg P, \Delta} \qquad \wedge R \frac{\Gamma \vdash P, \Delta \quad \Gamma \vdash Q, \Delta}{\Gamma \vdash P \wedge Q, \Delta} \qquad \vee R \frac{\Gamma \vdash P, Q, \Delta}{\Gamma \vdash P \vee Q, \Delta}$$

$$\neg L \frac{\Gamma \vdash P, \Delta}{\Gamma, \neg P \vdash \Delta} \qquad \wedge L \frac{\Gamma, P, Q \vdash \Delta}{\Gamma, P \wedge Q \vdash \Delta} \qquad \vee L \frac{\Gamma, P \vdash \Delta \quad \Gamma, Q \vdash \Delta}{\Gamma, P \vee Q \vdash \Delta}$$

$$\rightarrow R \frac{\Gamma, P \vdash Q, \Delta}{\Gamma \vdash P \rightarrow Q, \Delta} \qquad id \frac{}{\Gamma, P \vdash P, \Delta} \qquad TR \frac{}{\Gamma \vdash true, \Delta} \qquad WR \frac{\Gamma \vdash \Delta}{\Gamma \vdash P, \Delta}$$

$$\rightarrow L \frac{\Gamma \vdash P, \Delta \quad \Gamma, Q \vdash \Delta}{\Gamma, P \rightarrow Q \vdash \Delta} \qquad cut \frac{\Gamma \vdash C, \Delta \quad \Gamma, C \vdash \Delta}{\Gamma \vdash \Delta} \qquad \perp L \frac{}{\Gamma, false \vdash \Delta} \qquad WL \frac{\Gamma \vdash \Delta}{\Gamma, P \vdash \Delta}$$

$$\forall R \frac{\Gamma \vdash p(y), \Delta}{\Gamma \vdash \forall x\, p(x), \Delta} \;(y \notin \Gamma, \Delta, \forall x\, p(x)) \qquad \exists R \frac{\Gamma \vdash p(e), \Delta}{\Gamma \vdash \exists x\, p(x), \Delta} \;(\text{arbitrary term } e)$$

$$\forall L \frac{\Gamma, p(e) \vdash \Delta}{\Gamma, \forall x\, p(x) \vdash \Delta} \;(\text{arbitrary term } e) \qquad \exists L \frac{\Gamma, p(y) \vdash \Delta}{\Gamma, \exists x\, p(x) \vdash \Delta} \;(y \notin \Gamma, \Delta, \exists x\, p(x))$$

$$CER \frac{\Gamma \vdash C(Q), \Delta \quad \vdash P \leftrightarrow Q}{\Gamma \vdash C(P), \Delta} \qquad =R \frac{\Gamma, x = e \vdash p(e), \Delta}{\Gamma, x = e \vdash p(x), \Delta}$$

$$CEL \frac{\Gamma, C(Q) \vdash \Delta \quad \vdash P \leftrightarrow Q}{\Gamma, C(P) \vdash \Delta} \qquad =L \frac{\Gamma, x = e, p(e) \vdash \Delta}{\Gamma, x = e, p(x) \vdash \Delta}$$

Fig. 6.6 Proof rules of the dL sequent calculus considered in this chapter

Exercises

6.1. Prove the soundness of the following special purpose proof rule and use it to continue the proof in Fig. 6.3 similarly to the proof in Fig. 6.2:

$$\frac{\Gamma, \theta > 0 \vdash \Delta}{\Gamma, -(-\theta) > 0 \vdash \Delta}$$

6.2 (*). Since we are not adding the proof rule from Exercise 6.1 to the dL proof calculus, show how you can derive the same proof step using a creative combination of arithmetic and the other proof rules.

6.3. The sequent calculus proof in Fig. 6.2 proves the following dL formula

$$v^2 \leq 10 \wedge b > 0 \rightarrow b > 0 \wedge (\neg(v \geq 0) \vee v^2 \leq 10)$$

Its proof only used propositional sequent calculus rules and no arithmetic or dynamic axioms. What does that mean about the validity of the following formula with the same propositional structure?

$$x^5 = y^2 + 5 \wedge a^2 > c^2 \rightarrow a^2 > c^2 \wedge (\neg(z < x^2) \vee x^5 = y^2 + 5)$$

6.4 (Bouncing-ball sequent proof). Using just dL axioms and arithmetic, Sect. 5.4 showed a proof of a single-hop bouncing-ball formula:

$$0 \leq x \wedge x = H \wedge v = 0 \wedge g > 0 \wedge 1 \geq c \geq 0 \rightarrow$$
$$\left[\{x' = v, v' = -g\}; (?x = 0; v := -cv \cup ?x \geq 0)\right] (0 \leq x \wedge x \leq H) \quad (5.14^*)$$

What is the minimal change to make this proof a proof in the dL sequent calculus? Additionally conduct a sequent calculus proof for formula (5.14) that only applies proof rules and axioms at the top level.

6.5 (Proof practice). Give dL sequent calculus proofs for the following formulas:

$$x > 0 \rightarrow [x := x + 1 \cup x' = 2] x > 0$$
$$x > 0 \wedge v \geq 0 \rightarrow [x := x + 1 \cup x' = v] x > 0$$
$$x > 0 \rightarrow [x := x + 1 \cup x' = 2 \cup x := 1] x > 0$$
$$[x := 1; (x := x + 1 \cup x' = 2)] x > 0$$
$$[x := 1; x := x - 1; x' = 2] x \geq 0$$
$$x \geq 0 \rightarrow [x := x + 1 \cup (x' = 2; ?x > 0)] x > 0$$
$$x^2 \geq 100 \rightarrow [(?x > 0; x' = 2) \cup (?x < 0; x' = -2)] x^2 \geq 100$$

6.6. Could we have used the following proof rule for $\wedge$ instead of rule $\wedge R$? Is it sound? Does it have any advantages or disadvantages compared to rule $\wedge R$?

$$\frac{\Gamma \vdash P, \Delta \quad \Gamma, P \vdash Q, \Delta}{\Gamma \vdash P \wedge Q, \Delta}$$

6.7 (Propositional soundness). Prove soundness for the structural and propositional sequent proof rules considered in Fig. 6.1.

6.8 (Bi-implication). Prove that these proof rules for bi-implication are sound:

$$\leftrightarrow R \; \frac{\Gamma, P \vdash Q, \Delta \quad \Gamma, Q \vdash P, \Delta}{\Gamma \vdash P \leftrightarrow Q, \Delta}$$

$$\leftrightarrow L \; \frac{\Gamma, P \rightarrow Q, Q \rightarrow P \vdash \Delta}{\Gamma, P \leftrightarrow Q \vdash \Delta}$$

6.9. Without alluding to dynamic axiom $[\cup]$ or contextual equivalence CER, give a direct semantical soundness proof for the following sequent proof rules:

$$[\cup]R \; \frac{\Gamma \vdash [\alpha]P \wedge [\beta]P, \Delta}{\Gamma \vdash [\alpha \cup \beta]P, \Delta} \qquad [\cup]R2 \; \frac{\Gamma \vdash [\alpha]P, \Delta \quad \Gamma \vdash [\beta]P, \Delta}{\Gamma \vdash [\alpha \cup \beta]P, \Delta}$$

$$[\cup]L \; \frac{\Gamma, [\alpha]P \wedge [\beta]P \vdash \Delta}{\Gamma, [\alpha \cup \beta]P \vdash \Delta} \qquad [\cup]L2 \; \frac{\Gamma, [\alpha]P, [\beta]P \vdash \Delta}{\Gamma, [\alpha \cup \beta]P \vdash \Delta}$$

6.10 (dL sequent proof rules). Develop dynamic sequent calculus proof rules for the modalities similar to either the rules $[\cup]R$ and $[\cup]L$ that this chapter discussed briefly but did not pursue or similar to the rules $[\cup]R2$ and $[\cup]L2$ from Exercise 6.9. Prove soundness for these sequent calculus proof rules. You can use a general argument that soundness of the dynamic sequent proof rules follows from soundness of the dL axioms considered in Chap. 5, but you first need to prove soundness of those dL axioms (Exercise 5.10).

6.11. If we define the formula *true* as $1 > 0$ and the formula *false* as $1 > 2$, then are the proof rules $\top$R and $\bot$L derivable from the other proof rules?

6.12. Let $y(t)$ be the solution at time t of the differential equation $x' = f(x)$ with initial value $y(0) = x$. Show that the following sequent proof rule, which checks the evolution domain $q(x)$ at the end, is sound:

$$\frac{\Gamma \vdash \forall t \geq 0 \left([x := y(t)](q(x) \to p(x))\right), \Delta}{\Gamma \vdash [x' = f(x) \,\&\, q(x)]p(x), \Delta}$$

Would the following also be a sound axiom? Prove or disprove.

$$[x' = f(x) \,\&\, Q]P \leftrightarrow \forall t \geq 0 \left([x := y(t)](Q \to P)\right)$$

Is the following sequent proof rule sound, which checks the evolution domain $q(x)$ at the beginning and at the end?

$$\frac{\Gamma \vdash \forall t \geq 0 \left(q(x) \to [x := y(t)](q(x) \to p(x))\right), \Delta}{\Gamma \vdash [x' = f(x) \,\&\, q(x)]p(x), \Delta}$$

6.13 (*). Generalize the solution axiom schema $[']$ for differential equations from Chap. 5 to the case of systems of differential equations:

$$x'_1 = e_1, .., x'_n = e_n \,\&\, Q$$

First consider the easier case where $Q \equiv true$ and $n = 2$.

6.14 (MR monotonicity right rule). Prove that the following formulation of the monotonicity rule is sound. Either give a direct semantical soundness proof or derive rule MR from rule M$[\cdot]$ from Lemma 5.13.

$$\text{MR} \quad \frac{\Gamma \vdash [\alpha]Q, \Delta \quad Q \vdash P}{\Gamma \vdash [\alpha]P, \Delta}$$

6.15. Sect. 5.2 argued why the following proof rule is sound

$$\text{H;} \quad \frac{A \to [\alpha]E \quad E \to [\beta]B}{A \to [\alpha; \beta]B}$$

Prove that rule H; is indeed sound. Would the following be a sound axiom? Or can you find a counterexample?

$$[\alpha; \beta]B \leftrightarrow ([\alpha]E) \wedge (E \to [\beta]B)$$

6.16. By Sect. 6.5.1, quantifier elimination can be used to show the equivalence

$$\text{QE}(\exists x\,(ax+b=0)) \;\equiv\; (a\neq 0 \vee b=0) \qquad (6.7^*)$$

What is the result of applying quantifier elimination to $\exists x\,(ax^2+bx+c=0)$ instead?

6.17 (Derived propositional rules). Prove that the following rules are derived rules:

$$\text{cutR}\;\frac{\Gamma \vdash Q,\Delta \quad \Gamma \vdash Q \to P,\Delta}{\Gamma \vdash P,\Delta}$$

$$\text{cutL}\;\frac{\Gamma,Q \vdash \Delta \quad \Gamma \vdash P \to Q,\Delta}{\Gamma,P \vdash \Delta}$$

6.18 (Propositional completeness). A formula of propositional logic only uses the logical connectives $\wedge,\vee,\neg,\to$ and abstract atomic formulas such as p,q. For example, the propositional formula $p \wedge \neg q \to \neg(q \vee \neg p)$ is valid, whatever truth-values p and q have. Give an informal argument why every valid formula of propositional logic can be proved using the propositional sequent proof rules from Fig. 6.1.

References

[1] Peter B. Andrews. *An Introduction to Mathematical Logic and Type Theory: To Truth Through Proof.* 2nd. Dordrecht: Kluwer, 2002. DOI: 10.1007/978-94-015-9934-4.

[2] Saugata Basu, Richard Pollack, and Marie-Françoise Roy. *Algorithms in Real Algebraic Geometry.* 2nd. Berlin: Springer, 2006. DOI: 10.1007/3-540-33099-2.

[3] Jacek Bochnak, Michel Coste, and Marie-Francoise Roy. *Real Algebraic Geometry.* Vol. 36. Ergeb. Math. Grenzgeb. Berlin: Springer, 1998. DOI: 10.1007/978-3-662-03718-8.

[4] George Boolos. Don't eliminate cut. *Journal of Philosophical Logic* **13**(4) (1984), 373–378. DOI: 10.1007/BF00247711.

[5] George E. Collins. Quantifier elimination for real closed fields by cylindrical algebraic decomposition. In: *Automata Theory and Formal Languages.* Ed. by H. Barkhage. Vol. 33. LNCS. Berlin: Springer, 1975, 134–183. DOI: 10.1007/3-540-07407-4_17.

[6] George E. Collins and Hoon Hong. Partial cylindrical algebraic decomposition for quantifier elimination. *J. Symb. Comput.* **12**(3) (1991), 299–328. DOI: 10.1016/S0747-7171(08)80152-6.

[7] James H. Davenport and Joos Heintz. Real quantifier elimination is doubly exponential. *J. Symb. Comput.* **5**(1/2) (1988), 29–35. DOI: 10.1016/S0747-7171(88)80004-X.

[8]   Nathan Fulton, Stefan Mitsch, Jan-David Quesel, Marcus Völp, and André Platzer. KeYmaera X: an axiomatic tactical theorem prover for hybrid systems. In: *CADE*. Ed. by Amy Felty and Aart Middeldorp. Vol. 9195. LNCS. Berlin: Springer, 2015, 527–538. DOI: `10.1007/978-3-319-21401-6_36`.

[9]   Gerhard Gentzen. Untersuchungen über das logische Schließen I. *Math. Zeit.* **39**(2) (1935), 176–210. DOI: `10.1007/BF01201353`.

[10] Gerhard Gentzen. Untersuchungen über das logische Schließen II. *Math. Zeit.* **39**(3) (1935), 405–431. DOI: `10.1007/BF01201363`.

[11] Jacques Herbrand. Recherches sur la théorie de la démonstration. *Travaux de la Société des Sciences et des Lettres de Varsovie, Class III, Sciences Mathématiques et Physiques* **33** (1930), 33–160.

[12] Dejan Jovanović and Leonardo Mendonça de Moura. Solving non-linear arithmetic. In: *Automated Reasoning - 6th International Joint Conference, IJCAR 2012, Manchester, UK, June 26-29, 2012. Proceedings*. Ed. by Bernhard Gramlich, Dale Miller, and Ulrike Sattler. Vol. 7364. LNCS. Berlin: Springer, 2012, 339–354. DOI: `10.1007/978-3-642-31365-3_27`.

[13] André Platzer. Differential dynamic logic for hybrid systems. *J. Autom. Reas.* **41**(2) (2008), 143–189. DOI: `10.1007/s10817-008-9103-8`.

[14] André Platzer. *Logical Analysis of Hybrid Systems: Proving Theorems for Complex Dynamics*. Heidelberg: Springer, 2010. DOI: `10.1007/978-3-642-14509-4`.

[15] André Platzer. Logics of dynamical systems. In: *LICS*. Los Alamitos: IEEE, 2012, 13–24. DOI: `10.1109/LICS.2012.13`.

[16] André Platzer. The complete proof theory of hybrid systems. In: *LICS*. Los Alamitos: IEEE, 2012, 541–550. DOI: `10.1109/LICS.2012.64`.

[17] André Platzer. Differential game logic. *ACM Trans. Comput. Log.* **17**(1) (2015), 1:1–1:51. DOI: `10.1145/2817824`.

[18] André Platzer. A complete uniform substitution calculus for differential dynamic logic. *J. Autom. Reas.* **59**(2) (2017), 219–265. DOI: `10.1007/s10817-016-9385-1`.

[19] Abraham Seidenberg. A new decision method for elementary algebra. *Annals of Mathematics* **60**(2) (1954), 365–374. DOI: `10.2307/1969640`.

[20] Thoralf Skolem. Logisch-kombinatorische Untersuchungen über die Erfüllbarkeit oder Beweisbarkeit mathematischer Sätze nebst einem Theorem über dichte Mengen. *Videnskapsselskapets skrifter, 1. Mat.-naturv. klasse* **4** (1920), 1–36.

[21] Gilbert Stengle. A Nullstellensatz and a Positivstellensatz in semialgebraic geometry. *Math. Ann.* **207**(2) (1973), 87–97. DOI: `10.1007/BF01362149`.

[22] Alfred Tarski. *A Decision Method for Elementary Algebra and Geometry.* 2nd. Berkeley: University of California Press, 1951.

[23] Volker Weispfenning. Quantifier elimination for real algebra — the quadratic case and beyond. *Appl. Algebra Eng. Commun. Comput.* **8**(2) (1997), 85–101. DOI: `10.1007/s002000050055`.

Chapter 7
Control Loops & Invariants

Synopsis This chapter advances the analytical understanding of cyber-physical systems to cover control loops. While the syntax and semantics of hybrid programs from previous chapters already discussed loops, their logical characterization was so far limited to unfolding by the iteration axiom. That suffices for systems with a fixed finite number of control actions in a fixed finite number of repetitions of the control loop, but is not enough to understand and analyze the most interesting CPSs with unbounded time-horizons reaching an unbounded number of control decisions over time. This chapter uses the fundamental concept of invariants to handle loops and develops their operational intuition. CPS invariants are developed systematically based on inductive formulations of dynamic axioms for repetitions.

7.1 Introduction

Chap. 5 introduced rigorous reasoning for hybrid program models of cyber-physical systems, which Chap. 6 extended to a systematic and coherent reasoning approach for cyber-physical systems. Our understanding of the language exceeds our understanding of the reasoning principles, though, because we have not seen any credible ways of analyzing loops yet, despite the fact that loops are a perfectly harmless and common part of CPSs. In fact, computational thinking would argue that we do not truly understand an element of a programming language or a system model if we do not also understand ways of reasoning about them. This chapter sets out to make sure our analysis capabilities catch with on our modeling skills. This is, of course, all part of the agenda we set forth initially to study the language of cyber-physical systems gradually in layers that we master completely before advancing to the next challenge. The next challenge is control loops.

Chap. 3 demonstrated how important control is in CPS and that control loops are a very important feature for making this control happen. Without loops, CPS controllers are limited to short finite sequences of control actions, which are rarely sufficient to get our CPS anywhere. With loops, CPS controllers shine, because they

© Springer International Publishing AG, part of Springer Nature 2018
A. Platzer, *Logical Foundations of Cyber-Physical Systems*,
https://doi.org/10.1007/978-3-319-63588-0_7

can inspect the current state of the system, take action to control the system, let the physics evolve, and then repeat these steps in a loop over and over again to slowly get the state where the controller wants the system to be. Loops truly make feedback happen, by enabling a CPS to sense state and act in response to that over and over again. Think of programming a robot to drive on a highway. Would you be able to do that without some means of repetition or iteration as in repeated control? Probably not, because you would need to write a CPS program that monitors the traffic situation frequently and reacts in response to what the other cars do on the highway. There's no way of telling ahead of time, how often the robot will need to change its mind when it's driving a car on a highway.

A hybrid program's way of exercising repetitive control actions is the repetition operator *, which can be applied to any hybrid program α. The resulting hybrid program α^* repeats α any number of times, nondeterministically. That may be zero times or one time or 10 times or ....

Now, the flip side of the fact that control loops are responsible for a lot of the power of CPS is that they can also be tricky to analyze and fully understand. After all, it is easier to get a handle on what a system does in just one step than to understand what it will do in the long run when the CPS is running for any arbitrary amount of time. This is the CPS analogue of the fact that ultra-short-term predictions are often much easier than long-term predictions. It is easy to predict the weather a second into the future but much harder to predict next week's weather.[1]

The main insight behind the analysis of loops in CPS is to reduce the (complicated) analysis of their long-term global behavior to a simpler analysis of their local behavior for one control cycle. This principle significantly reduces the analytic complexity of loops in CPS. It leverages invariants, i.e., aspects of the system behavior that do not change as time progresses, so that our analysis can rely on them no matter how long the system already evolved. Invariants turn out also to lead to an important design principle for CPS, even more so than in programs. The significance of invariants in understanding CPS is not a coincidence, because the study of invariants (as with other mathematical structures) is also central to a large body of mathematics.

Since it is of central importance to develop a sense of how the parts of a proof fit together and what impact changes to preconditions or invariants have on a proof, this chapter will be very explicit about developing sequent calculus proofs to give you a chance to understand their structure. These proofs will also serve as a useful exercise to practice our skills on the sequent calculus reasoning for CPS that Chap. 6 developed. After some practice, subsequent chapters will often appeal in more intuitive ways to the canonical structure that a proof will have and focus on developing only its most crucial elements: invariants, because the remaining proof is relatively straightforward.

The most important learning goals of this chapter are:

---

[1] Of course, Nils Bohr already figured this out when he said that "prediction is very difficult, especially if it's about the future."

Modeling and Control: We develop a deeper understanding of control loops as a core principle behind CPS that ultimately underlies all feedback mechanisms in CPS control. This chapter also intensifyies our understanding of the dynamical aspects of CPS and how discrete and continuous dynamics interact.

Computational Thinking: This chapter extends the rigorous reasoning approach from Chap. 5 to systems with repetitions. This chapter is devoted to the development of rigorous reasoning techniques for CPS models with repetitive control loops or other loopy behavior, a substantially nontrivial problem in theory and practice. Without understanding loops, there is no hope of understanding the repetitive behavior of feedback control principles that are common to almost all CPSs. Understanding such behavior can be tricky, because so many things can change in the system and its environment over the course of the runtime of even just a few lines of code if that program runs repeatedly to control the behavior of a CPS. That is why the study of *invariants*, i.e., properties that do not change throughout the execution of the system, are crucial for their analysis. Invariants constitute the single most insightful and most important piece of information about a CPS. As soon as we understand the invariants of a CPS, we almost understand everything about it and will even be in a position to design the rest of the CPS around these invariants, a process known as the design-by-invariant principle. Identifying and expressing invariants of CPS models will be a part of this chapter as well.

The first part of the chapter shows a systematic development of invariance principles for loops from an axiomatic basis. The second part of the chapter focuses on loop invariants themselves along with their operational intuition.

Another aspect that this chapter reinforces is the important concept of global proof rules, which, just like Gödel's generalization rule G, for soundness reasons cannot keep the sequent context.

CPS Skills: We will develop a better understanding of the semantics of CPS models by understanding the core aspects of repetition and relating its semantics to corresponding reasoning principles. This understanding will lead us to develop a higher level of intuition for the operational effects involved in CPS by truly understanding what control loops fundamentally amount to.

7.2 Control Loops

Recall Quantum, the little acrophobic bouncing ball from Chap. 4:

$$\textbf{requires}(0 \leq x \wedge x = H \wedge v = 0)$$
$$\textbf{requires}(g > 0 \wedge 1 \geq c \geq 0)$$
$$\textbf{ensures}(0 \leq x \wedge x \leq H) \tag{4.25*}$$
$$\left(\{x' = v, v' = -g \,\&\, x \geq 0\};\right.$$
$$\left.\text{if}(x = 0)\, v := -cv\right)^*$$

rigorous reasoning for repetitions
identifying and expressing invariants
global vs. local reasoning
relating iterations to invariants
finitely accessible infinities
operationalize invariant construction
splitting & generalizations

control loops   semantics of control loops
feedback mechanisms   operational effects of control
dynamics of iteration

The contracts above have been augmented with the ones that we have identified in Chap. 4 by converting the initial contract specification into a logical formula in differential dynamic logic and then identifying the required assumptions to make it true in all states:

$$0 \leq x \wedge x = H \wedge v = 0 \wedge g > 0 \wedge 1 \geq c \geq 0 \rightarrow$$
$$\left[\left(\{x' = v, v' = -g \,\&\, x \geq 0\}; \; \text{if}(x = 0) \, v := -cv \right)^* \right] (0 \leq x \wedge x \leq H) \quad (4.23^*)$$

As we do not wish to be bothered by the presence of the additional if-then-else operator, which is not officially part of the minimal set of operators that differential dynamic logic dL provides, we rewrite (4.23) equivalently to:

$$0 \leq x \wedge x = H \wedge v = 0 \wedge g > 0 \wedge 1 \geq c \geq 0 \rightarrow$$
$$\left[\left(\{x' = v, v' = -g \,\&\, x \geq 0\}; \; (?x = 0; v := -cv \cup ?x \neq 0) \right)^* \right] (0 \leq x \wedge x \leq H) \quad (7.1)$$

In Chap. 4, we had an informal understanding why (7.1) is valid (true in all states), but no formal proof, albeit we proved a much simplified version of (7.1) in which we simply threw away the loop. Such ignorance is clearly not a correct way of understanding loops. Equipped with our refined understanding of what proofs are from Chap. 6, let's make up for that now by properly proving (7.1) in the dL calculus.

However, before going for a proof of this bouncing-ball property, however much Quantum may long for it, let us first take a step back and understand the rôle of loops in more general terms. Their semantics has been explored in Chap. 3 with unwinding-based reasoning in Chap. 5.

Quantum had a loop in which physics and its bouncing control alternated. Quantum desperately needs a loop for he doesn't know ahead of time how often he would bounce today. When falling from a great height, Quantum bounces quite a bit. Quantum also had a controller, albeit a rather impoverished one. All it can do is inspect the current height, compare it to the ground floor (at height 0) and, if $x = 0$, flip its velocity vector around after some casual damping by factor c. That is not a whole lot of flexibility for control choices, but Quantum was still rather proud to serve such an important rôle in controlling the ball's behavior. Indeed, without the control action, Quantum would never bounce back from the ground but would keep on falling forever—what a frightful thought for the acrophobic Quantum. On second thought Quantum would, actually, not even fall for very long without its controller, because of the evolution domain $x \geq 0$ for physics $x'' = -g \, \& \, x \geq 0$, which only allows physics to evolve for time zero if the ball is already at height 0, because gravity would otherwise try to pull it further down, except that the $x \geq 0$ constraint won't have it. So, in summary, without Quantum's control statement, it would simply fall and then lie flat on the ground without time being allowed to proceed. That would not sound very reassuring and certainly not as much fun as bouncing back up, so Quantum is really jolly proud of the controller.

This principle is not specific to the bouncing ball, but, rather, quite common in CPS. The controller performs a crucial task, without which physics would not evolve in the way that we want it to. After all, if physics did already always do what we want it to without any input from our side, we would not need a controller for it in the first place. Hence, control is crucial and understanding and analyzing its effect on physics is one of the primary responsibilities in CPS. After the implication in (7.1) is quickly consumed by the $\rightarrow$R proof rule, the trouble starts right away since Quantum needs to prove the safety of the loop.

7.3 Induction for Loops

This section develops induction principles for loops by systematically developing their intuition starting from the insights behind the iteration axiom.

7.3.1 Induction Axiom for Loops

Recall the loop semantics from Sect. 3.3.2 and its unwinding axiom from Sect. 5.3.7:

$$[\![\alpha^*]\!] = [\![\alpha]\!]^* = \bigcup_{n \in \mathbb{N}} [\![\alpha^n]\!] \qquad \text{with} \quad \alpha^{n+1} \equiv \alpha^n; \alpha \text{ and } \alpha^0 \equiv \,?true$$

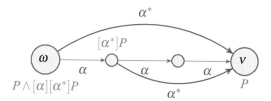

Lemma 5.7 ([*] **iteration axiom**). *The* iteration axiom *is sound:*

$$[*] \quad [\alpha^*]P \leftrightarrow P \wedge [\alpha][\alpha^*]P$$

Using the iteration axiom [*] from left to right, it "reduces" a safety property

$$[\alpha^*]P \tag{7.2}$$

of a loop α^* to the following equivalent dL formula:

$$P \wedge [\alpha][\alpha^*]P \tag{7.3}$$

The isolated left formula P and the $[\alpha]$ modality in the resulting formula (7.3) are simpler than the original (7.2) and could, thus, be analyzed using the other dL axioms. The only catch is that the postcondition $[\alpha^*]P$ of the $[\alpha]$ modality in (7.3) is as complicated as the original dL formula (7.2). While the iteration axiom [*] unpacked necessary conditions for the original repetition property (7.2), the true question of whether P always holds after repeating α any number of times remains, albeit nested within an extra $[\alpha]$. That does not look like a lot of progress in analyzing (7.2). In fact, it looks like using the iteration axiom [*] makes matters more complicated (unless perhaps a counterexample has been identified along the way). The iteration axiom [*] can still be useful to explicitly uncover the effect of one round of a loop.

Since (7.2) and (7.3) are equivalent, formula $[\alpha^*]P$ can only be true if P holds initially. So, if, in some state ω, we are trying to establish $\omega \in [[\alpha^*]P]$, then we only have a chance if the necessary condition $\omega \in [P]$ holds in the initial state ω. By the equivalent (7.3), $\omega \in [[\alpha^*]P]$ can also only hold if $\omega \in [[\alpha]P]$ since the loop in $[\alpha][\alpha^*]P$ may repeat 0 times (Exercise 7.2). So, we might as well establish the necessary condition $\omega \in [P \rightarrow [\alpha]P]$ since we already needed to assume $\omega \in [P]$. Showing the implication $P \rightarrow [\alpha]P$ in state ω is a little easier than showing $[\alpha]P$, because the implication assumes P. This shows $\mu \in [P]$ in any state μ after the *first* loop iteration, but since its α-successors will all also have to satisfy P for $\omega \in [[\alpha^*]P]$ to hold, we again need to show the same remaining condition $P \rightarrow [\alpha]P$, just in a different state μ.

If, instead, we manage to prove $P \rightarrow [\alpha]P$ *in all states* we get to by repeating α, not just the initial state ω, then we know P holds in all states after running α twice from ω, since we already know that P holds in all states μ after running α once from ω. By induction, no matter how often α is repeated, we know P is true afterwards if only P was true initially and $P \rightarrow [\alpha]P$ is always true after repeating α, i.e., $[\alpha^*](P \rightarrow [\alpha]P)$ is true in the current state, which is $\omega \in [[\alpha^*](P \rightarrow [\alpha]P)]$.

These thoughts lead to the induction axiom I expressing that a property P is always true after repeating HP α iff P is true initially and if, after any number of repetitions of α, P always holds after one more repetition of α if it held before.

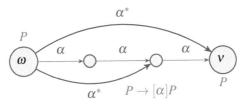

Lemma 7.1 (I induction axiom). *The induction axiom is sound:*

$$\text{I} \ [\alpha^*]P \leftrightarrow P \wedge [\alpha^*](P \to [\alpha]P)$$

Proof. Let $\omega \in [\![[\alpha^*]P]\!]$, then $\omega \in [\![P]\!]$ by choosing 0 iterations and $\omega \in [\![[\alpha^*][\alpha]P]\!]$ by choosing at least one iteration, which implies $\omega \in [\![[\alpha^*](P \to [\alpha]P)]\!]$. Conversely, let $\omega \in [\![P \wedge [\alpha^*](P \to [\alpha]P)]\!]$. Then consider a run of α^* from ω to ν with $n \in \mathbb{N}$ iterations, i.e., $(\omega, \nu) \in [\![\alpha^n]\!]$. The proof shows $\nu \in [\![P]\!]$ by induction on n (Fig. 7.1).

0. Case $n = 0$: Then $\nu = \omega$ satisfies $\nu \in [\![P]\!]$ by the first conjunct.
1. Case $n + 1$: By induction hypothesis for n, all states μ with $(\omega, \mu) \in [\![\alpha^n]\!]$ are assumed to satisfy $\mu \in [\![P]\!]$. Thus, $\mu \in [\![[\alpha]P]\!]$ by the second conjunct $\omega \in [\![[\alpha^*](P \to [\alpha]P)]\!]$ since $(\omega, \mu) \in [\![\alpha^n]\!] \subseteq [\![\alpha^*]\!]$. Hence, $\nu \in [\![P]\!]$ for all states ν with $(\mu, \nu) \in [\![\alpha]\!]$. Thus, $\nu \in [\![P]\!]$ for all states ν with $(\omega, \nu) \in [\![\alpha^{n+1}]\!]$. □

The $[\alpha^*]$ modality on the right-hand side of axiom I is necessary for soundness, because it would not be enough to merely show that the implication $P \to [\alpha]P$ is true in the current state. That is, the following formula would be an unsound axiom

$$[\alpha^*]P \leftrightarrow P \wedge (P \to [\alpha]P)$$

because its instance

$$[(x := x + 1)^*] x \leq 2 \leftrightarrow x \leq 2 \wedge (x \leq 2 \to [x := x + 1] x \leq 2)$$

is not true in a state ω with $\omega(x) = 0$, so it is also not valid. The $[\alpha^*]$ modality on the right-hand side of axiom I ensures that $P \to [\alpha]P$ is not just true in the current state, but true in all states reached after iterating the loop α^* any number of times.

7.3.2 Induction Rule for Loops

Even if axiom I has a pleasantly inductive flair to it, using it directly does not make matters any better compared to the iteration axiom $[^*]$. Using axiom I to prove a

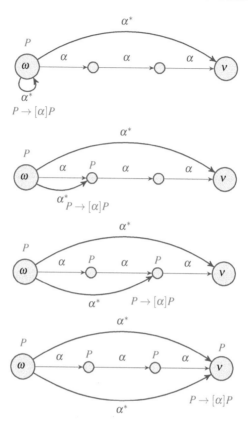

Fig. 7.1 Successively using induction axiom I at each state reached after running iterations of α^*

property of a loop (its left-hand side) will still reduce to proving a different property of a loop (its right-hand side). Why should the postcondition $P \to [\alpha]P$ be any easier to prove after the loop α^* than the original postcondition P?

The clou, however, is that the postcondition $P \to [\alpha]P$ in the right-hand side of induction axiom I can also be proved differently. Gödel's generalization rule G, which was already discussed in Sect. 5.6.3, provides a way of proving postconditions of arbitrary box modalities, even with loops, if the postcondition has a proof. Valid formulas (premise) are also true after all runs of any HP α (conclusion).

> **Lemma 5.12 (G Gödel generalization rule).** *The* Gödel rule *is sound:*
>
> $$G \; \frac{P}{[\alpha]P}$$

Generalization rule G can be used to prove $[\alpha^*](P \to [\alpha]P)$ by proving the post-condition $P \to [\alpha]P$. This leads to the induction rule, which reduces the proof that P always holds after repeating α (succedent of conclusion) provided that P was true initially (antecedent) to a proof of the induction step $P \to [\alpha]P$ (premise). The induction rule is a derived rule, i.e., it is proved from other axioms and proof rules.

> **Lemma 7.2 (ind induction rule).** *The loop induction rule ind is derived:*
>
> $$\text{ind } \frac{P \vdash [\alpha]P}{P \vdash [\alpha^*]P}$$

Proof. Derived rule ind derives from axiom I using rule G for the inductive step:

$$
\text{I} \cfrac{
  \text{$\wedge$R} \cfrac{
    \text{id} \cfrac{*}{P \vdash P}
    \qquad
    \text{G} \cfrac{
      \text{$\to$R} \cfrac{
        P \vdash [\alpha]P
      }{\vdash P \to [\alpha]P}
    }{P \vdash [\alpha^*](P \to [\alpha]P)}
  }{P \vdash P \wedge [\alpha^*](P \to [\alpha]P)}
}{P \vdash [\alpha^*]P}
$$

$\square$

The induction rule ind derives easily from the induction axiom I. Its premise expresses that P is inductive, i.e., true after all runs of α if P was true before. If P is inductive (premise), then P is always true after any number of repetitions of α^* (succedent of conclusion) if P is true initially (antecedent of the conclusion).

Loop induction rule ind requires the postcondition P to occur verbatim in the succedent's antecedent. But the rule does not directly apply for sequents $\Gamma \vdash [\alpha^*]P$ in which the antecedent Γ merely implies P but does not literally include it. The difference is easily overcome with a use of the cut rule, though, which, with a cut of P, can get the required formula P into the antecedent for the ind rule:

$$
\text{cut} \cfrac{
  \text{WR} \cfrac{\Gamma \vdash P, \Delta}{\Gamma \vdash P, [\alpha^*]P, \Delta}
  \qquad
  \text{WL,WR} \cfrac{
    \text{ind} \cfrac{P \vdash [\alpha]P}{P \vdash [\alpha^*]P}
  }{\Gamma, P \vdash [\alpha^*]P, \Delta}
}{\Gamma \vdash [\alpha^*]P, \Delta}
$$

Example 7.1. The only actual difference between the induction axiom I and the loop induction rule ind is that the latter already went a step further with the generalization rule G to discard the $[\alpha^*]$ modality, which makes rule ind more practical but also comes at a loss of precision. For example, the following simple dL formula is valid:

$$x \geq 0 \wedge v = 0 \to [(v := v+1; x' = v)^*] x \geq 0 \tag{7.4}$$

By induction axiom I, the valid formula (7.4) is equivalent to

$$x \geq 0 \wedge v = 0 \to x \geq 0 \wedge [(v := v+1; x' = v)^*](x \geq 0 \to [v := v+1; x' = v] x \geq 0) \tag{7.5}$$

Nevertheless, the induction step to which rule ind would reduce the proof of (7.4) is not valid:

$$x \geq 0 \to [v := v+1; x' = v] x \geq 0 \tag{7.6}$$

The reason why, unlike the formula (7.5) resulting from the induction axiom I, the induction step (7.6) resulting from rule ind is *not* valid is simply that rule ind discards the modality $[(v:=v+1;x':=v)^*]$ by Gödel generalization G. Discarding this modality misses out on its effect on the state, which changes the values of x and v. But since the change of x in (7.6) depends on the value of v, the postcondition $x \geq 0$ of (7.4) cannot possibly suffice for the induction step (7.6).

The formula for the induction step needs to be *strengthened* to retain information about the value of v always remaining nonnegative when discarding the repetition modality when rule ind tries to prove (7.5) by generalization. That is why the next section investigates ways of using the formula $x \geq 0 \land v \geq 0$ as an invariant to prove (7.4) even if its postcondition did not talk about v.

7.3.3 Loop Invariants

Even if the induction rule ind captures the core essentials of induction, that rule does not necessarily result in a successful proof. Unlike the induction axiom I, which is an equivalence, the premise of the induction rule ind does not have to be valid even if its conclusion is valid, because the Gödel generalization rule G, which is used to derive rule ind from axiom I, discards modality $[\alpha^*]$. Where axiom I was overly precise with its induction step after repetition, rule ind loses all information about the loop in the induction step. It can happen that the formula $P \to [\alpha]P$ is not valid in all states, but only true after repeating α any number of times (which is what the subformula $[\alpha^*](P \to [\alpha]P)$ in axiom I expresses). In other words, the truth of $P \to [\alpha]P$ might depend on a certain property that happens to always hold after repeating α, but does not follow from assumption P alone. But to establish such an auxiliary property to always hold after repeating α would also need a proof by induction just like P.

In fact, that phenomenon is quite familiar from mathematics. Some inductive proofs require a stronger formulation of the induction hypothesis for the proof to succeed. The proof of Fermat's Last Theorem is not an inductive proof assuming that $a^n + b^n \neq c^n$ for $n > 2$ has already been proved for all smaller natural numbers.

Fortunately, the monotonicity rule M[·], which was discussed in Sect. 5.6.4, already provides a way of suitably generalizing the postcondition of $[\alpha^*]$ to another formula for which the premise of induction rule ind will be proved successfully. If P implies Q (premise of M[·]), then $[\alpha]P$ implies $[\alpha]Q$ (conclusion).

Lemma 5.13 (M[·] monotonicity rule). *The monotonicity rules are sound:*

$$\text{M[·]} \quad \frac{P \to Q}{[\alpha]P \to [\alpha]Q} \qquad \text{M} \quad \frac{P \to Q}{\langle \alpha \rangle P \to \langle \alpha \rangle Q}$$

The monotonicity rule M[·] turns the bare-bones induction rule ind into the more useful loop invariant rule, which proves a safety property P of a loop α^* by proving that some loop invariant J is true initially (first premise), is inductive (second premise), and finally implies the original postcondition P (third premise).

Lemma 7.3 (Loop invariant rule). *The loop invariant rule is derived:*

$$\text{loop} \quad \frac{\Gamma \vdash J, \Delta \quad J \vdash [\alpha]J \quad J \vdash P}{\Gamma \vdash [\alpha^*]P, \Delta}$$

Proof. Rule loop is derived from the derived rule ind using a cut with $J \to [\alpha^*]J$ and weakening WL,WR (used without notice):

$$\text{cut} \frac{\text{→R} \dfrac{\text{ind} \dfrac{J \vdash [\alpha]J}{J \vdash [\alpha^*]J}}{\Gamma \vdash J \to [\alpha^*]J, \Delta} \qquad \text{→L} \dfrac{\Gamma \vdash J, \Delta \qquad \text{M}[\cdot] \dfrac{J \vdash P}{[\alpha^*]J \vdash [\alpha^*]P}}{\Gamma, J \to [\alpha^*]J \vdash [\alpha^*]P, \Delta}}{\Gamma \vdash [\alpha^*]P, \Delta}$$

□

First observe that the *inductive invariant J* occurs in all premises but not in the conclusion of rule loop. That means, whenever we apply the loop invariant rule to a desired conclusion, we get to choose what invariant J we want to use it for. Good choices of J will lead to a successful proof of the conclusion. Bad choices of J will stall the proof, because some of the premises cannot be proved.

The first premise of rule loop says that the initial state, about which we assume Γ (and that Δ does not hold), satisfies the invariant J, i.e., the invariant is initially true. The second premise of rule loop shows that the invariant J is *inductive*. That is, whenever J was true before running the loop body α, then J is always true again after running α. The third premise of rule loop shows that the invariant J is strong enough to imply the postcondition P that the conclusion was interested in.

Rule loop says that postcondition P holds after any number of repetitions of α if some invariant J holds initially (left premise), if that invariant J remains true after one iteration of α from any state where J was true (middle premise), and if that invariant J finally implies the desired postcondition P (right premise). If J is true after executing α whenever J has been true before (middle premise), then, if J holds in the beginning (left premise), J will continue to hold, no matter how often we repeat α in $[\alpha^*]P$, which is enough to imply $[\alpha^*]P$ if J implies P (right premise).

Taking a step back, these three premises correspond to the proof steps one would use to show that the contract of an ordinary program with a **requires**() contract Γ

(and not Δ), an **ensures**(P) contract, and a loop invariant J is correct. Now, we have this reasoning in a more general and formally more precisely defined context. We no longer need to appeal to intuition to justify why such a proof rule is fine, but can evoke a soundness proof for loop. We will also no longer be limited to informal arguments to justify invariance for a program but can do actual solid and rigorous formal proofs if we combine proof rule loop with the other proof rules from Chap. 6.

Invariants are crucial concepts for conventional programs and continue to be even more crucial for cyber-physical systems, where change is ubiquitous and any identification of aspects that remain unchanged over time is a blessing.

Of course, the search for suitable loop invariants J to be used with the loop invariant rule can be as much of a challenge as the search for invariants in mathematics. Yet, the fact that the difference between the equivalence in the induction axiom I and the induction steps of rules loop and ind is the absence of the $[\alpha^*]$ modality provides some guidance on what kind of information loop invariants J need. Loop invariants J may need to communicate something else that is also always true after running α^* and carries just information about the past behavior during α^* to imply that they are preserved after running α once more.

Example 7.2 (Stronger invariants). Consider an obvious example of a purely discrete loop to illustrate the rôle of loop invariants in proving the safety of loops:

$$x \geq 8 \land 5 \geq y \land y \geq 0 \to [(x:=x+y; y:=x-2\cdot y)^*]x \geq 0$$

This formula is valid. A proof with loop invariant J starts like this:

$$
\begin{array}{c}
\cfrac{\cfrac{x \geq 8 \land 5 \geq y \land y \geq 0 \vdash J \quad J \vdash [x:=x+y; y:=x-2\cdot y]J \quad J \vdash x \geq 0}{x \geq 8 \land 5 \geq y \land y \geq 0 \vdash [(x:=x+y; y:=x-2\cdot y)^*]x \geq 0}\;\text{loop}}{\vdash x \geq 8 \land 5 \geq y \land y \geq 0 \to [(x:=x+y; y:=x-2\cdot y)^*]x \geq 0}\;\to R
\end{array}
$$

A direct proof with the postcondition $x \geq 0$ as invariant J cannot succeed, because the induction step is not valid, since $x \geq 0$ is not guaranteed to be true after $x:=x+y$ if the inductive hypothesis only guarantees $x \geq 0$ about the previous state if y might be negative. The loop invariant J needs to imply the postcondition $x \geq 0$ but also contain additional information about the variable y that the change of x depends on.

The initial condition $x \geq 8 \land 5 \geq y \land y \geq 0$ also fails to be an invariant J since its induction step is not valid, because $5 \geq y$ is no longer guaranteed to be true after $x:=x+y; y:=x-2\cdot y$, e.g., if $x=8, y=0$ holds initially. The loop invariant J needs to be implied by the precondition, but may have to be weaker because the precondition itself does not have to remain true always when repeating the loop.

The loop invariant J, thus, has to be somewhere between the precondition (first premise) and the postcondition (third premise). It needs to involve bounds on both x and y, because the change of x depends on y and vice versa (second premise). The first assignment $x:=x+y$ obviously preserves $x \geq 0$ if also $y \geq 0$. The loop body obviously preserves this $y \geq 0$ if $x \geq y$. Indeed, the conjunction $x \geq y \land y \geq 0$ succeeds as loop invariant J:

$$\overset{\text{loop}}{\underset{\to\text{R}}{\dfrac{\dfrac{\overset{*}{\text{R}\,\dfrac{}{x\geq 8\wedge 5\geq y\wedge y\geq 0\vdash J}} \quad \dfrac{\overset{*}{\text{R}\,\dfrac{}{J\vdash x+y\geq x-y\wedge x-y\geq 0}}}{\dfrac{[:=]\,\dfrac{}{J\vdash [x:=x+y][y:=x-2\cdot y]J}}{[;]\,\dfrac{}{J\vdash [x:=x+y;\,y:=x-2\cdot y]J}}} \quad \overset{*}{\text{R}\,\dfrac{}{J\vdash x\geq 0}}}{x\geq 8\wedge 5\geq y\wedge y\geq 0\vdash [(x:=x+y;\,y:=x-2\cdot y)^*]x\geq 0}}{\vdash x\geq 8\wedge 5\geq y\wedge y\geq 0\to [(x:=x+y;\,y:=x-2\cdot y)^*]x\geq 0}}}$$

A similar proof uses the loop invariant $x\geq 0\wedge y\geq 0$ to prove Example 7.1.

> **Note 38 (Of loop invariants and relay races)** Loop invariants J are the proof analogue of a relay race. The initial state needs to show they have the baton J. Every state along the way after repeating α^* any number of times needs to wait to receive the baton J and then pass the baton J to the next state after running the next leg α of the relay race α^*. When the final state receives the baton J, that baton needs to carry enough information to meet the goal's safety condition P. Finding a loop invariant J is like designing the baton that makes all these passing phases work out as easily as possible.

7.3.4 Contextual Soundness Requirements

Since the loop rule derives via monotonicity rule M[·] from rule ind, which derives via Gödel's generalization G, it should not come as a surprise that it is crucial for soundness that the sequent context formulas Γ and Δ disappear from the middle and last premises of loop. It is equally soundness-critical that no context Γ,Δ carries over to the premise of rules G,M[·],M,ind. All those premises result from discarding the $[\alpha^*]$ modality, which ignores its effect. That is sound as long as no context Γ,Δ is preserved, which represents assumptions about the initial state before $[\alpha^*]$, which may no longer be true after $[\alpha^*]$. For the loop rule, information Γ,Δ about the initial state is only available to show that J is initially true (first premise), but no longer during the induction step (second premise) or use case (third premise).

Example 7.3 (No context). The context Γ,Δ cannot be kept in the ind rule without losing soundness:

$$\frac{x=0,x\leq 1\vdash [x:=x+1]x\leq 1}{x=0,x\leq 1\vdash [(x:=x+1)^*]x\leq 1}$$

This inference is unsound, because the premise is valid but the conclusion is not, since $x\leq 1$ will be violated after two repetitions. Even if $x=0$ is assumed initially (antecedent of conclusion), it cannot be assumed in the induction step (premise), because it is no longer true after iterating the loop any nonzero number of times. Almost the same counterexample shows that the middle premise of the loop rule cannot keep a context soundly. The following counterexample shows that the third premise of rule loop also cannot keep a context without losing soundness:

$$\frac{x=0 \vdash x \geq 0 \quad x \geq 0 \vdash [x:=x+1]x \geq 0 \quad x=0, x \geq 0 \vdash x=0}{x=0 \vdash [(x:=x+1)^*]x=0}$$

With some more thought, assumptions about *constant* parameters that cannot change during the HP α^* could be kept around without endangering soundness. This can be proved with the help of the vacuity axiom V from Sect. 5.6.2 (Exercise 7.8).

With Lemma 7.3, the loop invariant rule already has a simple and elegant soundness proof that simply derives it by monotonicity $M[\cdot]$ (Lemma 5.13) from the induction rule ind, which, in turn, is derived using Gödel's generalization rule G from the induction axiom I. Since loop invariants are such a fundamental concept, and since Example 7.3 just made us painfully aware how careful we need to be to keep CPS reasoning principles sound, we provide a second soundness proof directly from the semantics even if that proof is entirely redundant and more complicated than the first proof of Lemma 7.3.

Proof (of Lemma 7.3). In order to prove that rule loop is sound, we assume that all its premises are valid and need to show that its conclusion is valid, too. So let $\vDash \Gamma \vdash J, \Delta$ and $\vDash J \vdash [\alpha]J$ and $\vDash J \vdash P$. In order to prove that $\vDash \Gamma \vdash [\alpha^*]P, \Delta$, consider any state ω and show that $\omega \in [\![\Gamma \vdash [\alpha^*]P, \Delta]\!]$. If one of the formulas $Q \in \Gamma$ does not hold in ω (that is $\omega \notin [\![Q]\!]$) or if one of the formulas $Q \in \Delta$ holds in ω ($\omega \in [\![Q]\!]$), then there is nothing to show, because the formula that the sequent $\Gamma \vdash [\alpha^*]P, \Delta$ represents already holds in ω, either because one of the conjunctive assumptions Γ is not met in ω or because one of the other disjunctive succedents Δ already holds. Consequently, let all $Q \in \Gamma$ be true in ω and all $Q \in \Delta$ be false in ω or else there is nothing to show.

In that case, however, the first premise implies that $\omega \in [\![J]\!]$ because all its assumptions (which are the same Γ) are met in ω and all alternative succedents (which are the same Δ) do not already hold.[2]

In order to show that $\omega \in [\![[\alpha^*]P]\!]$, consider any run $(\omega, \nu) \in [\![\alpha^*]\!]$ from the initial state ω to some state ν and show that $\nu \in [\![\alpha]\!]$. According to the semantics of loops from Chap. 3, $(\omega, \nu) \in [\![\alpha^*]\!]$ if and only if, for some natural number $n \in \mathbb{N}$ that represents the number of loop iterations, there is a sequence of states $\mu_0, \mu_1, \ldots, \mu_n$ such that $\mu_0 = \omega$ and $\mu_n = \nu$ such that $(\mu_i, \mu_{i+1}) \in [\![\alpha]\!]$ for all $i < n$. The proof that $\mu_n \in [\![J]\!]$ is now by induction on n.

0. If $n = 0$, then $\nu = \mu_0 = \mu_n = \omega$, which implies by the first premise that $\nu \in [\![J]\!]$.
1. By induction hypothesis, $\mu_n \in [\![J]\!]$. By the second premise, $\vDash J \vdash [\alpha]J$, in particular for state μ_n we have $\mu_n \in [\![J \to [\alpha]J]\!]$, recalling the semantics of sequents. Combined with the induction hypothesis, this implies $\mu_n \in [\![[\alpha]J]\!]$, which means that $\mu \in [\![J]\!]$ for all states μ such that $(\mu_n, \mu) \in [\![\alpha]\!]$. Hence, $\mu_{n+1} \in [\![J]\!]$ because $(\mu_n, \mu_{n+1}) \in [\![\alpha]\!]$.

This implies, in particular, that $\nu \in [\![J]\!]$, because $\mu_n = \nu$. By the third premise, $\vDash J \vdash P$. In particular, $\nu \in [\![J \to P]\!]$, which with $\nu \in [\![J]\!]$ implies $\nu \in [\![P]\!]$. This con-

---

[2] In future soundness proofs, we will fast-forward to this situation right away, but it is instructive to see the full argument once.

cludes the soundness proof, since v was an arbitrary state such that $(\omega, v) \in [\![\alpha^*]\!]$, so $\omega \in [\![\alpha^*]\!]P]\!]$. $\qquad\qquad\qquad\qquad\qquad\qquad\qquad\qquad\qquad\qquad\qquad\qquad$ □

7.4 A Proof of a Happily Repetitive Bouncing Ball

Now that he understands the principles of how to prove loops in CPSs, Quantum is eager to put these skills to use. Quantum wants to relieve himself of his acrophobic fears once and for all by proving that he won't ever have to be afraid of excess heights $> H$ again nor of falling through the cracks in the ground to heights < 0.

Abbreviations have served Quantum well in trying to keep proofs on one page:

$$A \overset{\text{def}}{\equiv} 0 \leq x \wedge x = H \wedge v = 0 \wedge g > 0 \wedge 1 \geq c \geq 0$$

$$B_{(x,v)} \overset{\text{def}}{\equiv} 0 \leq x \wedge x \leq H$$

$$x''.. \overset{\text{def}}{\equiv} \{x' = v, v' = -g\}$$

Note the somewhat odd abbreviation for the differential equation just to condense notation. With these abbreviations, the bouncing-ball conjecture (7.1) turns into

$$A \to [(x''..; (?x = 0; v := -cv \cup ?x \neq 0))^*]B_{(x,v)} \qquad\qquad (7.1^*)$$

This formula is swiftly turned into the sequent at the top using proof rule $\to$R:

$$\to\text{R} \frac{A \vdash [(x''..; (?x = 0; v := -cv \cup ?x \neq 0))^*]B_{(x,v)}}{\vdash A \to [(x''..; (?x = 0; v := -cv \cup ?x \neq 0))^*]B_{(x,v)}}$$

Its premise leaves a loop to worry about, which gives Quantum a chance to practice what he learned in this chapter.

The first thing that Quantum will need for the proof of (7.1) is the appropriate choice for the invariant J to be used in the loop invariant proof rule loop. Quantum will use a dL formula $j_{(x,v)}$ for the invariant when instantiating J in the proof rule loop. But Quantum is still a little unsure about how exactly to define that formula $j_{(x,v)}$, not an unusual situation when trying to master the understanding of a CPS. Can you think of a good choice for the formula $j_{(x,v)}$ to help Quantum?

Before you read on, see if you can find the answer for yourself.

I don't know about you, but Quantum settles for the choice of using the post-condition as an invariant, because that is what he wants to show about the behavior:

$$j_{(x,v)} \overset{\text{def}}{\equiv} 0 \leq x \wedge x \leq H \qquad\qquad\qquad\qquad\qquad (7.7)$$

Because Quantum is so proud of his wonderful invariant $j_{(x,v)}$, he even uses it to perform a generalization with the newly acquired skill of the generalization proof rule MR in the inductive step to completely separate the proof about the differential equation and the proof about the bouncing dynamics.[3] Quantum conducts the proof in Fig. 7.2.

$$
\begin{array}{c}
\cfrac{\begin{array}{c}
{\scriptstyle [:=]}\cfrac{j(x,v),x=0 \vdash j(x,-cv)}{j(x,v),x=0 \vdash [v:=-cv]j(x,v)} \\
{\scriptstyle [?]}\cfrac{}{j(x,v) \vdash [?x=0][v:=-cv]j(x,v)} \\
{\scriptstyle [;]}\cfrac{}{j(x,v) \vdash [?x=0;v:=-cv]j(x,v)} \qquad {\scriptstyle [?]}\cfrac{j(x,v),x\neq 0 \vdash j(x,v)}{j(x,v) \vdash [?x\neq 0]j(x,v)} \\
{\scriptstyle \wedge R}\cfrac{}{j(x,v) \vdash [?x=0;v:=-cv]j(x,v) \wedge [?x\neq 0]j(x,v)} \\
{\scriptstyle [\cup]}\cfrac{}{j(x,v) \vdash [?x=0;v:=-cv \cup ?x\neq 0]j(x,v)}
\end{array}}{\begin{array}{c}
j(x,v) \vdash [x''..]j(x,v) \qquad \cfrac{}{j(x,v) \vdash [x''..][?x=0;v:=-cv \cup ?x\neq 0]j(x,v)} \\
{\scriptstyle MR}\cfrac{}{\cfrac{A \vdash j(x,v)}{} \qquad {\scriptstyle [;]}\cfrac{j(x,v) \vdash [x''..;(?x=0;v:=-cv \cup ?x\neq 0)]j(x,v)}{} \qquad j(x,v) \vdash B(x,v)}
\end{array}} \\[2pt]
{\scriptstyle loop}\cfrac{}{\begin{array}{c}
\cfrac{A \vdash [(x''..;(?x=0;v:=-cv \cup ?x\neq 0))^*]B(x,v)}{\vdash A \to [(x''..;(?x=0;v:=-cv \cup ?x\neq 0))^*]B(x,v)} {\scriptstyle \to R}
\end{array}}
\end{array}
$$

Fig. 7.2 Sequent calculus proof shape for bouncing ball (7.1)

The proof in Fig. 7.2 has five premises remaining to be proved. Quantum is pretty sure how to prove the first premise $(A \vdash j(x,v))$, corresponding to the initial condition, because $0 \le x \le H$ is true initially as $0 \le x = H$ follows from A. Quantum also knows how to prove the last premise $(j(x,v) \vdash B(x,v))$, because the invariant $j(x,v)$ from (7.7) is equal to the desired postcondition $B(x,v)$, so this is proved by the identity rule id.

But Quantum runs into unforeseen(?) trouble with the inductive step in the middle. While the third and fourth premise succeed, the second premise $j(x,v) \vdash [x''..]j(x,v)$ with the differential equation resists all proof attempts for the choice (7.7). That makes sense, because, even if the current height is bounded by $0 \le x \le H$ before the differential equation, there is no reason to believe it will remain bounded afterwards if this is all we know about the bouncing ball. If the ball were just below $x = H$, it would still ultimately exceed H if its velocity were too big.

Ah, right! We actually found that out about the bouncing ball in Chap. 4 already when we were wondering under what circumstances it might be safe to let a ball bounce around. As a matter of fact, everything we learned by the Principle of Cartesian Doubt about when it is safe to start a CPS is valuable information to preserve in the invariant. If it wasn't safe to start a CPS in a state, chances are, it wouldn't be safe either if we kept it running in such a state as we do in an inductive step.

Well, so Quantum found a (poor) choice of an invariant $j(x,v)$ in (7.7) that just cannot be proved because of the inductive step. What to do?, wonders Quantum.

Before you read on, see if you can find the answer for yourself.

---

[3] This is not necessary and Quantum might just as well not have used MR and gone for a direct proof using ['] right away instead. But it does save us some space on the page, and also showcases a practical use of proof rule MR.

There was trouble in the induction step, because $x \leq H$ could not be proved to be inductive. But Quantum does not despair. Quantum can demand a little less from the invariant and use the following weaker choice for $j_{(x,v)}$ instead of (7.7):

$$j_{(x,v)} \overset{\text{def}}{\equiv} x \geq 0 \qquad (7.8)$$

Armed with this new choice for an invariant, Quantum quickly gets to work constructing a new proof for (7.1). After frantically scribbling a couple of pages with sequent proofs, Quantum experiences a *déjà vu* and notices that his new proof has exactly the same form as the last sequent proof he began, just with a different choice for the logical formula $j_{(x,v)}$ to be used as the invariant when applying the loop rule with the choice (7.8) rather than (7.7) for $j_{(x,v)}$. Fortunately, Quantum already worked with an abbreviation last time he started a proof, so it is actually not surprising after all to see that the proof structure stays exactly the same and that the particular choice of $j_{(x,v)}$ only affects the premises, not the way the proof unraveled its program statements in the modalities.

Inspecting the five premises of the above sequent proof attempt in light of the improved choice (7.8) for the invariant, Quantum is delighted to find that the inductive step works out just fine. The height stays above ground always by construction with the evolution domain constraint $x \geq 0$ and is not changed in the subsequent discrete bouncing control. The initial condition $(A \vdash j_{(x,v)})$ also works out alright, because $0 \leq x$ was among the assumptions in A. Only this time, the last premise $(j_{(x,v)} \vdash B_{(x,v)})$ falls apart, because $x \geq 0$ is not at all enough to conclude the part $x \leq H$ of the postcondition. What's a ball to do to get himself verified these days?

Before you read on, see if you can find the answer for yourself.

Quantum takes the lesson from Cartesian Doubt to heart and realizes that the invariant needs to transport enough information about the state of the system to make sure the inductive step has a chance of holding true. In particular, the invariant desperately needs to preserve knowledge about the velocity, because how the height changes depends on the velocity (after all the differential equation reads $x' = v, \ldots$), so it would be hard to get a handle on height x without first understanding how velocity v changes, which it does in $v' = -g$ and at the bounce. Indeed, this is an entirely syntactic reason why neither (7.7) nor (7.8) could have worked out as invariants for the proof of (7.1). They only mention the height x, but how the height changes in the bouncing-ball HP depends on the velocity, which also changes. So unless the invariant preserves knowledge about v, it cannot possibly guarantee much about height x, except the fact $x \geq 0$ from the evolution domain constraint, which does not suffice to prove the postcondition $0 \leq x \leq H$.

Fine, so Quantum quickly discards the failed invariant choice from (7.7), which he is no longer quite so proud of, and also gives up on the weaker version (7.8), but instead shoots for a stronger invariant, which is surely inductive and strong enough to imply safety:

$$j_{(x,v)} \overset{\text{def}}{\equiv} x = 0 \wedge v = 0 \qquad (7.9)$$

This time, Quantum has learned his lesson and won't blindly set out to prove the property (7.1) from scratch again, but, rather, be clever about it and realize that he is still going to find the same shape of the sequent proof attempt above, just with, once again, a different choice for the invariant $j(x,v)$. So Quantum quickly jumps to conclusions and inspects the famous 5 premises of the above sequent proof attempt. This time, the postcondition is a piece of cake and the inductive step works like a charm (no velocity, no height, no motion). But the initial condition is giving Quantum quite a bit of a headache, because there is no reason to believe the ball would initially lie flat on the ground with velocity zero.

For a moment there, Quantum fancied the option of simply editing the initial condition A to include $x = 0$, because that would make this proof attempt work out just fine. But then he realized that this would mean that he would from now on be doomed to always start the day at speed zero on the ground, which would not lead to all that much excitement for a cheerful bouncing ball. That option would be safe, but a bit too much so for lack of motion.

What, then, is poor Quantum supposed to do to finally get a proof without losing all those exciting initial conditions?

> Before you read on, see if you can find the answer for yourself.

This time, Quantum thinks about the invariant question really hard and has a smart idea. Thinking back to where the idea of the loop invariants came from in the first place, they are replacements for the postcondition P that make $[\alpha^*](P \to [\alpha]P)$ provable despite discarding the $[\alpha^*]$ modality. They are stronger versions of the postcondition P, and need to at least imply that the postcondition P always holds after running α once, but, in fact, even need to imply they themselves continue to hold after α. For this to work out, their rôle is to capture whatever we still need to know about the previous runs of α^*.

If the loop invariant has to work for any number of loop iterations, it certainly has to work for the first few loop iterations. In particular, the loop invariant J is not unlike an intermediate condition of $\alpha; \alpha$. Quantum already identified an intermediate condition for the single-hop bouncing ball in Sect. 4.8.1. Maybe that will prove useful as an invariant, too:

$$j(x,v) \stackrel{\text{def}}{\equiv} 2gx = 2gH - v^2 \wedge x \geq 0 \qquad (7.10)$$

After all, an invariant is something like a permanent intermediate condition, i.e., an intermediate condition that keeps on working out alright for all future iterations. The bouncing ball is not yet sure whether this will work but it seems worth trying!

The shape of the proof in Fig. 7.2 again stays exactly the same, just with a different choice of $j(x,v)$, this time coming from (7.10). The remaining famous five premises are then proved easily. The first premise $A \vdash j(x,v)$ is proved using $x = H$ and $v = 0$:

$$0 \leq x \wedge x = H \wedge v = 0 \wedge g > 0 \wedge 1 \geq c \geq 0 \vdash 2gx = 2gH - v^2 \wedge x \geq 0$$

Expanding the abbreviations, the second premise $j_{(x,v)} \vdash [x''..]j_{(x,v)}$ is

$$2gx = 2gH - v^2 \wedge x \geq 0 \vdash [x' = v, v' = -g \,\&\, x \geq 0](2gx = 2gH - v^2 \wedge x \geq 0)$$

a proof that we have seen in previous chapters (Exercise 7.1). The third premise $j_{(x,v)}, x = 0 \vdash j_{(x,-cv)}$ is

$$2gx = 2gH - v^2 \wedge x \geq 0, x = 0 \vdash 2gx = 2gH - (-cv)^2 \wedge x \geq 0$$

which would be proved easily if we knew $c = 1$. Do we know $c = 1$? No, we do not know $c = 1$, because we only assumed $1 \geq c \geq 0$ in A. But we could prove this third premise easily if we edited the definition of the initial condition A to include $c = 1$. That is not the most general statement about bouncing balls, but let's happily settle for it till Exercise 7.5. Even then, however, we still need to augment $j_{(x,v)}$ to include $c = 1$ as well, since we otherwise would have lost this knowledge before we need it in the third premise. Having misplaced critical pieces of knowledge is a phenomenon you may encounter when you are conducting proofs. In such cases, you should trace where you lost the assumption in the first place and put it back in. But then you have also learned something valuable about your system, namely which assumptions are crucial for the correct functioning of which part of the system.

The fourth premise, $j_{(x,v)}, x \geq 0 \vdash j_{(x,v)}$ is proved splendidly whatever the abbreviations stand for simply using the identity rule id. In fact, Quantum could have noticed this earlier already but might have been distracted by his search for a good choice for the invariant $j_{(x,v)}$. This is but one indication of the fact that it may pay to take a step back from a proving effort and critically reflect on what all the pieces of the argument rely on exactly. Finally, the fifth premise $j_{(x,v)} \vdash B_{(x,v)}$, which is

$$2gx = 2gH - v^2 \wedge x \geq 0 \vdash 0 \leq x \wedge x \leq H$$

is proved by arithmetic as long as we know $g > 0$. This condition is already included in A. But we still managed to forget about that in our invariant $j_{(x,v)}$. So, again, the constant parameter assumption $g > 0$ should have been included in the invariant $j_{(x,v)}$, which, overall, should have been defined as

$$j_{(x,v)} \stackrel{\text{def}}{\equiv} 2gx = 2gH - v^2 \wedge x \geq 0 \wedge (c = 1 \wedge g > 0) \tag{7.11}$$

This is nearly the same definition as (7.10) except that assumptions about the system parameter choices are carried through. The last two conjuncts are trivially invariant, because neither c nor g changes while the little bouncing ball falls. As written, the loop invariant rule, unfortunately, still needs to have these constant assumptions included in the invariant, because it wipes the entire context Γ, Δ, which is crucial for soundness (Sect. 7.3.4). Exercise 7.8 investigates simplifications for this nuisance that will enable you to elide the trivial constant part $c = 1 \wedge g > 0$ from the invariant. Redoing the proof with the new loop invariant (7.11) will succeed, as will the proof in Sect. 7.5.

For the record, we now really have a full sequent proof of the undamped bouncing ball with repetitions. Quantum is certainly quite thrilled about this achievement!

Proposition 7.1 (Quantum is safe). *This dL formula has a proof and is, thus, valid:*

$$0 \le x \wedge x = H \wedge v = 0 \wedge g > 0 \wedge 1 = c \rightarrow$$
$$[(\{x' = v, v' = -g \,\&\, x \ge 0\}; (?x = 0; v := -cv \cup ?x \ne 0))^*](0 \le x \wedge x \le H) \quad (7.12)$$

Since invariants are a crucial part of a CPS design, you are encouraged to describe invariants in your hybrid programs. KeYmaera X will make use of the invariants annotated using the @invariant contract in hybrid programs to simplify your proof effort. But KeYmaera X solved Exercise 7.8 already, so it does not require a list of the constant expressions in the @invariant contracts. It is a good idea to rephrase (7.12) by explicitly including the invariant contract in the hybrid program for documentation as well as verification purposes:

$$0 \le x \wedge x = H \wedge v = 0 \wedge g > 0 \wedge 1 = c \rightarrow$$
$$[(\{x' = v, v' = -g \,\&\, x \ge 0\};$$
$$(?x = 0; v := -cv \cup ?x \ne 0))^* @\text{invariant}(2gx = 2gH - v^2 \wedge x \ge 0)] \quad (7.13)$$
$$(0 \le x \wedge x \le H)$$

Indeed, assumptions about constant parameters, which are trivially invariant, do not need to be listed, as the next section will explain.

7.5 Splitting Postconditions into Separate Cases

The invariant $j_{(x,v)}$ from formula (7.10) was not quite enough for proving the bouncing-ball property (7.12) since we need *constant parameter* assumptions from the modified invariant (7.11). Redoing the proof with the new invariant succeeds. But it would be easier if there was a way of reusing the old proof by threading the misplaced assumptions through to where we need them. Of course, we cannot simply add assumptions into the middle of a proof without losing soundness (Sect. 7.3.4). But Quantum wonders whether we might get away with doing that if it is merely a matter of adding assumptions about constant parameters such as $c = 1 \wedge g > 0$?

Indeed, there are two interesting insights about clever proof structuring that we can learn from this desire. One insight is an efficient way of proving the preservation of assumptions about constant parameters. The other is about modularly separating the reasoning into proofs for separate postconditions.

The dynamic axioms from Chap. 5 and the sequent proof rules from Chap. 6 decompose correctness analysis along the top-level operators, which, e.g., split the analysis into separate questions along the top-level operators in the hybrid programs. But it is also possible to split the reasoning along the postcondition to show

$[\alpha](P \wedge Q)$ by proving $[\alpha]P$ and $[\alpha]Q$ separately. If the HP α satisfies both the safety postcondition P and the safety postcondition Q, then it also satisfies the safety postcondition $P \wedge Q$, and vice versa, using the following result from Sect. 5.6.1.

Lemma 5.10 ($[]\wedge$ boxes distribute over conjunctions). *This axiom is sound:*

$$[]\wedge \quad [\alpha](P \wedge Q) \leftrightarrow [\alpha]P \wedge [\alpha]Q$$

The axiom $[]\wedge$ can decompose the box modality in the induction step along the conjunction in the loop invariant $j_{(x,v)} \wedge q$ to conduct separate proofs that $j_{(x,v)}$ is inductive (second premise in Fig. 7.3) and that the additional invariant q, which is defined as $c = 1 \wedge g > 0$, is inductive (third premise). In Fig. 7.3, bb denotes the loop body of the bouncing ball (7.12). Observe how the induction proof in the second premise is literally the same proof as the previous proof in Fig. 7.2, except that the missing assumption q is now available. The remaining proof that the additional loop invariant q is also inductive is isolated in the third premise.

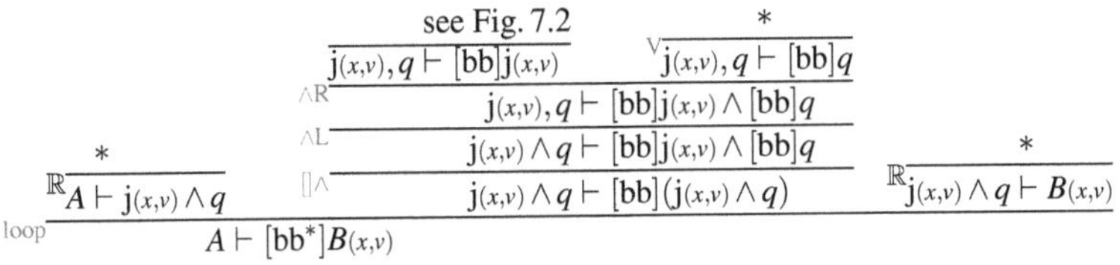

Fig. 7.3 Sequent calculus proof for bouncing ball (7.12) with split

There is an additional interesting twist in the proof in Fig. 7.3, though. The proof of its third premise establishes that formula q is inductive for the bouncing ball bb. This could be proved by successively decomposing the HP bb using the various dynamic axioms for sequential compositions, nondeterministic choices, differential equations, etc. While that proof would work, it is significantly more efficient to prove it in a single step using the axiom V (from Sect. 5.6.2) for postconditions of box modalities that do not change any of the variables in the postcondition, so that the postcondition is true after all runs of the HP if only it is true before.

Lemma 5.11 (V vacuous axiom). *The vacuous axiom is sound:*

$$V \quad p \rightarrow [\alpha]p \quad (FV(p) \cap BV(\alpha) = \emptyset)$$

where no free variable of p is bound (written) in α.

When used like this, the axioms $[]\wedge$ and V justify that constant parameter assumptions can be kept around without any harm to the proof (Exercise 7.8).

7.6 Summary

This chapter focused on developing and using the concept of invariants for CPS. Invariants enable us to prove properties of CPSs with loops, a problem of ubiquitous significance, because hardly any CPS get by without repeating some operations in a control loop. *Invariants constitute the single most insightful and most important piece of information about a CPS, because they tell us what we can rely on no matter how long a CPS runs.* Invariants are a fundamental force of computer science, and are just as important in mathematics and physics.

The axioms and proof rules investigated in this chapter are summarized in Fig. 7.4. While the loop invariant rule (loop) is the most practical approach for loops, the induction axiom I is an equivalence and explains the core principle of loop induction more directly. The loop invariant rule loop also derives directly from the induction axiom I by monotonicity rule M[·] and generalization rule G.

$$\text{I} \quad [\alpha^*]P \leftrightarrow P \wedge [\alpha^*](P \to [\alpha]P)$$

$$\text{G} \quad \frac{P}{[\alpha]P}$$

$$\text{M}[\cdot] \quad \frac{P \to Q}{[\alpha]P \to [\alpha]Q}$$

$$\text{loop} \quad \frac{\Gamma \vdash J, \Delta \quad J \vdash [\alpha]J \quad J \vdash P}{\Gamma \vdash [\alpha^*]P, \Delta}$$

$$\text{MR} \quad \frac{\Gamma \vdash [\alpha]Q, \Delta \quad Q \vdash P}{\Gamma \vdash [\alpha]P, \Delta}$$

$$[\,]\wedge \quad [\alpha](P \wedge Q) \leftrightarrow [\alpha]P \wedge [\alpha]Q$$

$$\text{V} \quad p \to [\alpha]p \quad (FV(p) \cap BV(\alpha) = \emptyset)$$

Fig. 7.4 Summary of proof rules for loops, generalization, monotonicity, and splitting boxes

The development that led to invariants has some interesting further consequences especially for finding bugs in CPSs by unrolling loops and disproving the resulting premises. But this bounded-model-checking principle is of limited use for ultimately verifying safety, because it only considers the system some finite number of steps in the future. This chapter focused on proving $[\alpha^*]P$ formulas, which were based on *invariants*, so properties that do not change. The discussion of proof techniques for proving $\langle\alpha^*\rangle P$ formulas will be postponed till Sect. 17.4, which will use *variants*, so properties that do change and steadily make progress toward the goal P.

In our effort to help the bouncing ball Quantum succeed with his proof, we saw a range of reasons why an inductive proof may not work out and what needs to be done to adapt the invariant.

7.7 Appendix

This appendix provides an alternative way of motivating the loop induction rule only from successively unwinding a loop with the iteration axiom [*] that works without using the more elegant induction axiom I.

7.7.1 Loops of Proofs

The iteration axiom [*] can be used to turn a safety property of a loop

$$A \to [\alpha^*]B \qquad\qquad (7.14)$$

into the following equivalent dL formula:

$$A \to B \wedge [\alpha][\alpha^*]B$$

What can we do to prove that loop? Investigating our proof rules from previous chapters, there is exactly one that addresses loops: the iteration [*] axiom again. Recall that, unlike sequent proof rules, axioms do not dictate where they can be used, so we might as well use them anywhere in the middle of the formula. Hence using axiom [*] on the inner loop yields

$$A \to B \wedge [\alpha](B \wedge [\alpha][\alpha^*]B)$$

Let's do that again because that was so much fun and use the [*] axiom on the only occurrence of $[\alpha^*]B$ to obtain

$$A \to B \wedge [\alpha](B \wedge [\alpha](B \wedge [\alpha][\alpha^*]B)) \qquad\qquad (7.15)$$

This is all very interesting but won't exactly get us any closer to a proof, because we could keep expanding the * star forever that way. How do we ever break out of this loop of never-ending proofs?

Before we get too disillusioned about our progress with axiom [*] so far, notice that (7.15) still allows us to learn something about α and whether it always satisfies B when repeating α. Since [*] is an equivalence axiom, formula (7.15) still expresses the same thing as (7.14), i.e., that postcondition B always holds after repeating α when A was true in the beginning. Yet, (7.15) explicitly singles out the first three runs of α. Let's make this more apparent with the derived axiom $[]\wedge$ for box splitting from Sect. 5.6.1. Using this valid equivalence turns (7.15) into

$$A \to B \wedge [\alpha]B \wedge [\alpha][\alpha](B \wedge [\alpha][\alpha^*]B)$$

Using $[]\wedge$ again gives us

$$A \rightarrow B \wedge [\alpha]B \wedge [\alpha]([\alpha]B \wedge [\alpha][\alpha][\alpha^*]B)$$

Using $[]\wedge$ once more gives

$$A \rightarrow B \wedge [\alpha]B \wedge [\alpha][\alpha]B \wedge [\alpha][\alpha][\alpha][\alpha^*]B \tag{7.16}$$

$$
\begin{array}{l}
{\scriptstyle \wedge R, \wedge R, \wedge R} \cfrac{A \vdash B \quad A \vdash [\alpha]B \quad A \vdash [\alpha][\alpha]B \quad A \vdash [\alpha][\alpha][\alpha][\alpha^*]B}{A \vdash B \wedge [\alpha]B \wedge [\alpha][\alpha]B \wedge [\alpha][\alpha][\alpha][\alpha^*]B} \\
{\scriptstyle []\wedge} \cfrac{}{A \vdash B \wedge [\alpha]B \wedge [\alpha]([\alpha]B \wedge [\alpha][\alpha][\alpha^*]B)} \\
{\scriptstyle []\wedge} \cfrac{}{A \vdash B \wedge [\alpha]B \wedge [\alpha][\alpha](B \wedge [\alpha][\alpha^*]B)} \\
{\scriptstyle []\wedge} \cfrac{}{A \vdash B \wedge [\alpha](B \wedge [\alpha](B \wedge [\alpha][\alpha^*]B))} \\
{\scriptstyle [^*]} \cfrac{}{A \vdash B \wedge [\alpha](B \wedge [\alpha][\alpha^*]B)} \\
{\scriptstyle [^*]} \cfrac{}{A \vdash B \wedge [\alpha][\alpha^*]B} \\
{\scriptstyle [^*]} \cfrac{}{A \vdash [\alpha^*]B}
\end{array}
$$

Fig. 7.5 Loops of proofs: iterating and splitting the box

Fig. 7.5 illustrates the proof construction so far.[4] Looking at it this way, (7.16) could be more useful than the original (7.14), because, even though the two formulas are equivalent, (7.16) explicitly singles out the fact that B has to hold initially, after doing α once, after doing α twice, and that $[\alpha^*]B$ has to hold after doing α three times. Even if we are not quite sure what to make of the latter $[\alpha][\alpha][\alpha][\alpha^*]B$, because it still involves a loop, we are quite certain how to understand and handle the first three:

$$A \rightarrow B \wedge [\alpha]B \wedge [\alpha][\alpha]B \tag{7.17}$$

If this formula is not valid, then, certainly, neither is (7.16) and, thus, neither is the original (7.14). Hence, if we find a counterexample to (7.17), we disproved (7.16) and (7.14). That can actually be rather useful!

However, if (7.17) is valid, we do not know whether (7.16) and (7.14) are, since they involve stronger requirements (B holds after any number of repetitions of α). What can we do then? Simply unroll the loop once more by using $[^*]$ on (7.15) to obtain

$$A \rightarrow B \wedge [\alpha](B \wedge [\alpha](B \wedge [\alpha](B \wedge [\alpha][\alpha^*]B))) \tag{7.18}$$

Or, equivalently, use axiom $[^*]$ on (7.16) to obtain the equivalent:

$$A \rightarrow B \wedge [\alpha]B \wedge [\alpha][\alpha]B \wedge [\alpha][\alpha][\alpha](B \wedge [\alpha][\alpha^*]B) \tag{7.19}$$

---

[4] Observe the $\wedge R, \wedge R, \wedge R$ at the top, which is not to be taken as an indication that the proof is stuttering, but merely meant as a notational reminder that the $\wedge R$ proof rule was actually used three times for that step. Because it will frequently simplify the notation, we will take the liberty of applying multiple rules at once like that without saying which derivation it was exactly. In fact, mentioning $\wedge R$ three times seems a bit repetitive, so we simply abbreviate this by writing $\wedge R$ even if we used the rule $\wedge R$three3 times and should have said $\wedge R, \wedge R, \wedge R$.

By sufficiently many uses of axiom $[]\wedge$, (7.18) and (7.19) are both equivalent to

$$A \to B \wedge [\alpha]B \wedge [\alpha][\alpha]B \wedge [\alpha][\alpha][\alpha]B \wedge [\alpha][\alpha][\alpha][\alpha][\alpha^*]B \qquad (7.20)$$

which we can again examine to see if we can find a counterexample to the first part:

$$A \to B \wedge [\alpha]B \wedge [\alpha][\alpha]B \wedge [\alpha][\alpha][\alpha]B$$

If yes, we disproved (7.14), otherwise we use axiom $[^*]$ once more.

> **Note 39 (Bounded model checking)** This process of iteratively unrolling a loop with the iteration axiom $[^*]$ and then checking the resulting (loop-free) conjuncts is called *Bounded Model Checking* and has been used with extraordinary success, e.g., in the context of finite-state systems [2]. The same principle can be useful to disprove properties of loops in differential dynamic logic by unwinding the loop, checking to see whether the resulting formulas have counterexamples and, if not, unrolling the loop once more. With certain computational refinements, this idea has found application in hybrid systems [1, 3, 5, 6] despite certain inevitable limits [9].

Suppose such a bounded model checking process has been followed to unroll the loop $N \in \mathbb{N}$ times. What can you conclude about the safety of the system?

If a counterexample is found or the formula can be disproved, then we are certain that the CPS is unsafe. If, instead, all but the last conjunct in the Nth unrolling of the loop are provable then the system will be safe for $N-1$ steps, but we cannot conclude anything about the safety of the system after more than $N-1$ steps. On the other hand, what we learn about the behavior of α from these iterations can still inform us about possible invariants.

7.7.2 Breaking Loops of Proofs

Proving properties of loops by unwinding them forever with axiom $[^*]$ is not a promising strategy, unless we find that the conjecture is not valid after a number of unwindings. Or unless we do not mind being busy with the proof forever for infinitely many proof steps (which would never get the acrophobic bouncing ball off the ground either with the confidence that a safety argument provides). One way or another, we will have to find a way to break the loop apart to complete our reasoning.

How can we prove the premises of Fig. 7.6? Sect. 7.7.1 investigated one way, which essentially amounts to Bounded Model Checking. Can we be more clever and prove the same premises in a different way? Preferably one that is more efficient and allows us to get the proof over with after finitely many steps?

There is not all that much we can do to improve the way we prove the first premise $(A \vdash B)$. We simply have to bite the bullet and do it, armed with all our knowledge of arithmetic from Chap. 6. But it's actually very easy at least for the

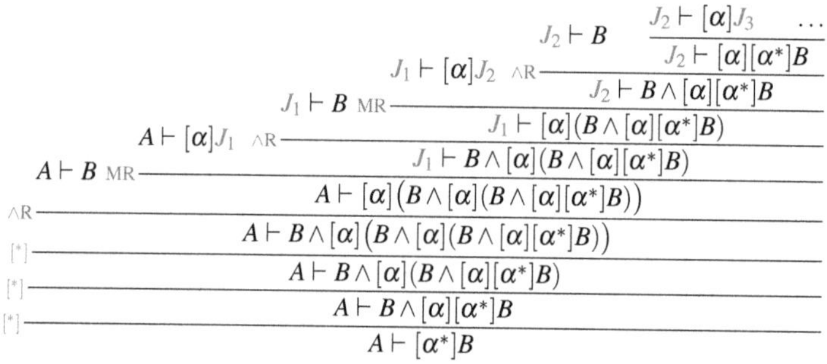

$$
\cfrac{
\cfrac{
\cfrac{
\cfrac{
A \vdash [\alpha]J_1
}{A \vdash B} {\scriptstyle MR}
}{A \vdash B} {\scriptstyle \wedge R}
}{\quad}{\scriptstyle [^*]}
}{\quad}{}
$$

$$
\cfrac{
  \cfrac{
    \cfrac{
      \cfrac{
        \cfrac{
          \cfrac{
            J_2 \vdash B \qquad \cfrac{J_2 \vdash [\alpha]J_3 \quad \cdots}{J_2 \vdash [\alpha][\alpha^*]B}
          }{J_2 \vdash B \wedge [\alpha][\alpha^*]B} {\scriptstyle \wedge R}
        }{J_1 \vdash [\alpha](B \wedge [\alpha][\alpha^*]B)}
        \quad
      }{J_1 \vdash B \wedge [\alpha](B \wedge [\alpha][\alpha^*]B)}
    }{A \vdash [\alpha]\bigl(B \wedge [\alpha](B \wedge [\alpha][\alpha^*]B)\bigr)}
  }{A \vdash B \wedge [\alpha]\bigl(B \wedge [\alpha](B \wedge [\alpha][\alpha^*]B)\bigr)}
  \quad [^*]
}{A \vdash B \wedge [\alpha](B \wedge [\alpha][\alpha^*]B)}
$$

Fig. 7.6 Loops of proofs: iterating and generalizing the box

bouncing ball. Besides, no dynamics have actually happened yet in the first premise, so if we despair in proving this one, the rest cannot become any easier either. For the second premise, there is not much that we can do either, because we will have to analyze the effect of the loop body α running once at least in order to be able to understand what happens if we run α repeatedly.

Yet, what's with the third premise $A \vdash [\alpha][\alpha]B$? We could just approach it as is and try to prove it directly using the dL proof rules. Alternatively, however, we could try to take advantage of the fact that it is the same hybrid program α that is running in the first and the second modality. Maybe they should have something in common that we can exploit as part of our proof?

How could that work? Can we possibly find something that is true after the first run of α and is all we need to know about the state for $[\alpha]B$ to hold? Can we characterize the intermediate state after the first α and before the second α? Suppose we manage to do that and identify a formula E that characterizes the intermediate state in this way. How do we use this intermediate condition E to simplify our proof?

Recall the intermediate condition contract version of the sequential composition proof rule from Chap. 4 that we briefly revisited in Chap. 5:

$$
\text{H;} \quad \frac{A \to [\alpha]E \quad E \to [\beta]B}{A \to [\alpha;\beta]B}
$$

Chap. 5 ended up dismissing the intermediate contract rule H; in favor of the more general axiom

$$
[;] \quad [\alpha;\beta]P \leftrightarrow [\alpha][\beta]P
$$

But, let us revisit rule H; just the same and see whether we can learn something from its way of using intermediate condition E. The first obstacle is that the conclusion of the H; rule does not match the form we need for $A \vdash [\alpha][\alpha]B$. That's not a problem in principle, because we can use axiom [;] backwards from right-hand side to left-hand side in order to turn $A \vdash [\alpha][\alpha]B$ back into

$$
A \vdash [\alpha;\alpha]B
$$

and then use rule H; to generalize with an intermediate condition E in the middle. However, this is what we generally want to stay away from, because using the axioms both forwards and backwards can get our proof search into trouble because we might loop around trying to find a proof forever without making any progress, by simply using axiom [;] forwards and then backwards and then forwards again and so on until the end of time. Such a looping proof does not strike us as useful. Instead, we'll adopt a proof rule that has some of the properties of H; but is more general. It is called *generalization* and allows us to prove any stronger postcondition Q for a modality, i.e., a postcondition that implies the original postcondition P.

Lemma 7.4 (MR monotonicity right rule). *This is a derived proof rule:*

$$\text{MR} \frac{\Gamma \vdash [\alpha]Q, \Delta \quad Q \vdash P}{\Gamma \vdash [\alpha]P, \Delta}$$

Proof. Rule MR can be derived from the monotonicity rule M[·] from Lemma 5.13:

$$\text{cut} \frac{\Gamma \vdash [\alpha]Q, \Delta \quad \text{M[·]} \dfrac{Q \vdash P}{\Gamma, [\alpha]Q \vdash [\alpha]P, \Delta}}{\Gamma \vdash [\alpha]P, \Delta}$$

$\square$

Because the proof rule MR is just a cut away from monotonicity rule M[·], we will also just say that we prove by M[·] even if we really also used it together with a cut as in rule MR.

If we apply rule MR on the third premise $A \vdash [\alpha][\alpha]B$ of our bounded-model-checking-style proof attempt with the intermediate condition E for Q that we assume we have identified, then we end up with

$$\text{MR} \frac{A \vdash [\alpha]E \quad E \vdash [\alpha]B}{A \vdash [\alpha][\alpha]B}$$

Let us try to use this principle to see whether we can find a way to prove

$$A \to B \wedge [\alpha](B \wedge [\alpha](B \wedge [\alpha](B \wedge [\alpha][\alpha^*]B))) \tag{7.18*}$$

Using rules $\wedge$R and MR a number of times for a sequence of intermediate conditions E_1, E_2, E_3 derives the proof in Fig. 7.7.

This particular derivation is still not very useful because it still has a loop in one of the premises, which is what we had originally started out with in (7.14) in the first place. But the derivation hints at a useful way we could possibly shortcut proofs. To lead to a proof of the conclusion, the above derivation requires us to prove the premises

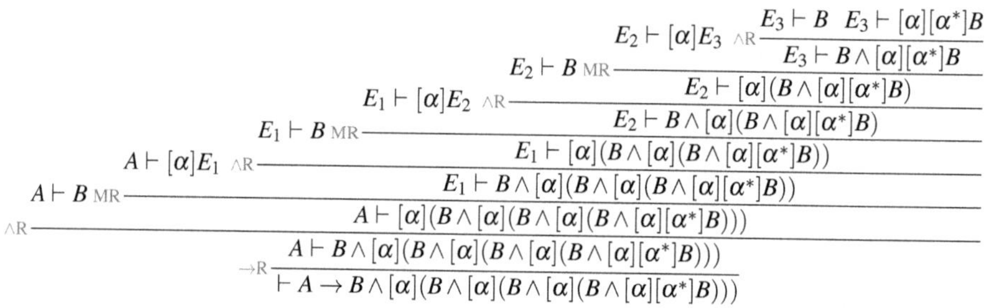

$$\cfrac{\cfrac{\cfrac{\cfrac{\cfrac{\cfrac{\cfrac{E_2 \vdash [\alpha]E_3 \;_{\wedge R} \cfrac{E_3 \vdash B \quad E_3 \vdash [\alpha][\alpha^*]B}{E_3 \vdash B \wedge [\alpha][\alpha^*]B}}{E_2 \vdash [\alpha](B \wedge [\alpha][\alpha^*]B)}}{E_1 \vdash [\alpha]E_2 \;_{\wedge R} \cfrac{}{E_2 \vdash B \wedge [\alpha](B \wedge [\alpha][\alpha^*]B)}}}{E_1 \vdash [\alpha](B \wedge [\alpha](B \wedge [\alpha][\alpha^*]B))}}{A \vdash [\alpha]E_1 \;_{\wedge R} \cfrac{}{E_1 \vdash B \wedge [\alpha](B \wedge [\alpha](B \wedge [\alpha][\alpha^*]B))}}}{A \vdash [\alpha](B \wedge [\alpha](B \wedge [\alpha](B \wedge [\alpha][\alpha^*]B)))}}{A \vdash B \wedge [\alpha](B \wedge [\alpha](B \wedge [\alpha](B \wedge [\alpha][\alpha^*]B)))}}{\vdash A \to B \wedge [\alpha](B \wedge [\alpha](B \wedge [\alpha](B \wedge [\alpha][\alpha^*]B)))}$$

(Proof tree, rules labeled MR, ∧R, ∧R, MR, ∧R, MR, ∧R, →R)

Fig. 7.7 Loops of proofs: intermediate generalizations

$$A \vdash [\alpha]E_1$$
$$E_1 \vdash [\alpha]E_2$$
$$E_2 \vdash [\alpha]E_3$$

as well as some other premises. What is an easy way to make that happen? What if all the intermediate conditions E_i were the same? Let's assume they are all the same condition E, that is, $E_1 \equiv E_2 \equiv E_3 \equiv E$. In that case, most of the resulting premises actually turn out to be one and the same premise:

$$E \vdash B$$
$$E \vdash [\alpha]E$$

except for the two left-most and the right-most premise. *Let us leverage this observation and develop a proof rule for which the same intermediate condition is used for all iterations of the loop.* Furthermore, we would even know the first premise

$$A \vdash [\alpha]E$$

if we could prove that the precondition A implies E:

$$A \vdash E$$

because we already have $E \vdash [\alpha]E$ as one of the premises.

7.7.3 Invariant Proofs of Loops

The condition $E \vdash [\alpha]E$ identified in the previous section seems particularly useful, because it basically says that whenever the system α starts in a state satisfying E, it will stay in E, no matter which of the states in E it was when the system started in the first place. It sounds like the system α^* cannot get out of E if it starts in E, since all that α^* can do is to repeat α some number of times. But every time we repeat α, the sequent $E \vdash [\alpha]E$ expresses that we cannot leave E that way. So no matter how often our CPS repeats α^*, it will still reside in E.

The other condition that the previous section identified as crucial is $E \vdash B$. And, indeed, if E does not imply the postcondition B that we have been interested in in the first place, then E is a perfectly true invariant of the system, but not really a very useful one as far as proving B goes.

What else could go wrong in a system that obeys $E \vdash [\alpha]E$, i.e., where this sequent is valid, because we found a proof for it? Indeed, the other thing that could happen is that E is an invariant of the system that implies safety, but our system just does not initially start in E; then we still don't know whether it's safe. Taking all three conditions together, we arrive exactly at the loop induction rule from Lemma 7.3.

7.7.4 Alternative Forms of the Induction Axiom

In the literature [4, 7, 8, 10], the induction axiom is classically presented as

$$\text{II} \quad [\alpha^*](P \to [\alpha]P) \to (P \to [\alpha^*]P)$$

instead of the slightly stronger and more intuitive form developed in Sect. 7.3.1:

$$\text{I} \quad [\alpha^*]P \leftrightarrow P \wedge [\alpha^*](P \to [\alpha]P)$$

The classical axiom II is equivalent to the sufficiency direction "←" of the equivalence axiom I just by propositional rephrasing, because both axioms need both P and $[\alpha^*](P \to [\alpha]P)$ to imply $[\alpha^*]P$. The derivation of the necessity direction "→" of the equivalence axiom I from II needs a more elaborate argument.

The proof first derives the backwards iteration axiom from either the "←" necessity direction of induction axiom I (or axiom II) with the help of others.

Lemma 7.5 ($\overleftarrow{*}$ backwards iteration axiom). *This axiom is derived:*

$$\overleftarrow{*} \quad [\alpha^*]P \leftrightarrow P \wedge [\alpha^*][\alpha]P$$

Proof. The sufficiency direction "←" of axiom $\overleftarrow{*}$ directly derives from the sufficiency direction "←" of the induction axiom I or its equivalent classical axiom II using monotonicity rule $M[\cdot]$, because postcondition $[\alpha]P$ is stronger than $P \to [\alpha]P$:

$$\frac{\frac{\frac{*}{[\alpha]P,P\vdash[\alpha]P}\,\mathrm{id}}{[\alpha]P\vdash P\to[\alpha]P}\to R}{\mathrm{M}[\cdot]\frac{}{[\alpha^*][\alpha]P\vdash[\alpha^*](P\to[\alpha]P)}}}{\land R\frac{\mathrm{id}\frac{*}{P\vdash P}\quad}{P,[\alpha^*][\alpha]P\vdash P\land[\alpha^*](P\to[\alpha]P)}}{\frac{\land L,!\frac{}{P\land[\alpha^*][\alpha]P\vdash[\alpha^*]P}}{\to R\frac{}{\vdash P\land[\alpha^*][\alpha]P\to[\alpha^*]P}}}$$

The necessity direction "$\to$" of $\overleftarrow{[^*]}$ is derived using $[^*]$,G,MR,$[]\land$ from axiom II or its equivalent sufficiency direction "$\leftarrow$" of axiom I as shown in Fig. 7.8.   □

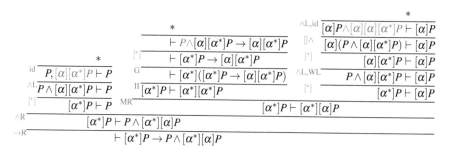

Fig. 7.8 Derivation of backwards unwinding axiom from alternative induction axiom

With the help of the backwards iteration axiom $\overleftarrow{[^*]}$, the proof of the necessity direction "$\to$" of axiom I is straightforward, because the only difference is the additional assumption P in the postcondition:

$$\land R\frac{\mathrm{id}\frac{*}{P\vdash P}\quad \overleftarrow{[^*]}\frac{\land L,WL\frac{\mathrm{M}[\cdot]\frac{\to R\frac{\mathrm{id}\frac{*}{[\alpha]P,P\vdash[\alpha]P}}{[\alpha]P\vdash P\to[\alpha]P}}{[\alpha^*][\alpha]P\vdash[\alpha^*](P\to[\alpha]P)}}{P\land[\alpha^*][\alpha]P\vdash[\alpha^*](P\to[\alpha]P)}}{[\alpha^*]P\vdash[\alpha^*](P\to[\alpha]P)}}{\to R\frac{[\alpha^*]P\vdash P\land[\alpha^*](P\to[\alpha]P)}{\vdash[\alpha^*]P\to P\land[\alpha^*](P\to[\alpha]P)}}$$

This completes the derivation of the induction axiom I from its classical formulation II using the other axioms.

This proof proved $[\alpha^*]P\to[\alpha^*][\alpha^*]P$, which has a stronger formulation.

Lemma 7.6 ($[^{**}]$ double iteration axiom). *This axiom is derived:*

$$[^{**}]\ \ [\alpha^*;\alpha^*]P\leftrightarrow[\alpha^*]P$$

Proof. The composition axiom [;] reduces the proof to two directions of which the direction "←" was already proved in the middle branch of Fig. 7.8:

$$
\begin{array}{c}
\cfrac{
\cfrac{
\cfrac{
\cfrac{*}{\wedge\mathrm{L,id}\ \overline{[\alpha^*]P\wedge[\alpha][\alpha^*][\alpha^*]P\vdash[\alpha^*]P}}}{
{}_{[^*]}\ [\alpha^*][\alpha^*]P\vdash[\alpha^*]P}
\qquad
\cfrac{*\ (\mathrm{Fig.}\ 7.8)}{[\alpha^*]P\vdash[\alpha^*][\alpha^*]P}}{
{}_{\leftrightarrow\mathrm{R}}\ \vdash[\alpha^*][\alpha^*]P\leftrightarrow[\alpha^*]P}}{
{}_{[;]}\ \vdash[\alpha^*;\alpha^*]P\leftrightarrow[\alpha^*]P}
\end{array}
$$

□

Derived axiom [**] is mostly meant to be used from left to right in order to collapse two subsequent loops into a single loop. Its diamond modality counterpart can also be useful to split a loop $\langle\alpha^*\rangle P$ into two separate loops $\langle\alpha^*;\alpha^*\rangle P$:

$$\langle^{**}\rangle\quad \langle\alpha^*;\alpha^*\rangle P\leftrightarrow\langle\alpha^*\rangle P$$

This splitting can be useful, e.g., to show that, on an empty soccer field, a robot can kick the ball into the goal in a control loop. After duplicating the control loop by axiom $\langle^{**}\rangle$, it is sufficient to show that the first control loop can navigate the robot close enough to the goal to have a chance to score, and then the second control loop can repeat until a goal is scored.

Exercises

7.1. Give a sequent proof for:

$$2gx=2gH-v^2\wedge x\geq 0\to[x'=v,v'=-g\ \&\ x\geq 0](2gx=2gH-v^2\wedge x\geq 0)$$

Does this property also hold if we remove the evolution domain constraint $x\geq 0$? That is, is the following formula valid?

$$2gx=2gH-v^2\wedge x\geq 0\to[x'=v,v'=-g](2gx=2gH-v^2\wedge x\geq 0)$$

7.2. Section 7.3.1 argued that both P and $[\alpha]P$ follow from (7.3). Show how $[\alpha]P$ is, indeed, implied by applying the iteration axiom [*] one more time to (7.3).

7.3 (Invariant candidates for bouncing balls). Could the bouncing ball use any of the following formulas as invariants to prove (7.1)? Explain why.

$$j_{(x,v)}\overset{\mathrm{def}}{\equiv}(x=0\vee x=H)\wedge v=0$$

$$j_{(x,v)}\overset{\mathrm{def}}{\equiv}0\leq x\wedge x\leq H\wedge v^2\leq 2gH$$

$$j_{(x,v)}\overset{\mathrm{def}}{\equiv}0\leq x\wedge x\leq H\wedge v\leq 0$$

7.4. Conduct a sequent proof for (7.12) without the monotonicity rule MR.

7.5 (Damped bouncing balls). Section 7.4 proved the bouncing-ball formula (7.12) in the dL sequent calculus for the case $c = 1$ of no damping at the bounce. Putting aside Quantum, actual bouncing balls are less perfect and only achieve damping coefficients satisfying $0 \leq c \leq 1$. Identify a suitable invariant and conduct a sequent calculus proof for this generalization.

7.6. Identify loop invariants proving the following dL formulas:

$$x > 1 \rightarrow [(x:=x+1)^*]x \geq 0$$
$$x > 5 \rightarrow [(x:=2)^*]x > 1$$
$$x > 2 \wedge y \geq 1 \rightarrow [(x:=x+y;y:=y+2)^*]x > 1$$
$$x > 2 \wedge y \geq 1 \rightarrow [(x' = y)^*]x > 1$$
$$x > 2 \wedge y \geq 1 \rightarrow [(x:=y;x' = y)^*]x \geq 1$$
$$x = -1 \rightarrow [(x:=2x+1)^*]x \leq 0$$
$$x = -1 \rightarrow [(\{x' = 2\})^*]x \geq -5$$
$$x = 5 \wedge c > 1 \wedge d > -c \rightarrow [(\{x' = c+d\})^*]x \geq 0$$
$$x = 1 \wedge u > x \rightarrow [(x:=2;\{x' = x^2+u\})^*]x \geq 0$$
$$x = 1 \wedge y = 2 \rightarrow [(x:=x+1;\{x' = y,y' = 2\})^*]x \geq 0$$
$$x \geq 1 \wedge v \geq 0 \rightarrow [(\{x' = v,v' = 2\})^*]x \geq 0$$
$$x \geq 1 \wedge v > 0 \wedge A > 0 \rightarrow [((a:=0 \cup a:=A);\{x' = v,v' = a\})^*]x \geq 0$$

7.7. Give a direct semantic soundness proof for rule ind and contrast its soundness proof with the soundness proof of rule loop to observe similarities and differences.

7.8 (Constant parameter assumptions). As Example 7.3 showed, it would be unsound for either the loop invariant rule or its core counterpart rule ind to keep the context Γ, Δ in the induction step (or in the third premise of rule loop). With adequate care, *some* select formulas from Γ and Δ can still be kept without losing soundness. These are the formulas from Γ and Δ that only refer to constant parameters that do not change during the HP α^*. Give a semantic argument why such a constant formula q in Γ or Δ can be kept soundly. Then show how q can be retained in the induction step with the help of the vacuous axiom V from Sect. 5.6.2.

7.9 (Far induction). The far induction axiom is for quicker inductions since its induction step takes two steps at once. Is it sound? Prove that it is or show a counterexample.

$$[\alpha^*]P \leftrightarrow P \wedge [\alpha^*](P \rightarrow [\alpha][\alpha]P)$$

7.10 (Unwound). The appendix motivated the loop invariant proof rule via systematic unwinding considerations in Sect. 7.7.2. We unwound loops in two different ways either directly in Fig. 7.6 or with intermediate generalizations in Fig. 7.7. Both approaches ultimately took us to the same inductive principle. But if unwinding is all that we are interested in, then which of the two ways of unwinding is more efficient? Which one produces feweer premises that are distractions in the argument? Which one has fewer choices of different intermediate conditions E_i in the first place?

7.11 (First arrival). Show that the first arrival axiom is sound, which says that if P is reachable by repeating α, then P is either *true* right away or one can repeat P to a state where P is not *true* yet but will become *true* after the next iteration of α:

$$\text{FA} \quad \langle \alpha^* \rangle P \to P \vee \langle \alpha^* \rangle (\neg P \wedge \langle \alpha \rangle P)$$

7.12 (*Dribbling basket balls). Identify a requirement on the initial state of the bouncing ball that allows it to move initially, so is more general than $v = 0$. Prove that this variation of the bouncing ball is safe.

References

[1]   Alessandro Cimatti, Sergio Mover, and Stefano Tonetta. SMT-based scenario verification for hybrid systems. *Formal Methods in System Design* **42**(1) (2013), 46–66. DOI: 10.1007/s10703-012-0158-0.

[2]   Edmund M. Clarke, Armin Biere, Richard Raimi, and Yunshan Zhu. Bounded model checking using satisfiability solving. *Form. Methods Syst. Des.* **19**(1) (2001), 7–34. DOI: 10.1023/A:1011276507260.

[3]   Andreas Eggers, Martin Fränzle, and Christian Herde. SAT modulo ODE: a direct SAT approach to hybrid systems. In: *Automated Technology for Verification and Analysis, 6th International Symposium, ATVA 2008, Seoul, Korea, October 20-23, 2008. Proceedings.* Ed. by Sung Deok Cha, Jin-Young Choi, Moonzoo Kim, Insup Lee, and Mahesh Viswanathan. Vol. 5311. LNCS. Berlin: Springer, 2008, 171–185. DOI: 10.1007/978-3-540-88387-6_14.

[4]   David Harel, Dexter Kozen, and Jerzy Tiuryn. *Dynamic Logic.* Cambridge: MIT Press, 2000.

[5]   Soonho Kong, Sicun Gao, Wei Chen, and Edmund M. Clarke. dReach: δ-reachability analysis for hybrid systems. In: *Tools and Algorithms for the Construction and Analysis of Systems - 21st International Conference, TACAS 2015, Held as Part of the European Joint Conferences on Theory and Practice of Software, ETAPS 2015, London, UK, April 11-18, 2015. Proceedings.* Ed. by Christel Baier and Cesare Tinelli. Vol. 9035. LNCS. Berlin: Springer, 2015, 200–205.

[6]   Carla Piazza, Marco Antoniotti, Venkatesh Mysore, Alberto Policriti, Franz Winkler, and Bud Mishra. Algorithmic algebraic model checking I: challenges from systems biology. In: *CAV.* Ed. by Kousha Etessami and Sriram K. Rajamani. Vol. 3576. LNCS. Berlin: Springer, 2005, 5–19. DOI: 10.1007/11513988_3.

[7]   André Platzer. The complete proof theory of hybrid systems. In: *LICS*. Los
      Alamitos: IEEE, 2012, 541–550. DOI: 10.1109/LICS.2012.64.

[8]   André Platzer. A complete uniform substitution calculus for differential dy-
      namic logic. *J. Autom. Reas.* **59**(2) (2017), 219–265. DOI: 10.1007/s108
      17-016-9385-1.

[9]   André Platzer and Edmund M. Clarke. The image computation problem in
      hybrid systems model checking. In: *HSCC*. Ed. by Alberto Bemporad, Anto-
      nio Bicchi, and Giorgio C. Buttazzo. Vol. 4416. LNCS. Springer, 2007, 473–
      486. DOI: 10.1007/978-3-540-71493-4_37.

[10] Vaughan R. Pratt. Semantical considerations on Floyd-Hoare logic. In: *17th
      Annual Symposium on Foundations of Computer Science, 25-27 October
      1976, Houston, Texas, USA*. Los Alamitos: IEEE, 1976, 109–121. DOI: 10
      .1109/SFCS.1976.27.

Chapter 8
Events & Responses

Synopsis Having already understood the analytical implications of control loops in cyber-physical systems via their logical characterizations with unwinding and invariants, this chapter investigates their impact on the important design paradigm of *event-triggered control systems*, also known as event-driven control systems. In such a system the controllers respond to certain events whenever they happen. The resulting event detection for the various events of interest is then executed in a control loop. A safe controller makes the appropriate response for each of the events of relevance. This direct response principle for the respective events provides systematic ways of designing event-triggered CPS controllers and leads to relatively simple safety arguments. But event-triggered systems are hard if not impossible to implement, because they require perfect event detection. That makes this chapter an ideal setting for a number of crucial modeling lessons for CPS.

8.1 Introduction

Chapter 3 already saw the importance of control and loops in CPS models, Chap. 5 presented a way of unwinding loops iteratively to relate repetition to runs of the loop body, and Chap. 7 finally explained the central inductive proof principle for loops using invariants.

That has been a lot of attention on loops, but there are even more things to be learned about loops. This is no coincidence, because loops or other forms of repetitions are one of the most difficult challenges in CPS [3–6]. The other difficult challenge comes from the differential equations. If differential equations are simple and there are no loops, CPSs suddenly become easy (they are even decidable [4]).

This chapter will focus on how these two difficult parts of CPS interact: how loops interface with differential equations. That interface is ultimately the connection between the cyber and the physical part, which, as we have known since Chap. 2, is fundamentally represented by the evolution domain constraints that determine when physics pauses to let cyber look and act.

© Springer International Publishing AG, part of Springer Nature 2018
A. Platzer, *Logical Foundations of Cyber-Physical Systems*,
https://doi.org/10.1007/978-3-319-63588-0_8

This and the next chapter focus on two important paradigms for making cyber interface with physics to form cyber-physical systems. The two paradigms play an equally important rôle in classical embedded systems. One paradigm is that of *event-triggered control*, where responses to events dominate the behavior of the system, and an action is taken whenever one of the events is observed. The other paradigm is *time-triggered control*, which uses periodic actions to influence the behavior of the system at certain frequencies. Both paradigms follow naturally from an understanding of hybrid program principles for CPS. Event-triggered control will be studied in this chapter, while time-triggered control will be pursued in the next chapter. Both chapters come with complementary lessons that are important in the design of virtually any CPS model and controller but are also important in simple embedded systems. Events and correct responses to events will be our challenge of choice for this chapter.

Based on the understanding of loops from Chap. 7, the most important learning goals of this chapter are:

Modeling and Control: This chapter provides a number of crucial lessons for modeling CPSs. We develop an understanding of one important design paradigm for control loops in CPS: event-triggered control. The chapter studies ways of developing models and controls corresponding to this feedback mechanism, which is based on issuing appropriate control responses for each of the relevant events of a system. This will turn out to be surprisingly subtle to model. The chapter focuses on CPS models with continuous sensing, i.e., we assume that sensor data is available and can be checked always.

Computational Thinking: This chapter uses the rigorous reasoning techniques for CPS loops with invariants from Chap. 7 based on the axiomatic reasoning approach from Chap. 5 to study CPS models with event-triggered control. As a running example, the chapter builds on the bouncing ball that has served us so well for conveying subtleties of hybrid system models in an intuitive example. This time, we add control decisions to the bouncing ball, turning it into a vertical ping-pong ball, which retains the intuitive simplicity of the bouncing ball, while enabling us to develop generalizable lessons about how to design event-triggered control systems correctly. While the chapter could hardly claim to show how to verify CPS models of any appropriate scale, the foundations laid in this chapter definitely carry significance for numerous practical applications, because the design of safe event-triggered controllers follows the same principles developed in a simple representative example here. It is easier to first see how cyber and physics interact in an event-triggered way for the significantly simpler and familiar phenomenon of bouncing balls, which provides the same generalizable lessons but in a simpler setting.

CPS Skills: This chapter develops an understanding of the precise semantics of event-triggered control, which can often be surprisingly subtle even if superficially simple. This understanding of the semantics will also guide our intuition of the operational effects caused by event-triggered control. Finally, the chapter shows a brief first glimpse of higher-level model-predictive control and design by invariant, even if a lot more can be said about that topic.

using loop invariants
design event-triggered control

modeling CPS
event-triggered control
continuous sensing
feedback mechanisms
control vs. physics

semantics of event-triggered control
operational effects
model-predictive control

8.2 The Need for Control

Having gotten accustomed to the little acrophobic bouncing ball Quantum since Chap. 4, this chapter will simply stick to that. Yet, Quantum asks for more action now, for he had so far no choice but to wait until he was down on the ground at height $x = 0$. When his patience paid off so that he finally observed height $x = 0$, then his *only* action was to make his velocity bounce back up. Frustrated by this limited menu of actions to choose from, Quantum begs for a ping-pong paddle. Thrilled at the opportunities opened up by flailing around with a ping-pong paddle, Quantum first performs some experiments to use it in all kinds of directions. But he never knew where he was going to land if he tried the ping-pong paddle sideways so he quickly gave up the thought of sideways actuation. The ball probably got so accustomed to his path of going up and down on the spot that he embraced the thought of keeping it that way. With the help of the ping-pong paddle, Quantum has high hopes to do the same, just faster without risking the terrified moments inflicted on him by his acrophobic attitude to heights. Setting aside all Münchausian concerns about how effective ping-pong paddles can be for the ball if the ball is using the paddle on itself in light of Newton's third law about opposing forces, let us investigate this situation regardless.[1] After all, the ping-pong-crazy bouncing ball Quantum still has what it takes to make control interesting: the dynamics of a physical system and decisions on when to react and how to react to the observed status of the system.

Chapter 7 proved the undamped bouncing ball with repetitions (shown in Fig. 8.1):

---

[1] If you find it hard to imagine a bouncing ball that uses a ping-pong paddle to pat itself on its top to propel itself back down to the ground again, just step back and consider the case where the ping-pong ball has a remote control to activate a device that moves the ping-pong paddle downwards. That will do just as well, but is less fun. Besides, Baron Münchhausen would surely be horribly disappointed if we settled for such a simple explanation for the need for control.

Fig. 8.1 Sample trajectory of a bouncing ball bouncing freely (plotted as position over time)

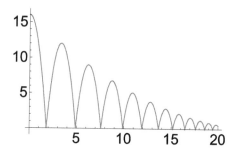

$$0 \leq x \wedge x = H \wedge v = 0 \wedge g > 0 \wedge 1 = c \rightarrow$$
$$[(\{x' = v, v' = -g \,\&\, x \geq 0\}; (?x = 0; v := -cv \cup ?x \neq 0))^*](0 \leq x \wedge x \leq H)$$
$$(7.12^*)$$

With this pretty complete understanding of bouncing balls, let's examine how to turn the simple bouncing ball into a fancy ping-pong ball using clever actuation of a ping-pong paddle. Quantum tried to actuate the ping-pong paddle. By making the ping-pong paddle solely move up and down, Quantum ultimately figured out that the ball would go back down pretty fast as soon as he got a pat on the top from the paddle. He also learned that the upwards direction turned out to be not just difficult but also rather dangerous. Moving the ping-pong paddle upwards from underneath the ball was rather tricky and only made the ball fly up even higher than before. Yet, that is what the acrophobic bouncing ball Quantum did not enjoy at all, so he only ever used the ping-pong paddle to push the ball downwards. Moving the ping-pong paddle sideways would make the bouncing ball leave its favorite path of going up and down on the same spot.

8.2.1 Events in Control

As a height that Quantum feels comfortable with, he chooses the magic number 5 and so he wants to establish $0 \leq x \leq 5$ to always hold as Quantum's favorite safety condition. The ball further installs the ping-pong paddle at a similar height so that he can actuate somewhere between height 4 and 5. He exercises great care to make sure he only moves the paddle downwards when the ball is underneath, never upwards when it is above, because that would take him frightfully high up. Thus, the effect of the ping-pong paddle will only be to reverse the ball's direction. For simplicity, Quantum figures that being hit by a ping-pong paddle might have a similar effect to being hit by the floor, except with a possibly different bounce factor $f \geq 0$ instead of the damping coefficient c.[2] So the paddle actuated this way is simply assumed to have the effect $v := -fv$. Since Quantum can decide to use the ping-pong paddle as

---

[2] The real story is a bit more complicated, but Quantum does not know any better yet.

Fig. 8.2 Sample trajectory of
a ping-pong ball (plotted as
position over time) with the
indicated ping-pong paddle
actuation range

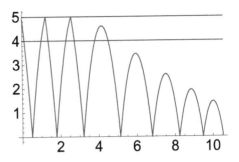

he sees fit (within the ping-pong paddle's reach between height 4 and 5), the ping-pong model is obtained from the bouncing-ball model by adding this additional (nondeterministic) choice to the HP. A sample trajectory for the ping-pong ball, where the ping-pong paddle is used twice is illustrated in Fig. 8.2. Observe how the use of the ping-pong paddle (here only at height $x = 5$) makes the ball bounce back faster.

Taking these thoughts into account, the ball devises a new and improved HP for ping-pong and conjectures its safety as expressed in the following dL formula:

$$0 \leq x \wedge x \leq 5 \wedge v \leq 0 \wedge g > 0 \wedge 1 \geq c \geq 0 \wedge f \geq 0 \rightarrow$$
$$[(\{x' = v, v' = -g \, \& \, x \geq 0\};$$
$$(?x = 0; v := -cv \cup ?4 \leq x \leq 5; v := -fv \cup ?x \neq 0))^*](0 \leq x \leq 5) \tag{8.1}$$

Having taken the *Principle of Cartesian Doubt* from Chap. 4 to heart, the aspiring ping-pong ball Quantum first scrutinizes conjecture (8.1) before setting out to prove it. What could go wrong?

For one thing, (8.1) allows the right control options of using the paddle by $?4 \leq x \leq 5; v := -fv$ but it also always allows the wrong choice $?x \neq 0$ when above ground. Remember that nondeterministic choices are just that: nondeterministic! So if Quantum is unlucky, the HP in (8.1) could run so that the middle choice is never chosen and, if the ball has a large downwards velocity v initially, it will jump back up higher than 5 even if it was below 5 initially. That scenario falsifies (8.1). A concrete counterexample can be constructed, e.g., from initial state ω with

$$\omega(x) = 5, \omega(v) = -10^{10}, \omega(c) = \frac{1}{2}, \omega(f) = 1, \omega(g) = 10$$

A less extreme scenario is shown in Fig. 8.3, where the first control at around time 3 works flawlessly but the second event is missed.

Despite this setback in his first control attempt, Quantum is thrilled by the extra prospect of a proper control decision for him to make. So Quantum "only" needs to figure out how to restrict the control decisions such that nondeterminism will only ever take one of the (possibly many) correct control choices, quite a common problem in CPS control. How can Quantum fix this bug in his control and turn

Fig. 8.3 Sample trajectory of a ping-pong ball (plotted as position over time) that misses one event to actuate the ping-pong paddle

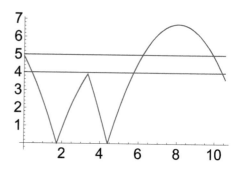

himself into a daring ping-pong ball? The problem with the controller in (8.1) is that it permits too many choices, some of which are unsafe. Restricting these choices and making them more deterministic is what it takes to ensure the ping-pong paddle is actuated as intended:

$$0 \leq x \wedge x \leq 5 \wedge v \leq 0 \wedge g > 0 \wedge 1 \geq c \geq 0 \wedge f \geq 0 \rightarrow$$
$$\left[(\{x' = v, v' = -g \,\&\, x \geq 0\}; \right.$$
$$\left. (?x = 0; v := -cv \cup ?4 \leq x \leq 5; v := -fv \cup ?x \neq 0 \wedge x < 4 \vee x > 5))^{*} \right] (0 \leq x \leq 5)$$

(8.2)

Recalling the if$(E)\,\alpha$ else β statement from Chap. 3, the same system can be modeled equivalently:

$$0 \leq x \wedge x \leq 5 \wedge v \leq 0 \wedge g > 0 \wedge 1 \geq c \geq 0 \wedge f \geq 0 \rightarrow$$
$$\left[(\{x' = v, v' = -g \,\&\, x \geq 0\}; \right.$$
$$\left. (?x = 0; v := -cv \cup ?x \neq 0; \text{if}(4 \leq x \leq 5)\,v := -fv))^{*} \right] (0 \leq x \leq 5)$$

Or, using if-then-else again, as the even shorter equivalent formula

$$0 \leq x \wedge x \leq 5 \wedge v \leq 0 \wedge g > 0 \wedge 1 \geq c \geq 0 \wedge f \geq 0 \rightarrow$$
$$\left[(\{x' = v, v' = -g \,\&\, x \geq 0\}; \right.$$
$$\left. \text{if}(x = 0)\,v := -cv \, \text{else if}(4 \leq x \leq 5)\,v := -fv)^{*} \right] (0 \leq x \leq 5)$$

(8.3)

Is conjecture (8.3) valid?

Before you read on, see if you can find the answer for yourself.

8.2.2 Event Detection

The problem with the controller in (8.3) is that, even though it exercises the appropriate control choice whenever the controller runs, the model does not ensure the controller will ever run at all when needed. The paddle control only runs after the differential equation stops, which can be almost any time. The differential equation is only guaranteed to stop when the ball bounces on the ground ($x = 0$), because its evolution domain constraint $x \geq 0$ would not be satisfied any longer on its way further down. Above ground, the differential equation model does not provide any constraints on how long it might evolve. Recall from Chap. 2 that the semantics of differential equations is nondeterministic in that the system can follow a differential equation *any amount of time*, as long as it does not violate the evolution domain constraints. In particular, the HP in (8.3) might miss the interesting *event* $4 \leq x \leq 5$ that the ping-pong ball's paddle control wanted to respond to. The system might simply skip over that region by following the differential equation $x' = v, v' = -g \& x \geq 0$ obliviously until the event $4 \leq x \leq 5$ has passed.

How can the HP from (8.3) be modified to make sure that the event $4 \leq x \leq 5$ will always be noticed and never missed?

> Before you read on, see if you can find the answer for yourself.

Essentially the only way to prevent the system from following a differential equation for too long is to restrict the evolution domain constraint, which is the predominant way to make cyber and physics interact. Indeed, that is what the evolution domain constraint $\ldots \& x \geq 0$ in (8.3) did in the first place. Even though this domain was introduced for different reasons (first principle arguments that light balls never fall through solid ground), its secondary effect was to make sure that the ground controller $?x = 0; v := -cv$ will never miss the right time to take action and reverse the direction of the ball from falling to climbing.

> **Note 40 (Evolution domains detect events)** Evolution domain constraints of differential equations in hybrid programs can detect events. That is, they can make sure the system evolution stops whenever an event happens on which the control wants to take action. Without such evolution domain constraints, the controller is not necessarily guaranteed to execute but may miss the event.

Following these thoughts further indicates that the evolution domain somehow ought to be augmented with more constraints that ensure the interesting event $4 \leq x \leq 5$ will never be missed accidentally. How can this be done? Should the event be conjoined to the evolution domain as follows?

$$0 \leq x \wedge x \leq 5 \wedge v \leq 0 \wedge g > 0 \wedge 1 \geq c \geq 0 \wedge f \geq 0 \rightarrow$$
$$\big[(\{x' = v, v' = -g \& x \geq 0 \wedge 4 \leq x \leq 5\};$$
$$\text{if}(x = 0) v := -cv \, \text{else if}(4 \leq x \leq 5) v := -fv)^* \big] (0 \leq x \leq 5)$$

> Before you read on, see if you can find the answer for yourself.

Of course not! This evolution domain would be entirely counterfactual and *require* the ball to always be at a height between 4 and 5, which is hardly the right physical model for any self-respecting bouncing ball. How could the ball ever fall to the ground and bounce back, this way? It couldn't.

Yet, on second thoughts, the way the event $x = 0$ got detected by the HP was not by including $\ldots \& x = 0$ in the evolution domain constraint, either, but by merely including the inclusive limiting constraint $\ldots \& x \geq 0$, which made sure the system could perfectly well evolve before the event domain $x = 0$, but that it just couldn't rush past the event $x = 0$. What would the inclusion of such an inclusive limiting constraint correspond to for the intended ping-pong paddle event $4 \leq x \leq 5$?

When the ball is hurled up into the sky, the last point at which action has to be taken to make sure not to miss the event $4 \leq x \leq 5$ is $x = 5$. The corresponding inclusive limiting constraint $x \leq 5$ thus should be somewhere in the evolution domain constraint. This is in direct analogy to the fact that the rôle of the evolution domain constraint $x \geq 0$ is to guarantee detection of the discrete event $x = 0$ in the discrete action that makes the ball bounce back up.

$$0 \leq x \wedge x \leq 5 \wedge v \leq 0 \wedge g > 0 \wedge 1 \geq c \geq 0 \wedge f \geq 0 \rightarrow$$
$$\big[\big(\{x' = v, v' = -g \,\&\, x \geq 0 \wedge x \leq 5 \}; \quad\quad\quad (8.4)$$
$$\mathsf{if}(x = 0)\, v := -cv\, \mathsf{else\, if}(4 \leq x \leq 5)\, v := -fv \big)^* \big] (0 \leq x \leq 5)$$

Is this the right model? Is dL formula (8.4) valid? Will its HP ensure that the critical event $4 \leq x \leq 5$ will not be missed?

> Before you read on, see if you can find the answer for yourself.

Formula (8.4) is valid. And, yet, (8.4) *is not at all the appropriate formula to consider!* It is absolutely crucial to understand why.

First, however, note that the HP in (8.4) allows the use of the ping-pong paddle anywhere in the height range $4 \leq x \leq 5$. Its evolution domain constraint enforces that this event $4 \leq x \leq 5$ will be noticed at the latest at height $x = 5$. So when exactly the ping-pong paddle is exercised in that range is nondeterministic (even if the control is written deterministically), because the duration of the differential equation is still chosen nondeterministically. This allows the ping-pong paddle to be controlled at the last height $x = 5$ or before it reaches height $x = 5$ as in Fig. 8.4.

Fig. 8.4 Sample trajectory of a ping-pong ball (plotted as position over time) with the indicated ping-pong paddle actuation range, sometimes actuating early, sometimes late

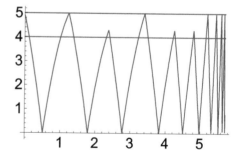

Notice that (8.4) does not make sure that the critical event $4 \leq x \leq 5$ will not be missed in the case of a ball that is climbing up above the lower trigger 4 but starts falling down again already before it exceeds the upper trigger 5 of the event. Such a possible behavior of the ping-pong ball was already shown in Fig. 8.2. Yet, this is not actually problematic, because missing out on the chance to actuate the ping-pong paddle in a situation where the paddle is not needed to ensure height control is just missing an opportunity for fun, not missing a critical control choice.

But there is a much deeper problem with (8.4). Formula (8.4) is perfectly valid. But why? Because all runs of the differential equation $x' = v, v' = -g \,\&\, x \geq 0 \wedge x \leq 5$ remain within the safety condition $0 \leq x \leq 5$ *by construction*. None of them are ever allowed to leave the region $x \geq 0 \wedge x \leq 5$, which, after all, is their evolution domain constraint. So formula (8.4) is trivially safe, because it says that a system that is constrained to not leave $x \leq 5$ cannot leave $x \leq 5$, which is a pretty trivial insight. A more careful proof involves that, every time around the loop, the postcondition holds trivially, because the differential equation's evolution constraint maintains it by definition, and the subsequent discrete control never changes the only variable x on which the postcondition depends. Hold on, the loop does not have to run but might be skipped over by zero iterations. Yet, in that case, the precondition ensures the postcondition, so, indeed, (8.4) is valid, but only quite trivially so.

Note 41 (Non-negotiability of physics) It is a good idea to make systems safe by construction; but not by changing the laws of physics, because physics is unpleasantly non-negotiable. If the only reason why a CPS model is safe is because we forgot to model all relevant behavior of the real physical system and modeled another universe instead, then correctness statements about those inadequate models do not apply to reality. We do not make this world any safer by writing CPS programs for another universe!

One common cause of counterfactual models is too restrictive evolution domain constraints that rule out physically realistic behavior.

That is what happened in (8.4). Quantum got so carried away with trying not to miss the event $4 \leq x \leq 5$ that he forgot to include a behavior in the model that takes place *after* the event has happened.

Contrast this with the rôle of the evolution domain constraint $\ldots \,\&\, x \geq 0$, which came into the system because of physics: to model the guaranteed bouncing back from the ground and to prevent the ball from falling through the ground. The constraint $x \geq 0$ models physical limitations of balls which cannot fall through solid soil. The evolution domain constraint $\ldots \,\&\, x \leq 5$ got added to the ping-pong HP for an entirely different reason. It came into play to model what our controller does, and inaptly so, because our feeble attempt ruled out physical behavior that could actually have happened in reality. There is no reason to believe that physics would be so kind to only evolve within $x \leq 5$ just because our controller model wants to respond to an event then. Remember never to do that. Ever!

> **Note 42 (Physical constraints versus control constraints)** Some constraints of system models are included for physical reasons; other constraints are added later to describe the controller. Take care to ensure not to accidentally limit the behavior of physics when all you meant to do is impose a constraint on your system controller. Physics will not listen to your desires! This applies to evolution domain constraints but also other aspects of your system model such as tests. It is fine to limit the force that the ping-pong paddle exerts, because that is for the controller to decide. But it is not a good idea for a controller to limit or change the values of gravity or damping coefficients, because that is rather hard to implement without first leaving the planet.

To belabor the point more formally, we could have told directly from a proof of formula (8.4) that its model is broken. Let us use the following abbreviations:

$$A \stackrel{\text{def}}{\equiv} 0 \le x \wedge x \le 5 \wedge v \le 0 \wedge g > 0 \wedge 1 \ge c \ge 0 \wedge f \ge 0$$

$$B(x) \stackrel{\text{def}}{\equiv} 0 \le x \wedge x \le 5$$

$$\{x'' = ..\& x \ge 0 \wedge x \le 5\} \stackrel{\text{def}}{\equiv} \{x' = v, v' = -g \,\&\, x \ge 0 \wedge x \le 5\}$$

$$ctrl \stackrel{\text{def}}{\equiv} \text{if}(x = 0)\, v := -cv\, \text{else if}\,(4 \le x \le 5)\, v := -fv$$

The proof of formula (8.4) is completely straightforward:

$$
\begin{array}{c}
* \\ \hline
{}^{\mathbb{R}} \dfrac{x \ge 0 \wedge x \le 5 \vdash B(x)}{x \ge 0 \wedge x \le 5 \vdash [ctrl]B(x)} \,{}_{\text{V}}
\end{array}
$$

$$
{}_{\text{loop}} \dfrac{{}^{\mathbb{R}}\dfrac{*}{A \vdash B(x)} \qquad {}_{\text{dW}}\dfrac{\;\;\dfrac{}{x\ge0\wedge x\le5\vdash[ctrl]B(x)}}{\dfrac{B(x)\vdash[\{x''=..\&x\ge0\wedge x\le5\}][ctrl]B(x)}{B(x)\vdash[\{x''=..\&x\ge0\wedge x\le5\};ctrl]B(x)}}\,{}_{[;]} \qquad {}^{\mathbb{R}}\dfrac{*}{B(x)\vdash B(x)}}{A \vdash [(\{x''=..\&x\ge0\wedge x\le5\};ctrl)^*]B(x)}
$$

In addition to the vacuous axiom V from Lemma 5.11 for unmodified postconditions, this sequent proof uses the differential weakening proof rule dW, which will later be explored in full in Lemma 11.2 of Chap. 11, but is already easy to understand right now.[3] The differential weakening proof rule dW proves any postcondition P of a differential equation that is implied by its evolution constraint Q:

$$\text{dW}\ \dfrac{Q \vdash P}{\Gamma \vdash [x' = f(x)\,\&\,Q]P, \Delta}$$

This is a beautiful and simple proof of (8.4) but there's a catch. Can you spot it?

Before you read on, see if you can find the answer for yourself.

---

[3] For solvable differential equations, rule dW can, of course, be derived from solution axiom ['] by appropriate generalization steps. But rule dW is sound for any other differential equation, which is why it will be explored in Chap. 11 of Part II, which focuses on advanced differential equations.

The above proof of (8.4) worked entirely by the differential weakening rule dW and the vacuous axiom V. The differential weakening rule dW discards the differential equation $\{x' = v, v' = -g\}$ and works entirely from the evolution domain constraint. The vacuous axiom V discards the controller *ctrl* since its postcondition $B(x)$ does not read any of the variables that the HP *ctrl* writes, which is just v.

Well, that is a remarkably efficient proof. But the fact that it *entirely discarded* the differential equation and controller everywhere shows that the property is independent of the differential equation and controller and, thus, holds for any other controller that does not assign to x and any other differential equation (bouncing ball under gravity or not) that shares the same evolution domain constraint $x \geq 0 \wedge x \leq 5$.

> **Note 43 (Irrelevance)** After having constructed a proof, we can go back and check which assumptions, which evolution domain constraints, which differential equations, and which parts of the controller were needed to establish it. This is not just useful to identify crucial versus irrelevant assumptions, but is also insightful to identify which part of a controller or dynamics can be changed without affecting the truth of the property. If almost every aspect of a controller and differential equation turns out to be irrelevant, we should be wary about the model. More generally, the set of facts and expressions on which a proof depends informs us how general or how unique its conclusion is. If a proof was independent of most aspects of a hybrid program, then it states a very broadly applicable general property, but also does not tell us any particularly deep fact that is unique to this specific hybrid program.

Consequently, the fact that the differential equations and controllers were irrelevant for the above proof of (8.4) confirms again that its physics model is broken, because our practical experience clearly demonstrates that safety of the ping-pong ball really depends on a clever use of the ping-pong paddle.

8.2.3 Dividing Up the World

Let's make up for this modeling mishap by developing a model that has both behaviors, the behaviors before and after the event, just in different continuous programs so that the decisive event in the middle cannot accidentally be missed.

$$0 \leq x \wedge x \leq 5 \wedge v \leq 0 \wedge g > 0 \wedge 1 \geq c \geq 0 \wedge f \geq 0 \rightarrow$$
$$\left[\left(\left(\{x' = v, v' = -g \,\&\, x \geq 0 \wedge x \leq 5\} \cup \{x' = v, v' = -g \,\&\, x > 5\}\right);\right.\right. \quad (8.5)$$
$$\left.\left.\mathsf{if}(x = 0)\,v := -cv\,\mathsf{else}\,\mathsf{if}(4 \leq x \leq 5)\,v := -fv\right)^*\right](0 \leq x \leq 5)$$

Instead of the single differential equation with a single evolution domain constraint in (8.4), the HP in (8.5) has a (nondeterministic) choice between two differential equations, here actually both the same, with two different evolution domain constraints. The left continuous system is restricted to the lower physics space $x \geq 0 \wedge x \leq 5$, the right continuous system is restricted to the upper physics space $x > 5$. Every time the loop repeats, there is a choice of either the lower physics equation or the upper physics equation. But the system can never stay in these differential equations for too long, because, e.g., when the ball is below 5 and speeding upwards very fast, then it cannot stay in the left differential equation above height 5, so it will have to stop evolving continuously and give the subsequent controller a chance to inspect the state and respond in case the event $4 \leq x \leq 5$ happened.

Now dL formula (8.5) has a much better model of events than the ill-advised (8.4). Is formula (8.5) valid?

> Before you read on, see if you can find the answer for yourself.

The model in (8.5) is, unfortunately, horribly broken. We meant to split the continuous evolution space into the regions before and after the event $4 \leq x \leq 5$. But we overdid it, because the space is now fractured into two disjoint regions, the lower physics space $x \geq 0 \wedge x \leq 5$ and the upper physics space $x > 5$. How can the ping-pong ball ever transition from one to the other? Certainly, as the ball moves upwards within the lower physics space $x \geq 0 \wedge x \leq 5$, it will have to stop evolving at $x = 5$ at the latest. But then even if the loop repeats, the ball still cannot continue in the upper physics space $x > 5$, because it is not quite there yet. Being at $x = 5$, it is an infinitesimal step away from $x > 5$. Of course, Quantum will only ever move continuously along a differential equation. There is no continuous motion that would take the ball from the region $x \geq 0 \wedge x \leq 5$ to the disjoint region $x > 5$ without leaving them. In other words, the HP in (8.5) has accidentally modeled that there will never ever be a transition from lower to upper physics space nor the other way around, because of an infinitesimal gap in between.

> **Note 44 (Connectedness and disjointness in evolution domains)** Evolution domain constraints need to be thought out carefully, because they determine the respective regions within which the system can evolve. Disjoint or unconnected evolution domain constraint regions often indicate that the model will have to be thought over again, because there cannot be any continuous transitions from one domain to the other if they are not connected. Even infinitesimal gaps in domain constraints can cause mathematical curiosities in a model that make it physically unrealistic.

Let's close the infinitesimal gap between $x \geq 0 \wedge x \leq 5$ and $x > 5$ by including the boundary $x = 5$ in both domains:

$$0 \leq x \wedge x \leq 5 \wedge v \leq 0 \wedge g > 0 \wedge 1 \geq c \geq 0 \wedge f \geq 0 \rightarrow$$
$$\left[\left(\left(\left(x' = v, v' = -g \,\&\, x \geq 0 \wedge x \leq 5\right) \cup \left(x' = v, v' = -g \,\&\, x \geq 5\right)\right);\right.\right. \quad (8.6)$$
$$\left.\left.\mathsf{if}(x = 0)\, v := -cv \,\mathsf{else}\,\mathsf{if}(4 \leq x \leq 5)\, v := -fv\right)^*\right](0 \leq x \leq 5)$$

Now there is a proper separation into lower physics $x \geq 0 \wedge x \leq 5$ and upper physics $x \geq 5$ but the system can be in either physics space at the switching boundary $x = 5$. This makes it possible for the ball to pass from lower physics into upper physics or back, but only at the boundary $x = 5$, which, in this case, is the only point that the two evolution domain constraints have in common.

In fact, it is generally a good idea to work with overlapping (and often closed) evolution domain constraints to minimize the likelihood of accidentally causing infinitesimal gaps in the domains of the model.

Now dL formula (8.6) has a much better model of events than the ill-advised (8.4). Is formula (8.6) valid?

> Before you read on, see if you can find the answer for yourself.

When the ball is jumping up from the ground, the model in (8.6) makes it impossible for the controller to miss the event $4 \leq x \leq 5$, because the only evolution domain constraint in the HP that applies at the ground is $x \geq 0 \wedge x \leq 5$. And that evolution domain stops being true above 5. Yet, suppose the ping-pong ball was flying up from the ground following the continuous program in the left choice and then stopped its evolution at height $x = 4.5$, which always remains perfectly within the evolution domain $x \geq 0 \wedge x \leq 5$ and is, thus, allowed. Then, after the sequential composition between the middle and last line of (8.6), the controller in the last line of (8.6) runs, notices that the formula $4 \leq x \leq 5$ for the event checking is true, and changes the velocity according to $v := -fv$, corresponding to the assumed effect of a pat with the paddle. That is actually its only choice in such a state, because the controller is deterministic, much unlike the differential equation. Consequently, the velocity has just become negative since it was positive before as the ball was climbing up. So the loop can repeat and the differential equation runs again. However, then the differential equation might evolve until the ball is at height $x = 4.25$, which will eventually happen since its velocity stays negative till the ground. If the differential equation stops then, the controller will run again, determine that $4 \leq x \leq 5$ is true still and so take action to change the velocity to $v := -fv$ again. That will, however, make the velocity positive again, since it was previously negative as the ball was in the process of falling. Hence, the ball will keep on climbing now, which, again, threatens the postcondition $0 \leq x \leq 5$. Will this falsify (8.6) or is it valid?

> Before you read on, see if you can find the answer for yourself.

On second thoughts, that alone still will not cause the postcondition to evaluate to *false*, because the only way the bouncing ball can evolve continuously from $x = 4.25$ is by the continuous program in the left choice of (8.6). And that differential equation is restricted to the evolution domain $x \geq 0 \wedge x \leq 5$, which causes the controller to run before leaving $x \leq 5$. That is, the event $4 \leq x \leq 5$ will again be noticed by the controller so that the ping-pong paddle pats the ball back down; see Fig. 8.5.

However, exactly the same reasoning applies also to the case where the ball successfully made it up to height $x = 5$, which is the height at which any climbing ball has to stop its continuous evolution, because it would otherwise violate the evolution domain $x \geq 0 \wedge x \leq 5$. As soon as that happens, the controller runs, notices that the event $4 \leq x \leq 5$ is true, and responds with the ping-pong paddle to cause $v := -fv$.

Fig. 8.5 Sample trajectory
of a ping-pong ball (plotted
as position over time) with
the controller firing multiple
times for the same event

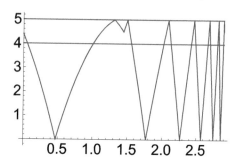

If, now, the loop repeats, yet the continuous evolution evolves for duration zero only, which is perfectly allowed, then the condition $4 \leq x \leq 5$ will still be true so that the controller again notices this "event" and responds with ping-pong paddle $v := -fv$. That will make the velocity positive, the loop can repeat, the continuous program on the right of the choice can be chosen since $x \geq 5$ holds true, and then the bouncing ball can climb and disappear into nothingness high up in the sky if only its velocity has been large enough. Such a behavior is shown in Fig. 8.6. The second illustration in Fig. 8.6 uses the artistic liberty of delaying the second ping-pong paddle use just a tiny little bit to make it easier to tell the two ping-pong paddle uses apart, even if that is not actually quite allowed by the HP model, because such behavior would actually be reflected by a third ping-pong paddle use as in Fig. 8.5.

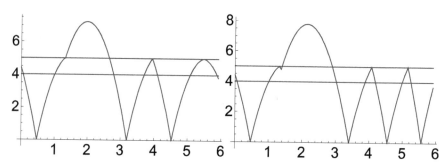

Fig. 8.6 Sample trajectory of a ping-pong ball (plotted as position over time) with the controller firing multiple times for the same event on the event boundary $x = 5$ between lower and upper physics

Ergo, (8.6) is not valid. What a pity! Poor Quantum would still have to be afraid of heights when following the control in (8.6). How can this problem be resolved?

Before you read on, see if you can find the answer for yourself.

8.2.4 Event Firing

The problem in (8.6) is that its left differential equation makes sure never to miss out on the event $4 \le x \le 5$ but its control may respond to it multiple times. Should each occasion of $4 \le x \le 5$ even be called a separate event? Quite certainly repeated responses to the same event according to control (8.6) cause trouble.

> **Note 45 (Multi-firing of events)** In event-triggered control, exercise care to ensure whether you want events to fire only once when they occur for the first time, or whether the system stays safe even if the same event is detected and responded to multiple times in a row. The latter systems are more robust.

One way of solving this problem is to change the condition in the controller to make sure it only responds to the $4 \le x \le 5$ event when the ball is on its way up, i.e., when its velocity is not negative ($v \ge 0$). That is what Quantum had in mind originally, but, in the great tradition of sophisticated systems, neglected to control it appropriately. The ping-pong paddle should only be actuated downwards when the ball is flying up.

These thoughts lead to the following variation:

$$0 \le x \wedge x \le 5 \wedge v \le 0 \wedge g > 0 \wedge 1 \ge c \ge 0 \wedge f \ge 0 \rightarrow$$
$$[(((\{x' = v, v' = -g \,\&\, x \ge 0 \wedge x \le 5\} \cup \{x' = v, v' = -g \,\&\, x \ge 5\}); \qquad (8.7)$$
$$\text{if}(x = 0)\,v := -cv\,\text{else if}(4 \le x \le 5 \wedge v \ge 0)\,v := -fv)^*](0 \le x \le 5)$$

Because the paddle action $v := -fv$ will disable the condition $v \ge 0$ for nonzero velocities, the controller in (8.7) can only respond once to the event $4 \le x \le 5$ to turn the upwards velocity into a downwards velocity, scaled by f (Exercise 8.1). Unlike in (8.6), this control decision cannot be immediately revertes inadvertently by the controller.

Is dL formula (8.7) valid?

Before you read on, see if you can find the answer for yourself.

Yes, formula (8.7) is valid. Finally! Note that it is still the case in (8.7) that, every time around the loop, there will be a nondeterministic choice to evolve within lower physics $x \ge 0 \wedge x \le 5$ or within upper physics $x \ge 5$. This choice is nondeterministic, so any outcome will be possible. If the left differential equation is chosen, the subsequent continuous evolution must be confined to $x \ge 0 \wedge x \le 5$ and stop before leaving that lower physics region to give the controller a chance to check for events and respond. If the right differential equation is chosen, the subsequent continuous evolution must be limited to $x \ge 5$ and must stop before leaving that upper physics region to give the controller a chance to inspect. In fact, the only way of leaving the upper physics space is downwards (with velocity $v < 0$), which, unlike in (8.6), will not trigger a response from the subsequent control in (8.7), because that controller checks for $v \ge 0$.

8.2.5 Event-Triggered Verification

How can dL formula (8.7) be proved, so that we have unquestionable evidence that it is, indeed, valid? The most critical element of a proof is finding a suitable invariant. What could be the invariant for proving (8.7)?

Before you read on, see if you can find the answer for yourself.

The postcondition

$$5 \geq x \geq 0 \qquad (8.8)$$

is an obvious candidate for an invariant. If it is true, it trivially implies the postcondition $0 \leq x \leq 5$ and it holds in the initial state. It is not inductive, though, because a state that satisfies only (8.8) could follow the second differential equation if it satisfied $x \geq 5$. In that case, if the velocity were positive, the invariant (8.8) would be violated immediately. Hence, at the height $x = 5$, the control has to make sure that the velocity is negative, so that the right differential equation in (8.7) has to stop immediately. Can (8.8) be augmented with a conjunction $v \leq 0$ to form an invariant?

$$5 \geq x \geq 0 \wedge v \leq 0$$

No, that would not work either, because the bounce on the ground immediately violates that invariant, since the whole point of bouncing is that the velocity will become positive again. In fact, the controller literally only ensures $v \leq 0$ at the event, which is detected at $x = 5$ at the latest that is the safety-critical decision point. Gathering these thoughts, it turns out that the dL formula (8.7) can be proved in the dL calculus using the invariant

$$5 \geq x \geq 0 \wedge (x = 5 \rightarrow v \leq 0) \qquad (8.9)$$

This invariant retains that the possible range of x is safe but is just strong enough to also remember the correct control choice at the event boundary $x = 5$. It expresses that the ball is either in lower physics space or at the boundary of both physics spaces. But if the ball is at the boundary of the physics spaces, then it is moving downwards. Invariant (8.9) follows the general principle of augmenting the expected postcondition with just enough information to guarantee safe control choices at all the critical handovers between the respective modes or decisions.

That is the reason why (8.9) is easily seen to be an invariant of (8.7). The invariant (8.7) is initially true, because the ball is initially in range and moving down. The invariant trivially implies the postcondition, because it consists of the postcondition plus an extra conjunction. The inductive step is most easily seen by considering cases. If the position before the loop body ran was $x < 5$, then the only physics possible to evolve is lower physics, which, by construction, implies the conjunct $5 \geq x \geq 0$ from its evolution domain constraint. The extra conjunct $x = 5 \rightarrow v \leq 0$ is true after the loop body has run, since, should the height actually be 5, which is the only case for which this extra conjunct is not already vacuously true, then the controller made sure to turn the velocity downwards by checking $4 \leq x \leq 5 \wedge$

$v \geq 0$ and negating the velocity. If the position before the loop body was $x \geq 5$ then the invariant (8.9) implies that the only position it could have had is $x = 5$ in which case either differential equation could be chosen. If the first differential equation is chosen, the reasoning for the induction step is as for the case $x < 5$. If the second differential equation is chosen, then the invariant (8.9) implies that the initial velocity is $v \leq 0$, which implies that the only possible duration that keeps the evolution domain constraint $x \geq 5$ of the upper physics true is duration 0, after which nothing has changed so the invariant still holds.

Observe how the scrutiny of a proof, which necessitated the transition from the broken invariant (8.8) to the provable invariant (8.9), has pointed us to subtleties with events and how ping-pong balls would become unsafe if they fired repeatedly. We discovered these issues by careful formal modeling with our "safety first" approach and a good dose of Cartesian Doubt. But had we not noticed it, the proof would not have let us get away with such oversights, because the (unreflected) invariant candidate (8.8) would not have worked, nor would the broken controller (8.6) have been provable. Of course, having a proof is not a replacement for exercising good judgment over a model to begin with.

Finally, recall that (global) invariants need to be augmented with the usual mundane assumptions about the unchanged variables, like $c \geq 0 \wedge g > 0 \wedge f \geq 0$, unless we use the more clever techniques from Sect. 7.5 that automatically preserve assumptions about constant parameters.

8.2.6 Event-Triggered Control Paradigm

The model that (8.7) and the other controllers in this section adhere to is called event-triggered control or sometimes also an event-triggered architecture.

> **Note 46 (Event-triggered control)** One common paradigm for controller design is *event-triggered control*, in which the controller runs in response to certain events that happen in the system. The controller might possibly run under other circumstances as well—when in doubt, the controller simply skips over without any effect if it does not want to change anything about the behavior of the system. But event-triggered controllers assume they will run for sure whenever certain events in the system happen.
>
> These events cannot be all too narrow, or else the system will not be implementable, though. For example, it is nearly impossible to build a controller that responds exactly at the point in time when the height of the bouncing ball is $x = 9.8696$. Chances are high that any particular execution of the system will have missed this particular height. Care must be taken in event-triggered design models also that the events do not inadvertently restrict the evolution of the system for the behavioral cases outside of events or after the events have happened. Those executions must still be verified.

Are we sure in model (8.7) that events are taken into account faithfully? That depends on what exactly we mean by an event like $4 \leq x \leq 5$. Do we mean that this event happens for the first time? Or do we mean every time this event happens? If multiple successive runs of the ping-pong ball's controller see this condition satisfied, do these count as the same or as separate instances of that event happening? Comparing the validity of (8.7) with the non-validity of (8.6) illustrates that these subtleties can have considerable impact on the system. Hence, a precise understanding of events and careful modeling is required.

The controller in (8.7) only takes an action for event $4 \leq x \leq 5$ when the ball is on the way up. Hence, the evolution domain constraint in the right continuous evolution is $x \geq 5$. If we wanted to model the occurrence of event $4 \leq x \leq 5$ also when the ball is on its way down, then we would have to have a differential equation with evolution domain $x \geq 4$ to make sure the system does not miss $4 \leq x \leq 5$ when the ball is on its way down either, without imposing that it would have to notice $x = 5$ already. This can be achieved by splitting the evolution domain regions appropriately, but was not necessary for (8.7) since the controller never responds to balls falling down, only those climbing up.

> **Note 47 (Subtleties with events)** Events are a slippery slope and great care needs to be exercised to use them without introducing an inadequate executional bias into the model.

There is a highly disciplined way of defining, detecting, and responding to general events in differential dynamic logic based on the there and back again axiom of differential dynamic logic [4]. That is, however, much more complicated than the simpler account shown here.

Finally, notice that the proof of (8.7) was almost independent of the differential equation and just a consequence of the careful choice of the evolution domain constraint to reflect the events of interest as well as getting the controller responses to these events right. That is, ultimately, the reason why the invariant (8.9) could be so simple. This also often contributes to making event-triggered controllers easier to get right.

> **Note 48 (Correct event-triggered control)** As long as a controller responds in the right ways to the right events, event-triggered controllers can be built rather systematically and are relatively easy to prove correct. But beware! You have to get the handling of events right, otherwise you only end up with a proof about counterfactual physics, which is not at all helpful since your actual CPS then follows an entirely different kind of physics.

8.2.7 Physics Versus Control Distinctions

> **Note 49 (Physics versus control)** Observe that some parts of hybrid program models represent facts and constraints from physics, and other parts represent controller decisions and choices. It is a good idea to keep the facts straight and remember which part of a hybrid program model comes from which. Especially, whenever a constraint is added because of a controller decision, it is good practice to carefully think through what happens if this is not the case. That is how we ended up splitting physics into different evolution domain constraints, for example.

Partitioning the hybrid program in the verified dL formula (8.7) into the parts that come from physics (typographically marked like physics) and the parts that come from control (typographically marked like control) leads to the following.

Proposition 8.1 (Event-triggered ping-pong is safe). *This* dL *formula is valid:*

$$0 \leq x \wedge x \leq 5 \wedge v \leq 0 \wedge g > 0 \wedge 1 \geq c \geq 0 \wedge f \geq 0 \rightarrow$$

$$\left[\left(\left(\{x' = v, v' = -g \,\&\, x \geq 0 \wedge x \leq 5\} \cup \{x' = v, v' = -g \,\&\, x \geq 5\} \right); \right. \right. \qquad (8.7^*)$$

$$\left. \left. \text{if}(x = 0)\, v := -cv \,\text{else}\, \text{if}(4 \leq x \leq 5 \wedge v \geq 0)\, v := -fv \right)^* \right] (0 \leq x \leq 5)$$

There could have been a second evolution domain constraint $x \geq 0$ for the physics in the second differential equation. But that evolution domain constraint was elided, because it is redundant in the presence of the evolution domain constraint $x \geq 5$ coming from the controller. Only controller constraints have been added compared to the initial physical model of the bouncing ball (7.12) that was entirely physics. This is a good indicator that the design was proper, because it did not alter the physics but merely added controller program parts, including control event detection by splitting differential equations into separate modes.

8.3 Summary

This chapter studied event-triggered control, which is one important principle for designing feedback mechanisms in CPSs and embedded systems. The chapter illustrated the most important aspects for a running example of a ping-pong ball. Even if the impoverished ping-pong ball went vertically and may not be the most exciting application of control in the world, the effects and pitfalls of control events were sufficiently subtle already to merit focusing on a simple intuitive case.

Event-triggered control assumes that all events are detected perfectly and right away. The event-triggered controller in (8.7) took some precautions by defining the event of interest for using the ping-pong paddle to be $4 \leq x \leq 5$. This may look like a big event in space to be noticed in practice, except when the ball moves too quickly, in which case the event $4 \leq x \leq 5$ is over rather quickly. However, the model still has $x \le 5$ as a hard limit in the evolution domain constraint to ensure that the event will never be missed in its entirety as the ball is rushing upwards.

Event-triggered control assumes permanent continuous sensing of the event of interest, because the hard limit of the event is ultimately reflected in the evolution domain constraint of the differential equation. This evolution domain constraint is checked permanently according to its semantics (Chap. 3). That gives event-triggered controllers quite simple mathematical models but also often makes them impossible to implement faithfully for lack of continuous-sensing capabilities.

Event-triggered control models can still be useful abstractions of the real world for systems that evolve slowly but sense quickly, because that is close enough to permanent sensing to still detect events quickly enough. Event-triggered control gives bad models for systems that change their state much more quickly than the sensors can catch up. Even in cases where event-triggered controllers are no good match for reality, they can still be helpful stepping stones for the analysis and design of the more realistic time-triggered controllers [1, 2] that the next chapter investigates. If a controller is not even safe when events are detected instantly and perfectly, it will not be safe when events may be discovered only sporadically with certain delay.

Exercises

8.1. Can the ping-pong paddle in (8.7) ever respond to the event $4 \le x \le 5$ twice in a row? What would happen if it did?

8.2. Is the ping-pong ball's loop invariant (8.9) also an invariant for just its two differential equations?

8.3. Are any of the following formulas invariants for proving (8.7)?

$$0 \le x \le 5 \wedge (x = 5 \rightarrow v \le 0) \wedge (x = 0 \rightarrow v \ge 0)$$
$$0 \le x < 5 \vee x = 5 \wedge v \le 0$$

8.4. Would the invariant (8.9) succeed in proving a variation of (8.7) in which the controller conjunction $\wedge v \ge 0$ is removed? If so explain why. If not, explain which part of the proof will fail and why.

8.5. Would a generalization of formula (8.7) be valid in which the assumption $v \le 0$ on the initial state is dropped? If yes, give a proof. If not, show a counterexample and explain how to fix this problem in a way that leads to a generalization of (8.7) that is still a valid formula.

8.6. Could we replace the two differential equations in (8.7) with a single differential equation and a disjunction of their evolution domain constraints to retain a valid formula?

$$0 \leq x \wedge x \leq 5 \wedge v \leq 0 \wedge g > 0 \wedge 1 \geq c \geq 0 \wedge f \geq 0 \rightarrow$$
$$[(\{x' = v, v' = -g \& (x \geq 0 \wedge x \leq 5) \vee x \geq 5\};$$
$$\text{if}(x = 0)\, v := -cv\, \text{else if}(4 \leq x \leq 5 \wedge v \geq 0)\, v := -fv)^*](0 \leq x \leq 5)$$

8.7. Conduct a sequent proof proving the validity of dL formula (8.7). In the sprit of proof irrelevance, carefully track which assumptions are used for which case?

8.8. The hybrid program in (8.4) was an inadequate model of physics because it terminated the world beyond height 5. Model (8.6) fixed this by introducing the same differential equation with the upper physics world and a nondeterministic choice. Would the following model have worked just as well? Would it be valid? Would it be an adequate model?

$$0 \leq x \wedge x \leq 5 \wedge v \leq 0 \wedge g > 0 \wedge 1 \geq c \geq 0 \wedge f \geq 0 \rightarrow$$
$$[(\{x' = v, v' = -g \& x \geq 0 \wedge (x = 5 \rightarrow v \leq 0)\};$$
$$\text{if}(x = 0)\, v := -cv\, \text{else if}(4 \leq x \leq 5)\, v := -fv)^*](0 \leq x \leq 5)$$

What about the evolution domain constraint $\ldots \& x \geq 0 \wedge x \neq 5$ instead?

8.9. What happens if we add an inner loop to (8.7)? Will the formula be valid? Will it be an adequate model of physics?

$$0 \leq x \wedge x \leq 5 \wedge v \leq 0 \wedge g > 0 \wedge 1 \geq c \geq 0 \wedge f \geq 0 \rightarrow$$
$$[((\{x' = v, v' = -g \& x \geq 0 \wedge x \leq 5\} \cup \{x' = v, v' = -g \& x \geq 5\})^*;$$
$$\text{if}(x = 0)\, v := -cv\, \text{else if}(4 \leq x \leq 5 \wedge v \geq 0)\, v := -fv)^*](0 \leq x \leq 5)$$

8.10. Modify the event-triggered controller such that its event detection also spots the event $4 \leq x \leq 5$ when descending at the latest at height 4 instead of always at height 5. Make sure this modified controller is safe and find a loop invariant for its proof.

8.11 (*). Design a variation of the event-triggered controller for the ping-pong ball that is allowed to use the ping-pong paddle within height $4 \leq x \leq 5$ but has a relaxed safety condition that accepts $0 \leq x \leq 2 \cdot 5$. Make sure to only force the use of the ping-pong paddle when necessary. Find an invariant and conduct a proof.

8.12 (2D ping-pong events). Design and verify the safety of a ping-pong controller that goes sideways with horizontal motion like in ordinary ping-pong matches.

8.13 (Robot chase). You are in control of a robot tailing another one in hot pursuit on a straight road. You can accelerate ($a := A$) or brake ($a := -b$). But so can the robot you're following! Your job is to design an event-triggered model whose controller makes sure the robots do not crash.

References

[1]   Sarah M. Loos. Differential Refinement Logic. PhD thesis. Computer Science Department, School of Computer Science, Carnegie Mellon University, 2016.

[2]   Sarah M. Loos and André Platzer. Differential refinement logic. In: *LICS*. Ed. by Martin Grohe, Eric Koskinen, and Natarajan Shankar. New York: ACM, 2016, 505–514. DOI: 10.1145/2933575.2934555.

[3]   André Platzer. Differential dynamic logic for hybrid systems. *J. Autom. Reas.* **41**(2) (2008), 143–189. DOI: 10.1007/s10817-008-9103-8.

[4]   André Platzer. The complete proof theory of hybrid systems. In: *LICS*. Los Alamitos: IEEE, 2012, 541–550. DOI: 10.1109/LICS.2012.64.

[5]   André Platzer. Differential game logic. *ACM Trans. Comput. Log.* **17**(1) (2015), 1:1–1:51. DOI: 10.1145/2817824.

[6]   André Platzer. A complete uniform substitution calculus for differential dynamic logic. *J. Autom. Reas.* **59**(2) (2017), 219–265. DOI: 10.1007/s108 17-016-9385-1.

Chapter 9
Reactions & Delays

Synopsis Time-triggered control systems are an important control paradigm. Event-triggered controllers focus on correct responses to appropriate events, which are assumed to be detected perfectly, which simplifies their design and analysis but makes them hard to implement. Time-triggered controllers, instead, focus on reacting to changes within certain reaction delays. Implementations become more straightforward using controllers that repeatedly execute within a certain maximum time period, or execute periodically with at least a certain frequency. While time-triggered control models can be easier to develop than event-triggered control models, the additional effects of reaction delays complicate the control logic and safety arguments.

9.1 Introduction

Chapter 7 explained the central proof principle for loops using invariants. Chapter 8 studied the important feedback mechanism of event-triggered control and made crucial use of invariants for rigorously reasoning about event-triggered control loops. Those invariants uncovered important subtleties with events that could be easily missed. In Chap. 8, we, in fact, already noticed these subtleties thanks to our "safety first" approach to CPS design, which guided us to exercise the scrutiny of Cartesian Doubt on the CPS model before even beginning a proof.

However, even if the final answer for the event-triggered controller for the ping-pong ball was rather clear and systematic, event-triggered control had an unpleasantly large number of modeling subtleties in store for us. Even in the end, event-triggered control has a rather high level of abstraction, because it assumes that all events are detected perfectly and right away with continuous sensing. The event-triggered model has $x \leq 5$ as a hard limit in the evolution domain constraint of the differential equation to ensure that the event $4 \leq x \leq 5$ will never ever be missed as the ball is rushing upwards.

As soon as we want to implement such a perfect event detection, it becomes clear that real controller implementations can usually only perform discrete sensing, i.e.,

© Springer International Publishing AG, part of Springer Nature 2018
A. Platzer, *Logical Foundations of Cyber-Physical Systems*,
https://doi.org/10.1007/978-3-319-63588-0_9

checking sensor data every once in a while at certain discrete points in time, whenever new measurements come from the sensor and when the controller has a chance to check whether the measurement is about to exceed height 5. Most controller implementations, thus, only end up checking for an event every once in a while, whenever the controller happens to run, rather than permanently as event-triggered controllers pretend.

This chapter, thus, focuses on the second important paradigm for making cyber interface with physics to form cyber-physical systems: the paradigm of *time-triggered control*, which uses periodic actions to affect the behavior of the system only at discrete points in time with certain frequencies. This is to be contrasted with the paradigm from Chap. 8 of *event-triggered control*, where responses to events dominate the behavior of the system and an action is taken whenever one of the events is observed. The two paradigms play an equally important rôle in classical embedded systems and both paradigms arise naturally from an understanding of the hybrid program principle for CPS.

Based on the understanding of loops from Chap. 7, the most important learning goals of this chapter are:

Modeling and Control: This chapter provides a number of crucial lessons for modeling CPSs and designing their controls. We develop an understanding of time-triggered control, which is an important design paradigm for control loops in CPS. This chapter studies ways of developing models and controls corresponding to this feedback mechanism, which is easier to implement but will turn out to be surprisingly subtle to control. Knowing and contrasting both event-triggered and time-triggered feedback mechanisms helps with identifying relevant dynamical aspects in CPS coming from events and reaction delays. This chapter focuses on CPS models assuming discrete sensing, i.e., sensing at (nondeterministically chosen) discrete points in time.

Computational Thinking: This chapter uses the rigorous reasoning approach from Chapters 5 and 7 to study CPS models with time-triggered control. As a running example, the chapter continues to develop the extension from bouncing balls to vertical ping-pong balls, now using time-triggered control. We again add control decisions to the bouncing ball, turning it into a ping-pong ball, which retains the intuitive simplicity of the bouncing ball, while enabling us to develop generalizable lessons about how to design time-triggered control systems correctly. The chapter will crucially study invariants and show a development of the powerful technique of design-by-invariant in a concrete example.

CPS Skills: This chapter develops an understanding of the semantics of time-triggered control. This understanding of the semantics will guide our intuition about the operational effects of time-triggered control and especially the impact it has on finding correct control constraints. An understanding of both is crucial for finding good tradeoffs to determine which parts of a model are faithfully understood as event-triggered and where time-triggered control is more accurate. Finally, the chapter studies some aspects of higher-level model-predictive control.

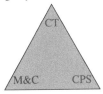

using loop invariants
design time-triggered control
design-by-invariant

modeling CPS
designing controls
time-triggered control
reaction delays
discrete sensing

semantics of time-triggered control
operational effect
finding control constraints
model-predictive control

9.2 Delays in Control

Event-triggered control is a useful and intuitive model matching our expectation that controllers react in response to certain critical conditions or events that necessitate intervention by the controller. However, one of its difficulties is that event-triggered control with its continuous-sensing assumption can be hard or impossible to implement in reality. On a higher level of abstraction, it is very intuitive to design controllers that react to certain events and change the control actuation in response to what events have happened. Closer to the implementation, this turns out to be difficult, because actual computer control algorithms do not actually run all the time, only sporadically every once in a while, albeit sometimes very often. Implementing event-triggered control faithfully would, in principle, require permanent continuous monitoring of the state to check whether an event has happened and respond appropriately. That is not particularly realistic, because fresh sensor data will only be available every once in a while, and controller implementations will only run at certain discrete points in time, causing delays in processing. Actuators may sometimes take time to get going. Think of the reaction time it takes you to turn the insight "I want to hit this ping-pong ball there" into action so that your ping-pong paddle will actually hit the ping-pong ball. Sometimes the ping-pong paddle acts early, sometimes late; see Fig. 9.1. Or think of the time it takes to react to the event "the car in front of me is turning on its red taillights" by appropriately applying the brakes.

Back to the drawing board. Let us reconsider the original dL formula (8.3) for the ping-pong ball (Fig. 9.1) from which we started out to design the event-triggered version in (8.7).

$$0 \leq x \wedge x \leq 5 \wedge v \leq 0 \wedge g > 0 \wedge 1 \geq c \geq 0 \wedge f \geq 0 \rightarrow$$
$$\left[\left(\{ x' = v, v' = -g \, \& \, x \geq 0 \}; \right. \right. \tag{8.3*}$$
$$\left. \left. \text{if}(x = 0) \, v := -cv \, \text{else if}(4 \leq x \leq 5) \, v := -fv \right)^* \right] (0 \leq x \leq 5)$$

Fig. 9.1 Sample trajectory of a ping-pong ball (plotted as position over time) with the indicated ping-pong paddle actuation range, sometimes actuating early, sometimes late

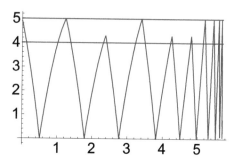

This simplistic formula (8.3) turned out not to be valid, because its differential equation was not guaranteed to be interrupted when the event $4 \leq x \leq 5$ happens. Consequently, (8.3) needs some other evolution domain constraint to make sure all continuous evolutions are stopped at some point for the control to have a chance to react to situation changes. Yet, it should not be something like $\ldots \& x \leq 5$ as in (8.7), because continuously monitoring for $x \leq 5$ requires permanent continuous sensing of the height, which is difficult to implement.

Note 50 (Physical versus controller events) The event $x = 0$ in the (physics) controller as well as the (physics) evolution domain constraint $x \geq 0$ for detecting the event $x = 0$ are perfectly justified in the bouncing-ball and ping-pong-ball models, because both represent physics. Physics is very well capable of keeping a ball above the ground, no matter how much checking for $x = 0$ it takes to make that happen. The ball just does not suddenly fall through the ground because physics looked the other way and forgot to check its evolution domain constraint $x \geq 0$! In our controller code, however, we need to exercise care when modeling events and their reactions. The controller implementations will not have the privilege of running all the time, which only physics possesses. Cyber happens every once in a while (even if it may execute quite quickly and quite frequently), while physics happens all the time. Controllers cannot sense and compute and act literally all the time.

How else could the continuous evolution of physics be interrupted to make sure the controller actually runs? By bounding the amount of time that physics is allowed to evolve before running the controller again. Before we can talk about time, the model needs to be changed to include some variable, let's call it t, that reflects the progress of time with a differential equation $t' = 1$:

$$0 \leq x \wedge x \leq 5 \wedge v \leq 0 \wedge g > 0 \wedge 1 \geq c \geq 0 \wedge f \geq 0 \rightarrow$$
$$\left[(\{x' = v, v' = -g, t' = 1 \,\&\, x \geq 0 \wedge t \leq 1\}; \right. \tag{9.1}$$
$$\left. \mathsf{if}(x = 0)\, v := -cv \,\mathsf{else}\,\mathsf{if}(4 \leq x \leq 5)\, v := -fv)^* \right](0 \leq x \leq 5)$$

Of course, the semantics of hybrid programs included some notion of time already, but it was inaccessible in the program itself because the duration r of differential equations was not a state variable that the model could read (Definition 3.2). No problem, (9.1) simply added a time variable t that evolves along the differential equation $t' = 1$ just like time itself does. In order to bound the progress of time by 1, the evolution domain includes $\ldots \& t \leq 1$ and declares that the clock variable t evolves with time as $t' = 1$.

Oops, that does not actually quite do it, because the HP in (9.1) restricts the evolution of the system so that it will never ever evolve beyond time 1, no matter how often the loop repeats. It imposes a global bound on the progress of time. That is not what we meant to say! Rather, we wanted the duration of each individual continuous evolution to be at most one second. The trick is to reset the clock t to zero by a discrete assignment $t := 0$ before the continuous evolution starts:

$$0 \leq x \wedge x \leq 5 \wedge v \leq 0 \wedge g > 0 \wedge 1 \geq c \geq 0 \wedge f \geq 0 \rightarrow$$
$$[(t:=0; \{x' = v, v' = -g, t' = 1 \& x \geq 0 \wedge t \leq 1\}; \tag{9.2}$$
$$\text{if}(x = 0) v := -cv \text{ else if}(4 \leq x \leq 5) v := -fv)^*](0 \leq x \leq 5)$$

In order to bound the duration by 1, the evolution domain includes $\ldots \& t \leq 1$ and the variable t is reset to 0 by $t := 0$ right before the differential equation. Hence, t represents a local clock measuring how long the evolution of the differential equation was. Its bound of 1 ensures that physics gives the controller a chance to react at least once per second. The system could stop the continuous evolution more often and earlier, because this model has no lower bound on t. Even if possible, it is inadvisable to constrain the model unnecessarily by lower bounds on the duration.

Before going any further, let's take a step back to notice an annoyance in the way the control in (9.2) was written. It is written in the style in which the original bouncing ball and the event-triggered ping-pong ball were phrased: continuous dynamics followed by control. That has the unfortunate effect that (9.2) lets physics happen *before* control does anything, which is not a very safe start. In other words, the initial condition would have to be modified to assume the initial control choice was fine. That would duplicate part of the control into the assumptions on the initial state. Instead, let's switch the statements from *plant*; *ctrl* to *ctrl*; *plant* to make sure control always happens before physics does anything:

$$0 \leq x \wedge x \leq 5 \wedge v \leq 0 \wedge g > 0 \wedge 1 \geq c \geq 0 \wedge f \geq 0 \rightarrow$$
$$[(\text{if}(x = 0) v := -cv \text{ else if}(4 \leq x \leq 5) v := -fv; \tag{9.3}$$
$$t := 0; \{x' = v, v' = -g, t' = 1 \& x \geq 0 \wedge t \leq 1\})^*](0 \leq x \leq 5)$$

Now that dL formula (9.3) has an upper bound on the time it takes between two subsequent control actions, is it valid? If so, which invariant can be used to prove it? If not, which counterexample shows its invalidity?

Before you read on, see if you can find the answer for yourself.

Even though (9.3) ensures a bound on how long it may take at most until the controller inspects the state and reacts, there is still a fundamental issue with (9.3).

Fig. 9.2 Sample trajectory of a time-triggered ping-pong ball (as position over time), missing the first event

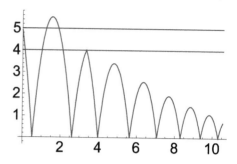

We can try to prove (9.3) and inspect the non-provable cases in the proof to find out what the issue is. Or we can just think about what could go wrong. The controller of (9.3) runs at the latest after one second (hence at least once per second) and then checks whether $4 \leq x \leq 5$. But if $4 \leq x \leq 5$ was not true when the controller ran last, there is no guarantee that this event will be detected reliably when the controller runs next. In fact, the ball might very well have been at $x = 3$ at the last controller run, then evolved continuously to $x = 6$ in a second and missed the event $4 \leq x \leq 5$ that it was supposed to detect (Exercise 9.2); see Fig. 9.2. Worse than that, the ping-pong ball has then not only missed the exciting event $4 \leq x \leq 5$ but already became unsafe.

Similarly, driving a car would be unsafe if you were to only open your eyes once a minute and monitor whether there is a car right in front of you. Too many things could happen in between that should prompt you to brake.

9.2.1 The Impact of Delays on Event Detection

How can this problem with formula (9.3) be solved? How can the CPS model make sure the controller does not miss its time to take action? Waiting until $4 \leq x \leq 5$ holds true is not guaranteed to be the right course of action for the controller.

Before you read on, see if you can find the answer for yourself.

The problem with (9.3) is that its controller is unaware of its own delay. It does not take into account how the ping-pong ball could move further before it gets a chance to react next. If the ball is already close to the ping-pong paddle's intended range of actuation, then the controller had better take action already if it is not sure whether it can still safely wait to take action till next time the time-triggered controller runs.

> **Note 51 (Delays may miss events)** Delays in controller reactions may cause events to be missed that it was supposed to monitor. When that happens, there is a discrepancy between an event-triggered understanding of a CPS and the real time-triggered implementation. Delays may make controllers miss events especially when slow controllers monitor events in relatively small regions for a fast-moving system. This relationship deserves special attention to make sure the impact of delays on a system controller cannot make it unsafe.
>
> It is often a good idea to first understand and verify an event-triggered design of a CPS controller to identify correct responses to the respective events and subsequently refine it to a time-triggered controller to analyze and verify that CPS in light of its reaction time. Discrepancies in this analysis hint at problems that event-triggered designs will likely experience at runtime and they indicate a poor event abstraction. Controllers need to be aware of their own delays to foresee what they might otherwise miss.

The ping-pong controller would be in trouble if $x > 5$ might already hold in its next control cycle after the continuous evolution, which would be outside the operating range of the ping-pong paddle (and already unsafe). Due to the evolution domain constraint, the continuous evolution can take at most 1 time unit, after which the ball will be at position $x + v - \frac{g}{2}$ as previous chapters already showed by solving the differential equation. Choosing gravity $g = 1$ to simplify the math, the controller would be in trouble in the next control cycle after 1 second, which would take the ball to position $x + v - \frac{1}{2} > 5$, if $x > 5\frac{1}{2} - v$ holds now.

9.2.2 Model-Predictive Control Basics

The idea is to make the controller now act based on how it predicts the state might evolve until the next control cycle (this is a very simple example of *model-predictive control* because the controller acts based on what its model predicts). Chap. 8 already discovered for the event-triggered case that the controller only wants to trigger the ping-pong paddle action if the ball is still flying up, not if it is already on its way down. Making (9.3) aware of the future in this way leads to

$$0 \leq x \wedge x \leq 5 \wedge v \leq 0 \wedge g = 1 \wedge 1 \geq c \geq 0 \wedge f \geq 0 \rightarrow$$
$$\left[\left(\text{if}(x = 0) \, v := -cv \, \text{else if}((x > 5\frac{1}{2} - v) \wedge v \geq 0) \, v := -fv; \right. \right. \tag{9.4}$$
$$\left. \left. t := 0; \{x' = v, v' = -g, t' = 1 \, \& \, x \geq 0 \wedge t \leq 1\})^* \right] (0 \leq x \leq 5)$$

Is conjecture (9.4) about the future-aware controller valid? If so, which invariant can be used to prove it? If not, which counterexample shows its invalidity?

Before you read on, see if you can find the answer for yourself.

The controller in formula (9.4) has been designed based on the prediction that the future may evolve for 1 time unit. If an action will no longer be possible in 1 time unit, because the event $x \le 5$ has passed in that future time instant, then the controller in (9.4) takes action right now already. That is a good start. The issue with that approach, however, is that there is no guarantee at all that the ping-pong ball will fly for exactly 1 time unit before the controller is asked to act again (and the postcondition is checked). The controller in (9.4) checks whether the ping-pong ball would be too far up after one time unit and does not intervene unless that is the case. Yet, what if the ball only flies for $\frac{1}{2}$ a time unit? Clearly, if the ball will be safe after 1 time unit, which is what the controller in (9.4) checks, it will also be safe after just $\frac{1}{2}$ a time unit, right?

Before you read on, see if you can find the answer for yourself.

Wrong! The ball may well be below height 5 again after 1 time unit but still could have been above 5 in between the current point of time and the time that is 1 time unit from now. Then the safety of the controller will be a mere rope of sand, because this incorrect controller will have a false sense of safety after having checked what happens 1 time unit from now, in complete ignorance of whether the behavior was actually safe until then. Such trajectories are shown in Fig. 9.3 from the same initial state and the same controller, just with different sampling periods. What a bad controller design if its behavior depends on the sampling period! But worse than that, such a bouncing ball will not be safe if it has been above 5 between two sampling points. After all, the bouncing ball follows a ballistic trajectory, which first climbs and then falls.

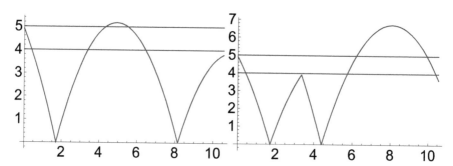

Fig. 9.3 Sample trajectory of a time-triggered ping-pong ball (as position over time), missing different events with different sampling periods

9.2.3 Design-by-Invariant

In order to get to the bottom of this, we need a quantity that tells us what the ball will do at all times, without mentioning the time variable explicitly, because we can hardly have the controller check its safety predictions at all times 0, 0.1, 0.25, 0.5, 0.786, ..., of which there are infinitely many anyhow.

Come to think of it, we were already investigating what we can say about a bouncing ball independently of the time when we were designing loop invariants for its proof in Sect. 7.4:

$$2gx = 2gH - v^2 \wedge x \geq 0 \wedge (c = 1 \wedge g > 0) \tag{7.11*}$$

This formula was proved to be an invariant of the bouncing ball, which means it holds true always while the bouncing ball is bouncing around. Invariants are the most crucial information about the behavior of a system that we can rely on all the time. Since (7.11) is only an invariant of the bouncing dynamics not the ping-pong ball, it, of course, only holds until the ping-pong paddle hits, which changes the control. But until the ping-pong paddle is used, (7.11) summarizes concisely all we need to know about the state of the bouncing ball at all times. Of course, (7.11) is an invariant of the bouncing ball, but it still needs to be true initially. The easiest way to make that happen is to assume (7.11) at the beginning of the ping-pong ball's life.[1]

Because (7.11) only conducted the proof of the bouncing ball invariant (7.11) for the case $c = 1$ to simplify the arithmetic, the ping-pong ball now adopts this assumption as well. To simplify the arithmetic and arguments, let us also adopt the assumption $f = 1$ in addition to $c = 1 \wedge g = 1$ for the proofs.

Substituting safety-critical height 5 for H in the invariant (7.11) for this instance of parameter choices leads to a condition when the energy exceeds safe height 5:

$$2x > 2 \cdot 5 - v^2 \tag{9.5}$$

as an indicator of the fact that the ball might end up climbing too high, because its energy would allow it to. Adding this condition (9.5) to the controller (9.4) leads to

$$2x = 2H - v^2 \wedge 0 \leq x \wedge x \leq 5 \wedge v \leq 0 \wedge g = 1 \wedge 1 = c \wedge 1 = f \geq 0 \rightarrow$$

$$\left[\left(\text{if}(x = 0) \, v := -cv \, \text{else if} \left((x > 5\tfrac{1}{2} - v \vee 2x > 2 \cdot 5 - v^2) \wedge v \geq 0 \right) v := -fv; \right. \right. \tag{9.6}$$

$$\left. \left. t := 0; \{x' = v, v' = -g, t' = 1 \, \& \, x \geq 0 \wedge t \leq 1\} \right)^* \right] (0 \leq x \leq 5)$$

The bouncing ball invariant (7.11) is now also assumed to hold in the initial state.

---

[1] Note that H is a variable that does not need to coincide with the upper height limit 5 as it did in the case of the bouncing ball, because the ping-pong ball has more control at its fingertips. In fact, the most interesting case is if $H > 5$ in which case the ping-pong ball will only stay safe *because of* its control. One way to think of H is as an indicator of the energy of the ball showing how high it might jump up if not for all its interaction with the ground and the ping-pong paddle.

Is dL formula (9.6) about the time-triggered controller valid? As usual, it is best to use an invariant or a counterexample for justification.

>   Before you read on, see if you can find the answer for yourself.

Formula (9.6) is "almost valid." But it is still not valid for a very subtle reason. It is great to have the help of proofs to catch those subtle issues. The controller in (9.6) takes action for two different conditions on the height x. However, the ping-pong paddle controller actually only runs in (9.6) if the ball is not at height $x = 0$, otherwise the ground performs the control action of reversing the direction of the ball. Now, if the ball is flat on the floor already ($x = 0$) yet its velocity so incredibly high that it will rush past height 5 in less than 1 time unit, then the ping-pong paddle controller will not even have had a chance to react before it is too late, because it does not execute on the ground according to (9.6); see Fig. 9.4.

Fig. 9.4 Sample trajectory of a time-triggered ping-pong ball (as position over time), failing to control on the ground

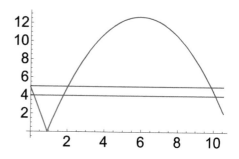

9.2.4 Sequencing and Prioritizing Reactions

Fortunately, these thoughts already indicate how the problem with multiple control actions can be fixed. We turn the nested if-then-else cascade into a sequential composition of two separate if-then statements that will ensure the ping-pong paddle controller runs even if the bouncing ball is still on the ground (Exercise 9.3).

$$2x = 2H - v^2 \wedge 0 \le x \wedge x \le 5 \wedge v \le 0 \wedge g = 1 \wedge 1 = c \wedge 1 = f \rightarrow$$

$$\left[\left(\text{if}(x = 0) \, v := -cv \, ; \, \text{if}((x > 5\tfrac{1}{2} - v \vee 2x > 2 \cdot 5 - v^2) \wedge v \ge 0) \, v := -fv; \right. \right. \quad (9.7)$$

$$\left. \left. t := 0; \{x' = v, v' = -g, t' = 1 \, \& \, x \ge 0 \wedge t \le 1\})^* \right] (0 \le x \le 5) \right.$$

Now, is formula (9.7) finally valid, please? If so, using which invariant? Otherwise, show a counterexample.

>   Before you read on, see if you can find the answer for yourself.

Yes, formula (9.7) is valid. What invariant can be used to prove formula (9.7)?
Formula (9.7) is valid, which, for $g = c = f = 1$, can be proved with this invariant:

$$2x = 2H - v^2 \wedge x \geq 0 \wedge x \leq 5 \tag{9.8}$$

This invariant instantiates the general bouncing ball invariant (7.11) for the present case of parameter choices and augments it with the desired safety constraint $x \leq 5$.

Yet, is the controller in (9.7) useful? That is where the problem lies now. The condition (9.5) that is the second disjunct in the controller of (9.7) checks whether the ping-pong ball could possibly *ever* fly up all the way to height 5. If this is ever true, it might very well be true long before the bouncing ball even approaches the critical control cycle where a ping-pong paddle action needs to be taken. In fact, if (9.5) is ever true, it will also be true at the very beginning. After all, the formula (7.11), from which condition (9.5) derived, is an invariant, so always true for the bouncing ball. What would that mean?

That would cause the controller in (9.7) to take action right away at the mere prospects of the ball ever being able to climb way up high, even if the ping-pong ball is still close to the ground and pretty far away from the last triggering height 5. That would make the ping-pong ball quite safe, since (9.7) is a valid formula. But it would also make it rather conservative and would not allow the ping-pong ball to bounce around nearly as much as it wants to. It would make the bouncing ball lie flat on the ground, because of an overly anxious ping-pong paddle. That is a horrendously acrophobic bouncing ball if it never even starts bouncing around in the first place. And the model would even require the (model) world to end, because there can be no progress beyond the point in time where the ball gets stuck on the ground. How can the controller in (9.7) be modified to resolve this problem?

Fig. 9.5 Sample trajectory of a time-triggered ping-pong ball (as position over time), stuck on the ground

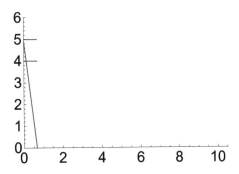

Before you read on, see if you can find the answer for yourself.

The idea is to restrict the use of the second if-then disjunct (9.5) in (9.7) to slow velocities, in order to make sure it only applies to the occasions that the first controller disjunct $x > 5\frac{1}{2} - v$ misses, because the ball will have been above height 5 in between. Only with slow velocities will the ball ever move so slowly that it is

Expedition 9.1 (Zeno paradox)

There is something quite surprising about how (9.7) may cause time to freeze. But, come to think of it, time did already freeze in mere bouncing balls!

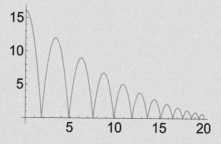

The duration between two hops on the ground of a bouncing ball keeps on decreasing rapidly. If, for simplicity, the respective durations are $1, \frac{1}{2}, \frac{1}{4}, \frac{1}{8}, \ldots$, then these durations sum to

$$\sum_{i=0}^{\infty} \frac{1}{2^i} = \frac{1}{1 - \frac{1}{2}} = 2$$

which shows that the bouncing-ball model will make the (model) world freeze almost to a complete stop, because it can never reach time 2 nor any time after. The bouncing ball model disobeys what is called *divergence of time*, i.e., that the real time keeps diverging to ∞. The reason this model prevents time from progressing beyond 2 is that the bouncing-ball model switches directions on the ground more and more frequently. This may be very natural for bouncing balls, but can cause subtleties and issues in other control systems if they switch infinitely often in finite time.

The name *Zeno paradox* comes from the Greek philosopher Zeno (ca. 490–430 BC) who found a paradox when fast runner Achilles gives the slow Tortoise a head start of 100 meters in a race: In a race, the quickest runner can never overtake the slowest, since the pursuer must first reach the point whence the pursued started, so that the slower must always hold a lead – recounted in Aristotle, Physics VI:9, 239b15.

Pragmatic solutions to counteract the Zeno paradox in bouncing balls add a statement to the model that makes the ball stop when the remaining velocity on the ground is too small. For example:

$$\text{if}(x = 0 \wedge -0.1 < v < 0.1)(v := 0; \{x' = 0\})$$

This statement switches to a differential equation that does not change position but, unlike the differential equation $x' = v, v' = -g \,\&\, x \geq 0$ for the bouncing ball, can be followed for any duration when $x = 0 \wedge v = 0$.

near its turning point to begin its descent and start falling down again before 1 time unit. And only then can the first condition miss that the ball is able to evolve above 5 within 1 time unit. When is a velocity slow in this respect?

For the ball to turn around and descend, it first needs to reach velocity $v = 0$ by continuity (during the flying phase) on account of the mean-value theorem. In gravity $g = 1$ the ball can reach velocity 0 within 1 time unit exactly when its velocity was $v < 1$ before the differential equation, because the velocity changes according to $v(t) = v - gt$. Consequently, adding a conjunct $v < 1$ to the second disjunct in the controller makes sure that the controller only checks for turnaround when it might actually happen during the next control cycle.

$$2x = 2H - v^2 \wedge 0 \leq x \wedge x \leq 5 \wedge v \leq 0 \wedge g = 1 \wedge 1 = c \wedge 1 = f \rightarrow$$

$$\left[\left(\text{if}(x=0)\, v := -cv; \text{if}\left((x > 5\tfrac{1}{2} - v \vee 2x > 2 \cdot 5 - v^2 \wedge v < 1 \right) \wedge v \geq 0 \right) v := -fv; \right.$$

$$\left. t := 0; \{x' = v, v' = -g, t' = 1 \,\&\, x \geq 0 \wedge t \leq 1\})^* \right] (0 \leq x \leq 5)$$

$$(9.9)$$

This dL formula is valid and provable with the same invariant (9.8) that was already used to prove (9.7). It has a much more aggressive controller than (9.7), though, so it is more fun for the ping-pong ball to bounce around with it.

> **Note 52 (Design by invariant)** Designing safe controller actions by following the system invariant is a great idea. After having identified an invariant for the bare-bones system (such as the bouncing ball), the remainder of the control actions can be designed safely by ensuring that each of them preserves the invariant. For example, the ping-pong paddle is used if the ball might violate the invariant. Some care is needed to avoid limiting the system unnecessarily. The reaction time determines which control cycle has the last chance to act to keep the invariant maintained. Of course, design-by-invariant does not extend to changing the laws of physics to please our controllers. But once the appropriate invariants have been identified for physics, the design of the controller can follow the objective of always maintaining the safety-critical invariants.

9.2.5 Time-Triggered Verification

The easiest way of proving that dL formula (9.9) is valid is to show that the invariant (9.8) holds after every line of code. Formally, this reasoning by lines corresponds to a number of uses of the generalization proof rule MR from Lemma 7.4 to show that the invariant (9.8) remains true after each line if it was true before. The first statement if$(x=0)\, v := -cv$ does not change the truth-value of (9.8), i.e.,

$$2x = 2H - v^2 \wedge x \geq 0 \wedge x \leq 5 \rightarrow [\text{if}(x=0)\, v := -cv](2x = 2H - v^2 \wedge x \geq 0 \wedge x \leq 5)$$

is valid, because, when $c = 1$, the statement can only change the sign of v and (9.8) is independent of signs, because the only occurrence of v satisfies $(-v)^2 = v^2$. Similarly, the second statement if$((x > 5\frac{1}{2} - v \vee 2x > 2 \cdot 5 - v^2 \wedge v < 1) \wedge v \geq 0) v := -fv$ does not change the truth-value of (9.8). That is the formula

$$2x = 2H - v^2 \wedge x \geq 0 \wedge x \leq 5 \rightarrow$$

$$[\text{if}((x > 5\frac{1}{2} - v \vee 2x > 2 \cdot 5 - v^2 \wedge v < 1) \wedge v \geq 0) v := -fv]$$

$$(2x = 2H - v^2 \wedge x \geq 0 \wedge x \leq 5)$$

is valid, because, at least for $f = 1$, the second statement can also only change the sign of v, which is irrelevant for the truth-value of (9.8). Finally, the relevant parts of (9.8) are a special case of (7.11), which has already been shown to be an invariant for the bouncing-ball differential equation and, thus, continues to be an invariant when adding a clock $t' = 1 \& t \leq 1$, which does not occur in (9.8). The additional invariant $x \leq 5$ that (9.8) has compared to (7.11) is easily taken care of using the corresponding knowledge about potential height H.

> **Note 53 (Time-triggered control)** One common paradigm for designing controllers is *time-triggered control*, in which controllers run periodically or pseudo-periodically with certain frequencies to inspect the state of the system. Time-triggered systems are closer to implementation than event-triggered control. They can be harder to build, however, because they invariably require the designer to understand the impact of delay on control decisions. That impact is important in reality, however, and, thus, effort invested in understanding the impact of time delays usually pays off in designing a safer system that is robust to bounded time delays.

Partitioning the hybrid program in the verified dL formula (9.9) into the parts that come from physics (typographically marked like physics) and the parts that come from control (typographically marked like control) leads to the following.

Proposition 9.1 (Time-triggered ping-pong is safe). *This dL formula is valid:*

$$2x = 2H - v^2 \wedge 0 \leq x \wedge x \leq 5 \wedge v \leq 0 \wedge g = 1 \wedge 1 = c \wedge 1 = f \rightarrow$$

$$[(\text{if}(x = 0) v := -cv; \text{if}((x > 5\frac{1}{2} - v \vee 2x > 2 \cdot 5 - v^2 \wedge v < 1) \wedge v \geq 0) v := -fv;$$

$$t := 0; \{x' = v, v' = -g, t' = 1 \& x \geq 0 \wedge t \leq 1\})^*](0 \leq x \leq 5)$$

$$(9.9^*)$$

Part of the differential equation, namely $t' = 1$, comes from the controller, because it corresponds to putting a clock on the controller and running it with at least the sampling frequency 1 (coming from the evolution domain constraint $t \leq 1$).

9.3 Summary

This chapter studied time-triggered control, which, together with event-triggered control from Chap. 8, is an important principle for designing feedback mechanisms in CPS and embedded systems. The chapter illustrated the most important aspects for a running example of a ping-pong ball. Despite or maybe even because of its simplicity, the ping-pong ball was an instructive source for the most important subtleties involved with time-triggered control decisions. Getting time-triggered controllers correct requires predictions about how the system state might evolve over short periods of time (one control cycle). The effects and subtleties of time-triggered actions in control were sufficiently subtle to merit focusing on a simple intuitive case.

Unlike event-triggered control, which assumes continuous sensing, the time-triggered control principle is more realistic by only assuming the availability and processing of sensor data at discrete instants of time (discrete sensing). Time-triggered system models avoid the modeling subtleties that events tend to cause for the detection of events. It is, thus, often much easier to get the models right and implementable for time-triggered systems than it is for event-triggered control. The price is that the burden of event-detection is then brought to the attention of the CPS programmer, whose time-triggered controller will now have to ensure it predicts and detects events early enough before it is too late to react to them. That is what makes time-triggered controllers more difficult to get correct, but is also crucial because important aspects of reliable event detection may otherwise be brushed under the rug, which does not help the final CPS become any safer either.

CPS design often begins by pretending the idealized world of event-triggered control (if the controller is not even safe when events are checked and responded to continuously, it is broken already) and then subsequently morphing the event-triggered controller into a time-triggered controller. This second step then often indicates additional subtleties that were missed in the event-triggered design. The additional insights gained in time-triggered controllers are crucial whenever the system reacts slowly or whenever it reacts quickly but needs a high precision in event detection to remain safe. For example, the reaction time for ground control decisions to reach a rover on Mars are so prohibitively large that they could hardly be ignored. Reaction times in a surgical robotics system that is running at, say, 55 Hz, are still crucial even if the system is moving slowly and reacting quickly, because the required precision of the system is in the sub-millimeter range [1]. But reaction times will have less of an impact for parking a slowly moving car somewhere in an empty football stadium.

Overall, the biggest issues with event-triggered control, besides sometimes being hard to implement, are the subtleties involved in properly modeling event detection without accidentally defying the laws of physics in pursuit of an event. But controlling event-triggered systems is reasonably straightforward as long as the events are chosen well. In contrast, finding a model is relatively easy in time-triggered control, but identifying appropriately safe controller constraints takes a lot more thought, leading, however, to important insights about the system at hand. It is possible to provide the best of both worlds by systematically reducing the safety proof of an (implementable) time-triggered controller to the (easier) safety proof of an event-triggered controller along with corresponding compatibility conditions [2, 3].

Exercises

9.1 (Time bounds). HP (9.3) imposes an upper bound on the duration of a continuous evolution. Can you impose an upper bound 1 and lower bound 0.5? Is there relevant safety-critical behavior in the system that is then no longer considered?

9.2. Give an initial state for which the controller in (9.3) would skip over the event without noticing it.

9.3. What would happen if the controller in (9.7) used the ping-pong paddle while the ball is still on the ground? To what physical phenomenon does that correspond?

9.4. The formula (9.9) with the time-triggered controller with reaction time at most 1 time unit is valid. Yet, if a ball is let loose ever so slightly above ground with a very fast negative velocity, couldn't it possibly bounce back and exceed the safe height 5 faster than the reaction time of 1 time unit? Does that mean the formula ought to have been falsifiable? No! Identify why and give a physical interpretation.

9.5. The controller in (9.9) runs at least once a second. How can you change the model and controller so that it runs at least twice a second? What changes can you make to the controller to reflect that increased frequency? How do you need to change (9.9) if the controller only runs reliably at least once every two seconds? Which of those changes are safety-critical, which are not?

9.6. What happens if we misread the binding precedences and think the condition $v < 1$ is added to both disjuncts in the controller in (9.9)?

$$2x = 2H - v^2 \wedge 0 \leq x \wedge x \leq 5 \wedge v \leq 0 \wedge g = 1 \wedge 1 = c \wedge 1 = f \rightarrow$$
$$\left[(\text{if}(x = 0)\, v := -cv; \text{if}((x > 5\frac{1}{2} - v \vee 2x > 2 \cdot 5 - v^2) \wedge v < 1 \wedge v \geq 0)\, v := -fv; \right.$$
$$\left. t := 0; \{x' = v, v' = -g, t' = 1 \& x \geq 0 \wedge t \leq 1\})^* \right](0 \leq x \leq 5)$$

Is the resulting formula still valid? Find an invariant or counterexample.

9.7. Conduct a sequent proof proving the validity of dL formula (9.9). Is it easier to follow a direct proof or is it easier to use the generalization rule MR for the proof?

9.8. The event-triggered controller we designed in Chap. 8 monitored the event $4 \leq x \leq 5$. The time-triggered controller in Sect. 9.2, however, ultimately only took the upper bound 5 into account. How and under what circumstances can you modify the controller so that it really only reacts to the event $4 \leq x \leq 5$ rather than under all circumstances where the ball is in danger of exceeding 5?

9.9. Devise a controller that reacts if the height changes by 1 when comparing the height before the continuous evolution to the height after. Can you make it safe? Can you implement it? Is it an event-triggered or a time-triggered controller? How does it compare to the controllers developed in this chapter?

9.10. The ping-pong ball proof relied on the parameter assumptions $g = c = f = 1$ for mere convenience of the resulting arithmetic. Develop a time-triggered model, controller, invariant, and proof for the general ping-pong ball without these unnecessarily strong simplifying assumptions.

9.11. Show that the ping-pong ball (9.9) can also be proved safe using just the invariant $0 \leq x \leq 5$ (possibly including assumptions on constants such as $g > 0$). Which assumptions on the initial state does this proof crucially depend on?

9.12 (*). Design a variation of the time-triggered controller for the ping-pong ball that is allowed to use the ping-pong paddle within height $4 \leq x \leq 5$ but has a relaxed safety condition that accepts $0 \leq x \leq 2 \cdot 5$. Make sure to only force the use of the ping-pong paddle when necessary. Find an invariant and conduct a proof.

9.13 (2D ping-pong time). Design and verify the safety of a ping-pong controller that goes sideways with horizontal motion like in ordinary ping-pong matches. What is the impact of reaction time?

9.14 (Robot chase). You are in control of a robot tailing another one in hot pursuit. You can accelerate ($a := A$), brake ($a := -b$), or coast ($a := 0$). But so can the robot you're following! Your job is to fill in the blanks with test conditions that make the robots not crash.

$$x \leq y \wedge v = 0 \wedge A \geq 0 \wedge b > 0 \rightarrow$$
$$\big[((c := A \cup c := -b \cup c := 0);$$
$$(? \underline{}; a := A \cup ? \underline{}; a := -b \cup ? \underline{}; a := 0);$$
$$t := 0; \{x' = v, v' = a, y' = w, w' = c, t' = 1 \& v \geq 0 \wedge w \geq 0 \wedge t \leq \varepsilon\})^*$$
$$\big] x \leq y$$

9.15 (Zeno's paradox of Achilles and the Tortoise). Hybrid systems make transparent the two different world models with which Zeno described the race of the fast runner Achilles against the slow Tortoise (Expedition 9.1). Achilles is at position a running with velocity v. The Tortoise is at position t crawling with velocity $w < v$. The model of successive motion uses separate differential equations, where Achilles first moves for duration s till he reaches the position t where the Tortoise was, which already moved on with its smaller velocity w for the same duration:

$$s := 0; (\{a' = v, s' = 1 \& a \leq t\}; ?a = t; \{t' = w, s' = -1 \& s \geq 0\}; ?s = 0)^*$$

Compare this to simultaneous motion in a combined differential equation system:

$$s := 0; \{a' = v, t' = w, s' = 1\}$$

Show that Achilles a will never reach Tortoise t in the first model despite $v > w$ if $a < t$ holds initially. For the second model prove that postcondition $a = t$ will eventually be true (with the help of a diamond modality). Then contrast both models with what happens when another Greek philosopher stumbles upon the race track, distracting Achilles with questions about other paradoxical models of motion.

References

[1] Yanni Kouskoulas, David W. Renshaw, André Platzer, and Peter Kazanzides. Certifying the safe design of a virtual fixture control algorithm for a surgical robot. In: *HSCC*. Ed. by Calin Belta and Franjo Ivancic. ACM, 2013, 263–272. DOI: 10.1145/2461328.2461369.

[2] Sarah M. Loos. Differential Refinement Logic. PhD thesis. Computer Science Department, School of Computer Science, Carnegie Mellon University, 2016.

[3] Sarah M. Loos and André Platzer. Differential refinement logic. In: *LICS*. Ed. by Martin Grohe, Eric Koskinen, and Natarajan Shankar. New York: ACM, 2016, 505–514. DOI: 10.1145/2933575.2934555.

Part II
Differential Equations Analysis

Overview of Part II on Differential Equations Analysis

This part of the book advances the study of cyber-physical systems to those whose dynamics can no longer be solved in closed form. If solutions of differential equations are no longer available or are too complicated, then indirect methods need to be used to analyze their properties. Just as induction is the crucial technique for understanding the behavior of loops in programs and in control systems from a local perspective, this part studies crucial generalizations of induction techniques to differential equations. The understanding gained so far in Part I for the intuition behind loop invariants will be a useful basis for extensions to differential equations. The primary remaining challenge is the development of a differential counterpart of induction, which is elusive in differential equations, because the very notion of a "next step" on which discrete induction is based is not at all meaningful in a continuous evolution. In addition to differential invariants as a sound generalization of induction to differential equations, this part studies differential cuts, which make it possible to prove and then use lemmas about the behavior of differential equations, and differential ghosts, which add new differential equations to the dynamics to enable additional invariants relating old and new variables. While the rôle of cuts for lemmas as well as ghost variables for additional state are well understood in discrete systems, both continue to play an arguably even more important rôle in the understanding of differential equations.

This Part II also provides a lightweight introduction to the meta-theory of differential equations by investigating the beginning of the provability theory for differential equations. While such a theory is not necessarily on the critical path for an understanding of practical invariant generation questions for cyber-physical systems, it still provides helpful intuition and insights about relationships between different invariants and different proof search approaches. It also serves as a relatively accessible, well-motivated, and intuitive segue into the study of proof theory, i.e., the theory of proofs or of proofs about proofs in the concrete setting of differential equations.

Chapter 10
Differential Equations & Differential Invariants

Synopsis This chapter leaves the realm of cyber-physical systems whose differential equations are solvable in closed form. Without closed-form solvable differential equations, the continuous dynamics of cyber-physical systems becomes much more challenging. The change is as noticeable and significant as the change from single-shot control systems to systems with an unbounded number of interactions in a control loop. All of a sudden, we can no longer pretend each differential equation could be replaced by an explicit representation of a function that describes the resulting state at time t along with a quantifier for t. Instead, differential equations have to be handled implicitly based on their actual dynamics as opposed to their solution. This leads to a remarkable shift in perspective opening up a new world of fascination in the continuous dynamical aspects of cyber-physical systems, and it begins by ascribing an entirely new meaning to primes in cyber-physical system models.

10.1 Introduction

So far, this textbook explored only one way to deal with differential equations: the $[']$ axiom schema from Lemma 5.3. Just like almost all other axioms, this axiom $[']$ is an equivalence, so it can be used to reduce a property of a more complex HP, in this case a differential equation, to a structurally easier logical formula.

$$[']\quad [x' = f(x)]p(x) \leftrightarrow \forall t{\geq}0\,[x := y(t)]p(x) \quad (y'(t) = f(y))$$

However, in order to use the $[']$ axiom for a differential equation $x' = f(x)$, we must first find a symbolic solution to the symbolic initial value problem (i.e., a function $y(t)$ such that $y'(t) = f(y)$ and $y(0) = x$). But what if the differential equation does not have such an explicit closed-form solution $y(t)$? Or what if $y(t)$ cannot be written down in first-order real arithmetic? Chapter 2 allows many more differential equations to be part of CPS models than just the ones that happen to have simple solutions. These are the differential equations we will look at in this chapter

© Springer International Publishing AG, part of Springer Nature 2018
A. Platzer, *Logical Foundations of Cyber-Physical Systems*,
https://doi.org/10.1007/978-3-319-63588-0_10

to provide rigorous reasoning techniques for them. In fact, the rigorous proofs for differential equations that this part of the textbook explores even simplify proofs of solvable differential equations and will ultimately make the solution axiom schema $[']$ superfluous.

You may have previously seen a whole range of methods for solving differential equations. These are indubitably useful for many common cases. But, in a certain sense, "most" differential equations are impossible to solve, because they have no explicit closed-form solution with elementary functions, for instance [18]:

$$x''(t) = e^{t^2}$$

Even if they do have solutions, the solution may no longer be in first-order real arithmetic. Example 2.5 showed that, for certain initial values, the solution of

$$x' = y, y' = -x$$

is $x(t) = \sin(t), y(t) = \cos(t)$, which is not expressible in real arithmetic (recall that both are infinite power series) and leads to undecidable arithmetic [6]. The sine function, for example, needs infinitely many powers, which does not give a finite term in first-order real arithmetic:

$$\sin(t) = t - \frac{t^3}{3!} + \frac{t^5}{5!} - \frac{t^7}{7!} + \frac{t^9}{9!} - \cdots$$

This chapter reinvestigates differential equations from a more fundamental perspective, which will lead to a way of proving properties of differential equations without using their solutions. It seeks unexpected analogies among the seemingly significantly different dynamical aspects of discrete dynamics and of continuous dynamics. The first and quite influential observation is that differential equations and loops have more in common than one might suspect.[1] Discrete systems may be complicated, but have a powerful ally: induction as a way of establishing truth for discrete dynamical systems by generically analyzing the one step that it performs (repeatedly like the body of a loop). What if we could use induction for differential equations? What if we could prove properties of differential equations directly by analyzing how these properties change along the differential equation rather than having to find a global solution first and inspecting whether it satisfies that property at all times? What if we could tame the analytic complexity of differential equations by analyzing the generic local "step" that a continuous dynamical system performs (repeatedly). The biggest conceptual challenge will, of course, be in understanding what exactly the counterpart of a step even is for continuous dynamical systems, because there is no such thing as a next step for a differential equation that evolves in continuous time.

This chapter is of central significance for the Foundations of Cyber-Physical Systems. The analytic principles begun in this chapter will be a crucial basis for analyzing all complex CPSs. The most important learning goals of this chapter are:

---

[1] In fact, discrete and continuous dynamics turn out to be proof-theoretically quite related [12].

Modeling and Control: This chapter will advance the core principles behind CPS by developing a deeper understanding of their continuous dynamical behavior. This chapter will also illuminate another facet of how discrete and continuous systems relate to one another, which ultimately leads to a fascinating view on understanding hybrid systems [12].

Computational Thinking: This chapter exploits the computational thinking principles in their purest form by seeking and exploiting surprising analogies between discrete dynamics and continuous dynamics, however different the two may appear at first sight. This chapter is devoted to rigorous reasoning about the differential equations in CPS models. Such rigorous reasoning is crucial for understanding the continuous behavior that CPSs exhibit over time. Without sufficient rigor in their analysis it can be impossible to understand their intricate behavior and spot subtle flaws in their control or say for sure whether and why a design is no longer faulty. This chapter systematically develops one reasoning principle for equational properties of differential equations that is based on *induction for differential equations* [8, 13]. It follows an axiomatic logical understanding of differential invariants via differential forms [14]. Subsequent chapters expand the same core principles developed in this chapter to the study of general invariant properties of differential equations. This chapter continues the *axiomatization* of differential dynamic logic dL [11, 12] pursued since Chap. 5 and lifts dL's proof techniques to systems with more complex differential equations. The concepts developed in this chapter form the differential facet illustrating the more general relation of *syntax* (which is notation), *semantics* (which carries meaning), and *axiomatics* (which internalizes semantic relations into universal syntactic transformations). These concepts and their relations jointly form the significant *logical trinity* of syntax, semantics, and axiomatics. This chapter studies the differential facet of this logical trinity. Finally, the verification techniques developed in this chapter are critical for verifying CPS models of appropriate scale and technical complexity.

CPS Skills: We will develop a deeper understanding of the semantics of the continuous dynamical aspects of CPS models and develop and exploit a significantly better intuition for the operational effects involved in CPS. In addition to exhibiting semantic nuances, this understanding is critical to rigorous reasoning for all but the most elementary cyber-physical systems.

10.2 A Gradual Introduction to Differential Invariants

This section provides a gradual development of the intuition behind differential invariants. Such an incremental development is useful to understand the working principles and to understand why differential invariants work the way they do. It can also support our intuition when designing systems or proofs for them.

discrete vs. continuous analogies
rigorous reasoning about ODEs
induction for differential equations
differential facet of logical trinity

understanding continuous dynamics   semantics of continuous dynamics
relate discrete+continuous   operational CPS effects

10.2.1 Global Descriptive Power of Local Differential Equations

Differential equations let the physics evolve continuously, possibly for longer periods of time. They describe such global behavior locally, however, just by the right-hand side of the differential equation.

> **Note 54 (Local descriptions of global behavior by differential equations)**
> The key principle behind the descriptive power of differential equations is that they describe the evolution of a continuous system over time using only a local description of the direction in which the system evolves at any point in space. The solution of a differential equation is a global description of how the system evolves. The differential equation itself is a local characterization. While the global behavior of a continuous system can be subtle, complex, and challenging, its local description as a differential equation is much simpler. This difference between local description and global behavior, which is fundamental to the descriptive power of differential equations, can be exploited for proofs.

Recall the semantics of differential equations from Chap. 3:

> **Definition 3.3 (Transition semantics of ODEs).**
>
> $$[\![x' = f(x) \,\&\, Q]\!] = \big\{(\omega, v) : \varphi(0) = \omega \text{ except at } x' \text{ and } \varphi(r) = v \text{ for a solution}$$
> $$\varphi : [0, r] \to \mathscr{S} \text{ of any duration } r \text{ satisfying } \varphi \models x' = f(x) \wedge Q\big\}$$
>
> where $\varphi \models x' = f(x) \wedge Q$, iff for all times $0 \le z \le r$: $\varphi(z) \in [\![x' = f(x) \wedge Q]\!]$
> with $\varphi(z)(x') \stackrel{\text{def}}{=} \frac{d\varphi(t)(x)}{dt}(z)$ and $\varphi(z) = \varphi(0)$ except at x, x'.

The solution φ describes the global behavior of the system, which is specified locally by the right-hand side $f(x)$ of the differential equation $x' = f(x)$.

Chap. 2 has shown a number of examples illustrating the descriptive power of differential equations, that is, examples in which the solution was very complicated even though the differential equation was rather simple. This is a strong property of differential equations: they can describe even complicated processes in simple ways. However, this representational advantage of differential equations does not carry over into the verification when verification is stuck with proving properties of differential equations only by way of their solutions, which, by the very nature of differential equations, are more complicated again.

This chapter, thus, investigates ways of proving properties of differential equations using the differential equations themselves, not their solutions. This leads to *differential invariants* [8, 13, 14], which can perform induction for differential equations just based on their local dynamics. In fact, loops and differential equations have a lot more in common [12] than meets the eye (Sect. 10.8.1).

10.2.2 Intuition for Differential Invariants

Just as inductive invariants are the premier technique for proving properties of loops, differential invariants [7–9, 13, 14] provide the primary inductive technique we use for proving properties of differential equations (without having to solve them). Recall the loop induction proof rule from Sect. 7.3.3

$$\text{loop} \quad \frac{\Gamma \vdash F, \Delta \quad F \vdash [\alpha]F \quad F \vdash P}{\Gamma \vdash [\alpha^*]P, \Delta}$$

The core principle behind loop induction is that the induction step for proving $[\alpha^*]P$ investigates the loop body as the local generator α and shows that it never changes the truth-value of the invariant F (see the middle premise $F \vdash [\alpha]F$ of proof rule loop from Sect. 7.3.3 or the only premise of the core essentials induction proof rule ind from Sect. 7.3.2). Let us try to establish the same inductive principle, just for differential equations. The first and third premise of the loop rule transfer easily to differential equations. The challenge is to figure out what the counterpart of the induction step $F \vdash [\alpha]F$ would be since, unlike loops, differential equations do not have a notion of "one step."

What does the local generator of a differential equation $x' = f(x)$ tell us about the evolution of a system? And how does it relate to the truth of a formula F all along the solution of that differential equation? That is, to the truth of the dL formula $[x' = f(x)]F$ expressing that all runs of $x' = f(x)$ lead to states satisfying F. Figure 10.1 depicts an example of a vector field for a differential equation (plotting the right-hand side of the differential equation as a vector at every point in the state space), a global solution (in red), and an unsafe region $\neg F$ (shown in blue). The safe region F is the complement of the blue unsafe region $\neg F$. Of course, it is quite impossible to draw the appropriate direction vector of the differential equation at literally every point in the state space in Fig. 10.1, so we have to settle for a few.

One way of proving that $[x' = f(x)]F$ is true in a state ω would be to compute the solution from that state ω, and check every point in time along the solution to

Fig. 10.1 Vector field and
one solution of a differential
equation that does not enter
the blue unsafe regions

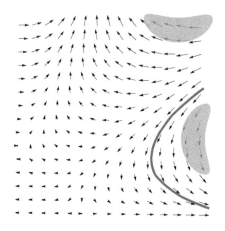

see whether it is in the safe region F or the unsafe region $\neg F$. Unfortunately, there
are uncountably infinitely many points in time to check. Furthermore, that only considers a single initial state ω, so proving validity of a formula would require considering all of the uncountably infinitely many possible initial states and computing
and following a solution in each of them. That is why this naïve approach does not
compute.

A similar idea can still be made to work when the symbolic initial-value problem
can be solved with a symbolic initial value x and a quantifier for time can be used,
which is what the solution axiom $[']$ does. Yet, even that only works when a solution
to the symbolic initial-value problem can be computed and the arithmetic resulting
from the quantifier for time can be decided. For polynomial solutions, this works
by Tarski's quantifier elimination (Sect. 6.5). But polynomial solutions come from
very simple systems only (the nilpotent linear differential equation systems from
Sect. 2.9.3).

Reexamining the illustration in Fig. 10.1, we suggest an entirely different way
of checking whether the system could ever lead to an unsafe state in $\neg F$ when
following the differential equation $x' = f(x)$. The intuition is the following. If there
were a vector in Fig. 10.1 that pointed from a safe state in F to an unsafe state $\neg F$
(in the blue region), then following the differential equation along that vector would
get the system into the unsafe region $\neg F$. If, instead, all vectors only pointed from
safe states to safe states in F, then, intuitively, following such a chain of vectors
would only lead from safe states to safe states. So if the system also started in a
safe state, it would stay safe forever. In fact, this also illustrates that we have some
leeway in how we show $[x' = f(x)]F$. We do not need to know where exactly the
system evolves to, just that it remains somewhere in F.

Let us make this intuition rigorous to obtain a sound proof principle that is perfectly reliable in order to be usable in CPS verification. What we need to do is to
find a way of characterizing how the truth of F changes when moving along the
differential equation. That will then enable us to show that the system only evolves
in directions in which the formula F stays true.

Fig. 10.2 One scenario for
the rotational dynamics and
relationship of direction vector (v, w) to radius r and angle
ϑ

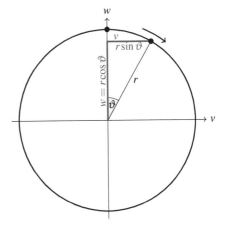

10.2.3 Deriving Differential Invariants

How can the intuition about directions of evolution of a logical formula F with
respect to differential equation $x' = f(x)$ be made rigorous? Let's develop step by
step.

Example 10.1 (Rotational dynamics). As a guiding example, consider a conjecture
about the rotational dynamics from Example 2.5 where v and w represent the coordinates of a direction vector rotating clockwise in a circle of radius r (Fig. 10.2):

$$v^2 + w^2 = r^2 \rightarrow [v' = w, w' = -v] \, v^2 + w^2 = r^2 \qquad (10.1)$$

The conjectured dL formula (10.1) is valid, because, indeed, if the vector (v, w)
is initially at distance r from the origin $(0,0)$, then it will always remain at that distance when rotating around the origin, which is what the dynamics does. That is, the
point (v, w) will always remain on the circle of radius r. But how can we prove that?
In this particular case, we could possibly investigate solutions, which are trigonometric functions (although the solutions indicated in Fig. 10.2 are not at all the only
solutions). With those solutions, we could perhaps find an argument why they stay
at distance r from the origin. But the resulting arithmetic will involve power series,
which makes it unnecessarily difficult. The argument for why the simple dL formula
(10.1) is valid should be an easy one. And it is, after we have discovered the right
proof principle as this chapter will do.

First, what is the direction in which a continuous dynamical system evolves?
The direction is exactly described by the differential equation, because the whole
point of a differential equation is to describe in which direction the state evolves at
every point in space. So the direction which a continuous system obeying $x' = f(x)$
follows from state ω is described by the time-derivative, which is exactly the value
$\omega[\![f(x)]\!]$ of term $f(x)$ in state ω. Recall that the term $f(x)$ can mention x and other
variables so its value $\omega[\![f(x)]\!]$ depends on the present state ω.

Fig. 10.3 Differential invariant F remains true in the direction of the dynamics

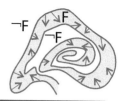

> **Note 55 ("Formulas that remain true in the direction of the dynamics")**
> Proving dL formula $[x' = f(x)]F$ does not really require us to answer where exactly the system evolves to but just how the evolution of the system relates to the formula F and the set of states ω in which F evaluates to *true*. It is enough to show that the system only evolves in directions in which formula F will stay *true* (Fig. 10.3).

A logical formula F is ultimately built from atomic formulas that are comparisons of (polynomial or rational) terms such as $e = 5$ or $v^2 + w^2 = r^2$. Let e denote such a (polynomial) term in the variable (vector) x that occurs in the formula F. The semantics of a polynomial term e in a state ω is the real number $\omega[\![e]\!]$ to which it evaluates. In which direction does the value of e evolve when following the differential equation $x' = f(x)$ for some time? That depends both on the term e that is being evaluated and on the differential equation $x' = f(x)$ that describes how the respective variables x evolve over time.

> **Note 56 (Directions)** Directions of evolutions are described by derivatives. After all, the differential equation $x' = f(x)$ states that the time-derivative of x is $f(x)$.

To find out how the value of a term changes, let's differentiate the term of interest and see what that tells us about how its value evolves over time. Wait, what do the resulting derivatives actually mean? That is a crucial question, but let us, nevertheless, take the inexcusable liberty of postponing this question till later and just develop a first intuition for now.

Example 10.2 (Differentiating terms in rotational dynamics). Which of the terms should be differentiated when trying to understand how the truth-value of the postcondition in (10.1) changes? Since that is not necessarily clear so far, let's rewrite formula (10.1) and consider the following equivalent (Exercise 10.2) dL formula, which only has a single interesting term to worry about:

$$v^2 + w^2 - r^2 = 0 \rightarrow [v' = w, w' = -v]\, v^2 + w^2 - r^2 = 0 \qquad (10.2)$$

Differentiating the only relevant term $v^2 + w^2 - r^2$ in the postcondition of (10.2) gives

$$(v^2 + w^2 - r^2)' = 2vv' + 2ww' - 2rr' \qquad (10.3)$$

Of course, differentiating $v^2 + w^2 - r^2$ does not just result in $2v + 2w - 2r$, because its value also depends on how the variables themselves change so on the derivative v' of v, etc. If only we knew what the symbols v', w', and r' mean in (10.3). The differential equation of (10.2) seems to indicate that v' equals w and w' equals $-v$. Would it be okay to replace the left-hand side w' of the differential equation with its right-hand side $-v$ in (10.3)? That would lead to

$$2vv' + 2ww' - 2rr' = 2vw + 2w(-v) - 2rr' \tag{10.4}$$

which clearly would be 0 if only r' were 0. Well, maybe we could consider r' to be 0, since r does not come with a differential equation, so r is not supposed to change, which is what the differential equation $r' = 0$ would tell us, too.

Lo and behold! This might lead to a possible proof because $2vw + 2w(-v)$ is indeed 0. We just do not know whether it is a proof yet. What proof rules should we have applied to prove (10.2)? Why are they sound proof rules? Was it okay to substitute the right-hand side of the differential equation for its left-hand side in (10.4)? Can we differentiate terms to find out how they change over time? What do the respective primed symbols v', w', r' mean? What is the meaning of the operator $(\cdot)'$ that we used on the term $v^2 + w^2 - r^2$ in (10.3)? How do we know that this operator makes the two sides of (10.3) equal? Or maybe even: do differential equations mind being substituted in?

These are a bunch of important questions on the road to turning the intuition of Example 10.2 into sound proof principles. Let's answer them one at a time.

10.3 Differentials

In order to clarify the intuition we followed for motivating differential invariant reasoning, we first add x' and $(e)'$ officially to the syntax since we used them in our reasoning in Example 10.2. The second step is to define their meaning. And the third step of the logical trinity is to develop axioms that can be proved sound with respect to the semantics and that enable correct syntactic reasoning about such primes.

10.3.1 Syntax of Differentials

The first step for understanding reasoning with differentiation is to ennoble the primes of x' and $(e)'$ and officially consider them as part of the language of differential dynamic logic by adding them to its syntax. For every variable x add a corresponding *differential symbol* x' that can be used like any other variable, but, in a differential equation $x' = f(x)$, of course, x' serves the special purpose of denoting the time-derivative of its associated variable x. For every term e, add the *differential term* $(e)'$. Formally, both really should have been part of differential dynamic logic all along, but our understanding only caught up with that fact in this chapter. Besides, it was easier to first suppress these primes and exclusively have them in differential equations in Part I.

Definition 10.1 (dL Terms). A *term e* of *(differential-form) differential dynamic logic* is defined by the grammar (where $e, \tilde{e}$ are terms, x is a variable with corresponding differential symbol x', and c a rational number constant):

$$e ::= x \mid x' \mid c \mid e + \tilde{e} \mid e - \tilde{e} \mid e \cdot \tilde{e} \mid e / \tilde{e} \mid (e)'$$

For emphasis, when primes are allowed, the logic is also called *differential-form* differential dynamic logic [14], but we will continue to just call it differential dynamic logic. The formulas and hybrid programs of (differential-form) differential dynamic logic are built as in Sects. 3.3 and 4.4. The semantics remains unchanged except that the new additions of differential terms $(e)'$ and differential symbols x' need to be outfitted with a proper meaning.

It is, of course, important to take care that division $e / \tilde{e}$ only makes sense in a context where the divisor $\tilde{e}$ is guaranteed not to be zero in order to avoid undefinedness. *We only allow division to be used in a context where the divisor is ensured not to be zero!*

10.3.2 Semantics of Differential Symbols

The meaning of a variable symbol x is defined by the state ω as $\omega(x)$, so its value $\omega[\![x]\!]$ in state ω is directly looked up from the state via $\omega[\![x]\!] = \omega(x)$. It is crucial to understand the significant subtleties and substantial challenges that arise when trying to give meaning to a differential symbol x' or anything else with a derivative connotation such as the differential term $(e)'$ of term e. The meaning of term e in state ω is $\omega[\![e]\!]$ and, thus, the meaning of the differential term $(e)'$ in state ω is written $\omega[\![(e)']\!]$. But now that we know how it's written, how is $\omega[\![(e)']\!]$ defined?

The first mathematical reflex may be to set out for a definition of x' and $(e)'$ in terms of a time-derivative $\frac{d}{dt}$ of something. But there is no time and, thus, no time-derivative in an isolated state ω. We cannot possibly define something like

$$\omega[\![(e)']\!] \stackrel{???}{=} \frac{d\omega[\![e]\!]}{dt}$$

because time t does not even occur anywhere on the right-hand side. In fact, it is entirely meaningless to ask for the rate of change of the value of anything over time in a single isolated state ω! For time-derivatives to make sense, we at least need a concept of time and the values understood as a function of time. That function needs to be defined on a big enough interval for derivatives to have a chance to become meaningful. And the function needs to be differentiable so that the time-derivatives even exist to begin with. In the presence of discrete state change, not every value will always have a time-derivative even if we were to keep its history around. None of this is the case when we try to define what the value $\omega[\![(e)']\!]$ of the syntactic term $(e)'$ would be in the state ω.

The next mathematical reflex may be to say that the meaning of x' and $(e)'$ depends on the differential equation. But the meaning of $(e)'$ in state ω is $\omega[\![(e)']\!]$, so there simply is no differential equation to speak of. Nothing can have a meaning that depends on something else outside, because that violates all principles of denotational semantics. Notice how useful it is that the principles of logic prompted us to be precise about the definition $\omega[\![(e)']\!]$ of the meaning of $(e)'$. Without the help of the mathematical rigor of logic, we might have just fallen for innocently writing down some primes and differential operators, and ultimately would have woken up surprised if this led us to "conclude" something that is not actually true.

While neither time-derivatives nor differential equations can come to the rescue to give x' or $(e)'$ a meaning, it is important to understand why the lack of having a value and a meaning would cause complications for the fabrics of logic. Denotational semantics defines the meaning of all expressions compositionally in a modular fashion and without reference to outside elements, such as the differential equation in which they also happen to occur. The meaning of terms is a function of the state, and not a function of the state and the context or purpose for which it happens to have been mentioned at the moment.

The mystery of giving meaning to differential symbols is resolved by declaring the state to be responsible for assigning a value not just to all variables $x \in \mathcal{V}$ but also to all differential symbols $x' \in \mathcal{V}'$. A *state* ω is a mapping $\omega : \mathcal{V} \cup \mathcal{V}' \to \mathbb{R}$ assigning a real number $\omega(x) \in \mathbb{R}$ to each variable $x \in \mathcal{V}$ and also a real number $\omega(x') \in \mathbb{R}$ to each differential symbol $x' \in \mathcal{V}'$. For example, when $\omega(v) = 1/2, \omega(w) = \sqrt{3}/2, \omega(r) = 5$ and $\omega(v') = \sqrt{3}/2, \omega(w') = -1/2, \omega(r') = 0$ the term $2vv' + 2ww' - 2rr'$ evaluates to

$$\omega[\![2vv' + 2ww' - 2rr']\!] = 2\omega(v) \cdot \omega(v') + 2\omega(w) \cdot \omega(w') - 2\omega(r) \cdot \omega(r') = 0$$

A differential symbol x' can have any arbitrary real value in a state ω. Along the solution $\varphi : [0, r] \to \mathcal{S}$ of a differential equation, however, we know precisely what value x' has. Or at least we do, if its duration r is nonzero so that we are not just talking about an isolated point $\varphi(0)$ again. At any point in time $z \in [0, r]$ along such a continuous evolution φ, the differential symbol x' has the same value as the time-derivative $\frac{d}{dt}$ of the value $\varphi(t)(x)$ of x over time t at the specific time z [8, 11, 14], because that is what we needed to make sense of the equation $x' = f(x)$.

Definition 3.3 (Transition semantics of ODEs).

$$[\![x' = f(x) \,\&\, Q]\!] = \{(\omega, v) : \varphi(0) = \omega \text{ except at } x' \text{ and } \varphi(r) = v \text{ for a solution}$$
$$\varphi : [0, r] \to \mathcal{S} \text{ of any duration } r \text{ satisfying } \varphi \models x' = f(x) \wedge Q\}$$

where $\varphi \models x' = f(x) \wedge Q$, iff for all times $0 \leq z \leq r$: $\varphi(z) \in [\![x' = f(x) \wedge Q]\!]$ with $\varphi(z)(x') \stackrel{\text{def}}{=} \frac{d\varphi(t)(x)}{dt}(z)$ and $\varphi(z) = \varphi(0)$ except at x, x'.

The value of differential symbol x' at time $z \in [0, r]$ along a solution $\varphi : [0, r] \to \mathcal{S}$ of a differential equation $x' = f(x) \,\&\, Q$ is equal to the analytic time-derivative at z:

Expedition 10.1 (Denotational semantics)

The whole paradigm of *denotational semantics*, initiated for programming languages by Dana Scott and Christopher Strachey [16], is based on the principle that the semantics of an expression of a programming language should be the mathematical object that it denotes. That is, a denotational semantics is a function assigning a mathematical object $\omega[\![e]\!]$ from a semantic domain (here $\mathbb{R}$) to each term e, depending on the state ω.

The *meaning of terms*, thus, is a function $[\![\cdot]\!] : \mathrm{Trm} \to (\mathscr{S} \to \mathbb{R})$ that maps each term $e \in \mathrm{Trm}$ to the function $[\![e]\!] : \mathscr{S} \to \mathbb{R}$ giving the real value $\omega[\![e]\!] \in \mathbb{R}$ that the term e has in each state $\omega \in \mathscr{S}$. In fact, this is exactly how the semantics of terms of dL has been defined in Chap. 2 in the first place. For classical logics such as first-order logic, this denotational semantics has always been the natural and dominant approach since Gottlob Frege [1].

Scott and Strachey [16], however, pioneered the idea of leveraging the denotational style of semantics to give meaning to programming languages. And, indeed, dL's hybrid programs have a denotational semantics. The meaning of an HP α is the reachability relation $[\![\alpha]\!] \subseteq \mathscr{S} \times \mathscr{S}$ that it induces on the states $\mathscr{S}$. Correspondingly, the (denotational) *meaning of hybrid programs* as defined in Chap. 3 is a function $[\![\cdot]\!] : \mathrm{HP} \to \wp(\mathscr{S} \times \mathscr{S})$ assigning a relation $[\![\alpha]\!] \subseteq \mathscr{S} \times \mathscr{S}$ in the powerset $\wp(\mathscr{S} \times \mathscr{S})$ of the product $\mathscr{S} \times \mathscr{S}$ to each HP α.

A crucial feature of denotational semantics, however, is *compositionality*. The meaning $[\![e + \tilde{e}]\!]$ of a compound such as $e + \tilde{e}$ should be a simple function of the meanings $[\![e]\!]$ and $[\![\tilde{e}]\!]$ of its pieces e and $\tilde{e}$. This compositionality is exactly the way the meaning of differential dynamic logic is defined. For example,

$$\omega[\![e + \tilde{e}]\!] = \omega[\![e]\!] + \omega[\![\tilde{e}]\!] \quad \text{for all states } \omega$$

With a point-wise understanding of $+$, this can be summarized as

$$[\![e + \tilde{e}]\!] = [\![e]\!] + [\![\tilde{e}]\!]$$

$$\varphi(z)(x') \stackrel{\text{def}}{=} \frac{\mathrm{d}\varphi(t)(x)}{\mathrm{d}t}(z) \tag{10.5}$$

Intuitively, the value $\varphi(z)(x')$ of x' is, thus, determined by considering how the value $\varphi(z)(x)$ of x changes along the solution φ when we change time z "only a little bit." Visually, it corresponds to the slope of the tangent of the value of x at time z; see Fig. 10.4. A subtlety poses the case of a solution of duration $r = 0$, in which case there still is no time-derivative to speak of. If $r = 0$, the more detailed explanation of Definition 3.3 in Sect. 3.3.2 ignores condition (10.5) leaving only the requirement that ω and ν agree except for the value of x' and that $\nu \in [\![x' = f(x) \wedge Q]\!]$.

Fig. 10.4 Semantics of differential symbol x' along differential equation

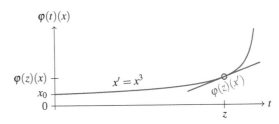

Now we finally figured out the answer to the question of what symbol x' means and what its value is. It all depends on the state. And nothing but the state! Along differential equations, we know a lot about the value of x', otherwise we know less.

The values assigned to x' by the states $\varphi(z)$ visited along a solution $\varphi : [0, r] \to \mathscr{S}$ of a differential equation $x' = f(x) \,\&\, Q$ will have a close relationship, namely (10.5) and $\varphi(z) \in [\![x' = f(x)]\!]$. But that relationship is by virtue of φ being a solution of a differential equation, so that the family of states $\varphi(z)$ for $z \in [0, r]$ have a unique link. It is perfectly consistent to have one state ω in which $\omega(x') = 1$ and another equally isolated state ν in which $\nu(x') = \sqrt{8}$. In fact, that is just what happens for the initial state ω and final state ν when following the differential equation $x' = x^3$ from $\omega(x) = 1$ for $\frac{1}{4}$ time units. If we do not know that ω and ν are the initial and final states of that differential equation or if we do not know that it was exactly for $\frac{1}{4}$ time units that we followed it, there is no reason to suspect much of a relationship between the values of $\omega(x')$ and $\nu(x')$.

Differential symbols x' have a meaning now as being interpreted directly by the state. Yet, what is the meaning of a differential term $(e)'$ such as $(v^2 + w^2 - r^2)'$?

> Before you read on, see if you can find the answer for yourself.

10.3.3 Semantics of Differential Terms

At this point it should no longer be a surprise that the first mathematical reflex of understanding differential terms $(e)'$ as time-derivatives will quickly fall short of its own expectations, because there still is no time-derivative in the isolated state ω that the value $\omega[\![(e)']\!]$ has at its disposal. Likewise, we still cannot ask any differential equations occurring somewhere else in the context, because that would break compositionality and would not explain the meaning in an isolated formula such as (10.3). Unfortunately, though, we cannot follow the same solution and ask the state to assign any arbitrary real value to each differential term. After all, there should be a close relationship of $\omega[\![(2x^2)']\!]$ and $\omega[\![(8x^2)']\!]$ namely that $4\omega[\![(2x^2)']\!] = \omega[\![(8x^2)']\!]$, and an arbitrary state would not respect this relationship if it were to remember arbitrary and unrelated real values for all possible differential terms. Thus, the structure and meaning of the term e should contribute to the meaning of $(e)'$.

The value of $(e)'$ is supposed to tell us something about how the value of e changes. But it is not and could not possibly be change over time to which this is referring, because there is no time or time-derivative to speak of in an isolated state ω. The trick is that we can still determine how the value of e will change, just not over time. We can tell just from the term e itself how its value will change locally depending on how its constituents change.

Recall that the *partial derivative* $\frac{\partial f}{\partial x}(\xi)$ of a function f with respect to the variable x at the point ξ characterizes how the value of f changes as the variable x changes at the point ξ, so when keeping all values of all variables at the point ξ, except for small local changes of the value of x. The term $2x^2$ will locally change according to the partial derivative of its value with respect to x, but the overall change will also depend on how x itself changes locally. The term $5x^2y$ also changes according to the partial derivative of its value with respect to x but it additionally changes according to its partial derivative with respect to y and overall also depends on how x and y themselves change locally.

The clou is that the state ω already has the values $\omega(x')$ of all differential symbols x' at its disposal, which, qua Definition 3.3, are reminiscent of the direction that x would be evolving to locally, if only state ω were part of a solution of a differential equation. The value $\omega(x')$ of differential symbol x' acts like the "local shadow" of the time-derivative $\frac{dx}{dt}$ at ω if only that derivative even existed at that point to begin with. But even if that time-derivative cannot exist at a general isolated state, we can still understand the value $\omega(x')$ that x' happens to have in that state as the direction that x would evolve in locally at that state. Likewise the value $\omega(y')$ of y' can be taken to indicate the direction that y would evolve in locally at that state. Now all it takes is a way to accumulate the change by summing it all up to lead to the meaning of differentials [14].

Definition 10.2 (Semantics of differentials). The semantics of differential term $(e)'$ in state ω is the value $\omega[\![(e)']\!]$ defined as

$$\omega[\![(e)']\!] = \sum_{x \in \mathcal{V}} \omega(x') \cdot \frac{\partial [\![e]\!]}{\partial x}(\omega)$$

The value $\omega[\![(e)']\!]$ is the sum of all (analytic) spatial partial derivatives at ω of the value $[\![e]\!]$ of e by each variable $x \in \mathcal{V}$ multiplied by the corresponding direction of evolution (*tangent*) described by the value $\omega(x')$ of differential symbol $x' \in \mathcal{V}'$.

That sum over all variables $x \in \mathcal{V}$ has finite support (only finitely many summands are nonzero), because term e only mentions finitely many variables x and the partial derivative with respect to variables x that do not occur in e is 0, so does not contribute to the sum. The spatial derivatives exist since the evaluation $\omega[\![e]\!]$ is a composition of smooth functions such as addition, multiplication, etc., so is itself smooth. Recall that the *partial derivative* with respect to variable $x \in \mathcal{V}$ of the value $[\![e]\!]$ of e at state $\omega \in \mathscr{S}$ represents how the value of $\omega[\![e]\!]$ changes with the value of x. It is defined as the limit of the corresponding difference quotient as the new value

10.3 Differentials
301

$\kappa \in \mathbb{R}$ that x has in state ω_x^κ converges to the value $\omega(x)$ that x has in state ω:

$$\frac{\partial [\![e]\!]}{\partial x}(\omega) = \lim_{\kappa \to \omega(x)} \frac{\omega_x^\kappa [\![e]\!] - \omega [\![e]\!]}{\kappa - \omega(x)}$$

Overall the (real) value of $(e)'$ depends not just on e itself and the values in the current state ω of the variables x that occur in e but also on the direction in which these variables are taken to evolve according to the values of the respective differential symbols x' in ω; see Fig. 10.5.

Fig. 10.5 Differential form semantics of differentials: their value depends on the point as well as on the direction of the vector field at that point

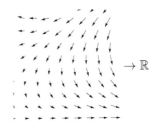

$\to \mathbb{R}$

Example 10.3 (Rotational dynamics). In state ω, the differential term $(v^2 + w^2 - r^2)'$ from the rotational dynamics has the semantics:

$$\omega[\![(v^2 + w^2 - r^2)']\!] = \omega(v') \cdot \omega[\![2v]\!] + \omega(w') \cdot \omega[\![2w]\!] - \omega(r') \cdot \omega[\![2r]\!]$$

Example 10.4. In a state ω, the differential term $(x^3 y + 2x + 1)'$ has the semantics:

$$\omega[\![(x^3 y + 2x + 5)']\!] = \omega(x') \cdot \omega[\![3x^2 y + 2]\!] + \omega(y') \cdot \omega[\![x^3]\!]$$

10.3.4 Derivation Lemma with Equations of Differentials

Observe one quite crucial byproduct of adopting differentials as first-class citizens in dL. Differentiation, the process of forming derivatives that we used in (10.2), was previously an amorphous operation without proper semantic counterparts. While it might have been clear how to differentiate a term, it was quite unclear what that really meant in a state. Using Definition 10.2, both sides of the equation (10.2) now have a precise semantics and, indeed, both sides always have the same value.

Differentiation has now simply become the perfectly meaningful use of equations of differential terms. For example, the use of Leibniz's product rule of differentiation simply corresponds to the use of the following equation:

$$(e \cdot k)' = (e)' \cdot k + e \cdot (k)' \tag{10.6}$$

Equations have a well-defined meaning on reals and both sides of the equation (10.6) have a semantics by Definition 10.2, which can be shown to agree. Equation (10.6) is an ordinary formula that is an equation of differential terms equating the differential $(e \cdot k)'$ of the product term $e \cdot k$ to the sum of terms $(e)' \cdot k$ and $e \cdot (k)'$. After establishing that the equation (10.6) is a valid formula, differentiating a product such as $x^3 \cdot y$ simply amounts to using the corresponding instance of (10.6) to justify

$$(x^3 \cdot y)' = (x^3)' \cdot y + x^3 \cdot (y)'$$

Corresponding equations of differentials hold for all other term operators.

Lemma 10.1 (Derivation lemma). *The following equations of differentials are valid formulas so sound axioms:*

$+'$ $(e+k)' = (e)' + (k)'$

$-'$ $(e-k)' = (e)' - (k)'$

$\cdot'$ $(e \cdot k)' = (e)' \cdot k + e \cdot (k)'$

$/'$ $(e/k)' = ((e)' \cdot k - e \cdot (k)')/k^2$

c' $(c())' = 0$ (for numbers or constants $c()$)

x' $(x)' = x'$ (for variable $x \in \mathcal{V}$)

Proof. We only consider the summation case of the proof, which is reported in full elsewhere [14].

$$\omega[\![(e+k)']\!] = \sum_x \omega(x') \frac{\partial[\![e+k]\!]}{\partial x}(\omega) = \sum_x \omega(x') \frac{\partial([\![e]\!] + [\![k]\!])}{\partial x}(\omega)$$

$$= \sum_x \omega(x') \left(\frac{\partial[\![e]\!]}{\partial x}(\omega) + \frac{\partial[\![k]\!]}{\partial x}(\omega) \right)$$

$$= \sum_x \omega(x') \frac{\partial[\![e]\!]}{\partial x}(\omega) + \sum_x \omega(x') \frac{\partial[\![k]\!]}{\partial x}(\omega)$$

$$= \omega[\![(e)']\!] + \omega[\![(k)']\!] = \omega[\![(e)' + (k)']\!]$$

□

This gives us a way of computing simpler forms for differentials of terms by applying the equations of Lemma 10.1 from left to right, which will, incidentally, lead us to the same result that differentiation would have, except now the result has been obtained by a chain of logical equivalence transformations on differentials each of which is individually grounded semantically with a soundness proof. It also becomes possible to selectively apply equations of differentials as needed in a proof without endangering soundness. Who would have figured that our study of differential equations would lead us down a path to study equations of differentials?

By axiom x', the differential $(x)'$ of a variable x is simply its corresponding differential symbol x', because they have the same semantics. The differential $(c())'$

of a constant symbol $c()$ is 0, because constant symbols do not change their value when the value of any variable changes, because no variables even occur. The differential of a division e/k uses a division, which is where we need to make sure not to accidentally divide by zero. Yet, in the definition of $(e/k)'$, the division is by k^2, which, fortunately, has the same roots that k already has, as $k = 0 \leftrightarrow k^2 = 0$ is valid for any term k. Hence, in any context in which e/k is defined, its differential $(e/k)'$ will also be defined.

Example 10.5. Computing the differential of a term like $v^2 + w^2$ is now easy just by using the respective equations from Lemma 10.1 in sequence as indicated:

$$(v^2 + w^2)' \stackrel{+'}{=} (v \cdot v)' + (w \cdot w)'$$

$$\stackrel{.'}{=} ((v)' \cdot v + v \cdot (v)') + ((w)' \cdot w + w \cdot (w)')$$

$$\stackrel{x'}{=} v' \cdot v + v \cdot v' + w' \cdot w + w \cdot w' = 2vv' + 2ww'$$

When r is a constant function symbol, an additional use of axiom c' also justifies

$$(v^2 + w^2 - r^2)' = 2vv' + 2ww'$$

10.3.5 Differential Lemma

Now that we have obtained a precise semantics of differential symbols x' and differentials $(e)'$ that is meaningful in any arbitrary state ω, no matter how isolated it may be, it is about time to come back to the question of what we can now learn from studying their values *along a differential equation*.

Along the solution φ of a differential equation, differential symbols x' do not have arbitrary values but, at all times z, are interpreted as time-derivatives of the value of x by Definition 3.3:

$$\varphi(z)[\![(x)']\!] = \varphi(z)(x') \stackrel{\text{def}}{=} \frac{\mathrm{d}\varphi(t)(x)}{\mathrm{d}t}(z) \tag{10.5*}$$

The key insight is that this equality of the value of differentials with analytic time-derivatives along a differential equation continues to hold not just for differentials of variables x but also for differentials $(e)'$ of arbitrary terms e.

The following central lemma [14], which is the differential counterpart of the substitution lemma, establishes the connection between the semantics of syntactic differentials of terms and semantic differentiation as an analytic operation to obtain analytic time-derivatives of the semantics of terms along differential equations. It will allow us to draw analytic conclusions about the behavior of a system along a differential equation from the values of differentials obtained syntactically.

Lemma 10.2 (Differential lemma). *Let* $\varphi \models x' = f(x) \wedge Q$ *for some solution* $\varphi : [0,r] \to \mathscr{S}$ *of duration* $r > 0$. *Then for all times* $0 \le z \le r$ *and all terms e defined all along* φ *with* $FV(e) \subseteq \{x\}$:

$$\varphi(z)[\![(e)']\!] = \frac{\mathrm{d}\,\varphi(t)[\![e]\!]}{\mathrm{d}t}(z)$$

Proof. Prior work reports the full proof [14], which is mostly by chain rule:

$$\frac{\mathrm{d}\varphi(t)[\![e]\!]}{\mathrm{d}t}(z) \overset{\text{chain}}{=} \sum_x \frac{\partial[\![e]\!]}{\partial x}(\varphi(z)) \frac{\mathrm{d}\varphi(t)(x)}{\mathrm{d}t}(z) = \sum_x \frac{\partial[\![e]\!]}{\partial x}(\varphi(z))\varphi(z)(x') = \varphi(z)[\![(e)']\!]$$

The proof uses that $\varphi(z)(x')$ equals $\frac{\mathrm{d}\varphi(t)(x)}{\mathrm{d}t}(z)$ along the solution φ of $x' = f(x)$. $\square$

In particular, $\varphi(z)[\![e]\!]$ is continuously differentiable in z. The same result applies to vectorial differential equations as long as all free variables of the term e have some differential equation so that their differential symbols agree with the time-derivatives.

Note 57 (Differential lemma clou) Lemma 10.2 shows that the analytic time-derivatives coincide with the values of differentials. The clou with Lemma 10.2 is that it equates precise but sophisticated analytic time-derivatives with purely syntactic differentials. The analytic time-derivatives on the right-hand side of Lemma 10.2 are mathematically precise and pinpoint exactly what we are interested in: the rate of change of the value of e along solution φ. But they are unwieldy for computers, because analytic derivatives are ultimately defined in terms of limit processes and also need a whole solution to be well-defined. The syntactic differentials on the left-hand side of Lemma 10.2 are purely syntactic (putting a prime on a term) and even their simplifications via the recursive use of the axioms from Lemma 10.1 are computationally tame.

Having said that, in order to be useful, the syntactic differentials need to be aligned with the intended analytic time-derivatives, which is exactly what Lemma 10.2 achieves. To wit, even differentiating polynomials and rational functions is much easier syntactically than by unpacking the meaning of analytic derivatives in terms of limit processes every time.

10.3.6 Differential Invariant Term Axiom

The differential lemma immediately leads to a first proof principle for differential equations. If the differential $(e)'$ is always zero along a differential equation, then e will always be zero if and only if it was zero initially. For emphasis, we use the backwards implication $P \leftarrow Q$ as alternative notation for the converse forward implication $Q \to P$.

Lemma 10.3 (Differential invariant term axiom). *This axiom is sound:*

$$\text{DI} \ \left(\left[x' = f(x)\right] e = 0 \leftrightarrow e = 0\right) \leftarrow \left[x' = f(x)\right] (e)' = 0$$

Proof. To prove that axiom DI is sound, we need to show the validity of the formula

$$\left[x' = f(x)\right] (e)' = 0 \rightarrow \left(\left[x' = f(x)\right] e = 0 \leftrightarrow e = 0\right)$$

Consider any state ω in which the assumption is true, so $\omega \in [\![[x' = f(x)] (e)' = 0]\!]$, and show that $\omega \in [\![[x' = f(x)] e = 0 \leftrightarrow e = 0]\!]$. Now, $\omega \in [\![[x' = f(x)] e = 0]\!]$ directly implies $\omega \in [\![e = 0]\!]$ when following the differential equation for duration 0. To show the converse implication, assume $\omega \in [\![e = 0]\!]$. If φ is a solution of $x' = f(x)$, then the assumption implies that $\varphi \models (e)' = 0$ since all restrictions of solutions are again solutions. Consequently, Lemma 10.2 implies

$$0 = \varphi(z)[\![(e)']\!] = \frac{d\,\varphi(t)[\![e]\!]}{dt}(z) \tag{10.7}$$

This implies that the term e always evaluates to zero along φ by the mean-value theorem (Lemma 10.4 below), since it initially started out 0 (by initial $\omega \in [\![e = 0]\!]$) and had 0 change over time by (10.7). Hold on, that use of Lemma 10.2 was, of course, predicated on having a solution φ of duration $r > 0$ (otherwise there are no time-derivatives to speak of). Yet, solutions of duration $r = 0$ also already satisfy $e = 0$ from the assumption $\omega \in [\![e = 0]\!]$. Strictly speaking [14], this proof requires that x' is not free in e. □

This proof uses the mean-value theorem [17, §10.10]:

Lemma 10.4 (Mean-value theorem). *If $g : [a, b] \rightarrow \mathbb{R}$ is continuous and differentiable in the open interval (a, b), then there is a $\xi \in (a, b)$ such that:*

$$g(b) - g(a) = g'(\xi)(b - a)$$

The only nuisance with axiom DI is that it never proves any interesting properties on its own. It reduces a proof of the postcondition $e = 0$ for a differential equation to the question of whether $e = 0$ is true initially but also to a proof of the postcondition $(e)' = 0$ for the same differential equation. This is similar to how the loop induction axiom I from Lemma 7.1 reduced the proof of postcondition P of a loop to another postcondition $P \rightarrow [\alpha]P$ of the same loop, so that we ultimately still needed the generalization rule G to get rid of the loop entirely. But just generalization rule G alone will not quite suffice for differential equations.

For Example 10.1, a use of axiom DI would lead to

$$\text{DI} \frac{v^2 + w^2 - r^2 = 0 \vdash [v' = w, w' = -v] v^2 + w^2 - r^2 = 0}{\vdash v^2 + w^2 - r^2 = 0 \rightarrow [v' = w, w' = -v] v^2 + w^2 - r^2 = 0}$$

where the top has $\vdash [v' = w, w' = -v] 2vv' + 2ww' - 2rr' = 0$ and the bottom rule is $\rightarrow$R.

Without knowing anything about v' and w' and r' in the postcondition, we have no chance of finishing this proof. Certainly the generalization rule G cannot succeed because the postcondition $2vv' + 2ww' - 2rr' = 0$ alone is not always true. In fact, it should not be valid, because whether a postcondition $e = 0$ is an invariant of a differential equation does not just depend on the differential $(e)'$ of the term in the postcondition, but also on the differential equation itself. What stands to reason is to use the right-hand sides of the differential equations for their left-hand sides; the two sides of the equation are supposed to be equal! The question is how to justify that that's sound.

10.3.7 Differential Substitution Lemmas

Lemma 10.2 shows that, along a differential equation, the value of the differential $(e)'$ of term e coincides with the analytic time-derivative of the value of term e. The value of a differential term $(e)'$ depends on the term itself as well as the value of its variables x and their corresponding differential symbols x'. Along a differential equation $x' = f(x)$, the differential symbols x' themselves actually have a simple interpretation: their values equal the right-hand side $f(x)$.

> The direction in which the value of a term e evolves as the system follows a differential equation $x' = f(x)$ depends on the differential $(e)'$ of the term e as well as on the differential equation $x' = f(x)$ that locally describes the evolution of its variable x over time.

What we need is a way of using the differential equation $x' = f(x)$ to soundly replace occurrences of the differential symbol x' from its left-hand side with the corresponding right-hand side $f(x)$ of the differential equation. Naïve replacement would be unsound, because that might violate the scope of the formula where x' equals $f(x)$. Discrete assignments $x := e$ were ultimately handled in axiom $[:=]$ from Lemma 5.2 by substituting the new value e for the variable x, and the axiom is already mindful of scoping challenges. The trick is to use the same assignments but for assigning terms to differential symbols x' instead of variables x. Since x' already always has the value $f(x)$ when following the differential equation $x' = f(x)$ along its solution φ, assigning $f(x)$ to x' by a discrete assignment $x' := f(x)$ has no effect.

> **Lemma 10.5 (Differential assignment).** *If $\varphi \models x' = f(x) \wedge Q$ for a solution $\varphi : [0, r] \to \mathscr{S}$ of any duration $r \geq 0$, then*
>
> $$\varphi \models P \leftrightarrow [x' := f(x)]P$$

Proof. The proof [14] is a direct consequence of the fact that the semantics of differential equations (Definition 3.3) requires that $\varphi(z) \in [\![x' = f(x)]\!]$ holds for all times z all along φ. Consequently, the assignment $x' := f(x)$ that changes the value of x' to be the value of $f(x)$ will have no effect, since x' already does have that value along the differential equation. Thus, P and $[x' := f(x)]P$ are equivalent along φ. $\qquad\square$

Using this equivalence at any state along a differential equation $x' = f(x)$ gives rise to a simple axiom characterizing the effect that a differential equation has on its differential symbol x'. Following a differential equation $x' = f(x)$ requires x' and $f(x)$ to always have the same value along the differential equation.

> **Lemma 10.6 (DE differential effect axiom).** *This axiom is sound:*
>
> $$\text{DE} \quad [x' = f(x) \,\&\, Q]P \leftrightarrow [x' = f(x) \,\&\, Q][x' := f(x)]P$$

  While axiom DE performs a no-op, its benefit is that it makes the effect that a differential equation has on the differential symbol available as a discrete assignment.
  The last ingredient is to use the assignment axiom $[:=]$ from Lemma 5.2 also for discrete assignments $x' := e$ to differential symbol x' instead of just for discrete assignments $x := e$ to variable x:

$$[:=] \quad [x' := e]p(x') \leftrightarrow p(e)$$

Let's continue the proof for Example 10.1:

$$
\begin{array}{ll}
{[:=]} & \dfrac{\vdash [v' = w, w' = -v]\, 2v(w) + 2w(-v) - 2rr' = 0}{\vdash [v' = w, w' = -v][v' := w][w' := -v]\, 2vv' + 2ww' - 2rr' = 0} \\[2ex]
\text{DE} & \dfrac{}{\vdash [v' = w, w' = -v]\, 2vv' + 2ww' - 2rr' = 0} \\[2ex]
\text{DI} & \dfrac{v^2 + w^2 - r^2 = 0 \vdash [v' = w, w' = -v]\, v^2 + w^2 - r^2 = 0}{} \\[2ex]
{\to}\text{R} & \dfrac{}{\vdash v^2 + w^2 - r^2 = 0 \to [v' = w, w' = -v]\, v^2 + w^2 - r^2 = 0}
\end{array}
$$

Oops, that did not make all differential symbols disappear, because r' is still around, since r did not have a differential equation in (10.2) to begin with. Stepping back, what we mean by a differential equation like $v' = w, w' = -v$ that does not mention r' is that r is not supposed to change. If r were supposed to change during a continuous evolution, then there would have to be a differential equation for r describing how exactly r changes.

Note 58 (Explicit change) Hybrid programs are *explicit change*. Nothing changes unless an assignment or differential equation specifies how (compare the semantics from Chap. 3 and the bound variables in Sect. 5.6.5). In particular, if a differential equation (system) $x' = f(x)$ does not mention z', then the variable z does not change during $x' = f(x)$, so $x' = f(x)$ and $x' = f(x), z' = 0$ are the same. Strictly speaking this equivalence only holds when z' itself also does not occur elsewhere in the program or formula, which is a condition that is usually met. The subtle nuance is that only $x' = f(x)$ will leave the value of z' untouched, but $x' = f(x), z' = 0$ will change z' to 0 by Definition 3.3.

Even if KeYmaera X has a rigorous treatment with uniform substitutions of free constant symbols, it suffices for our paper proofs to assume $z' = 0$ without further notice for variables z that do not change during a differential equation.

Since (10.2) does not have an r', Note 58 implies that instead of its differential equation $v' = w, w' = -v$ we could have used $v' = w, w' = -v, r' = 0$, which, with DE, would give rise to an extra $[r':=0]$, which we will assume implicitly from now on after showing its use explicitly just once.

$$
\begin{array}{ll}
 & \ast \\
\mathbb{R} & \overline{\vdash 2vw - 2wv - 0 = 0} \\
\text{G} & \overline{\vdash [v' = w, w' = -v]2v(w) + 2w(-v) - 0 = 0} \\
{[:=]} & \overline{\vdash [v' = w, w' = -v][v':=w][w':=-v][r':=0]2vv' + 2ww' - 2rr' = 0} \\
\text{DE} & \overline{\vdash [v' = w, w' = -v]2vv' + 2ww' - 2rr' = 0} \\
\text{DI} & \dfrac{v^2 + w^2 - r^2 = 0 \vdash [v' = w, w' = -v]v^2 + w^2 - r^2 = 0}{} \\
{\rightarrow}\text{R} & \overline{\vdash v^2 + w^2 - r^2 = 0 \rightarrow [v' = w, w' = -v]v^2 + w^2 - r^2 = 0}
\end{array}
$$

This is amazing, because we found out that the value of $v^2 + w^2 - r^2$ does not change over time along the differential equation $v' = w, w' = -v$. And we found that out without ever solving the differential equation, just by a few lines of simple but mathematically rigorous symbolic proof steps.

10.4 Differential Invariant Terms

In order to be able to use the above reasoning as part of a sequent proof efficiently, let's package up the argument in a simple proof rule. As a first shot, we stay with equations of the form $e = 0$, which gives us soundness for the following proof rule.

Lemma 10.7 (Differential invariant term rule). *The following special case of the differential invariants proof rule is sound, i.e., if its premise is valid then so is its conclusion:*

$$\text{dI} \frac{\vdash [x':=f(x)](e)' = 0}{e = 0 \vdash [x' = f(x)]e = 0}$$

Proof. We could prove soundness of this proof rule by going back to the semantics and lemmas we proved about it. The easier soundness proof is to prove that it is a derived rule, meaning that it can be expanded into a sequence of other axiom and proof rule applications that we have already seen to be sound:

$$\text{DI} \frac{\text{DE} \frac{\text{G} \frac{\vdash [x':=f(x)](e)' = 0}{\vdash [x' = f(x)\,\&\,Q][x':=f(x)](e)' = 0}}{\vdash [x' = f(x)\,\&\,Q](e)' = 0}}{e = 0 \vdash [x' = f(x)\,\&\,Q]e = 0}$$

This proof shows dI to be a derived rule because it starts with the premise of rule dI as the only open goal and ends with the conclusion of rule dI, using only proof rules we already know are sound. □

Notice that Gödel's generalization rule G was used to derive dI, so it would not be sound to retain a sequent context Γ, Δ in its premise (except, as usual, assumptions about constants). After all, its premise represents an induction step for a differential equation. Just like in loop invariants, we cannot assume the state considered in the induction step will still satisfy whatever we knew in the initial state.

This proof rule enables us to prove dL formula (10.2) easily in sequent calculus:

$$\to\text{R} \frac{\text{dI} \frac{[:=] \frac{\text{R} \frac{*}{\vdash 2vw + 2w(-v) - 0 = 0}}{\vdash [v':=w][w':= -v]\,2vv' + 2ww' - 0 = 0}}{v^2 + w^2 - r^2 = 0 \vdash [v' = w, w' = -v]\,v^2 + w^2 - r^2 = 0}}{\vdash v^2 + w^2 - r^2 = 0 \to [v' = w, w' = -v]\,v^2 + w^2 - r^2 = 0}$$

Taking a step back, this is an exciting development, because, thanks to differential invariants, the property (10.2) of a differential equation with a nontrivial solution has a very simple proof that we can easily check. The proof did not need to solve the differential equation, which has infinitely many solutions with combinations of trigonometric functions.[2] The proof only required deriving the postcondition and substituting in the differential equation.

---

[2] Granted, the solutions in this case are not quite so terrifying. They are all of the form

$$v(t) = a\cos t + b\sin t, \quad w(t) = b\cos t - a\sin t$$

But the special functions sin and cos still fall outside the decidable parts of arithmetic.

10.5 A Differential Invariant Proof by Generalization

So far, the differential invariant term proof rule dI works for

$$v^2 + w^2 - r^2 = 0 \rightarrow [v' = w, w' = -v]v^2 + w^2 - r^2 = 0 \qquad (10.2^*)$$

with an equation $v^2 + w^2 - r^2 = 0$ normalized to having 0 on the right-hand side. But it does not work for the original formula

$$v^2 + w^2 = r^2 \rightarrow [v' = w, w' = -v]v^2 + w^2 = r^2 \qquad (10.1^*)$$

because its postcondition is not of the form $e = 0$. Yet, the postcondition $v^2 + w^2 - r^2 = 0$ of (10.2) is trivially equivalent to the postcondition $v^2 + w^2 = r^2$ of (10.1), just by rewriting the polynomials on one side, which is a minor change. That is an indication that differential invariants can perhaps do more than what proof rule dI already knows about.

But before we pursue any further our discovery of what else differential invariants can do for us, let us first understand a very important proof principle.

> **Note 59 (Proof by generalization)** If you do not find a proof of a formula, it can sometimes be easier to prove a more general property from which the one you were looking for follows.

This principle, which may at first appear paradoxical, turns out to be very helpful. In fact, we have made ample use of Note 59 when proving properties of loops by induction. The loop invariant that needs to be proved is usually more general than the particular postcondition one is interested in. The desirable postcondition follows from having proved a more general inductive invariant.

Recall the monotonicity right rule MR from Lemma 7.4:

$$\text{MR} \quad \frac{\Gamma \vdash [\alpha]Q, \Delta \quad Q \vdash P}{\Gamma \vdash [\alpha]P, \Delta}$$

Instead of proving the desirable postcondition P of α (conclusion), proof rule MR makes it possible to prove the postcondition Q instead (left premise) and prove that Q is more general than the desired P (right premise). Generalization MR can help us prove the original dL formula (10.1) by first turning the postcondition into the form of the (provable) (10.2) and adapting the precondition using a corresponding cut with $v^2 + w^2 - r^2 = 0$, whose first premise $v^2 + w^2 = r^2 \vdash v^2 + w^2 - r^2 = 0$ is elided but is proved by ℝ:

$$\cfrac{\cfrac{\cfrac{\cfrac{*}{\vdash 2vw+2w(-v)-0=0}}{\vdash [v':=w][w':=-v]2vv'+2ww'-0} \;[:=]}{\cfrac{v^2+w^2-r^2=0\vdash [v'=w,w'=-v]v^2+w^2-r^2=0}{v^2+w^2=r^2\vdash [v'=w,w'=-v]v^2+w^2-r^2=0} \;\text{dI}} \quad \cfrac{\cfrac{*}{v^2+w^2-r^2=0\vdash v^2+w^2=r^2}}{} \;\mathbb{R}}{\cfrac{v^2+w^2=r^2\vdash [v'=w,w'=-v]v^2+w^2=r^2}{\vdash v^2+w^2=r^2\to [v'=w,w'=-v]v^2+w^2=r^2} \;\to\!\text{R}} \;\text{MR}}$$

This is a possible way of proving the original (10.1), but also unnecessarily complicated. Differential invariants can prove (10.1) directly once we generalize proof rule dI appropriately. For other purposes, however, it is still important to have the principle of generalization Note 59 in our repertoire of proof techniques.

10.6 Example Proofs

Of course, differential invariants are just as helpful for proving properties of other differential equations, of which this section lists a few.

Example 10.6. A simple proof shows the differential invariant illustrated in Fig. 10.6.

$$\cfrac{\cfrac{\cfrac{\cfrac{*}{\vdash 2x(-x^2)y+x^2(2xy)=0}}{\vdash [x':=-x^2][y':=2xy]\,2xx'y+x^2y'-0=0} \;[:=]}{x^2y-2=0\vdash [x'=-x^2,y'=2xy]x^2y-2=0} \;\text{dI}}{\vdash x^2y-2=0\to [x'=-x^2,y'=2xy]x^2y-2=0} \;\to\!\text{R}}$$

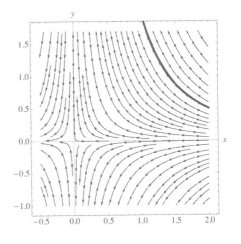

Fig. 10.6 Differential invariant (illustrated in thick red) of the indicated dynamics

Example 10.7 (Self-crossing). Another example is the invariant property illustrated in Fig. 10.7. It is proved easily using dI:

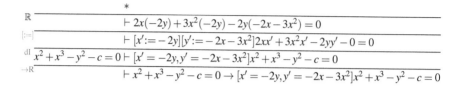

$$\mathbb{R} \; \overline{\qquad\qquad \ast \qquad\qquad\qquad \vdash 2x(-2y)+3x^2(-2y)-2y(-2x-3x^2)=0 \qquad\qquad\qquad}$$

$$[:=] \; \overline{\vdash [x':=-2y][y':=-2x-3x^2]2xx'+3x^2x'-2yy'-0=0}$$

$$\mathrm{dI} \; \overline{x^2+x^3-y^2-c=0 \vdash [x'=-2y,y'=-2x-3x^2]x^2+x^3-y^2-c=0}$$

$$\to\!\mathbb{R} \; \overline{\vdash x^2+x^3-y^2-c=0 \to [x'=-2y,y'=-2x-3x^2]x^2+x^3-y^2-c=0}$$

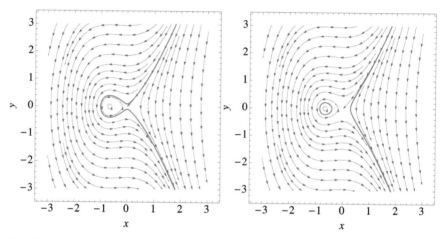

Fig. 10.7 Two differential invariants (illustrated in thick red) of the indicated self-crossing dynamics for Example 10.7 for different values of c

Example 10.8 (Motzkin). Another nice example is the Motzkin polynomial, which is an invariant of the following dynamics (see Fig. 10.8):

$$x^4y^2+x^2y^4-3x^2y^2+1=c \to$$
$$[x'=2x^4y+4x^2y^3-6x^2y,y'=-4x^3y^2-2xy^4+6xy^2]x^4y^2+x^2y^4-3x^2y^2+1=c$$

This dL formula is proved directly by dI, again after normalizing the equation to have right-hand side 0 (where .. abbreviates the antecedent):

$$\mathbb{R} \; \overline{\qquad\qquad\qquad\qquad \ast \qquad\qquad\qquad\qquad \vdash 0=0 \qquad\qquad\qquad\qquad\qquad\qquad}$$

$$[:=] \; \overline{\vdash [x':=2x^4y+4x^2y^3-6x^2y][y':=-4x^3y^2-2xy^4+6xy^2](x^4y^2+x^2y^4-3x^2y^2+1-c)'=0}$$

$$\mathrm{dI} \; \overline{.. \vdash [x'=2x^4y+4x^2y^3-6x^2y,y'=-4x^3y^2-2xy^4+6xy^2]x^4y^2+x^2y^4-3x^2y^2+1-c=0}$$

$$\to\!\mathbb{R} \; \overline{\vdash .. \to [x'=2x^4y+4x^2y^3-6x^2y,y'=-4x^3y^2-2xy^4+6xy^2]x^4y^2+x^2y^4-3x^2y^2+1-c=0}$$

The proof step [:=] is simple, but requires some space:

$$(x^4y^2 + x^2y^4 - 3x^2y^2 + 1 - c)' = (4x^3y^2 + 2xy^4 - 6xy^2)x' + (2x^4y + 4x^2y^3 - 6x^2y)y'$$

After substituting in the differential equation, this gives

$$(4x^3y^2 + 2xy^4 - 6xy^2)(2x^4y + 4x^2y^3 - 6x^2y)$$
$$+(2x^4y + 4x^2y^3 - 6x^2y)(-4x^3y^2 - 2xy^4 + 6xy^2)$$

which simplifies to 0 after expanding the polynomials, and, thus, leads to the equation $0 = 0$, which is easy arithmetic. Note that the arithmetic complexity is reduced when we hide unnecessary contexts as shown in Sect. 6.5.3.

(Thanks to Andrew Sogokon for the nice Example 10.8.)

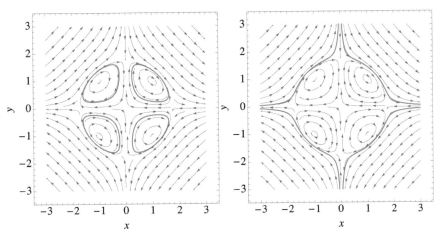

Fig. 10.8 Two differential invariants (illustrated in thick red) of the indicated dynamics for the Motzkin polynomial for Example 10.8 for different values of c

10.7 Summary

This chapter showed one form of differential invariants: the form where the differential invariants are terms whose value always stays 0 along all solutions of a differential equation. The next chapter will use the tools developed in this chapter to investigate more general forms of differential invariants and more advanced proof principles for differential equations. They all share the important discovery in this chapter: that properties of differential equations can be proved using the differential equation rather than its solution.

The most important technical insight of this chapter was that even very complicated behavior that is defined by mathematical properties of the semantics can be

captured by purely syntactical proof principles using differentials. The differential
lemma proved that the values of differentials of terms coincide with the analytic
derivatives of the values. The derivation lemma gave us the usual rules for computing derivatives as equations of differentials. The differential assignment lemma
allowed us the intuitive operation of substituting differential equations into terms.
Proving properties of differential equations using a mix of these simple proof principles is much more civilized and effective than working with solutions of differential
equations. The proofs are also computationally easier, because the proof arguments
are local and derivatives even decrease the polynomial degrees. The resulting axioms are summarized in Fig. 10.9 except the differential induction axiom DI since it
will be generalized in Chap. 11.

DE $[x' = f(x) \& Q]P \leftrightarrow [x' = f(x) \& Q][x' := f(x)]P$

$+'$ $(e+k)' = (e)' + (k)'$

$-'$ $(e-k)' = (e)' - (k)'$

$\cdot'$ $(e \cdot k)' = (e)' \cdot k + e \cdot (k)'$

$/'$ $(e/k)' = ((e)' \cdot k - e \cdot (k)')/k^2$

c' $(c())' = 0$ (for numbers or constants $c()$)

x' $(x)' = x'$ (for variable $x \in \mathcal{V}$)

Fig. 10.9 Axioms for differential invariant terms of differential equations without solutions

The principles begun in this chapter have significantly more potential, though,
and are not limited to proving only properties of the rather limited form $e = 0$. Subsequent chapters will make use of the results obtained and build on the differential
lemma, derivation lemma, and differential assignment lemma to develop more general proof principles for differential equations. But at least on open and connected
evolution domains, the differential invariance proof rule dI is pretty powerful, because it is able to prove all invariant terms, i.e., all terms that never change their
value along the differential equation (as Sect. 10.8.2 will explore). There also is
a way of deciding equational invariants of algebraic differential equations using a
higher-order generalization of differential invariants called differential radical invariants [2].

10.8 Appendix

This appendix discusses optional topics such as the relationship of differential equations to loops, the relationship to differential algebra, and the relationship of the differential invariant term proof rule to Sophus Lie's characterization of invariant functions.

10.8.1 Differential Equations Versus Loops

One way of developing an intuition for the purpose of differential invariants leads through a comparison of differential equations with loops. This perhaps surprising relation can be made completely rigorous and is at the heart of a deep connection equating discrete and continuous dynamics proof-theoretically [12]. This chapter will stay at the surface of this surprising connection but it still leverages the relation of differential equations to loops for our intuition.

To get started with relating differential equations to loops, compare

$$x' = f(x) \qquad \text{vs.} \qquad (x' = f(x))^*$$

How does the differential equation $x' = f(x)$ compare to the same differential equation in a loop $(x' = f(x))^*$ instead? Unlike the differential equation $x' = f(x)$, the repeated differential equation $(x' = f(x))^*$ can run the differential equation $x' = f(x)$ repeatedly any number of times. Albeit, on second thoughts, does that get the repetitive differential equation $(x' = f(x))^*$ to any more states than where the differential equation $x' = f(x)$ could evolve to?

Not really, because chaining lots of solutions of differential equations from a repetitive differential equation $(x' = f(x))^*$ together will still result in a single solution for the same differential equation $x' = f(x)$ that we could have followed all the way. This is precisely what a classical result about the continuation of solutions is about (Proposition 2.2).

> **Note 60 (Looping differential equations)** The loop $(x' = f(x))^*$ over a differential equation is equivalent to $x' = f(x)$, written $(x' = f(x))^* \equiv (x' = f(x))$, i.e., they have the same transition semantics:
>
> $$[\![(x' = f(x))^*]\!] = [\![x' = f(x)]\!]$$
>
> That is, differential equations "are their own loop".[3]

In light of Note 60, differential equations already have some aspects in common with loops. Like nondeterministic repetitions, differential equations might stop right

---

[3] Beware not to confuse this with the case for differential equations with evolution domain constraints, which is subtly different (Exercise 10.1).

away. Like nondeterministic repetitions, differential equations can evolve for longer or shorter durations and the choice of duration is nondeterministic. Like in nondeterministic repetitions, the outcome of the evolution of the system up to an intermediate state influences what happens in the future. And, in fact, in a deeper sense, differential equations actually really do correspond to loops executing their discrete Euler approximations [12].

With this rough relation in mind, let's advance the dictionary translating differential equation phenomena into loop phenomena and back. The local description of a differential equation as a relation $x' = f(x)$ of the state to its derivative corresponds to the local description of a loop by a repetition operator * applied to the loop body α. The global behavior of a global solution of a differential equation $x' = f(x)$ corresponds to the full global execution trace of a repetition α^*, but they are similarly unwieldy objects to handle. Because the local descriptions are so much more concise than the respective global behaviors, but still carry all information about how the system will evolve over time, we also say that the local relation $x' = f(x)$ is the *generator* of the global system solution and that the loop body α is the *generator* of the global behavior of repetition of the loop. Proving a property of a differential equation in terms of its solution corresponds to proving a property of a loop by unwinding it (infinitely often) using axiom [*] from Chap. 5. These comparisons are summarized in Table 10.1.

Table 10.1 Correspondence map between loops and differential equations

loop α^*	differential equation $x' = f(x)$
can repeat 0 times	can evolve for duration 0
repeat any number $n \in \mathbb{N}$ of times	evolve for any duration $r \in \mathbb{R}, r \geq 0$
effect depends on previous loop iteration	effect depends on the past solution
local generator is loop body α	local generator is $x' = f(x)$
full global execution trace	global solution $\varphi : [0, r] \to \mathscr{S}$
proof by unwinding iterations with axiom [*]	proof by global solution with axiom [']
proof by induction with loop invariant rule loop	proof by differential invariant

Now, Chap. 7 made the case that unwinding the iterations of a loop can be a rather tedious way of proving properties about the loop, because there is no good way of ever stopping unwinding, unless a counterexample can be found after a finite number of unwindings. This is where working with a global solution of a differential equation with axiom ['] is actually more useful, because the solution, if we can write it down in first-order real arithmetic, can be handled completely because of the quantifier $\forall t \geq 0$ over all durations. But Chap. 7 introduced induction with invariants as the preferred way of proving properties of loops, by, essentially, cutting the loop open and arguing that the generic state after any run of the loop body has the same characterization as the generic state before. After all these analogous correspondences between loops and differential equations, the obvious question is what the differential equation analogue of a proof concept would be that corresponds to proofs by induction for loops, which is the premier technique for proving loops.

Induction can be defined for differential equations using what are called *differential invariants* [8, 13, 14]. They have a similar principle to the proof rules for induction for loops. Differential invariants prove properties of the solution of the differential equation using only its local generator: the right-hand side of the differential equation.

Expedition 10.2 (Differential algebra)

Even though the following names and concepts are not needed for this textbook, let's take a brief scientific expedition to align the findings on equations of differentials with the algebraic structures from differential algebra [3, 15] in order to illustrate their systematic principle. The condition in axiom c' defines (rational) number symbols alias literals as *differential constants*, which do not change their value during continuous evolution. Their derivative is zero. The number symbol 5 will always have the value 5 and never change by anything other than 0. The condition in axiom $+'$ and the *Leibniz* or *product rule* from $\cdot'$ are the defining conditions for *derivation operators on rings*. The derivative of a sum is the sum of the derivatives (additivity or a homomorphic property with respect to addition, i.e., the operator $(\cdot)'$ applied to a sum equals the sum of the operator applied to each summand) according to axiom $+'$. Furthermore, the derivative of a product is the derivative of one factor times the other factor plus the one factor times the derivative of the other factor as in axiom $\cdot'$. The condition in axiom $-'$ is a derived rule for subtraction according to the identity $e - k = e + (-1) \cdot k$ and again expresses a homomorphic property, now with respect to subtraction rather than addition.

The equation in axiom x' uniquely defines the operator $(\cdot)'$ on the *differential polynomial algebra* spanned by the *differential indeterminates* $x \in \mathcal{V}$, i.e., the symbols x that have indeterminate derivatives x'. It says that we understand the differential symbol x' as the derivative of the symbol x for all state variables $x \in \mathcal{V}$. Axiom $/'$ canonically extends the derivation operator $(\cdot)'$ to the *differential field of quotients* by the usual *quotient rule*. As the base field $\mathbb{R}$ has no zero divisors[a], the right-hand side of axiom $/'$ is defined whenever the original division e/k can be carried out, which, as we assumed for well-definedness, is guarded by $k \neq 0$.

---

[a] In this setting, $\mathbb{R}$ has no zero divisors, because the formula $ab = 0 \to a = 0 \lor b = 0$ is valid, i.e., a product is zero only if a factor is zero.

Expedition 10.3 (Semantics of differential algebra)

The view of Expedition 10.2 sort of gave $(e)'$ a meaning, but, when we think about it, did not actually define it. Differential algebra studies the structural algebraic relations of, e.g., the derivative $(e+k)'$ to the derivatives $(e)'$ plus $(k)'$ and is incredibly effective at capturing and understanding them. But algebra— and differential algebra is no exception—is, of course, deliberately abstract about the question of what the individual pieces mean, because algebra is the study of structure, not the study of the meaning of the objects that are being structured in the first place. That is why we can learn all about the structure of derivatives and derivation operators from differential algebra, but have to go beyond differential algebra to complement it with a precise semantics that relates to what is needed to understand the mathematics of real CPSs.

10.8.2 Differential Invariant Terms and Invariant Functions

It is not a coincidence that the examples in this chapter were provable by differential invariant proof rule dI, because that proof rule can handle arbitrary invariant functions.

Despite the power that differential invariant terms offer, challenges lie ahead in proving properties. Theorem 10.1 from Expedition 10.4 gives an indication where challenges remain.

Example 10.9 (Generalizing differential invariants). This dL formula is valid

$$x^2 + y^2 = 0 \rightarrow [x' = 4y^3, y' = -4x^3]x^2 + y^2 = 0 \tag{10.9}$$

but cannot be proved directly using dI, because $x^2 + y^2$ is not an invariant function of the dynamics. In combination with generalization (MR to change the postcondition to the equivalent $x^4 + y^4 = 0$) and a cut (to change the antecedent to the equivalent $x^4 + y^4 = 0$), however, there is a proof using differential invariants dI:

$$
\begin{array}{ll}
& * \\
\mathbb{R} & \overline{\quad \vdash 4x^3(4y^3) + 4y^3(-4x^3) = 0 \quad} \\
{\scriptstyle[:=]} & \overline{\quad \vdash [x':=4y^3][y':=-4x^3]4x^3x' + 4y^3y' = 0 \quad} \\
{\scriptstyle \text{dI}} & \overline{\quad x^4 + y^4 = 0 \vdash [x' = 4y^3, y' = -4x^3]x^4 + y^4 = 0 \quad} \\
{\scriptstyle \text{cut,MR}} & \overline{\quad x^2 + y^2 = 0 \vdash [x' = 4y^3, y' = -4x^3]x^2 + y^2 = 0 \quad} \\
{\scriptstyle \rightarrow\text{R}} & \overline{\quad \vdash x^2 + y^2 = 0 \rightarrow [x' = 4y^3, y' = -4x^3]x^2 + y^2 = 0 \quad}
\end{array}
$$

The use of MR leads to another branch $x^4 + y^4 = 0 \vdash x^2 + y^2 = 0$ that is elided above. Similarly, the cut rule leads to another branch $x^2 + y^2 = 0 \vdash x^4 + y^4 = 0$ that is also elided. Both is proved easily using real arithmetic ($\mathbb{R}$).

Expedition 10.4 (Lie characterization of invariant functions)

The proof rule dI works by deriving the postcondition and substituting the differential equation in:

$$\text{dI} \quad \frac{\vdash [x':=f(x)](e)' = 0}{e = 0 \vdash [x' = f(x)]e = 0}$$

There is something quite peculiar about rule dI. Its premise is independent of the constant term in e. If, for any constant symbol c, the formula $e = 0$ is replaced by $e - c = 0$ in the conclusion, then the premise of rule dI stays the same, because $c' = 0$. Consequently, if dI proves

$$e = 0 \vdash [x' = f(x)]e = 0$$

then it also proves

$$e - c = 0 \vdash [x' = f(x)]e - c = 0 \tag{10.8}$$

for any constant c. This observation is the basis for a more general result, which simultaneously proves all formulas (10.8) for all c from the premise of dI.

On open connected domains, equational differential invariants are even a necessary and sufficient characterization of *invariant functions*, i.e., functions that are invariant along the dynamics of a system, because, whatever value c that function had in the initial state, the value will stay the same forever. The equational case of differential invariants is intimately related [10] to the seminal work by Sophus Lie on what are now called Lie groups [4, 5].

Theorem 10.1 (Lie's characterization of invariant terms). *Let* $x' = f(x)$ *be a differential equation system and let Q be a* domain, *i.e., a first-order formula of real arithmetic characterizing a connected open set. The following proof rule is a sound global equivalence rule, i.e., the conclusion is valid if and only if the premise is:*

$$\text{dI}_c \quad \frac{Q \vdash [x':=f(x)](e)' = 0}{\vdash \forall c\,(e = c \to [x' = f(x)\,\&\,Q]e = c)}$$

How could this happen? How could the original formula (10.9) be provable only after generalizing its postcondition to $x^4 + y^4 = 0$ and not before?

> **Note 61 (Strengthening induction hypotheses)** An important phenomenon we already encountered in Chap. 7 and other uses of induction is that, sometimes, the only way to prove a property is to strengthen the induction hypothesis. Differential invariants are no exception. It is worth noting, however, that the inductive structure in differential invariants includes their differential structure. And, indeed, the derivatives of $x^4 + y^4 = 0$ are different and more conducive to an inductive proof for Example 10.9 than those of $x^2 + y^2 = 0$ even if both have the same set of solutions.

Theorem 10.1 explains why $x^2 + y^2 = 0$ was doomed to fail as a differential invariant while $x^4 + y^4 = 0$ succeeded. All formulas of the form $x^4 + y^4 = c$ for all c are invariants of the dynamics in (10.9), because the proof succeeded. But $x^2 + y^2 = c$ is only an invariant for the lucky choice $c = 0$ and only equivalent to $x^4 + y^4 = 0$ for this case.

Exercises

10.1 (Repeating differential equations with domains). Note 60 explained that $(x' = f(x))^*$ is equivalent to $x' = f(x)$. Does the same hold for differential equations with evolution domain constraints? Are the hybrid programs $(x' = f(x) \& Q)^*$ and $x' = f(x) \& Q$ equivalent or not? Justify or modify the statement and justify the variation.

10.2. We argued that dL formulas (10.1) and (10.2) are equivalent and then went on to find a proof of (10.2). Continue this proof of (10.2) to a proof of (10.1) using the generalization rule MR and the cut rule.

10.3 (Derivation lemma proof). Prove the other cases of Lemma 10.1 where the term is a variable x or a subtraction $e - k$ or multiplication $e \cdot k$ or division e/k.

10.4 (Absence of solutions). What happens in the proof of Lemma 10.3 if there is no solution φ? Show that this is not a counterexample to axiom DI, but that the axiom is sound in that case, too.

10.5. Carry out the polynomial computations needed to prove Example 10.8 using proof rule dI.

10.6 (Rotation with angular velocity ω). Example 10.1 considered a rotation of vector (v, w) with angular velocity 1. Suppose the vector (v, w) is rotating with an arbitrary fixed angular velocity ω. Even if the vector rotates more quickly or slowly, it still always remains on the circle of radius r. Prove the resulting dL formula using differential invariants:

$$v^2 + w^2 = r^2 \rightarrow [v' = \omega w, w' = -\omega v \,\&\, \omega \neq 0] v^2 + w^2 = r^2$$

10.7. Prove the following dL formulas using differential invariants:

$$xy = c \to [x' = -x, y' = y, z' = -z] \, xy = c$$

$$4x^2 + 2y^2 = 1 \to [x' = 2y, y' = -4x] \, 4x^2 + 2y^2 = 1$$

$$x^2 + \frac{y^3}{3} = c \to [x' = y^2, y' = -2x] \, x^2 + \frac{y^3}{3} = c$$

$$x^2 + 4xy - 2y^3 - y = 1 \to [x' = -1 + 4x - 6y^2, y' = -2x - 4y] \, x^2 + 4xy - 2y^3 - y = 1$$

10.8 (Hénon-Heiles). Prove a differential invariant of a Hénon-Heiles system for the motion of a star at (x, y) flying in direction (u, v) around the center of the galaxy:

$$\frac{1}{2}(u^2 + v^2 + Ax^2 + By^2) + x^2 y - \frac{1}{3}\varepsilon y^3 = 0 \to$$

$$[x' = u, y' = v, u' = -Ax - 2xy, v' = -By + \varepsilon y^2 - x^2]$$

$$\frac{1}{2}(u^2 + v^2 + Ax^2 + By^2) + x^2 y - \frac{1}{3}\varepsilon y^3 = 0$$

References

[1]   Gottlob Frege. *Begriffsschrift, eine der arithmetischen nachgebildete Formelsprache des reinen Denkens*. Halle: Verlag von Louis Nebert, 1879.

[2]   Khalil Ghorbal and André Platzer. Characterizing algebraic invariants by differential radical invariants. In: *TACAS*. Ed. by Erika Ábrahám and Klaus Havelund. Vol. 8413. LNCS. Berlin: Springer, 2014, 279–294. DOI: 10.1007/978-3-642-54862-8_19.

[3]   Ellis Robert Kolchin. *Differential Algebra and Algebraic Groups*. New York: Academic Press, 1972.

[4]   Sophus Lie. *Vorlesungen über continuierliche Gruppen mit geometrischen und anderen Anwendungen*. Leipzig: Teubner, 1893.

[5]   Sophus Lie. Über Integralinvarianten und ihre Verwertung für die Theorie der Differentialgleichungen. *Leipz. Berichte* **49** (1897), 369–410.

[6]   André Platzer. Differential dynamic logic for hybrid systems. *J. Autom. Reas.* **41**(2) (2008), 143–189. DOI: 10.1007/s10817-008-9103-8.

[7]   André Platzer. Differential Dynamic Logics: Automated Theorem Proving for Hybrid Systems. PhD thesis. Department of Computing Science, University of Oldenburg, 2008.

[8]   André Platzer. Differential-algebraic dynamic logic for differential-algebraic programs. *J. Log. Comput.* **20**(1) (2010), 309–352. DOI: 10.1093/logcom/exn070.

[9]   André Platzer. *Logical Analysis of Hybrid Systems: Proving Theorems for Complex Dynamics*. Heidelberg: Springer, 2010. DOI: `10.1007/978-3-642-14509-4`.

[10] André Platzer. A differential operator approach to equational differential invariants. In: *ITP*. Ed. by Lennart Beringer and Amy Felty. Vol. 7406. LNCS. Berlin: Springer, 2012, 28–48. DOI: `10.1007/978-3-642-32347-8_3`.

[11] André Platzer. Logics of dynamical systems. In: *LICS*. Los Alamitos: IEEE, 2012, 13–24. DOI: `10.1109/LICS.2012.13`.

[12] André Platzer. The complete proof theory of hybrid systems. In: *LICS*. Los Alamitos: IEEE, 2012, 541–550. DOI: `10.1109/LICS.2012.64`.

[13] André Platzer. The structure of differential invariants and differential cut elimination. *Log. Meth. Comput. Sci.* **8**(4:16) (2012), 1–38. DOI: `10.2168/LMCS-8(4:16)2012`.

[14] André Platzer. A complete uniform substitution calculus for differential dynamic logic. *J. Autom. Reas.* **59**(2) (2017), 219–265. DOI: `10.1007/s10817-016-9385-1`.

[15] Joseph Fels Ritt. *Differential equations from the algebraic standpoint*. Vol. 14. Colloquium Publications. New York: AMS, 1932.

[16] Dana Scott and Christopher Strachey. *Towards a mathematical semantics for computer languages*. Tech. rep. PRG-6. Oxford Programming Research Group, 1971.

[17] Wolfgang Walter. *Analysis 1*. 3rd ed. Berlin: Springer, 1992. DOI: `10.1007/978-3-662-38453-4`.

[18] Eberhard Zeidler, ed. *Teubner-Taschenbuch der Mathematik*. Wiesbaden: Teubner, 2003. DOI: `10.1007/978-3-322-96781-7`.

Chapter 11
Differential Equations & Proofs

Synopsis Furthering the remarkable shift in perspective toward a more thorough investigation of the wonders of the continuous dynamics of cyber-physical systems, this chapter advances logical induction techniques for differential equations from differential invariant terms to differential invariant formulas. Its net effect will be that not just the real value of a term can be proved to be invariant during a differential equation but also the truth-value of a formula. Differential invariants can prove that, e.g., the sign of a term never changes even if its value changes. Continuing the axiomatization of the differential equation aspects of differential dynamic logic, this chapter exploits a differential equation twist of Gerhard Gentzen's cut principle to obtain differential cuts that prove and then subsequently use properties of differential equations. The chapter will also advance the intuitions behind the continuous operational effects involved in CPS.

11.1 Introduction

Chapter 10 introduced equational differential invariants of the form $e = 0$ for differential equations that are significantly more general than the ones supported by the solution axiom $[']$ from Chap. 5. Axiom $[']$ equivalently replaces properties of differential equations with universally quantified properties of solutions, but is limited to differential equations that have explicit closed-form solutions whose resulting arithmetic can be handled (mostly polynomials or rational functions). But axiom $[']$ at least works for any arbitrary postcondition. The equational differential invariant proof rule dI supports general differential equations, but is limited to equational postconditions of the form $e = 0$.

The goal of this chapter is to generalize the differential invariant proof rule to work for more general postconditions but retain the flexibility with the more complicated differential equations that differential invariants provide. Indeed, the principles developed in Chap. 10 generalize beautifully to logical formulas other than the limited form $e = 0$. While $[x' = f(x)]e = 0$ expresses that the value of term e

© Springer International Publishing AG, part of Springer Nature 2018
A. Platzer, *Logical Foundations of Cyber-Physical Systems*,
https://doi.org/10.1007/978-3-319-63588-0_11

never changes and remains 0 along the differential equation $x' = f(x)$, other logical formulas such as $[x' = f(x)]\, e \geq 0$ allow term e to change its value as long as its sign remains nonnegative so that $e \geq 0$ is invariant and its truth-value remains *true*.

This chapter will establish generalizations that make the differential invariant proof rule work for formulas F of more general forms. The core of the differential invariant proof rule is its use of the differential $(e)'$ of the involved terms to determine quantities that, along the differential equation, are locally equal to the rate of change of the term e over time. The tricky bit is that it is conceptually significantly more challenging to apply derivative-based invariant principles to formulas than to terms. While invariant terms already had enough surprises in store for us in Chap. 10, they ultimately ended up relating in simple, sound, and well-defined ways to the intuitive concept of time-derivatives of values as rates of change along differential equations. But what could possibly be the counterpart of the rate of change or time-derivative for a formula? Formulas are either *true* or *false*, which makes it difficult to understand what their rate of change should be. While derivatives of terms can, at least intuitively, be understood as the question of how a function changes its value in the reals $\mathbb{R}$ at close-by points, it is not at all clear how to understand a small change to a close-by value when the only possible values of the formula are boolean in the set $\{true, false\}$. We cannot just say "the truth-value of formula P changes just a little bit when the state changes its values just a little bit along the differential equation" in any particularly simple meaningful way.

Fortunately, these considerations already provide some intuitive guidance toward an answer. Even if there is no wiggle room in the set of truth-values $\{true, false\}$, we still want to use differential reasoning to argue that small changes of the points lead to close-by truth-values, so stay *true* if they were *true* initially, because there simply are no truth-values close to *true* other than *true* itself. Of course, the most subtle and most crucial part will be defining and justifying the differential $(F)'$ of a formula such that the shape of the differential invariant proof rule is sound:

$$\text{dI} \ \frac{Q \vdash [x':=f(x)](F)'}{F \vdash [x' = f(x)\,\&\,Q]F}$$

If, for example, the formula F in the conclusion is $e = 0$ and the evolution domain constraint Q is *true*, then Chap. 10 demonstrated that $(e)' = 0$ is a sound choice for the formula $(F)'$ in the premise of rule dI. This chapter investigates generalizations of rule dI that work for more general shapes of formula F than just $e = 0$. Differential invariants were originally introduced with another semantics [4, 5], but we follow an advanced axiomatic logical reading of differential invariants via differential forms [12] that also simplifies their intuitive understanding.

This chapter advances the capabilities of differential invariants that Chap. 10 started and continues to be of central significance for the Foundations of Cyber-Physical Systems in all but the most elementary CPSs. The most important learning goals of this chapter are:

Modeling and Control: This chapter continues the study of the core principles behind CPS by developing a deeper understanding of how continuous dynamical behavior affects the truth of logical formulas. The differential invariants developed in this chapter also have significance for developing models and controls using the design-by-invariant principle.

Computational Thinking: This chapter exploits computational thinking, continuing the surprising analogies between discrete dynamics and continuous dynamics discovered in Chap. 10. It is devoted to rigorous reasoning about differential equations in CPS models, which is crucial for understanding the continuous behavior that CPSs exhibit over time. This chapter systematically expands on the differential invariant terms for equational properties of differential equations developed in Chap. 10 and generalizes the same core principles to the study of general properties of differential equations. Computational thinking is exploited in a second way by generalizing Gentzen's cut principle, which is of seminal significance in discrete logic, to differential equations. This chapter continues the *axiomatization* of differential dynamic logic dL [9, 10] pursued since Chap. 5 and lifts dL's proof techniques to systems with more complex properties of more complex differential equations. The concepts developed in this chapter continue the differential facet illustrating the more general relation of *syntax* (which is notation), *semantics* (which carries meaning), and *axiomatics* (which internalizes semantic relations into universal syntactic transformations). These concepts and their relations jointly form the significant *logical trinity* of syntax, semantics, and axiomatics. Finally, the verification techniques developed in this chapter are critical for verifying CPS models of appropriate scale and technical complexity.

CPS Skills: The focus in this chapter is on reasoning about differential equations. As a beneficial side effect, we will develop better intuition for the operational effects involved in CPS by getting better tools for understanding how exactly state changes while the system follows a differential equation and what properties of the system will not change.

discrete vs. continuous analogy
rigorous reasoning about ODEs
beyond differential invariant terms
differential invariant formulas
cut principles for differential equations
axiomatization of ODEs
differential facet of logical trinity

understanding continuous dynamics   operational CPS effects
relate discrete+continuous   state changes along ODE

11.2 Recap: Ingredients for Differential Equation Proofs

Before studying differential invariant formulas in greater detail, we first recall the
semantics of differential equations from Chap. 3 and the semantics of differentials
from Chap. 10:

Definition 3.3 (Transition semantics of ODEs).

$$[\![x' = f(x)\,\&\,Q]\!] = \{(\omega, v) : \varphi(0) = \omega \text{ except at } x' \text{ and } \varphi(r) = v \text{ for a solution}$$
$$\varphi{:}[0,r] \to \mathscr{S} \text{ of any duration } r \text{ satisfying } \varphi \models x' = f(x) \wedge Q\}$$

where $\varphi \models x' = f(x) \wedge Q$, iff for all times $0 \le z \le r$: $\varphi(z) \in [\![x' = f(x) \wedge Q]\!]$
with $\varphi(z)(x') \stackrel{\text{def}}{=} \frac{d\varphi(t)(x)}{dt}(z)$ and $\varphi(z) = \varphi(0)$ except at x, x'.

Definition 10.2 (Semantics of differentials). The semantics of differential
term $(e)'$ in state ω is the value $\omega[\![(e)']\!]$ defined as

$$\omega[\![(e)']\!] = \sum_{x \in \mathscr{V}} \omega(x') \cdot \frac{\partial [\![e]\!]}{\partial x}(\omega)$$

Our approach for more general differential invariants will leverage the fact that
the following results from Chap. 10 already capture differential terms $(e)'$ and how
their values relate to the change of the value of the term e over time, as well as
the differential effects of differential equations on differential symbols. Equations
of differentials can be used to compute with differentials akin to the process of
forming derivatives which is called differentiation.

Lemma 10.1 (Derivation lemma). *The following equations of differentials are
valid formulas so sound axioms:*
+′ $(e + k)' = (e)' + (k)'$

−′ $(e - k)' = (e)' - (k)'$

·′ $(e \cdot k)' = (e)' \cdot k + e \cdot (k)'$

/′ $(e/k)' = ((e)' \cdot k - e \cdot (k)')/k^2$

c′ $(c())' = 0$ (for numbers or constants $c()$)

x′ $(x)' = x'$ (for variable $x \in \mathscr{V}$)

The value of a differential, at least along a differential equation, equals the analytic time-derivative.

> **Lemma 10.2 (Differential lemma).** *Let* $\varphi \models x' = f(x) \wedge Q$ *for some solution* $\varphi : [0, r] \to \mathscr{S}$ *of duration* $r > 0$. *Then for all times* $0 \leq z \leq r$ *and all terms* e *defined all along* φ *with* $FV(e) \subseteq \{x\}$:
>
> $$\varphi(z)[\![(e)']\!] = \frac{\mathsf{d}\,\varphi(t)[\![e]\!]}{\mathsf{d}t}(z)$$

Differential equations can be substituted in via their differential effect.

> **Lemma 10.5 (Differential assignment).** *If* $\varphi \models x' = f(x) \wedge Q$ *for a solution* $\varphi : [0, r] \to \mathscr{S}$ *of any duration* $r \geq 0$, *then*
>
> $$\varphi \models P \leftrightarrow [x' := f(x)]P$$

> **Lemma 10.6 (DE differential effect axiom).** *This axiom is sound:*
>
> $$\text{DE} \quad [x' = f(x)\,\&\,Q]P \leftrightarrow [x' = f(x)\,\&\,Q][x' := f(x)]P$$

These results are already more general and work for any postcondition P, not just normalized equations $e = 0$. Lemma 10.1 covers differentials of any polynomial (and rational) term. Lemma 10.2 relates their values to the change of value over time. Just the specific formulation of the differential invariant axiom needs to be generalized based on Lemma 10.2 to cover more general postconditions.

11.3 Differential Weakening

Just as the differential effect axiom DE perfectly internalizes the effect that differential equations have on the differential symbols, the differential weakening axiom internalizes the semantic effect of their evolution domain constraints (Definition 3.3). Of course, the effect of an evolution domain constraint Q is not to change the values of variables, but rather to limit the continuous evolution to always remain within the set of states $[\![Q]\!]$ where Q is true. There are multiple ways of achieving that [12] and you are invited to discover them.

One simple but useful way is the following *differential weakening* axiom, somewhat reminiscent of the way axiom DE is phrased but for domain Q.

> **Lemma 11.1 (DW differential weakening axiom).** *This axiom is sound:*
>
> $$\text{DW} \quad [x' = f(x)\,\&\,Q]P \leftrightarrow [x' = f(x)\,\&\,Q](Q \to P)$$

Since differential equations can never leave their evolution domain constraints (Fig. 11.1), any property P is true after the differential equation if and only if it is true whenever the evolution domain constraint Q is. The evolution domain constraint Q is always true throughout all evolutions of $x' = f(x) \& Q$ by Definition 3.3. We will see later that axiom DW justifies once and for all that the evolution domain constraint Q can be assumed soundly during any proof reasoning about differential equation $x' = f(x) \& Q$.

Fig. 11.1 Differential weakening axiom DW

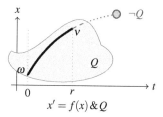

$$x' = f(x) \& Q$$

On its own, the differential weakening axiom DW has the same shortcoming as the differential effect axiom DE and differential invariant axiom DI. They reduce one property of a differential equation to another property of that differential equation. Following up with a generalization rule G after the differential weakening axiom DW leads to the following differential weakening sequent proof rule that can be quite useful.

Lemma 11.2 (dW differential weakening proof rule). *The differential weakening proof rule derives from axiom DW:*

$$\text{dW} \quad \frac{Q \vdash P}{\Gamma \vdash [x' = f(x) \& Q] P, \Delta}$$

The system $x' = f(x) \& Q$ will stop before it leaves Q, hence, if Q implies P (i.e., the region Q is contained in the region P), then P is always true after the continuous evolution, no matter what the actual differential equation $x' = f(x)$ does.

Of course, it is crucial for soundness that rule dW drops the context Γ, Δ, which could not soundly be available in the premise (Exercise 11.3). The context Γ contains information about the initial state, which is no longer guaranteed to remain true in the final state. As usual, keeping assumptions about constants around would be acceptable (Sect. 7.5). Yet, on its own, even rule dW cannot prove particularly interesting properties, because it only work when Q is rather informative. Differential weakening can, however, be useful to obtain partial information about the domains of differential equations or in combination with stronger proof rules (Sect. 11.8). If an entire system model is proved with just differential weakening dW, then this indicates that the model may have assumed overly strong evolution domain constraints, because its property would be true independently of the differential equations (Sect. 8.2.2).

11.4 Operators in Differential Invariants

This section develops ways of handling logical and arithmetical operators in differential invariants. Thanks to axiom DW, we will soon see that the evolution domain constraint Q can be assumed during the induction step. All differential invariant rules have the same shape but differ in how they define the *differential formula* $(F)'$ in the induction step depending on F:

$$\text{dI} \ \frac{Q \vdash [x' := f(x)](F)'}{F \vdash [x' = f(x) \& Q]F}$$

For the case where F is of the form $e = 0$ and Q is *true*, Chap. 10 justifies that this rule dI is sound when defining $(e = 0)'$ to be $(e)' = 0$. This leaves other shapes of the postcondition F and the evolution domain constraint Q to worry about. The differential invariant proof rules can also all be derived directly from corresponding differential induction axioms, given an appropriate definition and justification of $(F)'$. We first emphasize an intuitive gradual development, postponing the soundness proof until all cases of the rule have been developed.

11.4.1 Equational Differential Invariants

While Chap. 10 provided a way of proving postconditions of the form $e = 0$ for unsolvable differential equations, there are more general logical formulas that we would like to prove to be invariants of differential equations, not just the polynomial equations normalized such that they are single terms equaling 0. Direct proofs for postconditions of the form $e = k$ should work in almost the same way. In order to set the stage for the rest of this chapter, we develop an induction axiom and proof rule for differential equations with postcondition $e = k$ and simultaneously generalize it to the presence of evolution domain constraints Q using our newly discovered differential weakening principles captured in axiom DW.

Thinking back to the soundness proof for the case $e = 0$ in Sect. 10.3.6, the argument was based on the value of $\varphi(t)[\![e]\!]$ as a function of time t. The same argument can be made by considering the difference $\varphi(t)[\![e - k]\!]$ for postconditions of the form $e = k$. How does the inductive step for formula $e = k$ need to be defined to make a corresponding differential invariant proof rule sound? That is, for what premise is the following a sound proof rule when e and k are arbitrary terms?

$$\frac{\vdash ???}{e = k \vdash [x' = f(x)]e = k}$$

Before you read on, see if you can find the answer for yourself.

The following rule would make sense:

$$\frac{\vdash [x':=f(x)](e)' = (k)'}{e = k \vdash [x' = f(x)]e = k}$$

This rule for equational differential invariants captures the intuition that e always stays equal to k if it was initially (antecedent of conclusion) and the differential of term e is the same as the differential of k when using the right-hand side $f(x)$ of the differential equation $x' = f(x)$ for its left-hand side x'. For the case $Q \equiv true$, this rule fits the general shape of rule dI when we mnemonically define the "differential" of an equation $e = k$ as the formula

$$(e = k)' \overset{\text{def}}{\equiv} ((e)' = (k)')$$

This definition as the equation $(e)' = (k)'$ of the differentials of the two sides makes intuitive sense, because the truth-value of an equation $e = k$ does not change if the left- and right-hand side quantities have the same rate of change (Fig. 11.2).

Fig. 11.2 Equal rate of change from equal initial value (drawn slightly apart for visualization)

The way we justified the soundness of the $e = 0$ case of the differential invariant proof rule dI in Sect. 10.4 was by deriving it from a corresponding differential invariant axiom DI, which captured the fundamental induction principle for terms along differential equations in more elementary ways. Let us pursue the same approach for the invariant $e = k$.

$$\text{DI}_= \quad ([x' = f(x) \& Q]e = k \leftrightarrow [?Q]e = k) \leftarrow [x' = f(x) \& Q)](e)' = (k)'$$

This axiom expresses that, if $(e)' = (k)'$ always holds after the differential equation so that terms e and k always have the same rate of change, then e and k always have the same value after the differential equation if and only if they have the same value initially after the test $?Q$. The reason for the test $?Q$ is that the postcondition $e = k$ is vacuously true always after the differential equation $x' = f(x) \& Q$ when it starts outside its evolution domain constraint Q, because there is no evolution of the differential equation then. Correspondingly, the initial check $[?Q]e = k$ gets to assume the test $?Q$ passes, because otherwise there is nothing to show. Overall, axiom DI$_=$ expresses that two quantities that evolve with the same rate of change will always remain the same iff they start from the same value initially (Fig. 11.2).

Instead of going through a soundness proof for DI$_=$, however, we directly generalize the proof principles further and see whether differential invariants can prove even more formulas for us. We will later prove the soundness of the general differential invariant axiom, from which DI$_=$ derives as a special case.

11.4.2 Differential Invariant Proof Rule

Just as Sect. 11.4.1 did with axiom DI$_=$ for the case of equational postconditions $e = k$, this section provides induction axioms for postconditions of differential equations that are all of the form

$$\text{DI} \quad ([x' = f(x) \,\&\, Q]P \leftrightarrow [?Q]P) \leftarrow [x' = f(x) \,\&\, Q](P)'$$

This axiom expresses that, if a yet-to-be-defined differential formula $(P)'$ always holds after the differential equation so that P never changes its truth-value, then P is always true after the differential equation if and only if P was true initially after the test $?Q$. Since $[x' = f(x) \,\&\, Q]P$ always implies $[?Q]P$ by Definition 3.3 (using that $x' \notin \text{FV}(P) \cup \text{FV}(Q)$), only the converse implication needs the assumption $[x' = f(x) \,\&\, Q](P)'$. For each of the subsequently considered cases of $(P)'$, we only need to prove the validity of the following formula to prove the soundness of DI:

$$[x' = f(x) \,\&\, Q](P)' \rightarrow ([?Q]P \rightarrow [x' = f(x) \,\&\, Q]P)$$

For each case of this differential induction axiom DI, we obtain a corresponding differential invariant proof rule for free.

> **Lemma 11.3 (dI differential invariant proof rule).** *The differential invariant proof rule derives from axiom DI:*
>
> $$\text{dI} \quad \frac{Q \vdash [x' := f(x)](F)'}{F \vdash [x' = f(x) \,\&\, Q]F}$$

Proof. Proof rule dI derives from axiom DI as follows:

$$
\text{DI} \cfrac{
\text{[?]} \cfrac{
\to\text{R} \cfrac{
\text{id} \cfrac{*}{F, Q \vdash F}
}{
\cfrac{F \vdash Q \to F}{F \vdash [?Q]F}
}
\qquad
\text{DE} \cfrac{
\text{DW} \cfrac{
\text{G} \cfrac{
\to\text{R} \cfrac{
Q \vdash [x' := f(x)](F)'
}{
\vdash Q \to [x' := f(x)](F)'
}
}{
\vdash [x' = f(x) \,\&\, Q](Q \to [x' := f(x)](F)')
}
}{
\vdash [x' = f(x) \,\&\, Q][x' := f(x)](F)'
}
}{
\vdash [x' = f(x) \,\&\, Q](F)'
}
}{
F \vdash [x' = f(x) \,\&\, Q]F
}
$$

$\square$

The basic idea behind rule dI is that the premise of dI shows that the differential $(F)'$ holds within evolution domain Q when substituting the differential equations $x' = f(x)$ into $(F)'$. If F holds initially (antecedent of conclusion), then F itself always stays true (succedent of conclusion). Intuitively, the premise gives a condition showing that, within Q, the differential $(F)'$ along the differential constraints points inwards or transversally to F but never outwards to $\neg F$, as illustrated in

Fig. 11.3 Differential invariant F for safety

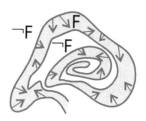

Fig. 11.3. Hence, if we start in F and, as indicated by $(F)'$, the local dynamics never points outside F, then the system always stays in F when following the dynamics.

Observe how useful it is that we have assembled an array of independent reasoning principles, differential effect DE, differential weakening DW, and generalization G, to combine and bundle the logically more elementary axiom DI to the more useful proof rule dI. Such modular combinations of reasoning principles are not just easier to prove sound, but also more flexible because they allow free variations in the argument structure. Recall, though, that the use of Gödel's generalization rule G for the derivation of dI implies it would be unsound to retain a sequent context Γ, Δ in its premise (except, as usual, assumptions about constants).

Example 11.1 (Rotational dynamics). Consider the system of rotational dynamics $v' = w, w' = -v$ from Example 10.1 on p. 293 once again. This dynamics is complicated in that the solution involves trigonometric functions, which are generally outside decidable classes of arithmetic. Yet, we can easily prove interesting properties about it using dI and decidable polynomial arithmetic. For instance, dI can directly prove formula (10.1), i.e., that $v^2 + w^2 = r^2$ is a differential invariant of the dynamics, using the following proof:

$$
\begin{array}{l}
\qquad\qquad\qquad * \\
\mathbb{R} \; \overline{\qquad \vdash 2vw + 2w(-v) = 0 \qquad} \\
[:=] \; \overline{\qquad \vdash [v':=w][w':=-v]\, 2vv' + 2ww' = 0 \qquad} \\
\text{dI} \; \overline{v^2 + w^2 = r^2 \vdash [v' = w, w' = -v]\, v^2 + w^2 = r^2} \\
\rightarrow\text{R} \; \overline{\qquad \vdash v^2 + w^2 = r^2 \rightarrow [v' = w, w' = -v]\, v^2 + w^2 = r^2}
\end{array}
$$

This proof is easier and more direct than the MR monotonicity proof in Chap. 10.

11.4.3 Differential Invariant Inequalities

The differential invariant axioms and proof rules considered so far give a good understanding of how to prove equational invariants. What about inequalities? How can they be proved?

> Before you read on, see if you can find the answer for yourself.

The primary question to generalize the differential invariant proof rule is again how to mnemonically define a "differential," which we do as follows:

$$(e \le k)' \stackrel{\text{def}}{\equiv} ((e)' \le (k)')$$

This gives the following differential invariant axiom, which we simply call DI again:

$$\big([x' = f(x) \,\&\, Q]e \le k \leftrightarrow [?Q]e \le k\big) \leftarrow [x' = f(x) \,\&\, Q]](e \le k)'$$

The only difference to the general axiom DI is the definition of the differential $(e \le k)'$ and its soundness proof. The intuition is that a quantity e with smaller or equal rate of change than that of k starting from a smaller or equal value initially will always remain smaller or equal (Fig. 11.4). Lemma 11.3 derives the corresponding case of the differential induction rule dI from this axiom:

$$\frac{Q \vdash [x' := f(x)](e \le k)'}{e \le k \vdash [x' = f(x) \,\&\, Q]e \le k}$$

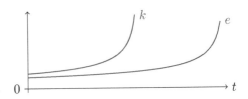

Fig. 11.4 Lesser or equal rate of change from lesser or equal initial value

Example 11.2 (Cubic dynamics). Similarly, differential induction can easily prove that $\frac{1}{3} \le 5x^2$ is an invariant of the cubic dynamics $x' = x^3$; see the proof in Fig. 11.5 for the dynamics in Fig. 11.6. To apply the differential induction rule dI, we form the derivative of the differential invariant $F \equiv \frac{1}{3} \le 5x^2$, which results in the dL formula $(F)' \equiv (\frac{1}{3} \le 5x^2)' \equiv 0 \le 5 \cdot 2xx'$. Now, the differential induction rule dI takes into account that the derivative of state variable x along the dynamics is known. Substituting the differential equation $x' = x^3$ into the inequality yields $[x' := x^3](F)' \equiv 0 \le 5 \cdot 2xx^3$, which is a valid formula and is closed by quantifier elimination with rule ℝ.

Differential invariants that are inequalities are not just a minor variation of equational differential invariants, because they can prove more. That is, it can be shown [11] that there are valid formulas that can be proved using differential invariant inequalities but *cannot* be proved just using equations as differential invariants. Sometimes, you need to be prepared to look for inequalities that you can use as differential invariants. The converse is not true. Everything that is provable using equational differential invariants is also provable using differential invariant inequalities [11], but you should still look for equational differential invariants if they give easier proofs.

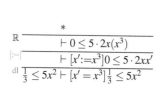

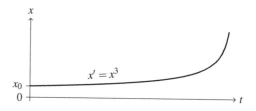

Fig. 11.5 Cubic dynamics proof   **Fig. 11.6** Cubic dynamics

Strict inequalities could also be used as differential invariants when defining their "differentials" mnemonically as

$$(e < k)' \stackrel{\text{def}}{\equiv} ((e)' < (k)')$$

However, we, instead, prefer a slightly relaxed definition that is also sound:

$$(e < k)' \stackrel{\text{def}}{\equiv} ((e)' \leq (k)')$$

The intuition is again that a quantity e that starts from a smaller initial value than k and has *no larger rate of change* than that of k will always remain smaller (Fig. 11.7). The cases $e \geq k$ and $e > k$ work analogously.

Fig. 11.7 Lesser or equal rate of change from lesser initial value

Example 11.3 (Rotational dynamics). An inequality property can be proved easily for the rotational dynamics $v' = w, w' = -v$ using the following proof:

$$\underset{\to R}{\dfrac{\underset{\text{dl}}{\dfrac{\underset{[:=]}{\dfrac{\mathbb{R} \dfrac{*}{\vdash 2vw + 2w(-v) \leq 0}}{\vdash [v':=w][w':=-v]\,2vv' + 2ww' \leq 0}}}{v^2 + w^2 \leq r^2 \vdash [v' = w, w' = -v]\,v^2 + w^2 \leq r^2}}}{\vdash v^2 + w^2 \leq r^2 \to [v' = w, w' = -v]\,v^2 + w^2 \leq r^2}}$$

Example 11.4 (Odd-order dynamics). The following proof easily proves a simple invariant with only even powers of a dynamics with only odd powers:

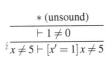

$$\frac{\ast \text{ (unsound)}}{\dfrac{\vdash 1 \neq 0}{{}^{t}x \neq 5 \vdash [x' = 1]x \neq 5}}$$

Fig. 11.8 Unsound attempt to use disequalities

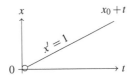

Fig. 11.9 Linear evolution of $x' = 1$

$$\text{dI} \; \frac{\text{[:=]} \; \dfrac{\mathbb{R} \; \dfrac{\ast}{\vdash 2x^6 + 14x^4 + 4x^2 \geq 0}}{\vdash [x':=x^5 + 7x^3 + 2x]\, 2xx' \geq 0}}{x^2 \geq 2 \vdash [x' = x^5 + 7x^3 + 2x]\, x^2 \geq 2}$$

Example 11.5 (Even-order dynamics). The following proof easily proves a simple invariant with only odd powers of a dynamics with only even powers:

$$\text{dI} \; \frac{\text{[:=]} \; \dfrac{\mathbb{R} \; \dfrac{\ast}{\vdash 2x^6 + 12x^4 + 10x^2 \geq 0}}{\vdash [x':=x^4 + 6x^2 + 5]\, 2x^2 x' \geq 0}}{x^3 \geq 2 \vdash [x' = x^4 + 6x^2 + 5]\, x^3 \geq 2}$$

Similar straightforward proofs work for any other appropriate sign condition on an odd power of a purely even dynamics or an even power of a purely odd dynamics, because the resulting arithmetic has only even powers and, thus, positive signs when added.

11.4.4 Disequational Differential Invariants

The case that is missing in differential invariant proof rules of atomic formulas is for postconditions that are disequalities $e \neq k$? How can they be proved?

> Before you read on, see if you can find the answer for yourself.

By analogy to the previous cases, one might expect the following definition:

$$(e \neq k)' \stackrel{?}{\equiv} ((e)' \neq (k)') \quad ???$$

It is crucial for soundness of differential invariants that $(e \neq k)'$ is *not* defined that way! In the counterexample in Fig. 11.8, variable x can reach $x = 0$ without its derivative ever being 0; again, see Fig. 11.9 for the dynamics. Of course, just because e and k start out different, does not mean they will always stay different if they evolve with different derivatives. *Au contraire*, it is because they evolve with different derivatives that they might catch each other (Fig. 11.10).

Fig. 11.10 Different rates of
change from different initial
values do not prove anything

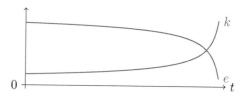

Instead, if e and k start out different and evolve with the same derivative, they
will always stay different. So the sound definition is slightly unexpected:

$$(e \neq k)' \overset{\text{def}}{\equiv} ((e)' = (k)')$$

11.4.5 Conjunctive Differential Invariants

The next case to consider is where the invariant that we want to prove is a conjunction $F \wedge G$. The crucial question then is again what a "differential" $(F \wedge G)'$ would
be that measures the rate of change in truth-values of the conjunction $F \wedge G$.

Before you read on, see if you can find the answer for yourself.

Of course, there aren't many changes of truth-values to speak of, because there
are only two: *true* and *false*. But, still, no change in truth-value is a good thing for
an invariant argument. An invariant should always stay *true* if it is *true* initially. To
show that a conjunction $F \wedge G$ is invariant it is perfectly sufficient to prove that both
are invariant. This can be justified separately, but is more obvious when recalling
how box distributes over conjunctions.

> **Lemma 5.10 ([]∧ boxes distribute over conjunctions).** *This axiom is sound:*
>
> $$[]\wedge \quad [\alpha](P \wedge Q) \leftrightarrow [\alpha]P \wedge [\alpha]Q$$

Consequently, the mnemonic "differential" for conjunction is the conjunction of
the differentials:

$$(A \wedge B)' \equiv (A)' \wedge (B)'$$

Soundness of this definition can be established by deriving it with the help of derived
axiom []∧ to split each of the postconditions into separate conjuncts; see Fig. 11.11.
The remaining premise in Fig. 11.11 is equivalent to the conjunction

$$\big([x' = f(x) \,\&\, Q](A)' \to [?Q]A \to [x' = f(x) \,\&\, Q]A\big)$$
$$\wedge \big([x' = f(x) \,\&\, Q](B)' \to [?Q]B \to [x' = f(x) \,\&\, Q]B\big)$$

Both conjuncts derive from axiom DI by induction hypothesis, because they have
simpler postconditions.

$$[]\wedge \frac{\vdash [x'=f(x)\,\&\,Q](A)'\wedge[x'=f(x)\,\&\,Q](B)' \to [?Q]A\wedge[?Q]B \to [x'=f(x)\,\&\,Q]A\wedge[x'=f(x)\,\&\,Q]B}{\vdash [x'=f(x)\,\&\,Q](A\wedge B)' \to [?Q](A\wedge B) \to [x'=f(x)\,\&\,Q](A\wedge B)}$$

Fig. 11.11 Soundness proof for conjunctive differential invariant axiom

Lemma 11.3 derives the corresponding case of the differential induction rule dI from this axiom, which enables us to prove conjunctions as in this example:

$$\mathrm{dI}\frac{[:=]\frac{\mathbb{R}\frac{*}{\vdash 2vw+2w(-v)\le 0 \wedge 2vw+2w(-v)\ge 0}}{\vdash [v':=w][w':=-v](2vv'+2ww'\le 0 \wedge 2vv'+2ww'\ge 0)}}{v^2+w^2\le r^2 \wedge v^2+w^2\ge r^2 \vdash [v'=w,w'=-v](v^2+w^2\le r^2 \wedge v^2+w^2\ge r^2)}$$

Of course, a manual proof using axiom $[]\wedge$ to conduct two separate proofs that the left conjunct is a differential invariant and that, separately, the right conjunct also is a differential invariant would have worked equally well. As the invariant $v^2+w^2\le r^2 \wedge v^2+w^2\ge r^2$ is equivalent to $v^2+w^2=r^2$, the above proof gives yet another proof of (10.1) when combined with a corresponding use of the generalization rule MR.

Example 11.6 (Bouncing ball's gravity). One of the major complications in the bouncing-ball proofs in Chap. 5 and Chap. 7 was the somewhat unwieldy arithmetic resulting from solving the differential equations. Its loop invariant can be proved more easily without solutions directly by differential invariants:

$$j_{(x,v)} \overset{\text{def}}{\equiv} 2gx = 2gH - v^2 \wedge x \ge 0 \tag{7.10*}$$

The only complication is that this conjunction is not a differential invariant for the bouncing ball's dynamics $x'=v, v'=-g\,\&\,x\ge 0$, because $x\ge 0$ is not inductive since the resulting induction step $v\ge 0$ obtained from the differential $(x\ge 0)'$ is not valid because the velocity is negative on the way down.

Just like the justification for $(A\wedge B)' \equiv (A)'\wedge(B)'$ did, the proof in Fig. 11.12 also uses the $[]\wedge$ axiom to split the postcondition and conduct independent proofs for independent questions. The trend for the conjunct $x\ge 0$ is potentially unsafe, because negative velocities would ultimately violate $x\ge 0$ if it wasn't for the evolution domain constraint keeping the ball above ground. Only the first conjunct is a differential invariant. The second conjunct can be proved by differential weakening (dW), because $x\ge 0$ is the evolution domain. Observe how the arithmetic in differential invariant reasoning is rather tame, because it is obtained by differentiation. This is quite unlike the arithmetic with solutions, which is obtained by integration.

$$\cfrac{\cfrac{\cfrac{*}{x{\ge}0 \vdash 2gv = -2v(-g)}}{\mathbb{R} \;\; x{\ge}0 \vdash [x':{=}v][v':{=}-g]2gx' = -2vv'}}{\mathrm{dI} \;\; \cfrac{2gx{=}2gH-v^2 \vdash [x''{=}-g\,\&\,x{\ge}0]2gx{=}2gH-v^2}{[]{\wedge}\;\; 2gx{=}2gH-v^2,\, x\ge 0 \vdash [x'' = -g\,\&\,x{\ge}0](2gx{=}2gH-v^2 \wedge x{\ge}0)}} \qquad \cfrac{\mathrm{id}\;\cfrac{*}{x{\ge}0 \vdash x{\ge}0}}{\mathrm{dW}\;\; \vdash [x''{=}-g\,\&\,x{\ge}0]x{\ge}0}$$

Fig. 11.12 Differential invariant proof for bouncing ball in gravity

11.4.6 Disjunctive Differential Invariants

The next case to consider is where the invariant that we want to prove is a disjunction $A \vee B$. Our other lemmas take care of how to handle differential effects and differential weakening, if only we define the correct "differential" $(A \vee B)'$. How?

> Before you read on, see if you can find the answer for yourself.

The "differential" of a conjunction is the conjunction of the differentials. So, by analogy, it might stand to reason to define the "differential" of a disjunction as the disjunction of the differentials.

$$(A \vee B)' \stackrel{?}{\equiv} (A)' \vee (B)' \quad ???$$

Let's give it a try:

$$\cfrac{\cfrac{\cfrac{\text{unsound}}{\vdash 2vw + 2w(-v) = 0 \vee 5v + rw \ge 0}}{[:=]\;\; \vdash [v':{=}w][w':{=}-v]2vv' + 2ww' = 0 \vee r'v + rv' \ge 0}}{{\oint}\;\; v^2 + w^2 = r^2 \vee rv \ge 0 \vdash [v' = w, w' = -v, r' = 5](v^2 + w^2 = r^2 \vee rv \ge 0)}}{}$$

That would be spectacularly wrong, however, because the formula at the bottom is not actually valid, so it does not deserve a proof, even if the formula at the top is valid. We have no business proving formulas that are not valid and if we ever could, we would have found a serious unsoundness in the proof rules.

For soundness of differential invariants, it is crucial that the "differential" $(A \vee B)'$ of a disjunction is defined, e.g., conjunctively as $(A)' \wedge (B)'$ instead of as $(A)' \vee (B)'$. From an initial state ω that satisfies $\omega \in [\![A]\!]$, and hence $\omega \in [\![A \vee B]\!]$, the formula $A \vee B$ is only sustained differentially if A itself is a differential invariant, not if B is. For instance, $v^2 + w^2 = r^2 \vee rv \ge 0$ is not an invariant of the above differential equation, because $rv \ge 0$ will be invalidated if we just follow the circle dynamics long enough. So if the disjunction was true because $rv \ge 0$ was true at the beginning, it does not stay invariant, even if the other disjunct $v^2 + w^2 = r^2$ is invariant.

Instead, splitting differential invariant proofs over disjunctions by the $\vee$L rule is the way to go, and, in fact, by axiom $[]\wedge$, also justifies the choice

$$(A \vee B)' \stackrel{\text{def}}{\equiv} (A)' \wedge (B)'$$

$$\rightarrow R \cfrac{\text{VL} \cfrac{\text{id} \cfrac{*}{A \vdash A, B}}{\text{VR} \cfrac{}{A \vdash A \vee B}} \quad \text{dI} \cfrac{\vdash [x':=f(x)](A)'}{A \vdash [x'=f(x)]A}}{\text{MR} \cfrac{}{A \vdash [x'=f(x)](A \vee B)} \quad \text{MR} \cfrac{\text{VR} \cfrac{\text{id} \cfrac{*}{B \vdash A, B}}{B \vdash A \vee B} \quad \text{dI} \cfrac{\vdash [x':=f(x)](B)'}{B \vdash [x'=f(x)]B}}{B \vdash [x'=f(x)](A \vee B)}}{A \vee B \vdash [x'=f(x)](A \vee B)} }{\vdash A \vee B \to [x'=f(x)](A \vee B)}$$

Soundness of the differential induction axiom with this definition of the differentials of disjunctions can be proved directly (Fig. 11.13). The proof uses that $[?Q](A \vee B)$ is indeed equivalent to $[?Q]A \vee [?Q]B$ because both are equivalent to $Q \to A \vee B$. From disjunct $[?Q]A$, the assumption $[x'=f(x) \& Q](A)'$ makes it possible to derive $[x'=f(x) \& Q]A$ from axiom DI by induction hypothesis (it has a simpler postcondition), from which $[x'=f(x) \& Q](A \vee B)$ derives by monotonicity rule M[·]. Similarly, disjunct $[?Q]B$ derives $[x'=f(x) \& Q]B$, from which $[x'=f(x) \& Q](A \vee B)$ also derives by monotonicity rule M[·].

11.5 Differential Invariants

Differential invariants are a general proof principle for proving invariants of differential equations. Summarizing what this chapter has discovered so far leads to a single axiom DI for differential invariants, from which the corresponding differential invariant proof rule dI derives.

> **Definition 11.1 (Differential).** The following definition generalizes the differential operator $(\cdot)'$ from terms to real-arithmetic formulas:
>
> $$(F \wedge G)' \equiv (F)' \wedge (G)'$$
> $$(F \vee G)' \equiv (F)' \wedge (G)'$$
> $$(e \geq k)' \equiv (e)' \geq (k)' \qquad \text{accordingly for } \leq, =$$
> $$(e > k)' \equiv (e)' \geq (k)' \qquad \text{accordingly for } <$$
> $$(e \neq k)' \equiv (e)' = (k)'$$
>
> The operation mapping F to $[x':=f(x)](F)'$ is also called the *Lie-derivative* of F with respect to $x' = f(x)$.

$$[\wedge] \cfrac{\vdash [x'=f(x) \& Q](A)' \wedge [x'=f(x) \& Q](B)' \to ([?Q]A \vee [?Q]B \to [x'=f(x) \& Q](A \vee B))}{\vdash [x'=f(x) \& Q](A \vee B)' \to [?Q](A \vee B) \to [x'=f(x) \& Q](A \vee B)}$$

Fig. 11.13 Soundness proof for disjunctive differential invariant axiom

By Definition 11.1, the "differential" $(F)'$ of formula F uses the differential $(e)'$ of the terms e that occur within F. It is possible to lift differential invariants to quantifiers [7], but for our purposes here, it is enough to assume quantifier elimination has been applied to first eliminate the quantifiers equivalently (Sect. 6.5).

Just like for the initial condition check $[?Q]P$, a minor twist on the DI axiom shows that the induction step $[x' = f(x) \& Q](P)'$ can also assume Q, because no evolution is possible if the system starts outside Q.

Lemma 11.4 (DI differential invariant axiom). *This axiom is sound:*

$$\text{DI} \ \left([x' = f(x) \& Q]P \leftrightarrow [?Q]P \right) \leftarrow \left(Q \rightarrow [x' = f(x) \& Q](P)' \right)$$

The general form of the differential invariant proof rule is derived as in Sect. 11.4.2.

Lemma 11.3 (dI differential invariant proof rule). *The differential invariant proof rule derives from axiom DI:*

$$\text{dI} \ \frac{Q \vdash [x':=f(x)](F)'}{F \vdash [x' = f(x) \& Q]F}$$

This proof rule enables us to easily prove (10.2) and all previous proofs as well. The following version dI' can be derived easily from the more fundamental, essential form dI similarly to how the most useful loop induction rule loop derives from the essential form ind. We do not use the version dI' in practice, because it is subsumed by a more general proof technique investigated in Sect. 11.8.

$$\text{dI'} \ \frac{\Gamma \vdash F, \Delta \quad Q \vdash [x':=f(x)](F)' \quad F \vdash \psi}{\Gamma \vdash [x' = f(x) \& Q]\psi, \Delta}$$

Proof (of Lemma 11.4). A detailed axiomatic proof of axiom DI is located elsewhere [12]. The proof of one implication was already in Sect. 11.4.2:

$$[x' = f(x) \& Q]P \rightarrow [?Q]P$$

The proof of the following implication will be by induction on the structure of P:

$$[x' = f(x) \& Q](P)' \rightarrow \left([?Q]P \rightarrow [x' = f(x) \& Q]P \right) \tag{11.1}$$

This proof directly implies the validity of the following direction, because the differential equation cannot run if Q is not initially true and then also fails test $?Q$:

$$\left(Q \rightarrow [x' = f(x) \& Q](P)' \right) \rightarrow \left([?Q]P \rightarrow [x' = f(x) \& Q]P \right)$$

The proof of the validity of (11.1) is by structural induction on P. The case of solutions of duration 0 follows directly from the assumption $[?Q]P$ by Definition 3.3 (using that $x' \notin \text{FV}(P) \cup \text{FV}(Q)$).

1. If P is of the form $e \geq 0$, so $(P)'$ is $(e)' \geq 0$, then consider a state ω satisfying $[x' = f(x) \& Q](e)' \geq 0$ and $[?Q]e \geq 0$. To show that $\omega \in [\![[x' = f(x) \& Q]e \geq 0]\!]$, consider any solution $\varphi : [0, r] \to \mathscr{S}$ with $\varphi \models x' = f(x) \wedge Q$ and $\varphi(0) = \omega$ except at x'. By Lemma 10.2, the function $h(t) \overset{\text{def}}{=} \varphi(t)[\![e]\!]$ is differentiable on $[0, r]$ if $r > 0$ and, provided $FV(e) \subseteq \{x\}$, its time-derivative is

$$\frac{\mathrm{d}h(t)}{\mathrm{d}t}(z) = \frac{\mathrm{d}\varphi(t)[\![e]\!]}{\mathrm{d}t}(z) = \varphi(z)[\![(e)']\!] \geq 0$$

for all times $z \in [0, r]$ by the assumption $\omega \in [\' \geq 0]\!]$. Since h is differentiable, there is some $0 < \xi < r$ by the mean-value theorem such that:

$$h(r) - h(0) = \underbrace{(r - 0)}_{>0} \underbrace{\frac{\mathrm{d}h(t)}{\mathrm{d}t}(\xi)}_{\geq 0} \geq 0 \qquad (11.2)$$
$$\underbrace{}_{\geq 0}$$

Since $h(0) \geq 0$ by $\omega \in [\![[?Q]e \geq 0]\!]$, this implies $h(r) \geq 0$. Hence, $\varphi(r) \in [\![e \geq 0]\!]$. Thus, $\omega \in [\![[x' = f(x) \& Q]e \geq 0]\!]$ since this proof works for any solution φ.
2. If P is of the form $e \geq k$ the above case applies to the equivalent $e - k \geq 0$, whose differential $(e)' - (k)' \geq 0$ is equivalent to the differential $(e)' \geq (k)'$.
3. If P is of the form $e = k$, a simple variation of the above proof applies. Alternatively, consider the equivalent $e \geq k \wedge k \geq e$, which has a differential $(e)' \geq (k)' \wedge (k)' \geq (e)'$ that is equivalent to the differential $(e)' = (k)'$. The minor twist is that this needs a shift in the well-founded induction to artificially consider conjunctions of inequalities smaller than equations.
4. If P is of the form $e > k$, a simple variation of the above proof applies, since its differential $(e)' \geq (k)'$ is equivalent to the differential of $e \geq 0$. The only additional thought is that the initial assumption $h(0) > 0$ implies $h(r) > 0$ by (11.2).
5. If P is of the form $A \wedge B$, then the derivation in Fig. 11.11 concludes the validity of (11.1) for postcondition $A \wedge B$ from the validity of (11.1) for the smaller postcondition A as well as the smaller postcondition B, which are both valid by induction hypothesis.
6. If P is of the form $A \vee B$, then the derivation in Fig. 11.13 concludes the validity of (11.1) for postcondition $A \vee B$ from the validity of (11.1) for the smaller postcondition A as well as the smaller postcondition B, which are both valid by induction hypothesis.   □

Generalizations to systems of differential equations are quite straightforward.

11.6 Example Proofs

So that we gain more experience with differential invariants, this section studies a few example proofs.

Example 11.7 (Quartic dynamics). The following simple dL proof uses rule dI to prove an invariant of a quartic dynamics:

$$\dfrac{\dfrac{\dfrac{*}{a \geq 0 \vdash 3x^2((x-3)^4+a) \geq 0} \; \mathbb{R}}{a \geq 0 \vdash [x':=(x-3)^4+a]3x^2x' \geq 0} \; [:=]}{x^3 \geq -1 \vdash [x'=(x-3)^4+a \,\&\, a \geq 0]x^3 \geq -1} \; \text{dI}$$

Rule dI directly makes the evolution domain constraint $a \geq 0$ available as an assumption in the premise, because the continuous evolution is never allowed to leave it.

Example 11.8 (Damped oscillator). Consider $x' = y, y' = -\omega^2 x - 2d\omega y$, which is the differential equation for the damped oscillator with the undamped angular frequency ω and the damping ratio d. See Fig. 11.14 for one example of an evolution along this continuous dynamics. Figure 11.14 shows an evolution of x over time t

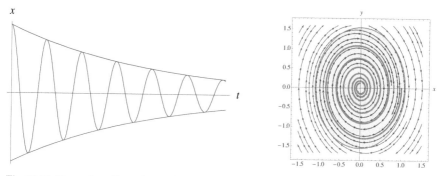

Fig. 11.14 Damped-oscillator time trajectory (**left**) and invariant in phase space (**right**)

on the left and a trajectory in the x, y state space on the right, which does not leave the green elliptic region $\omega^2 x^2 + y^2 \leq c^2$. General symbolic solutions of symbolic initial-value problems for this differential equation can become surprisingly difficult. A differential invariant proof, instead, is very simple:

$$\dfrac{\dfrac{\dfrac{*}{\omega \geq 0 \wedge d \geq 0 \vdash 2\omega^2 xy - 2\omega^2 xy - 4d\omega y^2 \leq 0} \; \mathbb{R}}{\omega \geq 0 \wedge d \geq 0 \vdash [x':=y][y':=-\omega^2 x - 2d\omega y]2\omega^2 xx' + 2yy' \leq 0} \; [:=]}{\omega^2 x^2 + y^2 \leq c^2 \vdash [x'=y, y'=-\omega^2 x - 2d\omega y \,\&\, (\omega \geq 0 \wedge d \geq 0)]\,\omega^2 x^2 + y^2 \leq c^2} \; \text{dI}$$

Observe that rule dI directly makes the evolution domain constraint $\omega \geq 0 \wedge d \geq 0$ available as an assumption in the premise, because the continuous evolution is never allowed to leave it.

11.7 Assuming Invariants

Let's make the dynamics more interesting and see what happens. Suppose there is a robot at a point with coordinates (x, y) that is facing in direction (v, w). Suppose the robot moves with constant (linear) velocity into direction (v, w). Suppose the direction (v, w) is simultaneously rotating as in Example 10.1 with an angular velocity ω as in Example 2.7 (Fig. 3.9). Then the resulting differential equations are:

$$x' = v, y' = w, v' = \omega w, w' = -\omega v$$

because the derivative of the x coordinate is the component v of the direction and

Fig. 11.15 Illustration of the Dubins dynamics of a point (x, y) moving in direction (v, w) along a dashed curve with angular velocity ω

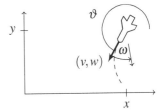

the derivative of the y coordinate is the component w of the direction. The angular velocity ω determines how fast the direction (v, w) rotates. Consider the conjecture

$$(x-1)^2 + (y-2)^2 \geq p^2 \rightarrow [x' = v, y' = w, v' = \omega w, w' = -\omega v](x-1)^2 + (y-2)^2 \geq p^2$$

(11.3)

This conjecture expresses that the robot at position (x, y) will always stay at distance $\geq p$ from the point $(1, 2)$ if it started there. Let's try to prove conjecture (11.3):

$$\dfrac{\dfrac{\vdash 2(x-1)v + 2(y-2)w \geq 0}{\vdash [x':=v][y':=w]2(x-1)x' + 2(y-2)y' \geq 0}}{{}^{\text{dI}}\,(x-1)^2 + (y-2)^2 \geq p^2 \vdash [x'=v, y'=w, v'=\omega w, w'=-\omega v](x-1)^2 + (y-2)^2 \geq p^2} {}_{[:=]}$$

Unfortunately, this differential invariant proof does not work. As a matter of fact, *fortunately* it does not work out, because conjecture (11.3) is not valid, so we will not be able to prove it with a sound proof technique. Conjecture (11.3) is too optimistic. Starting from a bad direction far far away, the robot will get too close to the point (1,2). Other directions may be fine.

Inspecting the above failed proof attempt, we could prove (11.3) if we knew something about the direction (v, w) that would allow the remaining premise to be proved. What could that be?

> Before you read on, see if you can find the answer for yourself.

Certainly, if we knew $v = w = 0$, the resulting premise would be proved. Yet, that case is pretty boring because it corresponds to the point (x, y) being stuck forever. A more interesting case in which the premise would easily be proved is if we knew

$x-1=-w$ and $y-2=v$. In what sense could we "know" $x-1=-w \wedge y-2=v$? Certainly, we would have to assume this compatibility condition for directions versus position is true in the initial state, otherwise we would not necessarily know the condition holds true where we need it. So let's modify (11.3) to include this assumption:

$$x-1=-w \wedge y-2=v \wedge (x-1)^2+(y-2)^2 \geq p^2 \rightarrow$$
$$[x'=v,y'=w,v'=\omega w,w'=-\omega v](x-1)^2+(y-2)^2 \geq p^2 \quad (11.4)$$

Yet, the place in the proof where we need to know $x-1=-w \wedge y-2=v$ for the above sequent proof to continue is in the middle of the inductive step. How can we make that happen?

> Before you read on, see if you can find the answer for yourself.

One step in the right direction is to check whether $x-1=-w \wedge y-2=v$ is a differential invariant of the dynamics, so it stays true forever if it is true initially:

$$\dfrac{\dfrac{\text{not valid}}{\vdash v=-(-\omega v) \wedge w=\omega w}}{\dfrac{\vdash [x':=v][y':=w][v':=\omega w][w':=-\omega v](x'=-w' \wedge y'=v')}{x-1=-w \wedge y-2=v \vdash [x'=v,y'=w,v'=\omega w,w'=-\omega v](x-1=-w \wedge y-2=v)}\, [:=]}\, \text{dI}$$

This prove does not quite work out, because the two sides of the equations are off by a factor of ω and, indeed, $x-1=-w \wedge y-2=v$ is not an invariant unless $\omega=1$. On second thoughts, that makes sense, because the angular velocity ω determines how quickly the robot turns, so if there is any relation between position and direction at all, it should somehow depend on the angular velocity ω.

Let's refine the conjecture to incorporate the angular velocity on the side of the equation where it was missing in the above proof and consider $\omega(x-1)=-w \wedge \omega(y-2)=v$ instead. That knowledge would still help the proof of (11.3), just with the same extra factor on both terms. So let's modify (11.4) to use this assumption on the initial state:

$$\omega(x-1)=-w \wedge \omega(y-2)=v \wedge (x-1)^2+(y-2)^2 \geq p^2 \rightarrow$$
$$[x'=v,y'=w,v'=\omega w,w'=-\omega v](x-1)^2+(y-2)^2 \geq p^2 \quad (11.5)$$

A simple proof shows that the new addition $\omega(x-1)=-w \wedge \omega(y-2)=v$ is a differential invariant of the dynamics, so it holds always if it holds at the beginning:

$$\dfrac{\dfrac{\dfrac{*}{\vdash \omega v=-(-\omega v) \wedge \omega w=\omega w}\,\text{R}}{\vdash [x':=v][y':=w][v':=\omega w][w':=-\omega v](\omega x'=-w' \wedge \omega y'=v')}\,[:=]}{\omega(x-1)=-w \wedge \omega(y-2)=v \vdash [x'=v,y'=w,v'=\omega w,w'=-\omega v](\omega(x-1)=-w \wedge \omega(y-2)=v)}\,\text{dI}$$

Now, how can this freshly proved invariant $\omega(x-1) = -w \wedge \omega(y-2) = v$ be made available in the previous proof? Perhaps we could prove (11.5) using the conjunction of the invariant we want with the additional invariant we need:

$$(x-1)^2 + (y-2)^2 \geq p^2 \wedge \omega(x-1) = -w \wedge \omega(y-2) = v$$

That does not work (eliding the antecedent in the conclusion just for space reasons):

$$\cfrac{\vdash 2(x-1)v + 2(y-2)w \geq 0 \wedge \omega v = -(-\omega v) \wedge \omega w = \omega w}{\cfrac{\vdash [x':=v][y':=w][v':=\omega w][w':=-\omega v](2(x-1)x' + 2(y-2)y' \geq 0 \wedge \omega x' = -w' \wedge \omega y' = v')}{{}^{\mathrm{dI}}\ .. \vdash [x'=v, y'=w, v'=\omega w, w'=-\omega v]((x-1)^2 + (y-2)^2 \geq p^2 \wedge \omega(x-1) = -w \wedge \omega(y-2) = v)}}$$

because the right conjunct in the premise is still proved beautifully but the left conjunct in the premise needs to know the invariant, while the differential invariant proof rule dI does not make the invariant F available in the antecedent of the premise.

In the case of loops, the invariant F can be assumed to hold before the loop body in the induction step (the other form loop of the loop invariant rule):

$$\text{ind}\ \ \frac{P \vdash [\alpha]P}{P \vdash [\alpha^*]P}$$

By analogy, we could augment the differential invariant proof rule dI similarly to include the invariant in the assumptions. Is that a good idea?

Before you read on, see if you can find the answer for yourself.

It looks tempting to suspect that rule dI could be improved by assuming the differential invariant F in the antecedent of the premise:

$$\text{dI}_{??}\ \ \frac{Q \wedge F \vdash [x':=f(x)](F)'}{F \vdash [x'=f(x) \& Q]F}\ \ \text{sound?}$$

After all, we really only care about staying safe when we are still safe since we start safe. Rule dI$_{??}$ would indeed easily prove the formula (11.5), which might make us cheer. But implicit properties of differential equations are a subtle business. Assuming F as in rule dI$_{??}$ would, in fact, be *unsound*, as the following simple counterexample shows, which "proves" an invalid property using the unsound proof rule dI$_{??}$:

$$\text{(unsound)}$$
$$\frac{\cfrac{v^2 - 2v + 1 = 0 \vdash 2vw - 2w = 0}{v^2 - 2v + 1 = 0 \vdash [v':=w][w':=-v](2vv' - 2v' = 0)}}{{}^{\oint}v^2 - 2v + 1 = 0 \vdash [v'=w, w'=-v]v^2 - 2v + 1 = 0}$$

Of course, $v^2 - 2v + 1 = 0$ does *not* stay true for the rotational dynamics, because v changes! And there are many other invalid properties that the unsound proof rule dI$_{??}$ would claim to "prove," for example.

$$
\begin{array}{c}
\text{(unsound)} \\[2pt]
\hline
-(x-y)^2 \geq 0 \vdash -2(x-y)(1-y) \geq 0 \\[2pt]
\hline
-(x-y)^2 \geq 0 \vdash [x':=1][y':=y](-2(x-y)(x'-y') \geq 0) \\[2pt]
\hline
-(x-y)^2 \geq 0 \vdash [x'=1,y'=y](-(x-y)^2 \geq 0)
\end{array}
$$

Assuming an invariant of a differential equation during its own proof is, thus, terribly incorrect, even though it has been suggested numerous times in the literature. There are some cases for which rule dI$_{??}$ or variations of it are still sound, but these are nontrivial [2, 3, 5, 8, 11]. The reason why assuming invariants for their own proof is problematic for the case of differential equations is subtle [5, 11]. In a nutshell, the proof rule dI$_{??}$ assumes more than it knows, so that the argument becomes cyclic. The antecedent only provides the invariant at a single point and Chap. 10 already explained that derivatives are not particularly well defined at a single point. That is one of the reasons why we had to exercise extraordinary care in our arguments to define precisely what derivatives and differentials were to begin with in Chap. 10. Unlike time-derivatives, differentials have meaning in isolated states.

11.8 Differential Cuts

Instead of these ill-guided attempts to assume invariants for their own proof, there is a complementary proof rule for *differential cuts* [4, 5, 8, 11] that can be used to strengthen assumptions about differential equations in a sound way.

> **Lemma 11.5 (dC differential cut proof rule).** *The differential cut proof rule is sound and derives from axiom DC, which will be considered subsequently:*
>
> $$
> \mathrm{dC} \ \frac{\Gamma \vdash [x'=f(x)\,\&\,Q]C, \Delta \quad \Gamma \vdash [x'=f(x)\,\&\,(Q \wedge C)]P, \Delta}{\Gamma \vdash [x'=f(x)\,\&\,Q]P, \Delta}
> $$

The differential cut rule works like a logical cut, but for differential equations. Recall the cut rule from Chap. 6, which can be used to prove a formula C in the left premise as a lemma and then assume it in the right premise:

$$
\mathrm{cut} \ \frac{\Gamma \vdash C, \Delta \quad \Gamma, C \vdash \Delta}{\Gamma \vdash \Delta}
$$

Similarly, differential cut rule dC proves a property C of a differential equation in the left premise and then assumes C to hold in the right premise, except that it assumes C to hold *during* a differential equation by restricting the behavior of the system. To prove the original postcondition P from the conclusion, rule dC restricts the system evolution in the right premise to the subdomain $Q \wedge C$ of Q, which changes the system dynamics but is a pseudo-restriction, because the left premise proves that C is an invariant anyhow (e.g., using rule dI). Note that rule dC is special in that it *changes the dynamics of the system* (it adds a constraint to the system evolution domain region), but it is still sound, because this change does not reduce the reachable set, thanks to the left premise; see Fig. 11.16

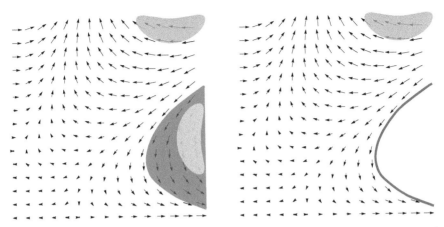

Fig. 11.16 If the solution of the differential equation can never leave region C and enter the red region $\neg C$ (**left**), then this unreachable region $\neg C$ can be cut out of the state space without changing the dynamics of the system by restricting it to C (**right**)

The benefit of rule dC is that C will (soundly) be available as an extra assumption for all subsequent dI uses in the right premise (see, e.g., the use of the evolution domain constraint in Example 11.8). In particular, the differential cut rule dC can be used to strengthen the right premise with more and more auxiliary differential invariants C that will be available as extra assumptions in the right premise, once they have been proven to be differential invariants in the left premise.

Example 11.9 (Increasingly damped oscillator). The damped oscillator in Example 11.8 was easily provable, but its proof crucially depended on having the damping coefficient $d \geq 0$ in the evolution domain constraint so that the induction step knew that the damping coefficient was not negative. In the following increasingly damped oscillator, the damping coefficient changes (albeit in arbitrary ways):

$$\omega^2 x^2 + y^2 \leq c^2 \wedge d \geq 0 \rightarrow [x' = y, y' = -\omega^2 x - 2d\omega y, d' = 7 \,\&\, \omega \geq 0]\, \omega^2 x^2 + y^2 \leq c^2$$

This makes the damped oscillator apply increasing damping, but the system still always stays in the ellipse (Fig. 11.17). A direct proof with a differential invariant will fail, because of the lack of knowledge about the damping coefficient d, which, after all, is now changing. But the indirect proof in Fig. 11.18 succeeds. It uses a differential cut with $d \geq 0$ to first prove, in the left branch, that d always remains nonnegative by a differential invariant argument, and then continues the right branch as in Example 11.8 using the new added evolution domain constraint $d \geq 0$.

Proposition 11.1 (Increasingly damped oscillation). *This* dL *formula is valid:*

$$\omega^2 x^2 + y^2 \leq c^2 \wedge d \geq 0 \rightarrow [x' = y, y' = -\omega^2 x - 2d\omega y, d' = 7 \,\&\, \omega \geq 0]\, \omega^2 x^2 + y^2 \leq c^2$$

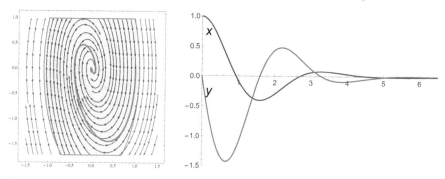

Fig. 11.17 Trajectory with vector field and evolution of an increasingly damped oscillator

$$
\mathbb{R} \; \frac{\ast}{\omega \geq 0 \vdash 7 \geq 0}
$$

$$
{}_{[:=]} \overline{\omega \geq 0 \vdash [d':=7]\, d' \geq 0}
$$

$$
{}_{dI} \overline{d \geq 0 \vdash [x'=y, y' = -\omega^2 x - 2d\omega y, d'=7 \,\&\, \omega \geq 0]\, d \geq 0} \qquad \text{proof as in Example 11.8}
$$

$$
{}_{dC} \frac{d \geq 0 \vdash \text{[above]}\, d \geq 0 \quad \omega^2 x^2 + y^2 \leq c^2 \vdash [x'=y, y'=-\omega^2 x - 2d\omega y, d'=7 \,\&\, \omega \geq 0 \wedge d \geq 0]\, \omega^2 x^2 + y^2 \leq c^2}{\omega^2 x^2 + y^2 \leq c^2, d \geq 0 \vdash [x'=y, y' = -\omega^2 x - 2d\omega y, d'=7 \,\&\, \omega \geq 0]\, \omega^2 x^2 + y^2 \leq c^2}
$$

Fig. 11.18 Differential cut proof for the increasingly damped oscillator

Example 11.10 (Robot formula). Proving the robot formula (11.5) in a sound way is now easy using a differential cut dC by $\omega(x-1) = -w \wedge \omega(y-2) = v$ after we abbreviate $(x-1)^2 + (y-2)^2 \geq p^2$ by A and $\omega(x-1) = -w \wedge \omega(y-2) = v$ by B:

$$
\mathbb{R} \; \frac{\ast}{B \vdash 2(x-1)v + 2(y-2)w \geq 0}
$$

$$
{}_{[:=]} \overline{B \vdash [x':=v][y':=w](2(x-1)x' + 2(y-2)y' \geq 0)}
$$

$$
{}_{dI} \; \frac{\lhd \quad A \vdash [x'=v, y'=w, v'=\omega w, w'=-\omega v \,\&\, \omega(x-1)=-w \wedge \omega(y-2)=v]\,(x-1)^2 + (y-2)^2 \geq p^2}{{}_{dC}\; A, B \vdash [x'=v, y'=w, v'=\omega w, w'=-\omega v]\,(x-1)^2 + (y-2)^2 \geq p^2}
$$

The first premise of the use of rule dC that is elided above (marked by $\lhd$) is proved:

$$
\mathbb{R} \; \frac{\ast}{\vdash \omega v = -(-\omega v) \wedge \omega w = \omega w}
$$

$$
{}_{[:=]} \overline{\vdash [x':=v][y':=w][v':=\omega w][w':=-\omega v](\omega x' = -w' \wedge \omega y' = v')}
$$

$$
{}_{dI} \overline{\omega(x-1)=-w \wedge \omega(y-2)=v \vdash [x'=v..](\omega(x-1)=-w \wedge \omega(y-2)=v)}
$$

Amazing. Now we have a proper sound proof of the quite nontrivial robot motion property (11.5). And it even is a surprisingly short proof.

It is not always enough to just do a single differential cut. Sometimes, you may want to do a differential cut with a formula C, then use C on the right premise of dC to prove a second differential cut with a formula D and then on its right premise have $C \wedge D$ available to continue the proof; see Fig. 11.19. For example, we could also have gotten a proof of (11.5) by first doing a differential cut with $\omega(x-1) = -w$, then continuing with a differential cut with $\omega(y-2) = v$, and then finally uising both to prove the postcondition (Exercise 11.6). Using this differential cut process repeatedly has turned out to be extremely useful in practice and even simplifies the invariant search, because it leads to several simpler properties to find and prove instead of a single complex property [6, 13, 14].

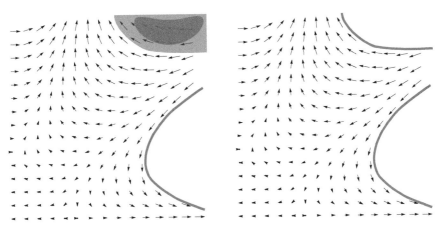

Fig. 11.19 If the solution of the differential equation can never leave region D and enter the top red region $\neg D$ (**left**), then this unreachable region $\neg D$ can also be cut out of the state space without changing the dynamics of the system by restricting it further to D (**right**)

It is straightforward to prove the differential cut rule dC sound from the semantics. The other differential equation proof rules were, however, proved sound by deriving them from corresponding axioms, which are, in turn, proved sound from the semantics. That approach also works for differential cuts.

Lemma 11.6 (DC differential cut axiom). *This axiom is sound:*

$$\text{DC} \quad \big([x' = f(x) \,\&\, Q]P \leftrightarrow [x' = f(x) \,\&\, Q \wedge C]P\big) \leftarrow [x' = f(x) \,\&\, Q]C$$

Proof. Any state that satisfies $[x' = f(x) \,\&\, Q]P$ also satisfies $[x' = f(x) \,\&\, Q \wedge C]P$, because every solution of $\varphi \models x' = f(x) \wedge Q \wedge C$ also solves $\varphi \models x' = f(x) \wedge Q$.

Conversely, consider an initial state ω satisfying the assumption $[x' = f(x) \,\&\, Q]C$. Thus, starting in ω, every solution φ that satisfies $\varphi \models x' = f(x) \wedge Q$ also satisfies C after the solution, so *all along* the solution, because every restriction of a solution is a solution. Thus, if solution φ starts in ω and satisfies $\varphi \models x' = f(x) \wedge Q$, it also satisfies $\varphi \models x' = f(x) \wedge Q \wedge C$, so that the assumption $\omega \in [\![x' = f(x) \,\&\, Q \wedge C]P]\!]$ implies $\omega \in [\![x' = f(x) \,\&\, Q]P]\!]$. $\qquad\square$

The differential cut rule dC derives directly from the differential cut axiom DC. Compared to rule dC, the axiom DC has the additional information that the right premise and conclusion of rule dC are, indeed, *equivalent* if the left premise is valid.

11.9 Differential Weakening Again

Observe how differential weakening from Sect. 11.3 can be useful in combination with differential cuts. For example, after having performed the differential cut illustrated in Fig. 11.16 and, then, subsequently, performing the differential cut illustrated in Fig. 11.19, all unsafe blue regions have been cut out of the state space, so that the system in Fig. 11.19(right) is trivially safe by differential weakening, because there are no more unsafe blue regions. That is, the ultimate evolution domain constraint $Q \wedge C \wedge D$ after the two differential cuts with C and with D trivially implies the safety condition F, i.e., $Q \wedge C \wedge D \vdash F$ is valid. But notice that it took the two differential cuts to make differential weakening useful. The original evolution domain constraint Q was not strong enough to imply safety, since there were still unsafe blue regions in the original system in Fig. 11.16(left) and even still in the intermediate system in Fig. 11.19(left) obtained after one differential cut with C.

If the system starts in an initial state where the evolution domain constraint is not satisfied, the system is stuck so cannot evolve for any duration, not even for duration 0. Any postcondition holds after *all* continuous evolutions of $x' = f(x) \& Q$ if there simply are *none*. In particular in a state where the evolution domain constraint Q is *false*, the differential invariant axiom DI proves $[x' = f(x) \& Q]$*false*, because the assumption Q in the induction step is not satisfied and the test $?Q$ in $[?Q]$*false* fails.

Such a proof, thus, is closed by a differential cut dC with *false* followed by a use of differential invariant axiom DI and differential weakening dW to show that the original postcondition follows from the augmented evolution domain constraint $Q \wedge false$. An easier proof is to use the monotonicity rule MR to prove the new postcondition *false*, which trivially implies the original postcondition P, and is proved by differential invariant axiom DI if the initial condition A implies $\neg Q$ so that the conjunction $A \wedge Q$ is a contradiction (even $(false)' \equiv true$ holds by Exercise 11.10):

$$
\mathrm{MR} \frac{
  \mathrm{DI} \frac{
    {}_{[?]} \frac{
      {}_{\to R} \frac{A, Q \vdash false}{A \vdash Q \to false}
    }{A \vdash [?Q]false}
  }{A \vdash [x' = f(x) \& Q]false}
  \qquad
  \mathrm{WR} \frac{A, Q \vdash}{A, Q \vdash [x' = f(x) \& Q](false)'}
  \qquad
  {}^{\mathbb{R}} \frac{*}{false \vdash P}
}{A \vdash [x' = f(x) \& Q]P}
$$

11.10 Differential Invariants for Solvable Differential Equations

The primary motivation for studying differential invariants, differential cuts, and differential weakening was the need for advanced induction techniques for advanced differential equations that have no closed-form solutions in decidable arithmetic. For such advanced differential equations, the solution axiom schema $[']$ cannot be used or leads to undecidable arithmetic. But differential invariant style reasoning is still helpful even for simpler differential equations that have (rational) solutions.

Example 11.11 (Differential cuts prove falling balls). Recall the dL formula for a falling ball that was a part of the bouncing-ball proof from Chap. 7:

$$2gx = 2gH - v^2 \wedge x \geq 0 \rightarrow [x'' = -g \,\&\, x \geq 0](2gx = 2gH - v^2 \wedge x \geq 0) \quad (11.6)$$

$$\text{where } \{x'' = -g \,\&\, x \geq 0\} \stackrel{\text{def}}{\equiv} \{x' = v, v' = -g \,\&\, x \geq 0\}$$

Chap. 7 proved dL formula (11.6) using the solution of the differential equations with the solution axiom schema $[']$. Yet, dL formula (11.6) can also be proved with a mix of differential invariants, differential cuts, and differential weakening, instead:

$$
\begin{array}{l}
^{\text{id}}\overline{x \geq 0 \wedge 2gx = 2gH - v^2 \vdash 2gx = 2gH - v^2 \wedge x \geq 0}^{\;*} \\[2pt]
{}^{\text{dW}}\lhd \overline{\quad 2gx = 2gH - v^2 \vdash [x'' = -g \,\&\, x \geq 0 \wedge 2gx = 2gH - v^2](2gx = 2gH - v^2 \wedge x \geq 0)} \\[2pt]
{}^{\text{dC}}\overline{2gx = 2gH - v^2 \vdash [x'' = -g \,\&\, x \geq 0](2gx = 2gH - v^2 \wedge x \geq 0)}
\end{array}
$$

The elided premise (marked $\lhd$) after dC is proved by differential invariants:

$$
\begin{array}{l}
^{\mathbb{R}}\overline{x \geq 0 \vdash 2gv = -2v(-g)}^{\;*} \\[2pt]
{}^{[:=]}\overline{x \geq 0 \vdash [x':=v][v':=-g]2gx' = -2vv'} \\[2pt]
{}^{\text{dI}}\overline{2gx = 2gH - v^2 \vdash [x'' = -g \,\&\, x \geq 0]2gx = 2gH - v^2}
\end{array}
$$

Note that differential weakening (dW) works for proving the postcondition $x \geq 0$, but dI would not work for proving $x \geq 0$, because its derivative is $(x \geq 0)' \equiv v \geq 0$, which is not an invariant of the bouncing ball since its velocity ultimately becomes negative when it is falling again under gravity.

The above proof is elegant and has notably easier arithmetic than the arithmetic required when working with solutions of the bouncing ball in Chap. 7.

> **Note 62 (Differential invariants lower degrees)** Differential invariant proof rule dI works by differentiation, which *lowers* polynomial degrees. The differential equation solution axiom $[']$ works with solutions, which ultimately integrate the differential equation and, thus, increase the degree. The computational complexity of the resulting arithmetic is, thus, often in favor of differential invariants even in cases where the differential equations can be solved so that the solution axiom $[']$ would be applicable.

Since the first conjunct of the postcondition in (11.6) is not needed for the proof of the second conjunct, a similar differential invariant proof can also be obtained using derived axiom $[]\wedge$ to split the postcondition instead of dC to nest it:

$$
\begin{array}{l}
{}^{\text{dI}}\cfrac{{}^{[:=]}\cfrac{{}^{\mathbb{R}}\overline{x \geq 0 \vdash 2gv = -2v(-g)}^{\;*}}{x \geq 0 \vdash [x':=v][v':=-g]2gx' = -2vv'}}{2gx = 2gH - v^2 \vdash [x'' = -g \,\&\, x \geq 0]2gx = 2gH - v^2} \qquad {}^{\text{dW}}\cfrac{{}^{\text{id}}\overline{x \geq 0 \vdash x \geq 0}^{\;*}}{2gx = 2gH - v^2 \vdash [x'' = -g \,\&\, x \geq 0]x \geq 0} \\[12pt]
{}^{[]\wedge}\overline{2gx = 2gH - v^2 \vdash [x'' = -g \,\&\, x \geq 0](2gx = 2gH - v^2 \wedge x \geq 0)}
\end{array}
$$

This is how it pays to pay attention to which parts of a postcondition hold by which principle. The second conjunct $x \geq 0$ follows from the evolution domain alone and, thus, holds by dW. The first conjunct is inductive and follows by dI.

Besides the favorably simple arithmetic coming from differential invariants, the other reason why the above proofs worked so elegantly is that the invariant was a clever choice that we came up with in a creative way in Chap. 4. There is nothing wrong with being creative. On the contrary! Please always be creative!

11.11 Summary

This chapter introduced very powerful proof rules for differential invariants, with which you can prove even complicated properties of differential equations in easy ways. Just like in the case of loops, where the search for invariants is nontrivial, differential invariants require some smarts (or good automatic procedures) to be found. Yet, once differential invariants have been identified, the proof follows easily.

The new proof rules and axioms that they are based on are summarized in Fig. 11.20. For convenience, the derivation axioms and axiom DE from Chap. 10 are included again. Differential invariants follow the intuition of proving properties of differential equations that get more true over time along the differential equation. Or they prove properties that at least do not get less true along a differential equation, so will remain true if they start true. The differential invariant proof rule determines locally whether a property remains true along a differential equation by inspecting the differential of the postcondition in the direction that the right-hand side of the differential equation indicates. Since the resulting premise is formed by differentiation and substitution, it is relatively easy to check whether the resulting real arithmetic is true.

If the postcondition of a differential equation is getting less true along the dynamics of the differential equation, however, then additional thoughts are needed. For example, differential cuts (axiom DC and corresponding rule dC) provide a way of enriching the dynamics with a property C that is first proved to be an invariant itself. The differential cut principle makes it possible to prove a sequence of additional properties of differential equations and then to use them subsequently in the proof. Differential cuts are powerful reasoning principles, because they can exploit additional implicit structure in the system by proving and then using lemmas about the behavior of the system. In particular, differential cuts can make into differential invariants properties that have not been differential invariants before by first restricting the domain to a smaller subset on which the property actually is an invariant. Indeed, differential cuts are a fundamental proof principle for differential equations, satisfying the *No Differential Cut Elimination* theorem [11], because some properties can only be proved with differential cuts, not without them. Yet another way properties of differential equations that are not differential invariants directly can be made inductive will be explored in the next chapter.

$$\text{dI } \frac{Q \vdash [x':=f(x)](F)'}{F \vdash [x' = f(x) \& Q]F} \qquad \text{dW } \frac{Q \vdash P}{\Gamma \vdash [x' = f(x) \& Q]P, \Delta}$$

$$\text{dC } \frac{\Gamma \vdash [x' = f(x) \& Q]C, \Delta \quad \Gamma \vdash [x' = f(x) \& (Q \wedge C)]P, \Delta}{\Gamma \vdash [x' = f(x) \& Q]P, \Delta}$$

$$\text{DW } [x' = f(x) \& Q]P \leftrightarrow [x' = f(x) \& Q](Q \to P)$$

$$\text{DI } \big([x' = f(x) \& Q]P \leftrightarrow [?Q]P\big) \leftarrow (Q \to [x' = f(x) \& Q](P)')$$

$$\text{DC } \big([x' = f(x) \& Q]P \leftrightarrow [x' = f(x) \& Q \wedge C]P\big) \leftarrow [x' = f(x) \& Q]C$$

$$\text{DE } [x' = f(x) \& Q]P \leftrightarrow [x' = f(x) \& Q][x' := f(x)]P$$

$$+' \ (e + k)' = (e)' + (k)'$$

$$-' \ (e - k)' = (e)' - (k)'$$

$$\cdot' \ (e \cdot k)' = (e)' \cdot k + e \cdot (k)'$$

$$/' \ (e/k)' = \big((e)' \cdot k - e \cdot (k)'\big)/k^2$$

$$c' \ (c())' = 0 \qquad\qquad \text{(for numbers or constants } c())$$

$$x' \ (x)' = x' \qquad\qquad \text{(for variable } x \in \mathscr{V})$$

Fig. 11.20 Axioms and proof rules for differential invariants and differential cuts of differential equations

11.12 Appendix: Proving Aerodynamic Bouncing Balls

This section studies a hybrid system with differential invariants. Remember the bouncing ball whose safety was proved in Chap. 7?

The little acrophobic bouncing ball has graduated from its study of loops and control and yearningly thinks back to its joyful time when it was studying continuous behavior. Caught up in nostalgia, Quantum the bouncing ball suddenly discovers that it unabashedly neglected the effect that air has on bouncing balls all the time. It sure is fun to fly through the air, so the little bouncing ball swiftly decides to make up for that oversight by including a proper aerodynamical model in its favorite differential equation. The effect that air has on the bouncing ball is air resistance and, it turns out, air resistance gets stronger the faster the ball is flying. After a couple of experiments, the little bouncing ball finds out that air resistance is quadratic in the velocity with an aerodynamic damping factor $r > 0$.

Now the strange thing with air is that air is always against the flying ball! Air always provides resistance, no matter in which direction the ball is flying. If the ball is hurrying upwards, the air holds it back and slows it down by decreasing its positive speed $v > 0$. If the ball is rushing back down to the ground, the air still holds the ball back and slows it down, only then that actually means *increasing* the negative velocity $v < 0$, because that corresponds to decreasing the absolute value $|v|$. How can that be modeled properly?

One way of modeling this situation would be to use the (discontinuous) sign function sign v that has value 1 for $v > 0$, value -1 for $v < 0$, and value 0 for $v = 0$:

$$x' = v, v' = -g - (\text{sign}\,v)rv^2 \,\&\, x \geq 0 \tag{11.7}$$

That, however, gives a differential equation with a discontinuous right-hand side [1]. Instead, the little bouncing ball has learned to appreciate the philosophy behind hybrid systems, which advocates for keeping the continuous dynamics simple and moving discontinuities and switching aspects to where they belong: the discrete dynamics. After all, switching and discontinuities are what the discrete dynamics is good at.

Consequently, the little bouncing ball decides to split modes and separate the upward-flying part $v \geq 0$ from the downward flying part $v \leq 0$ and offer the system a nondeterministic choice between the two:[1]

$$x \leq H \wedge v = 0 \wedge x \geq 0 \wedge g > 0 \wedge 1 \geq c \geq 0 \wedge r \geq 0 \rightarrow$$
$$\big[(\text{if}(x = 0)\,v := -cv;$$
$$(\{x' = v, v' = -g - rv^2 \,\&\, x{\geq}0 \wedge v{\geq}0\} \cup \{x' = v, v' = -g + rv^2 \,\&\, x{\geq}0 \wedge v{\leq}0\}))^*$$
$$\big](0 \leq x \leq H) \tag{11.8}$$

In pleasant anticipation of the new behavior that this *aerodynamic bouncing ball* model provides, the little bouncing ball is eager to give it a try. Before daring to bounce around with this model, though, the acrophobic bouncing ball first wants to be convinced that it would be safe to use, i.e., the model actually satisfies the height limit property in (11.8). So the bouncing ball first sets out on a proof adventure. After writing down several ingenious proof steps, the bouncing ball finds out that its previous proof does not carry over. For one thing, the nonlinear differential equations can no longer be solved quite so easily. That makes the solution axiom ['] rather useless. But, fortunately, the little bouncing ball brightens up again as it remembers that unsolvable differential equations were what differential invariants were good at. And the ball is rather keen on trying them in the wild, anyhow.

However, first things first. The first step of the proof after rule $\rightarrow$R is the search for an invariant for the loop induction proof rule loop. Yet, since the proof of (11.8) cannot work by solving the differential equations, we will also need to identify differential invariants for the differential equations. If we are lucky, maybe the same invariant could even work for both? Whenever we are in such a situation, we can search from both ends and either identify an invariant for the loop first and then try to adapt it to the differential equation, or, instead, look for a differential invariant first.

---

[1] Note that the reasons for splitting modes and offering a nondeterministic choice between them are not controller events as they have been in Chap. 8, but, rather, come from the physical model itself. The mechanism is the same, though, whatever the reason for splitting.

Since we know the loop invariant for the ordinary bouncing ball from (7.10), let's look at the loop first. The loop invariant for the ordinary bouncing ball was

$$2gx = 2gH - v^2 \wedge x \geq 0$$

We cannot really expect the equation in this invariant to work out for the aerodynamic ball (11.8) as well, because the whole point of the air resistance is that it slows the ball down. Since air resistance always works against the ball's motion, the height is expected to be less:

$$J_{x,v} \overset{\text{def}}{\equiv} 2gx \leq 2gH - v^2 \wedge x \geq 0 \tag{11.9}$$

In order to check right away whether this invariant that we suspect to be a loop invariant works for the differential equations as well, let's check for differential invariance:

$$
\mathbb{R} \cfrac{\ast}{\cfrac{g>0 \wedge r \geq 0, x \geq 0 \wedge v \geq 0 \vdash 2gv \leq 2gv + 2rv^3}{\cfrac{g>0 \wedge r \geq 0, x \geq 0 \wedge v \geq 0 \vdash 2gv \leq -2v(-g-rv^2)}{\overset{[:=]}{\cfrac{g>0 \wedge r \geq 0, x \geq 0 \wedge v \geq 0 \vdash [x':=v][v':=-g-rv^2](2gx' \leq -2vv')}{\overset{\text{dI}}{g>0 \wedge r \geq 0, 2gx \leq 2gH - v^2 \vdash [x'=v, v'=-g-rv^2 \,\&\, x \geq 0 \wedge v \geq 0]\, 2gx \leq 2gH - v^2}}}}}
$$

Note that for this proof to work, it is essential to keep the constants $g > 0 \wedge r \geq 0$ around, or at least $r \geq 0$. The easiest way of doing that is to perform a differential cut dC with $g > 0 \wedge r \geq 0$ and prove it to be a (trivial) differential invariant, because both parameters do not change, to make $g > 0 \wedge r \geq 0$ available in the evolution domain constraint for the rest of the proof.[2]

The differential invariant proof for the other ODE in (11.8) works as well:

$$
\mathbb{R} \cfrac{\ast}{\cfrac{g>0 \wedge r \geq 0, x \geq 0 \wedge v \leq 0 \vdash 2gv \leq 2gv - 2rv^3}{\cfrac{g>0 \wedge r \geq 0, x \geq 0 \wedge v \leq 0 \vdash 2gv \leq -2v(-g+rv^2)}{\overset{[:=]}{\cfrac{g>0 \wedge r \geq 0, x \geq 0 \wedge v \leq 0 \vdash [x':=v][v':=-g+rv^2]2gx' \leq -2vv'}{\overset{\text{dI}}{g>0 \wedge r \geq 0, 2gx \leq 2gH - v^2 \vdash [x'=v, v'=-g+rv^2 \,\&\, x \geq 0 \wedge v \leq 0]\, 2gx \leq 2gH - v^2}}}}}
$$

After this preparation, the rest of the proof of (11.8) is a matter of checking whether (11.9) is also a loop invariant. Except that the above two sequent proofs do not actually quite prove that (11.9) is a differential invariant, but only that its left conjunct $2gx \leq 2gH - v^2$ is. Would it work to add the right conjunct $x \geq 0$ and prove it to be a differential invariant?

Not exactly, because rule dI would lead to $[x':=v](x' \geq 0) \equiv v \geq 0$, which is obviously not always true for bouncing balls (except in the mode $x \geq 0 \wedge v \geq 0$). However, after proving the above differential invariant after a differential cut (elided use of the above proof is marked by ◁ in the next proof), a differential weakening argument by dW easily shows that the relevant part $x \geq 0$ of the evolution domain constraint always holds after the differential equation:

---

[2] Since this happens so frequently, KeYmaera X keeps constant parameter assumptions in the context using the vacuous axiom V as in Sect. 7.5.

$$\mathrm{dC}\frac{\mathrm{dW}\dfrac{\mathrm{id}\dfrac{*}{x\geq 0\wedge v\leq 0\wedge 2gx\leq 2gH-v^2\vdash 2gx\leq 2gH-v^2\wedge x\geq 0}}{\vartriangleleft\, 2gx\leq 2gH-v^2\vdash [x'=v,v'=-g+rv^2\,\&\,x\geq 0\wedge v\leq 0\wedge 2gx\leq 2gH-v^2](2gx\leq 2gH-v^2\wedge x\geq 0)}}{..\,2gx\leq 2gH-v^2\vdash [x'=v,v'=-g+rv^2\,\&\,x\geq 0\wedge v\leq 0](2gx\leq 2gH-v^2\wedge x\geq 0)}$$

From these pieces it now remains to prove that (11.9) is a loop invariant of (11.8). Without abbreviations, this proof will not fit on a page:

$$A_{x,v}\stackrel{\mathrm{def}}{\equiv} x\leq H\wedge v=0\wedge x\geq 0\wedge g>0\wedge 1\geq c\geq 0\wedge r\geq 0$$

$$B_{x,v}\stackrel{\mathrm{def}}{\equiv} 0\leq x\wedge x\leq H$$

$$x''\&v\geq 0\stackrel{\mathrm{def}}{\equiv}\{x'=v,v'=-g-rv^2\,\&\,x\geq 0\wedge v\geq 0\}$$

$$x''\&v\leq 0\stackrel{\mathrm{def}}{\equiv}\{x'=v,v'=-g+rv^2\,\&\,x\geq 0\wedge v\leq 0\}$$

$$J_{x,v}\stackrel{\mathrm{def}}{\equiv} 2gx\leq 2gH-v^2\wedge x\geq 0$$

$$\mathrm{loop}\frac{A_{x,v}\vdash J_{x,v}\;[:]\dfrac{\mathrm{MR}\dfrac{J_{x,v}\vdash [\mathrm{if}(x=0)\,v:=-cv]J_{x,v}\;[\cup]\dfrac{\wedge\mathrm{R}\dfrac{J_{x,v}\vdash [x''\&v\geq 0]J_{x,v}\quad J_{x,v}\vdash [x''\&v\leq 0]J_{x,v}}{J_{x,v}\vdash [x''\&v\geq 0]J_{x,v}\wedge [x''\&v\leq 0]J_{x,v}}}{J_{x,v}\vdash [x''\&v\geq 0\cup x''\&v\leq 0]J_{x,v}}}{J_{x,v}\vdash [\mathrm{if}(x=0)\,v:=-cv][x''\&v\geq 0\cup x''\&v\leq 0]J_{x,v}}}{J_{x,v}\vdash [\mathrm{if}(x=0)\,v:=-cv;(x''\&v\geq 0\cup x''\&v\leq 0)]J_{x,v}}\quad J_{x,v}\vdash B_{x,v}}{A_{x,v}\vdash [(\mathrm{if}(x=0)\,v:=-cv;(x''\&v\geq 0\cup x''\&v\leq 0))^*]B_{x,v}}$$

The first and last premise are proved by simple arithmetic using $g>0\wedge v^2\geq 0$. The third and fourth premise have been proved above by a differential cut with a subsequent differential invariant and differential weakening. That only leaves the second premise to worry about, which is proved as follows:

$$\wedge\mathrm{R}\frac{[\cup]\dfrac{[:]\dfrac{[?]\dfrac{\rightarrow\mathrm{R}\dfrac{[:=]\dfrac{J_{x,v},x=0\vdash J_{x,-cv}}{J_{x,v},x=0\vdash [v:=-cv]J_{x,v}}}{J_{x,v}\vdash x=0\rightarrow [v:=-cv]J_{x,v}}}{J_{x,v}\vdash [?x=0][v:=-cv]J_{x,v}}}{J_{x,v}\vdash [?x=0;v:=-cv]J_{x,v}}\quad [?]\dfrac{\rightarrow\mathrm{R}\dfrac{\mathrm{id}\dfrac{*}{J_{x,v},x\neq 0\vdash J_{x,v}}}{J_{x,v}\vdash x\neq 0\rightarrow J_{x,v}}}{J_{x,v}\vdash [?x\neq 0]J_{x,v}}}{J_{x,v}\vdash [?x=0;v:=-cv]J_{x,v}\wedge [?x\neq 0]J_{x,v}}}{\dfrac{J_{x,v}\vdash [?x=0;v:=-cv\cup ?x\neq 0]J_{x,v}}{J_{x,v}\vdash [\mathrm{if}(x=0)\,v:=-cv]J_{x,v}}}$$

This sequent proof first expands the if() with the axiom from Exercise 5.15 since if$(Q)\,\alpha$ is an abbreviation for $?Q;\alpha\cup ?\neg Q$. The resulting right premise is proved trivially by axiom (there was no state change in the corresponding part of the execution), the left premise is proved by arithmetic, because $2gH-v^2\leq 2gH-(-cv)^2$ since $1\geq c\geq 0$. This completes the sequent proof for the safety of the aerodynamic bouncing ball expressed in dL formula (11.8). That is pretty neat!

Proposition 11.2 (Aerodynamic Quantum is safe). *This* dL *formula is valid:*

$$x \leq H \land v = 0 \land x \geq 0 \land g > 0 \land 1 \geq c \geq 0 \land r \geq 0 \rightarrow$$
$$\big[\big(\text{if}(x = 0)\, v := -cv;$$
$$\quad (\{x' = v, v' = -g - rv^2 \,\&\, x \geq 0 \land v \geq 0\} \cup \{x' = v, v' = -g + rv^2 \,\&\, x \geq 0 \land v \leq 0\}) \big)^* $$
$$\big] (0 \leq x \leq H)$$

It is about time for the newly upgraded aerodynamic acrophobic bouncing ball to notice a subtlety in its (provably safe) model. The bouncing ball innocently split the differential equation (11.7) into two modes, one for $v \geq 0$ and one for $v \leq 0$, when developing the model (11.8). This seemingly innocuous step required more thought than the little bouncing ball put into it at the time. Of course, the single differential equation (11.7) could, in principle, switch between velocity $v \geq 0$ and $v \leq 0$ any arbitrary number of times during a single continuous evolution. The HP in (11.8) that splits the mode, however, enforces that the ground controller if$(x = 0)\, v := -cv$ will run in between switching from the mode $v \geq 0$ to the mode $v \leq 0$ or back. On its way up when gravity is just about to win out and pull the ball back down again, that is of no consequence, because the trigger condition $x = 0$ will not hold then anyhow, unless the ball really started the day without much energy ($x = v = 0$). On its way down, the condition may very well be true, namely when the ball is currently on the ground and just inverted its velocity. In that case, however, the evolution domain constraint $x \geq 0$ would have forced a ground controller action in the original system already anyhow.

So even if, in this particular model, the system could not in fact actually switch back and forth between the two modes too often in ways that would really matter, it is important to understand how to properly split modes in general, because that will be crucial for other systems. What the little bouncing ball should have done to become aerodynamical in a systematic way is to add an additional mini-loop around just the two differential equations, so that the system could switch modes repeatedly without requiring a discrete ground controller action to happen. This leads to the following dL formula with a systematic mode split, which is provably safe just the same (Exercise 11.7):

$$x \leq H \land v = 0 \land x \geq 0 \land g > 0 \land 1 \geq c \geq 0 \land r \geq 0 \rightarrow$$
$$\big[\big(\text{if}(x = 0)\, v := -cv;$$
$$\quad (\{x' = v, v' = -g - rv^2 \,\&\, x \geq 0 \land v \geq 0\} \cup \{x' = v, v' = -g + rv^2 \,\&\, x \geq 0 \land v \leq 0\})^* \big)^* $$
$$\big] (0 \leq x \leq H)$$

$$(11.10)$$

Exercises

11.1. Since ω does not change in this dL formula, its assumption $\omega \geq 0$ can be preserved soundly during the induction step for differential invariants (rule dI):

$$\omega \geq 0 \wedge x = 0 \wedge y = 3 \to [x' = y, y' = -\omega^2 x - 2\omega y]\omega^2 x^2 + y^2 \leq 9$$

Give a corresponding dL sequent calculus proof. How does the proof change if you do not preserve assumptions about constants in the context?

11.2 (Differential invariant practice). Prove the following formulas using differential invariants, differential cuts, and differential weakening as required:

$$xy^2 + x \geq 7 \to [x' = -2xy, y' = 1 + y^2]xy^2 + x \geq 7$$
$$x \geq 1 \vee x^3 \geq 8 \to [x' = x^4 + x^2](x \geq 1 \vee x^3 \geq 8)$$
$$x - x^2 y \geq 2 \wedge y \neq 5 \to [x' = -x^2, y' = -1 + 2xy]x - x^2 y \geq 2$$
$$x \geq 2 \wedge y \geq 22 \to [x' = 4x^2, y' = x + y^4]y \geq 22$$
$$x \geq 2 \wedge y = 1 \to [x' = x^2 y + x^4, y' = y^2 + 1]x^3 \geq 1$$
$$x = -1 \wedge y = 1 \to [x' = -6x^2 + 6xy^2, y' = 12xy - 2y^3] - 2xy^3 + 6x^2 y \geq 0$$
$$x \geq 2 \wedge y = 1 \to [x' = x^2 y^3 + x^4 y, y' = y^2 + 2y + 1]x^3 \geq 8$$
$$x = 1 \wedge y = 2 \wedge z \geq 8 \to [x' = x^2, y' = 4x, z' = 5y]z \geq 8$$
$$x^3 - 4xy \geq 99 \to [x' = 4x, y' = 3x^2 - 4y]x^3 - 4xy \geq 99$$

11.3 (Wrong differential weakening). Show that the following variation of the differential weakening rule dW would be unsound:

$$\frac{\Gamma, Q \vdash P, \Delta}{\Gamma \vdash [x' = f(x) \& Q]P, \Delta}$$

11.4 (Weak differentials of strong inequations). Prove that both of the following alternative definitions yield a sound differential invariant proof rule:

$$(e < k)' \equiv ((e)' < (k)')$$
$$(e < k)' \equiv ((e)' \leq (k)')$$

11.5 (Disequalities). We have defined

$$(e \neq k)' \equiv ((e)' = (k)')$$

Suppose you remove this definition so that you can no longer use the differential invariant proof rule for formulas involving $\neq$. Can you derive a proof rule to prove such differential invariants regardless? If so, how? If not, why not?

11.6. Prove dL formula (11.5) by first doing a differential cut with $\omega(x-1) = -w$, then continue with a differential cut with $\omega(y-2) = v$, and then finally use both to prove the original postcondition. Compare this proof to the proof in Sect. 11.8.

11.7 (Aerodynamic bouncing ball). The aerodynamic-bouncing-ball model silently imposed that no mode switching could happen without ground control being executed first. Even if that is not an issue for the bouncing ball, prove the more general formula (11.10) with its extra loop for more mode switching regardless. Compare the resulting proof to the sequent proof for (11.8).

11.8 (Generalizations). Sect. 5.6.4 explained how the proof of the dL formula $[x:=1;x' = x^2 + 2x^4]x^3 \geq x^2$ can be reduced by monotonicity rule M$[\cdot]$ to a proof of $[x:=1;x' = x^2 + 2x^4]x \geq 1$. Prove both formulas in the dL calculus. Is there a direct proof of the first formula using rule dI without first generalizing it to a proof of the second formula?

11.9 (Differential invariants assuming initial domains). The least that the proof rules for differential equations get to assume is the evolution domain constraint Q, because the system does not evolve outside it. Prove soundness for the following slightly stronger formulation of dI that assumes Q to hold initially:

$$\frac{\Gamma, Q \vdash F, \Delta \quad Q \vdash [x':=f(x)](F)'}{\Gamma \vdash [x' = f(x) \& Q]F, \Delta}$$

11.10 (Differentials of logical constants). Prove the following definitions to be sound for the differential invariant proof rule:

$$(true)' \equiv true$$
$$(false)' \equiv true$$

Show how you can use them to prove the formula

$$A \rightarrow [x' = f(x) \& Q]B$$

in the case where $A \rightarrow \neg Q$ is provable, i.e., where the system initially starts outside the evolution domain constraint Q. Can you derive both definitions from arithmetic definitions of the formulas *true* and *false*?

11.11 (Runaround robot). Identify differential cuts and differentials to prove the runaround robot control model from Exercise 3.9.

11.12 (Solutions without solution axiom schemata). Prove the following formula with differential cuts, differential invariants, and differential weakening and without using the solution axiom schema $[']$.

$$x = 6 \wedge v \geq 2 \wedge a = 1 \rightarrow [x' = v, v' = a]x \geq 5$$

Grab a big sheet of paper and then also similarly prove

$$x = 6 \wedge v \geq 2 \wedge a = 1 \wedge j \geq 0 \rightarrow [x' = v, v' = a, a' = j]x \geq 5$$

References

[1]   Jorge Cortés. Discontinuous dynamical systems: a tutorial on solutions, non-smooth analysis, and stability. *IEEE Contr. Syst. Mag.* **28**(3) (2008), 36–73.

[2]   Khalil Ghorbal and André Platzer. Characterizing algebraic invariants by differential radical invariants. In: *TACAS*. Ed. by Erika Ábrahám and Klaus Havelund. Vol. 8413. LNCS. Berlin: Springer, 2014, 279–294. DOI: 10.1007/978-3-642-54862-8_19.

[3]   Khalil Ghorbal, Andrew Sogokon, and André Platzer. Invariance of conjunctions of polynomial equalities for algebraic differential equations. In: *SAS*. Ed. by Markus Müller-Olm and Helmut Seidl. Vol. 8723. LNCS. Berlin: Springer, 2014, 151–167. DOI: 10.1007/978-3-319-10936-7_10.

[4]   André Platzer. Differential Dynamic Logics: Automated Theorem Proving for Hybrid Systems. PhD thesis. Department of Computing Science, University of Oldenburg, 2008.

[5]   André Platzer. Differential-algebraic dynamic logic for differential-algebraic programs. *J. Log. Comput.* **20**(1) (2010), 309–352. DOI: 10.1093/logcom/exn070.

[6]   André Platzer. *Logical Analysis of Hybrid Systems: Proving Theorems for Complex Dynamics*. Heidelberg: Springer, 2010. DOI: 10.1007/978-3-642-14509-4.

[7]   André Platzer. Quantified differential invariants. In: *HSCC*. Ed. by Marco Caccamo, Emilio Frazzoli, and Radu Grosu. New York: ACM, 2011, 63–72. DOI: 10.1145/1967701.1967713.

[8]   André Platzer. A differential operator approach to equational differential invariants. In: *ITP*. Ed. by Lennart Beringer and Amy Felty. Vol. 7406. LNCS. Berlin: Springer, 2012, 28–48. DOI: 10.1007/978-3-642-32347-8_3.

[9]   André Platzer. Logics of dynamical systems. In: *LICS*. Los Alamitos: IEEE, 2012, 13–24. DOI: 10.1109/LICS.2012.13.

[10] André Platzer. The complete proof theory of hybrid systems. In: *LICS*. Los Alamitos: IEEE, 2012, 541–550. DOI: 10.1109/LICS.2012.64.

[11] André Platzer. The structure of differential invariants and differential cut elimination. *Log. Meth. Comput. Sci.* **8**(4:16) (2012), 1–38. DOI: 10.2168/LMCS-8(4:16)2012.

[12] André Platzer. A complete uniform substitution calculus for differential dynamic logic. *J. Autom. Reas.* **59**(2) (2017), 219–265. DOI: 10.1007/s10817-016-9385-1.

[13] André Platzer and Edmund M. Clarke. Computing differential invariants of hybrid systems as fixedpoints. In: *CAV*. Ed. by Aarti Gupta and Sharad Malik. Vol. 5123. LNCS. Springer, 2008, 176–189. DOI: 10.1007/978-3-540-70545-1_17.

[14] André Platzer and Edmund M. Clarke. Computing differential invariants of hybrid systems as fixedpoints. *Form. Methods Syst. Des.* **35**(1) (2009). Spe-

cial issue for selected papers from CAV'08, 98–120. DOI: 10.1007/s107 03-009-0079-8.

Chapter 12
Ghosts & Differential Ghosts

Synopsis While working toward yet another fundamental reasoning technique for differential equations, this chapter describes somewhat surprising uses of additional auxiliary variables, called ghosts, in the modeling of and reasoning about cyber-physical systems. A *discrete ghost* is an extra variable introduced with an assignment into the proof (or model) for the sake of analyzing the model. A *differential ghost* is an extra variable that is added into the dynamics of a system with a quite arbitrarily made-up differential equation for the purposes of analyzing the system. What might at first sound counterproductive, because it increases the dimension of the system, will, upon closer inspection, turn out to be helpful for proving purposes, because the differential ghost variables provide additional quantities relative to whose (arbitrarily chosen) continuous evolution the behavior of the system can be understood. With a clever choice of the new differential equations for the differential ghosts, it can also become easier to understand the evolution of the original variables, because there is something else to relate to. Differential ghosts can make quite a surprising difference in our understanding of differential equations and can even explain how differential equations can be solved as part of an ordinary differential invariants proof.

12.1 Introduction

Chapters 10 and 11 equipped us with powerful tools for proving properties of differential equations without having to solve them. *Differential invariants* (dI) [3, 6] prove properties of differential equations by induction based on the right-hand side of the differential equation, rather than its much more complicated global solution. *Differential cuts* (dC) [3, 6] make it possible to first prove another property C of a differential equation and then change the dynamics of the system so that it will be restricted to never leave region C, which can, thus, be assumed about the system from then on. Differential cuts are a fundamental proof principle that can make inductive

© Springer International Publishing AG, part of Springer Nature 2018   363
A. Platzer, *Logical Foundations of Cyber-Physical Systems*,
https://doi.org/10.1007/978-3-319-63588-0_12

properties that are not otherwise invariants [6]. They do so by soundly changing the evolution domain constraints after a suitable proof of invariance.

Yet, not every true property of a differential equation can be proved even with the help of differential cuts [6]. There is yet another way of transforming the dynamics of the system to enable new proofs that were not possible before [6]. This transformation uses *differential ghosts* [6, 8] to soundly change the differential equations themselves instead of just changing their evolution domain constraints as differential cuts do. Of course, editing the differential equations should make us even more nervous about soundness than editing the evolution domain already did (in Chap. 11 before we knew how to do so soundly).

Differential ghosts are extra variables that are introduced into the differential equation system solely for the purposes of the proof. This existence just for analytic purposes is where the spooky name "ghost" comes from. Ghosts (or auxiliaries) refer to aspects of a model that do not exist in reality but are merely introduced for the sake of its analysis. Ghosts are not really present in the actual system, but just invented to make the story more interesting or, rather, the proof more conclusive.

In fact, ghost variables can also be useful for a proof when they remain entirely discrete variables that only change by discrete assignments, in which case they are called *discrete ghosts*. Such discrete ghosts are used to remember an intermediate state during the execution, which makes it possible to conduct a proof that relates the new value of the variable to the old value stored in the discrete ghost. Why would that be useful? Well, because it is sometimes easier to analyze the change of a variable than the value of the variable itself. In such cases, it is easier to show that the value of a variable increases compared to the discrete ghost and so stays above 10 if it started out above 10 than it is to directly show that the value always stays above 10.

Discrete ghosts and differential ghosts serve a similar intuitive purpose: they remember intermediate state values so that the relation of the values at intermediate states to the values at final states can be analyzed. The difference is that differential ghosts also update their value continuously at will along their very own special differential equation, which, if cleverly chosen, makes it particularly easy to conduct a proof. Discrete ghosts only receive a value during their own instantaneous discrete assignments but remain constant during differential equations. Ghosts give the proof a way of referring to how the state used to be that is no more. There are many reasons for introducing ghost state into a system, which will be investigated in this chapter.

One intuitive motivation for introducing a differential ghost is for proofs of properties that get less true over time, so that differential invariance techniques alone cannot prove them, because differential invariants prove properties that become more true over time (or at least not less true). If a postcondition such as $x > 0$ is getting less true over time, because the value of x continuously decreases, but the rate at which it decreases decreases too, then its value may still be above 0 always depending on its asymptotic behavior in the limit. In that case, it is useful to introduce a new differential ghost whose value relates to the change of the value of x in comparison to the present value of x. A particularly cleverly chosen differential ghost can serve as a counterweight to the change of truth of the original postcondition, and will then serve as an (evolving) point of reference, e.g., when the rate of change of the truth-value is changing over time. Similar differential ghosts are often needed to capture energy changes in systems with energy loss or energy gain. Technical details on differential ghosts are reported in prior work [6, 8].

The most important learning goals of this chapter are:

Modeling and Control: This chapter does not have much impact on modeling and control of CPS, because, after all, the whole point of ghosts and differential ghosts is that they are only added for the purposes of the proof. However, it can still sometimes be helpful to add such ghost and differential ghost variables into the original model right away. It is good style to mark such additional variables in the model and controller as ghost variables in order to retain the fact that they do not need to be included in the final system executable except for monitoring.

Computational Thinking: This chapter leverages computational thinking principles for the purpose of rigorous reasoning about CPS models by analyzing how extra dimensions can simplify or enable reasoning about lower-dimensional systems. From a state space perspective, extra dimensions are a horrible idea, because, e.g., the number of points on a gridded space grows exponentially in the number of dimensions (curse of dimensionality). From a reasoning perspective, however, the important insight of this chapter is that extra state variables sometimes help and may even make reasoning possible that is otherwise impossible [6]. One intuition why extra ghost state may help reasoning is that it can be used to consume the energy that a given dissipative system is leaking (similar to the reason dark matter has been speculated to exist) or produce the energy that a given system model consumes. The addition of such extra ghost state then enables invariants of generalized energy constants involving both original and ghost state that was not possible using only the original state. That is, ghost state may cause new energy invariants. This chapter continues the trend of generalizing important logical phenomena from discrete to continuous systems. The verification techniques developed in this chapter are critical for verifying some CPS models of appropriate scale and technical complexity but are not necessary for all CPS models. A secondary goal of this chapter is to develop more intuition and deeper understanding of differential invariants and differential cuts.

CPS Skills: The focus in this chapter is on reasoning about CPS models, but there is an indirect impact on developing better intuitions for operational effects in CPS by introducing the concept of relations of state to extra ghost state. A good grasp of such relations can substantially help with the intuitive understanding of CPS dynamics. The reason is that ghosts and differential ghosts enable extra invariants, which enable stronger statements about what we can rely on as a CPS evolves. They also enable relational arguments about how the change of some quantity over time relates to the change of another auxiliary quantity.

rigorous reasoning about ODEs
extra dimensions for extra invariants
higher-dimensional retreat
extra state enables reasoning
invent dark energy
intuition for differential invariants
states and proofs
verify CPS models at scale

none: ghosts are for proofs   relations of state
mark ghosts in models   extra ghost state
syntax of models   CPS semantics
solutions of ODEs

12.2 Recap

Recall the proof rules for differential invariants (dI), differential weakening (dW) and differential cuts (dC) for differential equations from Chap. 11:

> **Note 63 (Proof rules for differential equations)**
> $$\text{dI} \frac{Q \vdash [x':=f(x)](F)'}{F \vdash [x'=f(x)\&Q]F} \qquad \text{dW} \frac{Q \vdash P}{\Gamma \vdash [x'=f(x)\&Q]P,\Delta}$$
> $$\text{dC} \frac{\Gamma \vdash [x'=f(x)\&Q]C,\Delta \quad \Gamma \vdash [x'=f(x)\&(Q\wedge C)]P,\Delta}{\Gamma \vdash [x'=f(x)\&Q]P,\Delta}$$

12.3 A Gradual Introduction to Ghost Variables

This section provides a gradual introduction to various forms of ghost variables. It focuses on the motivation and intuition for what ghost variables are good for and how they can help us condcut a proof.

12.3.1 Discrete Ghosts

The discrete way of adding ghost variables is to introduce a new ghost variable y into a proof that remembers the value of any arbitrary term e for later usage. This can be useful in a proof in order to have a name, y, that recalls the value of e later on in the proof, especially when the value of e changes subsequently during the execution of an HP α in the remaining modalities. Such a discrete ghost y makes it possible to relate the value of e before and after the run of that hybrid program α.

> **Lemma 12.1 (iG discrete ghost rule).** *The following is a sound proof rule for introducing an auxiliary variable or (discrete)* ghost *y:*
>
> $$\text{iG} \;\frac{\Gamma \vdash [y := e]p, \Delta}{\Gamma \vdash p, \Delta} \quad (y \text{ new})$$

Proof. Rule iG derives from assignment axiom $[:=]$ from Chap. 5, which proves

$$p \leftrightarrow [y := e]p$$

because the new variable y does not occur in p (which can be thought of as a nullary predicate symbol here, which is why we write it lowercase following a principle we will explore in Chap. 18). $\qquad\square$

The discrete ghost rule iG directly derives from the assignment axiom $[:=]$ since it merely applies axiom $[:=]$ backwards to introduce a ghost variable y that was not there before. This is an example exploiting the flexibility of equivalence axioms to be used forwards as well as backwards. Of course, it is important to retain a forward momentum in the proof and not apply assignment axiom $[:=]$ to the premise of iG, which would make the carefully dreamt-up discrete ghost y go away again:

$$\text{iG} \frac{[:=]\dfrac{\Gamma \vdash p, \Delta}{\Gamma \vdash [y := e]p, \Delta}}{\Gamma \vdash p, \Delta}$$

Making ghosts go away would be a great goal for the Ghostbusters, but would not exactly make proof attempts productive. If we were really planning on eliminating the discrete ghost, then we should never have introduced it with rule iG in the first place. After observing that discrete ghosts are fancy names for backwards assignments and, thus, sound, the next question is what they could possibly be good for.

Discrete ghosts can be interesting when formula p in rule iG contains modalities that change variables of the term e so that y can remember the value that e had before that change. For example,

$$\text{iG} \frac{xy \geq 2 \vdash [c := xy][x' = x, y' = -y]\,xy \geq 2}{xy \geq 2 \vdash [x' = x, y' = -y]\,xy \geq 2}$$

This proof memorizes in the discrete ghost variable c the value that the interesting term xy had before the differential equation started. It is not obvious how to complete the proof, because substituting c away using the assignment axiom $[:=]$ would undo the pleasant effect that rule iG had, because the whole point of the new variable c

is that it does not occur elsewhere.[1] The only way the proof can make progress is by applying a proof rule to the differential equation, which is not top-level. We can either just keep the assignment around and directly use axioms at the postcondition, or we can first turn the assignment into an equation with this derived proof rule.

> **Lemma 6.5 ($[:=]_=$ equational assignment rule).** *This is a derived rule:*
>
> $$[:=]_= \frac{\Gamma, y = e \vdash p(y), \Delta}{\Gamma \vdash [x := e]p(x), \Delta} \quad (y \; new)$$

With that rule we can proceed as if nothing has happened:

$$
\begin{array}{l}
\mathbb{R} \dfrac{\ast}{\vdash 0 = xy + x(-y)} \\[6pt]
{\scriptstyle[:=]} \dfrac{}{\vdash [x':=x][y':=-y]0 = x'y + xy'} \\[6pt]
{\scriptstyle dI} \dfrac{xy \geq 2, c = xy \vdash [x'=x, y'=-y]c = xy}{} \quad \rhd \\[6pt]
{\scriptstyle MR} \dfrac{xy \geq 2, c = xy \vdash [x'=x, y'=-y]xy \geq 2}{} \\[6pt]
{\scriptstyle[:=]_=} \dfrac{xy \geq 2 \vdash [c := xy][x'=x, y'=-y]xy \geq 2}{} \\[6pt]
{\scriptstyle iG} \dfrac{xy \geq 2 \vdash [x'=x, y'=-y]xy \geq 2}{}
\end{array}
$$

The generalization step MR leads to a second premise that has been elided (marked by $\rhd$) and is proved, e.g., by vacuous axiom V, because the discrete ghost c starts out above 2 by antecedent and never changes its value during the differential equation. This particular property is also proved directly quite easily, but the proof technique of discrete ghosts is of more general interest beyond this demonstration. The next section provides one common source of such examples.

Notice that even the initial height H of the bouncing-ball model in Sect. 4.5 could have been considered a discrete ghost for the purpose of remembering the initial height via $H := x$ initially. The only reason why it is not a pure discrete ghost is because it is also used in the postcondition, so the safety conjecture cannot be stated without H. The variable H is part of the property, not just part of the proof.

12.3.2 Proving Bouncing Balls with Sneaky Solutions

Recall the dL formula for a falling ball in the bouncing-ball proof from Chap. 7:

$$2gx = 2gH - v^2 \wedge x \geq 0 \rightarrow [\{x' = v, v' = -g \, \& \, x \geq 0\}](2gx = 2gH - v^2 \wedge x \geq 0) \quad (11.6*)$$

---

[1] This potentially surprising phenomenon happens for other ghosts as well, because, the whole point of ghosts is to compute something that the original model and property do not depend on. Sufficiently sophisticated forms of dead-code elimination would get rid of ghosts, which would be counterproductive for the proof. In fact, dead-code elimination for compilers and ghosts for proofs are the same phenomenon, only backwards, because, applied from bottom to top, the discrete ghost rule iG introduces a variable that is dead code as opposed to eliminating it.

This formula was already proved twice: once in Chap. 7 using the solution axiom schema ['] and once in Sect. 11.10 on p. 350 using a mix of differential invariants with differential weakening, because the postcondition was cleverly constructed as an invariant of the bouncing ball in Chap. 4 already.

It's a good idea to be clever! But it also pays to be systematic and develop a rich toolbox of techniques for proving properties of differential equations. Is there a way to prove (11.6) without such a distinctively clever invariant that works as a differential invariant right away? Yes, of course, there is one, because (11.6) can even be proved using solution axiom [']. How many proofs does a formula need these days before we stop proving it?

Well, of course, every formula only needs one proof and then lives happily valid ever after. But it turns out that interesting things happen when we systematically try to understand how to make a proof happen that does not use the solution axiom ['] and, yet, still uses solution-based arguments. Can you conceive a way to use solutions for differential equations without invoking the actual solution axiom [']?

Before you read on, see if you can find the answer for yourself.

The solution of a differential equation should be invariant along the differential equation, because it describes an identity that always holds when following the differential equation. The solution for balls falling in gravity according to (11.6) is

$$x(t) = x + vt - \frac{g}{2}t^2$$
$$v(t) = v - gt$$

where x denotes the initial position and v_0 the initial velocity while $x(t)$ and $v(t)$ denote the position and velocity, respectively, after duration t. Now, the only trouble is that these equations cannot possibly be used as straight-out differential invariants, because $x(t)$ is not even allowed in the language we considered so far.[2] After the differential equation, the name for the position at that time is simply x and the name for velocity is just v. Obviously, $v = v - gt$ would not be a very meaningful equation as time goes on, so we somehow need to identify a new name for the old value that the position initially had before the differential equation, and likewise for the old velocity. This leads to the following rephrasing of the solution where x and v denote the variables after the differential equation at time t and x_0 and v_0 those before:

$$x = x_0 + v_0 t - \frac{g}{2}t^2$$
$$v = v_0 - gt$$
(12.1)

These equations are legitimate formulas and could possibly be differentially cut into (11.6) by dC, but there are still some nuanced subtleties with that approach.

---

[2] Function symbol applications like $x(t)$ will enter dL officially in Chap. 18, but that does not alter the considerations we are about to make.

Before you read on, see if you can find the answer for yourself.

Even if we mentioned that we intend v_0 to mean the initial value of velocity before the differential equation, there is no way that the proof will know this unless we do something about it. In particular, the proof will fail, because the resulting arithmetic is not true for all values of x_0 and v_0. Fortunately, there is a perfect proof rule that fits the task as if it had been made for this job. Before handling the differential equation, the proof rule iG can introduce a discrete ghost x_0 that remembers the initial value of x in the new discrete ghost variable x_0 for later reference. And the rule iG can be used again to remember the initial value of v in the discrete ghost v_0. From then on, the variables x_0 and v_0 really are the initial values of x and v before the ODE.

Now that the proof is equipped with a way of referring to the initial values, the next question is how exactly to go about differentially cutting the solutions (12.1) into the differential equations by dC. Maybe the most immediate suggestion would be to use rule dC with a conjunction of the two equations in (12.1) to get the solution into the system as quickly as possible. That will not work, however, because $x = x_0 + v_0 t - \frac{g}{2} t^2$ only is the correct solution for $x' = v$ after we have established that $v = v_0 - gt$ also is the correct solution of the differential equation $v' = -g$ on which x depends.

In retrospect, this ordering of differential cuts makes sense, because the differential equation $x' = v, v' = -g$ explicitly indicates that the change of x depends on v whose change, in turn, depends on g, which remains constant. Consequently, we first need to convey with a differential cut what the behavior of v is before we proceed to investigate the behavior of x, which, after all, depends on v.

Now, we are ready for a proof with solutions that does not use the solution axiom schema $[']$. In order to prevent us from accidentally being too clever and exploiting pure differential invariance principles again as in Sect. 11.10, we will pretend not to know anything about the specific precondition and postcondition and just call them A and $B_{(x,v)}$. Of course, we'll need a time variable $t' = 1$ in order to write a solution over time. Consider the following formulation of the dL formula (11.6):

$$A \vdash [\{x' = v, v' = -g, t' = 1 \,\&\, x \geq 0\}]B_{(x,v)} \qquad (12.2)$$

$$\text{where} \quad A \overset{\text{def}}{\equiv} 2gx = 2gH - v^2 \wedge x \geq 0$$

$$B_{(x,v)} \overset{\text{def}}{\equiv} 2gx = 2gH - v^2 \wedge x \geq 0$$

$$\{x'' = -g, t' = 1\} \overset{\text{def}}{\equiv} \{x' = v, v' = -g, t' = 1\}$$

The proof begins by introducing a discrete ghost v_0 to remember the initial velocity of the bouncing ball and then differentially cuts the solution $v = v_0 - tg$ into the system and proving it to be differentially invariant:

$$
\mathrm{iG}\,\dfrac{\mathrm{dC}\,\dfrac{\mathrm{dI}\,\dfrac{[:=]\,\dfrac{\mathbb{R}\,\dfrac{*}{x\ge 0\vdash -g=-1g}}{x\ge 0\vdash [v':=-g][t':=1]v'=-t'g}}{A\vdash [v_0:=v][x''=-g,t'=1\,\&\,x\ge 0]v=v_0-tg}\qquad A\vdash [v_0:=v][x''=-g,t'=1\,\&\,x\ge 0\wedge v=v_0-tg]B_{(x,v)}}{A\vdash [v_0:=v][x''=-g,t'=1\,\&\,x\ge 0]B_{(x,v)}}}{A\vdash [x''=-g,t'=1\,\&\,x\ge 0]B_{(x,v)}}
$$

Observe how the differential invariant rule dI made the sequent context A as well as the assignment $[v_0:=v]$ disappear, which is important for soundness, because both only hold in the initial state. That both are affected would be particularly easy to see if we had turned the assignment $v_0:=v$ into an equation $v_0=v$ with rule $[:=]_=$. Besides, $[v_0:=v]$ has to disappear from the induction step: if v_0 is the initial value of v then v_0 does not remain equal to v as the ball falls along $v'=-g$.

The left premise in the above proof proved by trivial arithmetic (rule $\mathbb{R}$). The right premise in the above proof is proved as follows by first introducing yet another discrete ghost x_0 with iG that remembers the initial position so that it can be referred to in the solution. The solution $x=x_0+v_0t-\frac{g}{2}t^2$ can then be differentially cut into the system by dC and is proved to be differentially invariant by dI using the new evolution domain $v=v_0-tg$:

$$
\mathrm{iG}\,\dfrac{\mathrm{dC}\,\dfrac{\mathrm{dI}\,\dfrac{[:=]\,\dfrac{\mathrm{id}\,\dfrac{*}{x\ge 0\wedge v=v_0-tg\vdash v=v_0-2\frac{g}{2}t}}{x\ge 0\wedge v=v_0-tg\vdash [x':=v][t':=1]x'=v_0t'-2\frac{g}{2}tt'}}{A\vdash [x_0:=x][v_0:=v][x''=-g,t'=1\,\&\,x\ge 0\wedge v=v_0-tg]x=x_0+v_0t-\frac{g}{2}t^2\,\triangleright}}{A\vdash [x_0:=x][v_0:=v][x''=-g,t'=1\,\&\,x\ge 0\wedge v=v_0-tg]B_{(x,v)}}}{A\vdash [v_0:=v][x''=-g,t'=1\,\&\,x\ge 0\wedge v=v_0-tg]B_{(x,v)}}
$$

The differential cut proof step (dC) has a second premise using the cut which is elided above (marked by $\triangleright$) and is proved directly by differential weakening (dW):

$$
\mathrm{dW}\,\dfrac{x\ge 0\wedge v=v_0-tg\wedge x=x_0+v_0t-\frac{g}{2}t^2\vdash B_{(x,v)}}{A\vdash [x_0:=x][v_0:=v][x''=-g,t'=1\,\&\,x\ge 0\wedge v=v_0-tg\wedge x=x_0+v_0t-\frac{g}{2}t^2]B_{(x,v)}}
$$

After expanding $B_{(x,v)}$, the resulting formula can be proved by real arithmetic, but it has a twist! First of all, the arithmetic can be simplified substantially using the equality substitution rule =R from Chap. 6 to replace v by v_0-tg and replace x by $x_0+v_0t-\frac{g}{2}t^2$ and use subsequent weakening (WL) to get rid of both equations after use. This simplification reduces the computational complexity of real arithmetic:

$$
\scriptsize
\wedge\mathrm{L}\,\dfrac{=\mathrm{R}\,\dfrac{=\mathrm{R}\,\dfrac{\mathrm{WL}\,\dfrac{\wedge\mathrm{R}\,\dfrac{\mathrm{WL}\,\dfrac{\vdash 2g(x_0+v_0t-\frac{g}{2}t^2)=2gH-(v_0-tg)^2}{x\ge 0\vdash 2g(x_0+v_0t-\frac{g}{2}t^2)=2gH-(v_0-tg)^2}\qquad \mathrm{id}\,\dfrac{*}{x\ge 0\vdash x\ge 0}}{x\ge 0\vdash 2g(x_0+v_0t-\frac{g}{2}t^2)=2gH-(v_0-tg)^2\wedge x\ge 0}}{x\ge 0,v=v_0-tg,x=x_0+v_0t-\frac{g}{2}t^2\vdash 2g(x_0+v_0t-\frac{g}{2}t^2)=2gH-(v_0-tg)^2\wedge x\ge 0}}{x\ge 0,v=v_0-tg,x=x_0+v_0t-\frac{g}{2}t^2\vdash 2gx=2gH-(v_0-tg)^2\wedge x\ge 0}}{x\ge 0,v=v_0-tg,x=x_0+v_0t-\frac{g}{2}t^2\vdash 2gx=2gH-v^2\wedge x\ge 0}}{x\ge 0\wedge v=v_0-tg\wedge x=x_0+v_0t-\frac{g}{2}t^2\vdash 2gx=2gH-v^2\wedge x\ge 0}
$$

Observe how this use of equality substitution and weakening helped simplify the arithmetic complexity of the formula substantially and even helped to eliminate a variable (v) right away. This can be useful to simplify arithmetic in many other cases as well. Both eliminating variables as well as applying and hiding equations right away can often simplify the complexity of handling real arithmetic. The arithmetic in the remaining left branch

$$2g\left(x_0 + v_0 t - \frac{g}{2}t^2\right) = 2gH - (v_0 - tg)^2$$

expands by polynomial arithmetic and cancels as indicated:

$$2g\left(x_0 + \cancel{v_0 t} - \cancel{\tfrac{g}{2}t^2}\right) = 2gH - v_0^2 + \cancel{2v_0 tg} + \cancel{t^2 g^2}$$

Those cancellations simplify the arithmetic, leaving the remaining condition

$$2gx_0 = 2gH - v_0^2 \tag{12.3}$$

Indeed, this relation characterizes exactly how H, which turns out to have been the maximal height, relates to the initial height x_0 and initial velocity v_0. In the case of initial velocity $v_0 = 0$, for example, the equation (12.3) collapses to $x_0 = H$, i.e., that H is the initial height in that case. Consequently, the computationally fastest way of proving the resulting arithmetic is to first prove by a differential cut dC that (12.3) is a trivial differential invariant (even by the vacuous axiom V), resulting in a proof of (11.6); see Exercise 12.3.

However, as we go through all proof branches again to check that we really have a proof, we notice a subtle but blatant oversight. Can you spot it, too?

The very first left-most branch with the initial condition for the differential invariant $v = v_0 = tg$ cannot, actually, be proved. The catch is that we silently assumed $t = 0$ to be the initial value for the new clock t, but our proof did not actually say so. Oh my, what can possibly be done about this glitch?

Before you read on, see if you can find the answer for yourself.

There are multiple approaches and, in fact, the most elegant approach will have to wait till the next section. But one feature that we have just learned about can be exploited again to talk about the change of time without having to assume that time t starts at 0. Discrete ghosts to the rescue! Even though we do not know the initial value of the differential ghost t, we can simply use a discrete ghost again to call it t_0 and get on with it. Will that work? Can you work it out? Or should we start a revision of the proof to find out?

$$
\begin{array}{l}
\mathbb{R} \; \dfrac{\ast}{x \geq 0 \vdash -g = -1g} \\[4pt]
{}^{[:=]}\dfrac{}{x \geq 0 \vdash [v' := -g][t' := 1]v' = 0 - (t' - 0)g} \\[4pt]
{}^{\mathrm{dI}}\dfrac{}{A \vdash [t_0 := t][v_0 := v][x'' = -g, t' = 1 \,\&\, x \geq 0]v = v_0 - (t - t_0)g \;\triangleright} \\[4pt]
{}^{\mathrm{dC}}\dfrac{}{A \vdash [t_0 := t][v_0 := v][x'' = -g, t' = 1 \,\&\, x \geq 0]B_{(x,v)}} \\[4pt]
{}^{\mathrm{iG}}\dfrac{}{A \vdash [v_0 := v][x'' = -g, t' = 1 \,\&\, x \geq 0]B_{(x,v)}}
\end{array}
$$

The proof continues similarly with this elided premise (marked $\triangleright$ above):

$$A \vdash [t_0 := t][v_0 := v][x'' = -g, t' = 1 \ \& \ x \geq 0 \land v = v_0 - (t - t_0)g]B_{(x,v)}$$

As this proof shows, everything works as expected as long as we realize that this requires a change of the invariants used for the differential cuts. The solution of the velocity to differentially cut in will be $v = v_0 - (t - t_0)g$ and the solution of the position to differentially cut in subsequently will be $x = x_0 + v_0(t - t_0) - \frac{g}{2}(t - t_0)^2$. With some thought you can also make sure to use the discrete ghosts for the initial values cleverly to initialize it at 0, which is significantly more convenient.

> **Note 64 (Ghost solutions)** Whenever there is a solution of a differential equation that we would like to make available to a proof without using the solution axiom schema ['], a differential cut and subsequent differential invariant can be used to cut the solution as an invariant into the system after a discrete ghost that remembers the initial values needed to express the solution. The tricky part is that solutions depend on time, and time may not be part of the differential equation system. If there is no time variable, however, an additional differential equation first needs to be added that pretends to be time.

For the case of the bouncing ball, this proof looks unnecessarily complicated, because the solution axiom ['] could have been used instead right away, instead. Yet, even if this particular proof was more involved, the arithmetic ended up being nearly trivial in the end (which Note 62 on p. 351 already observed to hold in general for differential invariant proofs). But the same proof technique of adding ghost variables as needed can be pretty useful in more complicated systems.

> **Note 65 (On the utility of ghosts)** Adding ghosts as needed can be useful in more complicated systems that do not have computable solutions, but in which other relations between initial (or intermediate) and final state can be proved. The same technique can also be useful for cutting in solutions when only part of a differential equation system admits a polynomial solution.

For example, the differential equation system $v_1' = \omega v_2, v_2' = -\omega v_1, v' = a, t' = 1$ is difficult, because it has non-polynomial solutions. Still, one part of this differential equation, the velocity $v' = a$, is easily solved. Yet, the solution axiom ['] is not applicable, because no real-arithmetic solution of the whole differential equation system exists (except when $\omega = 0$). Regardless, after suitable discrete ghosts, a differential cut with the solution $v = v_0 + at$ of $v' = a$ adds this precise knowledge about the time-dependent change of the variable v to the evolution domain for subsequent use.

The ghost solution technique is a useful technique to prove properties of differential equations by using solutions that are first proved to be solutions (with a differential cut) and then used in the remaining proof (e.g., by differential weakening). Unlike the solution axiom schema ['], the ghost solution approach also works if only part of a differential equation system can be solved, simply by cutting the required part of the solutions into the differential equations.

12.3.3 Differential Ghosts of Time

When we look back, the formula (12.2) that we now proved with ghost solutions
had a differential equation $t' = 1$ for time that was not actually part of the original
formula (11.6). Does that matter? Well, without having a time variable t, we could
hardly have even meaningfully written down the solutions (12.1). It does not seem
fair that this formula only has a proof by a differential cut with a solution if it already
has a differential equation $t' = 1$ for time to begin with! Even if $t' = 1$ is not in the
original differential equation, there should be a way to add it into the problem.

Indeed, come to think about it, every differential equation deserves a time variable if it really needs one. It does not actually change the system if we simply add a
new differential equation for a time variable into it. Of course, we had better make
sure that the variable is actually new and do not accidentally reuse a variable that
was already present. For example, squeezing $x' = 1$ into the differential equation
$x' = v, v' = -g$ for falling balls would seriously confuse the system dynamics, because x can hardly simultaneously follow $x' = v$ and the conflicting $x' = 1$. We also
cannot just squeeze in $g' = 1$ without significantly affecting the dynamics of bouncing balls, because gravity g was supposed to be a constant in $x' = v, v' = -g$ and
would suddenly be increasing over time in $x' = v, v' = -g, g' = 1$. But if we refrain
from making any of those silly mistakes and do not change the dynamics that was
already there but merely add a dynamics that was not there yet, then adding a new
time variable seems like a fine thing to do.

Now, if we want a way of adding a time variable into differential equation systems, we can add a proof rule just for exactly that purpose. The following proof rule
makes it possible to add a new differential equation for time:

$$\frac{\Gamma \vdash [x' = f(x), t' = 1 \& Q]P, \Delta}{\Gamma \vdash [x' = f(x) \& Q]P, \Delta} \quad (t \text{ fresh}) \tag{12.4}$$

A soundness justification for this proof rule would use that, as long as t is a fresh
variable that does not occur in the conclusion, then a proof of safety of the bigger
differential equation system with $t' = 1$ implies safety of the differential equation
without it. Indeed, this proof rule could have been used to prove the original falling
ball formula (11.6) from the above proof of (12.2). For once, the issue with this
proof rule is not one of soundness, but rather a matter of economy of reasoning
principles.

Proof rule (12.4) does a fine job of adding clocks but cannot do anything else. If
we ever want to add another differential equation into a differential equation system,
this narrow-minded proof rule is of no use. So before we end up wasting more time
with the special case of time as a motivation, let's proceed right away with the general case of differential ghosts, which are ghost variables with made-up differential
equations.

12.3.4 Constructing Differential Ghosts

Differential ghosts are ghost variables that are added into a differential equation system for the purpose of conducting a proof. The proof technique of differential ghosts is not limited to adding just the differential equation $t' = 1$ for time, but can add other differential equations $y' = g(x, y)$ into the differential equation system as well. In previous chapters, it has served us very well to first develop an axiomatic formulation and then proceed to package it up as the most useful proof rule subsequently. Let's proceed in the same way.

When we have a formula $[x' = f(x) \& Q]P$ about a differential equation (or a system) $x' = f(x) \& Q$, then we can add a new differential equation $y' = g(x, y)$ for a new variable y to obtain $[x' = f(x), y' = g(x, y) \& Q]P$. At what initial value does this new differential equation $y' = g(x, y)$ for the new differential ghost y start?

Before you read on, see if you can find the answer for yourself.

For the purposes of adding a differential ghost for a time variable $t' = 1$ for ghost solutions in Sect. 12.3.3 it would be best if the new variable were to start at 0. But for other use cases it might be much better to start the differential ghost elsewhere at a point that best fits the subsequent proof. Does it matter for soundness where the differential ghost y starts?

Since the differential ghost y is a new variable that was not in the original question and is merely added for the sake of the argument, it can also start at any initial state that we like. This phenomenon is somewhat similar to discrete ghosts, which can also soundly assume any arbitrary initial value by rule iG. This calls for the use of an existential quantifier for the initial value of the differential ghost y, because any initial value would justify the original formula. And, in fact, also vice versa, the original formula implies the existence of an initial value for the ghost y for which the bigger differential equation system $x' = f(x), y' = g(x, y) \& Q$ always stays in P. These thoughts lead to the following formulation of the differential ghost axiom:

$$\text{DG} \quad [x' = f(x) \& Q]P \leftrightarrow \exists y \, [x' = f(x), y' = g(x, y) \& Q]P \qquad (12.5)$$

Of course, y needs to be a new variable that does not occur in $[x' = f(x) \& Q]P$, since y is not much of a differential ghost if it was around before. Besides, it would be unsound to add a new differential equation for a variable that was used for a different purpose previously. If $x' = f(x) \& Q$ always stays in P, then there is an initial value for the differential ghost such that the augmented differential equation system $x' = f(x), y' = g(x, y) \& Q$ also always stays in P, and vice versa.

Certainly, the rule (12.4) for adding time can be derived from axiom DG when we use 1 for $g(x, y)$. In fact, when we cleverly instantiate by rule $\exists$R the existential quantifier for the ghost with 0, even the following improved rule can be derived:

$$\frac{\Gamma, t = 0 \vdash [x' = f(x), t' = 1 \& Q]P, \Delta}{\Gamma \vdash [x' = f(x) \& Q]P, \Delta} \quad (t \text{ fresh})$$

This rule is more helpful for ghost solutions because it makes sure the differential equation actually starts at time $t = 0$, which simplifies the arithmetic considerably.

But other differential equations $y' = g(x,y)$ can be added by axiom DG as well. How general can they be? And what are they good for? Is there a limit to what differential equations can be added?

Before you read on, see if you can find the answer for yourself.

During the soundness proof for axiom DG it will turn out that there is a limit to the differential equations that can be added soundly. But before proceeding with this crucial question, we take the inexcusable liberty of first exploring potential use cases of differential ghost axiom DG to sharpen our intuition and learn to appreciate what more general differential ghosts could be good for.

Example 12.1 (Matters get worse without differential ghosts). A guiding example for the use of differential ghosts is the following simple formula:

$$x > 0 \rightarrow [x' = -x]x > 0 \tag{12.6}$$

This formula is not susceptible to differential invariance proofs using just the rule dI, because the trend along $x' = -x$ for the truth-value of the postcondition $x > 0$ makes matters worse over time (the trend in Fig. 12.1 is negative even if its exponential solution $x_0 e^{-t}$ still just stays positive). The differential $-x \geq 0$ of $x > 0$ is invalid:

$$\frac{\displaystyle \frac{\text{not valid}}{\vdash -x \geq 0}}{\displaystyle \overset{[:=]}{} \frac{\vdash [x' := -x]x' \geq 0}{}}{\overset{\text{dI}}{} \; x > 0 \vdash [x' = -x]x > 0}$$

Fig. 12.1 Exponential decay along $x' = -x$ always makes matters worse for $x > 0$

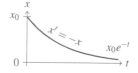

With significantly more thought, it can be shown that there is no indirect way of making (12.6) provable with the help of differential cuts either [6]. But even if along $x' = -x$ the postcondition $x > 0$ tends toward becoming *false*, the rate at which it is becoming *false* is slowing down as well, because the differential equation is $x' = -x$ (and not $x' = -x - 1$ where the extra offset -1 would indeed ultimately make x negative). While $x > 0$ is racing to the bottom, the rate at which x changes along $x' = -x$ simultaneously races to the bottom (toward 0). That begs the question of which of those two limit processes wins. If only we had a way to relate their progress to an extra quantity that could serve as a counterweight to the change of x and describe to what extent the rate at which x changes is being held up.

Suppose we had a differential ghost as an additional variable y with some differential equation that is still to be determined, which we can use as such a counterweight. What relationship of x and y would imply that x must be positive so that the postcondition $x > 0$ is *true*? If we think back to Lie's characterization of invariant terms in Sect. 10.8.2, then we recall that differential invariants are perfect at proving invariant terms that never change their value. The simplest equation of x and y that implies $x > 0$ is $xy^2 = 1$, because y^2 is surely nonnegative and, thus, x must be positive if its product with the nonnegative number y^2 is 1 (or any other positive number). And, in fact, $\exists y\, xy^2 = 1$ is even equivalent to $x > 0$.

Now, the only remaining question is by what differential equation should the differential ghost y change over time to preserve the invariant $xy^2 = 1$, which would imply the desired postcondition $x > 0$. This is the cool thing about differential ghosts: We get to choose their differential equations at our pleasure as $g(x,y)$ in axiom **DG**. Everywhere else the variables change according to their very own fixed differential equations that the original hybrid system comes with. But differential ghosts are different. We can make them change any way we want! We just need to make up our mind and choose cleverly to make our favorite proofs work out. How would we need y to change?

Before you read on, see if you can find the answer for yourself.

It may sound like a big open question. But it is completely systematic how we need a differential ghost to evolve if we want to make a formula such as $xy^2 = 1$ an invariant. The formula will be an invariant if its differential $x'y^2 + x2yy' = 0$ is proved along the differential equation. The differential equation already tells us that x' is $-x$, but we have not settled on our favorite differential equation for y' yet. Of course, the resulting formula $-xy^2 + 2xyy' = 0$ is best made *true* by a choice of $\frac{y}{2}$ for y', which we find by simply solving $-xy^2 + 2xyy' = 0$ for y'. That calls for the differential ghost $y' = \frac{y}{2}$ with which we set out for a proof of (12.6) using the dynamics illustrated in Fig. 12.2:

$$
\cfrac{
  \cfrac{
    \cfrac{
      \cfrac{
        \cfrac{
          \cfrac{*}{\mathbb{R}\;\; \vdash -xy^2 + 2xy\frac{y}{2} = 0}
        }{[:=]\;\; \vdash [x':=-x][y':=\frac{y}{2}]x'y^2 + x2yy' = 0}
      }{\text{dI}\;\; x > 0, xy^2 = 1 \vdash [x' = -x, y' = \frac{y}{2}]xy^2 = 1}
    }{\exists \text{R,cut}\;\; x > 0 \vdash \exists y\,[x' = -x, y' = \frac{y}{2}]xy^2 = 1}
  }{\text{MR}\;\; x > 0 \vdash \exists y\,[x' = -x, y' = \frac{y}{2}]x > 0}
}{\text{DG}\;\; x > 0 \vdash [x' = -x]x > 0}
$$

(with the left branch $\mathbb{R}\;\; \overline{xy^2 = 1 \vdash x > 0}$ and $*$ above the $\exists$R,cut step)

The monotonicity step **MR** replaces the original postcondition $x > 0$ with the desired postcondition $xy^2 = 1$ that implies $x > 0$. Recall that $\exists y\, xy^2 = 1$ is equivalent to $x > 0$, which makes it particularly easy to see[3] why this monotonicity step also works in the context $\exists y$. From the desired equation $xy^2 = 1$ it would be easy to read

---

[3] Just like modalities and universal quantifiers, existential quantifiers also satisfy the monotonicity rule that $\exists y\, P \to \exists y\, Q$ can be derived from $P \to Q$.

off the concrete witness $\sqrt{\frac{1}{x}}$ for the existential quantifier instantiation step $\exists R$ since $x > 0$ initially. But since existence of such a y is all that matters for the proof, it is enough to prove $\exists y\, xy^2 = 1$ from the antecedent $x > 0$, which is straightforward real arithmetic. Things get worse in the original variable x but we hold it up still against the trend with the help of the additional differential ghost variable y and its counterweight dynamics. Nice, this proof with a creative differential ghost gave us a surprising but systematic proof for exponential decay formula (12.6).

Fig. 12.2 Differential ghost y as counterweight for expo-nential decay along $x' = -x$

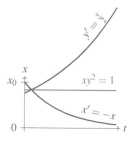

12.4 Differential Ghosts

Now that we have seen the benefit of dreaming up an entirely new differential equa-tion for a differential ghost for the sole purpose of doing a proof, it is about time to come back and see whether there are any limitations on the differential equations that can be added as differential ghosts for the purposes of a proof. What could possibly go wrong? Because the new variables are really new, adding new variables with new differential equations should not affect the original differential equations, because they cannot mention the differential ghosts (which would not otherwise be new).

The catch is that there is still a pretty subtle influence that additional differential equations can have on a preexisting differential equation system. If poorly chosen, then extra differential equations for differential ghosts can limit the duration of ex-istence of the solution of the joint differential equation system. It would not help to make a real system any safer if we were to compose it with a differential ghost that makes the imaginary world explode before the real system has a chance to run into an unsafe state. Blowing up the world does not make it any safer.

Example 12.2 (Nonexistent differential ghosts). It would be unsound to add a differ-ential ghost if its solution does not exist for at least as long as the original differen-tial equation system has a solution. Otherwise the following unsound proof attempt would reduce a conclusion that is not valid to a premise that is valid by adding a differential ghost $y' = y^2 + 1$ whose solution $y(t) = \tan t$ does not even exist for long enough to make $x \le 6$ *false* (as illustrated in Fig. 12.3):

$$\frac{\overset{\mathbb{R}}{\dfrac{x=0,y=0 \vdash [x'=1,y'=y^2+1]x\le 6}{x=0\vdash \exists y\,[x'=1,y'=y^2+1]x\le 6}}}{x=0\vdash [x'=1]x\le 6}$$

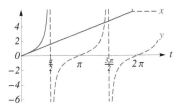

Fig. 12.3 Explosive differential ghosts that do not exist long enough would unsoundly limit the duration of solutions

When adding a differential ghost into a differential equation system it is, thus, crucial for soundness that the ghost differential equation has a solution that exists at least as long as the solutions of the rest of the differential equation system exist. The easiest way of making that happen is if the new differential equation added for the differential ghost y is linear in y, so of the form $y'=a(x)\cdot y+b(x)$. The terms $a(x)$ and $b(x)$ can be any terms of dL mentioning any number of variables any number of times, but they cannot mention y, because $y'=a(x)\cdot y+b(x)$ would not, then, be linear in y. This leads to the following (soundness-critical) correction for the formulation of the differential ghost axiom from (12.5) that we initially suspected in Sect. 12.3.4.

> **Lemma 12.2 (DG differential ghost axiom).** *The differential ghost axiom* DG *is sound:*
>
> $$\text{DG}\ [x'=f(x)\,\&\,Q]P \leftrightarrow \exists y\,[x'=f(x),y'=a(x)\cdot y+b(x)\,\&\,Q]P$$
>
> *where y is a new variable, not occurring in the left-hand side $[x'=f(x)\,\&\,Q]P$ or in $a(x)$ or in$b(x)$.*

Proof (Sketch). The proof, which is reported in full in prior work [8], uses a slight generalization of Corollary 2.1 from Sect. 2.9.2 to show that the new differential equation $y'=a(x)\cdot y+b(x)$ has a solution that exists at least as long as the solution of $x'=f(x)$ exists. The required Lipschitz condition follows from the fact that $a(x)$ and $b(x)$ in $y'=a(x)\cdot y+b(x)$ have continuous values over time and, thus, assume their maximum on the compact interval of existence of the solution of x. □

Axiom DG can be used to show that a property P holds after a differential equation if and only if it holds for some initial value y after an augmented differential equation with an extra $y'=a(x)\cdot y+b(x)$ that is linear in y so still has a solution that exists for sufficiently long. The case where $x'=f(x)$ is a (vectorial) differential equation system is similar, giving $y'=a(x)\cdot y+b(x)$ the opportunity to mention all variables other than the new y in $a(x)$ and $b(x)$.

A proof rule for differential ghosts derives from a direct use of axiom DG.

Lemma 12.3 (dG differential ghost rule). *This proof rule derives from DG:*

$$\text{dG} \quad \frac{\Gamma \vdash \exists y\,[x' = f(x), y' = a(x) \cdot y + b(x) \,\&\, Q]P, \Delta}{\Gamma \vdash [x' = f(x) \,\&\, Q]P, \Delta} \qquad (\textit{where } y \textit{ is new})$$

Proof. Proof rule dG is derived by a straightforward application of axiom DG. □

As illustrated in Example 12.1, it is almost always beneficial to subsequently re-place the postcondition P with a formula that makes use of the differential ghost y. The following rule dA was the first form [6] of differential ghosts and already bundles axiom DG up with others into a commonly useful form that adds a differential ghost while simultaneously replacing the postcondition to use the ghost.

Lemma 12.4 (dA differential auxiliaries rule). *The* differential auxiliaries *rule for introducing new auxiliary differential variables y derives from DG:*

$$\text{dA} \quad \frac{\vdash F \leftrightarrow \exists y\,G \qquad G \vdash [x' = f(x), y' = a(x) \cdot y + b(x) \,\&\, Q]G}{F \vdash [x' = f(x) \,\&\, Q]F}$$

Proof. Rule dA is derived from DG with transformations of the postcondition:

$$
\begin{array}{c}
\cfrac{
\cfrac{
\cfrac{\exists y\,G \vdash F}{G \vdash F}
\quad
\cfrac{F \vdash \exists y\,G \quad \Gamma, G \vdash [x' = f(x), y' = a(x) \cdot y + b(x)]G, \Delta}{\Gamma, F \vdash \exists y\,[x' = f(x), y' = a(x) \cdot y + b(x)]G, \Delta}\ \exists \text{R,cut}
}{
\Gamma, F \vdash \exists y\,[x' = f(x), y' = a(x) \cdot y + b(x)]F, \Delta
}\ \text{MR}
}{
\Gamma, F \vdash [x' = f(x)]F, \Delta
}\ \text{DG}
\end{array}
$$

□

By the right premise of rule dA, for any y, G is an invariant of the extended dynamics. Thus, G always holds after the evolution for some y (its value can be different than in the initial state), which still implies F by the left premise. Since y is fresh and its linear differential equation does not limit the duration of solutions of x on Q, this implies the conclusion. Since y is fresh, y does not occur in Q, and, thus, its solution does not leave Q, which would incorrectly restrict the duration of the evolution.

Intuitively, rule dA can help us to prove properties, because it may be easier to characterize how x changes in relation to an auxiliary differential ghost variable y with a suitable differential equation ($y' = a(x) \cdot y + b(x)$) compared to understanding the change of x in isolation. As usual, it would not be sound to keep the context Γ, Δ on the first premise of rule dA, because we have no reason to believe it would still hold after the differential equation, where we only know G (for some current value of y) according to the second premise but need to conclude that F also holds.

We conclude this section with a series of instructive examples that convey how differential ghosts are used as a powerful proof technique for differential equations.

In all examples the differential equation for the differential ghost is constructed entirely systematically as in Sect. 12.3.4 from the property that we want to be invariant.

Example 12.3 (Differential ghosts describe exponential growth). Exponential decay is not the only kind of dynamics that benefits from a differential ghost. Exponential growth is also proved just by flipping the sign of the differential ghost (Fig. 12.4).

$$
\mathrm{dA}\cfrac{\mathbb{R}\ \cfrac{}{\vdash x>0 \leftrightarrow \exists y\, xy^2 = 1}{}^{*} \qquad \mathrm{dI}\cfrac{[:=]\cfrac{\mathbb{R}\cfrac{}{\vdash xy^2 + 2xy(-\frac{y}{2}) = 0}{}^{*}}{\vdash [x':=x][y':=-\frac{y}{2}]x'y^2 + x2yy' = 0}}{xy^2 = 1 \vdash [x' = x, y' = -\frac{y}{2}]xy^2 = 1}}{x>0 \vdash [x' = x]x > 0}
$$

Fig. 12.4 Differential ghost y to balance exponential growth along $x' = x$

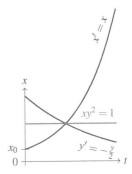

Example 12.4 (Exponential difference). It is equally easy to prove that x will never become zero along an exponential decay $x' = -x$. In this case, the condition on x and the additional differential ghost y that is equivalent to $x \neq 0$ is $\exists y\, xy = 1$, which requires the counterweight differential equation $y' = y$ to become invariant:

$$
\mathrm{dA}\cfrac{\mathbb{R}\ \cfrac{}{\vdash x>0 \leftrightarrow \exists y\, xy = 1}{}^{*} \qquad \mathrm{dI}\cfrac{[:=]\cfrac{\mathbb{R}\cfrac{}{\vdash -xy + xy = 0}{}^{*}}{\vdash [x':=-x][y':=y]x'y + xy' = 0}}{xy = 1 \vdash [x' = -x, y' = y]xy = 1}}{x \neq 0 \vdash [x' = -x]x \neq 0}
$$

Example 12.5 (Weak exponential decay). Proving that $x \geq 0$ is an invariant for exponential decay $x' = -x$ results in two cases, the case where $x = 0$ and the case where $x > 0$. Distinguishing between the two cases results in a successful proof. It is easier, however, to keep them together by rephrasing the desired invariant $x \geq 0$ with the equivalent $\exists y\,(y > 0 \land xy \geq 0)$ using a corresponding differential ghost $y' = y$:

$$
\dfrac{
  \dfrac{
    \dfrac{
      \dfrac{
        \dfrac{*}{\mathbb{R}\;\; \vdash -xy+xy \geq 0}
      }{[:=]\;\; \vdash [x':=-x][y':=y]x'y+xy' \geq 0}
    }{\mathrm{dl}\;\; \lhd\quad xy{\geq}0 \vdash [x'=-x,y'=y]xy{\geq}0}
  }{
    \dfrac{*}{\mathbb{R}\;\; \vdash x{\geq}0 \leftrightarrow \exists y(y{>}0\wedge xy{\geq}0)}\quad
    {[]\wedge}\;\; y{>}0\wedge xy{\geq}0 \vdash [x'=-x,y'=y](y{>}0\wedge xy{\geq}0)
  }
}{\mathrm{dA}\quad\quad x \geq 0 \vdash [x'=-x]x \geq 0}
$$

The $[]\wedge$ derived axiom leads to another premise (marked by $\lhd$), which proves with yet another differential ghost just like in Example 12.3, just carrying $x'=-x$ around:

$$
\dfrac{
  \dfrac{*}{\mathbb{R}\;\; \vdash y{>}0 \leftrightarrow \exists z\, yz^2 = 1}\quad
  \mathrm{dl}\;\;
  \dfrac{
    \dfrac{
      \dfrac{*}{\mathbb{R}\;\; \vdash yz^2 + 2yz(-\tfrac{z}{2}) = 0}
    }{[:=]\;\; \vdash [y':=y][z':=-\tfrac{z}{2}y]y'z^2 + y2zz' = 0}
  }{yz^2 = 1 \vdash [x'=-x,y'=y,z'=-\tfrac{z}{2}y]yz^2 = 1}
}{\mathrm{dA}\quad y > 0 \vdash [x'=-x,y'=y]y > 0}
$$

Example 12.6 (Exponential equilibrium). Proving that $x=0$ is an invariant for exponential decay $x'=-x$ succeeds by rephrasing it with the equivalent invariant $\exists y(y>0\wedge xy=0)$ and then following an analogue of Example 12.5. Alternatively, the splitting axiom $[]\wedge$ can directly reuse the proof of Example 12.5 to show that both $x \geq 0$ and $x \leq 0$ are invariants of $x'=-x$, which implies that their conjunction $x=0$ is invariant as well.

These examples are indicative of how proofs for other differential equations work. Systems with a different asymptotic behavior need a correspondingly shifted differential ghost.

Example 12.7 (Shifted exponentials). The following formula

$$x^3 > -1 \rightarrow [x'=-x-1]x^3 > -1$$

needs a differential ghost $y' = \tfrac{y}{2}$ with an inequality $(x+1)y^2 > 0$ instead of an equation in order for the second branch to be proved by $y^2 \geq 0$:

$$
\dfrac{
  \dfrac{*}{\mathbb{R}\;\; \vdash x^3{>}{-}1 \leftrightarrow \exists y(x{+}1)y^2{>}0}\quad
  \mathrm{dl}\;\;
  \dfrac{
    \dfrac{
      \dfrac{*}{\mathbb{R}\;\; \vdash -xy^2 + 2xy\tfrac{y}{2} + 2y\tfrac{y}{2} \geq 0}
    }{[:=]\;\; \vdash [x':=-x][y':=\tfrac{y}{2}]x'y^2 + (x{+}1)2yy' \geq 0}
  }{(x{+}1)y^2{>}0 \vdash [x'=-x,y'=\tfrac{y}{2}](x{+}1)y^2 > 0}
}{\mathrm{dA}\quad x^3 > -1 \vdash [x'=-x-1]x^3 > -1}
$$

Example 12.8 (Square resistance). The differential equations for differential ghosts may depend on previously existing variables, for example, when proving that $x > 0$ is an invariant along $x' = -x^2$ with the differential ghost $y' = \tfrac{x}{2}y$ (Fig. 12.5).

Fig. 12.5 Differential ghost y as counterweight for square resistance along $x' = -x^2$

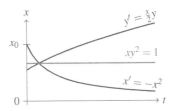

$$
\mathrm{dA} \dfrac{
\mathbb{R}\dfrac{*}{\vdash x > 0 \leftrightarrow \exists y\, xy^2 = 1}
\qquad
\mathrm{dI}\dfrac{
[:=]\dfrac{
\mathbb{R}\dfrac{*}{\vdash -x^2y^2 + 2xy(\frac{x}{2}y) = 0}
}{
\vdash [x':=-x^2][y':=\frac{x}{2}y]x'y^2 + x2yy' = 0
}
}{
xy^2 = 1 \vdash [x'=-x^2, y'=\frac{x}{2}y]\,xy^2 = 1
}
}{
x > 0 \vdash [x'=-x^2]\,x > 0
}
$$

Example 12.9 (Square activation). Proving that $x > 0$ is an invariant along $x' = x^2$ would also work with a differential ghost with a flipped sign. But that is unnecessarily difficult, because $x > 0$ is getting more true along $x' = x^2$ anyhow, so a direct differential invariant proof suffices.

$$
\mathrm{dI}\dfrac{
[:=]\dfrac{
\mathbb{R}\dfrac{*}{\vdash x^2 \ge 0}
}{
\vdash [x':=x^2]\,x' \ge 0
}
}{
x > 0 \vdash [x'=x^2]\,x > 0
}
$$

12.5 Substitute Ghosts

Ghosts even give us a, shockingly spooky, way of generating differential equations for differential ghosts on the fly as needed for proofs to work out. That might sound scary but is amazingly useful. To see how it works, invent your own differential ghost $y' = \bigcirc$ with a still-unspecified right-hand side $\bigcirc$, which is nothing but a *substitute ghost* or a common spooky cloud, and just keep "proving" as if nothing had happened:

$$
\mathrm{dA}\dfrac{
\mathbb{R}\dfrac{*}{\vdash x > 0 \leftrightarrow \exists y\, xy^2 = 1}
\qquad
\mathrm{dI}\dfrac{
[:=]\dfrac{
\dfrac{\text{could prove if } \bigcirc = \frac{y}{2}}{\vdash -xy^2 + 2xy\bigcirc = 0}
}{
\vdash [x':=-x][y':=\bigcirc]x'y^2 + x2yy' = 0
}
}{
xy^2 = 1 \vdash [x'=-x, y'=\bigcirc]\,xy^2 = 1
}
}{
x > 0 \vdash [x'=-x]\,x > 0
}
$$

The right premise could be proved if only ☁ were chosen to be $\frac{y}{2}$, in which case the premise $-xy^2 + 2xy$☁$= 0$ is quite easily proved. That, of course, was a bit too spooky for the soundness-loving truth-connoisseur. So let's instantiate the spooky cloud ☁ with its concrete choice $\frac{y}{2}$ and start all over with a proper proof:

$$
\dfrac{
\dfrac{
\;{}^{\mathbb{R}}\overline{\vdash x > 0 \leftrightarrow \exists y\, xy^2 = 1}\;
}{
{}_{\text{dA}}\;
}
\qquad
\dfrac{
{}^{\mathbb{R}}\dfrac{
\dfrac{*}{\vdash -xy^2 + 2xy\frac{y}{2} = 0}
}{
{}_{[:=]}\dfrac{\vdash [x':=-x][y':=\frac{y}{2}]x'y^2 + x2yy' = 0}{{}_{\text{dI}}\; xy^2 = 1 \vdash [x'=-x, y'=\frac{y}{2}]xy^2 = 1}
}
}{}
}{
x > 0 \vdash [x'=-x]\,x > 0
}
$$

Fortunately, this proper dL proof confirms the suspicion of a proof that we developed above. In that sense, all is fair in how we come up with a proof, even if we use spooky ghost arguments where ☁ is involved.[4] But in the end, it is crucial to conduct a proper proof with sound proof rules to ensure the conclusion is valid.

It can be shown [6] that there are properties such as this one that crucially need differential ghosts (alias differential auxiliaries) to be proved, which makes differential ghosts a powerful proof technique.

12.6 Limit Velocity of an Aerodynamic Ball

This section considers an insightful application of differential ghosts to prove asymptotic limit velocities. Safe position bounds were proved for the aerodynamic bouncing ball in Sect. 11.12: Unlike the original bouncing ball (4.6) from Sect. 4.2.1, the interesting twist is that its differential equation $x' = v, v' = -g + rv^2 \,\&\, x \geq 0 \wedge v \leq 0$ includes quadratic air resistance rv^2 working against the direction of motion. In preparation for the subsequent development, we assume positive aerodynamic resistance $r > 0$ instead of $r \geq 0$. The central argument in the proof of safety showed a velocity-dependent position bound by rule dI while the aerodynamic ball is falling:

$$
\dfrac{
{}^{\mathbb{R}}\dfrac{
\dfrac{
\dfrac{
\dfrac{*}{g > 0 \wedge r > 0, x \geq 0 \wedge v \leq 0 \vdash 2gv \leq 2gv - 2rv^3}
}{
g > 0 \wedge r > 0, x \geq 0 \wedge v \leq 0 \vdash 2gv \leq -2v(-g + rv^2)
}
}{
{}_{[:=]}\; g > 0 \wedge r > 0, x \geq 0 \wedge v \leq 0 \vdash [x':=v][v':=-g+rv^2]2gx' \leq -2vv'
}
}{
{}_{\text{dI}}\; g > 0 \wedge r > 0, 2gx \leq 2gH - v^2 \vdash [x'=v, v'=-g+rv^2 \,\&\, x \geq 0 \wedge v \leq 0]\,2gx \leq 2gH - v^2
}
}{}
}{}
$$

From every velocity bound (initially $v = 0$) the invariant $2gx \leq 2gH - v^2$ enables us to read off a corresponding position bound on the change of position x compared to fixed altitude H. But how fast could the aerodynamic bouncing ball be falling?

Before you read on, see if you can find the answer for yourself.

---

[4] Of course, ☁ is not quite as spooky as one might suspect. It can be made rigorous with function symbols that are subsequently substituted uniformly [8], as we discuss in Chap. 18.

Surely the original bouncing ball (4.6) from Sect. 4.2.1 could go arbitrarily fast if it is falling from a sufficiently high altitude, because its velocity permanently keeps increasing in absolute value along $x' = v, v' = -g \,\&\, x \geq 0$. But the aerodynamic ball is a different matter, because its aerodynamic resistance keeps increasing with the square of its velocity along $x' = v, v' = -g + rv^2 \,\&\, x \geq 0 \wedge v \leq 0$. Is there a maximal velocity for aerodynamic balls?

The higher the (absolute value of the) velocity v, the higher the air resistance rv^2. And indeed, the velocity v will not change anymore if the right-hand side of its differential equation has value 0:

$$v' = 0 \text{ iff } -g + rv^2 = 0 \text{ iff } v = \pm\sqrt{\frac{g}{r}}$$

Points where the right-hand side of the differential equation is 0 are called *equilibrium points*. In this case, of course, the position will still keep on changing, but the velocity is at an equilibrium when $v = \pm\sqrt{\frac{g}{r}}$. Recall how lucky we are for the division that the aerodynamic resistance coefficient r is not 0 (but then again, there would not otherwise be any limit to the velocity if r were 0). Consequently, we will prove that the (negative!) velocity is indeed always larger than the *limit velocity* $-\sqrt{g/r}$, so smaller than $\sqrt{g/r}$ in absolute value as shown in Fig. 12.6:

$$g > 0 \wedge r > 0 \wedge v > -\sqrt{\frac{g}{r}} \rightarrow [x' = v, v' = -g + rv^2 \,\&\, x \geq 0 \wedge v \leq 0] v > -\sqrt{\frac{g}{r}}$$

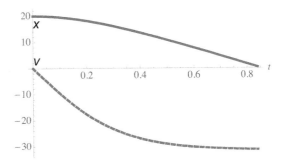

Fig. 12.6 Velocity of aerodynamic ball approaches limit velocity

Now the tricky bit is that, even though this dL formula is valid, it is not provable by differential invariants alone, because postcondition $v > -\sqrt{g/r}$ is getting less true over time. After all, the velocity keeps decreasing along $v' = -g + rv^2$ making $v > -\sqrt{g/r}$ less and less true as time progresses. The point is that, even if $v > -\sqrt{g/r}$ is getting less true, because the signed difference $v + \sqrt{g/r}$ between the two sides keeps decreasing forever, the rate at which the difference is decreasing is simultaneously shrinking, too. Indeed, $-\sqrt{g/r}$ is the limit velocity of the aerodynamic ball, because the ball will converge to velocity $-\sqrt{g/r}$ in the limit as time converges to infinity $t \to \infty$, but will never truly reach this asymptotic limit velocity in finite time. Differential cuts do not exactly help, because they either help establish a limit velocity but then also become less true over time so are not provable by differential invariants, or they are differential invariants but then do not provide asymptotic velocity bounds.

This calls for the help of a differential ghost to serve as counterweight to the change in value. What differential equation that differential ghost needs to balance the property is hard to predict. But the canonical property that the differential ghost y needs to satisfy to imply the postcondition $v > -\sqrt{g/r}$ is $\exists y\, y^2(v+\sqrt{g/r}) = 1$, because the two formulas are equivalent since $v > -\sqrt{g/r}$ iff $v+\sqrt{g/r} > 0$, which it has to be to obtain 1 when multiplied by some square $y^2 \geq 0$.

From this new postcondition $y^2(v+\sqrt{g/r}) = 1$ we can easily determine what the differential equation for the differential ghost y needs to be. All we need to do is first compute its differential like axiom DI would:

$$2yy'(v+\sqrt{g/r}) + y^2 v' = 0$$

Next, we substitute in the differential equation for v' like axiom DE would:

$$2yy'(v+\sqrt{g/r}) + y^2(-g+rv^2) = 0$$

Finally, we solve for y' to find out what differential equation makes it invariant:

$$y' = -{}^r\!/_2(v-\sqrt{g/r})y$$

This construction tells us all we need to complete the proof. The proof assumes fixed parameters $g > 0 \wedge r > 0$ everywhere to make $\sqrt{g/r}$ well-defined. The elided premise $v > -\sqrt{g/r} \leftrightarrow \exists y\, y^2(v+\sqrt{g/r})=1$ of dA (marked ◁) is proved by arithmetic.

$$
\begin{array}{ll}
\mathbb{R} & \dfrac{\quad * \quad}{\vdash -ry^2(v^2 - g/r) + y^2(-g+rv^2) = 0} \\[2ex]
& \overline{\vdash 2y(-{}^r\!/_2(v-\sqrt{g/r})y)(v+\sqrt{g/r}) + y^2(-g+rv^2) = 0} \\[2ex]
{[:=]} & \overline{\vdash [x':=v][v':=-g+rv^2][y':=-{}^r\!/_2(v-\sqrt{g/r})y]2yy'(v+\sqrt{g/r}) + y^2v' = 0} \\[2ex]
\text{dI} & \overline{◁\, y^2(v+\sqrt{g/r}) = 1 \vdash [x'=v,v'=-g+rv^2,y'=-{}^r\!/_2(v-\sqrt{g/r})y]\,y^2(v+\sqrt{g/r}) = 1} \\[2ex]
\text{dA} & \overline{v > -\sqrt{g/r} \vdash [x'=v,v'=-g+rv^2]v > -\sqrt{g/r}}
\end{array}
$$

Proposition 12.1 (Aerodynamic velocity limits). *This* dL *formula is valid:*

$$g > 0 \wedge r > 0 \wedge v > -\sqrt{\frac{g}{r}} \rightarrow [x'=v, v'=-g+rv^2 \,\&\, x \geq 0 \wedge v \leq 0]\, v > -\sqrt{\frac{g}{r}}$$

A similar construction always makes it possible to construct suitable differential ghosts [5, 6], but it is crucial for the proof that their solutions exist for long enough, which they do in the case of linear differential equations.

12.7 Axiomatic Ghosts

This section is devoted to yet another kind of ghosts: *axiomatic ghosts*. While irrelevant for simple systems, axiomatic ghosts are the way to go for systems that involve special functions such as sin, cos, tan, etc.

At a coordination level, the planar in-flight dynamics of an aircraft at x can be described by the following differential equation system [12]:

$$ x_1' = v\cos\vartheta \qquad\qquad x_2' = v\sin\vartheta \qquad\qquad \vartheta' = \omega \qquad (12.7) $$

That is, the linear velocity v of the aircraft changes both planar position coordinates x_1 and x_2 in the (planar) direction corresponding to the aircraft's current orientation ϑ. During curved flight, the angular velocity ω of the aircraft simultaneously changes the orientation ϑ of the aircraft (Fig. 12.7).

Fig. 12.7 Dubins aircraft
dynamics

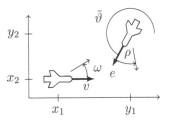

Unlike for straight-line flight with $\omega = 0$, the nonlinear dynamics in (12.7) is difficult to analyze [12] for curved flight with angular velocity $\omega \neq 0$. Solving (12.7) requires the Floquet theory of differential equations with periodic coefficients [13, Theorem 18.X] and yields mixed polynomial expressions with multiple trigonometric functions. Even more challenging is the verification of properties of the states that the aircraft reach by following these solutions, which requires proving that complicated formulas with mixed polynomial arithmetic and trigonometric functions hold true for all values of state variables and all possible evolution durations. However, quantified arithmetic with trigonometric functions is undecidable by Gödel's incompleteness theorem [2, 11].

To obtain polynomial dynamics, we *differentially axiomatize* [3] the trigonometric functions in the dynamics and reparametrize the state correspondingly. Instead of angular orientation ϑ and linear velocity v, we use the linear speed vector

$$ (v_1, v_2) \overset{\text{def}}{=} (v\cos\vartheta, v\sin\vartheta) \in \mathbb{R}^2 $$

which describes both the linear speed $\sqrt{v_1^2 + v_2^2} = v$ and the orientation of the aircraft in space; see Fig. 12.8. Substituting this coordinate change into differential equations (12.7) immediately yields $x_1' = v_1$ and $x_2' = v_2$. What differential equations the new axiomatic ghost variables v_1, v_2 obeys, is found by simple symbolic differentiation and substituting in differential equation system (12.7):

Fig. 12.8 Reparametrize for differential axiomatization

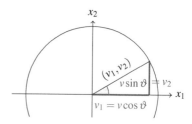

$$v_1' = (v\cos\vartheta)' = v'\cos\vartheta + v(-\sin\vartheta)\vartheta' = -(v\sin\vartheta)\omega = -\omega v_2$$
$$v_2' = (v\sin\vartheta)' = v'\sin\vartheta + v(\cos\vartheta)\vartheta' \quad = \quad (v\cos\vartheta)\omega = \quad \omega v_1$$

The middle equality holds when assuming constant linear velocity ($v' = 0$). Hence, equations (12.7) can be rephrased as the following differential equation:

$$x_1' = v_1 , x_2' = v_2 , v_1' = -\omega v_2 , v_2' = \omega v_1 \tag{12.8}$$
$$y_1' = e_1 , y_2' = u_2 , u_1' = -\rho u_2 , u_2' = \rho u_1 \tag{12.9}$$

Differential equation (12.8) expresses that position $x = (x_1, x_2)$ changes according to the linear speed vector (v_1, v_2), which in turn rotates according to ω. Simultaneous movement together with a second aircraft at $y \in \mathbb{R}^2$ having linear speed $(u_1, u_2) \in \mathbb{R}^2$ (also indicated with angle $\bar{\vartheta}$ in Fig. 12.7) and angular velocity ρ corresponds to the additional differential equation system (12.9). Differential equations capture simultaneous dynamics of multiple traffic agents succinctly using conjunction.

This *differential axiomatization* obtains polynomial differential equations. Their solutions still involve the same complicated nonlinear trigonometric expressions, so solutions still give undecidable arithmetic. But differential-invariant-type arguments work with the differential equations themselves and not with their solutions, so differential axiomatization actually helps when proving properties, because the solutions are still as complicated as they have always been, but the differential equations have become easier. The same technique helps when handling other special functions in other cases by differential axiomatization: introduce new ghost variables for the special functions and determine their differential equations by symbolic differentiation and substitution of the old differential equations.

12.8 Summary

The major lesson from this chapter is that it can sometimes be easier to relate a variable to its initial value or to other quantities than to understand its value in isolation. Ghosts, in their various forms, let us achieve that by adding auxiliary variables into the system dynamics, so that the values of the original variables of interest can be related to the values of the ghosts. Sometimes such ghosts are necessary to prove properties. Differential ghosts are especially useful for asymptotic properties or for proving properties whose trend alone makes them less true over time. The phenomenon that relations between state and ghost variables are sometimes easier to prove than just standalone properties of state variables applies frequently. This chapter shines a light on the power of relativity theory of differential equations in the sense of relating variables to one another. The ghost axioms and proof rules to introduce either discrete or continuous auxiliary variables are summarized in Fig. 12.9.

$$\text{DG} \quad [x' = f(x) \,\&\, Q]P \leftrightarrow \exists y\,[x' = f(x), y' = a(x) \cdot y + b(x) \,\&\, Q]P$$

$$\text{dG} \quad \frac{\Gamma \vdash \exists y\,[x' = f(x), y' = a(x) \cdot y + b(x) \,\&\, Q]P, \Delta}{\Gamma \vdash [x' = f(x) \,\&\, Q]P, \Delta}$$

$$\text{dA} \quad \frac{\vdash F \leftrightarrow \exists y\,G \quad G \vdash [x' = f(x), y' = a(x) \cdot y + b(x) \,\&\, Q]G}{F \vdash [x' = f(x) \,\&\, Q]F}$$

$$\text{iG} \quad \frac{\Gamma \vdash [y := e]p, \Delta}{\Gamma \vdash p, \Delta} \qquad\qquad (y \text{ new})$$

Fig. 12.9 Axioms and proof rules for ghosts and differential ghosts where y is new

This chapter also showcased a number of other useful proof techniques and even showed how properties of differential equations can be proved using solution-like arguments if only part of the differential equation system can be solved.

12.9 Appendix

This appendix provides a few additional cases where different kinds of ghost variables occur, e.g., for arithmetical purposes such as division or roots.

12.9.1 Arithmetic Ghosts

The easiest way to see why it sometimes makes sense to add variables into a system model is to take a look at divisions. Divisions are not officially part of real arithmetic, because divisions are definable indirectly. The point is that subtractions $b - c$ are definable as the term $b + (-1) \cdot c$ but divisions need a whole formula to be definable. For example, when a division b/c is ever mentioned in a term, then we can characterize a new variable q that remembers the value of b/c by indirectly characterizing q in terms of b and c by multiplication without $/$ and then subsequently using q wherever b/c first occurred:

$$q := \frac{b}{c} \quad \rightsquigarrow \quad q := *;\ ?qc = b \quad \rightsquigarrow \quad q := *;\ ?qc = b \wedge c \neq 0$$

where $q := *$ is the *nondeterministic assignment* that assigns an arbitrary real number to q. The first transformation (written $\rightsquigarrow$) characterizes $q = b/c$ indirectly by multiplying up c as $qc = b$. The second transformation then conscientiously remembers that divisions only make sense when we avoid dividing by zero. After all, divisions by zero excel at causing a lot of trouble. Divisions by zero won't stop for anything when causing trouble, not even when something that is as important and impactful as cyber-physical systems are concerned. The above transformation can be used when b/c occurs in the middle of a term, too:

$$x := 2 + \frac{b}{c} + e \rightsquigarrow q := *; \ ?qc = b; \ x := 2 + q + e \rightsquigarrow q := *; \ ?qc = b \wedge c \neq 0; \ x := 2 + q + e$$

Here q is called an *arithmetic ghost*, because q is an auxiliary variable that is only added to the program for the sake of defining the arithmetic quotient $\frac{b}{c}$. In similar ways we can define other functions such as square roots using an arithmetic ghost:

$$x := a + \sqrt{4y} \quad \rightsquigarrow \quad q := *; \ ?q^2 = 4y; \ x := a + q$$

But we should again scrutinize to make sure we realize that $4y$ should be nonnegative for the square root to make sense, and we could indeed add that into the test. We settle on not doing so, since non-negativity already follows from $q^2 = 4y$. Systematic transformations of divisions and square roots will also be considered in Chap. 20.

12.9.2 Nondeterministic Assignments & Ghosts of Choice

The HP statement $x := *$ that has been used in Sect. 12.9.1 is a *nondeterministic assignment* that assigns an arbitrary real number to x. Comparing with the syntax of hybrid programs from Chap. 3, however, it turns out that such a statement is not in the official language of hybrid programs:

$$\alpha, \beta \ ::= \ x := e \mid ?Q \mid x' = f(x) \& Q \mid \alpha \cup \beta \mid \alpha; \beta \mid \alpha^* \qquad (12.10)$$

What now?

One possible solution, which is the one taken in the implementation of the hybrid systems theorem prover KeYmaera [10] and its successor KeYmaera X [1], is to solve Exercise 5.13 and add the nondeterministic assignment $x := *$ as a statement to the syntax of hybrid programs.

$$\alpha, \beta \ ::= \ x := e \mid ?Q \mid x' = f(x) \& Q \mid \alpha \cup \beta \mid \alpha; \beta \mid \alpha^* \mid x := *$$

Consequently, the new syntactic construct of nondeterministic assignments needs a semantics to become meaningful:

7. $[\![x := *]\!] = \{(\omega, \nu) \ : \ \nu = \omega \text{ except for the value of } x, \text{ which can be any real}\}$

Nondeterministic assignments also need axioms or proof rules so that they can be understood and analyzed in proofs (Exercise 5.13). Those are reported in Fig. 12.10.

$\langle:*\rangle$ $\langle x:=*\rangle P \leftrightarrow \exists x P$

$[:*]$ $[x:=*]P \leftrightarrow \forall x P$

Fig. 12.10 Axioms for nondeterministic assignments

Axiom $\langle:*\rangle$ says that there is one way of assigning an arbitrary value to x so that P holds afterwards (i.e., $\langle x:=*\rangle P$ holds) if and only if P holds for some value of x (i.e., $\exists x P$ holds). And axiom $[:*]$ says that P holds for all ways of assigning an arbitrary value to x (i.e., $[x:=*]P$ holds) if and only if P holds for all values of x (i.e., $\forall x P$ holds), because x might have any such value after running $x:=*$, and because the $[\alpha]$ means that the postcondition needs to be true after all ways of running α.

An alternative approach for adding nondeterministic assignments $x:=*$ to hybrid programs is to reconsider whether we even have to add a new construct for $x:=*$ or whether it can be expressed in other ways. That is, to understand whether $x:=*$ is truly a new program construct or whether it can be defined in terms of the other hybrid program statements from (12.10). Is $x:=*$ definable by another hybrid program?

Before you read on, see if you can find the answer for yourself.

According to the proof rules $[:*]$ and $\langle:*\rangle$, nondeterministic assignments $x:=*$ can be expressed equivalently by suitable quantifiers. But that does not help at all in the middle of a program, where we can hardly write down a quantifier to express that the value of x now changes.

There is another way, though. Nondeterministic assignment $x:=*$ assigns any real number to x. One hybrid program that has essentially the same effect of giving x any arbitrary real value [4, Chapter 3] is

$$x:=* \quad \stackrel{\text{def}}{\equiv} \quad x' = 1 \cup x' = -1 \tag{12.11}$$

That is not the only definition of $x:=*$, though. An equivalent definition is [7]:

$$x:=* \quad \stackrel{\text{def}}{\equiv} \quad x' = 1; x' = -1$$

When working through the intended semantics of the left-hand side $x:=*$ shown in Case 7 above and the actual semantics of the right-hand side of (12.11) according to Chap. 3, it becomes clear that both sides of (12.11) have the same effect.[5]

---

[5] Observe a subtlety that, unlike the nondeterministic assignment, the differential equations also have an impact on the value of x', which is fine since most programs do not read x' any further, but needs extra care with an additional discrete ghost z otherwise: $z:=x'; \{x' = 1 \cup x' = -1\}; x':=z$

Hence, the above definition (12.11) captures the intended concept of giving x any arbitrary real value, nondeterministically. And, in particular, just like if-then-else, nondeterministic assignments do not really have to be added to the language of hybrid programs, because they can already be defined. Likewise, no proof rules have to be added for nondeterministic assignments, because there are already proof rules for the constructs used in the right-hand side of the definition of $x:=*$ in (12.11). Since the above proof rules $\langle :* \rangle, [:*]$ for $x:=*$ are particularly easy, though, it is usually more efficient to include them directly, which is what KeYmaera X does.

What may, at first sight, appear slightly spooky about (12.11), however, is that the left-hand side $x:=*$ is clearly an instant change in time where x changes its value instantaneously to some arbitrary new real number. That is not so for the right-hand side of (12.11), which involves two differential equations, which take time to follow.

The clue is that this passage of time is not observable in the state of the system. Consequently, the left-hand side of (12.11) really means the same as the right-hand side of (12.11). Remember from earlier chapters that time is not special. If a CPS wanted to refer to time, it would have a clock variable t with the differential equation $t' = 1$. With such an addition, however, the passage of time t would become observable in the value of variable t and, hence, a corresponding variation of the right-hand side of (12.11) would not be equivalent to $x:=*$ at all (indicated by $\neq$):

$$x:=* \quad \neq \quad \{x' = 1, t' = 1\} \cup \{x' = -1, t' = 1\}$$

The two sides differ, because the right side exposes the amount of time t it took to get the value of x to where it should be, which, secretly, records information about the absolute value of the change that x underwent from its old to its new value. That change is something that the left-hand side $x:=*$ knows nothing about.

12.9.3 Differential-Algebraic Ghosts

The transformation in Sect. 12.9.1 can eliminate all divisions, not just in assignments, but also in tests and all other hybrid programs, with the notable exception of differential equations. Eliminating divisions in differential equations turns out to be a little more involved.

The following elimination using a (discrete) arithmetic ghost q is correct:

$$x' = \frac{2x}{c} \& c \neq 0 \wedge \frac{x+1}{c} > 0 \quad \rightsquigarrow \quad q:=*; ?qc = 1; \{x' = 2xq \& c \neq 0 \wedge (x+1)q > 0\}$$

where the extra ghost variable q is supposed to remember the value of $\frac{1}{c}$.

The following attempt with a (discrete) arithmetic ghost q, however, would change the semantics rather radically:

$$x' = \frac{c}{2x} \& 2x \neq 0 \wedge \frac{c}{2x} > 0 \quad \rightsquigarrow \quad q:=*; ?q2x = 1; \{x' = cq \& 2x \neq 0 \wedge cq > 0\}$$

because q then only remembers the inverse of the *initial* value of $2x$, not the inverse of the value of $2x$ as x evolves along the differential equation $x' = \frac{c}{2x}$. That is q has a constant value during the differential equation but, of course, the quotient $\frac{c}{2x}$ changes over time as x does.

One way to proceed is to figure out how the value of the quotient $q = \frac{1}{2x}$ changes over time as x changes by $x' = \frac{c}{2x}$. By deriving what q stands for, that results in

$$q' = \left(\frac{1}{2x}\right)' = \frac{-2x'}{4x^2} = \frac{-2\frac{c}{2x}}{4x^2} = -\frac{c}{4x^3}$$

Alas, we are unlucky here, because that gives yet another division to keep track of.

The other and entirely systematic way to proceed is to lift nondeterministic assignments q to differential equations $q' = *$ with the intended semantics that q changes arbitrarily over time while following that nondeterministic differential equation:[6]

$$q' = \frac{b}{c} \quad \rightsquigarrow \quad q' = *\,\&\,qc = b \quad \rightsquigarrow \quad q' = *\,\&\,qc = b \wedge c \neq 0$$

While it is more complicated to give a semantics to $q' = *$, the idea behind the transformation is completely analogous to the case of discrete arithmetic ghosts:

$$x' = 2 + \frac{b}{c} + e \quad \rightsquigarrow \quad x' = 2 + q + e, q' = *\,\&\,qc = b$$
$$\rightsquigarrow \quad x' = 2 + q + e, q' = *\,\&\,qc = b \wedge c \neq 0$$

Variable q is a *differential-algebraic ghost* in the sense of being an auxiliary variable in the differential-algebraic equation for the sake of defining the quotient $\frac{b}{c}$.

Together with the reduction of divisions in discrete assignments from Sect. 12.9.1, along with the insight that divisions in tests and evolution domain constraints can always be rewritten to division-free form, this sketches a reduction showing that hybrid programs and differential dynamic logic do not need divisions [4]. The advantage of eliminating divisions this way is that differential dynamic logic does not need special precautions for divisions and that the handling of zero divisors is made explicit in the way the divisions are eliminated from the formulas. In practice, however, divisions are useful, yet great care has to be exercised to make sure that no inadvertent divisions by zero could ever cause troublesome singularities.

---

[6] The precise meaning of the nondeterministic differential equation $q' = *$ is reported elsewhere [4, Chapter 3]. It is the same as the differential-algebraic constraint $\exists d\, q' = d$, but differential-algebraic constraints have not been introduced in this textbook, either. Differential games also provide an elegant understanding [9]. The intuition of allowing arbitrary changes of the value of q over time is fine, though, for our purposes.

Note 66 (Divisions)

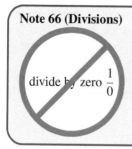

divide by zero $\frac{1}{0}$

Whenever dividing, exercise great care not to accidentally divide by zero, for that will cause quite some trouble. More often than not, this trouble corresponds to missing requirements in the system. For example $\frac{v^2}{2b}$ may be a good stopping distance when braking to a stop from initial velocity v, except when $b = 0$, which corresponds to having no brakes at all.

Exercises

12.1 (Conditions for discrete ghosts). Identify a minimal set of conditions necessary for proof rule iG from Sect. 12.3.1 to be sound. Show a counterexample for each of the remaining conditions to illustrate why it is necessary.

12.2. Augment the discrete ghost proofs in Sect. 12.3.1 to a full sequent proof of

$$xy - 1 = 0 \rightarrow [x' = x, y' = -y]xy = 1$$

12.3. Augment the proofs in Sect. 12.3.2 as described to obtain a full sequent proof of (11.6). Be advised to find a big sheet of paper, first.

12.4. For each of the following formulas provide a differential ghost proof:

$$x < 0 \rightarrow [x' = -x]x < 0$$
$$4x > -4 \rightarrow [x' = -x - 1]4x > -4$$
$$x > 0 \rightarrow [x' = -5x]x > 0$$
$$x > 2 \rightarrow [x' = -x + 2]4x > 2$$
$$x > 1 \rightarrow [x' = x + 1]x > 1$$
$$x > 4 \rightarrow [x' = x]x > 4$$
$$x^5 > 0 \rightarrow [x' = -2x]x^5 > 0$$
$$x > 0 \rightarrow [x' = x^2]x > 0$$
$$x > 0 \rightarrow [x' = -x^4]x > 0$$
$$x > 0 \rightarrow [x' = -x^5]x > 0$$

12.5 (Parachute). While traveling on a plane, your robot at height x with vertical velocity v finds a parachute that it can open ($r := p$ with parachute resistance p) or keep closed ($r = a$ with air resistance a). Of course, once deployed the parachute stays open. Your job is to fill in the blanks of a parachute controller with a test condition and a precondition that makes sure the robot opens the parachute early enough to land with bounded velocity.

$$g > 0 \wedge p > r = a > 0 \wedge x \geq 0 \wedge v < 0 \wedge \underline{\qquad\qquad} \rightarrow$$
$$\big[\big((?r = a \wedge \underline{\qquad\qquad} \cup r := p);$$
$$t := 0; \{x' = v, v' = -g + rv^2, t' = 1 \, \& \, t \leq \varepsilon \wedge x \geq 0 \wedge v < 0\}\big)^*$$
$$\big] (x = 0 \rightarrow v \geq m)$$

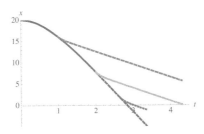

References

[1] Nathan Fulton, Stefan Mitsch, Jan-David Quesel, Marcus Völp, and André Platzer. KeYmaera X: an axiomatic tactical theorem prover for hybrid systems. In: *CADE*. Ed. by Amy Felty and Aart Middeldorp. Vol. 9195. LNCS. Berlin: Springer, 2015, 527–538. DOI: 10.1007/978-3-319-21401-6_36.

[2] Kurt Gödel. Über formal unentscheidbare Sätze der Principia Mathematica und verwandter Systeme I. *Monatshefte Math. Phys.* **38**(1) (1931), 173–198. DOI: 10.1007/BF01700692.

[3] André Platzer. Differential-algebraic dynamic logic for differential-algebraic programs. *J. Log. Comput.* **20**(1) (2010), 309–352. DOI: 10.1093/logcom/exn070.

[4] André Platzer. *Logical Analysis of Hybrid Systems: Proving Theorems for Complex Dynamics.* Heidelberg: Springer, 2010. DOI: 10.1007/978-3-642-14509-4.

[5] André Platzer. A differential operator approach to equational differential invariants. In: *ITP*. Ed. by Lennart Beringer and Amy Felty. Vol. 7406. LNCS. Berlin: Springer, 2012, 28–48. DOI: 10.1007/978-3-642-32347-8_3.

[6] André Platzer. The structure of differential invariants and differential cut elimination. *Log. Meth. Comput. Sci.* **8**(4:16) (2012), 1–38. DOI: 10.2168/LMCS-8(4:16)2012.

[7] André Platzer. Differential game logic. *ACM Trans. Comput. Log.* **17**(1) (2015), 1:1–1:51. DOI: 10.1145/2817824.

[8] André Platzer. A complete uniform substitution calculus for differential dynamic logic. *J. Autom. Reas.* **59**(2) (2017), 219–265. DOI: 10.1007/s10817-016-9385-1.

[9]   André Platzer. Differential hybrid games. *ACM Trans. Comput. Log.* **18**(3) (2017), 19:1–19:44. DOI: 10.1145/3091123.

[10] André Platzer and Jan-David Quesel. KeYmaera: a hybrid theorem prover for hybrid systems. In: *IJCAR*. Ed. by Alessandro Armando, Peter Baumgartner, and Gilles Dowek. Vol. 5195. LNCS. Berlin: Springer, 2008, 171–178. DOI: 10.1007/978-3-540-71070-7_15.

[11] Daniel Richardson. Some undecidable problems involving elementary functions of a real variable. *J. Symb. Log.* **33**(4) (1968), 514–520. DOI: 10.2307/2271358.

[12] Claire Tomlin, George J. Pappas, and Shankar Sastry. Conflict resolution for air traffic management: a study in multi-agent hybrid systems. *IEEE T. Automat. Contr.* **43**(4) (1998), 509–521. DOI: 10.1109/9.664154.

[13] Wolfgang Walter. *Ordinary Differential Equations*. Berlin: Springer, 1998. DOI: 10.1007/978-1-4612-0601-9.

Chapter 13
Differential Invariants & Proof Theory

Synopsis This advanced chapter studies some meta-properties of differential equations proving. It investigates aspects of the proof theory of differential equations, i.e., the theory of proofs about differential equations. While the primary focus in this chapter is on their theoretical significance, it also provides insights into the practical questions of what types of differential invariants to search for under which circumstances. The primary tool is the proof-theoretical device of relative deductive power, i.e., the question of whether all properties provable with technique $\mathscr{A}$ are also provable with technique $\mathscr{B}$. These results leverage appropriate insights about properties of real arithmetic and of differential equations.

13.1 Introduction

Chapters 10 and 11 equipped us with powerful tools for proving properties of differential equations without having to solve them. *Differential invariants* (dI) [10, 16] prove properties of differential equations by induction based on the right-hand side of the differential equation, rather than its much more complicated global solution. *Differential cuts* (dC) [10, 16] make it possible to prove another property C of a differential equation and then change the evolution domain of the dynamics of the system so that it is restricted to never leave that region C. Differential cuts turn out to be very useful when stacking inductive properties of differential equations on top of each other, so that easier properties are proved first and then assumed during the proof of the more complicated properties. In fact, in some cases, differential cuts are crucial for proving properties in the first place [5, 10, 14]. *Differential weakening* (dW) [10] proves simple properties that are entailed directly by the evolution domain, which becomes especially useful after the evolution domain constraint has been augmented sufficiently by way of a differential cut. *Differential ghosts* (dG) can prove properties by changing the dynamics of the system when adding a new differential equation for a new variable that was not there before. Differential ghosts are useful to, e.g., prove properties of systems with changing energy, where it helps

© Springer International Publishing AG, part of Springer Nature 2018
A. Platzer, *Logical Foundations of Cyber-Physical Systems*,
https://doi.org/10.1007/978-3-319-63588-0_13

397

to relate the change of state in the original system to auxiliary quantities that merely reflect a mathematical value for the sake of the argument, even if it is not part of the original system. In some cases, differential ghosts are crucial for proving properties, because they cannot be proved without them [14].

Just as in the case of loops, where the search for invariants is nontrivial, finding differential invariants also requires considerable smarts (or good automatic procedures [4, 7, 12, 17]). Once a differential invariant has been identified, however, the proof follows easily, which is a computationally attractive property.

Finding invariants of loops is very challenging. It can be shown to be the only fundamental challenge in proving safety properties of conventional discrete programs [8]. Likewise, finding invariants and differential invariants is the only fundamental challenge in proving safety properties of hybrid systems [9, 11, 13, 15]. A more delicate analysis even shows that just finding differential invariants is the only fundamental challenge for hybrid systems safety verification [13].

That is reassuring, because, at least, we know that the proofs will work[1] as soon as we find the right differential invariants. But it also tells us that we can expect the search for differential invariants (and invariants) to be quite challenging, because cyber-physical systems are extremely challenging. But it is worth the trouble, because CPSs are so important. Fortunately, differential equations also enjoy many pleasant properties that we can exploit to help us find differential invariants.

At the latest after this revelation, we fully realize the importance of studying and understanding differential invariants. So let us set out to develop a deeper understanding of differential invariants right away. The part of their understanding that this chapter develops is how various classes of differential invariants relate to each other in terms of what they can prove. There are properties that only differential invariants of the form $\mathscr{A}$ can prove, because differential invariants of the form $\mathscr{B}$ cannot ever succeed in proving them? Or are all properties provable by differential invariants of the form $\mathscr{A}$ also provable by differential invariants of the form $\mathscr{B}$?

These relations between classes of differential invariants tell us which forms of differential invariants we need to search for and which forms of differential invariants we don't need to bother considering. A secondary goal of this chapter besides this theoretical understanding is the practical development of better intuition about differential invariants and a more thorough appreciation of their effects. Some attention during the theoretical proofs will give us a generalizable understanding of which cases can or cannot be proved by which shape of differential invariants.

This chapter is based on prior work [14] and strikes a balance between comprehensive handling of the subject matter and core intuition. Many proofs in this chapter are simplified and only prove the core argument, while leaving out other aspects. Those—very important—further details of a comprehensive argument are beyond the scope of this textbook, however, and can be found elsewhere [14]. For example, this chapter will not study whether indirect proofs could conclude the same properties but will focus on the easier base case of direct proofs. With a more thorough analysis [14], it turns out that indirect proofs with the usual sequent calculus rules do

---

[1] Even if it may still be a lot of work to make the proofs work out in practice, at least they become possible, which is a good first step.

not change the results reported in this chapter, but the proofs become significantly more complicated and require a more precise choice of the sequent calculus formulation. In this chapter, we will also not always prove all statements conjectured in a theorem. The remaining proofs can be found in the literature [14].

> **Note 67 (Proof theory of differential equations)** The results in this chapter are part of the *proof theory* of differential equations, i.e., the theory of what can be proved about differential equations and with what techniques. They are proofs about proofs, because they prove relations between the provability of logical formulas with different proof calculi. That is, they relate the statements "formula P can be proved using $\mathscr{A}$" and "formula P can be proved using $\mathscr{B}$."

The most important learning goals of this chapter are:

Modeling and Control: This chapter helps in understanding the core argumentative principles behind CPS and sheds more light on the pragmatic question of how to tame their analytic complexity.

Computational Thinking: An important part of computer science studies questions about the *limits of computation* or, more generally, develops an understanding of *what can be done* and *what cannot be done*. Either in absolute terms (*computability theory* studies what is computable and what is not) or in relative terms (*complexity theory* studies what is computable in a characteristically quicker way or within classes of resource bounds on time and space). The answer is especially fundamental because it is independent of the model of computation, by the Church-Turing thesis [2, 20]. Often, the most significant understanding of a problem space starts with what cannot be done (the theorem of Rice [19] says that all nontrivial properties of programs are not computable) or what can be done (every problem that can be solved with a deterministic algorithm in polynomial time can also be solved with a nondeterministic algorithm in polynomial time, with the converse being the P versus NP [3] problem).

The primary purpose of this chapter is to develop such an understanding of the limits of what can and what cannot be done in the land of *proofs about differential equations*. Not all aspects of this deep question will be possible to answer in one chapter, but it will feature the beginning of the *proof theory of differential equations*, i.e., the theory of provability and proofs about differential equations. Proof theory is, of course, also of interest in other cases, but we will study it in the case that is most interesting and illuminating for cyber-physical systems: the case of proofs about differential equations.

The primary, scientific learning goals of this chapter are, thus, to develop a fundamental understanding of what can and cannot be proved in what ways about differential equations. This helps us in our search for differential invariants for applications, because such an understanding prevents us from asking the same analytic question again in equivalent ways (if two different classes of differential invariants prove the same properties and one of them already failed then there is no need to try the other) and guides our search toward the required

classes of differential invariants (by next choosing a class that can prove fundamentally more about properties of the requisite form).
The secondary, pragmatic learning goal is to practice inductive proofs about
differential equations using differential invariants and to develop an intuition for
which verification question to best address in which way. In these ways, both
fundamentally and pragmatically, the primary direct impact of this chapter is to
further our understanding of rigorous reasoning about CPS models as well as
how to verify CPS models of appropriate scale, in which more than one mode
of reasoning is often needed for the various parts and aspects of the system.
Finally this chapter has beneficial side effects informing differential invariant
search and deepening our intuition about differential equation proofs.

CPS Skills: This chapter serves no direct purpose in CPS Skills that the author
can think of, except indirectly via its impact on their analysis by informing
differential invariant search.

limits of computation
proof theory for differential equations
provability of differential equations
nonprovability of differential equations
proofs about proofs
relativity theory for proofs
inform differential invariant search
intuition for differential equation proofs

core argumentative principles   improved analysis
tame analytic complexity

13.2 Recap

Recall the following proof rules for differential equations from Chaps. 11 and 12:

Note 68 (Proof rules for differential equations)

$$\text{dI} \; \frac{Q \vdash [x':=f(x)](F)'}{F \vdash [x' = f(x) \,\&\, Q]F} \qquad \text{dW} \; \frac{Q \vdash P}{\Gamma \vdash [x' = f(x) \,\&\, Q]P, \Delta}$$

$$
\begin{array}{c}
\mathrm{dC}\ \dfrac{\Gamma \vdash [x' = f(x)\,\&\,Q]C, \Delta \quad \Gamma \vdash [x' = f(x)\,\&\,(Q \wedge C)]P, \Delta}{\Gamma \vdash [x' = f(x)\,\&\,Q]P, \Delta} \\[4mm]
\mathrm{dG}\ \dfrac{\Gamma \vdash \exists y\, [x' = f(x), y' = a(x)\cdot y + b(x)\,\&\,Q]P, \Delta}{\Gamma \vdash [x' = f(x)\,\&\,Q]P, \Delta}
\end{array}
$$

With cuts and generalizations, earlier chapters have already shown that the following can be proved:

$$
\text{cut,MR}\ \frac{A \vdash F \quad F \vdash [x' = f(x)\,\&\,Q]F \quad F \vdash B}{A \vdash [x' = f(x)\,\&\,Q]B} \tag{13.1}
$$

This proof step is useful for replacing a precondition A and a postcondition B with another invariant F that implies postcondition B (third premise) and is implied by precondition A (first premise) and is an invariant (second premise), which will now be done frequently in this chapter without further notice.

13.3 Comparative Deductive Study: Relativity Theory for Proofs

In order to find out what we can do when we have unsuccessfully searched for a differential invariant of one form, we need to understand what other form of differential invariants might work out better. If we have been looking for differential invariants of the form $e = 0$ with a term e without success and then move on to search for differential invariants of the form $e = k$, then we cannot expect to be any more successful than before, because $e = k$ can be rewritten as $e - k = 0$, which is of the first shape again. So we should, for example, try finding inequational differential invariants of the form $e \geq 0$, instead. In general, this begs the question of which generalizations would be silly (because differential invariants of the form $e = k$ cannot prove any more than those of the form $e = 0$) and which might be smart (because $e \geq 0$ might still succeed even if everything of the form $e = 0$ failed).

As a principled answer to questions like these, we study the relations of classes of differential invariants in terms of their relative deductive power. That is, we study whether some properties are only provable using differential invariants from the class $\mathscr{A}$, not using differential invariants from the class $\mathscr{B}$, or whether all properties provable with differential invariants from class $\mathscr{A}$ are also provable with class $\mathscr{B}$.

As a basis, we consider a propositional sequent calculus with logical cuts (which simplify glueing derivations together) and real arithmetic (denoted by proof rule $\mathbb{R}$) along the lines of what we saw in Chap. 6; see [14] for precise details. By $\mathscr{DI}$ we denote the proof calculus that, in addition, has general differential invariants (rule dI with arbitrary quantifier-free first-order formula F) but no differential cuts (rule dC) or differential ghosts (rule dG). For a set $\Omega \subseteq \{\geq, >, =, \wedge, \vee\}$ of operators, we denote by $\mathscr{DI}_\Omega$ the proof calculus where the differential invariant F in rule dI is further restricted to the set of formulas that uses only the operators in the set Ω. For

example, $\mathscr{DI}_{=,\wedge,\vee}$ is the proof calculus that allows only and/or-combinations of
equations to be used as differential invariants. Likewise, $\mathscr{DI}_{\geq}$ is the proof calculus
that only allows atomic weak inequalities $e \geq k$ to be used as differential invariants.

We consider classes of differential invariants and study their relations. If $\mathscr{A}$ and
$\mathscr{B}$ are two classes of differential invariants, we write $\mathscr{A} \leq \mathscr{B}$ if all properties provable using differential invariants from $\mathscr{A}$ are also provable using differential invariants from $\mathscr{B}$. We write $\mathscr{A} \not\leq \mathscr{B}$ otherwise, i.e., when there is a valid property that
can only be proven using differential invariants of $\mathscr{A} \setminus \mathscr{B}$. We write $\mathscr{A} \equiv \mathscr{B}$ for
equal deductive power if $\mathscr{A} \leq \mathscr{B}$ and $\mathscr{B} \leq \mathscr{A}$. We write $\mathscr{A} < \mathscr{B}$ for strictly more
deductive power of $\mathscr{B}$ if $\mathscr{A} \leq \mathscr{B}$ and $\mathscr{B} \not\leq \mathscr{A}$. Classes $\mathscr{A}$ and $\mathscr{B}$ are incomparable
if $\mathscr{A} \not\leq \mathscr{B}$ and $\mathscr{B} \not\leq \mathscr{A}$.

For example, the properties provable by differential invariants of the form $e = 0$
are the same as the properties provable by differential invariants of the form $e = k$.
That justifies $\mathscr{DI}_{=} \equiv \mathscr{DI}_{=0}$ where $\mathscr{DI}_{=0}$ denotes the class of properties provable with differential invariants of the form $e = 0$. Trivially, $\mathscr{DI}_{=} \leq \mathscr{DI}_{=,\wedge,\vee}$, because every property provable with differential invariants of the form $e = k$ is also
provable with differential invariants that additional are allowed to use conjunctions
and disjunctions. Likewise, $\mathscr{DI}_{\geq} \leq \mathscr{DI}_{\geq,\wedge,\vee}$. But the converses are not so clear,
because one might suspect that propositional connectives help.

13.4 Equivalences of Differential Invariants

Before we go any further, let us study whether there are straight out equivalence
transformations on formulas that preserve differential invariance. Every equivalence
transformation that we have for differential invariant properties helps us to structure
the proof search space and also helps to simplify the meta-proofs in the proof theory
of differential equations. For example, we should not expect $F \wedge G$ to be a differential invariant for proving a property when $G \wedge F$ was not. Neither would $F \vee G$ be
any better as a differential invariant than $G \vee F$.

> **Lemma 13.1 (Differential invariants and propositional logic).** *Differential
> invariants are invariant under propositional equivalences. That is, if $F \leftrightarrow G$
> is an instance of a propositional tautology then F is a differential invariant of
> $x' = f(x) \,\&\, Q$ if and only if G is.*

Proof. In order to prove this, we consider any property that is proved with F
as a differential invariant and show that the propositionally equivalent formula
G also works. Let F be a differential invariant of a differential equation system
$x' = f(x) \,\&\, Q$ and let G be a formula such that $F \leftrightarrow G$ is an instance of a propositional tautology. Then G is a differential invariant of $x' = f(x) \,\&\, Q$, because of the
following formal proof:

$$\frac{*}{F \vdash G} \quad \frac{\stackrel{[:=]}{Q \vdash [x':=f(x)](G)'}}{\stackrel{\text{dI}}{G \vdash [x' = f(x) \& Q]G}} \quad \frac{*}{G \vdash F}}{F \vdash [x' = f(x) \& Q]F}$$

cut,MR

The bottom proof step is easy to see using (13.1), which follows from rules cut and MR, because precondition F implies the new precondition G and postcondition F is implied by the new postcondition G propositionally. Subgoal $Q \vdash [x':=f(x)](G)'$ is provable, by the assumption that F is a differential invariant, so $Q \vdash [x':=f(x)](F)'$ provable. Note that $(G)'$ is ultimately a conjunction formed over the differentials of all atomic formulas of G. The set of atoms of G is identical to the set of atoms of F, because atoms are not changed by equivalence transformations with *propositional* tautologies. Furthermore, dL has a propositionally complete base calculus [14]. □

In all subsequent proofs, we can use propositional equivalence transformations by Lemma 13.1. In the following, we will also implicitly use equivalence reasoning for pre- and postconditions *à la* (13.1) as we have done in Lemma 13.1. Because of Lemma 13.1, we can, without loss of generality, work with arbitrary propositional normal forms for proof search.

13.5 Differential Invariants & Arithmetic

Depending on the reader's exposure to differential structures, it may come as a shock that not all logical equivalence transformations carry over to differential invariants. Differential invariance is not necessarily preserved under real-arithmetic equivalence transformations.

> **Lemma 13.2 (Differential invariants and arithmetic).** *Differential invariants are* not *invariant under equivalences of real arithmetic. That is, if $F \leftrightarrow G$ is an instance of a first-order real-arithmetic tautology, then F may be a differential invariant of $x' = f(x) \& Q$ yet G may not.*

Proof. There are two formulas that are equivalent in first-order real arithmetic but, for the same differential equation, one of them is a differential invariant by dI, the other one is not (because their differential structures differ). Since $5 \geq 0$, the formula $x^2 \leq 5^2$ is equivalent to $-5 \leq x \wedge x \leq 5$ in first-order real arithmetic. Nevertheless, $x^2 \leq 5^2$ is a differential invariant of $x' = -x$ by the following formal proof:

$$\frac{\stackrel{\mathbb{R}}{\overline{\vdash -2x^2 \leq 0}}}{\frac{\stackrel{[:=]}{\vdash [x':=-x]2xx' \leq 0}}{\stackrel{\text{dI}}{x^2 \leq 5^2 \vdash [x' = -x]x^2 \leq 5^2}}}$$

But the equivalent $-5 \leq x \wedge x \leq 5$ is not a differential invariant of $x' = -x$:

$$
{}^{\text{dl}} \dfrac{
{}^{[:=]} \dfrac{
\dfrac{\text{not valid}}{\vdash\, 0 \leq -x \wedge -x \leq 0}
}{
\vdash\, [x' := -x](0 \leq x' \wedge x' \leq 0)
}
}{
-5 \leq x \wedge x \leq 5 \vdash [x' = -x](-5 \leq x \wedge x \leq 5)
}
$$

$\square$

For proving the property in the proof of Lemma 13.2 we need to use (13.1) with the differential invariant $F \equiv x^2 \leq 5^2$ and cannot use $-5 \leq x \wedge x \leq 5$ directly. Both formulas are *true* for the exact same real values but their differential structure is different, because quadratic functions have different derivatives than linear functions.

By Lemma 13.2, we have to be explicit about using equivalences when investigating differential invariance, because some equivalence transformations affect whether a formula is a differential invariant. Not just the *elementary real-arithmetical equivalence* of having the same set of satisfying assignments matters, but also the differential structures that differential invariance depends on need to be compatible. Some equivalence transformations that preserve the set of solutions still destroy the differential structure. It is the equivalence of *real differential structures* that matters. Recall that differential structures are defined locally in terms of the behavior in neighborhoods of a point, not at the point itself.

Lemma 13.2 illustrates a notable point about differential equations. Many different formulas characterize the same set of satisfying assignments. But not all of them have the same differential structure. Quadratic polynomials have inherently different differential structure than linear polynomials even in cases where they happen to have the same set of solutions over the reals. The differential structure is finer-grained information. This is similar to the fact that two elementarily equivalent models of first-order logic can still be non-isomorphic. Both the set of satisfying assignments and the differential structure matter for differential invariance. In particular, there are many formulas with the same solutions but different differential structures. The formulas $x^2 \geq 0$ and $x^6 + x^4 - 16x^3 + 97x^2 - 252x + 262 \geq 0$ have the same solutions (all of $\mathbb{R}$), but very different differential structure; see Fig. 13.1.

The first two rows in Fig. 13.1 correspond to the polynomials from the two cases above. The third row is a structurally different degree 6 polynomial with again the same set of solutions ($\mathbb{R}$) but a rather different differential structure. Figure 13.1 illustrates that $(p)'$ can already have a very different characteristic even if the respective sets of satisfying assignments of $p \geq 0$ are identical.

We can, however, always normalize all atomic subformulas to have right-hand side 0, that is, of the form $p = 0, p \geq 0$, or $p > 0$. For instance, $p \leq q$ is a differential invariant if and only if $q - p \geq 0$ is, because $p \leq q$ is equivalent (in first-order real arithmetic) to $q - p \geq 0$. Moreover, for any variable x and term e, $[x' := e](p)' \leq (q)'$ is equivalent to $[x' := e](q)' - (p)' \geq 0$ in first-order real arithmetic, because the postcondition $(p)' \leq (q)'$ is equivalent to $(q)' - (p)' \geq 0$ in real arithmetic.

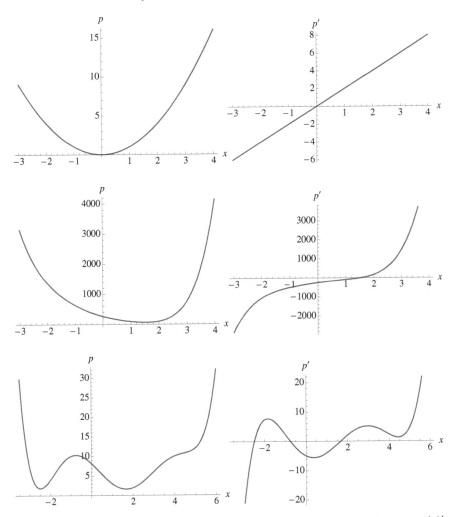

Fig. 13.1 Equivalent solutions ($p \geq 0$ on the left) with quite different differential structure $((p)'$ plotted on the right)

13.6 Differential Invariant Equations

Of course, we already know that $\mathscr{DI}_= \leq \mathscr{DI}_{=,\wedge,\vee}$ by definition, because every property provable without propositional logic in the differential invariants is also provable if we are allowed to use propositional logic. Indeed, for equational differential invariants $e = k$, alias differential invariant equations, propositional operators do not add to the deductive power [10, 14].

Proposition 13.1 (Equational deductive power). *The deductive power of differential induction with atomic equations is identical to the deductive power of differential induction with propositional combinations of polynomial equations. That is, each formula is provable with propositional combinations of equations as differential invariants iff it is provable with only atomic equations as differential invariants:*

$$\mathscr{DI}_= \equiv \mathscr{DI}_{=,\wedge,\vee}$$

How can we prove this positive statement about provability?

Before you read on, see if you can find the answer for yourself.

One direction is simple. Proving $\mathscr{DI}_= \leq \mathscr{DI}_{=,\wedge,\vee}$ is obvious, because every proof using a single differential invariant equation $e_1 = e_2$ also is a proof that is allowed to use a propositional combination of differential invariant equations: the propositional combination that just consists of the only conjunct $e_1 = e_2$ without making use of any of the propositional operators.

The other way around $\mathscr{DI}_= \geq \mathscr{DI}_{=,\wedge,\vee}$ is more difficult. If a formula can be proved using a differential invariant that is a propositional combination of equations, such as $e_1 = e_2 \wedge k_1 = k_2$, how can it possibly be proved using just a single equation?

Note 69 (Proofs of equal provability) A proof of Proposition 13.1 needs to show that every such provable property is also provable with a structurally simpler differential invariant. It effectively needs to transform proofs with propositional combinations of equations as differential invariants into proofs with just differential invariant equations. And, of course, the proof of Proposition 13.1 needs to prove that the resulting equations are provably differential invariants and still prove the same properties as before.

This is a general feature of proof theory. At the heart of the arguments, it often involves proof transformations. This explains why proof theory is a meta-theory conducting proofs about proofs: mathematical proofs about formal proofs.

Proof (of Proposition 13.1). Let $x' = f(x)$ be the (vectorial) differential equation to consider. We show that every differential invariant that is a propositional combination F of polynomial equations is expressible as a single atomic polynomial equation (the converse inclusion is obvious). We can assume F to be in negation normal form by Lemma 13.1 (recall that negations are resolved and $\neq$ can be assumed not to appear). Then we reduce F inductively to a single equation using the following transformations:

- If F is of the form $e_1 = e_2 \vee k_1 = k_2$, then F is equivalent to the single equation $(e_1 - e_2)(k_1 - k_2) = 0$. Furthermore, the formula in the induction step of dI, $[x':=f(x)](F)' \equiv [x':=f(x)]((e_1)' = (e_2)' \wedge (k_1)' = (k_2)')$ directly implies

$$[x':=f(x)]((e_1 - e_2)(k_1 - k_2))' = 0$$
$$\equiv [x':=f(x)]\left(((e_1)' - (e_2)')(k_1 - k_2) + (e_1 - e_2)((k_1)' - (k_2)') = 0\right)$$

which implies that the differential structure is compatible. So, the inductive step for $(e_1 - e_2)(k_1 - k_2) = 0$ will succeed if the inductive step for $e_1 = e_2 \vee k_1 = k_2$ succeeded. The converse implication does not hold, but also does not have to hold for this proof to work out, because we are merely saying that if the disjunction of equations is a differential invariant then the more complex single equation will also be, not vice versa.

- If F is of the form $e_1 = e_2 \wedge k_1 = k_2$, then F is equivalent to the single equation $(e_1 - e_2)^2 + (k_1 - k_2)^2 = 0$. Also, the formula in the induction step of rule dI, $[x':=f(x)](F)' \equiv [x':=f(x)]((e_1)' = (e_2)' \wedge (k_1)' = (k_2)')$ implies

$$[x':=f(x)]\left(((e_1 - e_2)^2 + (k_1 - k_2)^2)' = 0\right)$$
$$\equiv [x':=f(x)]\left(2(e_1 - e_2)((e_1)' - (e_2)') + 2(k_1 - k_2)((k_1)' - (k_2)') = 0\right)$$

Consequently propositional connectives of equations can successively be replaced by their equivalent arithmetic equations in pre- and postconditions, and the corresponding induction steps are still provable for the single equations.   □

Observe that the polynomial degree is increased quadratically by the reduction in Proposition 13.1, but, as a trade-off, the propositional structure is simplified. Consequently, differential invariant search for the equational case can either exploit propositional structure with lower-degree polynomials or suppress the propositional structure at the expense of higher degrees. This trade-off depends on the real-arithmetic decision procedure, but is often enough in favor of keeping propositional structure, because the proof calculus can still exploit the logical structure to decompose the verification question before invoking real arithmetic. There are cases, however, where such reductions are formidably insightful [12].

Equational differential invariants, thus, enjoy a lot of beautiful properties, including characterizing invariant functions [12] and generalizing to a decision procedure for algebraic invariants of algebraic differential equations [4].

13.7 Equational Incompleteness

Despite the fact that Proposition 13.1 confirms how surprisingly expressive single equations are, focusing exclusively on differential invariants with equations reduces the deductive power, because sometimes only differential invariant inequalities can prove properties.

> **Proposition 13.2 (Equational incompleteness).** *The deductive power of differential induction with equational formulas is strictly less than the deductive power of general differential induction, because some inequalities cannot be proven with equations.*
>
> $$\mathcal{DI}_= \equiv \mathcal{DI}_{=,\wedge,\vee} < \mathcal{DI}$$
> $$\mathcal{DI}_\geq \not\leq \mathcal{DI}_= \equiv \mathcal{DI}_{=,\wedge,\vee}$$
> $$\mathcal{DI}_> \not\leq \mathcal{DI}_= \equiv \mathcal{DI}_{=,\wedge,\vee}$$

How can such a proposition with a negative answer about provability be proved?

Before you read on, see if you can find the answer for yourself.

The proof strategy for the proof of Proposition 13.1 involved transforming dL proofs into other dL proofs to prove the inclusion $\mathcal{DI}_= \geq \mathcal{DI}_{=,\wedge,\vee}$. Can the same strategy prove Proposition 13.2? No, because we need to show the opposite! Proposition 13.2 conjectures $\mathcal{DI}_\geq \not\leq \mathcal{DI}_{=,\wedge,\vee}$, which means that there are true properties that are only provable using a differential invariant inequality $e_1 \geq e_2$ and not using any differential invariant equations or propositional combinations thereof.

For one thing, this means that we must find a property that a differential invariant inequality can prove. That ought to be easy enough, because Chap. 11 showed us how useful differential invariants are. But then a proof of Proposition 13.2 also requires a proof of why that very same formula cannot possibly ever be proved using only differential invariant equations or their propositional combinations. That is a proof about nonprovability. Proving provability in proof theory amounts to producing a proof (in dL's sequent calculus). Proving nonprovability most certainly does not mean it will be enough to write something down that is not a proof. After all, just because one proof attempt fails does not mean that other attempts will not be successful.

You have experienced this while you were working on proving the more challenging exercises of this textbook. The first proof attempt might have failed miserably and been impossible to ever complete. But, come the next day, you had a better idea with a different proof, and suddenly the same property turned out to be perfectly provable even if the first proof attempt failed.

How can we prove that *all* proof attempts do not work?

Before you read on, see if you can find the answer for yourself.

One way of showing that a logical formula cannot be proved is by giving a counterexample, i.e., a state that assigns values to the variables that falsify the formula. That, of course, does not help us prove Proposition 13.2, because a proof of Proposition 13.2 requires us to find a formula that can be proved with $\mathcal{DI}_\geq$ (so it cannot have any counterexamples, since it is perfectly valid), just cannot be proved with $\mathcal{DI}_{=,\wedge,\vee}$. Proving that a valid formula cannot be proved with $\mathcal{DI}_{=,\wedge,\vee}$ requires us to show that all proofs in $\mathcal{DI}_{=,\wedge,\vee}$ do not prove that formula.

Expedition 13.1 (Proving differences in set theory and linear algebra)

Recall what you know about sets. The way to prove that two sets M, N have the same "number" of elements is to come up with a pair of functions $\Phi : M \to N$ and $\Psi : N \to M$ between the sets and then prove that Φ, Ψ are inverses of each other, i.e., $\Phi(\Psi(y)) = y$ and $\Psi(\Phi(x)) = x$ for all $x \in M, y \in N$ to show that there is a bijection between the sets M and N. Proving that two sets M, N do not have the same "number" of elements works entirely differently, because that requires a proof for all pairs of functions $\Phi : M \to N$ and $\Psi : N \to M$ that there is an $x \in M$ such that $\Psi(\Phi(x)) \neq x$ or a $y \in N$ such that $\Phi(\Psi(y)) \neq y$. Since writing down every such pair of functions Φ, Ψ is a lot of work (an infinite amount of work if M and N are infinite), indirect criteria such as cardinality or countability are used instead, e.g., for proving that the reals $\mathbb{R}$ and rationals $\mathbb{Q}$ cannot possibly have the same number of elements, because $\mathbb{Q}$ is countable but $\mathbb{R}$ is not (by Cantor's diagonal argument [1, 18]).

Recall vector spaces from linear algebra. The way to prove that two vector spaces V, W are isomorphic is to think hard and construct a function $\Phi : V \to W$ and a function $\Psi : W \to V$ and then prove that Φ, Ψ are linear functions and inverses of each other. Proving that two vector spaces V, W are *not* isomorphic works entirely differently, because that requires a proof that all pairs of functions $\Phi : V \to W$ and $\Psi : W \to V$ are either not linear or not inverses of each other. Proving the latter literally is again a lot (usually an infinite amount) of work. Instead, indirect criteria are used. One proof that two vector spaces V, W are not isomorphic might show that the two have different dimensions and then prove that isomorphic vector spaces always have the same dimension, so V and W cannot possibly be isomorphic.

By analogy, proving non-provability leads to a study of indirect criteria about proofs of differential equations.

Note 70 (Proofs of different provability) Proving non-reducibility $\mathscr{A} \not\leq \mathscr{B}$ for classes of differential invariants requires an example formula P that is provable in $\mathscr{A}$ plus a proof that no proof using $\mathscr{B}$ proves P. The preferred way of doing that is to find an indirect criterion that all conclusions of all proofs in $\mathscr{B}$ possess but that P does not have, so that the proofs using $\mathscr{B}$ cannot possibly succeed in proving P.

Proof (of Proposition 13.2). Consider any positive term $a > 0$ (e.g., 5 or $x^2 + 1$ or $x^2 + x^4 + 2$). The following proof proves a formula by differential invariants with the weak inequality $x \geq 0$:

$$
\mathbb{R} \; \dfrac{*}{\vdash a \geq 0}
$$

$$
[:=] \; \dfrac{\vdash a \geq 0}{\vdash [x':=a]x' \geq 0}
$$

$$
\text{dl} \; \dfrac{\vdash [x':=a]x' \geq 0}{x \geq 0 \vdash [x'=a]x \geq 0}
$$

The same formula is not provable with an equational differential invariant, however. Any univariate polynomial p that is zero on all $x \geq 0$ is the zero polynomial and, thus, an equation of the form $p = 0$ cannot be equivalent to the half space $x \geq 0$. By the equational deductive power theorem (Proposition 13.1), the above formula then is not provable with any Boolean combination of equations as differential invariant either, because propositional combinations of equational differential invariants prove the same properties that single equational differential invariants do, and the latter cannot succeed in proving $x \geq 0 \rightarrow [x'=a]x \geq 0$.

The other parts of the theorem that involve generalizations of the non-provability argument to other indirect proofs using cuts, etc., are proved elsewhere [14].   □

It might be tempting to think that at least equational postconditions only need equational differential invariants for their proof. But that is not the case either [14]. So even if the property you care to prove involves only equations, you may still need to generalize your proof arguments to consider inequalities instead.

13.8 Strict Differential Invariant Inequalities

We show that, conversely, focusing on strict inequalities $p > 0$ also reduces the deductive power, because equations are obviously missing and there is at least one proof where this matters. That is, what are called strict barrier certificates do not prove (nontrivial) closed invariants.

> **Proposition 13.3 (Strict barrier incompleteness).** *The deductive power of differential induction with strict barrier certificates (formulas of the form $e > 0$) is strictly less than the deductive power of general differential induction:*
>
> $$\mathscr{DI}_> < \mathscr{DI}$$
> $$\mathscr{DI}_= \not\leq \mathscr{DI}_>$$

Proof. The following proof proves a formula by equational differential induction:

$$
\mathbb{R} \; \dfrac{*}{\vdash 2xy + 2y(-x) = 0}
$$

$$
[:=] \; \dfrac{\vdash 2xy + 2y(-x) = 0}{\vdash [x':=y][y':=-x]2xx' + 2yy' = 0}
$$

$$
\text{dl} \; \dfrac{\vdash [x':=y][y':=-x]2xx' + 2yy' = 0}{x^2 + y^2 = c^2 \vdash [x'=y, y'=-x]x^2 + y^2 = c^2}
$$

But the same formula is not provable with a differential invariant of the form $e > 0$. An invariant of the form $e > 0$ describes an open set and, thus, cannot be equivalent to the (nontrivial) closed set where $x^2 + y^2 = c^2$ holds true. The only sets that are both open and closed in (the Euclidean space) $\mathbb{R}^n$ are the empty set $\emptyset$ (described by the formula *false*) and the full space $\mathbb{R}^n$ (described by the formula *true*), both of which do not prove the property of interest, because *true* does not imply the postcondition and *false* does not hold initially. The other parts of the theorem are proved elsewhere [14]. □

Expedition 13.2 (Topology in real analysis)

The following proofs distinguish open sets from closed sets, which are concepts from real analysis (or topology). Roughly: A closed set is one whose boundary belongs to the set, for example the solid unit disk of radius 1. An open set is one for which no point of the boundary belongs to the set, for example the unit disk of radius 1 without the outer circle of radius 1.

closed solid disk   open disk
$x^2 + y^2 \leq 1$ $x^2 + y^2 < 1$
with boundary   without boundary

A set $O \subseteq \mathbb{R}^n$ is *open* iff, around every point of O, there is a small neighborhood that is contained in O. That is, for all points $a \in O$ there is an $\varepsilon > 0$ such that every point b of distance at most ε from a is still in O. A set $C \subseteq \mathbb{R}^n$ is *closed* iff its complement is open. Because $\mathbb{R}^n$ is what is called a complete metric space, a set $C \subseteq \mathbb{R}^n$ is closed iff every convergent sequence of elements in C converges to a limit in C (so C is closed under limits).

One takeaway message is that it makes sense to check whether the desired invariant is an open or a closed set and use differential invariants of the suitable type for the job. Of course, both $e = 0$ and $e \geq 0$ might still work for closed sets.

Beware, however, that openness and closedness depend on the ambient space. One proof in Chap. 12, for example, proved the strict inequality $x > 0$ to be an invariant of the differential equation $x' = -x$ by reducing it to a proof of invariance of the equation $xy^2 = 1$ with an additional differential ghost $y' = \frac{y}{2}$. Seemingly, this proves an open set to be an invariant by using a closed set, but the whole dimension of the state space changes due to the new variable y. And, indeed, the set of all x for which there is a y such that $xy^2 = 1$ is again the open set described by $x > 0$.

13.9 Differential Invariant Equations as Differential Invariant Inequalities

Weak inequalities $e \geq 0$, however, do subsume the deductive power of equational differential invariants $e = 0$. After some thought, this is somewhat obvious on the algebraic level but we will see that it also does carry over to the differential structure.

> **Proposition 13.4 (Equational definability).** *The deductive power of differen-tial induction with equations is subsumed by the deductive power of differential induction with weak inequalities:*
>
> $$\mathscr{DI}_{=,\wedge,\vee} \leq \mathscr{DI}_{\geq}$$

Proof. By Proposition 13.1, we only need to show that $\mathscr{DI}_{=} \leq \mathscr{DI}_{\geq}$, because Proposition 13.1 implies $\mathscr{DI}_{=,\wedge,\vee} = \mathscr{DI}_{=}$. Let $e = 0$ be an equational differential invariant of a differential equation $x' = f(x) \,\&\, Q$. Then we can prove the following:

$$
\cfrac{\quad\cfrac{*}{Q \vdash [x':=f(x)](e)' = 0}^{\;[:=],\mathbb{R}}\quad}{e = 0 \vdash [x' = f(x) \,\&\, Q]e = 0}{}^{\mathrm{dI}}
$$

Then, the inequality $-e^2 \geq 0$, which is equivalent to $e = 0$ in real arithmetic, also is a differential invariant of the same dynamics by the following dL proof:

$$
\cfrac{\quad\cfrac{*}{Q \vdash [x':=f(x)] - 2e(e)' \geq 0}^{\;[:=],\mathbb{R}}\quad}{-e^2 \geq 0 \vdash [x' = f(x) \,\&\, Q](-e^2 \geq 0)}{}^{\mathrm{dI}}
$$

The subgoal for the differential induction step is provable: if we can prove that Q implies $[x':=f(x)](e)' = 0$ according to the first sequent proof, then we can also prove that Q implies $[x':=f(x)] - 2e(e)' \geq 0$ for the second sequent proof, because the postcondition $(e)' = 0$ implies $-2e(e)' \geq 0$ in first-order real arithmetic. $\square$

Note that the differential invariant view of reducing properties of differential equations to differential properties in local states is crucial to make the last proof work. It is obvious that $(e)' = 0$ implies $-2e(e)' \geq 0$ holds in any single state. With-out differential invariance arguments, it is harder to relate this to the truth-values of corresponding properties along differential equations. By Proposition 13.4, differen-tial invariant search with weak inequalities can suppress equations. Note, however, that the polynomial degree is increased quadratically by the reduction in Proposi-tion 13.4. In particular, the polynomial degree is increased quartically by the reduc-tions in Proposition 13.1 and Proposition 13.4 one after another to turn propositional equational formulas into single weak inequalities. This quartic increase of the poly-nomial degree is likely a too-serious computational burden for practical purposes even if it is a valid reduction in theory.

13.10 Differential Invariant Atoms

Next we see that, with the notable exception of pure equations (Proposition 13.1), propositional operators do increase the deductive power of differential invariants.

> **Theorem 13.1 (Atomic incompleteness).** *The deductive power of differential induction with propositional combinations of inequalities exceeds the deductive power of differential induction with atomic inequalities.*
>
> $$\mathscr{DI}_{\geq} < \mathscr{DI}_{\geq,\wedge,\vee}$$
> $$\mathscr{DI}_{>} < \mathscr{DI}_{>,\wedge,\vee}$$

Proof. Consider any term $a \geq 0$ (e.g., 1 or $x^2 + 1$ or $x^2 + x^4 + 1$ or $(x-y)^2 + 2$). Then the formula $x \geq 0 \wedge y \geq 0 \to [x' = a, y' = y^2](x \geq 0 \wedge y \geq 0)$ is provable using a conjunction in the differential invariant:

$$
\dfrac{\dfrac{*}{\mathbb{R} \quad \vdash a \geq 0 \wedge y^2 \geq 0}}{\dfrac{[:=] \quad \vdash [x':=a][y':=y^2](x' \geq 0 \wedge y' \geq 0)}{{}^{\mathrm{dI}} \quad x \geq 0 \wedge y \geq 0 \vdash [x' = a, y' = y^2](x \geq 0 \wedge y \geq 0)}}
$$

By a sign argument similar to that in the proof of [10, Theorem 2] and [11, Theorem 3.3], no atomic formula is equivalent to $x \geq 0 \wedge y \geq 0$. Basically, no formula of the form $p(x,y) \geq 0$ for a polynomial p can be equivalent to $x \geq 0 \wedge y \geq 0$. This is because that would imply that $p(x,0) \geq 0 \leftrightarrow x \geq 0$ for all x, which, as $p(x,0)$ is a univariate polynomial with infinitely many roots (every $x \geq 0$), implies that $p(x,0)$ is the zero polynomial, which is not equivalent to $x \geq 0$, because the zero polynomial is also zero on $x < 0$. Similar arguments work for $p(x,y) > 0$ and $p(x,y) = 0$. Thus, the above property cannot be proven using an atomic differential invariant. The proof for a postcondition $x > 0 \wedge y > 0$ is similar.

The other—quite substantial—parts of the proof are proved elsewhere [14]. $\square$

Note that the formula in the proof of Theorem 13.1 is provable, e.g., using differential cuts (dC) with two atomic differential induction steps, one for $x \geq 0$ and one for $y \geq 0$. Yet, a similar, significantly more involved, argument can be made to show that the deductive power of differential induction with atomic formulas (even when using differential cuts) is still strictly less than the deductive power of general differential induction; see [10, Theorem 2]. This just needs another choice of differential equation and a more involved proof.

Consequently, in the case of inequalities, propositional connectives can be quite crucial when looking for differential invariants even in the presence of differential cuts.

13.11 Summary

Figure 13.2 summarizes the findings on provability relations of differential equations explained in this chapter and others reported in the literature [14]. This chapter considered the differential invariance problem, which, by a relative completeness argument [9, 13], is at the heart of hybrid systems verification. To better understand structural properties of hybrid systems, more than a dozen (16) relations between the deductive power of several (9) classes of differential invariants have been identified and analyzed. An understanding of these relations helps guide the search for suitable differential invariants and also provides an intuition for exploiting indirect criteria such as open/closedness of sets as a guide.

The results require a symbiosis of elements of logic with real-arithmetical, differential, semialgebraic, and geometrical properties. Future work includes investigating further this new field called *real differential semialgebraic geometry*, whose development has only just begun [5–7, 14].

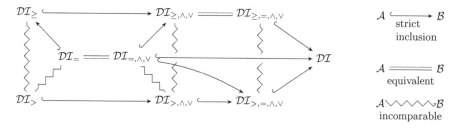

$\mathcal{DI}_\Omega$: properties verifiable using differential invariants built with operators from Ω

Fig. 13.2 Differential invariance chart (strict inclusions $\mathscr{A} < \mathscr{B}$, equivalences $\mathscr{A} \equiv \mathscr{B}$, and incomparabilities $\mathscr{A} \not\leq \mathscr{B}$, $\mathscr{B} \not\leq \mathscr{A}$ for classes of differential invariants are indicated)

13.12 Appendix: Curves Playing with Norms and Degrees

The proof of Lemma 13.2 showed a case where a formula with a higher-degree polynomial was needed to prove a property that a lower-degree polynomial could not prove. The conclusion from the proof of Lemma 13.2 is not that it is always better to use differential invariants of higher degrees, just because that worked in this particular proof.

For example, the following proof for an upper bound t on the supremum norm $\|(x,y)\|_\infty$ of the vector (x,y) defined as

$$\|(x,y)\|_\infty \leq t \overset{\text{def}}{\equiv} -t \leq x \leq t \wedge -t \leq y \leq t \tag{13.2}$$

is significantly easier for the curved dynamics:

$$
\dfrac{}{\mathbb{R}}
$$

$\mathbb{R}$	$\overset{*}{\overline{v^2+w^2\le 1\vdash -1\le v\le 1 \wedge -1\le w\le 1}}$
$[:=]$	$\overline{v^2+w^2\le 1\vdash [x':=v][y':=w][v':=\omega w][w':=-\omega v][t':=1](-t'\le x'\le t'\wedge -t'\le y'\le t')}$
dI	$\overline{\lessdot v^2+w^2\le 1\wedge x=y=t=0\vdash [x'=v,y'=w,v'=\omega w,w'=-\omega v,t'=1\,\&\,v^2+w^2\le 1]\|(x,y)\|_\infty\le t}$
dC	$v^2+w^2\le 1\wedge x=y=t=0\vdash [x'=v,y'=w,v'=\omega w,w'=-\omega v,t'=1]\|(x,y)\|_\infty\le t$

where the first premise of the differential cut (dC) above is elided (marked ◁) and proved as in Example 11.3. This proof shows that a point (x,y) starting with linear velocity at most 1 and angular velocity ω from the origin will not move further than the time t in supremum norm.

This simple proof is to be contrasted with the following proof attempt for a corresponding upper bound on the Euclidean norm $\|(x,y)\|_2$ defined as

$$\|(x,y)\|_2\le t\overset{\text{def}}{\equiv}x^2+y^2\le t^2 \tag{13.3}$$

for which a direct proof fails:

	$\overset{\text{not valid}}{\overline{v^2+w^2\le 1\vdash 2xv+2yw\le 2t}}$
$[:=]$	$\overline{v^2+w^2\le 1\vdash [x':=v][y':=w][v':=\omega w][w':=-\omega v][t':=1](2xx'+2yy'\le 2tt')}$
dI	$\overline{\lessdot v^2+w^2\le 1\wedge x=y=t=0\vdash [x'=v,y'=w,v'=\omega w,w'=-\omega v,t'=1\,\&\,v^2+w^2\le 1]\|(x,y)\|_2\le t}$
dC	$v^2+w^2\le 1\wedge x=y=t=0\vdash [x'=v,y'=w,v'=\omega w,w'=-\omega v,t'=1]\|(x,y)\|_2\le t$

An indirect proof is still possible but much more complicated. But the proof using the supremum norm (13.2) is much easier than the proof using the Euclidean norm (13.3) in this case. In addition, the arithmetic complexity decreases, because supremum norms are definable in linear arithmetic (13.2) unlike the quadratic arithmetic required for Euclidean norms (13.3). Finally, the simpler proof is, up to a factor of $\sqrt{2}$, just as good, because quantifier elimination easily proves that the supremum norm $\|\cdot\|_\infty$ and the standard Euclidean norm $\|\cdot\|_2$ are equivalent, i.e., their values are identical up to constant factors:

$$\forall x\forall y\,(\|(x,y)\|_\infty\le\|(x,y)\|_2\le\sqrt{n}\|(x,y)\|_\infty) \tag{13.4}$$

$$\forall x\forall y\,(\frac{1}{\sqrt{n}}\|(x,y)\|_2\le\|(x,y)\|_\infty\le\|(x,y)\|_2) \tag{13.5}$$

where n is the dimension of the vector space, here 2. That makes sense, because if, e.g., the coordinate with maximal absolute value is at most 1, then the Euclidean distance can be at most 1. And the extra factor of $\sqrt{2}$ is easily justified by Pythagoras' theorem. An illustration of the inclusion relationships of the unit discs in the various norms can be found in Fig. 13.3.

Fig. 13.3 *p*-norm inclusions

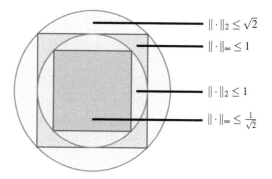

$\|\cdot\|_2 \leq \sqrt{2}$

$\|\cdot\|_\infty \leq 1$

$\|\cdot\|_2 \leq 1$

$\|\cdot\|_\infty \leq \frac{1}{\sqrt{2}}$

Exercises

13.1. Prove the norm relations (13.4) and (13.5). Use these relations in a sequent proof to relate the successful proof with a bound on the supremum norm $\|(x,y)\|_\infty$ to a corresponding result about a bound on the Euclidean norm $\|(x,y)\|_2$.

13.2. Prove the relation $\mathcal{DI}_> \leq \mathcal{DI}_{>,\wedge,\vee}$, i.e., that all properties provable using differential invariants of the form $p > q$ are also provable using propositional combinations of these formulas as differential invariants.

13.3. Prove the relation $\mathcal{DI}_\geq \equiv \mathcal{DI}_{\leq,\wedge,\vee}$.

13.4. Prove the relation $\mathcal{DI}_{\geq,\wedge,\vee} \equiv \mathcal{DI}_{\geq,=,\wedge,\vee}$.

13.5. Let $\mathcal{DI}_{true}$ denote the proof calculus in which only the formula *true* is allowed as a differential invariant. Prove the relation $\mathcal{DI}_{true} < \mathcal{DI}_=$.

13.6. Let $\mathcal{DI}_{false}$ denote the proof calculus in which only the formula *false* is allowed as a differential invariant. Prove the relation $\mathcal{DI}_{false} < \mathcal{DI}_>$.

13.7. Prove the relation $\mathcal{DI}_{=,\wedge,\vee} < \mathcal{DI}_{\geq,\wedge,\vee}$.

13.8. Prove the relation $\mathcal{DI}_{>,\wedge,\vee} < \mathcal{DI}_{>,=,\wedge,\vee}$.

13.9. What is the relationship of $\mathcal{DI}_{x=0}$ compared to $\mathcal{DI}_{x^2=0}$? That is, are there formulas that are only provable with invariants of the form $x = 0$ but not $x^2 = 0$, or vice versa?

References

[1]   Georg Cantor. Über eine elementare Frage der Mannigfaltigkeitslehre. *Jahresbericht der Deutschen Mathematiker-Vereinigung* **1** (1891), 75–78.

[2] Alonzo Church. A note on the Entscheidungsproblem. *J. Symb. Log.* **1**(1) (1936), 40–41.

[3] Stephen A. Cook. The complexity of theorem-proving procedures. In: *STOC*. Ed. by Michael A. Harrison, Ranan B. Banerji, and Jeffrey D. Ullman. New York: ACM, 1971, 151–158. DOI: `10.1145/800157.805047`.

[4] Khalil Ghorbal and André Platzer. Characterizing algebraic invariants by differential radical invariants. In: *TACAS*. Ed. by Erika Ábrahám and Klaus Havelund. Vol. 8413. LNCS. Berlin: Springer, 2014, 279–294. DOI: `10.1007/978-3-642-54862-8_19`.

[5] Khalil Ghorbal, Andrew Sogokon, and André Platzer. Invariance of conjunctions of polynomial equalities for algebraic differential equations. In: *SAS*. Ed. by Markus Müller-Olm and Helmut Seidl. Vol. 8723. LNCS. Berlin: Springer, 2014, 151–167. DOI: `10.1007/978-3-319-10936-7_10`.

[6] Khalil Ghorbal, Andrew Sogokon, and André Platzer. A hierarchy of proof rules for checking differential invariance of algebraic sets. In: *VMCAI*. Ed. by Deepak D'Souza, Akash Lal, and Kim Guldstrand Larsen. Vol. 8931. LNCS. Berlin: Springer, 2015, 431–448. DOI: `10.1007/978-3-662-46081-8_24`.

[7] Khalil Ghorbal, Andrew Sogokon, and André Platzer. A hierarchy of proof rules for checking positive invariance of algebraic and semi-algebraic sets. *Computer Languages, Systems & Structures* **47**(1) (2017), 19–43. DOI: `10.1016/j.cl.2015.11.003`.

[8] David Harel, Albert R. Meyer, and Vaughan R. Pratt. Computability and completeness in logics of programs (preliminary report). In: *STOC*. New York: ACM, 1977, 261–268.

[9] André Platzer. Differential dynamic logic for hybrid systems. *J. Autom. Reas.* **41**(2) (2008), 143–189. DOI: `10.1007/s10817-008-9103-8`.

[10] André Platzer. Differential-algebraic dynamic logic for differential-algebraic programs. *J. Log. Comput.* **20**(1) (2010), 309–352. DOI: `10.1093/logcom/exn070`.

[11] André Platzer. *Logical Analysis of Hybrid Systems: Proving Theorems for Complex Dynamics*. Heidelberg: Springer, 2010. DOI: `10.1007/978-3-642-14509-4`.

[12] André Platzer. A differential operator approach to equational differential invariants. In: *ITP*. Ed. by Lennart Beringer and Amy Felty. Vol. 7406. LNCS. Berlin: Springer, 2012, 28–48. DOI: `10.1007/978-3-642-32347-8_3`.

[13] André Platzer. The complete proof theory of hybrid systems. In: *LICS*. Los Alamitos: IEEE, 2012, 541–550. DOI: `10.1109/LICS.2012.64`.

[14] André Platzer. The structure of differential invariants and differential cut elimination. *Log. Meth. Comput. Sci.* **8**(4:16) (2012), 1–38. DOI: `10.2168/LMCS-8(4:16)2012`.

[15] André Platzer. Differential game logic. *ACM Trans. Comput. Log.* **17**(1) (2015), 1:1–1:51. DOI: `10.1145/2817824`.

[16]   André Platzer. A complete uniform substitution calculus for differential dynamic logic. *J. Autom. Reas.* **59**(2) (2017), 219–265. DOI: 10.1007/s108 17-016-9385-1.

[17]   André Platzer and Edmund M. Clarke. Computing differential invariants of hybrid systems as fixedpoints. *Form. Methods Syst. Des.* **35**(1) (2009). Special issue for selected papers from CAV'08, 98–120. DOI: 10.1007/s107 03-009-0079-8.

[18]   Willard Van Quine. On Cantor's theorem. *J. Symb. Log.* **2**(3) (1937), 120–124. DOI: 10.2307/2266291.

[19]   H. Gordon Rice. Classes of recursively enumerable sets and their decision problems. *Trans. AMS* **74**(2) (1953), 358–366. DOI: 10.2307/1990888.

[20]   Alan M. Turing. On computable numbers, with an application to the Entscheidungsproblem. *Proc. Lond. Math. Soc.* **42**(1) (1937), 230–265. DOI: 10.11 12/plms/s2-42.1.230.

Part III
Adversarial Cyber-Physical Systems

Overview of Part III on Adversarial Cyber-Physical Systems

This part fundamentally advances our understanding of cyber-physical system models by including an entirely new dynamical aspect of dynamical systems. The previous parts of this textbook thoroughly studied hybrid systems with their interacting discrete and continuous dynamics and elaborated their respective proof principles. There were choices in the hybrid system's evolution, but these were all resolved nondeterministically, so in an arbitrary, non-purposeful way.

Part III now explores what happens when there are different agents with different goals in the dynamics of a hybrid system, so that its choices may also be resolved differently by the different players at different times. Part III studies the *adversarial dynamics* of two players interacting on the discrete and continuous dynamics of hybrid systems, which leads to *hybrid games*. Unlike the hybrid systems models of cyber-physical systems that were studied in Parts I and II, the hybrid games of Part III provide mixed discrete, continuous, and adversarial dynamics. Hybrid games are for two players. Hybrid systems correspond to single-player hybrid games in which the only player is nondeterminism (or games in which the other player never has any choices to make). Adversarial dynamics gives more freedom in the overall dynamics of the system. It is important whenever multiple agents interact with possibly conflicting goals or possibly conflicting actions resulting from different perceptions of the world. Despite these significant generalizations of the system dynamics, the elementary understanding of cyber-physical systems from Parts I and II will continue to be generalizable quite seamlessly to cover hybrid games.

Chapter 14
Hybrid Systems & Games

Synopsis This chapter begins the study of an entirely new model of cyber-physical systems: that of *hybrid games*, which combine discrete, continuous, and adversarial dynamics. While hybrid systems with their discrete and continuous dynamics have served us well in the analysis of cyber-physical systems so far, other cyber-physical systems crucially require an understanding of additional dynamical effects. Adversarial dynamics is relevant whenever choices in the system can be resolved by different players. This happens frequently in CPSs with multiple agents who may or may not agree on a common goal or who, even if they share a common goal, may act differently based on a different perception of the world. This chapter discusses the far-reaching consequences of this insight and advances hybrid programs to a programming language for hybrid games.

14.1 Introduction

Hybrid systems have served us well throughout this textbook as a model of cyber-physical systems [1, 3, 7, 11]. But contrary to what we simply pretended in Parts I and II, hybrid systems and cyber-physical systems are not the same. Hybrid systems can also serve as models of other systems that are not cyber-physical per se, i.e., they are not built as a combination of cyber and computing capabilities with physical capabilities. Some biological systems can be understood as hybrid systems, because they combine discrete activation of genes and continuous biochemical reactions. Or physical processes can be understood as hybrid if things happen at very different speeds. Then, there is a slow process about which a continuous understanding is critical as well as a very fast process in which a discrete abstraction might be sufficient. Just think back to the bouncing ball where a discrete understanding of the event of the bounce was more suitable even if a continuous deformation occurs, but at a much faster pace than the continuous falling due to gravity. None of those examples is particularly cyber-physical. Nevertheless, they can be naturally modeled as hybrid systems, because their fundamental characteristic is the interaction of dis-

© Springer International Publishing AG, part of Springer Nature 2018
A. Platzer, *Logical Foundations of Cyber-Physical Systems*,
https://doi.org/10.1007/978-3-319-63588-0_14

crete and continuous dynamics, which is exactly what hybrid systems are good for. Hybrid systems are a mathematical model of dynamical systems with mixed discrete and continuous dynamics, whether cyber-physical or not. Hence, despite their good match, not *all* hybrid systems are cyber-physical systems.

One important point of this chapter is that the converse is not true either. Not all cyber-physical systems are hybrid systems! The reason for that is *not* that cyber-physical systems lack discrete and continuous dynamics, but, rather, that they involve also additional dynamical aspects. It is a pretty common phenomenon in cyber-physical systems that they involve several dynamical aspects, which is why they are best understood as *multi-dynamical systems*, i.e., systems with multiple dynamical features [4–7, 9, 10, 12].

In a certain sense, applications often have a $+1$ effect on dynamical aspects. Your analysis might start out focusing on some number of dynamical aspects only to observe during the elaboration of the analysis that there is another part of the system for which one more dynamical aspect is relevant than was originally anticipated. The bouncing ball is an example to which a preliminary analysis might first ascribe an entirely continuous dynamics, just to find out after a while that the singularity of bouncing back from the ground can be more easily understood by a discrete dynamics. Whenever you are analyzing a system, be prepared to find one more dynamical aspect around the corner! That is yet another reason why it is useful to have flexible and general analysis techniques grounded in logic that still work even after a new dynamical aspect has been found.

Of course, it is not going to be feasible to understand all multi-dynamical system aspects at once in this chapter. But this chapter is going to introduce one absolutely fundamental dynamical aspect: *adversarial dynamics* [9, 12]. Adversarial dynamics comes from multiple players that, in the context of a CPS, interact on a hybrid system and are allowed to make their respective choices arbitrarily, in pursuit of their goals. The combination of discrete, continuous, and adversarial dynamics leads to *hybrid games*. Unlike hybrid systems, hybrid games allow choices in the system dynamics to be resolved adversarially by different players with different objectives.

Hybrid games are necessary in situations where multiple agents actively compete. The canonical situation of a hybrid game would, thus, be when two teams of robots play robot soccer, moving around physically in space, controlled according to discrete computer decisions, and in active competition to score goals in opposite directions on the field. The robots in a robot soccer match can't agree on the direction in which they try to get the ball rolling. This leads to a mix of discrete, continuous, and adversarial dynamics for truly competitive reasons.

It turns out, however, that hybrid games also come up for reasons of *analytic competition*, that is, where possible competition is assumed only for the sake of a worst-case analysis. Consider a robot that is interacting with another robot, let's call it the *roguebot*. You are in control of the robot, but somebody else is controlling the roguebot. Your objective is to control your robot so that it will not run into the roguebot no matter what. That means you need to *find some* way of using your control choices for your robot so that it makes progress toward its goal but will remain safe *for all* possible control choices that the roguebot might follow. After all, you do not know exactly how the other roguebot is implemented and how it will react to your control decisions. That makes your robot play a hybrid game with the roguebot in which your robot is trying to safely avoid collisions. The roguebot might behave sanely and try to stay safe as well. But the roguebot's objectives might differ from yours, because its objective is not to get your robot to your goal. The roguebot rather wants to get to its own goal instead, which might cause unsafe interference whenever the roguebot takes an action in pursuit of its goal that is not in your robot's interest. If your robot caused a collision, because it chose an action that was incompatible with the roguebot's action, your robot would certainly be faulty and be sent back to the design table. And even when both robots perfectly agree on the same goal, their actions might still cause unintended interferences when their perception of the world differ. In that case the two robots could take conflicting actions despite pursuing the same goal, just because they each thought the state of the world was a little different. Just imagine a common goal of not colliding with a rule that whoever is further west moves even further to the west. Now if both robots think they are the one that is further west because their sensors tell them that, then they might still collide even if both really didn't mean to.

Alas, when you try to understand how you need to control your robot to stay safe, it can be instructive to think about what the worst-case action of a roguebot might be to make life difficult for you. When a test engineer is trying to demonstrate under which circumstance a simulation of your robot controller exhibits a faulty behavior, so that you can learn from the cases where your control does not work, they actually play a hybrid game with you. If your robot wins and stays safe, that is an indication of a good robot design at least in this scenario. But if the test engineer wins and shows an unsafe trace, then you still win even if you lose this particular simulation, because you learn more about the corner cases in your robot control design than when staring at simulation movies where everything is just fair-weather control.

This chapter is based on prior work [9], where more information can be found on logic and hybrid games. The most important learning goals of this chapter are:

Modeling and Control: We identify an important additional dynamical aspect, the aspect of *adversarial dynamics*, which adds an adversarial way of resolving the choices in the system dynamics. This dynamical aspect is important for understanding the core principles behind CPS, because multiple agents with possibly conflicting actions are featured frequently in CPS applications. Such conflicting actions might be chosen due to different goals or different perceptions of the world. It is helpful to learn under which circumstance adversarial dynamics is important for understanding a CPS and when it can be neglected without loss. CPSs in which all choices are resolved against you or all choices are resolved for you can already be described and analyzed in differential dynamic logic using its box and diamond modalities [7]. Adversarial dynamics is interesting in mixed cases, where some choices fall in your favor and others turn out against you. Another important goal of this chapter is to develop models and controls of CPS with adversarial dynamics corresponding to multiple agents.

Computational Thinking: This chapter follows fundamental principles from logic and computational thinking to capture the new phenomenon of adversarial dynamics in CPS models. We leverage core ideas from programming languages by extending syntax and semantics of program models and specification and verification logics with a new operator for duality to incorporate adversariality in a modular way into the realm of hybrid systems models. This leads to a compositional model of hybrid games with compositional operators. Modularity makes it possible to generalize our rigorous reasoning principles for CPS to hybrid games while simultaneously taming their complexity. This chapter introduces *differential game logic* dGL [9, 12] extending by adversarial dynamics the familiar differential dynamic logic, which has been used as the specification and verification language for CPS in Parts I and II. Computer science ultimately is about analysis such as worst-case analysis, expected-case analysis, or correctness analysis. Hybrid games enable analysis of CPSs at a more fine-grained level in between worst-case analysis and best-case analysis. In the dL formula $[\alpha]P$ all choices are resolved against us in the sense that $[\alpha]P$ is only true if P holds after all runs of α. In the dL formula $\langle\alpha\rangle P$ all choices are resolved in our favor in the sense that $\langle\alpha\rangle P$ is true if P holds after at least one run of α. Hybrid games can be used to attribute some but not all of the choices in a system to an opponent while leaving others to be resolved favorably. Finally, this chapter provides a perspective on advanced models of computation with alternating choices.

CPS Skills: We add a new dimension into our understanding of the semantics of a CPS model: the adversarial dimension corresponding to how a system changes state over time as multiple agents react to each other. This dimension is crucial for developing an intuition for the operational effects of multi-agent CPS. The presence of adversarial dynamics will cause us to reconsider the semantics of CPS models to incorporate the effects of multiple agents and their mutual reactions. This generalization, while crucial for understanding adversarial dynamics in CPS, also shines a helpful complementary light on the semantics of hybrid systems without adversariality by causing us to reflect on the rôle of choices.

14.2 A Gradual Introduction to Hybrid Games

This section gradually introduces the operations that hybrid games provide one step at a time. Its emphasis is on their motivation and an intuitive development starting from hybrid systems before subsequent sections provide a comprehensive view.

14.2.1 Choices & Nondeterminism

The first thing to remind ourselves about is that hybrid systems also already come with choices, and for good reasons, too.

fundamental principles of computational thinking
logical extensions
programming language modularity principles
compositional extensions
differential game logic
best/worst-case analysis
models of alternating computation

adversarial dynamics   multi-agent state change
conflicting actions   CPS semantics
multi-agent systems   reflections on choices
angelic/demonic choice

> **Note 71 (Choices in hybrid systems)** Hybrid systems involve choices. They manifest in hybrid programs as nondeterministic choices $\alpha \cup \beta$ whether to run HP α or HP β, in nondeterministic repetitions α^* where the choice is how often to repeat α, and in differential equations $x' = f(x) \,\&\, Q$ where the choice is how long to follow that differential equation. All those choices, however, have been resolved in one way, i.e., by the same entity or player: nondeterminism.

In which way the various choices are resolved depends on the context. In the box modality $[\alpha]$ of differential dynamic logic [1, 3, 7, 11], all nondeterminism is resolved in *all possible ways* so that the modal formula $[\alpha]P$ expresses that formula P holds for all ways in which the choices in HP α could be resolved. In the diamond modality $\langle \alpha \rangle$, instead, all nondeterminism is resolved in *some way* so that formula $\langle \alpha \rangle P$ expresses that formula P holds for at least one way of resolving the choices in HP α. The modality decides the mode of nondeterminism. The modal formula $[\alpha]P$ expresses that P holds necessarily after running α while $\langle \alpha \rangle P$ expresses that P is possible after α.

In particular, choices in α help $\langle \alpha \rangle P$, because what this formula calls for is *some* way of making P happen after α. If α has many possible behaviors, this is easier to satisfy. Choices in α hurt $[\alpha]P$, however, because this formula requires P to hold for *all* those choices. The more choices there are, the more difficult it is to make sure that P holds after every single combination of those choices.

> In differential dynamic logic, choices in α help uniformly (when they occur in $\langle \alpha \rangle P$) or make matters more difficult uniformly (when they occur in $[\alpha]P$).

That is why these various forms of choices in hybrid programs have been called *nondeterministic*. They are "unbiased." All possible resolutions of the choices in α can happen nondeterministically when running α. Which possibilities we care about (all or some) just depends on the modality around it. However, in each hybrid systems modality, all choices are uniformly resolved in one way, because we can only wrap one modality around the hybrid program. We cannot say that some choices within a modality are meant to help, others are meant to hinder.

By nesting other modalities in the postconditions, we can still express some limited form of alternation in how choices resolve:

$$[\alpha_1]\langle\alpha_2\rangle[\alpha_3]\langle\alpha_4\rangle P$$

This dL formula expresses that after all choices of HP α_1 there is a way of running HP α_2 such that for all ways of running HP α_3 there is a choice of running HP α_4 such that postcondition P is true. But that still only gives an opportunity for four rounds of alternation of choices in HPs and is not particularly concise even for that purpose. What we need is a more general way of ascribing actions to agents that allows an unbounded number of alternation of choices.

14.2.2 Control & Dual Control

Another way of looking at the choices that are to be resolved during the runs of a hybrid program α is that they can be resolved by one player. Let's call her *Angel*, because she helps us so much in making $\langle\alpha\rangle P$ formulas true. Whenever a choice is about to happen (by running the program statements $\alpha \cup \beta$, α^*, or $x' = f(x)\,\&\,Q$), Angel is called upon to see how the choice is supposed to be resolved this time. When playing $\alpha \cup \beta$, Angel chooses whether to play α or β. When playing α^*, Angel decides how often to play α. And when playing $x' = f(x)\,\&\,Q$, Angel decides how long to follow this differential equation within Q. Since Angel gets to choose, $\alpha \cup \beta$ is also called *angelic choice* and α^* is called *angelic repetition*.

From that perspective, it sounds easy enough to add a second player. Let's call him *Demon* as Angel's perpetual opponent.[1] Only so far, Demon will probably be quite bored after a while, when he realizes that he never actually gets to decide anything in the game, because Angel has all the fun in choosing how the game world unfolds and Demon just sits around idly and in apathy. So, to keep Demon engaged, we need to introduce some choices that fall under Demon's control.

One thing we could do to keep Demon interested in playing along in the hybrid game is to add a pair of shiny new controls especially for him. They might be called $\alpha \cap \beta$ for Demon's choice between α or β and $\alpha^\times$ for repetition of α under Demon's control. In fact, Demon might even demand an operation for continuous evolution under Demon's reign. But that would cause quite a lot of attention to Demon's controls, which might make him feel overly majestic. Let's not do that, because we don't want Demon to get any ideas.

---

[1] The names are quite arbitrary. But the responsibilities of such ontologically loaded names are easier to remember than those of neutral but boring player names such as player I and player II.

Instead, we will find it sufficient to add just a single operator to hybrid programs: the duality operator $\cdot^{\mathrm{d}}$ that can be used on any hybrid game α. What α^{d} does is to give all control that Angel had in game α to Demon, and, vice versa, all control that Demon had in α to Angel. The dual operator, thus, is a little bit like what happens when you turn a chessboard around by $180°$ in the middle of the game to play the game from the opponent's perspective. Whoever played the choices of player White previously will suddenly control Black, and whoever played Black now controls White (Fig. 14.1). Turning the game around twice as in $(\alpha^{\mathrm{d}})^{\mathrm{d}}$ restores the original game α. With just this single duality operator, Demon still gets his own set of controls ($\alpha \cap \beta$, $\alpha^{\times}$, and $\{x' = f(x) \,\&\, Q\}^{\mathrm{d}}$) by suitably nesting the operators, but we did not have to give him those controls specifically. Yet, now those extra controls are not special but simply an aspect of a more fundamental principle: duality.

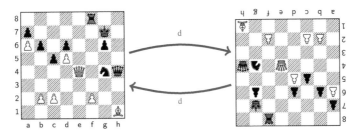

Fig. 14.1 Turning hybrid game α into the dual hybrid game α^{d} corresponds to turning a chessboard around by $180°$ so that the players control the choices in α^{d} that the opponent has in α

14.2.3 Demon's Derived Controls

Just as nondeterminism was in charge of all choices in a hybrid system, Angel has full control over all choices in each of the operators of hybrid games *except* when the operator $\cdot^{\mathrm{d}}$ comes into play. All choices within the scope of an odd number of $\cdot^{\mathrm{d}}$ belong to Demon, because $\cdot^{\mathrm{d}}$ makes the players switch sides. Demon's controls, i.e., direct controls for Demon, can be defined as derived operators with the duality operator $\cdot^{\mathrm{d}}$ from Angel's controls. Indeed, $(\alpha^{\mathrm{d}})^{\mathrm{d}}$, the dual of a dual, is the original game α, just like flipping a chessboard around twice results in the original chessboard. That is why it only matters whether a choice occurs within the scope of an odd number of $\cdot^{\mathrm{d}}$ (Demon's choice) or an even number of $\cdot^{\mathrm{d}}$ (Angel's choice).

Demonic choice $\alpha \cap \beta$ will play either hybrid game α or hybrid game β by Demon's choice. It is defined by $(\alpha^{\mathrm{d}} \cup \beta^{\mathrm{d}})^{\mathrm{d}}$. The choice for the $\cup$ operator belongs to Angel, yet since it is nested within $\cdot^{\mathrm{d}}$, that choice goes to Demon, except that the $\cdot^{\mathrm{d}}$ operators around hybrid games α and β restore the original ownership of controls. The hybrid game $(\alpha^{\mathrm{d}} \cup \beta^{\mathrm{d}})^{\mathrm{d}}$ corresponds to turning the chessboard around, thus, giving the choice between α^d and β^d that would have been Angel's to Demon, and then turning the chessboard in either α^d or β^d back again to either α or β.

Demonic repetition $\alpha^\times$ repeats hybrid game α as often as Demon chooses to. It is defined by $((\alpha^d)^*)^d$. The choice in the * operator belongs to Angel, but goes to Demon in a $\cdot^d$ context, while the choices in the α subgame underneath stay as they were originally thanks to the additional $\cdot^d$ operator that restores the game back to normal responsibilities. Again, $((\alpha^d)^*)^d$ corresponds to turning the chessboard around, thus giving the choice of repetition that would have been Angel's to Demon, yet turning the chessboard in α^d around again to play the original α.

The *dual differential equation* $\{x' = f(x) \,\&\, Q\}^d$ follows the same dynamics as $x' = f(x) \,\&\, Q$ except that, because of the duality operator, Demon now chooses the duration. He has to choose a duration during which Q holds all the time. Hence he loses when Q does not hold in the current state. Similarly, the *dual test* $?Q^d$ will make Demon lose the game immediately if the formula Q does not hold in the current state, just as the test $?Q$ will make Angel lose the game immediately if the formula Q does not hold currently. Dual assignment $(x := e)^d$ is equivalent to ordinary assignment $x := e$, because assignments never involve any choices to begin with, so it does not matter which player plays them.

Angel's control operators and Demon's control operators correspond to each other by duality:

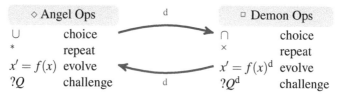

Because the double dual $(\alpha^d)^d$ is the same as the game α, we never have to use the duality operator $\cdot^d$ except in Demon's choice $\cap$, Demon's repetition $^\times$, or around differential equations and tests. But it is more systematic to just allow $\cdot^d$ everywhere.

14.3 Syntax of Differential Game Logic

Differential game logic (dGL) is a logic for studying properties of hybrid games [9]. The idea is to describe the game form, i.e., rules, dynamics, and choices of the particular hybrid game of interest, using a program notation and to then study its properties by proving the validity of logical formulas that refer to the existence of winning strategies for objectives of those hybrid games. This is analogous to how a differential dynamic logic formula $[\alpha]P$ separately describes the dynamics of the hybrid system as a hybrid program α and the property of interest in the modality's postcondition P.

14.3.1 Hybrid Games

Even though hybrid game forms only describe the *form* of the game with its dynamics and rules and choices, not the actual objective, they are still simply called hybrid games just for simplicity of terminology. The objective for a hybrid game is defined in the postcondition of the modal logical formula that refers to that hybrid game form. During a hybrid game the players can only lose by violating the rules of the game, never win. The proper winning condition is specified in the dGL formula. Hybrid games (HGs in Definition 14.1) and differential game logic formulas (Definition 14.2) are defined subsequently.

Definition 14.1 (Hybrid games). The *hybrid games of differential game logic* dGL are defined by the following grammar (α, β are hybrid games, x is a vector of variables, $f(x)$ is a vector of (polynomial) terms of the same dimension, and Q is a dGL formula or just a formula of first-order real arithmetic):

$$\alpha, \beta ::= x := e \mid x' = f(x) \& Q \mid ?Q \mid \alpha \cup \beta \mid \alpha; \beta \mid \alpha^* \mid \alpha^d$$

The only syntactical difference of hybrid games compared to hybrid programs for hybrid systems as in Chap. 3 is that, unlike hybrid programs, hybrid games also allow the dual operator α^d. This minor syntactic change will require us to reinterpret the meaning of the other operators in a much more flexible way in order to make sense of the presence of subgames within the games in which the players already interact. The basic principle is that whenever there used to be nondeterminism in the hybrid program semantics, there will now be a choice that is up to Angel in the hybrid game semantics. But don't be fooled! The parts of such a hybrid game may still be hybrid games, in which players interact, rather than just a single system running. So *all* operators of hybrid games still need to be carefully understood as games, not just the *duality operator* $\cdot^d$, because all operators can be applied to subgames that mention $\cdot^d$ or be part of a context that mentions a $\cdot^d$ duality.

The *atomic games* of dGL are assignments, continuous evolutions, and tests. In the *deterministic assignment game* (or discrete assignment game) $x := e$, the value of variable x changes instantly and deterministically to that of e by a discrete jump without any choices to resolve, just as it already was the case for the HP $x := e$. In the *continuous evolution game* (or continuous game) $x' = f(x) \& Q$, the system follows the differential equation $x' = f(x)$ where the duration is Angel's choice. But Angel is not allowed to choose a duration that would, at any time, take the state outside the region where *evolution domain constraint* formula Q holds. In particular, Angel is deadlocked and loses immediately if Q does not hold in the current state, because she cannot even evolve for duration 0 then without being outside Q.[2] The

---

[2] The most common case for Q is a formula of first-order real arithmetic, but any dGL formula will work (Definition 14.2). Evolution domain constraints turn out to be unnecessary, because they can be defined using hybrid games. In the ordinary differential equation $x' = f(x)$, the term x' denotes the time-derivative of x and $f(x)$ is a polynomial term that may mention x and other variables. More general forms of differential equations are possible [2, 3, 12], but will not be considered explicitly.

test game or *challenge* $?Q$ has no effect on the state, except that Angel loses the game immediately if dGL formula Q does not hold in the current state, because she failed the test she was supposed to pass. The test game $?Q$ challenges Angel and she loses immediately if she fails. Angel does not win just because she passes the challenge $?Q$, but at least the game continues. So passing challenges is a necessary condition to win games. Failing challenges, instead, immediately makes Angel lose. That makes tests $?Q$ in hybrid games the direct game counterpart of tests $?Q$ in hybrid programs. In order to properly track who won, we just need to get used to the notion of losing a game, instead of just aborting and discarding an HP execution.

The *compound games* of dGL are sequential, choice, repetition, and duals. The *sequential game* $\alpha;\beta$ is the hybrid game that first plays hybrid game α and then, when hybrid game α terminates without a player having lost already (so no challenge in α failed), continues by playing game β. When playing the *choice game* $\alpha \cup \beta$, Angel chooses whether to play hybrid game α or play hybrid game β. Like all the other choices, this choice is dynamic, i.e., every time $\alpha \cup \beta$ is played, Angel gets to choose again whether she wants to play α or β this time. She is not bound by whatever Angel chose last time. The *repeated game* α^* plays hybrid game α repeatedly and Angel chooses, after each play of α that terminates without a player having lost already, whether to play the game again or not, although she cannot choose to play indefinitely but has to stop repeating ultimately. Angel is allowed to stop α^* right away after zero iterations of α. Most importantly, the *dual game* α^d is the same as playing the hybrid game α with the rôles of the players swapped. That is Demon decides all choices in α^d that Angel has in α, and Angel decides all choices in α^d that Demon has in α. Players who are supposed to move but deadlock lose. Thus, while the test game $?Q$ causes Angel to lose if formula Q does not hold, the *dual test game* (or *dual challenge*) $(?Q)^d$ instead causes Demon to lose if Q does not hold.

For example, if α describes the game of chess, then α^d is chess where the players switch sides. If α, instead, describes a hybrid game where you are controlling your robot and a test engineer controls the roguebot, then α^d describes the dual game where you take control of the roguebot and the test engineer is stuck with your robot controls. If your test engineer is out for lunch, you can also play both robots. You just have to remember to play them both faithfully according to their objectives and can't cheat to make the test engineer's roguebot run away in terror just because that would make the job of your own robot easier. The real world isn't likely to make robot control so easy for you later. In fact, this pretend-play is another good way of understanding the intuition behind the duality operator. When playing α^d, you pretend to play for the other player in game α, respecting his objectives.

The dual operator $\cdot^d$ is the only syntactic difference of hybrid games compared to hybrid systems [1, 8], but a fundamental one [9], because it is the only operator where control passes from Angel to Demon or back. Without $\cdot^d$ all choices are resolved uniformly by Angel without interaction. The presence of $\cdot^d$ requires a thorough semantic generalization throughout the logic to cope with such flexibility.

Example 14.1 (Push-around cart). Consider a cart at position x moving along a straight line with velocity v that both Angel and Demon are pushing around simultaneously. Depending on whether they push or pull the cart, the two players will

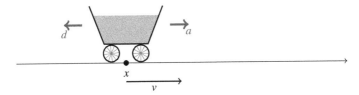

Fig. 14.2 Angel and Demon accelerating or braking by a and d, respectively, the cart at position x, which is moving with velocity x

each exert either an accelerating force or a braking force on x (Fig. 14.2):

$$((a:=1 \cup a:=-1); (d:=1 \cup d:=-1)^d; \{x'=v, v'=a+d\})^* \qquad (14.1)$$

First Angel chooses (by $\cup$) a positive or negative acceleration a, then Demon chooses a positive or negative acceleration d. This choice is Demon's because the choice $\cup$ occurs within the scope of a duality operator $\cdot^d$, so what used to be Angel's choice becomes Demon's choice. Recall that it does not matter who controls the assignments, because they come without a choice. Finally, the game follows the differential equation system $x'=v, v'=a+d$ in which the sum of both accelerations a and d chosen by Angel and Demon, respectively, take effect, because the sum of all forces acts as acceleration on the cart x with unit mass. The cart is a point x, so Demon can't cheat and use his force to make it fall over. The duration of the differential equation is Angel's choice. Finally the game repeats (*), as often as Angel wants, because Demon is bored and walks away from the cart if Angel ever decides to stop playing. Each round of this repetition, Angel does not know which choice Demon will use for d, because she chooses a first before the sequential composition (;). This is unlike the following hybrid game where Demon chooses first:

$$((d:=1 \cup d:=-1)^d; (a:=1 \cup a:=-1); \{x'=v, v'=a+d\})^* \qquad (14.2)$$

But Angel controls the duration of the differential equation in both (14.1) and (14.2), so she can still choose duration 0 if she does not like Demon's choice of d. He just might choose the same inopportune value for d during the next repetition, so Angel will ultimately have to accept some decision by Demon and evolve for a positive duration or else the cart will never move anywhere, which would be permitted but incredibly boring for everyone. Whichever player decides on the acceleration last, so Demon in (14.1) and Angel in (14.2), can decide to keep the velocity unchanged by playing the opposite acceleration value such that $a+d=0$.

Which choices and decisions are particularly clever ones for Angel and Demon is a separate question and depends on the objective of the hybrid game, which is what dGL formulas will be used for. Hybrid systems lack the ability to express that the choice of a in (14.1) and (14.2) is Angel's while the choice of d is Demon's. A hybrid system could also have choices (without $\cdot^d$ duals):

$$((d:=1 \cup d:=-1); (a:=1 \cup a:=-1); \{x'=v, v'=a+d\})^*$$

But then all choices are nondeterministic, so resolved by the same player and either all help (if in a diamond modality) or all hurt (if in a box modality). In hybrid game (14.2), however, the choice of the acceleration d helps Demon while the choice of the acceleration a helps Angel, as do the choice of the duration of the differential equation and the number of repetitions.

Demon's controls such as $\alpha \cap \beta$ and $\alpha^\times$ can be defined with the help of the duality operator $\cdot^d$ as in Sect. 14.2.3. In $\alpha^\times$, Demon chooses after each play of α whether to repeat the game, but cannot play indefinitely so he has to stop repeating ultimately. By duality, this follows from the fact that, in α^*, Angel also chooses after each play of α whether to repeat the game but she cannot play indefinitely.

Example 14.2 (Push-around cart). Demon's control operators rephrase (14.1) as

$$\big((a:=1 \cup a:=-1); (d:=1 \cap d:=-1); \{x' = v, v' = a+d\}\big)^*$$

Strictly speaking, $d:=1 \cap d:=-1$ is $((d:=1)^d \cup (d:=-1)^d)^d$. But that is equivalent to $(d:=1 \cup d:=-1)^d$, because deterministic assignments $x:=e$ are equivalent to dual assignments $(x:=e)^d$, since both involve no choice. Similarly, (14.2) is

$$\big((d:=1 \cap d:=-1); (a:=1 \cup a:=-1); \{x' = v, v' = a+d\}\big)^*$$

These were a lot of games but not a lot of purpose yet, which is where the dGL formulas come in. We consider them next.

14.3.2 Differential Game Logic Formulas

Hybrid games describe how the world can unfold when Angel and Demon interact according to their respective control choices. They explain the rules of the game, how Angel and Demon interact, and what the players can choose to do, but not who wins the game, nor what the respective objectives of the players are.[3] The actual winning conditions are specified by logical formulas of differential game logic.

We cannot continue the same understanding of modalities from Part I and Part II of this book, where the dL formula $[\alpha]P$ says that all runs of HP α satisfy P while the dL formula $\langle\alpha\rangle P$ says that at least one run of HP α satisfies P. It is not very meaningful to talk about all runs or some run of a hybrid game, because the whole point of games is that they provide a number of choices to the different players that may unfold differently in response to one another. Since the players have objectives, only some of those choices will manifest and be in their interest. What the players choose to do depends on what their opponent did before and vice versa. It is not particularly interesting if a player can lose a game by playing entirely stupidly. What is much more exciting is the question of whether the player can win if she plays in a clever way. And it is maximally compelling if a player even has a consistent way

---

[3] Except that players lose if they disobey the rules of the game by failing their respective challenges.

of always winning the game, no matter what the opponent is trying. Then the player has a *winning strategy*, i.e., a way to resolve her actions that will always win the game for all strategies that her opponent might try. This makes game play quite interactive; one has to find some choice for the player and consider all options for the opponent.

Modal formulas $\langle \alpha \rangle P$ and $[\alpha]P$ refer to hybrid games and the existence of winning strategies for Angel and Demon, respectively, in a hybrid game α with a winning condition specified by a logical formula P.

> **Definition 14.2 (dGL formulas).** The *formulas of differential game logic* dGL are defined by the following grammar (P, Q are dGL formulas, $e, \tilde{e}$ are terms, x is a variable, and α is a hybrid game):
>
> $$P, Q ::= e \geq \tilde{e} \mid \neg P \mid P \wedge Q \mid \exists x P \mid \langle \alpha \rangle P \mid [\alpha]P$$

Other operators $>, =, \leq, <, \vee, \rightarrow, \leftrightarrow, \forall x$ can be defined, e.g., $\forall x P \equiv \neg \exists x \neg P$.

The modal formula $\langle \alpha \rangle P$ expresses that Angel[4] has a winning strategy to achieve P in hybrid game α, i.e., Angel has a strategy to reach any of the states satisfying dGL formula P when playing hybrid game α, no matter what strategy Demon chooses. The modal formula $[\alpha]P$ expresses that Demon has a winning strategy to achieve objective P in hybrid game α, i.e., a strategy to reach any of the states satisfying P, no matter what strategy Angel chooses. The same game is played in $[\alpha]P$ as in $\langle \alpha \rangle P$ with the same choices resolved by the same players. The difference between the two dGL formulas is the player whose winning strategy they refer to. Both use the set of states where dGL formula P is true as the set of winning states for that player. The winning condition is defined by the modal formula; α only defines the hybrid game form, not when the game is won, which is what P does. Hybrid game α defines the rules of the game, including conditions on state variables that, if violated, cause the present player to lose for violation of the rules of the game. The dGL formulas $\langle \alpha \rangle P$ and $[\alpha] \neg P$ consider complementary winning conditions for Angel and Demon. Of course, the propositional logical connectives $\neg, \wedge, \vee, \rightarrow$ still mean what they always do and the quantifiers $\exists x P$ and $\forall x P$ quantify over the reals.

14.3.3 Examples

This section discusses some examples of hybrid games and states differential game logic formulas expressing properties of winning strategies for these games.

Example 14.3 (Push-around cart). Continuing Example 14.1, consider a dGL formula for the cart-pushing hybrid game from (14.2):

---

[4] It is easy to remember which modal operator is which. The formula $\langle \alpha \rangle P$ clearly refers to Angel's winning strategies because the diamond operator $\langle \cdot \rangle$ has wings. This is consistent with the fact that Angel takes charge of what nondeterminism used to do in dL, so $\langle \alpha \rangle P$ is where Angel's control $\cup, {}^*, x' = f(x)$ helps, just as nondeterminism helped in the diamond modality of dL.

$$v \geq 1 \rightarrow \left[\left((d := 1 \cup d := -1)^{\mathrm{d}}; (a := 1 \cup a := -1); \{x' = v, v' = a + d\} \right)^* \right] v \geq 0$$

This dGL formula expresses that Demon has a winning strategy to ensure that the cart's velocity v is nonnegative if it initially started at $v \geq 1$. That would have been trivial if we had considered hybrid game (14.1), in which Demon chooses d after Angel chose a such that the choice $d := -a$ would trivially ensure $v' = 0$. But Demon's choice $d := 1$ still makes sure that the velocity will never decrease, whether Angel subsequently chooses to also push ($a := 1$) or to slow the cart down ($a := -1$). For the same reason, Demon also has a winning strategy to achieve $x \geq 0$ if the cart initially starts with $v \geq 0$ at $x \geq 0$. That is, the following formula is valid:

$$x{\geq}0 \wedge v{\geq}0 \rightarrow \left[\left((d := 1 \cup d := -1)^{\mathrm{d}}; (a := 1 \cup a := -1); \{x' = v, v' = a + d\} \right)^* \right] x \geq 0$$

When replacing the box modality by a diamond modality, the formula

$$x \geq 0 \rightarrow \left\langle \left((d := 1 \cup d := -1)^{\mathrm{d}}; (a := 1 \cup a := -1); \{x' = v, v' = a + d\} \right)^* \right\rangle x \geq 0$$

expresses that Angel also has a winning strategy to achieve $x \geq 0$ in the same hybrid game starting from just the initial condition $x \geq 0$. But even if that dGL is valid as well, it is trivially valid, because Angel controls repetition (*) and can simply decide on 0 iterations which makes the game stay in the initial state where $x \geq 0$ already holds. The same would happen if Demon were to control the repetition with Demon's repetition $^\times$ instead of Angel's repetition * as long as Angel still controls the differential equation, because she can simply go for duration 0 every time:

$$x \geq 0 \rightarrow \left\langle \left((d := 1 \cup d := -1)^{\mathrm{d}}; (a := 1 \cup a := -1); \{x' = v, v' = a + d\} \right)^\times \right\rangle x \geq 0$$

Without that assumption $x \geq 0$ on the initial state, however,

$$\left\langle \left((d := 1 \cup d := -1)^{\mathrm{d}}; (a := 1 \cup a := -1); \{x' = v, v' = a + d\} \right)^* \right\rangle x \geq 0$$

is not valid, because, unless v is already nonnegative initially, Demon can always play $d := -1$, which will make it impossible for Angel to give it a positive velocity $a + d$. If Angel is stronger than Demon, the corresponding dGL formula is valid:

$$\left\langle \left((d := 1 \cup d := -1)^{\mathrm{d}}; (a := 2 \cup a := -2); \{x' = v, v' = a + d\} \right)^* \right\rangle x \geq 0$$

All that Angel needs to do to achieve $x \geq 0$ is to push really hard with $a := 2$ and continuously evolve for long enough. More subtly, even if Demon has the same strength, Angel, nevertheless, has a winning strategy to achieve $x^2 \geq 100$:

$$\left\langle \left((d := 2 \cup d := -2)^{\mathrm{d}}; (a := 2 \cup a := -2); \right. \right.$$
$$\left. \left. t := 0; \{x' = v, v' = a + d, t' = 1 \& t \leq 1\} \right)^* \right\rangle x^2 \geq 100 \tag{14.3}$$

Angel has no influence on Demon's decision on d. But all it takes for Angel is to play $a := 2$ if $v > 0$ and play $a := -2$ if $v < 0$ to ensure that the sign of v never changes, whatever Demon plays, and, thus, x will eventually either grow above 10

or shrink below -10. If $v = 0$ initially, then Angel first plays $a := d$ to mimic Demon in the first round and make v nonzero, which she can since Demon decides
first and Angel controls the differential equation's duration. Hence, (14.3) is valid,
too. If (14.3) did not have a time bound on the duration of the evolution, it would
be more obviously valid without repeating, because Angel could just mimic Demon
once and then follow the differential equation for a long time. The differential equation is Angel's choice, but she has an evolution domain constraint $t \leq 1$ to worry
about. Since clock $t' = 1$ was reset to $t := 0$ before, she cannot follow the differential
equation for more than 1 time unit without losing for violation of the rules of the
game. Thus, both players get to change their control variables at least once a second
but Angel controls when exactly. Every time they get to change control variables d
and a, Demon chooses first (before the sequential composition).

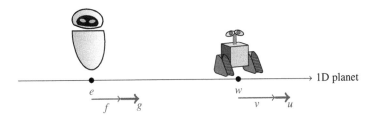

Fig. 14.3 Velocities v, f and accelerations u, g of two robots at w and e on a one-dimensional planet

Example 14.4 (WALL·E and EVE robot dance). Consider a game of the two robots
WALL·E and EVE moving on a rather flat one-dimensional planet (Fig. 14.3):

$$(w - e)^2 \leq 1 \wedge v = f \rightarrow \big\langle \big((u := 1 \cap u := -1);$$
$$(g := 1 \cup g := -1);$$
$$t := 0; \{w' = v, v' = u, e' = f, f' = g, t' = 1 \& t \leq 1\}^d \big)^\times$$
$$\big\rangle (w - e)^2 \leq 1$$

$$(14.4)$$

Despite the dimensionally somewhat impoverished planet, this dGL formula provides a canonical use case for a hybrid game. Robot WALL·E is at position w with
velocity v and acceleration u and plays the part of Demon. Robot EVE is at position
e with velocity f and acceleration g and plays the part of Angel.

The antecedent of (14.4) before the implication assumes that WALL·E and EVE
start close to one another (distance at most 1) and with identical velocities. The
objective of EVE, who plays Angel's part in (14.4), is to be close to WALL·E (i.e.,
$(w - e)^2 \leq 1$) as specified after the $\langle \cdot \rangle$ modality in the succedent. The hybrid game
proceeds as follows. Demon WALL·E controls how often the hybrid game repeats by
operator $^\times$. In each iteration, Demon WALL·E first chooses (with Demon's choice
operator $\cap$) to accelerate ($u := 1$) or brake ($u := -1$), then Angel EVE chooses (with

Angel's choice operator ∪) whether to accelerate ($g := 1$) or brake ($g := -1$). Every time that the $^\times$ loop repeats, the players get to make that choice again. They are not bound by what they chose in the previous iterations. Yet, depending on the previous choices, the state will have evolved differently, which influences indirectly what moves a player needs to choose to win. After this sequence of choices of u and g by Demon and Angel, respectively, a clock variable t is reset to $t := 0$. Then the hybrid game follows a differential equation system such that the time-derivative of WALL·E's position w is his velocity v and the time-derivative of v is his acceleration u; simultaneously, the time-derivative of EVE's position e is her velocity f and the time-derivative of f is her acceleration g. The time-derivative of clock variable t is 1, yet the differential equation is restricted to the evolution domain $t \leq 1$ so it can at most be followed for 1 time unit. Angel controls the duration of differential equations. Yet, this differential equation is within a dual game due to the operator $\cdot^d$ around it, so Demon actually controls the duration of the continuous evolution. Here, both WALL·E and EVE evolve continuously but Demon WALL·E decides how long. He cannot choose durations > 1, because that would make him violate the evolution domain constraint $t \leq 1$ and lose. So both players can change their control after at most one time unit, but Demon decides when exactly. Similar games can be studied for robot motion in higher dimensions using dGL.

The dGL formula (14.4) is valid, because Angel EVE has a winning strategy to get close to WALL·E by mimicking Demon's choices. Recall that Demon WALL·E controls the repetition $^\times$, so the fact that the hybrid game starts EVE off close to WALL·E is not sufficient for EVE to win the game. Mimicking by $g := u$ will also only work so easily because both start with the same initial velocity $v = f$. The hybrid game in (14.4) would be trivial if Angel were to control the repetition (because she would then win just by choosing not to repeat) or if Angel were to control the differential equation (because she would then win by always just evolving for duration 0). Hybrid games are most interesting when the choices are not already stacked in one player's favor. The analysis of (14.4) is more difficult if the first two lines in the hybrid game are swapped so that Angel EVE chooses g before Demon WALL·E chooses u, because she cannot play the copy strategy if Angel has to choose first.

Example 14.1 had a single differential equation system in which the controls of Angel and Demon mix via $x'' = a + d$, while Example 14.4 had a bigger differential equation system consisting of differential equations $w' = v, v' = u$ that belong to WALL·E and other differential equations $e' = f, f' = g$ that belong to EVE, which are joined together with time $t' = 1$. Both players evolve their respective variables together. The question of whether the resulting combined differential equation system is under Angel's or Demon's control is separate, and just depends on who gets to decide on the duration. This is in direct analogy to a loop body in which multiple operations by Angel and Demon might occur but still one of the two players needs to be responsible for deciding how often to repeat the loop itself, because the players might never come to an agreement if both were in charge of the same operator.

Example 14.5 (WALL·E and EVE and the world). The game in (14.4) accurately reflects the situation when WALL·E, who plays the part of Demon, is in control of time since the differential equation occurs within an odd number of $\cdot^{\mathrm{d}}$ operators. But this is not the only circumstance in which (14.4) is the right game to look at for EVE. Suppose there really is a third player, the external environment, which controls time. So, neither WALL·E nor EVE really gets to decide on how long differential equations are followed, nor on how often the loop repeats.

EVE can use a common modeling device to conservatively attribute the control of the differential equation to WALL·E, even if time is really under the control of a third player, the external environment. EVE's reason for this model would be that she is certainly not in control of time, so there is no reason to believe that time would help her. EVE, thus, conservatively cedes control of time to Demon, which corresponds to assuming that the third player of the external environment is allowed to collaborate with WALL·E to form an aggregate Demon player consisting of WALL·E and the environment. If, as the validity of the resulting formula (14.4) indicates, Angel EVE wins against the Demon team consisting of WALL·E and the world, then she wins no matter what WALL·E and the external world decide to do.

This answers what hybrid game EVE needs to analyze to find out when she has a winning strategy for all actions of WALL·E and the world. When WALL·E wants to analyze his winning strategies he cannot just use the $[\cdot]$ modality for the same hybrid game as in (14.4) anymore, because that hybrid game was formed by conservatively attributing the external world's control of time to Demon. However, Demon WALL·E can use the same modeling device to flip the world's differential equation control over to Angel's control by removing the $\cdot^{\mathrm{d}}$ to conservatively associate the environment with his opponent (and negate the postcondition to consider the opposite goal):

$$(w-e)^2 \leq 1 \wedge v = f \rightarrow \big[\big((u:=1 \cap u:=-1);$$
$$(g:=1 \cup g:=-1);$$
$$t:=0; \{w'=v, v'=u, e'=f, f'=g, t'=1 \,\&\, t \leq 1\}\big)^{\times}$$
$$\big] (w-e)^2 > 1$$

$$(14.5)$$

Observe how a three-player game of WALL·E, EVE, and the environment can be analyzed by combining the dGL formulas (14.4) and (14.5) propositionally, which then analyze the same game from different perspectives of different possible collaborations. The dGL formula expressing that neither (14.4) nor (14.5) is true, is true in exactly the states where WALL·E and EVE draw, because the external environment can choose the winner by helping either WALL·E or EVE. Here, (14.5) is unsatisfiable, because Demon needs to move first, so Angel can always mimic him to stay close.

The WALL·E and EVE examples were *games for analytic purposes*. WALL·E and EVE are not actually in adversarial competition with opposing objectives. They just did not know each other any better yet when they first met. And they still suffer some amount of uncertainty about each other's decisions, which can lead to a game situation for lack of better knowledge. The next example is one of *true adversarial competition* where the two players seriously complete.

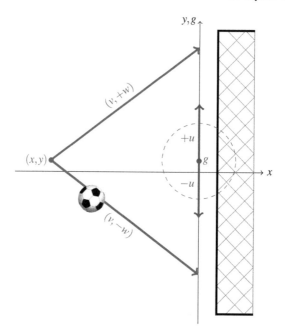

Fig. 14.4 Goalie g in robot soccer moves with velocity $\pm u$ up or down and, if within radius 1, can capture the ball (x,y) moving with velocity $(v,\pm w)$ sloped up or downards.

Example 14.6 (Goalie in robot soccer). Consider two robots engaged in a robot soccer match. Demon's robot is in possession of the ball and has a free kick toward the goal. Angel's robot is a goalie at position g who is trying hard to prevent Demon's robot from scoring a goal (Fig. 14.4). The ball is at position (x,y). Demon can either kick to roll the ball with vectorial velocity (v,w) into the left corner of the goal or with velocity $(v,-w)$ into its right corner. Angel's goalie robot can repeatedly move up or down near the goal line with linear velocity u or $-u$. She will capture the ball if the ball (v,w) is within radius 1 of the goalie $(0,g)$, i.e., if $x^2 + (y-g)^2 \le 1$. The two robots (and thus also the ball) are initially assumed to be at different x coordinates but the same y coordinate, with the ball being kicked toward the goal ($v > 0$):

$$x < 0 \wedge v > 0 \wedge y = g \rightarrow$$
$$\langle (w := +w \cap w := -w);$$
$$((u := +u \cup u := -u); \{x' = v, y' = w, g' = u\})^* \rangle x^2 + (y-g)^2 \le 1$$

(14.6)

Demon's robot only has one control decision, in line 2, which is the direction in which he kicks the ball. Once the ball is rolling, there's no turning back. Angel subsequently has a series of control decisions, both how often to repeat the subsequent control loop but also whether to move the goalie up or down (in line 3) and how long to follow the differential equation where the ball at position (x,y) rolls with velocity (v,w) in that direction and the goalie at position g moves with velocity u.

Whether dGL formula (14.6) is true depends on the relationship of the initial ball position x to the respective velocities v, w, u. The easiest case where it is true is $w = u$, in which case the vertical velocity w of the ball is identical to the goalie's velocity u, so that a mimic strategy $u := w$ will make Angel win and capture the ball. More generally,

$$\left(\frac{x}{v}\right)^2 (u-w)^2 \leq 1 \qquad (14.7)$$

implies that (14.6) is true. The time it takes for the ball to reach the goal line when moving with horizontal velocity v is $-x/v$. During that time the ball moves a distance of $-\frac{x}{v}w$ laterally in the y-direction, while the goalie moves a distance $-\frac{x}{v}u$. Since $y = g$ initially, the two positions will then be within capture distance 1 if (14.7) holds initially.

14.4 An Informal Operational Game Tree Semantics

Due to the subtleties and shift of perspective that hybrid games provide, the treatment of a proper semantics for differential game logic will be deferred to the next chapter. A graphical illustration of the choices that arise when playing hybrid games is depicted in Fig. 14.5. The nodes where Angel gets to decide are shown as diamonds $\diamond$, the nodes where Demon decides are shown as boxes $\diamond$. Circle nodes $\circ$ are shown when it depends on the remaining hybrid game which player gets to decide. Dashed edges - - - indicate Angel's actions to choose from, solid edges ——— indicates Demon's actions, while zigzag edges $\frown\!\!\frown$ indicate that a hybrid game is played and the respective players move as specified by that game.

The actions are the choice of real duration for $x' = f(x) \& Q$, the choice of playing the left or the right subgame for a choice game $\alpha \cup \beta$, and, after each round of a repetition, the choice of whether to stop or repeat in a repeated game α^*. The sequential game $\alpha; \beta$ has no actions to decide, except that game β starts after game α is done. There are no particular actions for the duality operator $\cdot^d$, which, however, flips the rôles of the players by flipping box nodes $\square$ that are under Demon's control with diamond nodes $\diamond$ that are under Angel's control. Mnemonically, $\cdot^d$ makes all nodes roll over by $45°$ so that boxes $\square$ turn into diamonds $\diamond$ and diamonds $\diamond$ turn into boxes $\square$. Assignments and tests also have no particularly interesting actions, except that Angel loses game $?Q$ unless Q is true in the current state.

The game tree action principles can be made rigorous in an operational game semantics [9], which conveys the intuition of interactive game play for hybrid games, relates to classical game theory and descriptive set theory, but is beyond the scope of this textbook, because Chap. 15 will investigate a significantly simpler denotational semantics. Observe how all choices involve at most two possibilities except differential equations, which have uncountably infinitely many choices, one option

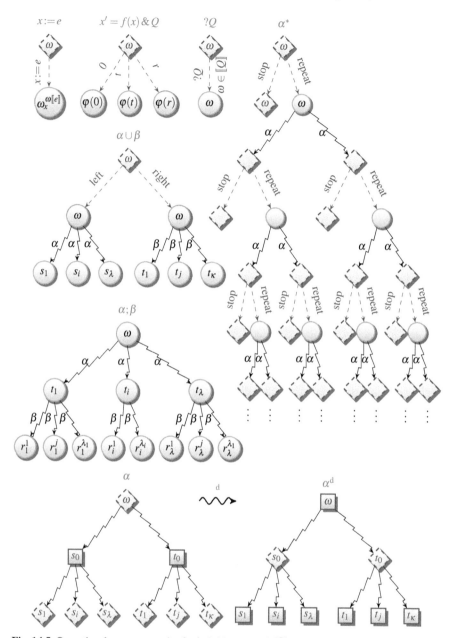

Fig. 14.5 Operational game semantics for hybrid games of dGL

for each nonnegative duration $r \in \mathbb{R}$. Of course, some of those durations may be a pretty bad idea if they would fail the evolution domain constraint, but that is up to the respective player to decide.

A *strategy* for a player in a hybrid game can be understood as a way of selecting a (state-dependent) action at each of the nodes of the game tree where that player has the choice, so an action at each diamond node for a strategy for Angel or an action at each box node for a strategy for Demon. A *winning strategy* is a strategy that leads to a winning state for all of the opponent's strategies.

As an example to illustrate some of the subtle nuances in defining an appropriate semantics for hybrid games, consider the discrete *filibuster formula*:

$$\langle (x:=0 \cap x:=1)^* \rangle x = 0 \tag{14.8}$$

It is Angel's choice whether to repeat (*), but every time Angel repeats, it is Demon's choice ($\cap$) whether to play $x:=0$ or $x:=1$. What is the truth-value of the dGL formula (14.8)?

The game in this formula never deadlocks, because each player always has at least one remaining move (here even two because Angel can stop or repeat and Demon can assign 0 or 1 to x). But it may appear that the game has perpetual checks, because no strategy helps either player win the game; see Fig. 14.6. Every time Angel chooses to repeat, hoping for an outcome of $x = 0$, Demon can stubbornly choose to play the right subgame $x:=1$ to make x one. That will not make Demon win either, because Angel is still in charge of deciding about repeating, which she will want to do to avoid the catastrophic outcome $x = 1$ that would make her lose. But next time around the loop, the situation is essentially unchanged, because Demon will still not want to give in and will, thus, cleverly play $x:=1$ again. How can that happen in this game and what can be done about it?

Before you read on, see if you can find the answer for yourself.

The mystery of the filibuster game can be solved when we remember that the game still ultimately has to stop in order that we may inspect who finally won the game. Angel is in charge of the * repetition and she can decide whether to stop or repeat. The filibuster game has no tests, so the winner only depends on the final state of the game, because both players are allowed to play arbitrarily without having to pass any tests in between. Angel wins a game play if $x = 0$ holds in the final state and Demon wins if $x \neq 0$ holds in the final state. What do the strategies indicated in Fig. 14.6 suggest? They postpone the end of the game, but if they did so indefinitely, there would never be a final state in which it could be evaluated who won. That is, indeed, not a way for anybody to win anything. Yet, Angel is in charge of the repetition *, so it is her responsibility to stop repeating eventually to evaluate who won. Consequently, the semantics of hybrid games allows the player in charge of a repetition to repeat as often as she wants, but she cannot repeat indefinitely. This will become apparent in the denotational semantics of hybrid games we will investigate in Chap. 15. Thus, (14.8) is *false* unless the winning condition $x = 0$ already holds initially, which allows Angel to just never repeat anything at all.

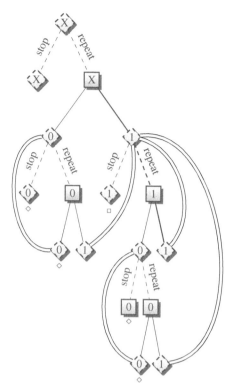

Fig. 14.6 The filibuster game formula $\langle(x:=0\cap x:=1)^*\rangle x=0$ looks as though it might be non-determined and not have a truth-value (unless $x=0$ initially) when the strategies follow the thick actions. Angel's action choices are illustrated by dashed edges from dashed diamonds, Demon's action choices by solid edges from solid squares, and double lines indicate identical states with the same continuous state and a subgame of the same structure of subsequent choices. States where Angel wins are marked by $\diamond$ and states where Demon wins by $\square$

The same phenomenon happens in the hybrid filibuster game:

$$\langle(x:=0;x'=1^d)^*\rangle x=0 \tag{14.9}$$

Both players can let the other one win. Demon can let Angel win by choosing to evolve his differential equation $x'=1^d$ for duration 0. And Angel can let Demon win by choosing to stop the repetition even if $x\neq 0$. Only because Angel will ultimately have to stop repeating does the formula in (14.9) have a proper truth-value and the formula is *false* unless $x=0$ already holds initially.

It is of similar importance that the players cannot decide to follow a differential equation forever (duration ∞), because that would make this game nondetermined:

$$\langle(x'=1^d;x:=0)^*\rangle x=0 \tag{14.10}$$

If players were allowed to evolve along a differential equation forever (duration ∞), then Demon would have an incentive to evolve along $x' = 1^d$ forever in the continuous filibuster (14.10). As soon as he stops the ODE, Angel can stop the loop and wins because of the subsequent assignment $x := 0$. But Angel cannot win without Demon stopping. Since Demon can evolve along $x' = 1^d$ for *any finite real amount of time* he wants, he will ultimately have to stop so that Angel wins and (14.10) is valid.

Table 14.1 Operators and (informal) meaning in differential game logic (dGL)

dGL	Operator	Meaning
$e = \tilde{e}$	equals	true iff values of e and $\tilde{e}$ are equal
$e \geq \tilde{e}$	greater or equal	true iff value of e greater-or-equal to $\tilde{e}$
$\neg P$	negation / not	true iff P is false
$P \wedge Q$	conjunction / and	true iff both P and Q are true
$P \vee Q$	disjunction / or	true iff P is true or if Q is true
$P \rightarrow Q$	implication / implies	true iff P is false or Q is true
$P \leftrightarrow Q$	bi-implication / equiv.	true iff P and Q are both true or both false
$\forall x\, P$	universal quantifier	true iff P is true for all values of variable x
$\exists x\, P$	existential quantifier	true iff P is true for some value of variable x
$[\alpha]P$	$[\cdot]$ modality / box	true iff Demon has winning strategy to achieve P in HG α
$\langle\alpha\rangle P$	$\langle\cdot\rangle$ modality / diamond	true iff Angel has winning strategy to achieve P in HG α

14.5 Summary

This chapter saw the introduction of differential game logic, summarized in Table 14.1, which extends the familiar differential dynamic logic with capabilities for modeling and understanding hybrid games. Hybrid games combine discrete dynamics, continuous dynamics, and adversarial dynamics, summarized in Table 14.2. Compared to hybrid systems, the new dynamical aspect of adversarial dynamics is captured entirely by the duality operator $\cdot^d$. Without it, hybrid games are single-player hybrid games, which are equivalent to hybrid systems. But the adversarial dynamics caused by the presence of the duality operator $\cdot^d$ also made us reflect on the semantics of all other composition operators for hybrid games.

After this chapter showed an informal and intuitive discussion of the actions that hybrid games allow, the next chapter gives a rigorous semantics to differential game logic and its hybrid games.

Table 14.2 Statements and effects of hybrid games (HGs)

HG Notation	Operation	Effect
$x := e$	assignment game	deterministically assigns value of e to variable x
$x' = f(x) \& Q$	continuous game	differential equation for x with term $f(x)$ within first-order constraint Q (evolution domain)
$?Q$	test / challenge	Angel loses unless formula Q holds at current state
$\alpha; \beta$	sequential game	HG β starts after HG α finishes
$\alpha \cup \beta$	choice game	Angel chooses between alternatives HG α or HG β
α^*	repeated game	Angel repeats HG α any finite number of times
α^d	dual game	swaps rôles of Angel and Demon in HG α

Exercises

14.1 (One-player games). Single-player hybrid games, i.e., ^{d}-free hybrid games, are just hybrid programs. For each of the following formulas, convince yourself that it has the same meaning whether you understand it as a differential dynamic logic formula with a hybrid system or as a differential game logic formula with a hybrid game (that happens to have only a single player):

$$\langle x := 0 \cup x := 1 \rangle x = 0$$
$$[x := 0 \cup x := 1]x = 0$$
$$\langle (x := 0 \cup x := 1); ?x = 1 \rangle x = 0$$
$$[(x := 0 \cup x := 1); ?x = 1]x = 0$$
$$\langle (x := 0 \cup x := 1); ?x = 0 \rangle x = 0$$
$$[(x := 0 \cup x := 1); ?x = 0]x = 0$$
$$\langle (x := 0 \cup x := 1)^* \rangle x = 0$$
$$[(x := 0 \cup x := 1)^*]x = 0$$
$$\langle (x := 0 \cup x := x + 1)^* \rangle x = 0$$
$$[(x := 0 \cup x := x + 1)^*]x = 0$$

14.2 (Single-player push-around carts). Hybrid game (14.2) was a single-player formulation in which all choices go to player Angel and Demon has nothing to do. Is the following dGL formula about it valid?

$$v \geq 1 \rightarrow \langle ((d := 1 \cup d := -1); (a := 1 \cup a := -1); \{x' = v, v' = a + d\})^* \rangle v \geq 0$$

Is the following dGL formula with a box modality valid, too?

$$v \geq 1 \rightarrow [((d := 1 \cup d := -1); (a := 1 \cup a := -1); \{x' = v, v' = a + d\})^*]v \geq 0$$

What does this imply about the required cleverness for appropriate control choices in similar hybrid games that we considered in Example 14.3? Even if we have not even fully considered a semantics let alone a proof calculus for hybrid games yet, can you still find a proof justifying the validity or a counterexample for the above two single-player hybrid game formulas?

14.3. In which states is the following dGL formula true and what is Demon's winning strategy in those states

$$[((\{x'=1\}\cup\{x'=-1\});(\{y'=1\}^d\cap\{y'=-1\}^d))^*]x<y$$

In which states is this variation true and what is Demon's winning strategy?

$$[((\{x'=1\}\cup\{x'=-1\});(\{y'=1\}^d\cap\{y'=-1\}^d))^*](x-y)^2<5$$

These dGL formulas have disconnected physics, where the duration of evolution of Angel's differential equation may have nothing to do with the duration of evolution of Demon's differential equation. Most games synchronize in time, however. The following dGL formula has different control choices for the different players but the differential equations are combined into a single differential equation system under Angel's control of time. In which states is the following formula true and what is Demon's winning strategy?

$$[((v:=1\cup v:=-1);(w:=1\cap w:=-1)\{x'=v,y'=w\})^*](x-y)^2<5$$

14.4. Consider the following dGL formulas and identify under which circumstance they are true:

$$\langle(x:=x+1;\{x'=x^2\}^d\cup x:=x-1)^*\rangle(0\le x<1)$$
$$\langle(x:=x+1;\{x'=x^2\}^d\cup(x:=x-1\cap x:=x-2))^*\rangle(0\le x<1)$$

14.5. Write down a valid formula that characterizes an interesting game between two robots and convince yourself whether it is valid or not.

14.6 (Robot simple chase game). The following dGL formula characterizes a one-dimensional game of chase between a robot at position x and another robot at position y, each with instant control of the velocity v among $a,-a,0$ for x (Angel's choice) and velocity w among $b,-b,0$ for y (Demon's subsequent choice). The game repeats any number of control rounds following Angel's choice (*). Angel is trying to get her robot x close to Demon's robot y. Under which circumstances is the formula true?

$$\langle(\ (v:=a\cup v:=-a\cup v:=0);$$
$$(w:=b\cap w:=-b\cap w:=0);$$
$$\{x'=v,y'=w\}\ )^*\rangle(x-y)^2\le 1$$

14.7 (Say when). For each of the following dGL formulas identify the set of states in which it is true and characterize this set by a formula of real arithmetic. For each case, briefly sketch the player's winning strategy when it is true and explain why the dGL formula is false in all other states:

$$\langle x := -1 \cup (x := 0 \cap x := y)\rangle x \geq 0$$

$$\langle \big((x := x+2 \cup (x := x-1; \{x' = -1\}^d))^* \big)\rangle 0 < x \leq 2$$

$$\langle x := x+2; x := x-1\rangle x \geq 0$$

$$\langle x := x-1 \cup (x := 0 \cap x := -y^2+1)\rangle x \geq 0$$

$$\langle x := y-1 \cup (\{x' = 1\}^d; x := x+2)\rangle x \geq 0$$

$$\langle x := -y \cup (x' = 2; \{x' = -1\}^d; x := x+2)\rangle x \geq 0$$

$$[(x := x \cap x' = -2)^*] x \geq 0$$

$$\langle (v := v \cap v := -v); (w := w \cup w := -w)\rangle v = w$$

$$\langle (v := v \cap v := -v); \{x' = v, y' = w\}\rangle x = y$$

$$\langle (v := v \cap v := -v); (w := w \cup w := -w); \{x' = v, y' = w\}\rangle x = y$$

$$\langle (x := x-1 \cap n := n-1; ?(n \geq 0)^d; x := x^2)^*\rangle x < 0$$

$$[(x := -x \cap x' = -x^2)^*] x \geq 0$$

$$\langle \big(x := 0 \cup ((x := x+1; \{x' = 1\}^d) \cup x := x-1) \big)^*\rangle 0 < x \leq 1$$

$$\langle \big(x := x^2 \cup (x := x+1 \cap x' = 2) \big)^*\rangle x > 0$$

$$\langle ((x := x+1; \{x' = x^2\}^d) \cup (x := x-1; \{x' = -1\}^d))^*\rangle 0 \leq x \leq 2$$

$$\langle ((x := x+1; \{x' = -1\}^d) \cup (x := x-1; \{x' = 1\}^d))^*\rangle 0 \leq x \leq 2$$

$$\langle ((x := x+1; \{x' = 1\}^d) \cup (x := x-1; \{x' = -1\}^d))^*\rangle 0 \leq x \leq 2$$

14.8 (* Robot chase). The following dGL formula characterizes a two-dimensional game of chase between a robot at position (x_1, x_2) facing in direction (d_1, d_2) and a robot at position (y_1, y_2) facing in direction (e_1, e_2). Angel has direct control over the angular velocity ω among $1, -1, 0$ for robot (x_1, x_2) and, subsequently, Demon has direct control over the angular velocity ρ among $1, -1, 0$ for robot (y_1, y_2). The game repeats any number of control rounds following Angel's choice (*). Angel is trying to get her robot close to Demon's robot. Is the following dGL formula valid? Can you identify some circumstances under which it is true? Or some circumstances under which it is false?

$$\bigg\langle \Big((\omega := 1 \cup \omega := -1 \cup \omega := 0);$$

$$(\rho := 1 \cap \rho := -1 \cap \rho := 0);$$

$$\{x_1' = d_1, x_2' = d_2, d_1' = -\omega d_2, d_2' = \omega d_1, y_1' = e_1, y_2' = e_2, e_1' = -\rho e_2, e_2' = \rho e_1\}^d$$

$$\Big)^* \bigg\rangle (x_1 - y_1)^2 + (x_2 - y_2)^2 \leq 1$$

14.9 (Goalies with resistance). Robot soccer balls have the irritating tendency to slow down after they have been kicked. Extend Example 14.6 with a model that takes the slowdown due to roll resistance and/or air resistance into account. Can you identify a condition under which the resulting formula is true? On a straight line, a point x of mass m with velocity v on flat terrain follows the differential equation

$$x' = v, v' = -av^2 - cgm$$

with gravity $g = 9.81\ldots$, a small roll resistance coefficient c and an even smaller aerodynamic coefficient a. What changes to say that no goal is scored?

References

[1] André Platzer. Differential dynamic logic for hybrid systems. *J. Autom. Reas.* **41**(2) (2008), 143–189. DOI: 10.1007/s10817-008-9103-8.

[2] André Platzer. Differential-algebraic dynamic logic for differential-algebraic programs. *J. Log. Comput.* **20**(1) (2010), 309–352. DOI: 10.1093/logcom/exn070.

[3] André Platzer. *Logical Analysis of Hybrid Systems: Proving Theorems for Complex Dynamics*. Heidelberg: Springer, 2010. DOI: 10.1007/978-3-642-14509-4.

[4] André Platzer. Stochastic differential dynamic logic for stochastic hybrid programs. In: *CADE*. Ed. by Nikolaj Bjørner and Viorica Sofronie-Stokkermans. Vol. 6803. LNCS. Berlin: Springer, 2011, 446–460. DOI: 10.1007/978-3-642-22438-6_34.

[5] André Platzer. A complete axiomatization of quantified differential dynamic logic for distributed hybrid systems. *Log. Meth. Comput. Sci.* **8**(4:17) (2012). Special issue for selected papers from CSL'10, 1–44. DOI: 10.2168/LMCS-8(4:17)2012.

[6] André Platzer. Dynamic logics of dynamical systems. *CoRR* **abs/1205.4788** (2012).

[7] André Platzer. Logics of dynamical systems. In: *LICS*. Los Alamitos: IEEE, 2012, 13–24. DOI: 10.1109/LICS.2012.13.

[8] André Platzer. The complete proof theory of hybrid systems. In: *LICS*. Los Alamitos: IEEE, 2012, 541–550. DOI: 10.1109/LICS.2012.64.

[9] André Platzer. Differential game logic. *ACM Trans. Comput. Log.* **17**(1) (2015), 1:1–1:51. DOI: 10.1145/2817824.

[10] André Platzer. Logic & proofs for cyber-physical systems. In: *IJCAR*. Ed. by Nicola Olivetti and Ashish Tiwari. Vol. 9706. LNCS. Berlin: Springer, 2016, 15–21. DOI: 10.1007/978-3-319-40229-1_3.

[11] André Platzer. A complete uniform substitution calculus for differential dynamic logic. *J. Autom. Reas.* **59**(2) (2017), 219–265. DOI: 10.1007/s10817-016-9385-1.

[12] André Platzer. Differential hybrid games. *ACM Trans. Comput. Log.* **18**(3) (2017), 19:1–19:44. DOI: 10.1145/3091123.

Chapter 15
Winning Strategies & Regions

Synopsis This chapter identifies a simple denotational semantics for hybrid games based on their winning regions, i.e., the set of states from which there is a winning strategy that wins the game for all strategies that the opponent might choose. Such a denotational semantics continues the successful trend in this book of understanding all operators in a compositional way. That is, the meaning of a compound hybrid game is a simple function of the meaning of its pieces. For repetitions in hybrid games, such a semantics will turn out to be surprisingly subtle, which will uncover a surprisingly rich complexity in hybrid games that is characteristically different from that of hybrid systems. This is the first indication that hybrid games come with their own unique sets of challenges beyond what hybrid systems already have in store for us.

15.1 Introduction

This chapter continues the study of hybrid games and their specification and verification logic, differential game logic [4], that Chap. 14 started. Chapter 14 saw the introduction of differential game logic with a primary focus on identifying and highlighting the new dynamical aspect of adversarial dynamics for modeling purposes. The meaning of hybrid games in differential game logic had been left informal, based on the intuition one relates to interactive gameplay and decisions in game trees. While it is possible to turn such a tree-type semantics into an operational semantics for hybrid games [4], the resulting development is technically rather involved. Even if such an operational semantics is informative and touches on interesting concepts from descriptive set theory, it is quite unnecessarily complicated.

This chapter will, thus, be devoted to developing a much simpler yet rigorous semantics, a denotational semantics for hybrid games. Chapter 14 already highlighted subtleties such as how never-ending game play ruins determinacy (i.e., that one player always has a winning strategy), simply because there never is a state in which the winner is declared. Especially the aspect of repetition and its interplay

© Springer International Publishing AG, part of Springer Nature 2018
A. Platzer, *Logical Foundations of Cyber-Physical Systems*,
https://doi.org/10.1007/978-3-319-63588-0_15

with differential equations will need careful attention now. The denotational semantics will make this subtle aspect crystal-clear.

This chapter is based on previous work [4], where more information can be found on logic and hybrid games. The most important learning goals of this chapter are:

Modeling and Control: We further our understanding of the core principles behind CPS for the adversarial dynamics resulting from multiple agents with possibly conflicting actions that occur in many CPS applications. This time, we devote attention to the nuances of their precise semantics. These observations will eventually uncover subtleties in the semantics of adversarial repetitions that makes them conceptually better behaved than the highly transfinite iterated winning-region construction. A byproduct of this development shows fixpoints in action, which play a prominent rôle in the understanding of other classes of models.

Computational Thinking: This chapter follows fundamental principles from computational thinking to capture the semantics of the new phenomenon of adversarial dynamics in CPS models. We leverage core ideas from programming languages by extending syntax and semantics of program models and specification and verification logics with the complementary operator of duality to incorporate adversariality in a modular way into the realm of hybrid systems models. This leads to a compositional model of hybrid games with compositional operators that each have a compositional semantics. Modularity makes it possible to generalize our rigorous reasoning principles for CPS to hybrid games while simultaneously taming their complexity. This chapter introduces the semantics of *differential game logic* dGL [4], which adds adversarial dynamics to the differential dynamic logic that has been used as the specification and verification language for CPS in the other parts of this textbook. Because of the fundamental rôle that alternation plays in hybrid games, this chapter also provides a perspective on advanced models of computation with alternating choices. Finally, the chapter will encourage us to reflect on the relationship of denotational and operational semantics. The former focuses on the mathematical object to which a syntactic expression refers. The latter instead has an emphasis on the actions that happen successively as the game unfolds.

CPS Skills: This chapter focuses on developing and understanding the semantics of CPS models with adversarial dynamics corresponding to how a system changes state over time as multiple agents react to each other. This understanding is crucial for developing an intuition for the operational effects of multi-agent CPSs. The presence of adversarial dynamics will cause us to reconsider the semantics of CPS models to incorporate the effects of multiple agents and their mutual reactions. This generalization, while crucial for understanding adversarial dynamics in CPS, also shines a helpful complementary light on the semantics of hybrid systems without adversariality by causing us to reflect on the meaning of choices. The semantics of hybrid games genuinely generalizes the semantics of hybrid systems from earlier chapters.

fundamental principles of computational thinking
logical extensions
PL modularity principles
compositional extensions
differential game logic
denotational vs. operational semantics

adversarial dynamics
adversarial semantics
adversarial repetitions
fixpoints

CPS semantics
multi-agent operational effects
mutual reactions
complementary hybrid systems

15.2 Semantics of Differential Game Logic

What is the most elegant way of defining a semantics for differential game logic? How can a semantics be defined at all? First of all, the dGL formulas P that are used in the postconditions of dGL modal formulas $\langle\alpha\rangle P$ and $[\alpha]P$ define the winning conditions for the hybrid game α. When playing the hybrid game α, we, thus, need to know the set of states in which the winning condition P is satisfied, because that is the region that the respective player wants to reach. That set of states in which P is true is denoted $[\![P]\!]$, which defines the semantics of dGL formula P. Recall that $\omega \in [\![P]\!]$ indicates that state ω is among the set of states in which P is true. The state ω in a hybrid game is still just a mapping that assigns real numbers to all variables, just as in hybrid programs, because that is what is needed to make sense of terms such as $x \cdot y + 2$ and formulas such as $x^2 \geq x \cdot y + 2$ in the hybrid game. A *state* ω is a mapping from variables to $\mathbb{R}$. The *set of states* is denoted $\mathscr{S}$.

15.2.1 Limits of Reachability Relations

The semantics of hybrid games is more subtle than that of hybrid systems. The semantics of a hybrid program α is simply a reachability relation $[\![\alpha]\!] \subseteq \mathscr{S} \times \mathscr{S}$ where $(\omega, \nu) \in [\![\alpha]\!]$ indicates that final state ν is reachable from initial state ω by running HP α. That made it possible to define the semantics of the dL formula $\langle\alpha\rangle P$ via

$$[\![\langle\alpha\rangle P]\!] = \{\omega \in \mathscr{S} : \nu \in [\![P]\!] \text{ for some } \nu \text{ with } (\omega, \nu) \in [\![\alpha]\!]\} \quad \text{for HP } \alpha \quad (15.1)$$

This approach does not suffice for hybrid games. First of all, the reachability relation $(\omega, \nu) \in [\![\alpha]\!]$ is only defined when α is a hybrid program, not when it is a hybrid game. And it is not even clear whether a reachability relation is all that it takes to understand the semantics of a hybrid game, because mere reachability information about states hardly retains enough information to represent the interactive aspects of gameplay in which some choices are better than others for the respective players. But the deeper reason is that the shape (15.1) is too restrictive. Criteria of this shape would require Angel to single out a single state ω that satisfies the winning condition $v \in [\![P]\!]$ and then get to that state v by playing hybrid game α from ω. Yet, all that Demon then has to do to spoil this plan is lead the play into a different state (even one in which Angel would also have won) but which is different from the projected state v. More generally, winning into a single state is really difficult.

15.2.2 Set-Valued Semantics of Differential Game Logic Formulas

Winning by leading the play into one of several states that satisfy the winning condition is more feasible. If we know the whole set of states $[\![P]\!]$ where postcondition P is true as the winning condition, then the hybrid game α uniquely determines the set of states from which Angel has a winning strategy in the game α to reach a state in $[\![P]\!]$. This *winning region* in hybrid game α for Angel's winning condition $[\![P]\!]$ will be denoted $\varsigma_\alpha([\![P]\!])$. More generally, for any set of states $X \subseteq \mathscr{S}$, will $\varsigma_\alpha(X)$ denote the set of states from which Angel has a winning strategy in the hybrid game α to reach a state in Angel's winning condition X. Correspondingly, $\delta_\alpha(X)$ will denote the set of states from which Demon has a winning strategy in the hybrid game α to reach a state in Demon's winning condition X. Both sets will be defined in Sect. 15.2.3.

For a subset $X \subseteq \mathscr{S}$ the complement $\mathscr{S} \setminus X$ is denoted $X^\complement$. The notation ω_x^d from (2.9) on p. 49 still denotes the state that agrees with state ω except for the interpretation of variable x, which is changed to $d \in \mathbb{R}$. The value of term e in state ω is denoted by $\omega[\![e]\!]$ as in Definition 2.4. The denotational semantics of dGL formulas will be defined in Definition 15.1 by simultaneous induction along with the denotational semantics, $\varsigma_\alpha(\cdot)$ and $\delta_\alpha(\cdot)$, of hybrid games, defined in Definition 15.2, because dGL formulas are defined by simultaneous induction with hybrid games. The *(denotational) semantics of a hybrid game* α defines for each set of Angel's winning states $X \subseteq \mathscr{S}$ the *winning region*, i.e., the set of states $\varsigma_\alpha(X)$ from which Angel has a winning strategy to achieve X (whatever strategy Demon chooses). The *winning region* of Demon, i.e., the set of states $\delta_\alpha(X)$ from which Demon has a winning strategy to achieve X (whatever strategy Angel chooses) is defined later as well.

Definition 15.1 (dGL semantics). The *semantics of a* dGL *formula* P is the subset $[\![P]\!] \subseteq \mathscr{S}$ of states in which P is true. It is defined inductively as follows:

1. $[\![e \geq \tilde{e}]\!] = \{\omega \in \mathscr{S} : \omega[\![e]\!] \geq \omega[\![\tilde{e}]\!]\}$
   That is, the set of states in which $e \geq \tilde{e}$ is true is the set in which the value of e is greater than or equal to the value of $\tilde{e}$.

2. $[\![\neg P]\!] = ([\![P]\!])^\complement$
   That is, the set of states in which $\neg P$ is true is the complement of the set of states in which P is true.
3. $[\![P \wedge Q]\!] = [\![P]\!] \cap [\![Q]\!]$
   That is, the set of states in which $P \wedge Q$ is true is the intersection of the set of states in which P is true with the set of states in which Q is true.
4. $[\![\exists x P]\!] = \{\omega \in \mathscr{S} : \omega_x^r \in [\![P]\!] \text{ for some } r \in \mathbb{R}\}$
   That is, the states in which $\exists x P$ is true are those which only differ in the real value of x from a state in which P is true.
5. $[\![\langle \alpha \rangle P]\!] = \varsigma_\alpha([\![P]\!])$
   That is, the set of states in which $\langle \alpha \rangle P$ is true is Angel's winning region to achieve $[\![P]\!]$ in hybrid game α, i.e., the set of states from which Angel has a winning strategy in hybrid game α to reach a state where P holds.
6. $[\![[\alpha] P]\!] = \delta_\alpha([\![P]\!])$
   That is, the set of states in which $[\alpha] P$ is true is Demon's winning region to achieve $[\![P]\!]$ in hybrid game α, i.e., the set of states from which Demon has a winning strategy in hybrid game α to reach a state where P holds.

A dGL formula P is *valid*, written $\vDash P$, iff it is true in all states, i.e., $[\![P]\!] = \mathscr{S}$.

The semantics $\varsigma_\alpha(X)$ and $\delta_\alpha(X)$ of Angel's and Demon's winning regions for winning condition X in hybrid game α will be defined next.

15.2.3 Winning-Region Semantics of Hybrid Games

Definition 15.1 uses the winning regions $\varsigma_\alpha(\cdot)$ and $\delta_\alpha(\cdot)$ for Angel and Demon, respectively, in the hybrid game α. Rather than taking a detour to understand those by operational game semantics (as in Chap. 14), the winning regions of hybrid games can be defined directly, giving a denotational semantics to hybrid games.[1] The winning regions for Angel are illustrated in Fig. 15.1, for Demon in Fig. 15.2.

Definition 15.2 (Semantics of hybrid games without repetition). The *semantics of a hybrid game* α is a function $\varsigma_\alpha(\cdot)$ that, for each set of Angel's winning states $X \subseteq \mathscr{S}$, gives the *winning region*, i.e., the set of states $\varsigma_\alpha(X)$ from which Angel has a winning strategy to achieve X (whatever strategy Demon chooses). It is defined inductively as follows:

---

[1] The semantics of a hybrid game is not merely a reachability relation between states as for hybrid systems [3], because the adversarial dynamic interactions and nested choices of the players have to be taken into account. For brevity, the informal explanations sometimes say "win the game" when really they mean "have a winning strategy to win the game". The semantics of differential equations could be augmented to ignore the initial value of the differential symbol x' as in Part II. This is not pursued for simplicity, because considering $x' := *; x' = f(x) \& Q$ has the same effect.

1. $\varsigma_{x:=e}(X) = \{\omega \in \mathscr{S} : \omega_x^{\omega[\![e]\!]} \in X\}$

   That is, an assignment $x := e$ wins a game into X from any state ω whose modification $\omega_x^{\omega[\![e]\!]}$ that changes the value of x to $\omega[\![e]\!]$ is in X.

2. $\varsigma_{x'=f(x)\,\&\,Q}(X) = \{\varphi(0) \in \mathscr{S} : \varphi(r) \in X \text{ for some solution } \varphi : [0,r] \to \mathscr{S}$
   of any duration $r \in \mathbb{R}$ satisfying $\varphi \models x' = f(x) \wedge Q\}$

   That is, Angel wins the differential equation $x' = f(x)\,\&\,Q$ into X from any state $\varphi(0)$ from which there is a solution φ of $x' = f(x)$ of any duration r that remains in Q all the time and leads to a final state $\varphi(r) \in X$.

3. $\varsigma_{?Q}(X) = [\![Q]\!] \cap X$

   That is, Angel wins into X for a challenge $?Q$ from the states that satisfy Q to pass the challenge and are already in X, because challenges $?Q$ do not change the state. Angel only achieves winning condition X in game $?Q$ in the states in X that also satisfy the test formula Q.

4. $\varsigma_{\alpha \cup \beta}(X) = \varsigma_\alpha(X) \cup \varsigma_\beta(X)$

   That is, Angel wins a game of choice $\alpha \cup \beta$ into X whenever she wins game α into X or wins β into X (by choosing a subgame for which she has a winning strategy).

5. $\varsigma_{\alpha;\beta}(X) = \varsigma_\alpha(\varsigma_\beta(X))$

   That is, Angel wins a sequential game $\alpha;\beta$ into X whenever she has a winning strategy in game α to achieve $\varsigma_\beta(X)$, i.e., to make it to one of the states from which she has a winning strategy in game β to achieve X.

6. $\varsigma_{\alpha^*}(X)$ will be defined later.

7. $\varsigma_{\alpha^{\mathrm{d}}}(X) = (\varsigma_\alpha(X^{\complement}))^{\complement}$

   That is, Angel wins α^{d} to achieve X in exactly the states in which she does not have a winning strategy in game α to achieve the opposite $X^{\complement}$.

Since the players switch sides in a dual game α^{d}, Angel's winning region $\varsigma_{\alpha^{\mathrm{d}}}(X)$ from which she has a winning strategy to achieve X in the dual game α^{d} is the same as the complement $(\varsigma_\alpha(X^{\complement}))^{\complement}$ of the set $\varsigma_\alpha(X^{\complement})$ where Angel would have a winning strategy in the game α to achieve the complement region $X^{\complement}$ where she loses the dual game α^{d}. The winning region $\varsigma_\alpha(X^{\complement})$ corresponds to Angel simulating Demon's controls in α^{d} by playing Angel's controls in α but for Demon's objective $X^{\complement}$ instead of Angel's objective X. The complement of this region then is the winning region $\varsigma_{\alpha^{\mathrm{d}}}(X)$ where Angel has a winning strategy in the dual game α^{d} to achieve X, because she would not have had a winning strategy to achieve $X^{\complement}$ when simulating Demon with pretend-play in game α.

After having defined the winning region $\varsigma_\alpha(X)$ from which Angel has a winning strategy to achieve X in the hybrid game α, the next question is how to define the winning region $\delta_\alpha(X)$ from which Demon has a winning strategy to achieve X in the hybrid game α. Together, these define the functions used in the semantics of dGL formulas (Definition 15.1). For discrete assignments $x := e$, the winning region $\varsigma_{x:=e}(X)$ for Angel is the same as the winning region $\delta_{x:=e}(X)$ for Demon in the same game with the same winning condition X, because there are no choices to resolve in a discrete assignment. But for differential equations, the winning regions are very different, because Angel is in control of the duration of the differential equation, so Demon only has a chance if the differential equation starts in X (because Angel could follow it for duration 0 from the evolution domain) and stays in X all the time (because Angel could follow it for any other duration within the evolution domain). Likewise, since Angel gets to decide how to resolve a choice $\alpha \cup \beta$, Demon can only win if he wins both subgames.

Definition 15.3 (Semantics of hybrid games without repetition, continued).
The *winning region* of Demon, i.e., the set of states $\delta_\alpha(X)$ from which Demon has a winning strategy to achieve X (whatever strategy Angel chooses) is defined inductively as follows:

1. $\delta_{x:=e}(X) = \{\omega \in \mathscr{S} : \omega_x^{\omega[\![e]\!]} \in X\}$
   That is, an assignment $x:=e$ wins a game into X from any state ω whose modification $\omega_x^{\omega[\![e]\!]}$ that changes the value of x to $\omega[\![e]\!]$ is in X.

2. $\delta_{x'=f(x)\,\&\,Q}(X) = \{\varphi(0) \in \mathscr{S} : \varphi(r) \in X \text{ for all durations } r \in \mathbb{R} \text{ and all solutions } \varphi : [0,r] \to \mathscr{S} \text{ satisfying } \varphi \models x' = f(x) \wedge Q\}$
   That is, Demon wins the differential equation $x' = f(x)\,\&\,Q$ into X from any state $\varphi(0)$ from which all solutions φ of $x' = f(x)$ of any duration r that remain within Q all the time lead to states $\varphi(r) \in X$ in the end.

3. $\delta_{?Q}(X) = (\![Q]\!)^{\complement} \cup X$
   That is, Demon wins into X for a challenge $?Q$ from the states which violate Q so that Angel fails her challenge $?Q$ or that are already in X, because challenges $?Q$ do not change the state. Demon achieves the winning condition X in game $?Q$ in the states in X (whether or not Q holds) as well as in the states in which Angel fails test formula Q.

4. $\delta_{\alpha \cup \beta}(X) = \delta_\alpha(X) \cap \delta_\beta(X)$
   That is, Demon wins a game of choice $\alpha \cup \beta$ into X whenever he wins α into X *and* wins β into X (because Angel might choose either subgame).

5. $\delta_{\alpha;\beta}(X) = \delta_\alpha(\delta_\beta(X))$
   That is, Demon wins a sequential game $\alpha;\beta$ into X whenever he has a winning strategy in game α to achieve $\delta_\beta(X)$, i.e., to make it to one of the states from which he has a winning strategy in game β to achieve X.

6. $\delta_{\alpha^*}(X)$ will be defined later.

7. $\delta_{\alpha^d}(X) = (\delta_\alpha(X^{\complement}))^{\complement}$
   That is, Demon wins α^d to achieve X in exactly the states in which he does not have a winning strategy in game α to achieve the opposite $X^{\complement}$.

Strategies do not occur explicitly in the dGL semantics, because the semantics is based on the existence of winning strategies, not on the strategies themselves. Just like the semantics of dL, the semantics of dGL is *compositional*, i.e., the semantics of a compound dGL formula is a simple function of the semantics of its pieces. Likewise, the semantics of a compound hybrid game is a simple function of the semantics of its pieces. Also observe how the existence of a strategy in hybrid game

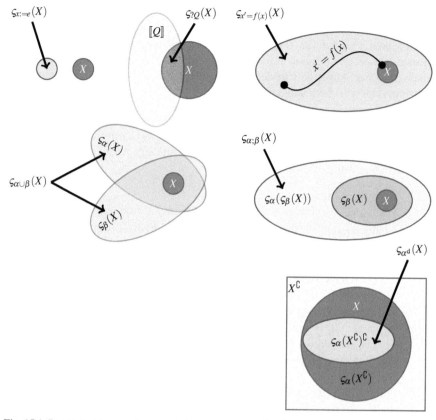

Fig. 15.1 Denotational semantics of hybrid games as Angel's winning region

α to achieve X is independent of any game and dGL formula surrounding α, but just depends on the remaining game α itself and on the goal X.

Even if we will only prove the following monotonicity property of winning regions in Chap. 16 after having defined a semantics of repetition, we already state it now, because it provides useful intuition. The semantics is monotone [4], i.e., larger sets of winning states have larger winning regions, because it is easier to win into larger sets of winning states (Fig. 15.3).

> **Lemma 15.1 (Monotonicity).** *The* dGL *semantics is* monotone, *that is, both* $\varsigma_\alpha(X) \subseteq \varsigma_\alpha(Y)$ *and* $\delta_\alpha(X) \subseteq \delta_\alpha(Y)$ *for all* $X \subseteq Y$.

Note the big qualitative difference in the denotational semantics style of defining the winning region $\varsigma_\alpha(X)$ in Definition 15.2 compared to the operational semantics captured (informally) in Sect. 14.4. The denotational semantics directly associates with a hybrid game α and a winning condition X the set of states from which player Angel has a winning strategy in game α to achieve X. This results in a simple in-

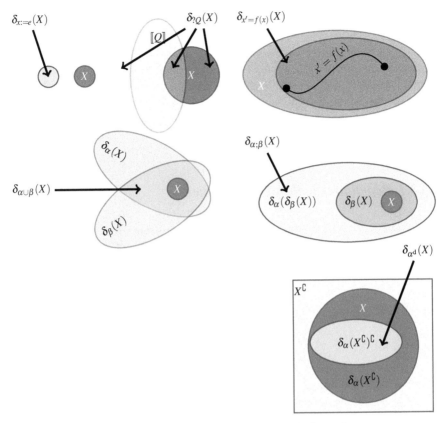

Fig. 15.2 Denotational semantics of hybrid games as Demon's winning region

Fig. 15.3 Monotonicity: it is easier to win into larger sets of winning states $Y \supseteq X$

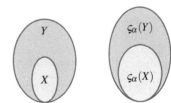

ductive definition of $\varsigma_\alpha(X)$ based on the structure of α. The game trees from the operational semantics in Sect. 14.4 give a more direct operational intuition of how a game can be played by moving along the edges of its game graph. But rigorously defining the structure of such an (infinite) graph and what it means to have a winning strategy in it is technically more involved and complicates subsequent analysis, compared to the more convenient denotational semantics. There are other circumstances where an operational semantics is more useful, so it is good to be familiar with both styles to choose the best fit for any question at hand.

15.3 Semantics of Repetition in Hybrid Games

Before going any further we need to define a semantics for repetition, which will turn out to be surprisingly subtle. The final answer in Sect. 15.3.4 is not quite so complicated, but it takes considerable deliberation to get there. Since the insights along the way are of general interest and nicely illuminate interesting complexities of hybrid games, we do not mind taking a careful route toward understanding the rôle of repetition in hybrid games.

15.3.1 Repetitions with Advance Notice

Definition 15.2 is still missing a definition for the semantics of repetition in hybrid games. With $\alpha^{n+1} \equiv \alpha^n; \alpha$ and $\alpha^0 \equiv \,?true$, the semantics of repetition in hybrid systems was

$$[\![\alpha^*]\!] = \bigcup_{n \in \mathbb{N}} [\![\alpha^n]\!]$$

The obvious counterpart for the semantics of repetition in hybrid games would be

$$\varsigma_{\alpha^*}(X) \stackrel{?}{=} \bigcup_{n < \omega} \varsigma_{\alpha^n}(X) \tag{15.2}$$

where ω is the first infinite ordinal (if you have never seen ordinals before, just read $n < \omega$ as n is in the natural numbers, i.e., as $n \in \mathbb{N}$). Would that give the intended meaning to repetition? Would Angel be forced to stop in order to win if the game of repetition were played this way? Yes, she would, because, even though there is no bound on the number of repetitions that she can choose, for each natural number n, the resulting game $\varsigma_{\alpha^n}(X)$ is finite.

Would this definition capture the intended meaning of repeated game play?

Before you read on, see if you can find the answer for yourself.

The issue is that each way of playing a repetition according to (15.2) would require Angel to choose a natural number $n \in \mathbb{N}$ of repetitions and *expose this number to Demon* when playing α^n so that he would know how often Angel decided to repeat before he even has to make a move.

That would lead to what is called the *advance notice semantics* [5] for α^*, which requires the players to announce the number of times that game α will be repeated when the loop begins. The advance notice semantics defines $\varsigma_{\alpha^*}(X)$ as $\bigcup_{n<\omega} \varsigma_{\alpha^n}(X)$ and defines $\delta_{\alpha^*}(X)$ as $\bigcap_{n<\omega} \delta_{\alpha^n}(X)$. When playing α^*, Angel, thus, announces to Demon the number of repetitions $n < \omega$ when the game α^* starts and Demon announces the number of repetitions when the game $\alpha^\times$ starts. This advance notice makes it easier for Demon to win loops α^* and easier for Angel to win loops $\alpha^\times$, because the opponent announces an important feature of their strategy immediately, as opposed to revealing whether or not to repeat the game once more one iteration at a time as we had meant with the operational game trees in Chap. 14.

If we gave repetition an advance notice semantics, then that would be a big disadvantage for the player controlling repetitions. The following formula, for example, is valid in dGL, but would not be valid in the advance notice semantics (Fig. 15.4):

$$x = 1 \wedge a = 1 \rightarrow \langle ((x := a; a := 0) \cap x := 0)^* \rangle x \neq 1 \tag{15.3}$$

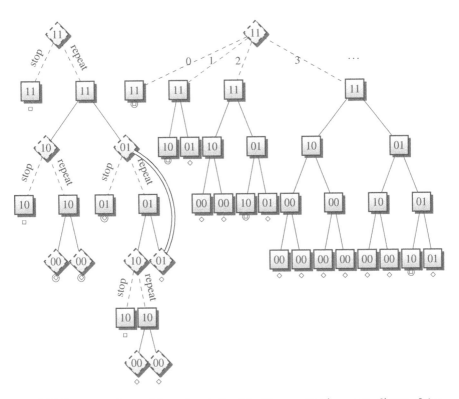

Fig. 15.4 Game trees for $x = 1 \wedge a = 1 \rightarrow \langle \alpha^* \rangle x \neq 1$ with game $\alpha \equiv (x := a; a := 0) \cap x := 0$ (notation: x, a). **(left)** valid in dGL by strategy "repeat once and repeat once more if $x = 1$, then stop" **(right)** false in advance notice semantics by the strategy "$n - 1$ choices of $x := 0$ followed by $x := a; a := 0$ once", where n is the number of repetitions Angel announced

The game starts with x and a both being 1 and asks whether Angel has a winning strategy to reach $x \neq 1$ with a repetition that she controls but Demon gets to choose whether 0 or a is put into x. The catch is that whenever the value of x is copied over to x in Demon's left choice, then a is zeroed out, so this only helps him once.

If, in the advance notice semantics, Angel announces when the repetition starts that she has chosen n repetitions of the game, then Demon wins with flying colors by choosing the right choice $x:=0$ option $n-1$ times followed by the left choice of $x:=a;a:=0$ in the last repetition. This strategy would not work in the dGL seman-tics, where Angel is free to decide whether to repeat α^* *after* each repetition based on the resulting state of the game. Inspecting the state makes a big difference to deciding whether to stop. If x has the value 0, then Angel decides to stop, otherwise she repeats. If Demon played the right choice $x:=0$, Angel stops. If he played the left choice $x:=a;a:=0$, then Angel decides to repeat but will stop after the next iteration, regardless of which option Demon chose. The winning strategy for (15.3) indicated with $\oslash$ in Fig. 15.4(left) shows that this dGL formula is valid.

Of course, there are also formulas that would be valid in the advance notice semantics but are not valid in dGL, for example, the dual of formula (15.3):

$$x = 1 \wedge a = 1 \rightarrow [((x:=a;a:=0) \cap x:=0)^*]x = 1$$

Just as an advance notice semantics would make it easy for Demon to win α^* games with repetitions under Angel's control, it would also make it easy for Angel to win $\alpha^\times$ games with repetitions under Demon's control.

The advance notice semantics misses out on the existence of perfectly reasonable winning strategies, because it is just not interactive enough for proper hybrid game play. The dGL semantics is more general and gives the player in charge of repetition more control to inspect the state before having to decide on whether to repeat again. If you ever really need parts of a game where the number of repetitions is announced to the other player ahead of time, then it is easy to model them (Exercise 15.2).

> Despite being built in direct analogy to the semantics of repetition in hybrid systems, the advance notice semantics is inappropriate for hybrid games, be-cause it is very difficult for a CPS to predict ahead of time exactly how many iterations of a control cycle it will take to get to the goal.

For hybrid systems, it does not matter whether the number of iterations for a rep-etition is chosen ahead of time or afterwards, because there are no surprises during its evolution. All choices are consistently resolved by nondeterminism. This corre-sponds to all choices being resolved by Angel, which means she can always choose every choice in the best possible way. But for games, Demon can have a number of surprises in store for Angel, so that she will have to wait and see to decide how often to repeat.

15.3.2 Repetitions as Infinite Iterations

The trouble with the semantics in Sect. 15.3.1 is that Angel's move for the repetition reveals too much to Demon, because Demon can inspect the remaining game α^n to find out how long the game will be played before he even has to make his first move.

Let's try to undo this. Instead of considering a choice over all n-fold repetitions α^n that reveals the chosen number n, we could consider a semantics that iterates n

times the winning region of α for any arbitrary finite number n (see Fig. 15.5):

$$\varsigma_{\alpha^*}(X) \overset{?}{=} \bigcup_{n<\omega} \varsigma_\alpha^n(X) \tag{15.4}$$

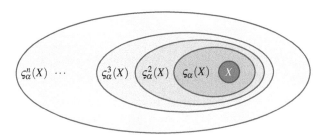

Fig. 15.5 Iteration $\varsigma_\alpha^n(X)$ of $\varsigma_\alpha(\cdot)$ from winning condition X

This semantics is called the *ω-semantics* and also denoted $\varsigma_\alpha^\omega(X)$. All we need to do then is to define the iteration of the winning-region construction. For any winning condition $X \subseteq \mathscr{S}$, the *n-times iterated winning region* $\varsigma_\alpha^n(X)$ of α is defined by induction on n as

$$\varsigma_\alpha^0(X) \overset{\text{def}}{=} X$$
$$\varsigma_\alpha^{\kappa+1}(X) \overset{\text{def}}{=} X \cup \varsigma_\alpha(\varsigma_\alpha^\kappa(X))$$

The only states from which a repetition can win without repeating are the ones that start at the goal X already ($\varsigma_\alpha^0(X) = X$). The states from which a repetition can win into the set X with up to $\kappa+1$ repetitions are those that start in X as well as all the states for which there is a winning strategy in the hybrid game α to achieve a state in $\varsigma_\alpha^\kappa(X)$. That is, the construction successively applies $\varsigma_\alpha(\cdot)$ while retaining the winning condition X:

$$\varsigma_\alpha^0(X) = X$$
$$\varsigma_\alpha^1(X) = X \cup \varsigma_\alpha(X)$$
$$\varsigma_\alpha^2(X) = X \cup \varsigma_\alpha(X \cup \varsigma_\alpha(X))$$
$$\varsigma_\alpha^3(X) = X \cup \varsigma_\alpha(X \cup \varsigma_\alpha(X \cup \varsigma_\alpha(X)))$$
$$\varsigma_\alpha^4(X) = X \cup \varsigma_\alpha(X \cup \varsigma_\alpha(X \cup \varsigma_\alpha(X \cup \varsigma_\alpha(X))))$$
$$\vdots$$

Does this give the right semantics for repetition of hybrid games? Does it match the existence of winning strategies that we were hoping to define?

Before you read on, see if you can find the answer for yourself.

The surprising answer is *no* for a very subtle but also very fundamental reason. The existence of winning strategies for α^* does *not* coincide with the ωth iteration of α.

Would the following dGL formula be valid with the semantics from (15.4)?

$$\langle (\underbrace{x:=1;x'=1^d}_{\beta} \cup \underbrace{x:=x-1}_{\gamma})^* \rangle (0 \leq x < 1) \tag{15.5}$$

Before you read on, see if you can find the answer for yourself.

As usual, $[a,b)$ denotes the interval from a inclusive to b exclusive. Using the abbreviations indicated in (15.5) such as $\alpha \equiv \beta \cup \gamma$, it is easy to see that $\varsigma_\alpha^n([0,1)) = [0,n+1)$ for all $n \in \mathbb{N}$ by a simple inductive proof:

$$\varsigma_{\beta \cup \gamma}^0([0,1)) = [0,1)$$

$$\varsigma_{\beta \cup \gamma}^{n+1}([0,1)) = [0,1) \cup \varsigma_{\beta \cup \gamma}(\varsigma_{\beta \cup \gamma}^n([0,1))) \overset{\text{IH}}{=} [0,1) \cup \varsigma_{\beta \cup \gamma}([0,n+1))$$

$$= [0,1) \cup \varsigma_\beta([0,n+1)) \cup \varsigma_\gamma([0,n)) = [0,1) \cup \emptyset \cup [1,n+2) = [0,n+1+1)$$

Consequently, the ω-semantics from (15.4) consists of all nonnegative reals:

$$\bigcup_{n<\omega} \varsigma_\alpha^n([0,1)) = \bigcup_{n<\omega} [0,n+1) = [0,\infty) \tag{15.6}$$

Hence, the ω-semantics from (15.4) indicates that the hybrid game (15.5) can be won only from initial states in $[0,\infty)$, that is, for those that satisfy $0 \leq x$.

Unfortunately, this is complete nonsense! True, the hybrid game in dGL formula (15.5) can be won from all initial states that satisfy $0 \leq x$. But it can also be won from all other initial states! The only twist is that Angel may need an unbounded number of iterations to win it from initial states with $x < 0$, because Demon can increase the value of x arbitrarily far during his differential equation. In fact, there are cases where the ω-semantics is minuscule compared to the true winning region and arbitrarily far away from the truth [4].

For the formula (15.5), the ω-semantics misses out on Angel's perfectly reasonable winning strategy "first choose $x:=1;x'=1^d$ and then always choose $x:=x-1$ until stopping at $0 \leq x < 1$." This winning strategy wins from *every* initial state in $\mathbb{R}$, which is a much bigger set than the set of nonnegative reals from (15.6).

This winning strategy justifies that the dGL formula (15.5) is valid. Yet, is there a direct way to see that (15.6) is not the final answer for (15.5) without putting the winning-region computations aside and constructing a separate ingenious winning strategy, which would undermine the whole point of using winning regions for the semantics?

Before you read on, see if you can find the answer for yourself.

The crucial observation comes from a closer inspection of what exactly we did to arrive at (15.6). The fact (15.6) shows that the hybrid game in (15.5) can be won from all nonnegative initial values with at most ω (that is "first countably infinitely many") steps. The induction step proving $\varsigma_\alpha^n([0,1)) = [0, n+1)$ for all $n \in \mathbb{N}$ showed that if, for whatever reason (by inductive hypothesis really), $[0, n)$ is in the winning region, then $[0, n+1)$ also is in the winning region by simply applying $\varsigma_\alpha(\cdot)$ to $[0, n)$.

How about doing exactly that again? For whatever reason (i.e., by the above argument), $[0, \infty)$ is in the winning region. Doesn't that mean that $\varsigma_\alpha([0, \infty))$ should again be in the winning region by exactly the same inductive argument above?

Before you read on, see if you can find the answer for yourself.

Note 72 (+1 argument) Whenever a set Z is in the winning region $\varsigma_{\alpha^*}(X)$ of repetition, then $\varsigma_\alpha(Z)$ is in the winning region $\varsigma_{\alpha^*}(X)$ as well, because it is just one round away from Z and α^* can simply repeat once more. That is,

$$\text{if } Z \subseteq \varsigma_{\alpha^*}(X) \text{ then } \varsigma_\alpha(Z) \subseteq \varsigma_{\alpha^*}(X)$$

Fig. 15.6 Winning regions $\varsigma_\alpha(Z)$ of sets $Z \subseteq \varsigma_{\alpha^*}(X)$ are already included in $\varsigma_{\alpha^*}(X)$ since $\varsigma_\alpha(Z)$ is just one more round away from Z

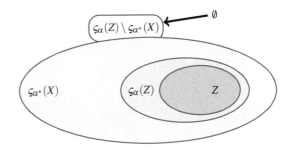

Applying Note 72, which is illustrated in Fig. 15.6, to the situation at hand works as follows. The fact (15.6) explains that at least $[0, \infty) \subseteq \varsigma_{(\beta \cup \gamma)^*}([0, 1))$ is in the winning region of repetition. By Note 72, the winning region $\varsigma_{(\beta \cup \gamma)^*}([0, 1))$ also contains the one-step winning region $\varsigma_{\beta \cup \gamma}([0, \infty)) \subseteq \varsigma_{(\beta \cup \gamma)^*}([0, 1))$ of $[0, \infty)$. Computing what that is gives

$$\varsigma_{\beta \cup \gamma}([0, \infty)) = \varsigma_\beta([0, \infty)) \cup \varsigma_\gamma([0, \infty)) = \mathbb{R} \cup [0, \infty) = \mathbb{R}$$

Beyond that, the winning region cannot contain anything else, because $\mathbb{R}$ is the whole state space already (since there is only one variable in this hybrid game) and it is kind of hard to add anything to that. Indeed, trying to use the winning-region construction once more on $\mathbb{R}$ does not change the result:

$$\varsigma_{\beta \cup \gamma}(\mathbb{R}) = \varsigma_\beta(\mathbb{R}) \cup \varsigma_\gamma(\mathbb{R}) = \mathbb{R} \cup \mathbb{R} = \mathbb{R}$$

This result, then, coincides with what the ingenious winning strategy above told us as well: formula (15.5) is valid, because there is a winning strategy for Angel from every initial state. However, the repeated $\varsigma_{\beta \cup \gamma}(\cdot)$ winning-region construction seems more systematic than an ingenious guess of a smart winning strategy. So it gives a more constructive and explicit semantics.

Let's recap. It took us more than infinitely many steps to find the winning region of the hybrid game described in (15.5). After infinitely many iterations to arrive at $\varsigma_\alpha^\omega([0,1)) = \bigcup_{n<\omega} \varsigma_\alpha^n([0,1)) = [0,\infty)$, it took us one more step to arrive at

$$\varsigma_{(\beta \cup \gamma)^*}([0,1)) = \varsigma_\alpha^{\omega+1}([0,1)) = \mathbb{R}$$

where we denote the number of steps we took overall by $\omega + 1$, since it was one more step than (first countably) infinitely many (i.e., ω many); see Fig. 15.7 for an illustration. More than infinitely many steps to get somewhere are plenty. Even worse: there are cases where even $\omega + 1$ is not enough iterations to get to the semantics of repetition. The number of iterations needed to find $\varsigma_{\alpha^*}(X)$ could in general be much larger than just a little more than first countably infinitely many [4].

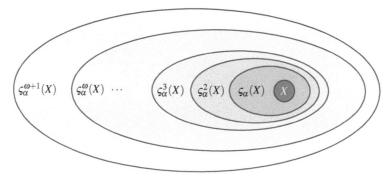

Fig. 15.7 Iteration $\varsigma_\alpha^{\omega+1}(X)$ of $\varsigma_\alpha(\cdot)$ from winning condition $X = [0,1)$ stops when applying $\varsigma_\alpha(\cdot)$ to the ωth infinite iteration $\varsigma_\alpha^\omega(X)$

The existence of the above winning strategy is found at level $\varsigma_\alpha^{\omega+1}([0,1)) = \mathbb{R}$. Even though any particular use of the winning strategy in any game play uses only some finite number of repetitions of the loop, the argument that it will always work requires $> \omega$ many iterations of $\varsigma_\alpha(\cdot)$, because Demon can change x to an arbitrarily big value, so that ω many iterations of $\varsigma_\alpha(\cdot)$ are needed to conclude that Angel has a winning strategy for any positive value of x. There is no smaller upper bound on the number of iterations it takes Angel to win. Angel cannot even promise ω as a bound on the repetition count, which is what the ω-semantics would effectively require her to do. But strategies do converge after $\omega + 1$ iterations for (15.5).

> The ω-semantics is inappropriate, because it can be arbitrarily far away from characterizing the winning region of hybrid games.

15.3.3 Inflationary Semantics of Repetition

Despite the quite discouraging fact that infinitely many iterations of the winning-region construction $\varsigma_\alpha(\cdot)$ do not suffice to accurately describe the winning region of repetition α^*, there still is a way of rescuing the situation if we simply keep iterating. We just need to repeat the construction more than infinitely often, leading us into the wonderful world of ordinals. Even if we will ultimately discard this higher iteration semantics with ordinals in favor of a simpler semantics of repetition in Sect. 15.3.4, we still learn interesting subtle nuances about hybrid games by pursuing a little more iteration at first.

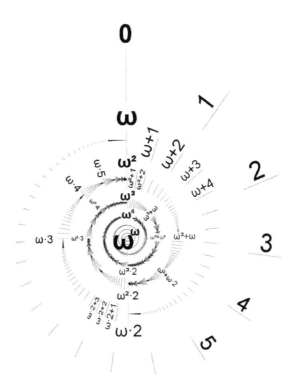

Fig. 15.8 Illustration of infinitely many ordinals up to ω^ω, including $0 < 1 < 2 < \cdots < \omega < \omega+1 < \cdots < \omega \cdot 2 < \omega \cdot 2+1 < \cdots < \omega^2 < \omega^2+1 < \cdots < \omega^2+\omega < \omega^2+\omega+1 < \cdots$

The key to understanding *ordinals* is that each ordinal κ always has a *successor ordinal* $\kappa+1$ but every set of ordinals also has a least upper bound λ, called the *limit ordinal* if it is not already a successor ordinal. For example, ω is the first infinite ordinal, and the smallest ordinal that is bigger than all natural numbers. But ω also has a successor ordinal $\omega+1$, which, in turn, has a successor ordinal $\omega+2$, and all those have a least upper bound $\omega \cdot 2$ (Fig. 15.8). All ordinals starting at ω are called *transfinite ordinals*, because there are infinitely many smaller ordinals.

When we apply the winning-region construction $\varsigma_\alpha(\cdot)$ for each successor ordinal $\kappa + 1$, but take the union of all previous winning regions at limit ordinals λ such as ω, the semantics of repetition can be defined using transfinite iteration (Fig. 15.9):

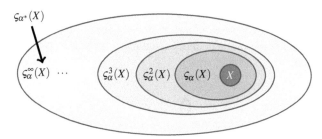

Fig. 15.9 Transfinite iteration $\varsigma_\alpha^\infty(X)$ of $\varsigma_\alpha(\cdot)$ from winning condition X results in winning region $\varsigma_{\alpha^*}(X)$ of repetition

$$\varsigma_\alpha^0(X) \stackrel{\text{def}}{=} X$$

$$\varsigma_\alpha^{\kappa+1}(X) \stackrel{\text{def}}{=} X \cup \varsigma_\alpha(\varsigma_\alpha^\kappa(X)) \qquad \kappa + 1 \text{ is a successor ordinal}$$

$$\varsigma_\alpha^\lambda(X) \stackrel{\text{def}}{=} \bigcup_{\kappa < \lambda} \varsigma_\alpha^\kappa(X) \qquad \lambda \neq 0 \text{ is a limit ordinal}$$

The semantics of repetition is the union of all winning regions for all ordinals:

$$\varsigma_{\alpha^*}(X) = \varsigma_\alpha^\infty(X) \stackrel{\text{def}}{=} \bigcup_{\kappa \text{ ordinal}} \varsigma_\alpha^\kappa(X) \qquad (15.7)$$

> **Note 73 (Infinite iterations infinitely often)** Unfortunately, hybrid games require rather big infinite ordinals until this inflationary style of computing their winning regions stops [4]. That translates into an infinite amount of work and then some more, infinitely often, to compute the winning region. Hardly the sort of thing we would like to wait for to find out who wins a game.

This semantics for repetition from (15.7) provides the correct answer if we do not mind the highly transfinite number of iterations it needs. Unfortunately, even the pretty infinite ordinal ω^ω is not enough for all hybrid games [4, Theorem 3.8]. The semantics we pursue in Sect. 15.3.4 is much easier in this respect, it just does not provide the same insights about the iterative complexities of hybrid games.

With this refined understanding of iteration, look back at dGL formula (15.5) and observe what the above argument about the winning-region computation terminating at $\omega + 1$ implies about bounds on how long it takes Angel to win the game in (15.5). Since the winning region only terminates at $\omega + 1$, she could not win with

Expedition 15.1 (Ordinal numbers)

Ordinals extend natural numbers. Natural numbers are inductively defined as the (smallest) set $\mathbb{N}$ containing 0 and the successor $n+1$ of every number $n \in \mathbb{N}$ that is in the set. Natural numbers are totally ordered: given any two different natural numbers, one number is going to be strictly smaller than the other one. For every finite set of natural numbers there is a smallest natural number that's bigger than all of them. Ordinals extend this beyond infinity. They just refuse to stop after all natural numbers have been written down. Taking all those (countably infinitely many) natural numbers $\{0, 1, 2, 3, \dots\}$, there is a smallest ordinal that's bigger than all of them. This ordinal is ω, the first[a] infinite ordinal: $0 < 1 < 2 < 3 < \cdots < \omega$ Unlike the ordinals $1, 2, 3, \dots$ from the natural numbers, the ordinal ω is a *limit ordinal*, because it is not the successor of any other ordinal. The ordinals $1, 2, 3, \dots$ are *successor ordinals*, because each of them is the successor $n+1$ of another ordinal n. The ordinal 0 is special, because it is not a successor ordinal of any ordinal or natural number.

Ordinals are keen on ensuring that every ordinal has a successor and for every set of ordinals there is a bigger ordinal. So, ω must have a successor, which is the successor ordinal $\omega + 1$, the successor of which is $\omega + 2$, etc.:

$$0 < 1 < 2 < 3 < 4 < \cdots < \omega < \omega+1 < \omega+2 < \omega+3 < \omega+4 < \cdots$$

Of course, in ordinal land, there ought to be an ordinal that's bigger than even all of those ordinals as well. It's the limit ordinal $\omega + \omega = \omega \cdot 2$, at which point we have counted to countable infinity twice already and will keep on finding bigger ordinals, because even $\omega \cdot 2$ will have a successor, namely $\omega \cdot 2 + 1$:

$$0 < 1 < 2 < \cdots < \omega < \omega+1 < \omega+2 < \cdots \omega \cdot 2 < \omega \cdot 2 + 1 < \omega \cdot 2 + 2 < \cdots$$

Now the set of all these will have a bigger ordinal $\omega \cdot 2 + \omega = \omega \cdot 3$, which again has successors and so on. That happens infinitely often so that $\omega \cdot n$ will be an ordinal for any natural number $n \in \mathbb{N}$. All those infinitely many ordinals will still have a limit ordinal that's bigger than all of them, which is $\omega \cdot \omega = \omega^2$. That one again has a successor $\omega^2 + 1$ and so on (Fig. 15.8):

$$0 < 1 < 2 < \cdots \omega < \omega+1 < \omega+2 < \cdots \omega \cdot 2 < \omega \cdot 2 + 1 < \cdots \omega \cdot 3 < \omega \cdot 3 + 1 < \cdots$$
$$\omega^2 < \omega^2 + 1 < \cdots \omega^2 + \omega < \omega^2 + \omega + 1 < \cdots \omega^\omega < \cdots \omega^{\omega^\omega} < \cdots \omega_1^{CK} < \cdots \omega_1 \cdots$$

The first infinite ordinal is ω, the Church-Kleene ordinal ω_1^{CK} is the first nonrecursive ordinal, and ω_1 is the first uncountable ordinal. Every ordinal κ is either a successor ordinal, i.e., the smallest ordinal $\kappa = \iota + 1$ greater than some ordinal ι, or a limit ordinal, i.e., the supremum of all smaller ordinals. Depending on the context, 0 is considered a limit ordinal or separate.

---

[a] For a moment read "$\omega = \infty$" as infinity, but you will realize in an instant that this naïve view does not go far enough, because there will be ample reason to distinguish different infinities.

any finite bound $n \in \mathbb{N}$ on the number of repetitions it takes her to win. Even though she will surely win in the end according to her winning strategy, she has no way of saying how long that would take. Not that Angels will ever do this, but suppose she were to brag to impress Demon by saying she could win (15.5) within $n \in \mathbb{N}$ repetitions, then it would be impossible for her to keep that promise. No matter how big a bound $n \in \mathbb{N}$ she were to choose, Demon could still always spoil it from any negative initial state by evolving his differential equation $x' = 1^d$ for much longer than n time units so that it takes Angel more than n rounds to decrease the resulting value down to the interval $[0, 1)$ again.

This illustrates the dual of the discussion on the advance notice semantics in Sect. 15.3.1, which showed that Demon could make Angel win faster than she announced just to make her lose in the final round. In (15.5), Demon can always make Angel win later than she promised even if she ultimately will still win. This is the sense in which $\omega + 1$ is the best bound on the number of rounds it takes Angel to win the hybrid game in (15.5). Consequently, a variation of the advance notice semantics based on Angel announcing that she will repeat at most $n \in \mathbb{N}$ times (as opposed to exactly $n \in \mathbb{N}$ times) does not capture the semantics of repetition appropriately.

Expedition 15.2 (Ordinal arithmetic)

Ordinals support addition, multiplication, and exponentiation, which can be defined by induction on the second argument quite similarly to how they are defined for natural numbers. The only oddity is that these operations are non-commutative. The constructions distinguish the case of successor ordinals, which are direct successors of a smaller ordinal compared to limit ordinals, which are the least upper bounds, over all smaller ordinals:

$$\iota + 0 = \iota$$
$$\iota + (\kappa + 1) = (\iota + \kappa) + 1 \qquad \text{for successor ordinals } \kappa + 1$$
$$\iota + \lambda = \bigsqcup_{\kappa < \lambda} \iota + \kappa \qquad \text{for limit ordinals } \lambda$$
$$\iota \cdot 0 = 0$$
$$\iota \cdot (\kappa + 1) = (\iota \cdot \kappa) + \iota \qquad \text{for successor ordinals } \kappa + 1$$
$$\iota \cdot \lambda = \bigsqcup_{\kappa < \lambda} \iota \cdot \kappa \qquad \text{for limit ordinals } \lambda$$
$$\iota^0 = 1$$
$$\iota^{\kappa+1} = \iota^\kappa \cdot \iota \qquad \text{for successor ordinals } \kappa + 1$$
$$\iota^\lambda = \bigsqcup_{\kappa < \lambda} \iota^\kappa \qquad \text{for limit ordinals } \lambda$$

where $\bigsqcup$ denotes the supremum or least upper bound. Carefully note ordinal oddities like the noncommutativity coming from $2 \cdot \omega = 4 \cdot \omega$ and $\omega \cdot 2 < \omega \cdot 4$.

15.3.4 Characterizing Winning Repetitions Implicitly

Section 15.3.3 culminated in a semantics of repetition defined as the union of all winning regions for all ordinals by an explicit (albeit wildly infinite) construction (15.7). Is there a more immediate way of characterizing the winning region $\varsigma_{\alpha^*}(X)$ of repetition implicitly rather than by explicit construction? This thought will lead to a beautiful illustration of Bertrand Russell's enlightening bon mot:

> The advantages of implicit definition over construction are roughly those of theft over honest toil. — Bertrand Russell (slightly paraphrased)

The iterated winning-region construction (15.7) describes the semantics of repetition by iterating from below, i.e., starting from $\varsigma_\alpha^0(X) = X$ and adding states. Could the semantics of repetition be characterized more indirectly but more concisely from above? With an implicit characterization instead of an explicit construction?

The +1 argument (Note 72) implies $\varsigma_\alpha(Z) \subseteq \varsigma_{\alpha^*}(X)$ for any set $Z \subseteq \varsigma_{\alpha^*}(X)$. In particular, the set $Z \stackrel{\text{def}}{=} \varsigma_{\alpha^*}(X)$ itself satisfies

$$\varsigma_\alpha(\varsigma_{\alpha^*}(X)) \subseteq \varsigma_{\alpha^*}(X) \tag{15.8}$$

After all, repeating α once more from the winning region $\varsigma_{\alpha^*}(X)$ of repetition of α cannot give us any states that did not already have a winning strategy in α^*, because α^* could have just been repeated one more time itself. Consequently, if a set $Z \subseteq \mathscr{S}$ claims to be the winning region $\varsigma_{\alpha^*}(X)$ of repetition, it at least has to satisfy

$$\varsigma_\alpha(Z) \subseteq Z \tag{15.9}$$

because, by (15.8), the true winning region $\varsigma_{\alpha^*}(X)$ does satisfy (15.9). Thus, strategizing along α from Z does not give anything that Z does not already know about.

Is there anything else that such a set Z needs to satisfy to qualify for being the winning region $\varsigma_{\alpha^*}(X)$ of repetition? Is there only one choice for Z? Or many? If there are multiple choices, which Z is it? Does such a Z always exist, even?

Before you read on, see if you can find the answer for yourself.

One such Z always exists, even though it may be rather boring. The empty set $Z \stackrel{\text{def}}{=} \emptyset$ looks like it would satisfy (15.9) because it is rather hard to win a game that requires Angel to enter the empty set of states $\emptyset$ to win.

On second thoughts, $\varsigma_\alpha(\emptyset) \subseteq \emptyset$ does not actually always hold for all hybrid games α. It is violated for states from which Angel can make sure Demon violates the rules of the game α by losing a challenge or failing to comply with evolution domain constraints. When Q is a nontrivial formula like $x > 0$ Demon fails $?Q^d$ sometimes:

$$\varsigma_{?Q^d}(\emptyset) = (\varsigma_{?Q}(\emptyset^\complement))^\complement = ([\![Q]\!] \cap \mathscr{S})^\complement = ([\![Q]\!])^\complement = [\![\neg Q]\!] \not\subseteq \emptyset$$

Yet, then the set of states $[\![\neg Q]\!]$ that make Demon violate the rules satisfies (15.9):

$$\varsigma_{?Q^d}([\![\neg Q]\!]) = (\varsigma_{?Q}([\![\neg Q]\!]^{\complement}))^{\complement} = (\varsigma_{?Q}([\![Q]\!]))^{\complement} = ([\![Q]\!] \cap [\![Q]\!])^{\complement} = [\![\neg Q]\!] \subseteq [\![\neg Q]\!]$$

But even in cases where the empty set $\emptyset$ satisfies (15.9), it may be too small. Likewise, even if the set of states where Demon violates the rules immediately satisfies (15.9), this set may still be too small. Angel is still in charge of repetition and can decide how often to repeat and whether to repeat at all. The winning region $\varsigma_{\alpha^*}(X)$ of repetition of α should at least also contain the winning condition X, because the winning condition X is particularly easy to reach when already starting in X because Angel can then simply decide to stop fooling around and just repeat zero times. Consequently, if a set $Z \subseteq \mathscr{S}$ claims to be the winning region $\varsigma_{\alpha^*}(X)$, then it has to satisfy (15.9) and also satisfy

$$X \subseteq Z \tag{15.10}$$

Both conditions (15.9), (15.10) together can be summarized in one condition.

Note 74 (Pre-fixpoint) Every candidate Z for the winning region $\varsigma_{\alpha^*}(X)$ satisfies the *pre-fixpoint* condition:

$$X \cup \varsigma_\alpha(Z) \subseteq Z \tag{15.11}$$

Again: what is this set Z that satisfies (15.11)? Is there only one choice? Or many? If there are multiple choices, which Z is the right one for the semantics of repetition? Does such a Z always exist, even?

Before you read on, see if you can find the answer for yourself.

One such Z certainly exists. The empty set does not qualify unless $X = \emptyset$ (and even then $\emptyset$ actually only works if Demon cannot be tricked into violating the rules of the game). The set X itself is too small as well, unless the game has no incentive to start repeating, because $\varsigma_\alpha(X) \subseteq X$. But the full state space $Z \overset{\text{def}}{=} \mathscr{S}$ always satisfies (15.11) trivially, so (15.11) definitely has a solution. Now, the whole space is a little too big to call it Angel's winning region independently of the hybrid game α. Even if the full space may very well be the winning region for a particularly Demonophobic Angel-friendly hybrid game like (15.5), the full state space is hardly the right winning region for any arbitrary hybrid game α^*. It definitely depends on the hybrid game α and the winning condition P whether Angel has a winning strategy for $\langle\alpha\rangle P$ or not. For example for Demon's favorite game where he always wins, the winning region $\varsigma_{\alpha^*}(X)$ of Angel had better be $\emptyset$, not $\mathscr{S}$. Thus, the largest solution Z of (15.11) hardly qualifies.

So which solution Z of (15.11) do we define to be $\varsigma_{\alpha^*}(X)$ now?

Before you read on, see if you can find the answer for yourself.

Among the many sets Z that solve (15.11), the largest one is not informative, because the largest Z simply degrades to the full state space $\mathscr{S}$. So smaller solutions Z are preferable. Which one? How do multiple solutions relate to each other? Suppose

Y, Z are both solutions of (15.11). That is

$$X \cup \varsigma_\alpha(Y) \subseteq Y \qquad\qquad (15.12)$$
$$X \cup \varsigma_\alpha(Z) \subseteq Z \qquad\qquad (15.13)$$

Then, by the monotonicity lemma (Lemma 15.1)

$$X \cup \varsigma_\alpha(Y \cap Z) \overset{\text{mon}}{\subseteq} X \cup (\varsigma_\alpha(Y) \cap \varsigma_\alpha(Z)) \overset{(15.12),(15.13)}{\subseteq} Y \cap Z \qquad (15.14)$$

Hence, by (15.14), the intersection $Y \cap Z$ of solutions Y and Z of (15.11) also is a solution of (15.11).

> **Lemma 15.2 (Intersection closure).** *For any two solutions Y, Z of the prefix condition (15.11), the intersection $Y \cap Z$ is a solution of (15.11) as well.*

Whenever there are two solutions Z_1, Z_2 of (15.11), their intersection $Z_1 \cap Z_2$ solves (15.11) as well. When there's yet another solution Z_3 of (15.11), the intersection $Z_1 \cap Z_2 \cap Z_3$ also solves (15.11). Similarly the intersection of any larger family of solutions solves (15.11). If we keep on intersecting solutions, we will arrive at smaller and smaller solutions until, some fine day, there's not going to be a smaller one. This yields the smallest solution Z of (15.11), which is $\varsigma_{\alpha^*}(X)$.

> **Note 75 (Semantics of repetitions)** Among the many Z that solve (15.11), $\varsigma_{\alpha^*}(X)$ is defined to be the smallest Z that solves prefix condition (15.11):
>
> $$\varsigma_{\alpha^*}(X) = \bigcap \{Z \subseteq \mathscr{S} : X \cup \varsigma_\alpha(Z) \subseteq Z\} \qquad (15.15)$$

In other words, the winning region $\varsigma_{\alpha^*}(X)$ is the smallest set Z that already contains the winning condition X and the set of states $\varsigma_\alpha(Z)$ from which Angel can win into Z with one more round of game α. Hence, adding to Z the set of states $\varsigma_\alpha(Z)$ where one more round would win does not change the set Z, as illustrated in Fig. 15.10.

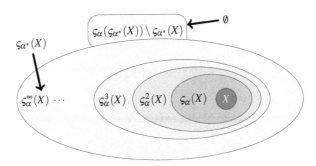

Fig. 15.10 Illustration of denotational semantics of winning region of hybrid game repetitions

The fact that $\varsigma_{\alpha^*}(X)$ is defined as the *smallest* of all these sets makes sure that Angel only wins games by a well-founded number of repetitions. That is, she only wins a repetition if she ultimately stops repeating, not by postponing termination forever with a filibuster [4].

The characterization in terms of iterated winning regions from Sect. 15.3.3 leads to the same set $\varsigma_{\alpha^*}(X)$, but the (least pre-fixpoint or) least fixpoint characterization (15.15) is easier to describe and reason with. Understanding why the two styles of definition of the semantics lead to the same result will take some thought.

The set on the right-hand side of (15.15) is an intersection of solutions, thus, a solution by Lemma 15.2 (or its counterpart for arbitrary families of solutions). Hence $\varsigma_{\alpha^*}(X)$ itself satisfies the prefix condition (15.11):

$$X \cup \varsigma_\alpha(\varsigma_{\alpha^*}(X)) \subseteq \varsigma_{\alpha^*}(X) \tag{15.16}$$

Also compare this with where we came from when we argued for (15.8). Could it be the case that the inclusion in (15.16) is strict, i.e., not equals? No this cannot happen, because $\varsigma_{\alpha^*}(X)$ is the smallest such set. That is, by (15.16), the set $Z \stackrel{\text{def}}{=} X \cup \varsigma_\alpha(\varsigma_{\alpha^*}(X))$ satisfies $Z \subseteq \varsigma_{\alpha^*}(X)$ and, thus, by Lemma 15.1:

$$X \cup \varsigma_\alpha(Z) \stackrel{\text{mon}}{\subseteq} X \cup \varsigma_\alpha(\varsigma_{\alpha^*}(X)) = Z$$

Thus, the set $Z \stackrel{\text{def}}{=} X \cup \varsigma_\alpha(\varsigma_{\alpha^*}(X))$ satisfies the condition $X \cup \varsigma_\alpha(Z) \subseteq Z$ from (15.15). Since $\varsigma_{\alpha^*}(X)$ is the smallest such set by (15.15), it is a subset of Z:

$$\varsigma_{\alpha^*}(X) \subseteq Z = X \cup \varsigma_\alpha(\varsigma_{\alpha^*}(X))$$

Consequently, with (15.16), this implies that both inclusions hold, so $\varsigma_{\alpha^*}(X) = Z$. Thus, the least pre-fixpoint $\varsigma_{\alpha^*}(X)$ satisfies not just the *pre-fixpoint inclusion* (15.11) but it even satisfies the *fixpoint equation*:

$$X \cup \varsigma_\alpha(\varsigma_{\alpha^*}(X)) = \varsigma_{\alpha^*}(X)$$

Note 76 (Semantics of repetitions, fixpoint formulation) The semantics or winning region $\varsigma_{\alpha^*}(X)$ of repetition is a *fixpoint* solving the equation

$$X \cup \varsigma_\alpha(Z) = Z \tag{15.17}$$

It is the *least fixpoint*, i.e., the smallest set Z solving the equation (15.17). That is, it satisfies

$$\varsigma_{\alpha^*}(X) = \bigcap\{Z \subseteq \mathscr{S} : X \cup \varsigma_\alpha(Z) = Z\} \tag{15.18}$$

Due to the seminal fixpoint theorem of Knaster-Tarski [6], the least fixpoint semantics $\varsigma_{\alpha^*}(X)$ from (15.15) alias (15.18) gives the same set of states as the inflationary semantics (15.7), because the semantics is monotone (Lemma 15.1). That is, after iterating the winning-region construction in Sect. 15.3.3 for all (large enough) ordinals starting from X, the result will be the least fixpoint of (15.17).

Lemma 15.3 (Transfinite inflation leads to a least fixpoint).

$$\varsigma_{\alpha^*}(X) \stackrel{\text{def}}{=} \bigcap\{Z \subseteq \mathscr{S} : X \cup \varsigma_\alpha(Z) \subseteq Z\} = \varsigma_\alpha^\infty(X) \stackrel{\text{def}}{=} \bigcup_{\kappa \text{ ordinal}} \varsigma_\alpha^\kappa(X)$$

But the iterated winning-region constructions go significantly transfinite [4], way beyond the first infinite ordinal ω.

The situation for Demon's winning region for repetition is analogously. The difference is that Angel controls repetition α^*, so Demon only has a winning strategy to achieve X if he starts in X (because Angel might repeat 0 times) and has a winning strategy to stay in X all the time. Postponing termination forever will make Demon win if only he stays in X, because Angel is in charge of repetition and will ultimately have to stop repeating. Consequently, the winning region for Demon for Angel's repetition is the largest fixpoint.

Note 77 (Demon's winning region for repetition)

$$\delta_{\alpha^*}(X) = \bigcup\{Z \subseteq \mathscr{S} : X \cap \delta_\alpha(Z) = Z\} = \bigcup\{Z \subseteq \mathscr{S} : Z \subseteq X \cap \delta_\alpha(Z)\}$$

The winning region $\delta_{\alpha^*}(X)$ is the largest set Z that is contained in the winning condition X and in the set of states $\delta_\alpha(Z)$ where Demon has a winning strategy to remain in Z for one more round of game α. This set is the largest fixpoint, because Demon does not mind repeating indefinitely, since he knows that Angel will ultimately have to stop repeating at some point anyhow. He only needs to make sure not to have left the winning condition X, because he cannot know how often Angel will choose to repeat.

15.4 Semantics of Hybrid Games

The semantics of hybrid games from Sect. 15.2.3 was still pending a definition of the winning regions $\varsigma_\alpha(\cdot)$ and $\delta_\alpha(\cdot)$ for Angel and Demon, respectively, in the hybrid game α. Rather than taking a detour to understand those by an operational game tree semantics (as in Chap. 14), or in terms of transfinitely iterated winning-region constructions (Sect. 15.3.3), the winning regions of all hybrid games can be defined directly (Sect. 15.3.4), giving a denotational semantics to hybrid games.

The only difference of the following semantics compared to the previous Definition 15.2 is the new case of repetition α^* illustrated in Fig. 15.10.

Definition 15.4 (Semantics of hybrid games). The *semantics of a hybrid game* α is a function $\varsigma_\alpha(\cdot)$ that, for each set of Angel's winning states $X \subseteq \mathscr{S}$, gives the *winning region*, i.e., the set of states $\varsigma_\alpha(X)$ from which Angel has a winning strategy to achieve X (whatever strategy Demon chooses). It is defined inductively as follows:

1. $\varsigma_{x:=e}(X) = \{\omega \in \mathscr{S} : \omega_x^{\omega[\![e]\!]} \in X\}$
   That is, an assignment $x := e$ wins a game into X from any state ω whose modification $\omega_x^{\omega[\![e]\!]}$ that changes the value of x to $\omega[\![e]\!]$ is in X.

2. $\varsigma_{x'=f(x)\,\&\,Q}(X) = \{\varphi(0) \in \mathscr{S} : \varphi(r) \in X \text{ for some solution } \varphi : [0,r] \to \mathscr{S}$ of any duration $r \in \mathbb{R}$ satisfying $\varphi \models x' = f(x) \land Q\}$
   That is, Angel wins the differential equation $x' = f(x)\,\&\,Q$ into X from any state $\varphi(0)$ from which there is a solution φ of $x' = f(x)$ of any duration r that remains in Q all the time and leads to a final state $\varphi(r) \in X$.

3. $\varsigma_{?Q}(X) = [\![Q]\!] \cap X$
   That is, Angel wins into X for a challenge $?Q$ from the states which satisfy Q to pass the challenge and are already in X, because challenges $?Q$ do not change the state. Angel only achieves winning condition X in game $?Q$ in the states in X that also satisfy the test formula Q.

4. $\varsigma_{\alpha \cup \beta}(X) = \varsigma_\alpha(X) \cup \varsigma_\beta(X)$
   That is, Angel wins a game of choice $\alpha \cup \beta$ into X whenever she wins α into X or wins β into X (by choosing a subgame she has a winning strategy for).

5. $\varsigma_{\alpha;\beta}(X) = \varsigma_\alpha(\varsigma_\beta(X))$
   That is, Angel wins a sequential game $\alpha;\beta$ into X whenever she has a winning strategy in game α to achieve $\varsigma_\beta(X)$, i.e., to make it to one of the states from which she has a winning strategy in game β to achieve X.

6. $\varsigma_{\alpha^*}(X) = \bigcap\{Z \subseteq \mathscr{S} : X \cup \varsigma_\alpha(Z) \subseteq Z\}$
   That is, Angel wins a game of repetition α^* into X from the smallest set of states Z that includes both X and the set of states $\varsigma_\alpha(Z)$ from which Angel can achieve Z in one more round of game α.

7. $\varsigma_{\alpha^d}(X) = (\varsigma_\alpha(X^\complement))^\complement$
   That is, Angel wins α^d to achieve X in exactly the states in which she does not have a winning strategy in game α to achieve the opposite $X^\complement$.

Definition 15.5 (Semantics of hybrid games, continued). The *winning region* of Demon, i.e., the set of states $\delta_\alpha(X)$ from which Demon has a winning strategy to achieve X (whatever strategy Angel chooses) is defined inductively:

1. $\delta_{x:=e}(X) = \{\omega \in \mathscr{S} : \omega_x^{\omega[\![e]\!]} \in X\}$
   That is, an assignment $x := e$ wins a game into X from any state ω whose modification $\omega_x^{\omega[\![e]\!]}$ that changes the value of x to $\omega[\![e]\!]$ is in X.

2. $\delta_{x'=f(x)\,\&\,Q}(X) = \{\varphi(0) \in \mathscr{S} : \varphi(r) \in X \text{ for all durations } r \in \mathbb{R} \text{ and all}$ solutions $\varphi : [0,r] \to \mathscr{S}$ satisfying $\varphi \models x' = f(x) \land Q\}$

That is, Demon wins the differential equation $x' = f(x) \& Q$ into X from any state $\varphi(0)$ from which all solutions φ of $x' = f(x)$ of any duration r that remain within Q all the time lead to states $\varphi(r) \in X$ in the end.

3. $\delta_{?Q}(X) = (\llbracket Q \rrbracket)^{\complement} \cup X$

That is, Demon wins into X for a challenge $?Q$ from the states which violate Q so that Angel fails her challenge $?Q$ or that are already in X, because challenges $?Q$ do not change the state. Demon achieves the winning condition X in game $?Q$ in the states in X (whether or not Q holds) as well as in the states in which Angel fails test formula Q.

4. $\delta_{\alpha \cup \beta}(X) = \delta_\alpha(X) \cap \delta_\beta(X)$

That is, Demon wins a game of choice $\alpha \cup \beta$ into X whenever he wins α into X *and* wins β into X (because Angel might choose either subgame).

5. $\delta_{\alpha;\beta}(X) = \delta_\alpha(\delta_\beta(X))$

That is, Demon wins a sequential game $\alpha;\beta$ into X whenever he has a winning strategy in game α to achieve $\delta_\beta(X)$, i.e., to make it to one of the states from which he has a winning strategy in game β to achieve X.

6. $\delta_{\alpha^*}(X) = \bigcup\{Z \subseteq \mathscr{S} : Z \subseteq X \cap \delta_\alpha(Z)\}$

That is, Demon wins a game of repetition α^* into X from the biggest set of states Z that is included both in X and in the set of states $\delta_\alpha(Z)$ from which Demon can achieve Z in one more round of game α.

7. $\delta_{\alpha^{\mathrm{d}}}(X) = (\delta_\alpha(X^{\complement}))^{\complement}$

That is, Demon wins α^{d} to achieve X in exactly the states in which he does not have a winning strategy in game α to achieve the opposite $X^{\complement}$.

The semantics of dGL is still *compositional*, i.e., the semantics of a compound dGL formula is a simple function of the semantics of its pieces, and the semantics of a compound hybrid game is a function of the semantics of its pieces.

The semantics of $\varsigma_{\alpha^*}(X)$ is a least fixpoint, which results in a well-founded repetition of α, i.e., Angel can repeat any number of times but she ultimately needs to stop at a state in X in order to win. The semantics of $\delta_{\alpha^*}(X)$ is a greatest fixpoint, for which Demon needs to achieve a state in X after *every* number of repetitions, because Angel might choose to stop at any time, but Demon still wins if he only postpones Angel's victory forever, because Angel ultimately has to stop repeating.

Thus, for the formula $[\alpha^*]P$, Demon already has a winning strategy if he only has a strategy that is not losing by preventing P indefinitely, because Angel eventually has to stop repeating anyhow and will then end up in a state not satisfying P, which makes her lose. For Demon's repetition $[\alpha^{\times}]P$ the situation is dual, so Demon will ultimately have to stop repeating and get to state P in finite time. But Angel is happy to postpone Demon's victory forever, because Demon will eventually have to stop since he is in charge of Demon's repetition $\alpha^{\times}$.

15.5 Summary

This chapter saw the introduction of a proper formal semantics for differential game logic and hybrid games. This resulted in a simple denotational semantics, where the meaning of all formulas and hybrid games is a simple function of the meaning of its pieces. The only possible outlier was the semantics of repetition, which turned out to be somewhat subtle and ultimately required higher-ordinal iterations of winning-region constructions. This led to an insightful appreciation for the complexities, challenges, and flexibilities of hybrid games. But the final word on the semantics of repetition was a simpler implicit characterization via fixpoints. The next chapter will leverage their semantic basis for the next leg in the logical trinity: axiomatics. That will enable us to succinctly reason about hybrid games and whether our player of interest has a winning strategy.

The concepts that we touched upon in this chapter are of independent interest. Fixpoints play a huge rôle in many areas of science [1, 2, 7]. Ordinals are also of more general interest. Differences between operational and denotational semantics are more broadly impactful beyond CPS.

Exercises

15.1. Use the semantics of differential game logic to explain why the following formulas are valid and then give a corresponding winning strategy:

$$\langle x:=x^2; (x:=x+1 \cap x:=x+2)\rangle x > 0$$
$$\langle x:=x^2 \cup (x:=x+1 \cap x:=x+2)\rangle x > 0$$
$$\langle x:=x^2 \cup (x:=x+1 \cap x'=2)\rangle x > 0$$
$$[(x:=x^2 \cup x:=-x^2); (x:=x+1 \cap x:=x-1)]x^2 \geq 1$$
$$\langle (x:=x^2 \cap x:=-x^2); \{x'=1\}; (x:=x+1 \cap x:=x-1)\rangle x^2 \geq 1$$
$$[(x:=x^2 \cup ?x < 0; x:=-x); \{x'=1\}; (x:=x+1 \cap x:=0)]x^2 \geq 1$$

15.2 (Modeling advance notice semantics). The advance notice semantics from Sect. 15.3.1 was discarded in favor of the more general semantics of repetition in Sect. 15.3.4, which allows the player controlling repetition to decide arbitrarily each round based on observing the state. Suppose, however, that you have a game where you do want to allow Angel to repeat α any (finite) number of times but you require that she announces the number of repetitions of α ahead of time, just like in the advance notice semantics. Construct a hybrid game that requires Angel to disclose the intended number of repetitions of α to Demon ahead of time even in the semantics of Definition 15.4.

Hint: you can use additional variables.

15.3. The formula (15.5) was shown to need $\omega + 1$ iterations of the winning region construction to terminate with the following answer justifying the validity of (15.5):

$$\varsigma_{\alpha^*}([0,1)) = \varsigma_\alpha^{\omega+1}([0,1)) = \varsigma_\alpha([0,\infty)) = \mathbb{R}$$

What happens if the winning region construction is used once more to compute $\varsigma_\alpha^{\omega+2}([0,1))$? How often does the winning region construction need to be iterated to justify validity of

$$\langle (x := x+1; x' = 1^d \cup x := x-1)^* \rangle (0 \le x < 1)$$

15.4. Explain how often you will have to repeat the winning-region construction from Sect. 15.3.3 to show that the following dGL formulas are valid:

$$\langle (x := x+1; x' = 1^d \cup x := x-1)^* \rangle (0 \le x < 1)$$
$$\langle (x := x-1; y' = 1^d \cup y := y-1; z' = 1^d \cup z := z-1)^* \rangle (x < 0 \land y < 0 \land z < 0)$$

15.5 (* Clockwork ω). How often does the winning-region construction from Sect. 15.3.3 need to be iterated to justify validity of

$$\langle (?y < 0; x := x-1; y' = 1^d \cup ?z < 0; y := y-1; z' = 1^d \cup z := z-1)^* \rangle x < 0$$

Give a winning strategy for Angel. Do the answers change for the following formula?

$$\langle (?y < 0; x := x-1; y' = 1^d \cup ?z < 0; y := y-1; z' = 1^d \cup z := z-1)^* \rangle$$
$$(x < 0 \land y < 0 \land z < 0)$$

15.6. Can you find dGL formulas for which the winning-region construction takes even longer to terminate? How far can you push this?

15.7 (Monotonicity). Prove Lemma 15.1 by induction on the structure of α.

15.8 (Doubly recursive inflationary semantics*). The inflationary semantics of repetition was defined as

$$\varsigma_\alpha^0(X) \overset{\text{def}}{=} X$$
$$\varsigma_\alpha^{\kappa+1}(X) \overset{\text{def}}{=} X \cup \varsigma_\alpha(\varsigma_\alpha^\kappa(X)) \qquad \kappa+1 \text{ is a successor ordinal}$$
$$\varsigma_\alpha^\lambda(X) \overset{\text{def}}{=} \bigcup_{\kappa < \lambda} \varsigma_\alpha^\kappa(X) \qquad \lambda \ne 0 \text{ is a limit ordinal}$$

Show that we could also have modified the successor ordinal case $\kappa + 1$ with two recursive calls without changing the final outcome:

$$\varsigma_\alpha^{\kappa+1}(X) \overset{\text{def}}{=} \varsigma_\alpha^\kappa(X) \cup \varsigma_\alpha(\varsigma_\alpha^\kappa(X))$$

References

[1]   Edmund M. Clarke, Orna Grumberg, and Doron A. Peled. *Model Checking*. Cambridge: MIT Press, 1999.

[2]   Andrzej Granas and James Dugundji. *Fixed Point Theory*. Berlin: Springer, 2003. DOI: 10.1007/978-0-387-21593-8.

[3]   André Platzer. The complete proof theory of hybrid systems. In: *LICS*. Los Alamitos: IEEE, 2012, 541–550. DOI: 10.1109/LICS.2012.64.

[4]   André Platzer. Differential game logic. *ACM Trans. Comput. Log.* **17**(1) (2015), 1:1–1:51. DOI: 10.1145/2817824.

[5]   Jan-David Quesel and André Platzer. Playing hybrid games with KeYmaera. In: *IJCAR*. Ed. by Bernhard Gramlich, Dale Miller, and Ulrike Sattler. Vol. 7364. LNCS. Berlin: Springer, 2012, 439–453. DOI: 10.1007/978-3-642-31365-3_34.

[6]   Alfred Tarski. A lattice-theoretical fixpoint theorem and its applications. *Pacific J. Math.* **5**(2) (1955), 285–309.

[7]   Eberhard Zeidler. *Nonlinear Functional Analysis and Its Applications*. Vol. II/A. Berlin: Springer, 1990. DOI: 10.1007/978-1-4612-0985-0.

Chapter 16
Winning & Proving Hybrid Games

Synopsis This chapter begins the development of the logical characterization of the dynamics of hybrid games, which proves from which states which player can win which game. It investigates compositional reasoning principles with dynamic axioms for adversarial dynamical systems, where each axiom captures how the existence of a winning strategy for a more complex hybrid game relates to the existence of corresponding winning strategies for simpler game fragments. These dynamic axioms enable rigorous reasoning for adversarial CPS models and axiomatize differential game logic, which turns the specification logic dGL into a verification logic for CPS. This is the cornerstone for lifting hybrid systems reasoning techniques to hybrid games.

16.1 Introduction

This chapter continues the study of hybrid games and their logic, differential game logic [11], whose syntax was introduced in Chap. 14 and whose semantics was developed in Chap. 15. This chapter furthers the development of differential game logic to the third leg of the logical trinity: its axiomatics. It will focus on the development of rigorous reasoning techniques for hybrid games as models of CPSs with adversarial dynamics. Without such analysis and reasoning techniques, a logic that only comes with syntax and semantics can be used as a specification language with a precise meaning, but it is not very helpful for actually analyzing and verifying hybrid games. It is the logical trinity of syntax, semantics, and axiomatics that gives logics the power to serve as well-founded specification and verification languages with a (preferably concise) syntax, an unambiguous semantics, and actionable analytic reasoning principles. Thus, this chapter is the hybrid games analogue of Chap. 5, where we investigated dynamic axioms for dynamical systems but did not know about adversarial dynamics yet. Indeed, after the logical sophistication we achieved throughout the textbook, this chapter will settle for a (Hilbert-type) proof calculus predominantly with axioms as in Chap. 5 as opposed to the more easily automatable

© Springer International Publishing AG, part of Springer Nature 2018
A. Platzer, *Logical Foundations of Cyber-Physical Systems*,
https://doi.org/10.1007/978-3-319-63588-0_16

sequent calculus from Chap. 6. A sequent calculus can be built around the axioms of dGL in the same way that the sequent calculus of Chap. 6 was built around the dL axioms from Chap. 5, resulting in the same proof-structuring advantages.

Playing hybrid games is fun. Winning hybrid games is even more fun. But the most fun comes from proving that you'll win a hybrid game. Only don't tell your opponent that you have proved that you have a winning strategy, because he might not want to play this game with you any more.

This chapter is based on [11], where more information can be found on logic and hybrid games. The most important learning goals of this chapter are:

Modeling and Control: We advance our understanding of the core principles behind CPS with hybrid games by understanding analytically and semantically how discrete, continuous, and the adversarial dynamics resulting, e.g., from multiple agents are integrated and interact in CPS. Fixpoints in the semantics of repetitions of games will provide one important aspect for the subsequent development of rigorous reasoning techniques.

Computational Thinking: This chapter is devoted to the development of rigorous reasoning techniques for CPS models involving adversarial dynamics, which is critical to getting CPSs with such interactions right. Hybrid games provide even more subtle interactions than hybrid systems did, which makes it even more challenging to say for sure whether and why a design is correct without sufficient rigor in its analysis. After Chap. 15 captured the semantics of differential game logic and hybrid games compositionally, this chapter exploits the compositional meaning to develop compositional reasoning principles for hybrid games. This chapter systematically develops one reasoning principle for each of the operators of hybrid programs, resulting in a compositional verification approach. A compositional semantics is de facto a necessary but not a sufficient condition for the existence of compositional reasoning principles. Despite the widely generalized semantics of hybrid games compared to hybrid systems, this chapter will strive to generalize reasoning techniques for hybrid systems to hybrid games as smoothly as possible. This leads to a modular way of integrating adversariality into the analysis of hybrid systems models while simultaneously taming their complexity. This chapter provides an *axiomatization* of differential game logic dGL [11] to lift dGL from a specification language to a verification language for CPS with adversarial dynamics.

CPS Skills: We will develop a deep understanding of the semantics of CPS models with adversariality by carefully relating their semantics to their reasoning principles and aligning them in perfect unison. This understanding will also enable us to develop better intuition for the operational effects involved in CPS.

In our quest to develop rigorous reasoning principles for hybrid games, we will strive to identify compositional reasoning principles that align in perfect unison with the compositional semantics of hybrid games developed in Chap. 15. This enterprise will be enlightening and, for the most part, quite successful. And, in fact, the reader is encouraged to start right away with the development of a proof calculus for differential game logic and later compare it with the one that this textbook develops.

rigorous reasoning for adversarial dynamics
compositional reasoning from compositional semantics
modular addition of adversarial dynamics
axiomatization of dGL

analytical & semantical interaction
of discrete+continuous+adversarial
fixpoints

CPS semantics
align semantics and reasoning
operational CPS effects

The part, where this will turn out to be rather difficult is repetition, which is why the textbook take a scenic detour through the characterization of its semantics.

16.2 Semantical Considerations

Before submerging completely into the development of rigorous reasoning techniques for hybrid games as models for CPS with adversarial dynamics, however, it will be wise to take a short detour by investigating some simple properties of their semantics. This section discusses simple but important meta-properties of the semantics of hybrid games that we will make use of subsequently, but which are also of independent interest.

16.2.1 Monotonicity

As Chap. 15 already conjectured, the semantics is monotone [11], i.e., larger sets of winning states have larger winning regions. It is easier to win into larger sets of winning states (Fig. 15.3 on p. 457), which can be proved by inspection of Definition 15.4 on p. 474.

> **Lemma 15.1 (Monotonicity).** *The* dGL *semantics is* monotone, *that is, both* $\varsigma_\alpha(X) \subseteq \varsigma_\alpha(Y)$ *and* $\delta_\alpha(X) \subseteq \delta_\alpha(Y)$ *for all* $X \subseteq Y$.

Proof. The proof is a simple check of Definition 15.4 based on the observation that X only occurs with an even number of negations in the semantics. It is proved by induction on the structure of the hybrid game α. So when proving Lemma 15.1 for a hybrid game α, we assume that it has already been proved for all subgames of α.

1. $\varsigma_{?Q}(X) = [\![Q]\!] \cap X \subseteq [\![Q]\!] \cap Y = \varsigma_{?Q}(Y)$, because $X \subseteq Y$.

2. The cases of discrete assignments and differential equations are equally simple.
3. $\varsigma_{\alpha \cup \beta}(X) = \varsigma_\alpha(X) \cup \varsigma_\beta(X) \subseteq \varsigma_\alpha(Y) \cup \varsigma_\beta(Y) = \varsigma_{\alpha \cup \beta}(Y)$, because monotonicity is already assumed to hold for the subgames α and β of $\alpha \cup \beta$ by induction hypothesis.
4. $\varsigma_\beta(X) \subseteq \varsigma_\beta(Y)$ by induction hypothesis for the subgame β, because $X \subseteq Y$. Hence, $\varsigma_{\alpha;\beta}(X) = \varsigma_\alpha(\varsigma_\beta(X)) \subseteq \varsigma_\alpha(\varsigma_\beta(Y)) = \varsigma_{\alpha;\beta}(Y)$ by induction hypothesis for the subgame α, because $\varsigma_\beta(X) \subseteq \varsigma_\beta(Y)$.
5. $\varsigma_{\alpha^*}(X) = \bigcap\{Z \subseteq \mathscr{S} : X \cup \varsigma_\alpha(Z) \subseteq Z\} \subseteq \bigcap\{Z \subseteq \mathscr{S} : Y \cup \varsigma_\alpha(Z) \subseteq Z\} = \varsigma_{\alpha^*}(Y)$ if $X \subseteq Y$.
6. $X \subseteq Y$ implies $X^\complement \supseteq Y^\complement$, hence $\varsigma_\alpha(X^\complement) \supseteq \varsigma_\alpha(Y^\complement)$, so $\varsigma_{\alpha^d}(X) = (\varsigma_\alpha(X^\complement))^\complement \subseteq (\varsigma_\alpha(Y^\complement))^\complement = \varsigma_{\alpha^d}(Y)$.

The proof showing $\delta_\alpha(X) \subseteq \delta_\alpha(Y)$ when $X \subseteq Y$ is left for Exercise 16.6.   □

While monotonicity is of independent interest, it also implies that the least fixpoint in $\varsigma_{\alpha^*}(X)$ and greatest fixpoint in $\delta_{\alpha^*}(X)$ are well defined at all [4, Lem. 1.7].

16.2.2 Determinacy

Every particular match played in a hybrid game is won by exactly one player, because hybrid games are *zero-sum* (one player's loss is another player's win) and there are no *draws* (the outcome of a particular game play is never inconclusive because every final state is won by one of the players). This is a simple property of each individual match. All we need to do for one particular match is to wait until the players are done playing, which will happen eventually, and then check the winning condition in the final state.

Hybrid games satisfy a much stronger property: *determinacy*, i.e., from any initial situation, one of the players always has a winning strategy to force a win, regardless of how the other player chooses to play. Determinacy is quite a strong property indicating that for every state, there is a player who can force a win, so there is a winning strategy that will make that player win every single match in the given hybrid game from that initial state, no matter what the opponent does.

If, from the same initial state, both Angel and Demon had a winning strategy for opposing winning conditions, then something would be terribly inconsistent. It cannot possibly happen that Angel has a winning strategy in hybrid game α to get to a state where $\neg P$ and, from the same initial state, Demon supposedly also has a winning strategy in the same hybrid game α to get to a state where P holds. After all, a winning strategy is a strategy that makes that player win no matter what strategy the opponent follows. If both players had such winning strategies for winning conditions $\neg P$ and P, respectively, then their strategies would take the final state simultaneously to $\neg P$ and to P, which is impossible. Hence, for any initial state, at most one player can have a winning strategy for complementary winning conditions. This argues for the validity of $\neg([\alpha]P \land \langle \alpha \rangle \neg P)$, which can be proved (Theorem 16.1 below).

So hybrid games are *consistent*, because it cannot happen that both players have a winning strategy for complementary winning conditions in the same state. But maybe no one has a winning strategy, i.e., both players can let the other player win, but cannot win strategically themselves (recall, e.g., the filibuster example from Chap. 14, which first appeared as if no player has a winning strategy but then turned out to make Demon win, because Angel needs to stop her repetition eventually). For hybrid games at least one (in fact, exactly one) player has a winning strategy for complementary winning conditions from any initial state [11]. This property, called *determinacy*, is important to be able to assign classical truth-values to formulas, because their modalities refer to the existence of winning strategies. If it is not clear which player has a winning strategy then we cannot say whether formulas of the form $\langle\alpha\rangle P$ and $[\alpha]P$ are true.

If Angel has no winning strategy to achieve $\neg P$ in hybrid game α, then Demon has a winning strategy to achieve P in the same hybrid game α, and vice versa.

> **Theorem 16.1 (Consistency & determinacy).** *Hybrid games are* consistent *and* determined, *i.e.,* $\vDash \neg\langle\alpha\rangle\neg P \leftrightarrow [\alpha]P$.

Proof. The proof shows by induction on the structure of α that $\varsigma_\alpha(X^\complement)^\complement = \delta_\alpha(X)$ for all $X \subseteq \mathscr{S}$, which implies the validity of $\neg\langle\alpha\rangle\neg P \leftrightarrow [\alpha]P$ using $X \stackrel{\text{def}}{=} [\![P]\!]$. For the most part, the proof only expands Definition 15.4 and Definition 15.5 directly.

1. $\varsigma_{x:=e}(X^\complement)^\complement = \{\omega \in \mathscr{S} : \omega_x^{\omega[e]} \notin X\}^\complement = \varsigma_{x:=e}(X) = \delta_{x:=e}(X)$
2. $\varsigma_{x'=f(x)\,\&\,Q}(X^\complement)^\complement = \{\varphi(0) \in \mathscr{S} : \varphi(r) \notin X$ for some $0 \le r \in \mathbb{R}$ and some (differentiable) $\varphi : [0,r] \to \mathscr{S}$ such that $\frac{\mathrm{d}\varphi(t)(x)}{\mathrm{d}t}(\zeta) = \varphi(\zeta)[\![f(x)]\!]$ and $\varphi(\zeta) \in [\![Q]\!]$ for all $0 \le \zeta \le r\}^\complement = \delta_{x'=f(x)\,\&\,Q}(X)$, because the set of states from which there is no winning strategy for Angel to reach a state in $X^\complement$ prior to leaving $[\![Q]\!]$ along $x' = f(x)\,\&\,Q$ is exactly the set of states from which $x' = f(x)\,\&\,Q$ always stays in X (until leaving $[\![Q]\!]$ in case that ever happens).
3. $\varsigma_{?Q}(X^\complement)^\complement = ([\![Q]\!] \cap X^\complement)^\complement = ([\![Q]\!])^\complement \cup (X^\complement)^\complement = \delta_{?Q}(X)$
4. $\varsigma_{\alpha\cup\beta}(X^\complement)^\complement = (\varsigma_\alpha(X^\complement) \cup \varsigma_\beta(X^\complement))^\complement = \varsigma_\alpha(X^\complement)^\complement \cap \varsigma_\beta(X^\complement)^\complement = \delta_\alpha(X) \cap \delta_\beta(X) = \delta_{\alpha\cup\beta}(X)$
5. $\varsigma_{\alpha;\beta}(X^\complement)^\complement = \varsigma_\alpha(\varsigma_\beta(X^\complement))^\complement = \varsigma_\alpha(\delta_\beta(X)^\complement)^\complement = \delta_\alpha(\delta_\beta(X)) = \delta_{\alpha;\beta}(X)$
6. $\varsigma_{\alpha^*}(X^\complement)^\complement = \left(\bigcap\{Z \subseteq \mathscr{S} : X^\complement \cup \varsigma_\alpha(Z) \subseteq Z\}\right)^\complement =$
   $\left(\bigcap\{Z \subseteq \mathscr{S} : (X \cap \varsigma_\alpha(Z)^\complement)^\complement \subseteq Z\}\right)^\complement = \left(\bigcap\{Z \subseteq \mathscr{S} : (X \cap \delta_\alpha(Z^\complement))^\complement \subseteq Z\}\right)^\complement$
   $= \bigcup\{Z \subseteq \mathscr{S} : Z \subseteq X \cap \delta_\alpha(Z)\} = \delta_{\alpha^*}(X)$ [1]
7. $\varsigma_{\alpha^d}(X^\complement)^\complement = (\varsigma_\alpha((X^\complement)^\complement)^\complement)^\complement = \delta_\alpha(X^\complement)^\complement = \delta_{\alpha^d}(X)$ □

---

[1] The penultimate equation follows from a μ-calculus [7] equivalence that the greatest fixpoint $\nu Z.\Upsilon(Z)$ of $\Upsilon(Z)$ is the same as the complement $\neg\mu Z.\neg\Upsilon(\neg Z)$ of the least fixpoint $\mu Z.\neg\Upsilon(\neg Z)$ of the dual $\neg\Upsilon(\neg Z)$. Applicability of this equation uses the insights from Chap. 15 that least pre-fixpoints are fixpoints and greatest post-fixpoints are fixpoints for monotone functions.

The determinacy direction of Theorem 16.1 is $\models \neg\langle\alpha\rangle\neg P \to [\alpha]P$, which is propositionally equivalent to $\models \langle\alpha\rangle\neg P \vee [\alpha]P$, implying that from all initial states, either Angel has a winning strategy to achieve $\neg P$ or Demon has a winning strategy to achieve P. The consistency direction of Theorem 16.1 is $\models [\alpha]P \to \neg\langle\alpha\rangle\neg P$, i.e., $\models \neg([\alpha]P \wedge \langle\alpha\rangle\neg P)$, which implies that there is no state from which both Demon has a winning strategy to achieve P and, simultaneously, Angel has a winning strategy to achieve $\neg P$.

16.3 Dynamic Axioms for Hybrid Games

This section develops axioms for decomposing hybrid games [11], which will make it possible to reason rigorously about hybrid games. We continue the compositionality principles of logic such that each axiom describes one operator on hybrid games in terms of simpler hybrid games. The major twist compared to the dynamic axioms for dynamical systems described by hybrid programs from Chap. 5 is that the dynamic axioms now need to capture the existence of winning strategies in hybrid games and handle the subtle challenges of interactive game play.

What gives us hope to identify reasonable axioms is that the semantics of dGL is well behaved, because the meaning of each hybrid game is a function of the meaning of its subgames.

16.3.1 Dynamic Axioms for Determinacy

The easiest way to get started with an axiomatization for the operators of differential game logic is to internalize the insights from the semantical results in Sect. 16.2 as logical axioms.

Consistency and determinacy (Theorem 16.1) showed that $\models \neg\langle\alpha\rangle\neg P \leftrightarrow [\alpha]P$ is valid. That is, if Angel has no winning strategy to achieve $\neg P$ then Demon has a winning strategy to achieve P in the same hybrid game α, and vice versa. This insight helpfully related box and diamond modalities, but Theorem 16.1 is not yet available in our proofs, because it is about validity or truth, not proof.

All it takes to use Theorem 16.1 in proofs is to internalize it as an axiom.

> **Lemma 16.1 ([·] determinacy axiom).** *The* determinacy axiom *is sound:*
>
> $$[·] \quad [\alpha]P \leftrightarrow \neg\langle\alpha\rangle\neg P$$

Proof. Soundness of axiom [·], i.e., that each of its instances is valid, directly follows from Theorem 16.1. □

After we adopt [·] as an axiom and give a soundness proof for it, we can, from now on, just use the determinacy principle by referring to axiom [·]. We do not need to worry on a case-by-case basis whether it can be used in a proof, because we settled its soundness question once and for all. Of course, Theorem 16.1 sort of says the same thing as axiom $[\cdot]$ does, but proofs do not come with a mechanism for applying external mathematical theorems, while they very much come with a mechanism for applying logical axioms.

16.3.2 Monotonicity

Transliterating Theorem 16.1 into the axiomatization of dGL was straightforward, almost copy-and-paste. What is the axiomatic counterpart of Lemma 15.1?

> Before you read on, see if you can find the answer for yourself.

Lemma 15.1 says that $\varsigma_\alpha(X) \subseteq \varsigma_\alpha(Y)$ if $X \subseteq Y$. What is the logical counterpart of $\varsigma_\alpha(X)$ and of $\varsigma_\alpha(Y)$?

Of course the logical counterpart of $\varsigma_\alpha(X)$ cannot possibly be $\langle\alpha\rangle X$, because that is not even a syntactically well-formed formula when $X \subseteq \mathscr{S}$ is a set of states. But for a logical formula P, the dGL formula $\langle\alpha\rangle P$ corresponds to $\varsigma_\alpha([\![P]\!])$, because $[\![\langle\alpha\rangle P]\!] = \varsigma_\alpha([\![P]\!])$ by Definition 15.1. Likewise, when Q is another logical formula, then $\langle\alpha\rangle Q$ corresponds to $\varsigma_\alpha([\![Q]\!])$. What does the inclusion $\varsigma_\alpha([\![P]\!]) \subseteq \varsigma_\alpha([\![Q]\!])$ correspond to?

> Before you read on, see if you can find the answer for yourself.

Since $\varsigma_\alpha([\![P]\!]) \subseteq \varsigma_\alpha([\![Q]\!])$ is $[\![\langle\alpha\rangle P]\!] \subseteq [\![\langle\alpha\rangle Q]\!]$, this inclusion of sets of states of truth is equivalent to the validity of the dGL formula $\langle\alpha\rangle P \to \langle\alpha\rangle Q$. Now Lemma 15.1 does not imply that $\langle\alpha\rangle P \to \langle\alpha\rangle Q$ is valid. Lemma 15.1 only implies $\vDash \langle\alpha\rangle P \to \langle\alpha\rangle Q$ under the assumption that $[\![P]\!] \subseteq [\![Q]\!]$. What is a corresponding rigorous reasoning principle in the dGL proof calculus?

> Before you read on, see if you can find the answer for yourself.

The logical internalization of the monotonicity principle from Lemma 15.1 as a proof principle is the following proof rule.

Lemma 16.2 (M monotonicity rule). *The monotonicity rules are sound:*

$$\text{M}\ \frac{P \to Q}{\langle\alpha\rangle P \to \langle\alpha\rangle Q} \qquad \text{M}[\cdot]\ \frac{P \to Q}{[\alpha]P \to [\alpha]Q}$$

Proof. This proof rule is sound, i.e., validity of all premises (here just one) implies validity of the conclusion, which directly follows from Lemma 15.1. If the premise $P \to Q$ is valid, then $[\![P]\!] \subseteq [\![Q]\!]$, which implies $[\![\langle\alpha\rangle P]\!] \subseteq [\![\langle\alpha\rangle Q]\!]$ by Lemma 15.1, which says that the conclusion $\langle\alpha\rangle P \to \langle\alpha\rangle Q$ is valid. Rule $\text{M}[\cdot]$ is similar. ☐

This lemma is identical to Lemma 5.13 on p. 163, except that the new lemma applies to arbitrary hybrid games α, not just to hybrid programs as Lemma 5.13 did.

Of course, Lemma 15.1 cannot be internalized as the following formula:

$$(P \to Q) \to (\langle\alpha\rangle P \to \langle\alpha\rangle Q) \tag{16.1}$$

The formula (16.1) only assumes the implication $P \to Q$ to be true in the current state, while rule M assumes the implication $P \to Q$ is valid, so true in all states, including the final states that Angel is trying to achieve to win $\langle\alpha\rangle P$.

The validity from Theorem 16.1 gave rise to an axiom for dGL while the conditional validity from Lemma 15.1 leads to a proof rule with a premise for the assumption and a conclusion.

16.3.3 Dynamic Axioms for Assignments

The semantics of hybrid games is a set-valued semantics, giving the set of states from which Angel has a winning strategy to achieve set $X \subseteq \mathscr{S}$ as the winning region $\varsigma_\alpha(X) \subseteq \mathscr{S}$. But except for the style of definition, assignments $x:=e$ still have the same semantics that they had in hybrid systems, because assignments have a deterministic result and involve no choices by any player whatsoever. Consequently, Angel has a winning strategy in the discrete assignment game $x:=e$ to achieve $p(x)$ iff $p(e)$ is true, because the assignment $x:=e$ exactly has the effect of changing the value of the variable x to the value of e.

Lemma 16.3 ($\langle:=\rangle$ assignment axiom). *The* assignment axiom *is sound:*

$$\langle:=\rangle \quad \langle x:=e\rangle p(x) \leftrightarrow p(e)$$

16.3.4 Dynamic Axioms for Differential Equations

Unlike discrete assignments, differential equations involve a choice, namely Angel's choice of duration. Recall that the winning-region semantics of differential equations from Definition 15.4 is the set of all states from which there is a solution of the differential equation to the winning condition:

$$\varsigma_{x'=f(x)}(X) = \{\varphi(0) \in \mathscr{S} : \varphi(r) \in X \text{ for some solution } \varphi : [0,r] \to \mathscr{S} \text{ of any}$$
$$\text{duration } r \in \mathbb{R} \text{ satisfying } \varphi \models x' = f(x)\}$$

A schematical illustration of what this region looks like is as follows:

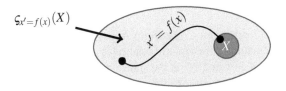

If we have a solution $y(\cdot)$ of the initial value problem $y'(t) = f(y), y(0) = x$, then Angel has a winning strategy for $\langle x' = f(x) \rangle p(x)$ iff there is a duration $t \geq 0$ such that $p(x)$ holds after assigning the solution $y(t)$ to x.

Lemma 16.4 ($\langle ' \rangle$ solution axiom). *The* solution axiom schema *is sound:*

$$\langle ' \rangle \quad \langle x' = f(x) \rangle p(x) \leftrightarrow \exists t \geq 0 \, \langle x := y(t) \rangle p(x) \qquad (y'(t) = f(y))$$

where $y(\cdot)$ solves the symbolic initial value problem $y'(t) = f(y), y(0) = x$.

While differential equations are games that provide a choice for Angel, they, at least, do not give any choices to the other player Demon. That is why there is only one quantifier, the existential quantifier for time, because it is up to Angel to choose her favorite time t to reach $p(x)$. The soundness proof for axiom $\langle ' \rangle$ is essentially the same as the correctness argument for the solution axiom $[']$ for hybrid programs from Lemma 5.3, based on the assumption that $y(\cdot)$ is a solution of the differential equation.

The solution axiom schema $\langle ' \rangle$ inherits the same shortcomings that solution axiom schema $[']$ for hybrid systems already had. It only works for simple differential equations for which Angel happens to have a solution. More complicated differential equations need the induction techniques for differential equations from Part II, which continue to work in hybrid games and generalize to differential games [13].

As stated, the axiom schema $\langle ' \rangle$ also does not support differential equations with evolution domain constraints. While a corresponding generalization is quite straightforward, hybrid games ultimately turn out to provide a more elegant approach for evolution domains (Sect. 16.6). For convenience, we state the evolution domain constraint version of axiom $\langle ' \rangle$ regardless.

Lemma 16.5 ($\langle ' \rangle$ solution with domain axiom). *This axiom is sound:*

$$\langle ' \rangle \quad \langle x' = f(x) \,\&\, q(x) \rangle p(x) \leftrightarrow \exists t \geq 0 \left((\forall 0 \leq s \leq t \, q(y(s))) \wedge \langle x := y(t) \rangle p(x) \right)$$

where $y(\cdot)$ solves the symbolic initial value problem $y'(t) = f(y), y(0) = x$.

16.3.5 Dynamic Axioms for Challenge Games

Test games or challenge games $?Q$ require Angel to pass the test Q or else she will lose the game prematurely for violating the rules of the game. Angel only achieves winning condition X in game $?Q$ in the states in X that also satisfy the test formula Q, because she will otherwise lose the game for violation of the rules. Recall the semantics of test games from Definition 15.4:

$$\varsigma_{?Q}(X) = [\![Q]\!] \cap X \tag{16.2}$$

An illustration of the winning region in (16.2) is:

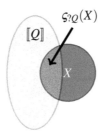

Correspondingly, if Angel wants to win $\langle ?Q \rangle P$ then she will have to be in a state where the postcondition P is already true, because tests do not change the state so P can only be true after $?Q$ if P was already true before. Furthermore, that initial state will also have to satisfy the test condition Q or else she will lose for having violated the rules of the game by failing her test.

> **Lemma 16.6 ($\langle ? \rangle$ test axiom).** *The* test axiom *is sound:*
>
> $$\langle ? \rangle \quad \langle ?Q \rangle P \leftrightarrow Q \wedge P$$

Proof. The axiom is sound iff each of its instances is valid, i.e., true in all states. The equivalence is valid iff the set of all states $[\![\langle ?Q \rangle P]\!]$ where its left-hand side is true is equal to the set of states $[\![Q \wedge P]\!]$ where its right-hand side is true. Indeed, $[\![\langle ?Q \rangle P]\!] = \varsigma_{?Q}([\![P]\!]) = [\![Q]\!] \cap [\![P]\!] = [\![Q \wedge P]\!]$. $\qquad\Box$

16.3.6 Dynamic Axioms for Choice Games

Proving the existence of winning strategies in a choice game $\alpha \cup \beta$ is more difficult, because this hybrid game involves a choice by Angel and may involve further choices by both players in the respective subgames α and β. Recall the semantics of choice games from Definition 15.4, which is a union of the semantics for the subgames:

$$\varsigma_{\alpha \cup \beta}(X) = \varsigma_\alpha(X) \cup \varsigma_\beta(X) \tag{16.3}$$

Let us illustrate what (16.3) means:

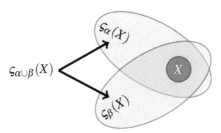

According to the winning region semantics (16.3), the states from which there is a winning strategy in the game $\alpha \cup \beta$ for Angel to achieve X is the union of the set of states from which Angel has a winning strategy in the left subgame α to achieve X and the set of states from which Angel has a winning strategy in the right subgame β to achieve X. Consequently, $\langle \alpha \cup \beta \rangle P$ is true, i.e., Angel has a winning strategy to achieve P in $\alpha \cup \beta$, iff Angel has a winning strategy to achieve P in α or a winning strategy to achieve P in β.

> **Lemma 16.7 ($\langle \cup \rangle$ axiom of choice).** *The* axiom of game of choice *is sound:*
>
> $$\langle \cup \rangle \quad \langle \alpha \cup \beta \rangle P \leftrightarrow \langle \alpha \rangle P \vee \langle \beta \rangle P$$

Proof. The axiom is sound iff each of its instances is valid, i.e., true in all states. The equivalence is valid iff the set of all states $[\![\langle \alpha \cup \beta \rangle P]\!]$ where its left-hand side is true is equal to the set of states $[\![\langle \alpha \rangle P \vee \langle \beta \rangle P]\!]$ where its right-hand side is true.
$$[\![\langle \alpha \cup \beta \rangle P]\!] = \varsigma_{\alpha \cup \beta}([\![P]\!]) = \varsigma_\alpha([\![P]\!]) \cup \varsigma_\beta([\![P]\!]) = [\![\langle \alpha \rangle P]\!] \cup [\![\langle \beta \rangle P]\!] = [\![\langle \alpha \rangle P \vee \langle \beta \rangle P]\!]$$
$\square$

Proving existence of a winning strategy for Angel in a game of choice under Angel's control in $\langle \alpha \cup \beta \rangle P$ merely amounts to proving the disjunction $\langle \alpha \rangle P \vee \langle \beta \rangle P$.

For Demon's choice $\alpha \cap \beta$, Angel has to invest more work to prove that she has a winning strategy for it, because her opponent Demon gets to make the choice. Consequently, Angel only has a winning strategy if she has a winning strategy for both subgames that Demon might choose:

$$\langle \alpha \cap \beta \rangle P \leftrightarrow \langle \alpha \rangle P \wedge \langle \beta \rangle P \tag{16.4}$$

Even if this formula is valid, it will not be adopted as an axiom, because (16.4) can be derived easily from the choice axiom $\langle \cup \rangle$ together with the duality axiom $\langle {}^d \rangle$, which we will explore later. After all, Demon's choice $\alpha \cap \beta$ is built with a derived operator that is defined as the double dual $(\alpha^d \cup \beta^d)^d$ from Angel's choice.

16.3.7 Dynamic Axioms for Sequential Games

The next case to consider is a proof of existence of winning strategies in a sequential game $\alpha; \beta$. Recall the semantics of sequential games from Definition 15.4, which is a composition of the winning regions:

$$\varsigma_{\alpha;\beta}(X) = \varsigma_{\alpha}(\varsigma_{\beta}(X)) \tag{16.5}$$

An illustration of what (16.5) means is the following:

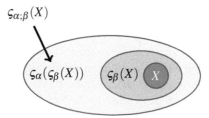

Thus, the set of states from which Angel has a winning strategy for $\alpha; \beta$ is the composition, so the winning region in which she has a strategy in α to reach the winning region for β. The formula characterizing from which states Angel has a winning strategy in the game β to achieve postcondition P is the dGL formula $\langle \beta \rangle P$. Consequently, the formula characterizing from which states Angel has a winning strategy in the game α to reach $\langle \beta \rangle P$ is $\langle \alpha \rangle \langle \beta \rangle P$. By (16.5), that formula is exactly equivalent to $\langle \alpha; \beta \rangle P$ characterizing the states from which Angel has a winning strategy in game $\alpha; \beta$ to achieve P.

Lemma 16.8 ($\langle ; \rangle$ composition axiom). *The composition axiom is sound:*

$$\langle ; \rangle \quad \langle \alpha; \beta \rangle P \leftrightarrow \langle \alpha \rangle \langle \beta \rangle P$$

Proof. The semantics of the composition of the modal operators $\langle \alpha \rangle$ and $\langle \beta \rangle$ exactly corresponds to the semantics of the modal operator $\langle \alpha; \beta \rangle$ for the sequential composition game: $[\![\langle \alpha; \beta \rangle P]\!] = \varsigma_{\alpha;\beta}([\![P]\!]) = \varsigma_{\alpha}(\varsigma_{\beta}([\![P]\!])) = \varsigma_{\alpha}([\![\langle \beta \rangle P]\!]) = [\![\langle \alpha \rangle \langle \beta \rangle P]\!]$ □

16.3.8 Dynamic Axioms for Dual Games

So far, all the axioms for hybrid games looked conspicuously familiar. Such a structural similarity may be somewhat surprising, because the new axioms of this chapter allow hybrid *games*, which have an entirely new semantics with adversarial dynamics compared to the hybrid systems from Part I.

But then again, hybrid systems are special cases of hybrid games, the ones that do not need the other player, because HPs do not mention the duality operator so that control never passes to Demon. Every axiom for hybrid games also holds for hybrid systems, because hybrid systems are special cases of hybrid games. In retrospect it is, thus, not quite so surprising that the reasoning principles for hybrid games have a lot in common with reasoning principles for hybrid systems, even if they need new soundness proofs, because hybrid games have a more general semantics.

For the duality operator in the dual game α^d, however, we will run out of luck trying to find inspiration from generalizations of corresponding reasoning principles for hybrid systems, because the whole point is that the duality operator is the only difference between hybrid systems and hybrid games. Hybrid systems cannot yet know how to handle α^d, because α^d is a hybrid game but not a hybrid system.

Recall the semantics of dual games from Definition 15.4:

$$\varsigma_{\alpha^d}(X) = \varsigma_\alpha(X^\complement)^\complement \tag{16.6}$$

An illustration of what (16.6) means is the following:

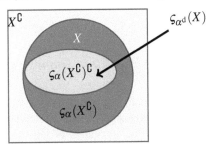

Now, how does that turn into a logical axiom? The complement $X^\complement$ corresponds to negation $\neg P$ of the postcondition P. Hence, the logical internalization of $\varsigma_\alpha(\llbracket P \rrbracket^\complement)$ corresponds to $\langle\alpha\rangle\neg P$ and its complement $\varsigma_\alpha(\llbracket P \rrbracket^\complement)^\complement$ corresponds to $\neg\langle\alpha\rangle\neg P$.

Lemma 16.9 ($\langle^d\rangle$ duality axiom). *The* duality axiom *is sound:*

$$\langle^d\rangle \quad \langle\alpha^d\rangle P \leftrightarrow \neg\langle\alpha\rangle\neg P$$

Proof. $\llbracket\langle\alpha^d\rangle P\rrbracket = \varsigma_{\alpha^d}(\llbracket P\rrbracket) = \varsigma_\alpha(\llbracket P\rrbracket^\complement)^\complement = \varsigma_\alpha(\llbracket\neg P\rrbracket)^\complement = (\llbracket\langle\alpha\rangle\neg P\rrbracket)^\complement = \llbracket\neg\langle\alpha\rangle\neg P\rrbracket$ □

Example 16.1 (Demon's choice). Since Demon's choice $\alpha \cap \beta$ is $(\alpha^d \cup \beta^d)^d$, the duality axiom $\langle^d\rangle$ and the axiom for Angel's choice $\langle\cup\rangle$ can be used to derive the axiom (16.4) for Demon's choice:

$$
\begin{array}{c}
* \\
\hline
\langle\alpha\rangle P \wedge \langle\beta\rangle P \leftrightarrow \langle\alpha\rangle P \wedge \langle\beta\rangle P \\
\hline
\neg(\neg\langle\alpha\rangle\neg\neg P \vee \neg\langle\beta\rangle\neg\neg P) \leftrightarrow \langle\alpha\rangle P \wedge \langle\beta\rangle P \\
\hline
\neg(\langle\alpha^d\rangle\neg P \vee \langle\beta^d\rangle\neg P) \leftrightarrow \langle\alpha\rangle P \wedge \langle\beta\rangle P \\
\hline
\neg\langle\alpha^d \cup \beta^d\rangle\neg P \leftrightarrow \langle\alpha\rangle P \wedge \langle\beta\rangle P \\
\hline
\langle(\alpha^d \cup \beta^d)^d\rangle P \leftrightarrow \langle\alpha\rangle P \wedge \langle\beta\rangle P \\
\hline
\langle\alpha \cap \beta\rangle P \leftrightarrow \langle\alpha\rangle P \wedge \langle\beta\rangle P
\end{array}
$$

Having proved this formula once, we can, from now on, just use the corresponding derived axiom for Demon's choice instead of reproving it every time:

$$\langle\cap\rangle \quad \langle\alpha\cap\beta\rangle P \leftrightarrow \langle\alpha\rangle P \wedge \langle\beta\rangle P$$

$$[\cap] \quad [\alpha\cap\beta]P \leftrightarrow [\alpha]P \vee [\beta]P$$

The derived axiom $[\cap]$ for Demon's winning strategy in Demon's choice can be derived directly from derived axiom $\langle\cap\rangle$:

$$
\begin{array}{c}
\ast \\
\hline
[\alpha]P \vee [\beta]P \leftrightarrow [\alpha]P \vee [\beta]P \\
\hline
{}^{[\cdot]}\quad \neg\langle\alpha\rangle\neg P \vee \neg\langle\beta\rangle\neg P \leftrightarrow [\alpha]P \vee [\beta]P \\
\hline
\neg(\langle\alpha\rangle\neg P \wedge \langle\beta\rangle\neg P) \leftrightarrow [\alpha]P \vee [\beta]P \\
\hline
{}^{\langle\cap\rangle}\quad \neg\langle\alpha\cap\beta\rangle\neg P \leftrightarrow [\alpha]P \vee [\beta]P \\
\hline
{}^{[\cdot]}\quad [\alpha\cap\beta]P \leftrightarrow [\alpha]P \vee [\beta]P
\end{array}
$$

16.3.9 Dynamic Axioms for Repetition Games

The remaining challenge is axioms for repetition games α^*. Repetitions in hybrid games turned out to be semantically significantly more subtle than repetitions in hybrid systems (Chap. 15). Recall the semantics of repetition games from Definition 15.4, where we finally settled on defining it as a least fixpoint of the winning regions of α, because iteration went quite transfinite:

$$\varsigma_{\alpha^*}(X) = \bigcap\{Z \subseteq \mathscr{S} : X \cup \varsigma_\alpha(Z) \subseteq Z\} = \bigcap\{Z \subseteq \mathscr{S} : X \cup \varsigma_\alpha(Z) = Z\} \quad (16.7)$$

The second equation uses that the least pre-fixpoint was also a least fixpoint (Note 76 on p. 472). This semantics (16.7) is best illustrated as follows:

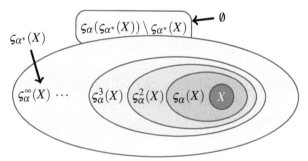

By the second equation of (16.7), $\varsigma_{\alpha^*}(X)$ is a fixpoint of $X \cup \varsigma_\alpha(Z) = Z$, so

$$\varsigma_{\alpha^*}(X) = X \cup \varsigma_\alpha(\varsigma_{\alpha^*}(X)) \quad (16.8)$$

How can (16.8) be internalized as a syntactic reasoning principle in logic?

Before you read on, see if you can find the answer for yourself.

As usual, the set of states $X \subseteq \mathscr{S}$ does not fit into a logical formula, but its logical counterpart is a logical formula P, whose semantics $[\![P]\!]$ will be some set of states. Consequently, the left-hand side of (16.8) corresponds to the logical formula $\langle \alpha^* \rangle P$ expressing that Angel has a winning strategy in the repeated hybrid game α^* to achieve P. What does the right-hand side of (16.8) correspond to?

Since the set X is internalized by the logical formula P, $\varsigma_{\alpha^*}(X)$ corresponds to the logical formula $\langle \alpha^* \rangle P$, because $[\![\langle \alpha^* \rangle P]\!] = \varsigma_{\alpha^*}([\![P]\!])$. Consequently, $X \cup \varsigma_\alpha(\varsigma_{\alpha^*}(X))$ corresponds to the logical formula $P \vee \langle \alpha \rangle \langle \alpha^* \rangle P$. This leads to the following axiom.

> **Lemma 16.10 ($\langle^* \rangle$ iteration axiom).** *The* iteration axiom *is sound:*
>
> $$\langle^* \rangle \quad \langle \alpha^* \rangle P \leftrightarrow P \vee \langle \alpha \rangle \langle \alpha^* \rangle P$$

Proof. The proof is a direct consequence of the fact that the winning region of repetition is a fixpoint (Note 76). Since $[\![\langle \alpha^* \rangle P]\!] = \varsigma_{\alpha^*}([\![P]\!])$ is a fixpoint, we have $[\![\langle \alpha^* \rangle P]\!] = [\![P]\!] \cup \varsigma_\alpha([\![\langle \alpha^* \rangle P]\!])$. Thus, $[\![P \vee \langle \alpha \rangle \langle \alpha^* \rangle P]\!] = [\![P]\!] \cup [\![\langle \alpha \rangle \langle \alpha^* \rangle P]\!] = [\![P]\!] \cup \varsigma_\alpha([\![\langle \alpha^* \rangle P]\!]) = [\![\langle \alpha^* \rangle P]\!]$. □

This axiom $\langle^* \rangle$ is identical to the iteration axiom for hybrid systems (which is the diamond version of Lemma 5.7), except that its soundness justification is completely different. But, once proved sound, the reasoning with axiom $\langle^* \rangle$ works in the same way. Does the axiom $\langle^* \rangle$ say all there is to say about repetition in hybrid games?

Before pursuing this question, first observe that the iteration axiom $\langle^* \rangle$ for Angel's winning strategy in Angel's repetition implies a corresponding iteration axiom for Demon's winning strategy in Demon's repetition.

Example 16.2 (Demon's repetition). Since Demon's repetition $\alpha^\times$ is $((\alpha^d)^*)^d$, the duality axiom $\langle^d \rangle$ and determinacy axiom $[\cdot]$ turn Angel's iteration axiom $\langle^* \rangle$ into a corresponding iteration axiom for Demon's winning strategy

$$[^\times] \quad [\alpha^\times]P \leftrightarrow P \vee [\alpha][\alpha^\times]P$$

This derived axiom $[^\times]$ can be proved easily:

$$
\begin{array}{c}
\ast \\[2pt]
\hline
P \vee [\alpha][\alpha^\times]P \leftrightarrow P \vee [\alpha][\alpha^\times]P \\[2pt]
\hline
P \vee [\alpha][((\alpha^d)^*)^d]P \leftrightarrow P \vee [\alpha][\alpha^\times]P \\[2pt]
\hline
{\scriptstyle \langle d \rangle, [\cdot]} \quad P \vee \langle \alpha^d \rangle \langle (\alpha^d)^* \rangle P \leftrightarrow P \vee [\alpha][\alpha^\times]P \\[2pt]
\hline
{\scriptstyle \langle^* \rangle} \quad \langle (\alpha^d)^* \rangle P \leftrightarrow P \vee [\alpha][\alpha^\times]P \\[2pt]
\hline
{\scriptstyle \langle d \rangle, [\cdot]} \quad [((\alpha^d)^*)^d]P \leftrightarrow P \vee [\alpha][\alpha^\times]P \\[2pt]
\hline
[\alpha^\times]P \leftrightarrow P \vee [\alpha][\alpha^\times]P
\end{array}
$$

The derivation of the diamond modality for Demon's repetition is correspondingly

$$\langle^\times \rangle \quad \langle \alpha^\times \rangle P \leftrightarrow P \wedge \langle \alpha \rangle \langle \alpha^\times \rangle P$$

16.3.10 Proof Rules for Repetition Games

The iteration axiom $[^*]$ was established to be sound in Sect. 5.3.7, but Chap. 7 identified a significantly more useful approach of proving properties of loops by induction. Similarly, one might wonder whether the iteration axiom $\langle^*\rangle$ really already captures all there is to say about repetition in hybrid games.

Taking a step back, axiom $\langle^*\rangle$ expresses that $\langle\alpha^*\rangle P$ is a fixpoint of (16.8), which follows from (16.7), but does not convey that, among all the possible fixpoints, $\langle\alpha^*\rangle P$ is the least fixpoint. How can this be rendered in a logical proof principle?

> Before you read on, see if you can find the answer for yourself.

Since $\langle\alpha^*\rangle P$ is the least fixpoint, the set of all states in which it is true is a subset of any other fixpoint. The logical internalization is that if Q is a logical formula whose semantics also satisfies the fixpoint condition from (16.7), then the set of states where $\langle\alpha^*\rangle P$ is true is smaller, that is $[\![\langle\alpha^*\rangle P]\!] \subseteq [\![Q]\!]$, which means that $\langle\alpha^*\rangle P \to Q$ is valid. Since it is, here, a little more convenient to work with the pre-fixpoint condition from (16.7), saying that the logical formula Q is a pre-fixpoint amounts to assuming that $P \vee \langle\alpha\rangle Q \to Q$ is valid.

Lemma 16.11 (FP fixpoint rule). *The* fixpoint rule *is sound:*

$$\text{FP} \quad \frac{P \vee \langle\alpha\rangle Q \to Q}{\langle\alpha^*\rangle P \to Q}$$

Proof. The proof is a direct consequence of the fact that the winning region of repetition is the least fixpoint (Note 76). Assume the premise $P \vee \langle\alpha\rangle Q \to Q$ is valid, i.e., $[\![P \vee \langle\alpha\rangle Q]\!] \subseteq [\![Q]\!]$. That is, $[\![P]\!] \cup \varsigma_\alpha([\![Q]\!]) = [\![P]\!] \cup [\![\langle\alpha\rangle Q]\!] = [\![P \vee \langle\alpha\rangle Q]\!] \subseteq [\![Q]\!]$. Thus, Q is a pre-fixpoint of $Z = [\![P]\!] \cup \varsigma_\alpha(Z)$. By monotonicity (Lemma 15.1), $[\![\langle\alpha^*\rangle P]\!] = \varsigma_{\alpha^*}([\![P]\!])$ is the least fixpoint [8, Appendix A]. Hence, $[\![\langle\alpha^*\rangle P]\!] \subseteq [\![Q]\!]$, which implies that $\langle\alpha^*\rangle P \to Q$ is valid.   □

Together with the iteration axiom $\langle^*\rangle$, the fixpoint proof rule FP is in most direct correspondence with the semantics of repetition in hybrid games, which is defined as a least fixpoint. The iteration axiom $\langle^*\rangle$ expresses that $\langle\alpha^*\rangle P$ is a fixpoint while rule FP expresses that it is the least fixpoint.

Admittedly, though, the fixpoint rule FP can be a bit unwieldy to use. Fortunately, the old familiar loop invariant rule, generalized to hybrid games, can be derived from the fixpoint rule FP and even vice versa [11, Lemma 4.1].

Corollary 16.1 (ind loop invariant rule). *The* loop invariant *proof rule is derived:*

$$\text{ind} \quad \frac{P \to [\alpha]P}{P \to [\alpha^*]P}$$

Proof. The proof uses contraposition, i.e., that $A \to B$ is equivalent to $\neg B \to \neg A$ in classical logic or similarly simple propositional rewriting at the steps that are not marked:

$$
\cfrac{
\cfrac{
\cfrac{
\cfrac{
\cfrac{
\cfrac{\vdash P \to [\alpha]P}{\vdash P \to P \wedge [\alpha]P}
}{\vdash P \to P \wedge \neg\langle\alpha\rangle\neg P}\;{}^{[\cdot]}
}{\vdash \neg P \vee \langle\alpha\rangle\neg P \to \neg P}
}{\vdash \langle\alpha^*\rangle\neg P \to \neg P}\;{}^{\mathrm{FP}}
}{\vdash P \to \neg\langle\alpha^*\rangle\neg P}
}{\vdash P \to [\alpha^*]P}\;{}^{[\cdot]}
$$

This proof shows that rule ind is a derived rule, since its conclusion can be proved from rule FP using the other axioms and propositional reasoning.   □

Based on our improved understanding of structuring proofs with sequents from Chap. 6, it is easy to see that a corresponding sequent formulation of rule ind is derived from rules $\to$R,cut:

$$
\text{ind}\ \ \frac{P \vdash [\alpha]P}{P \vdash [\alpha^*]P}
$$

Example 16.3 (Invariants and fixpoints for Demon's repetition). Since Demon's repetition $\alpha^\times$ is $((\alpha^{\mathrm d})^*)^{\mathrm d}$, the duality axiom $\langle^d\rangle$ and determinacy axiom $[\cdot]$ turn Demon's invariant rule ind for Angel's repetition into a corresponding invariant rule for Angel's winning strategy in Demon's repetition:

$$
\text{ind}^\times\ \ \frac{P \to \langle\alpha\rangle P}{P \to \langle\alpha^\times\rangle P}
$$

Likewise, Demon's counterpart for Demon's repetition of the fixpoint rule FP derives by duality axiom $\langle^d\rangle$ and determinacy axiom $[\cdot]$:

$$
\text{FP}^\times\ \ \frac{P \vee [\alpha]Q \to Q}{[\alpha^\times]P \to Q}
$$

The proofs of correctness for rules $\text{ind}^\times$ and $\text{FP}^\times$ are explored in Exercise 16.5.

16.4 Example Proofs

This section shows how the dGL axioms can be used to prove the existence of winning strategies for some hybrid games.

Example 16.4. The dual filibuster game formula from Chap. 14 is proved easily in the dGL calculus by going back and forth between players [11] using abbreviations $\cap,^\times$:

$$
\begin{array}{c}
*\\
\mathbb{R}\;\overline{\;x=0\vdash 0=0\vee 1=0\;}\\
\langle:=\rangle\;\overline{\;x=0\vdash \langle x:=0\rangle x=0\vee \langle x:=1\rangle x=0\;}\\
\langle\cup\rangle\;\overline{\;x=0\vdash \langle x:=0\cup x:=1\rangle x=0\;}\\
\langle{}^d\rangle\;\overline{\;x=0\vdash \neg\langle (x:=0\cup x:=1)^{\mathrm d}\rangle\neg x=0\;}\\
\overline{\;x=0\vdash \neg\langle x:=0\cap x:=1\rangle\neg x=0\;}\\
[\cdot]\;\overline{\;x=0\vdash [x:=0\cap x:=1]x=0\;}\\
\mathrm{ind}\;\overline{\;x=0\vdash [(x:=0\cap x:=1)^{*}]x=0\;}\\
[\cdot]\;\overline{\;x=0\vdash \neg\langle (x:=0\cap x:=1)^{*}\rangle\neg x=0\;}\\
\langle{}^d\rangle\;\overline{\;x=0\vdash \langle (x:=0\cap x:=1)^{*\mathrm d}\rangle x=0\;}\\
\overline{\;x=0\vdash \langle (x:=0\cup x:=1)^{\times}\rangle x=0\;}
\end{array}
$$

Example 16.5 (Push-around cart). Recall the following dGL formula about the push-around cart game from Example 14.3:

$$x\geq 0\wedge v\geq 0\rightarrow \big[\big((d:=1\cup d:=-1)^{\mathrm d};\,(a:=1\cup a:=-1);\,\{x'=v,v'=a+d\}\big)^{*}\big]x\geq 0$$

Using Demon's choice and the fact that the duals of assignments are the assignments themselves, recall that this dGL formula is equivalent to

$$x\geq 0\wedge v\geq 0\rightarrow \big[\big((d:=1\cap d:=-1);\,(a:=1\cup a:=-1);\,\{x'=v,v'=a+d\}\big)^{*}\big]x\geq 0$$

$$
\begin{array}{l}
[:=]\;\dfrac{J\vdash [\{x'=v,v'=1+1\}]J\wedge[\{x'=v,v'=-1+1\}]J}{J\vdash [a:=1][\{x'=v,v'=a+1\}]J\wedge[a:=-1][\{x'=v,v'=a+1\}]J}\\[4pt]
[\cup]\;\dfrac{}{J\vdash [a:=1\cup a:=-1][\{x'=v,v'=a+1\}]J}\\[4pt]
[\cdot]\;\dfrac{}{J\vdash [(a:=1\cup a:=-1);\{x'=v,v'=a+1\}]J}\\[4pt]
[:=]\;\dfrac{}{J\vdash [d:=1][(a:=1\cup a:=-1);\{x'=v,v'=a+d\}]J}\\[4pt]
\mathrm{VR,WR}\;\dfrac{}{J\vdash [d:=1][(a:=1\cup a:=-1);\{x'=v,v'=a+d\}]J\vee[d:=-1]\dots}\\[4pt]
[\cap]\;\dfrac{}{J\vdash [d:=1\cap d:=-1][(a:=1\cup a:=-1);\{x'=v,v'=a+d\}]J}\\[4pt]
[\cdot]\;\dfrac{}{J\vdash [(d:=1\cap d:=-1);(a:=1\cup a:=-1);\{x'=v,v'=a+d\}]J}\\[4pt]
\mathrm{ind}\;\dfrac{}{J\vdash [((d:=1\cap d:=-1);(a:=1\cup a:=-1);\{x'=v,v'=a+d\})^{*}]x\geq 0}
\end{array}
$$

Choosing the loop invariant $J\overset{\mathrm{def}}{\equiv}x\geq 0\wedge v\geq 0$ will complete this proof, because both remaining differential equation properties can be proved by solving them:

$$[\,'\,],[:=]\;\dfrac{x\geq 0\wedge v\geq 0\vdash \forall t\geq 0\,(x+vt+t^2\geq 0\wedge v+2t\geq 0)}{J\vdash [\{x'=v,v'=1+1\}]J}$$

$$[\,'\,],[:=]\;\dfrac{x\geq 0\wedge v\geq 0\vdash \forall t\geq 0\,(x+vt\geq 0\wedge v\geq 0)}{J\vdash [\{x'=v,v'=0\}]J}$$

They can also both be proved directly by differential invariants from Part II.

Proposition 16.1 (Push-around carts are safe). *This* dGL *formula is valid:*

$$x\geq 0\wedge v\geq 0\rightarrow \big[\big((d:=1\cap d:=-1);\,(a:=1\cup a:=-1);\,\{x'=v,v'=a+d\}\big)^{*}\big]x\geq 0$$

$$
\begin{array}{ll}
 & * \\
\hline
\mathbb{R} & J \vdash \forall t{\geq}0\,\big((w+vt+\tfrac{1}{2}t^2-e-ft-\tfrac{1}{2}t^2)^2 \leq 1 \land v+t=f+t\big) \\
\hline
[']{,}\langle{:=}\rangle & J \vdash \langle t:=0\rangle[\{w''=1,e''=1\}]J \\
\hline
\lor\mathrm{R,WR} & J \vdash \langle t:=0\rangle[\{w''=1,e''=1\}]J \lor \langle t:=0\rangle[\{w''=1,e''=-1\}]J \\
\hline
[\cdot] & J \vdash \langle t:=0\rangle\neg\langle\{w''=1,e''=1\}\rangle\neg J \lor \langle t:=0\rangle\neg\langle\{w''=1,e''=-1\}\rangle\neg J \\
\hline
\langle{}^d\rangle & J \vdash \langle t:=0\rangle\langle\{w''=1,e''=1\}^{\mathrm{d}}\rangle J \lor \langle t:=0\rangle\langle\{w''=1,e''=-1\}^{\mathrm{d}}\rangle J \\
\hline
\langle;\rangle & J \vdash \langle t:=0;\{w''=1,e''=1\}^{\mathrm{d}}\rangle J \lor \langle t:=0;\{w''=1,e''=-1\}^{\mathrm{d}}\rangle J \\
\hline
\langle{:=}\rangle & J \vdash \langle g:=1\rangle\langle t:=0;\{w''=1,e''=g\}^{\mathrm{d}}\rangle J \lor \langle g:=-1\rangle\langle t:=0;\{w''=1,e''=g\}^{\mathrm{d}}\rangle J \\
\hline
\langle\cup\rangle & J \vdash \langle g:=1 \cup g:=-1\rangle\langle t:=0;\{w''=1,e''=g\}^{\mathrm{d}}\rangle J \\
\hline
\langle;\rangle & J \vdash \langle(g:=1 \cup g:=-1);t:=0;\{w''=1,e''=g\}^{\mathrm{d}}\rangle J \\
\hline
\langle{:=}\rangle & J \vdash \langle u:=1\rangle\langle(g:=1 \cup g:=-1);t:=0;\{w''=u,e''=g\}^{\mathrm{d}}\rangle J \qquad \triangleright \\
\hline
\land\mathrm{R} & J \vdash \langle u:=1\rangle\langle(g:=1 \cup g:=-1);t:=0;\{w''=u,e''=g\}^{\mathrm{d}}\rangle J \land \langle u:=-1\rangle\ldots \\
\hline
\langle\cap\rangle & J \vdash \langle u:=1 \cap u:=-1\rangle\langle(g:=1 \cup g:=-1);t:=0;\{w''=u,e''=g\}^{\mathrm{d}}\rangle J \\
\hline
\langle;\rangle & J \vdash \langle(u:=1 \cap u:=-1);(g:=1 \cup g:=-1);t:=0;\{w''=u,e''=g\}^{\mathrm{d}}\rangle J \\
\hline
\mathrm{ind}^{\times} & J \vdash \langle\big((u:=1 \cap u:=-1);(g:=1 \cup g:=-1);t:=0;\{w''=u,e''=g\}^{\mathrm{d}}\big)^{\times}\rangle(w-e)^2 \leq 1
\end{array}
$$

Fig. 16.1 Proof of the two-robot dance

Example 16.6 (WALL·E and EVE robot dance). Recall the following dGL formula about a robot dance from Example 14.4:

$$
\begin{aligned}
(w-e)^2 \leq 1 \land v=f \rightarrow \big\langle\big(&(u:=1 \cap u:=-1);\\
&(g:=1 \cup g:=-1);\\
&t:=0;\{w'=v,v'=u,e'=f,f'=g,t'=1\,\&\,t\leq 1\}^{\mathrm{d}}\big)^{\times}\\
\big\rangle\,&(w-e)^2 \leq 1
\end{aligned}
\tag{14.4}
$$

With loop invariant $J \stackrel{\text{def}}{\equiv} (w-e)^2 \leq 1 \land v=f$, the proof of dGL formula (14.4) is shown in Fig. 16.1. The (crucial) branch proving that a winning strategy also exists when Demon chooses $u:=-1$ is elided (marked $\triangleright$) but quite similar. When conducting a proof for the existence of Angel's winning strategy (a $\langle\cdot\rangle$ formula), note how Angel's choices turn into conjunctions while Demon's choices turn into disjunctions.

After we finish the proof in Fig. 16.1, we see that the rules $\lor$R,WR can already discard the right disjunct earlier than they did. This is a common phenomenon in hybrid games. In hindsight, proofs make it easy to see the best options. But proofs also become simpler if we identify clever actions early.

Proposition 16.2 (Robot dance is safe). *This dGL formula is valid:*

$$
\begin{aligned}
(w-e)^2 \leq 1 \land v=f \rightarrow \big\langle\big(&(u:=1 \cap u:=-1);\\
&(g:=1 \cup g:=-1);\\
&t:=0;\{w'=v,v'=u,e'=f,f'=g,t'=1\,\&\,t\leq 1\}^{\mathrm{d}}\big)^{\times}\\
\big\rangle\,&(w-e)^2 \leq 1
\end{aligned}
$$

These proofs illustrate that proving properties of hybrid games is entirely analogous to proving properties of hybrid systems. All we need to pay attention to is to *only* use the axioms of differential game logic and not accidentally those of differential dynamic logic from Part I. Of course, the two logics share most axioms. For example, we still prove box properties of loops by a proof rule with loop invariants.

16.5 Axiomatization

The axiomatization for differential game logic [11] that we just developed gradually is summarized in Fig. 16.2.

$[\cdot]$ $[\alpha]P \leftrightarrow \neg\langle\alpha\rangle\neg P$

$\langle:=\rangle$ $\langle x:=e\rangle p(x) \leftrightarrow p(e)$

$\langle'\rangle$ $\langle x'=f(x)\rangle p(x) \leftrightarrow \exists t{\geq}0\,\langle x:=y(t)\rangle p(x)$ $(y'(t)=f(y))$

$\langle?\rangle$ $\langle ?Q\rangle P \leftrightarrow Q \wedge P$

$\langle\cup\rangle$ $\langle\alpha\cup\beta\rangle P \leftrightarrow \langle\alpha\rangle P \vee \langle\beta\rangle P$

$\langle;\rangle$ $\langle\alpha;\beta\rangle P \leftrightarrow \langle\alpha\rangle\langle\beta\rangle P$

$\langle*\rangle$ $\langle\alpha^*\rangle P \leftrightarrow P \vee \langle\alpha\rangle\langle\alpha^*\rangle P$

$\langle d\rangle$ $\langle\alpha^d\rangle P \leftrightarrow \neg\langle\alpha\rangle\neg P$

M $\dfrac{P \to Q}{\langle\alpha\rangle P \to \langle\alpha\rangle Q}$

FP $\dfrac{P \vee \langle\alpha\rangle Q \to Q}{\langle\alpha^*\rangle P \to Q}$

ind $\dfrac{P \to [\alpha]P}{P \to [\alpha^*]P}$

Fig. 16.2 Differential game logic axiomatization

The determinacy axiom $[\cdot]$ describes the duality of winning strategies for complementary winning conditions of Angel and Demon, i.e., that Demon has a winning strategy to achieve P in hybrid game α if and only if Angel does not have a counter strategy, i.e., winning strategy to achieve $\neg P$ in the same game α. The determinacy axiom $[\cdot]$ internalizes Theorem 16.1. Axiom $\langle:=\rangle$ is the assignment axiom. In the differential equation axiom $\langle'\rangle$, $y(\cdot)$ is the unique [14, Theorem 10.VI] solution of the symbolic initial value problem $y'(t)=f(y),y(0)=x$. The duration t how long to follow solution y is for Angel to decide, hence existentially quantified. It goes without saying that variables such as t are fresh in Fig. 16.2.

Axioms $\langle ? \rangle$, $\langle \cup \rangle$, and $\langle ; \rangle$ are as in differential dynamic logic [10] except that their meaning is quite different, because they refer to winning strategies of hybrid games instead of reachability relations of systems. The challenge axiom $\langle ? \rangle$ expresses that Angel has a winning strategy to achieve P in the test game $?Q$ exactly from those positions that are already in P (because $?Q$ does not change the state) and that satisfy Q for otherwise she would fail the test and lose the game immediately. The axiom of choice $\langle \cup \rangle$ expresses that Angel has a winning strategy in a game of choice $\alpha \cup \beta$ to achieve P iff she has a winning strategy in either hybrid game α or in β, because she can choose which one to play. The sequential game axiom $\langle ; \rangle$ expresses that Angel has a winning strategy in a sequential game $\alpha ; \beta$ to achieve P iff she has a winning strategy in game α to achieve $\langle \beta \rangle P$, i.e., to get to a position from which she has a winning strategy in game β to achieve P. The "$\leftarrow$" direction of the iteration axiom $\langle * \rangle$ characterizes $\langle \alpha^* \rangle P$ as a pre-fixpoint. It expresses that, if the game is already in a state satisfying P or if Angel has a winning strategy for game α to achieve $\langle \alpha^* \rangle P$, i.e., to get to a position from which she has a winning strategy for game α^* to achieve P, then, either way, Angel has a winning strategy to achieve P in game α^*. The "$\rightarrow$" direction of $\langle * \rangle$ can already be derived by other axioms [11]. The dual axiom $\langle ^d \rangle$ characterizes dual games. It says that Angel has a winning strategy to achieve P in dual game α^d iff Angel does not have a winning strategy to achieve $\neg P$ in game α. Combining dual game axiom $\langle ^d \rangle$ with the determinacy axiom $[\cdot]$ yields $\langle \alpha^d \rangle P \leftrightarrow [\alpha]P$, i.e., that Angel has a winning strategy to achieve P in α^d iff Demon has a winning strategy to achieve P in α. Similar reasoning derives $[\alpha^d]P \leftrightarrow \langle \alpha \rangle P$.

Monotonicity rule M is the generalization rule of monotonic modal logic **C** [2] and logically internalizes monotonicity (Lemma 15.1). It expresses that, if the implication $P \rightarrow Q$ is valid, then, wherever Angel has a winning strategy in a hybrid game α to achieve P, she also has a winning strategy to achieve Q, because Q holds wherever P does. So rule M expresses that easier objectives are easier to win. Fixpoint rule FP characterizes $\langle \alpha^* \rangle P$ as a *least* pre-fixpoint. It says that, if Q is another formula that is a pre-fixpoint, i.e., that holds in all states that satisfy P or from which Angel has a winning strategy in game α to achieve that condition Q, then Q also holds wherever $\langle \alpha^* \rangle P$ does, i.e., in all states from which Angel has a winning strategy in game α^* to achieve P.

The proof rule FP and the induction rule ind are equivalent in the sense that one can be derived from the other in the dGL calculus [11]. How the loop induction rule ind derives from the fixpoint rule FP was shown in Corollary 16.1.

16.5.1 Soundness

Summarizing the individual lemmas that established the soundness of the axioms allows us to conclude that the dGL proof calculus is sound [11]. In analogy to Sect. 6.2.2, we write $\vdash_{dGL} P$ iff dGL formula P can be *proved* with dGL rules from dGL axioms. Likewise, we write $\Gamma \vdash_{dGL} P$ iff dGL formula P can be proved from the set of formulas Γ. In particular, $\vdash_{\mathsf{dGL}} P$ iff $\emptyset \vdash_{\mathsf{dGL}} P$. We write $\vDash P$ iff P is valid, i.e., true in all states (Definition 15.1). The two notions are intimately related by soundness.

> **Theorem 16.2 (Soundness of dGL).** *The* dGL *axiomatization in Fig. 16.2 is sound, i.e., all provable formulas are valid. That is,*
>
> $$\vdash_{\mathsf{dGL}} P \ implies \ \vDash P$$

Proof. An axiomatization or proof calculus is sound iff *all* provable formulas are valid. There are a lot of provable formulas, so this might call for a lot of work. But, as in Part I and Part II, by far the best way of establishing soundness of a proof calculus is by exploiting logical compositionality principles to show that each axiom and proof rule is sound individually. An axiom is sound iff each of its instances is a valid formula. A proof rule is sound iff the validity of all its premises implies the validity of its conclusion. Once all axioms and all proof rules are sound, every formula that has a proof will be valid, because a proof must end with axioms (which are sound so only have valid instances) and must have used proof rules in between (which, if sound, make the conclusion valid since the premises were valid). Most axioms have been proved sound already in their respective lemmas (such as Lemmas 16.7 and 16.11). The full proof can be found in prior work [11].   □

This gives us a sound proof approach for CPSs that are as challenging as hybrid games. What exactly did we prove the axioms sound for again? What does sound mean and entail exactly?

> **Note 78 (The miracle of soundness)** Soundness of the dGL proof calculus means that all dGL formulas that are provable using the dGL calculus are valid, a *conditio sine qua non* for logic, i.e., a condition without which logic could not be. It would not make sense to prove a formula if that proof would not even entail the formula's validity, i.e., that it is true in all states.
> For a proof calculus to be sound, every formula that it proves with any proof has to be valid. Fortunately, proofs are composed from axioms by proof rules. So all we need to do to ensure that a proof calculus is sound is to prove its few axioms to be sound and then everything we ever derive from them by sound proof rules is correct as well, no matter how big and complicated. A proof is a long combination of many simple arguments, each of which just involves one of the axioms or proof rules. Once each of those finitely many axioms and proof rules is proved to be sound, all those infinitely many proofs that can be conducted in the dGL proof calculus become sound as well. That is compositionality in its finest form for the soundness argument. *It is soundness that ultimately links semantics and axiomatics in perfect unison[a] so that axiomatic proof coincides with semantic truth*, an important aspect of the logical trinity.
> One subtlety is that a proof might use many instances of the same finite axiom list. Then the soundness proof for the axioms has to work for any instance.

This aspect is often left implicit in soundness arguments, although a rigorous treatment can be given by distinguishing axioms from axiom schemes [11, 12].

---

[a] In search of perfection, completeness is another important aspect in achieving perfect unison, which, incidentally, holds for differential game logic as well [11].

16.5.2 Completeness

Soundness is the most crucial condition for any proof calculus of any logic, and is especially crucial for something as impactful and safety-critical as cyber-physical systems. By soundness, every formula with a proof is valid. The most intriguing condition, however, is the converse: whether the calculus is *complete*, i.e., can prove all formulas that are valid. That would be very exciting, because we would then know that whenever a formula is valid, there's a proof for it, so if we have not found it yet, we just need to look a little harder.

In particular, even if we deleted all axioms and proof rules, then the resulting empty proof calculus would be sound, just not at all useful, because we could not prove anything with it. Completeness considers the question of whether the proof calculus comes with all the axioms and proof rules that it needs in order to be able to do "all" proofs in it. Certainly, if we were to delete all axioms and proof rules that handle the $\cup$ operator, then the calculus would become quite incomplete, because we could no longer prove any interesting properties of hybrid games with a choice. But even if every operator has a corresponding axiom, it is not clear whether those axioms are enough to prove every valid property about them. Repetitions, for example, come with two reasoning principles, the iteration axiom $\langle * \rangle$ and the fixpoint rule FP, which serve different purposes.

Sadly, absolute completeness would be too good to be true for something as expressive as differential game logic, because Gödel's second incompleteness theorem shows that every system extending first-order natural number arithmetic of addition and multiplication is incomplete [3]. While differential game logic does not directly provide natural numbers, it still characterizes them indirectly as the set of all values of x for which repeatedly subtracting 1 can lead to 0:

$$\langle (x := x - 1)^* \rangle x = 0$$

And, indeed, thinking back to previous chapters, there were quite a few challenges in proving properties of hybrid systems as well as hybrid games. The primary challenge in Part I was the need to find invariants for loops. The primary challenge in Part II was in finding differential invariants for differential equations. It can be shown that these are essentially the only fundamental challenges in CPS verification [10, 11].

Differential game logic supports relative completeness, i.e., its axiomatization can prove every valid dGL formula from elementary tautologies. The dGL axiomatization is complete relative to any differentially expressive[2] logic [11].

> **Theorem 16.3 (Relative completeness of dGL).** *The dGL calculus is a complete axiomatization of hybrid games relative to any differentially expressive logic L, i.e., every valid dGL formula is provable in the dGL calculus from L tautologies. That is,*
>
> $$\vDash P \ \textit{implies} \ L \vdash_{\mathsf{dGL}} P$$

In fact, the name axiomatization is reserved for proof calculi that do not just come with soundness guarantees but provide completeness guarantees as well. Indeed, the differential dynamic logic axiomatization has soundness and completeness guarantees relative to any differentially expressive logic as well [9, 10, 12]. So we were justified in calling it an axiomatization in Part I and Part II.

The rôle of the differentially expressive logic L relative to which completeness is proved is particularly intuitive for differential dynamic logic. By the first relative completeness theorem [9, 10], differential dynamic logic is complete relative to properties of differential equations, so if a dL formula is valid then it can be proved using the dL axioms from elementary valid properties of differential equations. Of course, we need to be able to prove the safety of the differential equations (e.g., using differential invariants) in order to be able to understand a hybrid system. But by the first relative completeness theorem, it is enough to worry about differential equations, because the dL axioms will then be able to prove the hybrid system, too. By the second relative completeness theorem [10], differential dynamic logic is also complete relative to purely discrete dynamics, so if a dL formula is valid then it can also be proved using the dL axioms from elementary valid properties of discrete systems. Again one needs to be able to master loops (by finding appropriate loop invariants), but the dL axioms can then prove the whole hybrid system, too. In fact, dL is complete relative to any differentially expressive logic [12], of which the purely continuous as well as the purely discrete fragment are two canonical examples.

These insights give rise to a relative decision procedure for differential dynamic logic [10], which decides differential dynamic logic from an oracle for L. Such a relative decision procedure is an algorithm that accepts any (without loss of generality fully quantified) dL formula as input and will correctly output "valid" or "not valid" by asking a finite number of questions to an oracle for L.

---

[2] A logic L that is closed under first-order connectives is *differentially expressive* (for dGL) if every dGL formula P has an equivalent $P^\flat$ in L and all equivalences of the form $\langle x' = f(x) \rangle G \leftrightarrow (\langle x' = f(x) \rangle G)^\flat$ for formulas G in L are provable in its calculus.

16.6 There and Back Again Game

Quite unlike in hybrid systems and (poor test[3]) differential dynamic logic [9, 10], every hybrid game containing a differential equation $x' = f(x) \,\&\, Q$ with evolution domain constraints Q can be replaced equivalently by a hybrid game without evolution domain constraints. Evolution domains are definable in hybrid games [11] and can, thus, be removed equivalently.

> **Lemma 16.12 (Evolution domain reduction).** *Evolution domains of differential equations are definable as hybrid games: For every hybrid game there is an equivalent hybrid game that has no evolution domain constraints, i.e., all continuous evolutions are of the form $x' = f(x)$.*

Proof. For notational convenience, assume vectorial differential equation $x' = f(x)$ to contain a clock $x_0' = 1$ and that t_0 and z are fresh variables. Then a differential equation $x' = f(x) \,\&\, Q(x)$ with evolution domain is equivalent to the hybrid game:

$$t_0 := x_0;\, x' = f(x);\, (z := x; z' = -f(z))^d;\, ?(z_0 \geq t_0 \to Q(z)) \qquad (16.9)$$

See Fig. 16.3 for an illustration. Suppose the current player is Angel. The idea be-

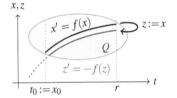

Angel plays forward game, reverts flow and time x_0; Demon checks Q in backwards game until initial t_0

Fig. 16.3 "There and back again game": Angel evolves x forwards in time along $x' = f(x)$, Demon checks evolution domain backwards in time along $z' = -f(z)$ on a copy z of the state vector x

hind (16.9) is that the fresh variable t_0 remembers the initial time x_0, and Angel then evolves forward along $x' = f(x)$ for any amount of time (Angel's choice). Afterwards, the opponent Demon copies the state x into a fresh variable (vector) z that he can evolve backwards along $(z' = -f(z))^d$ for any amount of time (Demon's choice). The original player Angel must then pass the challenge $?(z_0 \geq t_0 \to Q(z))$, i.e., Angel loses immediately if Demon was able to evolve backwards and leave region $Q(z)$ while satisfying $z_0 \geq t_0$, which checks that Demon did not evolve backward for longer than Angel evolved forward, i.e., to before the initial time. Otherwise, when Angel passes the test, the extra variables t_0, z become irrelevant (they are fresh) and the game continues from the current state x that Angel chose originally (by selecting a duration for the evolution that Demon could not invalidate).   □

---

[3] *Poor test* means that each test $?Q$ uses only first-order formulas Q. If modalities are used within Q, then $?Q$ is a *rich test*.

From now on, Lemma 16.12 can eliminate all evolution domain constraints equivalently in hybrid games. While evolution domain constraints are fundamental parts of standard hybrid systems [1, 5, 6, 9], they turn out to be mere convenience notation for hybrid games. In that sense, hybrid games are more fundamental than hybrid systems, because they feature elementary operators. In theory, we never have to worry about evolution domains any more, because they are just part of the other operators for hybrid games. In practice, it still helps to handle evolution domain constraints directly, because axioms like DW for differential weakening and DI for differential invariants are conceptually easier than the reduction in (16.9).

16.7 Summary

This chapter developed an axiomatization for differential game logic [11]. The resulting axioms, summarized in Fig. 16.2 on p. 498, coincide with corresponding axioms for hybrid systems. But they needed entirely new soundness justification, because, due to the interactive game play features caused by the presence of the duality operator, the semantics of hybrid games is significantly more general than that of hybrid systems. The simple syntactic reasoning principles of differential game logic are substantially more succinct than the corresponding subtleties with purely semantical arguments. Just contrast the simplicity of the axiomatization with the enormous (more than infinite) number of iterations needed in semantical arguments about winning regions for repetition from Chap. 15.

This dGL axiomatization provides a strong foundation for hybrid games. But Parts I and II also studied other reasoning principles for hybrid systems that we have not yet considered for hybrid games. It is not necessarily obvious, e.g., whether Gödel's generalization rule is sound for dGL, which is what Chap. 17 will explore:

$$G \frac{P}{[\alpha]P}$$

Exercises

16.1 (Diamond proofs). Use the dGL axioms to prove the following formulas:

$$\langle x := -x \cup (x := x + 1 \cap x := x + 2) \rangle x > 0$$
$$\langle (x := -x + 1 \cup x := x + 1) \cap x := 2 \rangle x > 0$$
$$\langle (x := x \cup x := -x); (x := x + 1 \cap x := x + 2) \rangle x > 0$$
$$\langle x := x^2 \cup (x := x + 1 \cap x' = 2) \rangle x > 0$$
$$\langle x := -x \cup (x' = 1 \cap x' = 2) \rangle x \geq 0$$
$$\langle x := -x \cup (x := x + 2 \cap x' = 2) \rangle x \geq 0$$

$$\langle(x:=-x\cup(x:=x+2\cap x'=2))^*\rangle x\geq 0$$

16.2. Explain how determinacy relates to the two possible understandings of the filibuster example discussed in (14.8) on p. 441.

16.3 (Box modalities). Show that the dGL axioms for box modalities in Fig. 16.4 are derived by duality (with duality axiom $\langle^d\rangle$ and determinacy axiom $[\cdot]$) from the dGL axioms for diamond modalities.

$\langle\cdot\rangle$ $\langle\alpha\rangle P\leftrightarrow\neg[\alpha]\neg P$

$[:=]$ $[x:=e]p(x)\leftrightarrow p(e)$

$[']$ $[x'=f(x)]p(x)\leftrightarrow\forall t\geq 0\,[x:=y(t)]p(x)$ $(y'(t)=f(y))$

$[?]$ $[?Q]P\leftrightarrow(Q\rightarrow P)$

$[\cup]$ $[\alpha\cup\beta]P\leftrightarrow[\alpha]P\wedge[\beta]P$

$[;]$ $[\alpha;\beta]P\leftrightarrow[\alpha][\beta]P$

$[^*]$ $[\alpha^*]P\leftrightarrow P\wedge[\alpha][\alpha^*]P$

$[^d]$ $[\alpha^d]P\leftrightarrow\neg[\alpha]\neg P$

$M[\cdot]$ $\dfrac{P\rightarrow Q}{[\alpha]P\rightarrow[\alpha]Q}$

Fig. 16.4 Differential game logic derived axioms for box modalities

---

16.4 (Demon's controls). Show that the dGL axioms for Demon's controls in Fig. 16.5 are derived from the definition of Demon's control operators (with the help of the duality axiom $\langle^d\rangle$ and determinacy axiom $[\cdot]$).

$\langle\cap\rangle$ $\langle\alpha\cap\beta\rangle P\leftrightarrow\langle\alpha\rangle P\wedge\langle\beta\rangle P$

$[\cap]$ $[\alpha\cap\beta]P\leftrightarrow[\alpha]P\vee[\beta]P$

$\langle^\times\rangle$ $\langle\alpha^\times\rangle P\leftrightarrow P\wedge\langle\alpha\rangle\langle\alpha^\times\rangle P$

$[^\times]$ $[\alpha^\times]P\leftrightarrow P\vee[\alpha][\alpha^\times]P$

Fig. 16.5 Differential game logic derived axioms for Demon's controls

---

16.5 (Demon's repetition). Use the duality axiom $\langle^d\rangle$ and determinacy axiom $[\cdot]$ to show that the following proof rules for Demon's repetition are derived rules:

$$\text{ind}^\times \; \frac{P \to \langle\alpha\rangle P}{P \to \langle\alpha^\times\rangle P} \qquad \text{FP}^\times \; \frac{P \vee [\alpha]Q \to Q}{[\alpha^\times]P \to Q}$$

16.6 (Demon's monotonicity). Prove the second part of Lemma 15.1, i.e., that the inclusion $\delta_\alpha(X) \subseteq \delta_\alpha(Y)$ holds for all hybrid games α and all sets $X \subseteq Y$.

16.7 (Box proofs). Use the dGL axioms and proof rules to prove the following formulas:

$$[x := -x^2 \cup (x := x \cap x := -x)]x \leq 0$$
$$[x := -x^2 \cup (x' = 2 \cup x' = -1)^d]x \leq 0$$
$$[(x := x^2 \cap x := -x^2) \cup (\{x' = 1\}^d \cap \{x' = -2\})]x \leq 0$$
$$x \geq 0 \to [(x := x + 1 \cup \{x' = -2\}; \{x' = 1\}^d)^*]x \geq 0$$
$$x \geq 0 \to [(x := x + 1 \cup (t := 0; \{x' = -5, t' = 1 \,\&\, t \leq 1\} \cap \{x' = 2\}))^*]x \geq 0$$

16.8 (Unsound axioms). Not all hybrid systems axioms can be used for hybrid games. Prove that the following perfectly valid hybrid systems axiom is unsound for hybrid games by giving a counterexample, i.e., an instance of the axiom that is not a valid dGL formula:

$$[]\wedge \; [\alpha](P \wedge Q) \leftrightarrow [\alpha]P \wedge [\alpha]Q$$

16.9 (What about other hybrid systems axioms?). Chaps. 5 and 7 and Exercises 5.22 and 7.11 identified several other useful axioms and proof rules for hybrid systems, summarized in Fig. 16.6. All hybrid games axioms are sound for hybrid systems, but not vice versa. Identify which of the axioms and proof rules listed in Fig. 16.6 are also sound for hybrid games and which ones are not.

K $[\alpha](P \to Q) \to ([\alpha]P \to [\alpha]Q)$ $\qquad$ M$_{[\cdot]}$ $\dfrac{P \to Q}{[\alpha]P \to [\alpha]Q}$

$\overleftarrow{\text{M}}$ $\langle\alpha\rangle(P \vee Q) \to \langle\alpha\rangle P \vee \langle\alpha\rangle Q$ $\qquad$ M $\langle\alpha\rangle P \vee \langle\alpha\rangle Q \to \langle\alpha\rangle(P \vee Q)$

I $[\alpha^*]P \leftrightarrow P \wedge [\alpha^*](P \to [\alpha]P)$ $\qquad$ ind $\dfrac{P \to [\alpha]P}{P \to [\alpha^*]P}$

B $\langle\alpha\rangle \exists x P \to \exists x \langle\alpha\rangle P$ $\qquad (x \notin \alpha) \qquad$ $\overleftarrow{\text{B}}$ $\exists x \langle\alpha\rangle P \to \langle\alpha\rangle \exists x P$

V $p \to [\alpha]p$ $\qquad (\text{FV}(p) \cap \text{BV}(\alpha) = \emptyset) \qquad$ VK $p \to ([\alpha]true \to [\alpha]p)$

G $\dfrac{P}{[\alpha]P}$ $\qquad$ M$_{[\cdot]}$ $\dfrac{P \to Q}{[\alpha]P \to [\alpha]Q}$

R $\dfrac{P_1 \wedge P_2 \to Q}{[\alpha]P_1 \wedge [\alpha]P_2 \to [\alpha]Q}$ $\qquad$ M$_{[\cdot]}$ $\dfrac{P_1 \wedge P_2 \to Q}{[\alpha](P_1 \wedge P_2) \to [\alpha]Q}$

FA $\langle\alpha^*\rangle P \to P \vee \langle\alpha^*\rangle(\neg P \wedge \langle\alpha\rangle P)$

$\overleftarrow{[*]}$ $[\alpha^*]P \leftrightarrow P \wedge [\alpha^*][\alpha]P$ $\qquad\qquad$ $[*]$ $[\alpha^*]P \leftrightarrow P \wedge [\alpha][\alpha^*]P$

Fig. 16.6 More hybrid systems axioms, some of which are sound for hybrid games

16.10 (Robot simple chase game). Following up on the one-dimensional game of chase between a robot at position x and another robot at position y from Exercise 14.6, now fill in the blanks with a suitable precondition and prove that Demon then has a winning strategy to avoid capture:

$$\underline{\hspace{2cm}} \to \big[\big(\, (v:=a \cup v:=-a \cup v:=0);$$
$$(w:=b \cap w:=-b \cap w:=0);$$
$$\{x'=v, y'=w\}\,\big)^*\big]\,(x-y)^2 \geq 1$$

References

[1] Rajeev Alur, Costas Courcoubetis, Thomas A. Henzinger, and Pei-Hsin Ho. Hybrid automata: an algorithmic approach to the specification and verification of hybrid systems. In: *Hybrid Systems*. Ed. by Robert L. Grossman, Anil Nerode, Anders P. Ravn, and Hans Rischel. Vol. 736. LNCS. Berlin: Springer, 1992, 209–229. DOI: 10.1007/3-540-57318-6_30.

[2] Brian F. Chellas. *Modal Logic: An Introduction*. Cambridge: Cambridge Univ. Press, 1980. DOI: 10.1017/CBO9780511621192.

[3] Kurt Gödel. Über formal unentscheidbare Sätze der Principia Mathematica und verwandter Systeme I. *Monatshefte Math. Phys.* **38**(1) (1931), 173–198. DOI: 10.1007/BF01700692.

[4] David Harel, Dexter Kozen, and Jerzy Tiuryn. *Dynamic Logic*. Cambridge: MIT Press, 2000.

[5] Thomas A. Henzinger. The theory of hybrid automata. In: *LICS*. Los Alamitos: IEEE Computer Society, 1996, 278–292. DOI: 10.1109/LICS.1996.561342.

[6] Thomas A. Henzinger, Peter W. Kopke, Anuj Puri, and Pravin Varaiya. What's decidable about hybrid automata? In: *STOC*. Ed. by Frank Thomson Leighton and Allan Borodin. New York: ACM, 1995, 373–382. DOI: 10.1145/225058.225162.

[7] Dexter Kozen. Results on the propositional μ-calculus. *Theor. Comput. Sci.* **27**(3) (1983), 333–354. DOI: 10.1016/0304-3975(82)90125-6.

[8] Dexter Kozen. *Theory of Computation*. Berlin: Springer, 2006.

[9] André Platzer. Differential dynamic logic for hybrid systems. *J. Autom. Reas.* **41**(2) (2008), 143–189. DOI: 10.1007/s10817-008-9103-8.

[10] André Platzer. The complete proof theory of hybrid systems. In: *LICS*. Los Alamitos: IEEE, 2012, 541–550. DOI: 10.1109/LICS.2012.64.

[11] André Platzer. Differential game logic. *ACM Trans. Comput. Log.* **17**(1) (2015), 1:1–1:51. DOI: 10.1145/2817824.

[12] André Platzer. A complete uniform substitution calculus for differential dynamic logic. *J. Autom. Reas.* **59**(2) (2017), 219–265. DOI: 10.1007/s10817-016-9385-1.

[13]   André Platzer. Differential hybrid games. *ACM Trans. Comput. Log.* **18**(3) (2017), 19:1–19:44. DOI: 10.1145/3091123.

[14]   Wolfgang Walter. *Ordinary Differential Equations.* Berlin: Springer, 1998. DOI: 10.1007/978-1-4612-0601-9.

Chapter 17
Game Proofs & Separations

Synopsis The primary purpose of this chapter is to compare the proof principles of hybrid games versus those of hybrid systems. Having established reasoning principles for hybrid games in the previous chapter, our attention shifts to contrasting and identifying what the actual difference really is. Despite being rooted in a different semantics, hybrid game axioms are surprisingly close to those for hybrid systems. But there are also some major soundness-critical discrepancies to notice. These findings are important for correctly reasoning about hybrid games, but also shine a complementary light on reasoning principles for hybrid systems by highlighting which ones crucially depend on the absence of adversarial dynamics.

17.1 Introduction

This chapter continues the study of hybrid games and their logic, differential game logic [4]. After Chap. 14 introduced hybrid games and Chap. 15 developed their winning-region semantics, Chap. 16 achieved major breakthroughs in their understanding by studying the axioms of hybrid games. The resulting simple axioms made it surprisingly easy to prove correctness properties of hybrid games with dGL in ways that were quite similar to how we have already successfully proved properties of hybrid systems with dL in this book.

Of course, it should make us wonder why two logics that are based on such different conditions (hybrid systems versus hybrid games) end up being so surprisingly close in their axioms. And, indeed, upon closer inspection, we will find notable differences that we definitely need to respect when analyzing hybrid games. The chapter starts out with a comparison of the axioms of hybrid systems versus hybrid games and inspects what we have missed so far when considering hybrid game axioms. We will find a surprising logical robustness that even two semantically quite different logics end up having, for the most part, quite similar axioms.

This chapter is based on prior work [4], where more information can be found on logic and hybrid games. The most important learning goals of this chapter are:

© Springer International Publishing AG, part of Springer Nature 2018
A. Platzer, *Logical Foundations of Cyber-Physical Systems*,
https://doi.org/10.1007/978-3-319-63588-0_17

Modeling and Control: While the primary learning objectives in this chapter come from computational thinking, modeling and control observations still find out in passing that continuous and adversarial dynamics also mix in the form of differential games to which differential game logic generalizes [5]. A coverage of those findings is beyond the scope of this textbook, though.

Computational Thinking: This chapter solidifies our understanding of rigorous reasoning techniques for CPS models involving adversarial dynamics. Its primary purpose is to identify which hybrid systems reasoning principles are still sound for hybrid games and which ones crucially depend on the absence of adversariality. This delineation is critical to ensure that no incorrect arguments enter our proofs for CPSs with adversarial interactions. This refined understanding of what is sound and what is not also leads to a new appreciation for the robustness of logic. Finally, these findings shine a complementary light on what is specific to hybrid systems and what is more general.

CPS Skills: We will develop a complementary understanding of CPS models and how they are impacted by the presence or absence of adversariality. Being rooted in a syntactic characterization of the difference of hybrid systems and hybrid games, this understanding will make it easier to pinpoint what the exact nuances in different CPS operations and arguments are.

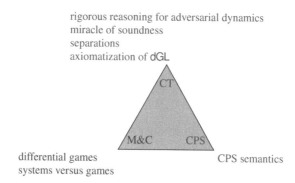

rigorous reasoning for adversarial dynamics
miracle of soundness
separations
axiomatization of dGL

differential games   CPS semantics
systems versus games

17.2 Recap: Hybrid Games

Recall a result from Chap. 16 and, in Fig. 17.1, the axiomatization of differential game logic [4] that Chap. 16 discussed.

Theorem 16.1 (Consistency & determinacy). *Hybrid games are* consistent *and* determined, *i.e.,* $\models \neg\langle\alpha\rangle\neg P \leftrightarrow [\alpha]P$.

$[\cdot]$ $[\alpha]P \leftrightarrow \neg\langle\alpha\rangle\neg P$

$\langle:=\rangle$ $\langle x:=e\rangle p(x) \leftrightarrow p(e)$

$\langle'\rangle$ $\langle x'=f(x)\rangle p(x) \leftrightarrow \exists t{\geq}0\, \langle x:=y(t)\rangle p(x)$ $\qquad (y'(t)=f(y))$

$\langle?\rangle$ $\langle ?Q\rangle P \leftrightarrow Q \wedge P$

$\langle\cup\rangle$ $\langle\alpha\cup\beta\rangle P \leftrightarrow \langle\alpha\rangle P \vee \langle\beta\rangle P$

$\langle;\rangle$ $\langle\alpha;\beta\rangle P \leftrightarrow \langle\alpha\rangle\langle\beta\rangle P$

$\langle*\rangle$ $\langle\alpha^*\rangle P \leftrightarrow P \vee \langle\alpha\rangle\langle\alpha^*\rangle P$

$\langle d\rangle$ $\langle\alpha^{d}\rangle P \leftrightarrow \neg\langle\alpha\rangle\neg P$

M $\quad \dfrac{P \to Q}{\langle\alpha\rangle P \to \langle\alpha\rangle Q}$

FP $\quad \dfrac{P \vee \langle\alpha\rangle Q \to Q}{\langle\alpha^*\rangle P \to Q}$

ind $\quad \dfrac{P \to [\alpha]P}{P \to [\alpha^*]P}$

Fig. 17.1 Differential game logic axiomatization (repeated)

17.3 Separating Axioms

Parts I and II of this textbook identified a number of useful axioms for hybrid systems. Chapter 16 did the same for hybrid games, albeit at a faster pace, because the earlier parts of this book already prepared us well for the typical challenges when developing and using axioms. When we compare the axioms of differential game logic dGL (Fig. 17.1) to those of differential dynamic logic dL, we notice that they share significant similarities. But the dGL axioms already had more involved soundness justifications, most notably in axiom $[\cdot]$, which is a simple observation for hybrid systems (all runs satisfy P iff it is not the case that there is a run satisfying $\neg P$) but already needs the full determinacy theorem Theorem 16.1 as justification for hybrid games.

Without any doubt, the axioms of differential game logic in Fig. 17.1 are sound for hybrid systems as well, because every hybrid system is a (single-player) hybrid game. In fact, except of course the duality axiom $\langle d\rangle$, they all look surprisingly close to the axioms for hybrid systems from Chaps. 5 and 7. Many look almost identical when comparing dL axioms in Fig. 5.4 on p. 160 to the box modality formulation of the dGL axioms in Fig. 16.4 on p. 505. If they look so close, couldn't we have arrived at the dGL axioms more quickly by inferring them from dL axioms?

Well not quite, because all axioms for hybrid games are sound for hybrid systems, since all hybrid systems are hybrid games, but not the other way around! We need to pay more attention and conduct more refined proofs to justify that axioms are *even* sound for hybrid games, not just hybrid systems, because hybrid games can exhibit more behaviors. Of course, we could still have used hybrid systems axioms as an inspiration for possible hybrid games axioms, precisely because of the fact that an axiom can only work for hybrid games if it is, at least, sound for hybrid systems. But once we list hybrid system axioms, we need to scrutinize them very carefully to ensure they continue to be sound for hybrid games, still. In fact, the best preparation for this chapter is to do exactly that by first solving Exercise 16.9.

Before you read on, see if you can find the answer for yourself.

In order to understand the fundamental difference between hybrid systems and hybrid games, it is instructive to investigate separating axioms, i.e., axioms of hybrid systems that are not sound for hybrid games. Some of these axioms that are sound for hybrid systems but not for hybrid games are summarized in Fig. 17.2.

$\not{K}$ $[\alpha](P \to Q) \to ([\alpha]P \to [\alpha]Q)$

$M_{[\cdot]} \dfrac{P \to Q}{[\alpha]P \to [\alpha]Q}$

$\overleftarrow{\not{M}}$ $\langle\alpha\rangle(P \vee Q) \to \langle\alpha\rangle P \vee \langle\alpha\rangle Q$

M $\langle\alpha\rangle P \vee \langle\alpha\rangle Q \to \langle\alpha\rangle(P \vee Q)$

$\not{I}$ $[\alpha^*]P \leftrightarrow P \wedge [\alpha^*](P \to [\alpha]P)$

ind $\dfrac{P \to [\alpha]P}{P \to [\alpha^*]P}$

$\not{B}$ $\langle\alpha\rangle\exists x P \to \exists x \langle\alpha\rangle P$ $\qquad (x\not\in\alpha)$

$\overleftarrow{\text{B}}$ $\exists x \langle\alpha\rangle P \to \langle\alpha\rangle\exists x P$

$\not{V}$ $p \to [\alpha]p$ $\qquad (\text{FV}(p) \cap \text{BV}(\alpha) = \emptyset)$

VK $p \to ([\alpha]true \to [\alpha]p)$

$\not{G}$ $\dfrac{P}{[\alpha]P}$

$M_{[\cdot]} \dfrac{P \to Q}{[\alpha]P \to [\alpha]Q}$

$\not{R}$ $\dfrac{P_1 \wedge P_2 \to Q}{[\alpha]P_1 \wedge [\alpha]P_2 \to [\alpha]Q}$

$M_{[\cdot]} \dfrac{P_1 \wedge P_2 \to Q}{[\alpha](P_1 \wedge P_2) \to [\alpha]Q}$

$\not{FA}$ $\langle\alpha^*\rangle P \to P \vee \langle\alpha^*\rangle(\neg P \wedge \langle\alpha\rangle P)$

$\overleftarrow{\not{M}}$ $[\alpha^*]P \leftrightarrow P \wedge [\alpha^*][\alpha]P$

$[*]$ $[\alpha^*]P \leftrightarrow P \wedge [\alpha][\alpha^*]P$

Fig. 17.2 Separating axioms: The axioms and rules on the left are sound for hybrid systems but not for hybrid games. The related axioms or rules on the right are sound for hybrid games

Detailed counterexamples showing that the axioms on the left of Fig. 17.2 are unsound for hybrid games are reported in previous work [4], but let us investigate the intuition for the difference causing their unsoundness in hybrid games. Kripke's modal modus ponens K from Lemma 5.9 is unsound for hybrid games: even if Demon can play robot soccer so that his robots score a goal every time they pass the ball (they just never try to even pass the ball) and Demon can also play robot soccer so that his robots always pass the ball (somewhere in some random direction), that does not mean Demon has a strategy to always score goals in robot soccer, because that is significantly more difficult to achieve. The problem with axiom K for hybrid games is that Demon's strategies in its two assumptions can be incompatible, which is something that cannot happen in hybrid systems where both box modalities refer to all runs of HP α.

A concrete counterexample illustrating why K is unsound for hybrid games is

$$[x:=0\cap(x:=1\cup x:=-1)](x\neq 0 \to x>0) \to$$
$$([x:=0\cap(x:=1\cup x:=-1)]x\neq 0 \to [x:=0\cap(x:=1\cup x:=-1)]x>0)$$

The first assumption is *true*, because Demon can play left ($x:=0$), which trivially satisfies the postcondition $x\neq 0 \to x>0$. The second assumption is *true*, because Demon can play right ($x:=1\cup x:=-1$), which satisfies $x\neq 0$ whichever way Angel decides. But there is no winning strategy that enables Demon to achieve $x>0$, because playing left makes x zero and playing right enables Angel to play right ($x:=-1$), too.

As Chap. 16 showed, the closely related monotonicity rule M$[\cdot]$ is sound also for hybrid games. The difference of monotonicity rule M$[\cdot]$ to the unsound Kripke axiom K is that it requires the implication $P \to Q$ in the premise to be valid, so true in all states, not just in the states that some of Demon's winning strategies reaches as axiom K requires. The converse monotonicity axiom $\overleftarrow{\text{M}}$, however, is also unsound for hybrid games: just because Angel EVE has a strategy to be close to WALL·E or far away does not mean EVE either has a strategy to always end up close to WALL·E or a strategy to always be far away. It is a mere triviality to be either close or far, because if EVE isn't close to WALL·E then she's far away. Period. But consistently staying close may be about as challenging as consistently always staying far away. The other direction of the monotonicity axiom M is still sound, because if there is a winning strategy for Angel to achieve P in hybrid game α then she also has a winning strategy to achieve the easier $P \vee Q$, because P implies $P \vee Q$.

The induction axiom I from Lemma 7.1 is unsound for hybrid games: just because Demon has a strategy for his soccer robots (e.g., power down) that, no matter how often α^* repeats, Demon still has a strategy such that his robots do not run out of battery for just one more control cycle (one control cycle does not need a lot of battery), that does not mean he has a strategy to keep his robots' batteries nonempty all the time, because that would require quite a revolution in battery designs. The problem is that one more round may be possible for Demon with the appropriate control choices even if the winning condition cannot be sustained forever. The loop induction rule ind (Corollary 16.1) is sound for hybrid games, because its premise requires that $P \to [\alpha]P$ be valid so true in all states, not just true for one particular winning strategy of Demon in the hybrid game α^*.

The Barcan axiom B, which provides a way of commuting modalities with like-minded quantifiers [1], is unsound for hybrid games: just because the winner of a robot soccer tournament who satisfies P can be chosen for x after the robot game α does not mean it is possible to predict this winner x before the game α. By contrast, the converse Barcan axiom $\overleftarrow{\text{B}}$ [1] is sound for hybrid games since, if x is known before the game α, selecting the winner for x can still be postponed until after the game, because that is much easier. The reason why both Barcan axioms are sound for hybrid systems is that all choices are nondeterministic in hybrid systems, so there is no opponent that will take an unexpected turn, which is why predicting x ahead of time is possible.

The vacuous axiom V from Lemma 5.11, in which no free variable of p is bound by α, is unsound for hybrid games. Even if p does not change its truth-value during α does not mean it is possible for Demon to reach any final state at all without being tricked into violating the rules of the game along the way by Angel. With an additional assumption ($[\alpha]true$) implying that Demon has a winning strategy to reach any final state at all (in which *true* holds, which imposes no condition), the possible vacuous axiom VK is still sound for hybrid games. Similarly, Gödel's rule G from Lemma 5.12 is unsound for hybrid games: even if P holds in all states, Demon may still fail to win $[\alpha]P$ if he loses prematurely since Angel tricks Demon into violating the rules during the hybrid game α. The following counterexample is an instance of Gödel's rule G with a valid premise but a conclusion that is equivalent to *false*:

$$\frac{true}{[?false^d]true}$$

The monotonicity rule M$[\cdot]$ is again similar to Gödel's G but sound for hybrid games, because its assumption at least implies that Demon has a winning strategy to get to P at all, which then implies by the premise that he also has a winning strategy to get to the easier Q. Likewise, the regularity rule R is unsound for hybrid games: just because Demon's soccer robots have a strategy to focus all robots on strong defense and another strategy to, instead, focus them all on strong offense that does not mean he has a strategy to win robot soccer even if simultaneously strong defense and strong offense together might imply victory (premise), because offensive and defensive strategies are in conflict. Demon cannot possibly send all his robots both into offense and into defense at the same time, because they won't know which way to go. They have to choose. A special instance of the monotonicity rule M$[\cdot]$ is the closest rule that is still sound, because its assumption requires Demon to achieve both P_1 and P_2 at the same time with the same strategy, which, by the premise, implies Q.

The first-arrival axiom FA, which is the dual of the induction axiom I, is unsound for hybrid games: just because Angel's robot has a strategy to ultimately capture Demon's faster robot with less battery does not mean she either starts with capture or has a strategy to repeat her control cycle so that she exactly captures Demon's robot during the next control cycle, as Demon might save up his energy and speed up just when Angel expected to catch him. Having a better battery, Angel will still ultimately win even if Demon speeds ahead, but not in the round she thought she would be able to predict.

Another way of understanding why several hybrid systems axioms summarized in Fig. 17.2 are not sound for hybrid games is that hybrid games can turn box modalities into diamond modalities by duality and vice versa. After all, the duality axiom $\langle^d\rangle$ together with the determinacy axiom $[\cdot]$ derive

$$\langle\alpha^d\rangle P \leftrightarrow [\alpha]P$$
$$[\alpha^d]P \leftrightarrow \langle\alpha\rangle P$$

Consequently, if an axiom such as K were sound for hybrid games, then it would also be sound for the hybrid game α^d instead of α, which, by axioms $\langle {}^d \rangle, [\cdot]$, turns its box modalities into diamond modalities, but the resulting pure diamond formulation of K is not even sound for hybrid systems:

$$\langle \alpha \rangle (P \to Q) \to (\langle \alpha \rangle P \to \langle \alpha \rangle Q)$$

> **Note 79 (One game's boxes are another game's diamonds)** If a hybrid systems axiom is not also sound when replacing box modalities with diamond modalities and vice versa, then it cannot possibly be sound for hybrid games.

This principle does not explain all cases listed in Fig. 17.2, though! Not even the backwards iteration axiom $\overleftarrow{[*]}$ from Lemma 7.5 on p. 239 is sound for hybrid games, however innocently similar the backwards iteration axiom $\overleftarrow{[*]}$ may be to the (sound) forward iteration axiom $[*]$. The only difference between the unsound $\overleftarrow{[*]}$ and the sound axiom $[*]$ is whether α or the repetition α^* comes first. But that makes a significant difference for hybrid games, because in $[\alpha^*][\alpha]P$ Demon will observe when Angel stopped the repetition α^* but the winning condition P is only checked after one final round of α. Consequently, the right-hand side of the unsound $\overleftarrow{[*]}$ gives Demon one round of early notice about when Angel is going to stop the game, which she will not do in the left-hand side of $\overleftarrow{[*]}$. For example, because of inertia, Demon's robot can easily make sure that it is still moving for one round even though he turned its power off. But that does not mean that the Robot will always keep on moving when its power is off. The following easier instance of hybrid systems axiom $\overleftarrow{[*]}$ is not valid, so the axiom is unsound for hybrid games:

$$[(x := a; a := 0 \cap x := 0)^*] x = 1 \leftrightarrow$$
$$x = 1 \wedge [(x := a; a := 0 \cap x := 0)^*][x := a; a := 0 \cap x := 0] x = 1$$

If $a = 1$ initially, then the right-hand side is true by Demon's winning strategy of always playing $x := 0$ in the repetition but playing $x := a; a := 0$ afterwards. The left-hand side is not true, because all that Angel needs to do is repeat sufficiently often at which point Demon will have caused x to be 0, because he cannot predict when Angel will stop. By the sequential composition axiom $[;]$, the two formulas from axioms $[*]$ and $\overleftarrow{[*]}$ are equivalent to the following two formulas, respectively:

$$[\alpha^*]P \leftrightarrow P \wedge [\alpha; \alpha^*]P \qquad\qquad \text{from } [*] \text{ by } [;]$$
$$[\alpha^*]P \leftrightarrow P \wedge [\alpha^*; \alpha]P \qquad\qquad \text{from } \overleftarrow{[*]} \text{ by } [;]$$

From a hybrid systems perspective, the HP $\alpha; \alpha^*$ is equivalent to the HP $\alpha^*; \alpha$, but that does not extend to hybrid games! Hybrid game $\alpha^*; \alpha$ corresponds to Angel announcing the end of the game one round before the game is over, which makes it easier for Demon to win. Unrolling loops in the beginning is acceptable in hybrid games, but unrolling them in the end may change their semantics! Unrolling loops at the end as in $\overline{[*]}$ is not sound for hybrid games, because it requires predicting the end of the game prematurely.

17.4 Repetitive Diamonds – Convergence Versus Iteration

More fundamental differences between hybrid systems and hybrid games also exist in terms of convergence rules, even if these have not played a prominent rôle in this textbook. These differences are discussed in detail elsewhere [4]. In a nutshell, Harel's convergence rule [2] is not separating, because it is sound for dGL, just unnecessarily, and, furthermore, not even particularly useful for hybrid games [4]. The hybrid version of Harel's convergence rule [3] for dL makes it possible to prove diamond properties of loops. It reads as follows (where v does not occur in α):

$$ \text{con} \ \frac{p(v) \wedge v > 0 \vdash \langle \alpha \rangle p(v-1)}{\Gamma, \exists v\, p(v) \vdash \langle \alpha^* \rangle \exists v \leq 0\, p(v), \Delta} \quad (v \notin \alpha) $$

The convergence rule con uses a *variant* $p(v)$, which is the diamond counterpart of an invariant of the induction rule loop for box modalities of repetitions. Just as an invariant expresses what never changes as a loop executes (Chap. 7), a variant expresses what does change and make progress toward a goal when a loop executes. The dL proof rule con expresses that the variant $p(v)$ holds for some nonpositive real number $v \leq 0$ after repeating α sufficiently often if $p(v)$ holds for any real number at all in the beginning (antecedent) and, by premise, $p(v)$ can decrease after some execution of α by 1 (or another positive real constant) if $v > 0$. This rule can be used to show positive progress (by 1) with respect to $p(v)$ by executing α. The variant $p(v)$ is an abstract progress measure that can decrease by at least 1 unless already at the goal and will, thus, eventually reach the goal (for a nonpositive distance $v \leq 0$).

Just as the induction rule ind is often used with a separate premise for the initial and postcondition check (loop from Chap. 7), rule con is often used in the following derived form that we simply also call con since it will be easy enough for us now to disambiguate which of the two versions of the rule we are referring to:

$$ \text{con} \ \frac{\Gamma \vdash \exists v\, p(v), \Delta \quad \vdash \forall v > 0\, (p(v) \to \langle \alpha \rangle p(v-1)) \quad \exists v \leq 0\, p(v) \vdash Q}{\Gamma \vdash \langle \alpha^* \rangle Q, \Delta} \quad (v \notin \alpha) $$

The following sequent proof shows how convergence rule con with $x < n+1$ for $p(n)$ can be used to prove a simple dL liveness property of a discrete HP:

$$
\cfrac{
  \cfrac{
    \mathbb{R}\cfrac{*}{x \geq 0 \vdash \exists n\, x < n+1}
    \quad
    \cfrac{
      \cfrac{
        \cfrac{
          \cfrac{
            \mathbb{R}\cfrac{*}{x < n+1 \wedge n > 0 \vdash x - 1 < n}
          }{x < n+1 \wedge n > 0 \vdash \langle x := x-1 \rangle x < n - 1 + 1}(:=)
        }{\vdash x < n+1 \wedge n > 0 \to \langle x := x-1 \rangle x < n - 1 + 1}{\to}R
      }{\vdash \forall n > 0\, (x < n+1 \to \langle x := x-1 \rangle x < n - 1 + 1)}\forall R
    \quad
    \mathbb{R}\cfrac{*}{\exists n \leq 0\, x < n+1 \vdash x < 1}
  }{x \geq 0 \vdash \langle (x := x-1)^* \rangle x < 1}\text{con}
}{\vdash x \geq 0 \to \langle (x := x-1)^* \rangle x < 1}{\to}R
$$

Let's compare how dGL proves diamond properties of repetitions based on the iteration axiom $\langle * \rangle$. In addition to the iteration axiom $\langle * \rangle$, the following proofs for diamond repetitions employ a clever use of the uniform substitution proof rule US, which concludes that any substitution instance of a provable formula is provable, too. That is, if ϕ has a proof, then the instance $\sigma(\phi)$ that is obtained by performing any (admissible) uniform substitution σ on ϕ is valid, too:

$$\text{US} \quad \frac{\phi}{\sigma(\phi)}$$

Uniform substitutions replace function symbols with suitable terms and predicate symbols by logical formulas. For example, a uniform substitution σ may substitute an abstract predicate symbol p such that $p(x)$ is replaced with the dL formula $\langle (x := x - 1)^* \rangle (0 \leq x < 1)$ of the (same) free variable x. Uniform substitution will be explored in Chap. 18 of Part IV, but its intuition can already be easily understood, which is all we need right now. For the time being, all that is important to know about it is that rule US substitutes formulas for predicate symbols and has appropriate implementations that check and ensure soundness. Using it from the conclusion to its premise, rule US can be used to abstract formulas by predicate symbols. And all we need to know about predicate symbols is that they are indeed symbolic in the sense that unlike for a concrete logical formula such as $0 \leq x < 1$ we do not know a priori when exactly $p(x)$ is *true*.

Example 17.1 (Non-game system). The same simple non-game dGL formula

$$x \geq 0 \rightarrow \langle (x := x - 1)^* \rangle 0 \leq x < 1$$

above is provable without con, as shown in Fig. 17.3, where $\langle \alpha^* \rangle 0 \leq x < 1$ is short for $\langle (x := x - 1)^* \rangle (0 \leq x < 1)$. Note that, as in the subsequent proofs, the extra assumption for cut near the bottom of the proof in Fig. 17.3 is provable by $\langle * \rangle, \forall R$:

$$
\cfrac{
  \cfrac{
    *
  }{
    \langle * \rangle \vdash 0 \leq x < 1 \vee \langle x := x - 1 \rangle \langle \alpha^* \rangle 0 \leq x < 1 \rightarrow \langle \alpha^* \rangle 0 \leq x < 1
  }
}{
  \forall R \; \vdash \forall x (0 \leq x < 1 \vee \langle x := x - 1 \rangle \langle \alpha^* \rangle 0 \leq x < 1 \rightarrow \langle \alpha^* \rangle 0 \leq x < 1)
}
$$

The rôle of the predicate symbol in Fig. 17.3 is to have $p(x)$ serve as an abstract formula standing for $\langle (x := x - 1)^* \rangle (0 \leq x < 1)$. Since the premise of rule US can be proved for the abstract predicate $p(x)$, its conclusion for the concrete formula $\langle (x := x - 1)^* \rangle (0 \leq x < 1)$ in place of $p(x)$ is valid by rule US as well.

Example 17.2 (Choice game). The dGL formula

$$x = 1 \wedge a = 1 \rightarrow \langle (x := a; a := 0 \cap x := 0)^* \rangle x \neq 1$$

is provable as shown in Fig. 17.4, where $\beta \cap \gamma$ is short for $x := a; a := 0 \cap x := 0$ and $\langle (\beta \cap \gamma)^* \rangle x \neq 1$ is short for $\langle (x := a; a := 0 \cap x := 0)^* \rangle x \neq 1$.

Example 17.3 (2-Nim-type game). The dGL formula

$$
\begin{array}{ll}
\mathbb{R} & \overline{\hspace{10cm}}^{*} \\
& \overline{\forall x\,(0 \le x < 1 \vee p(x-1) \to p(x)) \to (x \ge 0 \to p(x))} \\
\langle := \rangle & \overline{\forall x\,(0 \le x < 1 \vee \langle x := x-1 \rangle p(x) \to p(x)) \to (x \ge 0 \to p(x))} \\
\text{US} & \overline{\forall x\,(0 \le x < 1 \vee \langle x := x-1 \rangle \langle \alpha^{*} \rangle 0 \le x < 1 \to \langle \alpha^{*} \rangle 0 \le x < 1) \to (x \ge 0 \to \langle \alpha^{*} \rangle 0 \le x < 1)} \\
\langle^{*} \rangle, \forall R, \text{cut} & \overline{\hspace{2cm} x \ge 0 \to \langle \alpha^{*} \rangle 0 \le x < 1}
\end{array}
$$

Fig. 17.3 dGL Angel proof for non-game system Example 17.1 $x \ge 0 \to \langle (x := x-1)^{*} \rangle 0 \le x < 1$

$$
\begin{array}{ll}
\mathbb{R} & \overline{\hspace{10cm}}^{*} \\
& \overline{\forall x\,(x \ne 1 \vee p(a,0) \wedge p(0,a) \to p(x,a)) \to (\mathit{true} \to p(x,a))} \\
\langle ; \rangle, \langle := \rangle & \overline{\forall x\,(x \ne 1 \vee \langle \beta \rangle p(x,a) \wedge \langle \gamma \rangle p(x,a) \to p(x,a)) \to (\mathit{true} \to p(x,a))} \\
\langle \cup \rangle, \langle^{d} \rangle & \overline{\forall x\,(x \ne 1 \vee \langle \beta \cap \gamma \rangle p(x,a) \to p(x,a)) \to (\mathit{true} \to p(x,a))} \\
\text{US} & \overline{\forall x\,(x \ne 1 \vee \langle \beta \cap \gamma \rangle \langle (\beta \cap \gamma)^{*} \rangle x \ne 1 \to \langle (\beta \cap \gamma)^{*} \rangle x \ne 1) \to (\mathit{true} \to \langle (\beta \cap \gamma)^{*} \rangle x \ne 1)} \\
\langle^{*} \rangle, \forall R, \text{cut} & \overline{\hspace{3cm} \mathit{true} \to \langle (\beta \cap \gamma)^{*} \rangle x \ne 1} \\
\mathbb{R} & \overline{\hspace{3cm} x = 1 \wedge a = 1 \to \langle (\beta \cap \gamma)^{*} \rangle x \ne 1}
\end{array}
$$

Fig. 17.4 dGL Angel proof for demonic choice game Example 17.2
$x = 1 \wedge a = 1 \to \langle (x := a; a := 0 \cap x := 0)^{*} \rangle x \ne 1$

$$
x \ge 0 \to \langle (x := x-1 \cap x := x-2)^{*} \rangle 0 \le x < 2
$$

is provable as shown in Fig. 17.5, where $\beta \cap \gamma$ is short for $x := x-1 \cap x := x-2$ and $\langle (\beta \cap \gamma)^{*} \rangle 0 \le x < 2$ is short for $\langle (x := x-1 \cap x := x-2)^{*} \rangle 0 \le x < 2$.

$$
\begin{array}{ll}
\mathbb{R} & \overline{\hspace{10cm}}^{*} \\
& \overline{\forall x\,(0 \le x < 2 \vee p(x-1) \wedge p(x-2) \to p(x)) \to (\mathit{true} \to p(x))} \\
\langle := \rangle & \overline{\forall x\,(0 \le x < 2 \vee \langle \beta \rangle p(x) \wedge \langle \gamma \rangle p(x) \to p(x)) \to (\mathit{true} \to p(x))} \\
\langle \cup \rangle, \langle^{d} \rangle & \overline{\forall x\,(0 \le x < 2 \vee \langle \beta \cap \gamma \rangle p(x) \to p(x)) \to (\mathit{true} \to p(x))} \\
\text{US} & \overline{\forall x\,(0 \le x < 2 \vee \langle \beta \cap \gamma \rangle \langle (\beta \cap \gamma)^{*} \rangle 0 \le x < 2 \to \langle (\beta \cap \gamma)^{*} \rangle 0 \le x < 2) \to (\mathit{true} \to \langle (\beta \cap \gamma)^{*} \rangle 0 \le x < 2)} \\
\langle^{*} \rangle, \forall R, \text{cut} & \overline{\hspace{3cm} \mathit{true} \to \langle (\beta \cap \gamma)^{*} \rangle 0 \le x < 2} \\
\mathbb{R} & \overline{\hspace{3cm} x \ge 0 \to \langle (\beta \cap \gamma)^{*} \rangle 0 \le x < 2}
\end{array}
$$

Fig. 17.5 dGL Angel proof for 2-Nim-type game Example 17.3
$x \ge 0 \to \langle (x := x-1 \cap x := x-2)^{*} \rangle 0 \le x < 2$

Example 17.4 (Hybrid game). The dGL formula

$$
\langle (x := 1; x' = 1^{d} \cup x := x-1)^{*} \rangle 0 \le x < 1
$$

is provable as shown in Fig. 17.6, where the notation $\langle (\beta \cup \gamma)^{*} \rangle 0 \le x < 1$ is short for $\langle (x := 1; x' = 1^{d} \cup x := x-1)^{*} \rangle (0 \le x < 1)$: The proof steps for β use in $\langle' \rangle$ that $t \mapsto x+t$ is the solution of the differential equation, so the subsequent use of $\langle := \rangle$ substitutes 1 in for x to obtain $t \mapsto 1+t$. Recall from Chap. 16 that the winning regions for this formula need $> \omega$ iterations to converge. It is still provable easily.

$$*$$

$$
\begin{array}{rl}
\mathbb{R} & \dfrac{}{\forall x\,(0 \le x < 1 \lor \forall t \ge 0\, p(1+t) \lor p(x-1) \to p(x)) \to (true \to p(x))} \\[4pt]
\langle := \rangle & \dfrac{}{\forall x\,(0 \le x < 1 \lor \langle x:=1\rangle \neg \exists t \ge 0\, \langle x:=x+t\rangle \neg p(x) \lor p(x-1) \to p(x)) \to (true \to p(x))} \\[4pt]
\langle ' \rangle & \dfrac{}{\forall x\,(0 \le x < 1 \lor \langle x:=1\rangle \neg \langle x'=1\rangle \neg p(x) \lor p(x-1) \to p(x)) \to (true \to p(x))} \\[4pt]
\langle ; \rangle,\langle {}^d \rangle & \dfrac{}{\forall x\,(0 \le x < 1 \lor \langle \beta\rangle p(x) \lor \langle \gamma\rangle p(x) \to p(x)) \to (true \to p(x))} \\[4pt]
\langle \cup \rangle & \dfrac{}{\forall x\,(0 \le x < 1 \lor \langle \beta \cup \gamma\rangle p(x) \to p(x)) \to (true \to p(x))} \\[4pt]
\text{US} & \dfrac{}{\forall x\,(0 \le x < 1 \lor \langle \beta \cup \gamma\rangle \langle (\beta\cup\gamma)^*\rangle 0 \le x < 1 \to \langle (\beta\cup\gamma)^*\rangle 0 \le x < 1) \to (true \to \langle(\beta\cup\gamma)^*\rangle 0 \le x < 1)} \\[4pt]
\langle * \rangle,\forall\text{R,cut} & \dfrac{}{true \to \langle (\beta\cup\gamma)^*\rangle 0 \le x < 1}
\end{array}
$$

Fig. 17.6 dGL Angel proof for hybrid game Example 17.4
$\langle(x:=1; x'=1^d \cup x:=x-1)^*\rangle 0 \le x < 1$

A downside of the approach of using uniform substitution rule US with the iteration axiom $\langle * \rangle$ to prove diamond properties of loops is that the resulting arithmetic (marked $\mathbb{R}$) mixes real arithmetic with predicate symbols, which is quite challenging. This is a reason to still take note of the convergence rule con despite its limitations.

17.5 Summary

This chapter solidified our understanding of rigorous reasoning principles for hybrid games by developing an appreciation for the axiomatic differences of hybrid systems versus hybrid games. While the previous chapter emphasized the aspects of surprising similarities of hybrid systems and hybrid games reasoning, this chapter now carefully emphasized the differences. We have explored intuitive reasons, which make it easier to remember which axioms can carry over from hybrid systems to hybrid games. But it is, of course, crucial for soundness in our arguments to understand precisely which hybrid systems axioms continue to be sound for hybrid games.

The sophisticated differential equation reasoning principles from Part II that prove properties of differential equations without the need for explicit closed-form solutions carry over to hybrid games, because they do not involve any game aspects. More importantly, though, differential invariants generalize to differential games that directly combine continuous and adversarial dynamics by allowing both players to provide continuous-time input on which the differential equation depends [5]. The idea is to give both players the ability to provide input controls during the continuous system while following a differential game.

17.6 Appendix: Relating Differential Game Logic and Differential Dynamic Logic

Now that we have come to appreciate the value of soundness, couldn't we have known about that, for the most part, before the soundness result of Theorem 16.2? Most dGL axioms look rather familiar when we compare them to the dL axioms from Chap. 5. Does that not mean that these same axioms are already trivially sound? Why did we go to the (admittedly rather minor) trouble of proving Theorem 16.2?

> Before you read on, see if you can find the answer for yourself.

It is not quite so easy. After all, we could have given the same syntactical operator $\cup$ an entirely different meaning for hybrid games than before for hybrid systems. Maybe we could have been silly and flipped the meaning of ; and $\cup$ around just to confuse everybody. The fact of the matter is, of course, that we did not. The operator $\cup$ still means choice, just for hybrid games rather than hybrid systems. So can we deduce the soundness of the dGL axioms in Fig. 17.1 from the soundness of the corresponding dL axioms from Chap. 5 and focus on the new axioms, only?

Before we do anything of the kind, we first need to convince ourselves that the dL semantics really coincide with the more general dGL semantics in case there are no games involved. How can that be done? Maybe by proving the validity of all formulas of the following form

$$\underbrace{\langle\alpha\rangle P}_{\text{in dL}} \leftrightarrow \underbrace{\langle\alpha\rangle P}_{\text{in dGL}} \tag{17.1}$$

for dual-free hybrid games α, i.e., those that do not mention ^{d} (not even indirectly hidden in the abbreviations $\cap, ^{\times}$).

> Before you read on, see if you can find the answer for yourself.

The problem with (17.1) is that it is not directly a formula in any logic, because the $\leftrightarrow$ operator can hardly be applied meaningfully to two formulas from different logics. Well, of course, every dL formula is a dGL formula, so the left-hand side of (17.1) could be embedded into dGL. But then (17.1) would become well-defined but is only stating a mere triviality. Everything is equivalent to itself, which is not a gigantic insight to write home about.

Instead, a proper approach would be to rephrase the well-intended but ill-fated (17.1) semantically:

$$\underbrace{\omega \in [\![\langle\alpha\rangle P]\!]}_{\text{in dL}} \text{ iff } \underbrace{\omega \in [\![\langle\alpha\rangle P]\!]}_{\text{in dGL}} \tag{17.2}$$

which is equivalent to

$$\underbrace{\left(v \in [\![P]\!] \text{ for some } v \text{ with } (\omega, v) \in [\![\alpha]\!]\right)}_{\text{statement about reachability in dL}} \text{ iff } \underbrace{\omega \in \varsigma_\alpha([\![P]\!])}_{\text{winning in dGL}}$$

Equivalence (17.2) can be shown. In fact, Exercise 3.15 in Chap. 3 already developed an understanding of the dL semantics based on sets of states, preparing for (17.2).

The trouble is that, besides requiring a proof itself, the equivalence (17.2) will still not quite justify soundness of the dGL axioms in Fig. 17.1 that look innocuously like dL axioms. Equivalence (17.2) is for dual-free hybrid games α. But even if the top-level operator in axiom $\langle \cup \rangle$ is not ^{d}, that dual operator can still occur within α or β, which can only be made sense of with a game semantics.

Consequently, we are much better off proving soundness for the dGL axioms according to their actual semantics, like in Theorem 16.2, as opposed to trying half-witted ways out that only make soundness matters worse.

Exercises

17.1 (Good and bad axioms). Prove each of the axioms on the left of Fig. 17.2 to be unsound for hybrid games. For each of the axioms, provide a concrete dGL formula that is an instance of that axiom but not a valid formula. For the unsound proof rules on the left of Fig. 17.2 give an instance where the premise is valid but the conclusion is not. Then go on to show a way of using each of the reasoning principles on the right of Fig. 17.2 for a hybrid game.

17.2. Prove the following dGL formula with the iteration and uniform substitution technique as in Example 17.2

$$\left\langle \left(x := x^2 \cup (x := x + 1 \cap x' = 2) \right)^* \right\rangle x > 0$$

17.3 (***). The following formula was proved using dGL's hybrid games proof rules in Fig. 17.3
$$x \geq 0 \rightarrow \langle (x := x - 1)^* \rangle 0 \leq x < 1$$

Try to see whether you can prove it using the convergence rule con instead.

References

[1] Ruth C. Barcan. The deduction theorem in a functional calculus of first order based on strict implication. *J. Symb. Log.* **11**(4) (1946), 115–118.

[2] David Harel, Albert R. Meyer, and Vaughan R. Pratt. Computability and completeness in logics of programs (preliminary report). In: *STOC.* New York: ACM, 1977, 261–268.

[3] André Platzer. Differential dynamic logic for hybrid systems. *J. Autom. Reas.* **41**(2) (2008), 143–189. DOI: 10.1007/s10817-008-9103-8.

[4] André Platzer. Differential game logic. *ACM Trans. Comput. Log.* **17**(1) (2015), 1:1–1:51. DOI: 10.1145/2817824.

[5] André Platzer. Differential hybrid games. *ACM Trans. Comput. Log.* **18**(3) (2017), 19:1–19:44. DOI: 10.1145/3091123.

Overview of Part IV on Comprehensive CPS Correctness

This part shifts perspective yet again and investigates techniques that lead to comprehensive correctness arguments for cyber-physical systems. Based on rigorous reasoning principles for elementary cyber-physical systems from Part I and rigorous reasoning principles for continuous dynamics with unsolvable differential equations from Part II, this part now explores the most fundamental remaining elements that make it possible to give a pervasive correctness result for a CPS. Without rigorous axiomatizations of hybrid systems as in Part I, and of differential equations in Part II, and/or of hybrid games as in Part III, it is very difficult to reason soundly about CPSs. But even with the help of such sound axiomatizations, there is still some remaining potential for error in the subtle world of cyber-physical systems.

Part IV provides a number of unrelated approaches that help safeguard different aspects of correctness analysis results for CPSs. First, this part provides a completely axiomatic approach for hybrid systems based on uniform substitutions that enables a simple and correct implementation of differential dynamic logic reasoning with an extremely parsimonious logical framework. Uniform substitutions provide a convenient framework for conducting flexible proofs about CPSs that can be implemented in a simple straightforward way. This framework treats axioms as data in the object logic and reduces the required mechanism for sound theorem proving to just the uniform substitution algorithm. This makes it easy to implement simple but powerful theorem provers for hybrid systems from an extraordinarily small soundness-critical core.

Part IV also investigates a logical way to tame the subtle relationship of CPS models to CPS implementations in a provably correct way. Since the nuances of cyber-physical systems provide ample opportunity for subtle discrepancies, the relationship of CPS models to CPS implementations is rather nontrivial. It is quite important for the comprehensive success of a CPS analysis and design effort to identify the relevant parts of physics on the appropriate level of abstraction. But that leaves open the question of how to justify that the physical model is adequate. Techniques from the logical foundations of model safety transfer can synthesize provably correct monitor conditions that, if checked to hold at runtime, are provably guaranteed to imply that offline safety verification results about CPS models apply to the present run of the actual CPS implementation. This crucial link is needed to make safety results for CPS models transfer to CPS implementations. The link can be characterized and tamed elegantly in differential dynamic logic using the diamond modality that played a less prominent rôle in Parts I and II but already became significantly more relevant in Part III.

Finally, Part IV considers logical elements of reasoning techniques for the real arithmetic to which the differential dynamic logic axiomatizations reduce CPS correctness. Real-arithmetic verification is pervasive in CPSs and comes up in all CPS verification. Part IV explains virtual substitutions, which provide a systematic logical approach that is practically significant for real-arithmetic formulas at least of up to polynomial degree 3. Techniques for higher degrees are beyond the scope of this textbook, but one simple technique will be explained nevertheless.

Chapter 18
Axioms & Uniform Substitutions

Synopsis This chapter explores a succinct approach for soundly implementing rigorous reasoning for hybrid systems. Unlike previous chapters, this chapter is not concerned with identifying new reasoning principles for cyber-physical systems, but, rather, focuses on how they can best be implemented correctly. Uniform substitutions are identified as a simple concept based upon which differential dynamic logic proof systems can be implemented quite easily. Uniform substitutions uniformly instantiate predicate symbols by formulas. Since all reasoning can be reduced to finding the appropriate sequence of uniform substitutions, this makes it possible to implement theorem provers with a small soundness-critical core.

18.1 Introduction

The logic and reasoning principles for hybrid systems (Part I), differential equations (Part II), and hybrid games (Part III) identified in previous chapters are conducive to quite simple correctness arguments. Proof principles decouple the question of what a correct argument is from the question of how to find it. Soundness even of the biggest and most complicated proofs directly follows from the soundness of each of the proof steps. Every proof step uses one of a small set of dL axioms and proof rules, which can each be proved sound individually quite easily. The transfer of soundness was already rooted in Definition 6.2 on p. 179, which defined a proof rule to be sound iff the validity of all premises implies the validity of the conclusion. For axioms, Definition 5.1 on p. 146 defined an axiom to be sound iff all its instances are valid. Since proofs only consist of axioms composed with proof rules, this implies that the conclusion of every (completed) proof is valid.

The remaining challenge for soundness of proofs is to ensure that all axioms and proof rules are also implemented correctly in a theorem prover. The primary obstacle is that the reasoning principles identified so far were considered as *axiom schemata*, i.e., they stand for an infinite family of formulas of the same shape. That is easily said, but still needs some form of implementation. Moreover, a fair number of the

© Springer International Publishing AG, part of Springer Nature 2018
A. Platzer, *Logical Foundations of Cyber-Physical Systems*,
https://doi.org/10.1007/978-3-319-63588-0_18

axiom schemata have soundness-critical side conditions that need to be respected to guarantee soundness. That these soundness-critical side conditions cannot be elided is most obvious in the vacuous axiom schema from Lemma 5.11:

$$V \quad p \to [\alpha]p \quad (FV(p) \cap BV(\alpha) = \emptyset)$$

Of course, every use of axiom schema V needs to ensure that the same formula p is used in the precondition and the postcondition. But without checking that no free variable of p is written to in the hybrid program α, it would be quite unsound to conclude that p always holds after running HP α if p was true initially. After all, if α changes a variable that p reads, its truth-value may change. It is only thanks to this side condition that the following invalid formula is not provable by axiom V:

$$x \geq 0 \to [x' = -5]x \geq 0$$

The differential equation solution axiom schema from Lemma 5.3 has even more complicated side conditions:

$$['] \quad [x' = f(x)]p(x) \leftrightarrow \forall t \geq 0\,[x := y(t)]p(x) \quad (y'(t) = f(y))$$

The soundness-critical side conditions for axiom schema $[']$ are:

1. The variable t needs to be fresh and cannot have occurred already, because it is supposed to represent the independent variable for time.
2. The function of time $y(\cdot)$ needs to solve the differential equation $y(t)' = f(y(t))$ and needs to be defined at all times that the quantifier for t quantifies over, because the continuous dynamics of the differential equation can only be equivalently replaced by a discrete assignment when $y(\cdot)$ is the correct solution of the differential equation.
3. The solution $y(\cdot)$ needs to solve the symbolic initial value condition $y(0) = x$ for the variable x, because we usually do not have a specific numerical initial value when using axiom schema $[']$.
4. The solution $y(\cdot)$ needs to cover all solutions parametrically, e.g., when the solution has different shapes for different choices of the initial value x.
5. The postcondition $p(x)$ cannot have differential symbol x' as a free variable, because it receives the value $f(x)$ after the differential equation but retains the initial value after the discrete assignment to x.[1]

A correct implementation of axiom schema $[']$, thus, amounts to an algorithm accepting every formula of this shape after checking all the required side conditions. Fortunately, Part II already provided a substantially more elegant way of proving properties of differential equations by induction that also makes the solution axiom schema $[']$ superfluous [8], because it can be replaced by appropriate differential cuts to augment the evolution domain with the solution after a suitable differential ghost has shifted the dynamics into the time domain $t' = 1$ (Chap. 12). But the fact

---

[1] Of course, this is easily fixed by adding an assignment $x' := f(x)$ after the assignment to x.

remains that axiom schemata have a tendency to require a somewhat unwieldy set of side conditions that are soundness-critical and, thus, need to be enforced for every reasoning step. Compared to verification algorithms that do not even benefit from a similar logical foundation, it is still substantially easier to devise correct implementations of individual axiom schemata and then glue them together with correct implementations of proof rules. But this chapter will find a more straightforward way that is even easier to get correct.

The primary observation to make this happen comes from a shift in perspective that distinguishes between axioms and axiom schemata. An *axiom* is a single valid formula that is adopted as a basis for reasoning in a proof calculus. An *axiom schema* stands for an infinite family of formulas of the same shape (subject to the required side conditions) and, thus, needs to be implemented with an algorithm. Implementing an axiom is trivial, because an axiom is just a single formula in the object logic. The only downside is that the only formula that an axiom enables us to prove is literally that formula in verbatim, which we are rarely interested in proving.

Consequently, the missing element for a reasoning system that is based on axioms is a mechanism for instantiating them. Church's uniform substitutions [2] provide such a mechanism for first-order logic. Uniform substitutions make it possible to instantiate predicate symbols by formulas and check the required conditions to ensure that that instantiation is sound. Generalizing uniform substitutions from first-order logic to differential dynamic logic leads to the corresponding mechanism to implement flexible dL proving parsimoniously with uniform substitution as essentially the only proof rule [7, 8].

Differential dynamic logic provides sound reasoning principles. Uniform substitutions make it easy to implement them correctly. Uniform substitutions are the secret for simple sound hybrid systems provers such as KeYmaera X [3]. This chapter has a major impact, enabling the 1,700 lines of soundness-critical code in KeYmaera X compared to the 66,000 lines of soundness-critical code[2] in its predecessor KeYmaera [9], which implements a schematic sequent calculus for dL [5].

The most important learning goals of this chapter are:

Modeling and Control: We will eventually see how the shift in perspective to axioms gives us an opportunity to reflect on the significance of the local meaning of differentials in hybrid systems.

Computational Thinking: This chapter investigates the relationship and fundamental difference of axioms versus axiom schemata. This philosophical distinction leads to a significant algorithmic impact on the style of implementing hybrid systems reasoning. This chapter explores the local meaning of axioms, which is the axiomatic counterpart of how generic points are understood as a nondegenerate generalization of concrete points in algebraic geometry. The fundamental concept of uniform substitutions will be explored, which makes it possible to use axioms as if they were axiom schemata without the need for any additional mechanisms or side condition checking. This purely axiomatic

---

[2] These numbers are to be taken with a grain of salt because the two provers were implemented in different programming languages.

reconsideration of the proof calculus for differential dynamic logic will lead to a new level of appreciation for what the axioms of differential dynamic logic already offered throughout this book without us noticing.

CPS Skills: We identify techniques for a parsimonious straightforward implementation of CPS reasoning. These techniques enable a modular implementation of the logic and the prover mostly independently in parallel, which reduces complexity and makes it easier to advance the reasoning techniques.

axiom vs. axiom schema
algorithmic impact of philosophical difference
local meaning of axioms
generic axioms like generic points
uniform substitution

local meaning of differentials   parsimonious CPS reasoning implementation
                                                                  modular implementation of logic ∥ prover

18.2 Axioms Versus Axiom Schemata

Recall the axiom $[\cup]$ for hybrid programs with a nondeterministic choice $\alpha \cup \beta$ from Lemma 5.1 on p. 144:

$$[\cup] \quad [\alpha \cup \beta]P \leftrightarrow [\alpha]P \wedge [\beta]P \tag{18.1}$$

The innocent way of reading (18.1) is as an axiom schema $[\cup]$. An axiom schema is meant to stand for the infinite family of formulas that have the shape of that axiom schema, so α, β are schema variables or placeholders for arbitrary HPs and P is a placeholder for an arbitrary dL formula. The left-hand side of axiom schema $[\cup]$ applies for any dL formula of the form $[\alpha \cup \beta]P$, so to any box modality of any HP that begins with a nondeterministic choice as the top-level operator and has any HPs as subprograms and any dL formula as a postcondition. For example, the left-hand side of axiom schema $[\cup]$ fits dL formula $[x := x + 1 \cup x' = x^2]x \geq 0$, which implies that the axiom schema $[']$ justifies the following equivalence:

$$[x := x + 1 \cup x' = x^2]x \geq 0 \leftrightarrow [x := x + 1]x \geq 0 \wedge [x' = x^2]x \geq 0 \tag{18.2}$$

Of course, this is not the only dL formula that needs to be recognized to be of the shape that axiom schema $[\cup]$ indicates. Here are a few more:

$$[x' = x^2 \cup x := x+1]x \geq 0 \leftrightarrow [x' = x^2]x \geq 0 \wedge [x := x+1]x \geq 0$$
$$[x' = 5 \cup x' = -x]x^2 \geq 5 \leftrightarrow [x' = 5]x^2 \geq 5 \wedge [x' = -x]x^2 \geq 5$$
$$[v := v+1; x' = v \cup x' = 2]x \geq 5 \leftrightarrow [v := v+1; x' = v]x \geq 5 \wedge [x' = 2]x \geq 5$$

A direct implementation of axiom schema $[\cup]$ consists of an algorithm that takes a dL formula as an input and decides whether that formula is of the form of schema $[\cup]$. Of course, it is crucially important that literally the same postcondition is used for all three modalities of axiom schema $[\cup]$. And it is important that the same HP is used in the left part of the nondeterministic choice $\alpha \cup \beta$ and in the first modality $[\alpha]$ on the right-hand side, and that the same HP is used in the right part of $\alpha \cup \beta$ that is also used in the second modality $[\beta]$ on the right-hand side.[3] Axiom schema $[\cup]$ does not even have any side conditions yet, but it already comes with a few tedious conditions to check (if implementing it in an imperative programming language) or match correctly (in a functional programming language with pattern matching).

A more conscious way of reading (18.1) is as an axiom $[\cup]$ that literally only refers to one dL formula:

$$[\alpha \cup \beta]P \leftrightarrow [\alpha]P \wedge [\beta]P \tag{18.3}$$

Of course, we will still have to make sure that (18.3) actually is a syntactically well-formed dL formula, which it is not presently. The only formula that such an axiom $[\cup]$ ever proves is (18.3). That alone is not so useful, but an axiom is easily implemented, just by copying the dL formula (18.3) from axiom $[\cup]$ into the prover.

Alonzo Church's seminal observation is that the sole operation it takes to make more use of an axiom is to provide a uniform substitution mechanism that replaces parts of formulas with other formulas [2]. The trick is to identify when such a replacement is sound. Of course, Church did not know about differential dynamic logic yet, so he settled for first-order logic. But with a sufficiently generalized notion of uniform substitutions for differential dynamic logic [8], we can prove the dL formula (18.3) from axiom $[\cup]$ and then use a uniform substitution to prove (18.2) from (18.3). All this takes is the uniform substitution that substitutes $x := x+1$ for α and substitutes $x' = x^2$ for β and simultaneously also substitutes $x \geq 0$ for P uniformly everywhere in (18.3).

Now, the one crucial missing piece is a precise definition of this uniform substitution mechanism. The other crucial element is a precise understanding of whether and what the uniform substitution mechanism needs to check to ensure that all its replacements preserve soundness. And the final missing element is the question of what precise form the syntactic expressions α, β, and P take in (18.1) if it is to be taken literally as an axiom. Then the same process needs to be repeated with an axiomatic reinterpretation of all other dL axioms to find out how they can all be read as axioms instead of as significantly more complicated axiom schemata.

---

[3] If the formula that is used in place of P has a modality, then the textual description of the places where these occur is, of course, slightly more complex, but they are still in the same places of the expression tree corresponding to the formula.

Admittedly, on a sheet of paper, it is more convenient to work with axiom schemata, because we are now already so well trained to pay attention to make no incorrect reasoning steps by checking all required side conditions. But for precision purposes in a formal verification tool it is substantially easier to work with axioms instead, because the uniform substitution mechanism only needs to be understood and implemented once and because the axioms can be implemented by copy-and-paste. And even when working on a sheet of paper it may be easier to just remember a single uniform substitution mechanism instead of a diverse list of side conditions.

18.3 What Axioms Want

If the axiom of nondeterministic choice $[\cup]$ is internalized as an axiom, not as an axiom schema, then what syntactic elements of differential dynamic logic do the parts of its formula (18.1) correspond to? Suddenly, α and β need to be concrete HPs in the syntax of dL as opposed to schematic variables or placeholders for concrete HPs. Likewise the postcondition P needs to be a concrete dL formula. In fact, revisiting the differential dynamic logic axiom schemata from Chap. 5, there are three different cases of postconditions:

$[:=]$ $[x := e] p(x) \leftrightarrow p(e)$

$[\cup]$ $[\alpha \cup \beta] P \leftrightarrow [\alpha] P \wedge [\beta] P$

V $p \to [\alpha] p$ $(FV(p) \cap BV(\alpha) = \emptyset)$

The postcondition p of the vacuous axiom schema V cannot have any variable free that is bound by the HP α. But anything that is not written to by HP α can still be mentioned in p, which is the whole point of this axiom compared to Gödel's generalization proof rule G. In comparison, the postcondition $p(x)$ of the assignment axiom schema $[:=]$ should be allowed to mention variable x despite the fact that it is written to in the HP $x := e$. That is why the postcondition $p(x)$ mentions x explicitly. The postcondition on the left-hand side of axiom schema $[:=]$ can have the argument x free in the same places that the formula $p(e)$ on the right-hand side of that axiom schema has term e. Its postcondition $p(x)$ can still mention other free variables besides x, because no other variable is written to in the discrete assignment $x := e$. The postcondition P of the axiom schema of nondeterministic choice $[\cup]$, instead, can have any free variables without reservation, because the axiom is correct whether or not the HPs $\alpha \cup \beta$, α, or β modify the values of free variables of P.

Predicate Symbols

Predicate symbols explain all three cases of postconditions with one joint mechanism. The postcondition p in axiom V has a predicate symbol p with 0 arguments, so it has no special permission to have its truth-value depend on any particular free variables. The postcondition $p(x)$ in axiom $[:=]$ has a predicate symbol p with variable x as its only argument, so its truth-value can depend on the value of x since $[x:=e]$ binds no other variables, x is the only variable that needs explicit permission to be mentioned in the context $[x:=e]p(x)$. When reading the postcondition P in axiom $[\cup]$ as $p(\bar{x})$ for a predicate symbol p that receives the vector $\bar{x}$ of all variables as argument, so its truth-value can depend on the values of all variables, then all cases of postconditions are covered by corresponding predicate symbols that only differ in the number of their arguments.

It is conceptually easier to read the axioms $[:=],[\cup],$V as axioms, so concrete dL formulas, with predicate symbols as postconditions instead of placeholders for formulas. The concrete dL formula $p(x)$ from axiom $[:=]$ literally tells us that its truth-value depends on variable x and apparently nothing else. The formula p from axiom V directly indicates that its truth-value does not depend on the values of any variables. And the case $p(\bar{x})$, which is how we read P in axiom $[\cup]$, indicates that its truth-value may depend on the values of all variables $\bar{x}$. We no longer need to keep in mind what other dL formulas the respective postconditions might stand for, but we see the concrete dL formula explicitly.

Separately, we can then worry about what formulas are acceptable as drop-in replacements for predicate symbols. We can find out which replacements are fine once and for all, and independently of the particular axiom at hand. This separation of concerns is liberating because it enables us to understand the soundness of an axiom via the validity of its dL formula independently of the soundness of the mechanism that generalizes and replaces syntactic elements of the axioms with other concrete dL expressions. For example, the concrete instance (18.2) can be obtained from the concrete dL formula (18.3) of axiom $[\cup]$ by the uniform substitution

$$\sigma = \{\alpha \mapsto x:=x+1, \ \beta \mapsto x' = x^2, \ P \mapsto x \geq 0\}$$

This substitution σ substitutes HP $x:=x+1$ for α and HP $x' = x^2$ for β and dL formula $x \geq 0$ for P alias $p(\bar{x})$. Of course, this will require us to better understand the substitution process itself and the rôle of the HPs α and β. But let us first stay on the topic of how to interpret predicate symbols.

Unlike a formula such as $x^2 > 5$, which comes with a fixed interpretation of when it is *true*, namely exactly when the square of the value of x exceeds 5, a predicate symbol p does not have a fixed meaning, but is subject to our interpretation. That is what makes it a *symbol*, because it *stands for something*. Certainly, a predicate symbol can take different truth-values depending on its argument. So for example, depending on the value of its argument e, the formula $p(e)$ in axiom $[:=]$ will be *true* or *false*. But if two terms e and $\tilde{e}$ are evaluated to the same real value, then $p(e)$ and $p(\tilde{e})$ will, of course, either both be *true* consistently, or both be *false* consistently. Likewise, the predicate symbol p with 0 arguments in axiom V may be either *true* or *false*. But since it does not take any arguments at all, its truth-value does not depend on the values of any variables, so is independent of the state, and will either be *true* consistently everywhere or *false* consistently everywhere. Indeed, if the assumption

p of axiom V holds then the arity 0 predicate symbol p is *true*, which makes it *true* everywhere, even after running HP α, because its truth-value visibly does not depend on the values of any variables. If, instead p is *false*, then the assumption of axiom V is not met, so its implication is trivially *true*.

Function Symbols

Predicate symbols capture the different cases of formulas in dL axioms. Similarly, in the assignment axiom $[:=]$, the term e needs to be a concrete dL term, but one that can take on any value, because that is what a schema variable placeholder in the corresponding axiom schema $[:=]$ would be able to do. A function symbol with 0 arguments plays that rôle, because a function symbol can be evaluated to any real value, but will then have the same value in all states since it has no variables in its 0 arguments, just as predicate symbols can evaluate to any truth-value.

The following concrete dL formula can, then, be used as assignment axiom $[:=]$

$$[x:=c()]p(x) \leftrightarrow p(c()) \tag{18.4}$$

with predicate symbol p of arity 1 and function symbol $c()$ of arity 0. For example, the concrete instance

$$[x:=x^2-1]x \geq 0 \leftrightarrow x^2-1 \geq 0 \tag{18.5}$$

can be obtained from (18.4) by the uniform substitution

$$\sigma = \{c() \mapsto x^2-1,\ p(\cdot) \mapsto (\cdot \geq 0)\} \tag{18.6}$$

This substitution σ substitutes the term x^2-1 for the arity 0 function symbol $c()$ and substitutes the greater-or-equal zero comparison formula for the arity 1 predicate symbol p. To indicate that every occurrence of the predicate symbol p of any argument is affected and substituted with the corresponding ≥ 0 comparison, the substitution substitutes $p(\cdot)$ with a dL formula in which the dot $\cdot$ marks where the argument goes in the resulting dL formula. So for any argument e, the formula $p(e)$ will be replaced with $\sigma(e) \geq 0$. Of course, the substitution σ will also need to be applied to the argument e of $p(e)$, not just to the predicate symbol p, which is why $\sigma(e) \geq 0$ is substituted for $p(e)$ and not just $e \geq 0$ when forming (18.5) from (18.4) by (18.6). The result of applying the substitution σ to e is denoted $\sigma(e)$ and will be defined properly later.

Program Constant Symbols

Finally, we return to the rôle that the HPs α and β play in axiom $[\cup]$. On the one hand, both need to be concrete HPs for axiom $[\cup]$ to become a concrete dL formula. On the other hand, neither α nor β has a concrete specific behavior, because the axiom [∪] works for whatever HPs α and β do. Consequently, the HPs we use for α and β in axiom [∪] are what we call *program constant symbols* and can have any arbitrary behavior. Just as predicate symbols do not have a fixed interpretation but might be *true* of any argument, and just as function symbols f do not have a fixed interpretation but might have any real value as a function of the argument's value, so program constant symbols do not have a fixed interpretation but might have any arbitrary behavior. Depending on its interpretation, a program constant symbol might possibly transition from any initial state to any final state, because its behavior is not described explicitly as it would be in the case of a specific differential equation or discrete assignment.

18.4 Differential Dynamic Logic with Interpretations

After having realized what syntactic elements dL axioms need so that they can be faithfully represented as concrete axioms instead of axiom schemata, the first thing we do is officially add those elements to the syntax of differential dynamic logic [8]. Of course, we could have added them right away when introducing hybrid programs in Chap. 3 and differential dynamic logic in Chap. 4, but that would have been a distraction, because we did not need them until now.

18.4.1 Syntax

Differential dynamic logic dL is as usual, except that *function symbols*, *predicate symbols*, and *program constant symbols* are added. Function symbols are usually written f, g, h, predicate symbols p, q, r, and program constant symbols are written a, b, c. Each function and predicate symbol expects a fixed number of terms as arguments, called its *arity*. When f is a function symbol of arity n, then $f(e_1, \ldots, e_n)$ is now also allowed as a term for any n terms $e_1, \ldots, e_n$. Likewise, when p is a predicate symbol of arity n, then $p(e_1, \ldots, e_n)$ is now a formula for any n terms $e_1, \ldots, e_n$. But $f(e_1, \ldots, e_{n-1})$ is not a term, because the function symbol f of arity n has not even received sufficiently many arguments. When we have a function that can, say, add two numbers that we pass as arguments, then we cannot just call this function with one argument or with seven arguments, but need to provide exactly two.

Function symbols are essentially a more liberal generalization of built-in term operators, such as $+$, which has arity 2, is written infix as $e_1 + e_2$ instead of as $+(e_1, e_2)$, and always means addition. Function symbols can have a different number of arguments, but also always expect exactly the same number of arguments as indicated by their arity. Function symbols of arity 0 are also called *constant symbols*, because their value does not depend on any arguments. The use of a function symbol c of arity 0 is sometimes written as $c()$ with empty parentheses for emphasis. In fact, we already allowed rational numbers as constant symbols of arity 0 when originally defining terms. The meaning of a rational number constant is, of course, also fixed. The meaning of the rational number constant 1 is always 1 and the meaning of the rational number constant $1/2$ is always the real number 0.5.

By contrast, function symbols are more general, because they are actually meant as symbols. That is, they do not have a fixed meaning once and for all time, but are symbolic, so their meaning is subject to interpretation. Similarly, predicate symbols are symbols so their meaning depends on our interpretation, and likewise for program constant symbols.

Definition 18.1 (Terms). A *term e* is defined by augmenting the grammar from Definition 2.2 on p. 42 with the following case (where $e_1, \ldots, e_n$ are n terms and f is a function symbol of arity n):

$$e ::= f(e_1, \ldots, e_n) \mid \ldots$$

Definition 18.2 (Hybrid program). *Hybrid programs* are defined by augmenting the grammar from Definition 3.1 on p. 76 with the following case (where a is any program constant symbol):

$$\alpha, \beta ::= a \mid \ldots$$

Definition 18.3 (dL formula). The *formulas of differential dynamic logic* (dL) are defined by augmenting the grammar from Definition 4.1 on p. 111 with the following case (where $e_1, \ldots, e_n$ are terms and p is an arity n predicate symbol):

$$P ::= p(e_1, \ldots, e_n) \mid \ldots$$

For emphasis, we might call the resulting logic *differential dynamic logic with interpretations*, but continue to just call it dL, because we just neglected to consider these extensions until now, since they were not necessary for our understanding yet.

This extension of the syntax of dL makes it possible to phrase all axioms we saw before as axioms with concrete dL formulas (instead of as axiom schemata that represent their infinitely many instances subject to side conditions). For example, the axiom schemata we considered as motivating examples above turn into

$$[:=] \quad [x := c()]p(x) \leftrightarrow p(c())$$

$$[\cup] \quad [a \cup b]p(\bar{x}) \leftrightarrow [a]p(\bar{x}) \wedge [b]p(\bar{x})$$

$$V \quad p \rightarrow [a]p$$

18.4.2 Semantics

The semantics of function symbols, predicate symbols, and program constant symbols is actually easy, but it comes with a twist compared to all other definitions of semantics we saw anywhere else in this textbook. The whole point of function symbols, predicate symbols, and program constant symbols is that they are symbolic, so they do not come with a fixed interpretation. Consequently, unlike for the binary $+$ operator, which always means addition, the semantics of a term mentioning a function symbol f of arity 2 depends on how we interpret the symbol f, which may be addition or multiplication or any other reasonable function from two reals to a real.

In order to be able to evaluate to a real number any term in any state, we fix an *interpretation* I that assigns a (sufficiently smooth[4]) n-ary function $I(f) : \mathbb{R}^n \to \mathbb{R}$ to every function symbol f of arity n. Given such an interpretation I, we can easily evaluate every term in any state ω just by looking up in the interpretation I the corresponding function $I(f)$ for every function symbol f in the term and using the variable values from the state ω.

Definition 18.4 (Semantics of terms). The *value of term* e in state $\omega \in \mathscr{S}$ for interpretation I is a real number denoted $\omega[\![e]\!]$ and is defined by augmenting Definition 2.4 with the following case:

$$\omega[\![f(e_1,\ldots,e_n)]\!] = I(f)\big(\omega[\![e_1]\!],\ldots,\omega[\![e_n]\!]\big) \quad \text{if } f \text{ is a function symbol of arity } n$$

That is, in state ω, a function symbol application is evaluated to the result of the function $I(f)$ applied to the real values $\omega[\![e_i]\!]$ to which the respective argument terms e_i are evaluated in the state ω.

As predicate symbols have no fixed interpretation either, the interpretation I also assigns an n-ary relation $I(p) \subseteq \mathbb{R}^n$ to every predicate symbol p of arity n. With such an interpretation, the set of states in which a formula is true can be defined easily.

Definition 18.5 (dL semantics). The *semantics* of a dL formula P for interpretation I is the set of states $[\![P]\!] \subseteq \mathscr{S}$ in which P is true, and is defined by augmenting Definition 4.2 with the following case:

12. $[\![p(e_1,\ldots,e_n)]\!] = \big\{\omega \ : \ (\omega[\![e_1]\!],\ldots,\omega[\![e_n]\!]) \in I(p)\big\}$
   That is, a predicate symbol application is true in the set of states ω in which the arguments terms e_i are evaluated to a tuple of real numbers that is in the relation $I(p)$.

A formula P is *valid*, written $\vDash P$, iff it is true in all states of all interpretations I, i.e., $[\![P]\!] = \mathscr{S}$, so $\omega \in [\![P]\!]$ for all states ω and all interpretations I.

---

[4] Functions that are continuously differentiable are smooth enough for our purposes.

Finally, the interpretation I also assigns a reachability relation $I(a) \subseteq \mathscr{S} \times \mathscr{S}$ to every program constant symbol a. As usual, $(\omega, v) \in [\![a]\!]$ indicates that final state v is reachable from initial state ω in the HP a.

> **Definition 18.6 (Transition semantics of HPs).** Each HP α is interpreted semantically as a binary reachability relation $[\![\alpha]\!] \subseteq \mathscr{S} \times \mathscr{S}$ over states for each interpretation ω, and is defined by augmenting Definition 3.2 with the case:
>
> 7. $[\![a]\!] = I(a)$
>   That is, the reachability relation for program constant symbol a is an arbitrary state transition relation determined by the interpretation I.

With this extension of the semantics, it is now easy to see that the dL formula in the V axiom is valid. In fact, this is the easiest possible proof of the soundness of the vacuous axiom V (Lemma 5.11).

> **Lemma 18.1 (V vacuous axiom).** *The vacuous axiom is sound:*
>
> $$\text{V} \quad p \rightarrow [a]p$$

Proof. The truth of an arity 0 predicate symbol p just depends on the interpretation I but not on the state ω since p does not have any variables. Consequently, either p is interpreted to be *true* by I, in which case $[a]p$ is *true* as well, because if p holds in all states then it also holds in all states reachable after running HP a. Or p is interpreted to be *false* by I, in which case the assumption p is *false* and the implication $p \rightarrow [a]p$ is vacuously *true*. $\qquad\square$

Likewise the equivalence of the dL formula in the assignment axiom $[:=]$ is easily seen to be valid (Lemma 5.2).

> **Lemma 18.2 ($[:=]$ assignment axiom).** *The assignment axiom is sound:*
>
> $$[:=] \quad [x:=c()]p(x) \leftrightarrow p(c())$$

Proof. Predicate symbol p is *true* of x after assigning the new value $c()$ to x (so $[x:=c()]p(x)$) iff predicate symbol p is *true* of the new value $c()$ (so $p(c())$). $\qquad\square$

18.5 Uniform Substitution

A *uniform substitution* σ substitutes function symbols with terms, predicate symbols with formulas, and program constant symbols with hybrid programs, and it does so uniformly, e.g., it uses the same HP as replacement for program constant symbol b in all places.[5] The result of applying the uniform substitution σ to dL

---

[5] Replacing the same program constant symbol b with different HPs in different places would be very illogical and break all structure there ever was. Let's don't ever be so silly!

formula ϕ is denoted $\sigma(\phi)$. Similarly, $\sigma(\theta)$ denotes the result of applying the uniform substitution σ to term θ and $\sigma(\alpha)$ denotes the result of applying the uniform substitution σ to HP α. They will all be defined rigorously in Sect. 18.5.3.

The substitution σ defines a term σf as a replacement for each arity 0 function symbol f. The substitution σ also defines a dL formula σp as a replacement for each arity 0 predicate symbol p. It also defines a hybrid program σa for each program constant symbol a. Applying the substitution σ will replace every occurrence of program constant symbol a uniformly with the HP σa and every occurrence of arity 0 function symbol f with σf and every occurrence of arity 0 predicate symbol p with the corresponding replacement σp. For function and predicate symbols with arguments, the reserved function symbol $\cdot$ is used as a placeholder to indicate where the argument goes. For an arity 1 function symbol f, the substitution defines a term whose occurrences of function symbol $\cdot$ indicate where the argument of f is placed. For an arity 1 predicate symbol p, the substitution defines a dL formula whose occurrences of function symbol $\cdot$ indicate where the argument of p goes.

The notation for describing a uniform substitution σ that substitutes term e_1 for arity 1 function symbol f and substitutes term e_2 for arity 0 function symbol c, and that substitutes dL formula ϕ_1 for arity 1 predicate symbol p and substitutes dL formula ϕ_2 for arity 0 predicate symbol q and substitutes hybrid program α for program constant symbol a is

$$\sigma = \{f(\cdot) \mapsto e_1, c \mapsto e_2, p(\cdot) \mapsto \phi_1, q \mapsto \phi_2, a \mapsto \alpha\} \qquad (18.7)$$

The occurrences of reserved arity 0 function symbol $\cdot$ in the term e_1 and in the formula ϕ_1, respectively, indicate where the arguments of f and of p, respectively, go in the replacement. We have already seen examples of uniform substitutions in Sect. 18.3. The uniform substitution σ in (18.7) replaces arity 1 function symbol f and predicate symbol p, arity 0 function symbol c and predicate symbol q, and program constant symbol a but leaves all other symbols alone. The *domain* of substitution σ is the set of all symbols it replaces, so $\{f, c, p, q, a\}$ for (18.7).

18.5.1 Uniform Substitution Rule

Church's uniform substitution proof rule US says that the result $\sigma(\phi)$ of applying a uniform substitution σ to a valid formula ϕ is valid, too. Its generalization to differential dynamic logic is sound as well [8]. The intuition is that, if a formula ϕ is valid, so true in all states with any interpretation of its predicate, function, and program constant symbols, then it is also valid after substituting concrete formulas in for its predicate symbols, etc., because the predicate symbol very well may be interpreted to have the same truth-value as its substitute formula. The tricky part is the correct handling of arguments of the predicate symbols and of variables in the replacements, because variables may have different values in different subformulas.

Theorem 18.1 (Uniform substitution). *The proof rule US is sound:*

$$\text{US} \ \frac{\phi}{\sigma(\phi)}$$

So if formula ϕ has a proof, then its uniform substitution instance $\sigma(\phi)$ has a proof, too, just by applying the uniform substitution proof rule US. *The uniform substitution mechanism checks that it does not introduce a free variable in a context in which it is bound in $\sigma(\phi)$.* If the uniform substitution σ applied to ϕ were to introduce a free variable x into a context in which x has been bound, then $\sigma(\phi)$ is not defined, because it *clashes*, and the proof rule US is not applicable to ϕ.

Before proceeding with an exact definition of the uniform substitution mechanism constructing $\sigma(\phi)$ in Sect. 18.5.3, we explore a number of representative examples to gain intuition for the rôle of rule US in proving.

The formula $(\neg\neg p) \leftrightarrow p$, for example, is valid (in classical logic). When we pick any dL formula ψ, then also valid will be the formula that results from $(\neg\neg p) \leftrightarrow p$ by uniformly substituting all occurrences of arity 0 predicate symbol p with this formula ψ. For example, the uniform substitution $\sigma = \{p \mapsto [x' = x^2]x \geq 0\}$ proves

$$\text{US}\frac{(\neg\neg p) \leftrightarrow p}{(\neg\neg[x' = x^2]x \geq 0) \leftrightarrow [x' = x^2]x \geq 0}$$

Any other formula could have been used as a replacement for p (consistently everywhere) as well and rule US would have proved the result from $(\neg\neg p) \leftrightarrow p$.

Substitutions are more subtle when working, e.g., from the formula $(\forall x\, p) \leftrightarrow p$. This formula expresses for an arity 0 predicate symbol p that p is true for all x if and only if p is true in the current state, which makes apparent sense, because the arity 0 predicate symbol p quite visibly does not mention any variables that its truth-value would depend on. In fact, it is precisely this absence of the mention of x that the validity of the formula $(\forall x\, p) \leftrightarrow p$ depends on. We cannot possibly soundly replace p with $x \geq 0$, because that would lead to

$$\text{clash}\frac{(\forall x\, p) \leftrightarrow p}{\forall x\, (x \geq 0) \leftrightarrow x \geq 0}$$

which is unsound, because not all values of x are nonnegative (left) just because the present value of x is nonnegative (right) in the current state. Indeed, the uniform substitution mechanism will clash when applying $\sigma = \{p \mapsto x \geq 0\}$ to $(\forall x\, p) \leftrightarrow p$, because σ would introduce the free variable x in the replacement for p in a context $\forall x\, p$ in which x refers to a bound variable, such that the variable x in the replacements for the two occurrences of p would possibly refer to two different values. The requirement that the replacement for p does not have x as a free variable is quite consistent with our original reason why the premise $(\forall x\, p) \leftrightarrow p$ was valid at all.

Variables other than x can be mentioned free in the replacement for p, though, because they are not bound anywhere where p occurs. For example, the uniform substitution $\sigma = \{p \mapsto y \geq 0\}$ enables rule US to prove

$$\text{US} \frac{(\forall x\, p) \leftrightarrow p}{\forall x\, (y \geq 0) \leftrightarrow y \geq 0}$$

18.5.2 Examples

The primary, but not the only, use case of the uniform substitution proof rule US is that it makes it possible to instantiate axioms with specific dL formulas. The following examples will, thus, have an axiom as premise, which is proved in the dL calculus just by mentioning its name. The primary focus will be on demonstrating how uniform substitutions work, when they clash, and why that is soundness-critical.

How Uniform Substitutions Handle Arguments

Rule US proves, for example, (18.5) from (18.4) with uniform substitution (18.6):

$$\text{US} \frac{[x := c()]p(x) \leftrightarrow p(c())}{[x := x^2 - 1]x \geq 0 \leftrightarrow x^2 - 1 \geq 0}$$

Intuitively, this uniform substitution replaces all occurrences of function symbol $c()$ with $x^2 - 1$ while also replacing all occurrences of predicate symbol p with a greater-or-equal-to-zero comparison. Of course, in addition to substituting $(\cdot \geq 0)$ for $p(\cdot)$, the uniform substitution is also used on all arguments e of p in any subformula $p(e)$. So σ uniformly replaces every occurrence of $p(e)$ with $\sigma(e) \geq 0$. In particular, σ replaces $p(x)$ with $x \geq 0$ but $p(c())$ with $x^2 - 1 \geq 0$.

The uniform substitution $\sigma = \{c() \mapsto x^2 - 1, \ p(\cdot) \mapsto (\cdot \geq x)\}$, instead, clashes for the same formula, because the replacement for $p(\cdot)$ would introduce the free variable x in a context $[x := x^2 - 1]\_$ in which x is bound:

$$\text{clash} \notin \frac{[x := c()]p(x) \leftrightarrow p(c())}{[x := x^2 - 1]x \geq x \leftrightarrow x^2 - 1 \geq x} \tag{18.8}$$

It is crucial for soundness that this substitution clashes, because the premise is valid (axiom $[:=]$) but the conclusion is not, because the postcondition $x \geq x$ of the assignment is valid, but the right-hand side $x^2 - 1 \geq x$ is not. This makes sense, because all free occurrences of x in the postcondition are affected by the assignment $x := x^2 - 1$, so the substitution $\{p(\cdot) \mapsto (\cdot \geq x)\}$ does not select *all* occurrences of x for the $\cdot$ placeholder. In contrast, the uniform substitution $\sigma = \{c() \mapsto x^2 - 1, \ p(\cdot) \mapsto (\cdot \geq \cdot)\}$ gives the perfectly acceptable result

$$\text{US}\frac{[x:=c()]p(x) \leftrightarrow p(c())}{[x:=x^2-1]x \ge x \leftrightarrow x^2-1 \ge x^2-1}$$

Likewise, the uniform substitution $\sigma = \{c() \mapsto x^2 - 1, \ p(\cdot) \mapsto (2(\cdot) \ge \cdot)\}$ results in

$$\text{US}\frac{[x:=c()]p(x) \leftrightarrow p(c())}{[x:=x^2-1]2x \ge x \leftrightarrow 2(x^2-1) \ge x^2-1}$$

In comparison, the uniform substitution $\sigma = \{c() \mapsto x^2 - 1, \ p(\cdot) \mapsto (\cdot \ge y)\}$ is acceptable, because, even if the replacement for $p(\cdot)$ introduces the free variable y, it only introduces it in the context $[x:=x^2-1]\_$ in which y is not bound either:

$$\text{US}\frac{[x:=c()]p(x) \leftrightarrow p(c())}{[x:=x^2-1]x \ge y \leftrightarrow x^2-1 \ge y}$$

Observe how the explicit argument x in the subformula $p(x)$ of the premise makes it possible for the substitute $x \ge y$ to mention x instead of placeholder $\cdot$ in its replacement $(\cdot \ge y)$. But as (18.8) demonstrated, even such a mention of x as an argument does not give license to use variable x anywhere else in the replacements. Of course, the argument x in $p(x)$ only indicates an explicit license for a possible dependence on x, not that $p(x)$ has to depend on x. For example, this uniform substitution $\sigma = \{c() \mapsto 2x + 1, \ p(\cdot) \mapsto (y^2 \ge y)\}$ does not use the $\cdot$ argument placeholder:

$$\text{US}\frac{[x:=c()]p(x) \leftrightarrow p(c())}{[x:=2x+1]y^2 \ge y \leftrightarrow y^2 \ge y}$$

Uniform substitutions can also have predicates in which the argument placeholder $\cdot$ appears in more deeply nested positions. For example, the uniform substitution $\sigma = \{c() \mapsto x^2, \ p(\cdot) \mapsto [(y:=\cdot+y)^*](\cdot \ge y)\}$ is acceptable, because it does not introduce any free variables in a context in which they are bound:

$$\text{US}\frac{[x:=c()]p(x) \leftrightarrow p(c())}{[x:=x^2][(y:=x+y)^*](x \ge y) \leftrightarrow [(y:=x^2+y)^*](x^2 \ge y)}$$

How Uniform Substitutions Handle Constant Predicate Symbols

Without the original formula mentioning x as an argument in $p(x)$, the uniform substitution cannot use x in a context in which x is bound. For example, the uniform substitution $\sigma = \{a \mapsto x' = 5, \ p \mapsto (x \le 5)\}$ clashes, because the replacement for p introduces the free variable x into the context $[x' = 5]\_$ in which x is bound, which is the context that results from applying σ to $[a]p$:

$$\text{clash} \notin \frac{p \to [a]p}{x \le 5 \to [x'=5]x \le 5}$$

It is crucial for soundness that this substitution clashes, because the premise is valid (axiom V), but the conclusion is not, because x does not stay below 5 forever when following the differential equation $x' = 5$. This is precisely what the side condition of axiom schema V from Lemma 5.11 prevents, too. But unlike axiom schemata, rule US does not need special purpose knowledge about how to prevent such incorrect uses for the particular case of axiom V. It provides a generic mechanism.

In contrast, the uniform substitution $\sigma = \{a \mapsto x' = 5, \ p \mapsto (y \leq 5)\}$ works fine, because it only introduces free variable y in the resulting context $[x' = 5]\_$ in which y is not bound anyhow:

$$\text{US} \frac{p \to [a]p}{y \leq 5 \to [x' = 5]\, y \leq 5}$$

The uniform substitution $\sigma = \{a \mapsto (v := v + 1; \{x' = v, v' = -b\}), \ p \mapsto (y \leq b)\}$ works fine, because its function and predicate symbols only introduce free variables y and b in a context in which they are possibly read but never written:

$$\text{US} \frac{p \to [a]p}{y \leq b \to [v := v + 1; \{x' = v, v' = -b\}]\, y \leq b}$$

Telling the respective good and bad cases of axiom instantiation attempts apart is what uniform substitutions achieve without having to provide any side conditions that are specific to the particular formulas or axiom schemata at hand. Uniform substitutions provide a uniform answer, once and for all, to the question of which instantiations of formulas are sound because they preserve validity.

How Uniform Substitutions Handle Program Constant Symbols

When working from the assignment axiom $[:=]$ with its postcondition $p(x)$ or from the vacuous axiom V with its postcondition p, the uniform substitution rule US needs to check for capture of other variables, which is important for soundness. When working from the nondeterministic choice axiom $[\cup]$ with its postcondition $p(\bar{x})$, instead, then this postcondition has explicit permission to mention all variables $\bar{x}$, such that any dL formula can be accepted as replacement. The uniform substitution $\sigma = \{a \mapsto v := -cv, \ b \mapsto x'' = -g, \ p(\bar{x}) \mapsto 2gx \leq 2gH - v^2\}$ yields

$$\text{US} \frac{[a \cup b]p(\bar{x}) \leftrightarrow [a]p(\bar{x}) \wedge [b]p(\bar{x})}{[v := -cv \cup x'' = -g]2gx \leq 2gH - v^2 \leftrightarrow [v := -cv]2gx \leq 2gH - v^2 \wedge [x'' = -g]2gx \leq 2gH - v^2}$$

As usual, $x'' = -g$ is short for $\{x' = v, v' = -g\}$.

18.5.3 Uniform Substitution Application

A uniform substitution can replace any number of function, predicate, or program constant symbols simultaneously. The notation $\sigma f(\cdot)$ denotes the replacement for $f(\cdot)$ according to σ, i.e., the value $\sigma f(\cdot)$ of function σ at $f(\cdot)$. By contrast, $\sigma(\phi)$ denotes the result of applying σ to ϕ which we defined now (likewise for $\sigma(\theta)$ and $\sigma(\alpha)$). The notation $f \in \sigma$ signifies that σ replaces function symbol f, i.e., $\sigma f(\cdot) \neq f(\cdot)$, so f is in the domain of σ. Likewise, the notation $p \in \sigma$ signifies that σ replaces predicate symbol p, and correspondingly $a \in \sigma$ means that σ replaces program constant symbol a.

$$\sigma(x) = x \qquad \text{for variable } x \in \mathcal{V}$$
$$\sigma(f(e)) = (\sigma(f))(\sigma(e)) \stackrel{\text{def}}{=} \{\cdot \mapsto \sigma(e)\}(\sigma f(\cdot)) \quad \text{for function symbol } f \in \sigma$$
$$\sigma(g(e)) = g(\sigma(e)) \qquad \text{for function symbol } g \notin \sigma$$
$$\sigma(e + \tilde{e}) = \sigma(e) + \sigma(\tilde{e})$$
$$\sigma(e \cdot \tilde{e}) = \sigma(e) \cdot \sigma(\tilde{e})$$
$$\sigma((e)') = (\sigma(e))' \qquad \text{if } \sigma \text{ is } \mathcal{V}\text{-admissible for } e$$
$$\sigma(e \geq \tilde{e}) \equiv \sigma(e) \geq \sigma(\tilde{e}) \qquad \text{likewise for } >, =, <, \leq$$
$$\sigma(p(e)) \equiv (\sigma(p))(\sigma(e)) \stackrel{\text{def}}{=} \{\cdot \mapsto \sigma(e)\}(\sigma p(\cdot)) \quad \text{for predicate symbol } p \in \sigma$$
$$\sigma(q(e)) \equiv q(\sigma(e)) \qquad \text{for predicate symbol } q \notin \sigma$$
$$\sigma(\neg \phi) \equiv \neg \sigma(\phi)$$
$$\sigma(\phi \wedge \psi) \equiv \sigma(\phi) \wedge \sigma(\psi) \qquad \text{likewise for } \vee, \rightarrow, \leftrightarrow$$
$$\sigma(\forall x \phi) \equiv \forall x \sigma(\phi) \qquad \text{if } \sigma \text{ is } \{x\}\text{-admissible for } \phi$$
$$\sigma(\exists x \phi) \equiv \exists x \sigma(\phi) \qquad \text{if } \sigma \text{ is } \{x\}\text{-admissible for } \phi$$
$$\sigma([\alpha]\phi) \equiv [\sigma(\alpha)]\sigma(\phi) \qquad \text{if } \sigma \text{ is } \mathrm{BV}(\sigma(\alpha))\text{-admissible for } \phi$$
$$\sigma(\langle\alpha\rangle\phi) \equiv \langle\sigma(\alpha)\rangle\sigma(\phi) \qquad \text{if } \sigma \text{ is } \mathrm{BV}(\sigma(\alpha))\text{-admissible for } \phi$$
$$\sigma(a) \equiv \sigma a \qquad \text{for program constant symbol } a \in \sigma$$
$$\sigma(b) \equiv b \qquad \text{for program constant symbol } b \notin \sigma$$
$$\sigma(x := e) \equiv x := \sigma(e)$$
$$\sigma(x' = e \,\&\, Q) \equiv x' = \sigma(e) \,\&\, \sigma(Q) \qquad \text{if } \sigma \text{ is } \{x, x'\}\text{-admissible for } e, Q$$
$$\sigma(?Q) \equiv ?\sigma(Q)$$
$$\sigma(\alpha \cup \beta) \equiv \sigma(\alpha) \cup \sigma(\beta)$$
$$\sigma(\alpha; \beta) \equiv \sigma(\alpha); \sigma(\beta) \qquad \text{if } \sigma \text{ is } \mathrm{BV}(\sigma(\alpha))\text{-admissible for } \beta$$
$$\sigma(\alpha^*) \equiv (\sigma(\alpha))^* \qquad \text{if } \sigma \text{ is } \mathrm{BV}(\sigma(\alpha))\text{-admissible for } \alpha$$

Fig. 18.1 Recursive application of uniform substitution σ

Figure 18.1 defines the result $\sigma(\phi)$ of applying to a dL formula ϕ the *uniform substitution* σ that uniformly replaces all occurrences of a function f by a term (instantiated with its respective argument of f) and all occurrences of a predicate p by a formula (instantiated with its argument) as well as of a program constant symbol a by a program. Each case in Fig. 18.1 applies the uniform substitution recursively.[6] In each case, the uniform substitution application mechanism checks that the substitution is admissible for the bound variables of the operator, i.e., σ will not introduce

---

[6] This makes the uniform substitution a *homomorphism*, because the substitution of an addition is the addition of substitutions: $\sigma(e + \tilde{e}) = \sigma(e) + \sigma(\tilde{e})$ and accordingly for all other operators.

free variables in the scope of an operator in which they are bound (which will be defined in Definition 18.7 below).

For example, for the case $\sigma(\forall x\,\phi)$, the set of bound variables that σ needs to be admissible for ϕ is $\{x\}$, because if σ were to introduce free variable x while forming $\sigma(\phi)$, then x would be incorrectly captured by quantifier $\forall x$. Suppose the substitution σ were to replace an arity 0 predicate symbol p that occurs in ϕ with the formula $x \geq 0$; then within the scope of the quantifier of $\forall x\,\phi$, this formula $x \geq 0$ refers to a different variable called x, namely the one bound by the universal quantifier $\forall x$ and no longer the free variable x. That is why such a uniform substitution is not defined, because it is not admissible. This is crucial for soundness, e.g., for the formula $p \leftrightarrow \forall x\,p$, because the substitution would otherwise replace p inconsistently with the same formula $x \geq 0$ but referring to different values of x in different places, since one of the newly introduced occurrences of x in the resulting $x \geq 0 \leftrightarrow \forall x\,(x \geq 0)$ is in the scope of a quantifier binding x. In the case of a modal formula $\sigma([\alpha]\phi)$, the bound variables that are taboo and cannot be introduced as free variables when forming the substituted postcondition $\sigma(\phi)$ are the bound variables $\mathrm{BV}(\sigma(\alpha))$ of the substituted HP $\sigma(\alpha)$. In the case of a differential equation $\sigma(x' = e\,\&\,Q)$, the bound variables $\{x, x'\}$ are taboo and cannot be introduced as free variables when forming $\sigma(e)$ or $\sigma(Q)$, since the differential equation changes the values of both.

Arguments are put in for the placeholder $\cdot$ recursively by uniform substitution $\{\cdot \mapsto \sigma(\theta)\}$ in Fig. 18.1, which is well-defined since it replaces the placeholder function symbol $\cdot$ of arity 0 by the readily substituted argument $\sigma(\theta)$. Recall the definition of the free variables $\mathrm{FV}(P)$ as well as the bound variables $\mathrm{BV}(P)$ of formula P from Sects. 5.6.5 and 5.6.6.

Definition 18.7 (Admissible uniform substitution). A uniform substitution σ is *U-admissible for formula* ϕ (or term θ or HP α, respectively) with respect to the variables $U \subseteq \mathcal{V}$ iff $\mathrm{FV}(\sigma|_{\Sigma(\phi)}) \cap U = \emptyset$, where $\sigma|_{\Sigma(\phi)}$ is the restriction of substitution σ that only replaces symbols that occur in ϕ, and $\mathrm{FV}(\sigma) = \bigcup_{f \in \sigma} \mathrm{FV}(\sigma f(\cdot)) \cup \bigcup_{p \in \sigma} \mathrm{FV}(\sigma p(\cdot))$ is the set of *free variables* that σ introduces for function or predicate symbols.

A uniform substitution σ is *admissible for* ϕ (or θ or α, respectively) iff the bound variables U of each operator of ϕ are not free in the substitution on its arguments, i.e., σ is U-admissible. These admissibility conditions are listed explicitly in Fig. 18.1, which defines the result $\sigma(\phi)$ of applying σ to ϕ. For each case in Fig. 18.1, the taboo set U whose U-admissibility is required of σ is exactly the set of variables that are bound by its top-level operator.

The substitution σ is said to *clash* and its result $\sigma(\phi)$ (or $\sigma(\theta)$ or $\sigma(\alpha)$) is not defined if σ is not admissible, in which case rule US is not applicable either. All the admissibility conditions in Fig. 18.1 are easily summarized:

If you bind a free variable, you go to logic jail!

Note that the free variables $\mathrm{FV}(\sigma)$ of a substitution σ are only defined as the union of the free variables of the replacements for its function symbols f and pred-

18 Axioms & Uniform Substitutions

icate symbols p, not the program constant symbols, because programs may already read the full state and change it to a new state. Likewise, replacements of predicate symbols $p(\bar{x})$ with all variables $\bar{x}$ as arguments are disregarded in the free variable determination, because they apparently already have explicit permission to depend on the values of all variables and, thus, do not introduce any new free variables.

Finally, observe that σ is already U-admissible for formula ϕ if the sufficient condition $\mathrm{FV}(\sigma) \cap U = \emptyset$ holds. The only reason for Definition 18.7 to restrict the admissibility check to the restriction $\sigma|_{\Sigma(\phi)}$ of the substitution to the symbols that actually occur in the affected formula ϕ is that there is no need for the substitution to clash if σ introduces free variables for function or predicate symbols that do not even occur in ϕ. For example, $\sigma = \{p(\cdot) \mapsto (\cdot \leq y), q \mapsto (x \leq 5)\}$ is $\{x\}$-admissible for $\phi \stackrel{\text{def}}{=} (x > 2 \wedge p(y))$, because the dangerous predicate symbol q with its free variable x that would not be $\{x\}$-admissible does not even occur in ϕ, so the substitution is restricted to $\sigma|_{\Sigma(\phi)} = \{p(\cdot) \mapsto (\cdot \leq y)\}$, whose only free variable is y. Neither the original substitution σ nor its restriction $\sigma|_{\Sigma(\phi)}$ are $\{y\}$-admissible for ϕ, because both have y as a free variable. The original substitution σ also would not be $\{x\}$-admissible for $\psi \stackrel{\text{def}}{=} (x > 2 \wedge p(y) \wedge q)$, because its replacement for the predicate symbol q that occurs in ψ has x as a free variable.

For example, this uniform substitution $\sigma = \{a \mapsto x' = 5,\ p \mapsto (y \leq 5)\}$ succeeds:

$$\text{US} \frac{p \to [a]p}{y \leq 5 \to [x' = 5]y \leq 5}$$

It uses the uniform substitution mechanism in Fig. 18.1 and also $y \notin \mathrm{BV}(x' = 5)$:

$$\sigma(p \to [a]p) \equiv \sigma(p) \to \sigma([a]p) \equiv \sigma(p) \to [\sigma(a)]\sigma(p)$$
$$\equiv \sigma p \to [\sigma a]\sigma p \equiv y \leq 5 \to [x' = 5]y \leq 5$$

In addition to the previous examples, we consider a few very insightful ones. The uniform substitution $\sigma = \{p(\cdot) \mapsto (\cdot \geq 0),\ q \mapsto (y < 0)\}$ works fine, because it only introduces free variable y in the context $\forall x\_$ in which y is not bound:

$$\text{US} \frac{\forall x\,(p(x) \vee q) \leftrightarrow (\forall x\,p(x)) \vee q}{\forall x\,(x \geq 0 \vee y < 0) \leftrightarrow (\forall x\,(x \geq 0)) \vee y < 0}$$

The application of uniform substitution according to Fig. 18.1 is straightforward:

$$\sigma(\forall x\,(p(x) \vee q) \leftrightarrow (\forall x\,p(x)) \vee q) \equiv \sigma(\forall x\,(p(x) \vee q)) \leftrightarrow \sigma((\forall x\,p(x)) \vee q)$$
$$\equiv \forall x\,(\sigma(p(x) \vee q)) \leftrightarrow \sigma(\forall x\,p(x)) \vee \sigma(q) \equiv \forall x\,(\sigma(p(x)) \vee \sigma(q)) \leftrightarrow \forall x\,\sigma(p(x)) \vee \sigma(q)$$
$$\equiv \forall x\,(x \geq 0 \vee y < 0) \leftrightarrow (\forall x\,(x \geq 0)) \vee y < 0$$

This substitution uses that x is not free in the replacement for q.

In contrast, uniform substitution $\sigma = \{p(\cdot) \mapsto (\cdot \geq 0),\ q \mapsto (x < 0)\}$ clashes as its replacement for q introduces free variable x in a context $\forall x\_$ in which it is bound:

$$\text{clash}_\notin \frac{\forall x\,(p(x) \vee q) \leftrightarrow (\forall x\, p(x)) \vee q}{\forall x\,(x \geq 0 \vee x < 0) \leftrightarrow (\forall x\,(x \geq 0)) \vee x < 0}$$

This is soundness-critical, because the left formula is valid (every number is either greater-or-equal or smaller than 0) but the right formula is not, because it is equivalent to $x < 0$, which imposes a condition on the present value of x. The uniform substitution application $\sigma(\forall x\,(p(x) \vee q))$ clashes, because σ is not $\{x\}$-admissible for $p(x) \vee q$ on account of the replacement $x < 0$ for q having free variable x which would already be bound by the $\forall x$ quantifier. Of course, this makes sense, because a disjunction can only be pulled outside the scope of a quantifier if it does not actually use the quantified variable. That is precisely what the premise expresses. Indeed, the premise can be proved from other quantifier axioms.

Observe that it is crucial for soundness that even an occurrence of $p(x)$ in a context where x is bound does not permit free variable x to be mentioned in the replacement except in the places of the $\cdot$ placeholder. For example, uniform substitution $\sigma = \{c() \mapsto 0,\ p(\cdot) \mapsto (\cdot \geq x)\}$ clashes when used on the assignment axiom $[:=]$, because the replacement for $p(\cdot)$ would introduce the extra free variable x in a context $[x:=0]\_$ in which x is bound:

$$\text{clash}_\notin \frac{[x:=c()]p(x) \leftrightarrow p(c())}{[x:=0]x \geq x \leftrightarrow 0 \geq x}$$

The premise is valid (axiom $[:=]$) but the conclusion is not, because the postcondition $x \geq x$ of the assignment is valid, but the right-hand side $0 \geq x$ is not. The reason is that the free variable x in the replacement $(\cdot \geq x)$ for $p(\cdot)$ would refer to the variable bound by $x:=0$ in the substitute for $p(x)$ but would refer to a free variable x in the substitute for $p(c())$.

Uniform substitutions also need to pay attention when substituting in the argument. For example, $\sigma = \{c() \mapsto y^2,\ p(\cdot) \mapsto [(y:=\cdot+y)^*](\cdot \geq y)\}$ clashes when applied to the assignment axiom $[:=]$ while substituting the replacement y^2 for $c()$ for the argument placeholder $\cdot$ in the replacement for $p(c())$, since that would introduce free variable y into a context $[(y:=\cdot+y)^*](\cdot \geq y)$ where it is bound:

$$\text{clash}_\notin \frac{[x:=c()]p(x) \leftrightarrow p(c())}{[x:=y^2][(y:=x+y)^*](x \geq y) \leftrightarrow [(y:=y^2+y)^*](y^2 \geq y)}$$

This is, of course, crucial for soundness because the left loop always adds to y the *same* value in each round (the square of the initial value of y) while the right loop, instead, always adds to y the square of the *most recent* value of y.

18.5.4 Uniform Substitution Lemmas

The key to understanding why the rule US is sound is the uniform substitution lemma that relates the syntactic change that a uniform substitution makes to a corresponding semantic reinterpretation called *adjoint interpretation*. The idea is that instead of syntactically replacing a predicate symbol p with another formula when forming the result $\sigma(\phi)$ of a uniform substitution σ, one might just as well modify the interpretation of the predicate symbol p. The uniform substitute $\sigma(\phi)$ of a formula is true in state ω in an interpretation I iff the formula ϕ itself is true in ω in its adjoint interpretation $\sigma_\omega^* I$. The semantic modification of adjoint interpretations has the same effect as the syntactic uniform substitution but on the semantics.

For example, recall that to prove (18.5) from (18.4) we used US with substitution

$$\sigma = \{c() \mapsto x^2 - 1, \ p(\cdot) \mapsto (\cdot \geq 0)\} \qquad (18.6^*)$$

$$\text{US} \frac{[x := c()]p(x) \leftrightarrow p(c())}{[x := x^2 - 1]x \geq 0 \leftrightarrow x^2 - 1 \geq 0}$$

Instead of syntactically substituting $(\cdot \geq 0)$ for $p(\cdot)$ everywhere, we could have reinterpreted predicate symbol p in a different way, namely such that $\sigma_\omega^* I(p)$ holds true iff its argument is greater-or-equal 0. And instead of syntactically substituting $x^2 - 1$ for $c()$ everywhere, we could have reinterpreted function symbol $c()$ such that $\sigma_\omega^* I(c())$ has the value that $x^2 - 1$ has in state ω. In the so-modified adjoint interpretation $\sigma_\omega^* I$ the original $[x := c()]p(x) \leftrightarrow p(c())$ now has exactly the same meaning that the substituted formula $[x := x^2 - 1]x \geq 0 \leftrightarrow x^2 - 1 \geq 0$ has in I.

Since the exact details of this construction are inconsequential for the purposes of this textbook, we refer to previous work [8] for a precise construction of the adjoint interpretation $\sigma_\omega^* I$ for I, ω in this way. The only important point is that adjoint interpretations enable the following uniform substitution lemma, whose proof can be found in previous work [8, Lemma 24].

> **Lemma 18.3 (Uniform substitution for formulas).** *The uniform substitution σ and its adjoint interpretation $\sigma_\omega^* I$ for I, ω have the same semantics for all formulas ϕ:*
> $$\omega \in I[\![\sigma(\phi)]\!] \text{ iff } \omega \in \sigma_\omega^* I[\![\phi]\!]$$

18.5.5 Soundness

Equipped with the uniform substitution lemma, which equates the semantics of a uniform substitute with the semantics of the original in an adjoint interpretation, it is now easy to establish the soundness of proof rule US (Theorem 18.1). Of course, the uniform substitution proof rule US is only applicable if its uniform substitution is defined, so respects its admissibility conditions.

Theorem 18.1 (Uniform substitution). *The proof rule US is sound:*

$$\text{US} \quad \frac{\phi}{\sigma(\phi)}$$

Proof. The proof [8] uses that truth of the substituted formula is equivalent to truth of the original formula in the adjoint interpretation to conclude that validity of the premise in all interpretations implies validity in the adjoint interpretation so validity of the conclusion. Let the premise ϕ of rule US be valid, i.e., $\omega \in I[\![\phi]\!]$ for all states ω and for all interpretations I of the program, predicate, and function symbols. To show that the conclusion is valid, consider any state ω and any interpretation I and show that $\omega \in I[\![\sigma(\phi)]\!]$. By Lemma 18.3, the uniformly substituted formula $\sigma(\phi)$ is true in state ω of interpretation I iff the original formula ϕ is true in state ω of the adjoint interpretation $\sigma_\omega^* I$ that has already been modified according to the substitution σ, that is $\omega \in I[\![\sigma(\phi)]\!]$ iff $\omega \in \sigma_\omega^* I[\![\phi]\!]$. Now $\omega \in \sigma_\omega^* I[\![\phi]\!]$ holds, because $\omega \in I[\![\phi]\!]$ for all states ω and interpretations I, including for state ω and interpretation $\sigma_\omega^* I$, by premise. $\qquad\square$

The other missing ingredient for the uniform substitution proof rule US is the exact definition of the free and bound variables, which is needed in the definition of admissibility (Definition 18.7). Those were already reported in Sect. 5.6.6. The only addition is the definition of free and bound variables for the newly added function and predicate symbols as well as program constant symbols. For function and predicate symbols this is just a matter of asking the argument terms:

$$FV(f(e_1,\ldots,e_k)) = FV(e_1) \cup \cdots \cup FV(e_k)$$
$$FV(p(e_1,\ldots,e_k)) = FV(e_1) \cup \cdots \cup FV(e_k)$$
$$BV(p(e_1,\ldots,e_k)) = \emptyset$$

The interpretations of a program constant symbol a can read and write any variable from the set $\mathcal{V}$ of all variables but is not guaranteed to write any particular variable so has no must-bound variables:

$$FV(a) = \mathcal{V}$$
$$BV(a) = \mathcal{V}$$
$$MBV(a) = \emptyset$$

18.6 Axiomatic Proof Calculus for dL

A purely axiomatic formulation of the differential dynamic logic axiomatization [8] is shown in Fig. 18.2. The axioms listed in Fig. 18.2 are axioms, so concrete dL for-

$$\boxed{\begin{array}{ll}
\text{[:=]} \;\; [x:=c()]p(x) \leftrightarrow p(c()) & \qquad \text{G} \;\; \dfrac{p(\bar{x})}{[a]p(\bar{x})} \\[2.5ex]
\text{[?]} \;\; [?q]p \leftrightarrow (q \to p) & \qquad \forall \;\; \dfrac{p(x)}{\forall x\, p(x)} \\[2.5ex]
\text{[∪]} \;\; [a \cup b]p(\bar{x}) \leftrightarrow [a]p(\bar{x}) \wedge [b]p(\bar{x}) & \\[1ex]
\text{[;]} \;\; [a;b]p(\bar{x}) \leftrightarrow [a][b]p(\bar{x}) & \qquad \text{MP} \;\; \dfrac{p \to q \quad p}{q} \\[2.5ex]
\text{[*]} \;\; [a^*]p(\bar{x}) \leftrightarrow p(\bar{x}) \wedge [a][a^*]p(\bar{x}) & \\[1ex]
\langle \cdot \rangle \;\; \langle a \rangle p(\bar{x}) \leftrightarrow \neg [a]\neg p(\bar{x}) & \\[1ex]
\text{K} \;\; [a](p(\bar{x}) \to q(\bar{x})) \to ([a]p(\bar{x}) \to [a]q(\bar{x})) & \\[1ex]
\text{I} \;\; [a^*]p(\bar{x}) \leftrightarrow p(\bar{x}) \wedge [a^*](p(\bar{x}) \to [a]p(\bar{x})) & \\[1ex]
\text{V} \;\; p \to [a]p &
\end{array}}$$

Fig. 18.2 Differential dynamic logic axioms and proof rules

mulas, and not axiom schemata that stand for an infinite collection of formulas. The axioms are *sound*, i.e., valid dL formulas. Besides the cases we already discussed so far, these axioms are formed by using program constant symbols a and b as concrete hybrid programs and by using $p(\bar{x})$ as a concrete formula for the postconditions that have no admissibility requirement. During uniform substitution with rule US, those program constant symbols a and b and the formulas $p(\bar{x})$ and $q(\bar{x})$ in the axioms can, in turn, be substituted with arbitrary HPs and dL formulas, respectively.

The only exception is the test axiom [?], which might have been phrased as either of the following two dL formulas:

$$[?q]p \leftrightarrow (q \to p) \tag{18.9}$$
$$[?q(\bar{x})]p(\bar{x}) \leftrightarrow (q(\bar{x}) \to p(\bar{x})) \tag{18.10}$$

It looks as if the second formulation (18.10) would be more flexible, because its explicit mention of the list of all variables in $p(\bar{x})$ and $q(\bar{x})$ make it obvious that the axiom can be instantiated with any arbitrary dL formulas for the test $?q(\bar{x})$ and postcondition $p(\bar{x})$. However, the first formulation (18.9) is sufficient, because any arbitrary dL formulas can already be substituted in for the arity 0 predicate symbols p and q as well, since no variables are bound anywhere in (18.10), so its intersection with any arbitrary set of free variables of any substitution will always be empty.

Soundness of the axioms and proof rules in Fig. 18.2 follows from soundness of the corresponding axiom schemata and proof rule schemata in Chap. 5 and Chap. 7 from Part I of this textbook. The concrete axioms in Fig. 18.2 are instances of the previous axiom schemata, even if their soundness would have been easier to prove directly [8]. Implementing the axioms of Fig. 18.2 in a theorem prover is now straightforward, because each axiom is just a single concrete dL formula that the prover needs to remember. It can be shown that the uniform substitution proof rule US can prove all instances of these axioms that are required for completeness [8].

18.7 Differential Axioms

The axiomatic approach discussed in this chapter is not limited to logically internalizing the CPS reasoning principles from Part I but works equally well elsewhere, including the proof principles for differential equations from Part II. The key ingredient enabling such an approach for differential equations is the differential forms that we have already gotten to know in Part II. A purely axiomatic formulation of the differential equation axioms and axioms for differentials of dL [8] is shown in Fig. 18.3. These axioms are special instances of the axiom schemata from Part II, which explains their soundness.

DW $[x' = f(x)\,\&\,q(x)]p(x) \leftrightarrow [x' = f(x)\,\&\,q(x)](q(x) \rightarrow p(x))$

DI $\big([x' = f(x)\,\&\,q(x)]p(x) \leftrightarrow [?q(x)]p(x)\big) \leftarrow \big(q(x) \rightarrow [x' = f(x)\,\&\,q(x)](p(x))'\big)$

DC $\big([x' = f(x)\,\&\,q(x)]p(x) \leftrightarrow [x' = f(x)\,\&\,q(x) \wedge r(x)]p(x)\big) \leftarrow [x' = f(x)\,\&\,q(x)]r(x)$

DE $[x' = f(x)\,\&\,q(x)]p(\bar{x}) \leftrightarrow [x' = f(x)\,\&\,q(x)][x' := f(x)]p(\bar{x})$

DG $[x' = f(x)\,\&\,q(x)]p(x) \leftrightarrow \exists y\,[x' = f(x), y' = a(x) \cdot y + b(x)\,\&\,q(x)]p(x)$

DS $[x' = c()\,\&\,q(x)]p(x) \leftrightarrow \forall t{\geq}0\,\big((\forall 0{\leq}s{\leq}t\,q(x+c()s)) \rightarrow [x := x+c()t]p(x)\big)$

$+'$ $(f(\bar{x}) + g(\bar{x}))' = (f(\bar{x}))' + (g(\bar{x}))'$

$-'$ $(f(\bar{x}) - g(\bar{x}))' = (f(\bar{x}))' - (g(\bar{x}))'$

$\cdot'$ $(f(\bar{x}) \cdot g(\bar{x}))' = (f(\bar{x}))' \cdot g(\bar{x}) + f(\bar{x}) \cdot (g(\bar{x}))'$

$/'$ $(f(\bar{x})/g(\bar{x}))' = \big((f(\bar{x}))' \cdot g(\bar{x}) - f(\bar{x}) \cdot (g(\bar{x}))'\big)/g(\bar{x})^2$

c' $(c())' = 0$ (for numbers or constants $c()$)

x' $(x)' = x'$ (for variable $x \in \mathcal{V}$)

Fig. 18.3 Differential equation axioms and differential axioms

The structural advantages of uniform substitutions are exploited in the axioms listed in Fig. 18.3. Since the arity 1 function symbols a and b in the differential ghost axiom DG receive argument x, their respective replacements also have special permission to depend on x. Their uniform substitution replacements can, thus, also have free variable x and any other variable but *not* the new differential ghost y, because y is bound by $y' = a(x) \cdot y + b(x)$. The replacements for $a(x)$ and $b(x)$ in axiom DG can, thus, overall mention any variable other than the differential ghost y. It is crucial for soundness (Chap. 12) that the replacements for $a(x)$ and $b(x)$ do not have free variable y, because the new differential equation $y' = a(x) \cdot y + b(x)$ is not otherwise guaranteed to have a solution of sufficient duration when $y' = a(x) \cdot y + b(x)$ is not actually linear since the replacements for $a(x)$ or $b(x)$ secretly depend on y. Unlike for another variable z, the axiom DG needs to provide special permission in the form $a(x)$ and $b(x)$ to depend on x, because x is bound by $x' = f(x)$.

Observe how much easier it is to establish the soundness of the concrete axiom DG with its concrete mentions of free variables compared to establishing what precise relationships of variable occurrences are soundly acceptable in the schematic instances of DG. The uniform substitution mechanism in the form of rule US takes care of these generalization and instantiation questions once and for all, as opposed to on a case-by-case basis for each axiom schema again.

Looking through the axioms in Fig. 18.3, it is also important that x' is not free in the postcondition $p(x)$ of the differential invariant axiom DI, because x' is guaranteed to equal $f(x)$ in $[x' = f(x) \& q(x)]p(x)$ but not in $[?q(x)]p(x)$. Indeed, uniform substitution maintains this during instantiation, because x' is bound by $x' = f(x) \& q(x)$ so cannot occur in the replacements for the postcondition $p(x)$ without special permission. This is unlike for axiom DW, where the postcondition $p(x)$ also disallows a mention of x' for simplicity even if it would have been perfectly sound, because all occurrences of $p(x)$ are in the scope of $[x' = f(x) \& q(x)]$.

Similarly, it is important for the solution axiom DS to not have x' in the postcondition $p(x)$, because otherwise an additional assignment $[x' := c()]p(x)$ would be needed instead of $p(x)$ on the right-hand side to propagate the effect that the differential equation $x' = c() \& q(x)$ has on x'. Of course, it is even more important for the replacement of the arity 0 constant symbol $c()$ in the differential equation to not have x as a free variable, because $x + c()t$ would not otherwise be the correct solution of the differential equation $x' = c()$. The constant differential equation axiom DS is weaker than the full solution axiom schema $[']$, because it only works for differential equations with constant (symbolic) right-hand side. But axiom DS can be used along with a differential ghost DG to introduce time $t' = 1$ and with differential cuts DC to introduce and then prove by DI the solutions of other solvable differential equations [8] similar to the approach discussed in Chap. 12.

The arity 0 function symbol $c()$ in axiom c' cannot be substituted with formulas that mention variables, because, similarly to a quantifier, the differential operator $(\dots)'$ does not accept the introduction of variables. The reason why the differential operator $(\dots)'$ does not allow any new variables to be introduced during uniform substitution is that the value of $(xy)'$ equals the value of $x'y + xy'$ and depends on x, x', y, y', which is why it is important to know all free variables of any $(\dots)'$ term.

In particular, the arity 0 function symbol $c()$ in axiom c' can be replaced by constant terms like $5 \cdot 2$ or $5 + b()$ for an arity 0 (constant) function symbol such as $b()$ for braking force, but not by a term like $5 + x$ with a new variable whose differential would indeed depend on the value of variables x and x'.

This is to be contrasted with axiom $+'$ whose occurrence of $f(\bar{x})$ and $g(\bar{x})$ can be replaced by any arbitrary terms, because they already mention all variables $\bar{x}$, so no new variables remain to be introduced during uniform substitution. Indeed, differentials act as specified in axioms $+', -', \cdot', /'$ on the arithmetic operations for any terms $f(\bar{x}), g(\bar{x})$, but the differential 0 as specified in axiom c' only applies to terms that are actually constant and cannot have any free variables. Uniform substitutions make it very easy to distinguish between the two cases just by the syntactic expressions used in the respective concrete axiom formulas.

18.8 Summary

The main insight of this chapter is that uniform substitutions provide a simple and modular way of implementing differential dynamic logic reasoning for hybrid systems. Based on a straightforward recursive implementation of uniform substitution, the soundness-critical part of a theorem prover reduces to mere copy-and-paste of the concrete formulas adopted as axioms. The resulting proof calculus continues to be sound and complete relative to any differentially expressive logic [8], including first-order logic of differential equations [5, 6] and discrete dynamic logic [6].

> **Theorem 18.2 (Axiomatization of dL).** *The uniform substitution* dL *calculus listed in Figs. 18.2 and 18.3 is a* sound and complete axiomatization *of hybrid systems relative to* any *differentially expressive logic L, i.e., every valid* dL *formula is provable in the* dL *calculus from L tautologies.*

This succinct approach explains the small soundness-critical core of the uniform substitution prover KeYmaera X [3] and why it was relatively easy to cross-verify it [1] both in Isabelle/HOL [4] and in Coq [10]. In fact, with a minor generalization of the set of symbols that uniform substitutions can instantiate, it is easy to derive the contextual equivalence rewriting rules (Lemma 6.2) from uniform substitution [8] as well, which are featured prominently to apply axioms in context. The Appendix explores that all other proof rules of dL are not schematic but *axiomatic proof rules* consisting of concrete dL formulas instantiated by uniform substitutions [8].

18.9 Appendix: Uniform Substitution of Rules and Proofs

Uniform substitutions are not limited to be used on axioms, but can also be used on proof rules

$$\frac{\phi_1 \quad \cdots \quad \phi_n}{\psi}$$

or entire proofs that conclude ψ from the premises ϕ_1 to ϕ_n (likewise for sequents). We just need to use the same uniform substitution on the premises that we use for the conclusion. By Lemma 18.3, using the same uniform substitution everywhere semantically corresponds to fixing and using the same interpretation I everywhere.

An inference or proof rule is *locally sound* iff its conclusion is valid in any interpretation I in which all its premises are valid. All locally sound proof rules are sound, because if all premises are valid in all interpretations, then local soundness makes the conclusion valid in each of the interpretations. But locally sound proof rules can also be soundly substituted uniformly, which preserves local soundness.

Theorem 18.3 (Uniform substitution of rules). *All uniform substitution instances (with $FV(\sigma) = \emptyset$) of locally sound inferences are* locally sound:

$$\frac{\phi_1 \quad \cdots \quad \phi_n}{\psi} \text{ locally sound} \quad \text{implies} \quad \frac{\sigma(\phi_1) \quad \cdots \quad \sigma(\phi_n)}{\sigma(\psi)} \text{ locally sound}$$

The idea behind the proof of Theorem 18.3 [8] is that, by Lemma 18.3, the truth of the right premises $\sigma(\phi_i)$ in a state ω and interpretation I is equivalent to the truth of the corresponding left premises ϕ_i in ω and adjoint interpretation $\sigma_\omega^* I$. By local soundness of the left inference, if all premises ϕ_i are valid in interpretation $\sigma_\omega^* I$, then so is its conclusion ψ, which, by Lemma 18.3, implies that the substituted conclusion $\sigma(\psi)$ is valid in I. The assumption that $FV(\sigma) = \emptyset$ is used to ensure that the same argument works in one adjoint interpretation $\sigma_\omega^* I$ regardless of the state ω. If $n = 0$ so that ψ has a proof, then this theorem also holds when $FV(\sigma) \neq \emptyset$, since soundness and local soundness are equivalent notions for $n = 0$ premises.

Theorem 18.3 explains how all proof rules of dL (except US) are *axiomatic proof rules* that are merely pairs of concrete dL formulas. For example, generalization rule

$$G \frac{p(\bar{x})}{[a]p(\bar{x})}$$

is a pair of concrete dL formulas. Rule G can be instantiated with Theorem 18.3 to:

$$\frac{x^2 \geq 0}{[x := x + 1; (x' = x \cup x' = -2)]x^2 \geq 0}$$

using the uniform substitution

$$\sigma = \{a \mapsto x := x + 1; (x' = x \cup x' = -2),\ p(\bar{x}) \mapsto x^2 \geq 0\}$$

All of a sudden, the only proof rule that needs an implementation as an algorithm is the uniform substitution mechanism itself that is used in rule US and Theorem 18.3. All other axioms and axiomatic proof rules are just concrete data.

Exercises

18.1. Give the result of applying the uniform substitution rule US with substitution $\sigma = \{a \mapsto \{x'' = -g \,\&\, x \geq 0\},\ b \mapsto ?(x = 0); v := -cv,\ p(\bar{x}) \mapsto 2gx \leq 2gH - v^2\}$ to the following formulas, respectively:

$$[a \cup b]p(\bar{x}) \leftrightarrow [a]p(\bar{x}) \wedge [b]p(\bar{x})$$
$$[a;b]p(\bar{x}) \leftrightarrow [a][b]p(\bar{x})$$
$$[a^*]p(\bar{x}) \leftrightarrow p(\bar{x}) \wedge [a][a^*]p(\bar{x})$$
$$\langle a \rangle p(\bar{x}) \leftrightarrow \neg[a]\neg p(\bar{x})$$
$$[a^*]p(\bar{x}) \leftrightarrow p(\bar{x}) \wedge [a^*](p(\bar{x}) \rightarrow [a]p(\bar{x}))$$

18.2 (Clash or not). The uniform substitution proof rule US checks that the substitution σ has no replacements that would introduce a free variable in a context where that variable is bound. List the conclusion that rule US produces when being applied with the given substitution on the following examples or explain why and how US clashes, and explain whether it is soundness-critical that US clashes:

$$[x := c()]p(x) \leftrightarrow p(c()) \qquad \sigma = \{c() \mapsto 0,\ p(\cdot) \mapsto (\cdot = x)\}$$
$$[x := c()]p(x) \leftrightarrow p(c()) \qquad \sigma = \{c() \mapsto y+1,\ p(\cdot) \mapsto [y := 1; (y := \cdot)^*]y \leq 1\}$$
$$[x' = c()]p(x) \leftrightarrow \forall t \geq 0 [x := x + t \cdot c()]p(x) \quad \sigma = \{c() \mapsto -x,\ p(\cdot) \mapsto (\cdot \geq 0)\}$$

18.3 (Make it clash). Let p an arity 0 predicate symbol. Give a uniform substitution σ for which it is necessary for soundness that US clashes when being applied to

$$p \rightarrow [a]p$$

Can you also give a uniform substitution that clashes when applied to the following?

$$[a;b]p(\bar{x}) \leftrightarrow [a][b]p(\bar{x})$$

18.4. Give the result of applying uniform substitution rule US with substitution $\sigma = \{c() \mapsto x \cdot y^2 + 1,\ p(\cdot) \mapsto (y + \cdot \geq z)\}$ on the following formulas or explain why and how US clashes:

$$[u := c()]p(u) \leftrightarrow p(c())$$
$$[x := c()]p(x) \leftrightarrow p(c())$$
$$[y := c()]p(y) \leftrightarrow p(c())$$
$$[z := c()]p(z) \leftrightarrow p(c())$$
$$[u := c()]p(u) \leftrightarrow \forall u\, (u = c() \rightarrow p(u))$$
$$[x := c()]p(x) \leftrightarrow \forall x\, (x = c() \rightarrow p(x))$$
$$[y := c()]p(y) \leftrightarrow \forall y\, (y = c() \rightarrow p(y))$$
$$[z := c()]p(z) \leftrightarrow \forall z\, (z = c() \rightarrow p(z))$$

If σ clashes, give a "similar" uniform substitution that would not clash.

18.5. What is the result of applying rule US to the $[:=]$ axiom with the substitution $\sigma = \{c() \mapsto x+y,\ p(\cdot) \mapsto [z := \cdot + 1; (z := z + \cdot)^* \cdot + 1 \geq 0]\}$? For each of the following formulas, either say which uniform substitution proves it by rule US from axiom [:=], or explain why no such uniform substitution exists.

$$[x:=-x]x^2 \geq 2x \leftrightarrow (-x)^2 \geq 2(-x)$$
$$[x:=y+1][(z:=z+x)^*]x^2 \geq z \leftrightarrow [(z:=z+y+1)^*](y+1)^2 \geq z$$
$$[x:=2x][(z:=z+x)^*]x^2 \geq z \leftrightarrow [(z:=z+2x)^*](2x)^2 \geq z$$
$$[x:=2x][(z:=z+x;z:=z+x)^*]x^2 \geq z \leftrightarrow [(z:=z+2x;z:=z+x)^*](2x)^2 \geq z$$
$$[x:=z+1][(z:=z+x)^*]x^2 \geq z \leftrightarrow [(z:=z+z+1)^*](z+1)^2 \geq z$$
$$[x:=z][x'=2x]x \geq 0 \leftrightarrow [z'=2z]z \geq 0$$
$$[x:=z+1][x'=2x]x \geq 0 \leftrightarrow [z'=2z]z \geq 0$$

18.6 (Local soundness). Theorem 18.3 shows that locally sound proof rules can be uniformly substituted. Show that all proof rules of dL except US are locally sound.

18.7 (Renaming). Uniform substitutions are perfect for substituting formulas for predicate symbols, terms for function symbols, and programs for program constant symbols. Yet, no matter how many uniform substitutions we try to use on the assignment axiom [:=], it will always be the variable x that it assigns to. The only other axioms that even mention variable names are the differential variable axiom x', the quantifier generalization rule $\forall$, and the differential equation axioms (which resolve this question by generalizing to systems of differential equations, however).

In order to prove instances of these axioms with other variable names, it would be helpful to have a proof rule for renaming. Renaming can be done in at least two different ways. Uniform renaming renames a variable x to a variable y uniformly everywhere. Bound renaming only renames one bound occurrence of a variable x to y (and, of course, consistently renames occurrences of x within the scope of this bound occurrence to y), which can be used to prove $\forall x\, p(x) \leftrightarrow \forall y\, p(y)$. Uniform renaming, for example, proves

$$\text{UR}\frac{x \geq 0 \wedge \forall x\,(x^2 \geq 0) \to [x:=x+1]x > 0}{y \geq 0 \wedge \forall y\,(y^2 \geq 0) \to [y:=y+1]y > 0}$$

Bound renaming, instead, proves

$$\text{BR}\frac{x \geq 0 \wedge \forall x\,(x^2 \geq 0) \to [x:=x+1]x > 0}{x \geq 0 \wedge \forall x\,(x^2 \geq 0) \to [y:=x+1]y > 0}$$

Give a precise construction defining both styles of renaming proof rules and carefully identify all requirements for soundness. If you are up for a challenge, prove both rules sound.

References

[1]   Brandon Bohrer, Vincent Rahli, Ivana Vukotic, Marcus Völp, and André Platzer. Formally verified differential dynamic logic. In: *Certified Programs and Proofs - 6th ACM SIGPLAN Conference, CPP 2017, Paris, France, January 16-17, 2017*. Ed. by Yves Bertot and Viktor Vafeiadis. New York: ACM, 2017, 208–221. DOI: `10.1145/3018610.3018616`.

[2]   Alonzo Church. *Introduction to Mathematical Logic*. Princeton: Princeton University Press, 1956.

[3]   Nathan Fulton, Stefan Mitsch, Jan-David Quesel, Marcus Völp, and André Platzer. KeYmaera X: an axiomatic tactical theorem prover for hybrid systems. In: *CADE*. Ed. by Amy Felty and Aart Middeldorp. Vol. 9195. LNCS. Berlin: Springer, 2015, 527–538. DOI: `10.1007/978-3-319-21401-6_36`.

[4]   Tobias Nipkow, Lawrence C. Paulson, and Markus Wenzel. *Isabelle/HOL — A Proof Assistant for Higher-Order Logic*. Vol. 2283. LNCS. Berlin: Springer, 2002.

[5]   André Platzer. Differential dynamic logic for hybrid systems. *J. Autom. Reas.* **41**(2) (2008), 143–189. DOI: `10.1007/s10817-008-9103-8`.

[6]   André Platzer. The complete proof theory of hybrid systems. In: *LICS*. Los Alamitos: IEEE, 2012, 541–550. DOI: `10.1109/LICS.2012.64`.

[7]   André Platzer. A uniform substitution calculus for differential dynamic logic. In: *CADE*. Ed. by Amy Felty and Aart Middeldorp. Vol. 9195. LNCS. Berlin: Springer, 2015, 467–481. DOI: `10.1007/978-3-319-21401-6_32`.

[8]   André Platzer. A complete uniform substitution calculus for differential dynamic logic. *J. Autom. Reas.* **59**(2) (2017), 219–265. DOI: `10.1007/s108 17-016-9385-1`.

[9]   André Platzer and Jan-David Quesel. KeYmaera: a hybrid theorem prover for hybrid systems. In: *IJCAR*. Ed. by Alessandro Armando, Peter Baumgartner, and Gilles Dowek. Vol. 5195. LNCS. Berlin: Springer, 2008, 171–178. DOI: `10.1007/978-3-540-71070-7_15`.

[10] The Coq development team. *The Coq proof assistant reference manual*. Version 8.0. LogiCal Project. 2004.

Chapter 19
Verified Models & Verified Runtime Validation

Synopsis This chapter provides an important twist on cyber-physical systems analysis. Without any doubt, formal verification provides crucial safety information for CPSs with exhaustive coverage of all the infinitely many possible behaviors that no finite amount of testing could ever provide. The catch is that the safety result then covers *all* behavior of the verified CPS models, but only provides safety guarantees for the actual CPS implementation to the extent that this implementation fits the model. Obtaining good enough models of physics is a nontrivial challenge in and of itself. This chapter provides a systematic way to transfer the safety guarantees about a CPS model to safety results for the actual implementation with the help of provably correct runtime compliance monitors. When run on the CPS implementation, these runtime monitors validate the actual execution in a verified way against the verified models that were proved safe previously.

19.1 Introduction

Since cyber-physical systems provide so many interesting control challenges of subtle interactions with the uncertainties and complexities of the physical world, it is quite nontrivial to get them right. Due to the large number of ways in which the behavior of the relevant systems can interact, full coverage is best achieved with the support of formal verification and validation techniques. In order to have the benefit of full coverage of safety for all possible behaviors, it is, of course, necessary to provide a model of the system under scrutiny, including a model not just of its controllers but also the relevant part of physics.

Models of reality come with certain inevitable challenges. The world is a complicated place, which implies that our models of the world will either also be exceedingly complex or else focus on certain simpler fragments. The trick is to focus exclusively on the relevant aspects of reality and devise a model of the physical dynamics that makes use of simplifying abstractions, including nondeterministic overapproximations, as much as possible. Just recall how hybridness and nonde-

© Springer International Publishing AG, part of Springer Nature 2018
A. Platzer, *Logical Foundations of Cyber-Physical Systems*,
https://doi.org/10.1007/978-3-319-63588-0_19

terminism helped simplify the bouncing-ball model in Chap. 4 and Sect. 11.12 by giving it more behavior than realistically possible but in ways that make it easier to describe. What is relevant and how do we make sure that the model covers reality?

When we are done with a proof of safety, we have the most exhaustive guarantee for the particular question about the particular model in the differential dynamic logic formula that we proved. While this proves that model to be safe, it only verifies the actual CPS implementation to the extent that this implementation fits the model. How can we establish that it really does? More generally, how can we have the model's safety result transfer to the safety of the actual implementation?

As we have already seen in Part I, we can hardly squeeze an actual self-driving car into a logical modality in a formula and expect to prove any meaningful properties about this mix of a syntactic expression and a physical object. Besides, any such attempt would still be missing out on the physical model of relevance for the car's motion and the behavior of its environment.

Instead, we will take a more subtle route and produce a monitor program to be run on the CPS implementation that will check all the time whether the present behavior fits what the model in the safety proof assumed, which ensures that the safety result about the CPS model applies to the present reality. But that monitor will be accompanied by a proof that it performs this checking in provably correct ways such that a successful response from the monitor program implies that the specific behavior of the concrete implementation is safe.

While a substantially more detailed technique can be found in previous work [5], this chapter emphasizes a simple intuitive approach to overcoming the challenges of model mismatches. The most important learning goals of this chapter are:

Modeling and Control: The crucial lesson of this chapter is that there are *inevitable differences* of models compared to reality, because it is infeasible or even impossible to model all complexities of reality exactly. Fortunately, it also is not necessary to model all of reality to make predictions about a cyber-physical system affecting only a part of the world! But even then, there are nontrivial challenges in making sure that the models provide an adequate level of detail. This chapter investigates systematic ways of ensuring that the real system complies with the model, or, vice versa, the model fits reality. Another sideline will be a few insights about the impact that safety considerations have on architectural design, because clever system architectures simplify the safety argument by helping to reduce safety to a smaller subsystem.

Computational Thinking: Relationships between truth and proof are of fundamental significance in logic and are the backbone of soundness and completeness considerations. A unique twist in cyber-physical systems is the fundamental challenge that, despite all soundness in the proof calculus, there can still be discrepancies between proofs in a model and truth in reality. While a sound proof makes perfect guarantees about the system behavior in the model, these guarantees only apply to the real system *if* accurate models of the system can be obtained. For CPS, this includes the daunting task of finding models not only of the controllers but also of the physical dynamics. This problem is fundamental and cannot be overcome by shifting to more precise models. Seemingly, the problem might be sidestepped by working with CPS data and experiments, but those have *no* predictive power without again assuming a corresponding model of reality, which leads to a vicious circle of assumptions either way.

This chapter gives a high-level account of a technique called *ModelPlex* [5], which provides a way of cutting through this Gordian knot of models and assumptions by combining offline proof with verifiably correct online monitoring. By turning a theorem prover upside down, ModelPlex generates provably correct monitor conditions that, if checked to hold at runtime, are provably guaranteed to imply that the offline safety verification results about the CPS model apply to the present run of the actual CPS implementation so that it is provably safe. This results in a correct-by-construction approach leveraging dynamic contracts to transfer proofs about models to CPS implementations.

CPS Skills: This chapter provides ways of taming CPS complexity by making it possible to isolate safety-critical parts and by providing disciplined ways of working with simplified models without losing the connection to the real CPS. It introduces the important pragmatic concept of runtime validation with online monitors that check the behavior of the real CPS implementation at runtime. Their primary purpose is to check for deviations of predictions about the behavior of the system compared to the actual observed runs and stop the system safely whenever potentially dangerous deviations are detected.

proof in a model vs. truth in reality
tracing assumptions
turning provers upside down
correct-by-construction
dynamic contracts
proofs for CPS implementations

models vs. reality   tame CPS complexity
inevitable differences   runtime validation
model compliance   online monitor
architectural design   prediction vs. run

19.2 Fundamental Challenges with Inevitable Models

In differential dynamic logic [7–9], models can be expressed directly as a hybrid program with its discrete controller actions and continuous differential equations that interact according to the program operators. After specifying the correctness properties of interest in differential dynamic logic, this logic provides rigorous reasoning techniques for proving the correctness properties as we saw in Part I for elementary CPS and in Part II for advanced CPS with sophisticated continuous dynamics. Once we are done with such a proof, we have achieved a major advancement of our understanding of the system and have obtained a rigorous safety argument why the controllers keep the CPS safe according to the dL formula.

Indubitably, such a rigorous safety result provides a lot of confidence in the correct design of the system, especially since it is even accompanied by an undeniable proof as a safety certificate. The more nuanced subtlety, however, is that we need to make sure we have phrased the dL formula correctly. This formula contains the preconditions, controller, physical model, and postconditions. For the sake of illustration, consider a typical dL formula of this shape:

$$A \rightarrow [(ctrl; plant)^*]B \tag{19.1}$$

Asking René Descartes for help with his skepticism, what could still go wrong even after we have a proof of (19.1)? What would happen if we wrote down the wrong postcondition B? If we ask the wrong question, we will get a perfect answer but for a question we are not interested in. It is, thus, of paramount importance that we review postcondition B carefully to make sure it really expresses all safety-critical properties of significance for the system.

What would happen if we used the wrong precondition A? Unlike in postcondition B, we cannot have forgotten a crucial condition in the precondition A, because the dL formula (19.1) would otherwise just not have a proof. What might happen, still, is that the precondition A on the initial state is overly conservative, so that we cannot turn the CPS on safely in as many circumstances as we would like. That is a pity but at least not unsafe, except when A is never true, because it contains a contradiction, in which case (19.1) is vacuously true, because its assumption is impossible. It is, thus, helpful to prove satisfiability of all preconditions as a sanity check.

Note 80 (Impossible assumptions) Whenever you make an assumption in your model or safety property, it is a good idea to check whether that assumption is possible. It is an even better idea to prove that it is at least satisfiable, so there is a state in which it is true.

What else could have gone wrong in phrasing the safety conjecture (19.1)? We could have described the controller *ctrl* incorrectly. Or, rather, there could have been an important discrepancy between what our controller model *ctrl* says it does and what an actual software implementation or low-level microcontroller really does. Differential refinement logic [3, 4] is an extension of differential dynamic logic that provides a systematic approach for relating more abstract controllers with safety proofs to more concrete controllers with additional efficiency properties that inherit safety for free. As we saw in Part I, more abstract models are often easier to verify than models containing full detail. If we were to choose a very different implementation platform or, say, the additional semantic ambiguities of a low-level C program, we would still need some way of establishing a guaranteed link between what was verified in (19.1) and the code that is really running on the CPS in the end.

Finally, and most critically, we could have gotten the physical model *plant* wrong in (19.1). What would happen then? Well, then our CPS would be in trouble. If we accidentally wrote down the physical model of a train and prove the dL formula in (19.1), we cannot expect it to control a rocket satisfactorily, because their physical dynamics are so different on account of the noticeable absence of rails in space. But even if we got the basic principles of the physical model right, yet missed some of its crucial aspects, then our proof of (19.1) would have limited predictive power for reality.

For the controller *ctrl*, the savior might be to move it closer to the implementation with increasingly fine-grained details, possibly absorbing the verification complexity shock with refinement proof techniques [4]. Maybe the same approach would help for the physical model *plant*? Well it would. And then again it would not! On the one hand, safety results about increasingly higher-fidelity models improve the range of behaviors where the safety results apply to reality. On the other hand, even higher-fidelity models are still just that: models of reality. Even a model with Schrödinger's equation of quantum mechanics [10] is still only a model of reality, and not even a very useful one for describing and predicting the motion of cars on the road, because quantum mechanics is more relevant for particles that are significantly smaller than cars. A model with Einstein's field equations of general relativity [2] is also still a model of reality, and not any more useful for describing a car, because that is more relevant for fast objects close to the speed of light, which is forbidden for cars on most highways.

Be that as it may, some models are pretty useful for making predictions about reality. This attitude [6] has been well-captured by George Box's [1] slogan.

Note 81 (George Box) All models are wrong but some are useful.

So, we will still continue to use models, because they enable predictions, but will from now on be more aware of the fact that models come with certain tradeoffs of analyzability and accuracy.

Is there a way to sidestep the issue by just not using any models in the first place? In fact, what can we even do without a model? We can run experiments and gather sample data about the behavior of a system. While that is certainly quite useful, too, it is important to understand that we will still need models to enable any predictions at all. Generality comes from the use of models! Unless we fix a model of how the behavior at the setup of the experiment relates to the behavior in other circumstances, the data alone will not provide any predictive power. Is the altitude of where your car drives relevant for its operation? Probably not. Yet, does the slope of the road affect its behavior? Quite likely. Does the weather have any influence? Unless we at least make some such dependence and independence assumptions in a model, there is no way that we can ever conjecture with any amount of certainty that our car will need at least 16 m to brake to a standstill from 35 mph (or roughly 50 km/h) next time, too, no matter where else we have already tried.

So it looks like, for better or for worse, we are stuck with the use of models at least for the physics. Is there anything at all that we can do to make sure our guarantees about the CPS models transfer to the actual CPS implementations? That is what this chapter investigates.

19.3 Runtime Monitors

To begin with, assume that we have put the lessons from Parts I and II to good use by having come up with a proof of a dL formula, e.g., of the illustrative shape:

$$A \to [(ctrl; plant)^*]B \tag{19.1*}$$

Now all that remains is to figure out a way to make sure that this safety result about the CPS model $(ctrl; plant)^*$ transfers to the actual CPS implementation. If offline results alone do not check whether all parts of (19.1) fit the real CPS, then let us investigate runtime monitors that address this question online when the CPS runs.

Checking that the initial condition A applies for the real CPS is straightforward as long as all its quantities can be measured. In that case, all it takes is to evaluate at runtime whether the formula A is true in the initial state and only permit the CPS to be turned on if it is. That's still relatively straightforward (if every relevant quantity is physically measurable).

Monitoring the postcondition B to see whether it is true would be equally straightforward. But that is not actually useful at all, because if B is ever false, then the system is already hopelessly unsafe by definition and there is nothing that can be done about it anymore, since we cannot just go back in time and do things differently. Just think of a postcondition B expressing that the controlled car should have positive distance to other cars. Once that condition evaluates to *false*, the cars have collided and all hope for a happy ending is lost. Well, maybe matters get better if we work with a postcondition B that has an extra margin such as a distance of 1 m at least? That does not really help either, because once that safety margin is violated, the car might already be going so fast that a collision is unavoidable regardless.

So, the most challenging aspect in runtime monitoring is to find out what precise condition should be monitored, and to identify what exactly one knows about the system behavior if that monitor condition is checked successfully. Proofs play a crucial rôle in identifying what needs to be monitored, and they are certainly fundamental in proving what correctness properties the resulting monitors come with, meaning what one knows about the CPS if the monitors all evaluate to *true*.

Continuing down the list, how can we monitor the controller *ctrl* in (19.1) to see whether the real CPS fits it? For various reasons, the controller implementation in the CPS may have slight discrepancies compared to its higher-level control model *ctrl*. For example, the controller might have been implemented in a low-level language such as C, or might use preexisting legacy code, or might have been synthesized from a Stateflow/Simulink model, or it might be running low-level microcontroller machine code. More fundamentally, recall that the controller model *ctrl* includes a model of the discrete actions of agents in the environment, such as the nondeterministic choice of acceleration or braking for the car in front, where we may not have access to the actual implementation. All these factors contribute to potential deviations of the actual controllers compared to the controller model *ctrl*. What can we monitor to check whether the real CPS fits to the model *ctrl*?

> Before you read on, see if you can find the answer for yourself.

The most important impact of the controller *ctrl* is to decide how to set control variables for the physical dynamics *plant* after suitable computations based on certain measurements of sensor inputs. All these final control variable decisions in the real controller (let's call it γ_{ctrl}) should be monitored and checked for compatibility with what the controller model *ctrl* permits. Of course, due to nondeterminism, the real controller implementation γ_{ctrl} can reach different decisions than the model *ctrl*, but it should only ever reach decisions that the verified controller model *ctrl* at least allows. If, in any circumstance, the real controller implementation γ_{ctrl} ever decides to assign values to control variables that the verified model *ctrl* does not allow, then these are potentially unsafe and should be rejected for safety reasons.

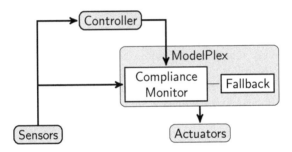

Fig. 19.1 ModelPlex monitors sit between controller and actuator to check the controller's decisions for compliance with the model based on sensor data with veto leading to a safe fallback action

Such a *controller monitor* inspects each and every resulting decision by the controller implementation for compliance with the verified controller model *ctrl* based on the current sensor data and vetoes the decision if *ctrl* does not allow it; see Fig. 19.1. Of course, the resulting controller monitor cannot just reject a control decision, but must also override it with a safe fallback action to execute on *plant* instead. Figuring out such a safe fallback action is not always obvious either. But at least it is an easier problem, because that safe fallback action just needs to keep the system in safe stasis without doing anything particularly useful. In a car, for example, this last resort action might consist of applying emergency brakes, cutting the engine's power, and asking the human to investigate. In an aircraft, it might be flying in a loitering circle until the problematic situation is resolved. In a quadrotor drone, it might be hovering in place.

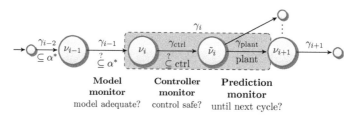

Model   Controller   Prediction
monitor   monitor   monitor
model adequate?   control safe?   until next cycle?

Fig. 19.2 Use of ModelPlex monitors along a system run

Let us assume that we have already found and proved safe such a safe fallback action, since that is an easier problem. That leaves open the question of how we can best monitor and determine whether an observed controller action of the real controller implementation γ_{ctrl} fits the verified model *ctrl*. While you are invited to think about this challenge already, we will postpone it and first consider another challenge.

Proceeding further, how can we monitor the physical model *plant* in (19.1) to see whether the real CPS physics fits it? That is even more subtle, because, no matter what we try, the real physical world does not come with any source code that we can run or read to try to find out whether it fits the model *plant*. Instead, the only chance is to try out the physical system and observe what it does to see whether it fits the model *plant*. Interestingly, that is already quite well aligned with the approach we settled on for the real controller γ_{ctrl}, just for completely different reasons.

Somewhat similar to the controller monitor, which models just the responsibilities of the controller for compliance with *ctrl*, the *model monitor* models the whole system for compliance with the model *ctrl*; *plant* that includes the physical model.[1] The model monitor will inspect the data from the current state v_i and the data from the previous state v_{i-1} when it ran last to check whether the transition from v_{i-1} to v_i can be explained by the model *ctrl*; *plant*. If that transition fits the model *ctrl*; *plant*, then since all repetitions of said model from a safe initial state satisfying A (which we checked at runtime initially) are safe (satisfying B by the offline proof), then the concrete run of the real CPS implementation ending in v_i is also safe. If the transition v_{i-1} to v_i does not fit *ctrl*; *plant*, however, then the safety proof of the CPS model (19.1) does not apply to the current execution, so the model monitor vetoes and initiates a safe fallback action.

---

[1] The reason for monitoring the whole control loop body *ctrl*; *plant* instead of just separately the physics *plant* is that this provides better guarantees [5] and also works if the HP is of any arbitrary form other than $(ctrl;plant)^*$.

19.4 Model Compliance

Based on these general runtime monitoring principles, the primary remaining question is how to actually check a concrete system execution for compliance with the verified model $(ctrl; plant)^*$ from (19.1). Of course, it is relatively easy to check whether the initial state satisfies the precondition A, or at least it is easy when all its relevant quantities can be measured appropriately. But the same is not true for the hybrid program $(ctrl; plant)^*$, because of all the nondeterminism and differential equations that it involves. It, thus, takes a more clever approach to determine whether the real system execution fits the hybrid program $(ctrl; plant)^*$. This section motivates and intuitively develops such an approach with the simple example of a bouncing ball.

Example 19.1 (Bouncing-ball monitors). As a simple guiding example, recall the familiar acrophobic bouncing ball Quantum that has been with us ever since Chap. 4:

$$0 \leq x \wedge x = H \wedge v = 0 \wedge g > 0 \wedge 1 \geq c \geq 0 \rightarrow$$
$$\left[\left(\{ x' = v, v' = -g \,\&\, x \geq 0 \}; \; (?x = 0; v := -cv \cup ?x \neq 0) \right)^* \right] (0 \leq x \wedge x \leq H) \tag{4.24}$$

This formula has been proved in Proposition 7.1 (with the additional assumption $c = 1$ that Exercise 7.5 showed can be removed again). This led to a perfect proof of safety for Quantum, if only Quantum actually fits the hybrid model in the formula (4.24) that was proved in dL's sequent calculus.

You might already have noticed in earlier chapters that Quantum is easily startled. Even before reading Expedition 4.5 on p. 123, Quantum was already a natural born Cartesian skeptic. His level of scrutiny and skeptical doubt only increased after reading Chap. 19. Since Quantum really wants to get things right, he checks the initial condition of formula (4.24) and then figures out how to check whether the real execution fits the hybrid program model in (4.24).

Figuring that the differential equations in (4.24) are surely the best possible differential equations that could ever exist (since they are meant to describe bouncing balls, after all), Quantum first only worries about the discrete controller part of (4.24). Is there a logical formula characterizing that a controller implementation switching from position x and velocity v to new position x^+ and velocity v^+ fits the discrete controller in (4.24)?

Before you read on, see if you can find the answer for yourself.

A run of an actual discrete controller implementation changing the position from x to x^+ and the velocity from v to v^+ (leaving all other variables alone) only faithfully fits the controller in (4.24) if the following logical formula evaluates to *true*:

$$\left(x = 0 \wedge v^+ = -cv \;\vee\; x > 0 \wedge v^+ = v \right) \wedge x^+ = x \tag{19.2}$$

This formula represents a controller monitor if no physical motion ever happens. Certainly, the conjunct $x^+ = x$ needs to be *true* for an execution to faithfully run the discrete controller of (4.24), because the discrete dynamics of the bouncing ball does not affect the ball's altitude but merely its velocity. Furthermore, it is either the case that the ball runs the first control branch so is presently on the ground ($x = 0$) and the new velocity v^+ after the discrete ground controller is the damped version $-cv$ of the previous velocity v, or it is the case that the ball is still in the air ($x > 0$ because of the second control branch $?x \neq 0$ together with the evolution domain constraint $x \geq 0$) and then the velocity is unaltered ($v^+ = v$). Every run satisfying controller monitor (19.2) fits an execution of the discrete controller (4.24).

In retrospect, it is reasonably obvious how controller monitor (19.2) relates to the controller in HP (4.24), with a disjunction corresponding to the controller's choices and conjunctions accumulating the conditions of tests and effects of assignments. Of course, the particular discrete controller in (4.24) is deterministic and, hence, so is the monitor (19.2). But the same principle applies for controllers with ample nondeterminism, in which case the resulting monitor condition is more flexible.

To validate the controller model, Quantum takes out his favorite oscilloscope and plenty of other measuring devices and quickly evaluates controller monitor (4.24) for a number of trial bounces in a high-fidelity simulation environment (Fig. 19.3).

Fig. 19.3 Sample run of a bouncing ball (plotted as height over time) that ultimately lies down flat

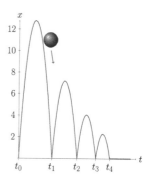

The controller implementation holds up pretty well with respect to controller monitor (19.2) for quite a while, except that the ball ultimately dares to just lie flat on the ground. If the implemented controller turns $v = -3$ to $v^+ = 0$, then this is clearly a violation of (19.2) and flagged as such at time t_4. Quantum was initially tempted to dismiss t_4 as a measurement error but decided after retrying a few times that there must be an actual controller model mismatch. Indeed, looking back at (4.24), its HP does not yet include the refinements from Sect. 4.2.3 that enable bouncing balls to ultimately deflate and lie flat when their energy is insufficient to jump back up.

The controller implementation is not the only aspect of reality with the potential for a mismatch between model and reality. In fact, much more challenging than the controller's implementation is the implementation of physics that is also known as the real world. Despite dozens of centuries of progress, mankind is still sometimes at a loss when it comes to explaining physical phenomena, let alone the behavior of other agents acting in the environment. Consequently, there is even more that might go wrong if we start the analysis with a completely inadequate physical model. At the same time, even the best physical model is still just that: a model.

Example 19.2 (Bouncing-ball model monitors). When Quantum tries to devise a clever monitor to check the continuous physics behind HP (4.24), he discovers even more challenges than with the controllers. It is quite nontrivial to run a differential equation for an unspecified nondeterministic amount of time and check whether the resulting state coincides with the state that has been measured in a run of the real system implementation experimentally. Granted, the particular double integrator ODE in (4.24) is still of the relatively tame kind, but, for the most part, this only simplifies the illustration, not the question.

Is there a logical formula characterizing that a physical motion evolving from position x and velocity v to new position x^+ and velocity v^+ can be explained by the differential equation in (4.24)?

Before you read on, see if you can find the answer for yourself.

For just the differential equations $x' = v, v' = -g$, a corresponding monitor can be read off from the invariants (7.10) that we used to prove the bouncing ball:

$$2g(x^+ - x) = v^2 - (v^+)^2 \tag{19.3}$$

The change $v^2 - (v^+)^2$ in the squares of the velocities equals $-2g$ times the change $x - x^+$ in positions. This monitor condition can also easily be read off quite directly from the solutions of the differential equation (expressing the new position x^+ and velocity v^+ as a function of the old) by eliminating time t, which is unobservable in the HP of (4.24):

$$
\begin{aligned}
v^+ &= v - gt & \overset{g\neq0}{\equiv} \quad & t = \tfrac{v-v^+}{g} \\
x^+ &= x + vt - \tfrac{g}{2}t^2 & \overset{above}{\equiv} \quad & x^+ = x + v\tfrac{v-v^+}{g} - \tfrac{g}{2}\tfrac{v^2-2vv^+-(v^+)^2}{g^2} \\
& & \equiv \quad & 2g(x^+ - x) = 2v^2 - 2vv^+ - v^2 + 2vv^+ - (v^+)^2
\end{aligned}
$$

Indeed, the resulting equation (19.3) is the crucial invariant of the dynamics, but it does not characterize the system behavior sufficiently, because equation (19.3) also holds for physically impossible situations such as when the new velocity v^+ is chosen to be bigger than the previous velocity v, which is impossible under gravity $v' = -g$. The reason is that (19.3) is an invariant equation that is always true during the differential equation but neglects the fact that the differential equation must evolve forward. So (19.3) holds whether the differential equation evolves forward or backward in time, because it is unaware of the directionality in the system. Since $g > 0$, this directionality is easily expressed by an additional conjunct saying that the velocity never increases:

$$2g(x^+ - x) = v^2 - (v^+)^2 \wedge v^+ \le v \tag{19.4}$$

The only challenge remaining to adequately represent all assumptions of the plant model in the bouncing ball is to add the domain constraint for the initial $x \geq 0$ and final position $x^+ \geq 0$:

$$2g(x^+ - x) = v^2 - (v^+)^2 \wedge v^+ \leq v \wedge x \geq 0 \wedge x^+ \geq 0 \qquad (19.5)$$

Quantum can now validate trial runs of the physical motion in gravity for compatibility with the plant model by testing whether monitor (19.5) evaluates to *true*. In particular, if the observed change in the ball's position and change in velocity while falling according to gravity are always compatible with (19.5), then this supports the hypothesis that the differential equation and evolution domain constraints in HP (4.24) describe the reality of falling balls well.[2] The combination into an overall model monitor respecting the discrete controller and continuous motion is natural [5], essentially by substituting the plant monitor (19.5) into the controller monitor (19.2).

19.5 Provably Correct Monitor Synthesis

Reading off the appropriate monitor conditions from the models by an educated guess as in Sect. 19.4 is one thing. But justifying their correctness rigorously and making sure that no subtle but critical conditions are left out is another. While the relationship of controller monitor (19.2) to the discrete control model in (4.24) is reasonably transparent after suitable transformations of the evolution domains and tests, it is non-obvious whether (19.5) correctly checks *all* correctness-critical conditions of the plant model.

Since correctness in CPS is so important, it is also important to get the correctness monitors themselves correct. After all, the controller and model monitors are supposed to be last-resort mechanisms that, when things do not go according to plan, avert potential catastrophes by interfering before it is too late. It would not help if the monitor conditions monitored the wrong expressions and ended up missing critical safety hazards that they were meant to spot.

The question is: what makes a runtime monitor condition correct? How do we tell a correct runtime monitor apart from a well-intended but flawed monitor? These questions are not unlike the ones we asked out for cyber-physical systems themselves in Chap. 4. But the big difference is that it is a more narrow problem to ask whether a runtime monitor adequately represents the conditions of a model, because we already have both the runtime monitor as well as the model it is supposed to check compatibility with. Contrast this with Chap. 4 where we were given a model and were still trying to identify its appropriate safety condition. So when is a runtime monitor correct?

---

[2] Of course, falling balls cannot deny a certain resemblance to the apples that were falling from trees in Isaac Newton's time. So one model might describe two related scenarios.

Before you read on, see if you can find the answer for yourself.

The specification that correct runtime monitors need to obey is that they imply that there, indeed, is a run of the model that explains the observed transition of the previous state to the present state. If the monitor condition evaluates to *false*, then an alarm about a possible model violation is raised and a safe fallback action is initiated (such as applying emergency brakes and cutting power). But if it evaluates to *true*, then the monitor had better be right about the fact that the observed behavior fits to the model. In other words, if the monitor condition is *true* for the old position x and new position x^+, then there really needs to be a run of the corresponding model that, indeed, leads from position x to x^+ as observed (and accordingly for previous velocity v to new velocity v^+ and other variables).

Solve Exercise 19.1 now to demonstrate that the runtime monitors are correct.

19.5.1 Logical State Relations

Let us call the hybrid program of the relevant model α and let $\chi(x,x^+)$ be a run-time monitor formula. The runtime monitor formula $\chi(x,x^+)$ is a formula in which the variable (vector) x is meant to refer to the previous position (and velocity or other relevant state variables) while variable (vector) x^+ is meant to refer to the new position (and velocity).

> **Definition 19.1 (Correctness of runtime monitors).** The *runtime monitor* formula $\chi(x,x^+)$ is called *correct* for the hybrid program model α with bound variables $\mathrm{BV}(\alpha) \subseteq \{x\}$ iff the following dL formula is valid:
>
> $$\chi(x,x^+) \rightarrow \langle\alpha\rangle x = x^+$$

If x is a vector of variables $(x_1,\ldots,x_n)$ and x^+, thus, is the vector of variables $(x_1^+,\ldots,x_n^+)$, then the vectorial equation $x = x^+$ means the same as the conjunction

$$\bigwedge_{i=1}^{n} x_i = x_i^+$$

Example 19.3 (Correctness of controller monitor). Continuing Example 19.1, let $\chi(x,v,x^+,v^+)$ denote the controller monitor formula (19.2) for the discrete controller of (4.24).

$$\chi(x,v,x^+,v^+) \stackrel{\mathrm{def}}{\equiv} \left(x=0 \wedge v^+ = -cv \ \vee \ x>0 \wedge v^+ = v\right) \wedge x^+ = x \qquad (19.2*)$$

The correctness of controller monitor (19.2) can be proved in the dL calculus:

$$\frac{\chi(x,v,x^+,v^+) \vdash (x=0 \rightarrow x{=}x^+ \wedge -cv{=}v^+) \vee (x \neq 0 \rightarrow x{=}x^+ \wedge v{=}v^+)}{}$$

$$(?) \quad \chi(x,v,x^+,v^+) \vdash \langle ?x=0 \rangle (x{=}x^+ \wedge -cv{=}v^+) \vee \langle ?x \neq 0 \rangle (x{=}x^+ \wedge v{=}v^+)$$

$$(:=) \quad \chi(x,v,x^+,v^+) \vdash \langle ?x = 0 \rangle \langle v := -cv \rangle (x{=}x^+ \wedge v{=}v^+) \vee \langle ?x \neq 0 \rangle (x{=}x^+ \wedge v{=}v^+)$$

$$(;) \quad \chi(x,v,x^+,v^+) \vdash \langle ?x = 0; v := -cv \rangle (x{=}x^+ \wedge v{=}v^+) \vee \langle ?x \neq 0 \rangle (x{=}x^+ \wedge v{=}v^+)$$

$$(\cup) \quad \chi(x,v,x^+,v^+) \vdash \langle ?x = 0; v := -cv \cup ?x \neq 0 \rangle (x{=}x^+ \wedge v{=}v^+)$$

While this simple notion of correctness of runtime monitors is particularly easy to understand, there are also improved ways of establishing the correctness of controller monitors that additionally guarantee that the controller will never violate the evolution domain constraints of the plant [5].

Example 19.4 (Correctness of plant monitor). Continuing Example 19.2, this time, let $\chi(x,v,x^+,v^+)$ denote the differential equation monitor formula (19.4) for the differential equation of (4.24) without its evolution domain constraint. Because $g > 0$ holds, this monitor $\chi(x,v,x^+,v^+)$ is correct for the differential equation:

$$\mathbb{R} \quad \frac{}{g > 0, \chi(x,v,x^+,v^+) \vdash \exists t \geq 0 \left(-\tfrac{g}{2}t^2 + vt + x = x^+ \wedge v - gt = v^+\right)}$$

$$(:=) \quad g > 0, \chi(x,v,x^+,v^+) \vdash \exists t \geq 0 \langle x := -\tfrac{g}{2}t^2 + vt + x \rangle \langle v := v - gt \rangle (x{=}x^+ \wedge v{=}v^+)$$

$$(') \quad g > 0, \chi(x,v,x^+,v^+) \vdash \langle x' = v, v' = -g \rangle (x{=}x^+ \wedge v{=}v^+)$$

In fact, it can even be shown that the differential equation monitor condition (19.4) is perfect, because it is *true* if and only if the differential equation can reach position x^+ and velocity v^+:

$$g > 0 \rightarrow \left(\langle x' = v, v' = -g \rangle (x = x^+ \wedge v = v^+) \leftrightarrow 2g(x^+ - x) = v^2 - (v^+)^2 \wedge v^+ \leq v\right)$$

Because this equivalence is provable for $g > 0$, the differential equation monitor (19.4) will never raise false alarms. In a similar way (19.5) is a provably correct runtime monitor for the plant of (4.24) including the evolution domain constraint. That is, the following dL formula is provable:

$$g > 0 \wedge 2g(x^+ - x) = v^2 - (v^+)^2 \wedge v^+ \leq v \wedge x \geq 0 \wedge x^+ \geq 0 \rightarrow$$
$$\langle x' = v, v' = -g \,\&\, x \geq 0 \rangle (x = x^+ \wedge v = v^+)$$

Again, we can prove that this plant monitor is perfect because, given $g > 0$, it is equivalent to the reachability of x^+ and v^+ along the continuous plant:

$$g > 0 \rightarrow \big(2g(x^+ - x) = v^2 - (v^+)^2 \wedge v^+ \leq v \wedge x \geq 0 \wedge x^+ \geq 0$$
$$\leftrightarrow \langle x' = v, v' = -g \,\&\, x \geq 0 \rangle (x = x^+ \wedge v = v^+)\big)$$

19.5.2 Model Monitors

The above runtime monitors (19.2) for the discrete controller, (19.4) for the differential equation and (19.5) for the full plant are all useful, but do not cover all details of the HP model (4.24) of the bouncing ball yet. We can, nevertheless, follow essentially the same approach again for a runtime monitor of the full HP model. The only tricky part is the need to deal with the loop. The most natural way of monitoring the correct execution of a control loop, however, is to separately check each round of the control loop. Consequently, all we need to do is unwind the loop once with the iteration axiom $\langle * \rangle$ and just find a runtime monitor for the loop body α instead of for the full loop α^*. *Then this runtime monitor can be used for each round of the control loop in the controller implementation.* Such a runtime monitor for just the loop body is, in fact, also much more useful than a runtime monitor for the full loop α^*, because it can take quite a while before we finally know whether the entire loop execution fits the model. We would much prefer to already check during each run of the loop body whether everything still operates according to the model.

Example 19.5 (Correctness of model monitor). A runtime monitor for the HP's loop body can be constructed by substituting the respective monitors into one another:

$$x^+ > 0 \wedge 2g(x^+ - x) = v^2 - (v^+)^2 \wedge v^+ \leq v \wedge x \geq 0$$
$$\vee\, x^+ = 0 \wedge c^2 2g(x^+ - x) = c^2 v^2 - (v^+)^2 \wedge v^+ \geq -cv \wedge x \geq 0 \qquad (19.6)$$

Correctness of the model monitor (19.6) for one iteration of the full loop body of the HP model (4.24) can be proved in dL.

Proposition 19.1 (Correct bouncing-ball model monitor). *Formula* (19.6) *is a correct model monitor for* (4.24). *That is this* dL *formula is valid:*

$$g > 0 \wedge 1 \geq c \geq 0 \rightarrow$$
$$\big(x^+ > 0 \wedge 2g(x^+ - x) = v^2 - (v^+)^2 \wedge v^+ \leq v \wedge x \geq 0$$
$$\vee\, x^+ = 0 \wedge c^2 2g(x^+ - x) = c^2 v^2 - (v^+)^2 \wedge v^+ \geq -cv \wedge x \geq 0$$
$$\rightarrow \langle \{x' = v, v' = -g\, \&\, x \geq 0\}; (?x = 0; v := -cv \cup ?x \neq 0) \rangle (x = x^+ \wedge v = v^+) \big)$$

In fact, equivalence instead of implication can again be proved as well, which demonstrates that the model monitor is exact and does not produce any false alarms.

19.5.3 Correct-by-Construction Synthesis

So far, we have settled for educated guesses to produce runtime monitor formulas and subsequently proved them correct in the dL proof calculus. That is perfectly acceptable except that we then always have to get creative to produce the monitor formulas in the first place.

It would be better if we could construct the runtime monitor formulas system-atically. In fact, it would be even more helpful if we had a way of constructing the runtime monitor formulas in a correct-by-construction approach such that we simultaneously generate the runtime monitor and a proof of its correctness. Surprisingly, this is perfectly possible as well. All we need to do is to exploit the rigorous reasoning principles of differential dynamic logic for a purpose other than safety verification. The starting point is one crucial observation. What is the easiest formula satisfying correctness criterion Definition 19.1?

> Before you read on, see if you can find the answer for yourself.

By far the easiest and most obviously correct formula satisfying runtime monitoring correctness (Definition 19.1) for model α is the dL formula $\langle\alpha\rangle x = x^+$ itself. Of course, when choosing $\chi(x,x^+) \stackrel{\text{def}}{\equiv} \langle\alpha\rangle x = x^+$, then the correctness condition is trivially valid because every formula implies itself, even $\langle\alpha\rangle x = x^+$ does:

$$\chi(x,x^+) \to \langle\alpha\rangle x = x^+$$

True beyond any doubt. What could be wrong with that?

> Before you read on, see if you can find the answer for yourself.

Well, nobody can argue against the validity of $\langle\alpha\rangle x = x^+ \to \langle\alpha\rangle x = x^+$. But we hardly learn anything at all about what constitutes a faithful execution of the controller model α if this is how we choose $\chi(x,x^+)$. Indeed, the HP α itself is a pretty perfect model of what it means to fit to its model. Yet, the problem is that if α is quite a complicated HP with ample nondeterminism, then it can be rather time-consuming to exhaustively execute all its different choices just to find out whether there is one combination that explains the present state transition in the concrete controller implementation. For that reason, it is better to find a simpler formula that also implies $\langle\alpha\rangle x = x^+$ but is easier to evaluate at runtime, for example, a formula that is pure real arithmetic. How else might we construct such a simpler monitor formula $\chi(x,x^+)$ that provably implies $\langle\alpha\rangle x = x^+$ as well?

> Before you read on, see if you can find the answer for yourself.

The idea to construct such a monitor $\chi(x,x^+)$ implying $\langle\alpha\rangle x = x^+$ is about as easy as it is far-reaching. The clou is that we already have a pretty powerful and rigorously correct transformation technique at our disposal: the axioms and proof rules of differential dynamic logic from Parts I and II. All we need to do is to apply them to $\langle\alpha\rangle x = x^+$ and find out how that simplifies the formula.

Example 19.6 (Synthesis of controller monitor). The controller monitor (19.2) from Example 19.1 for the bouncing-ball controller (4.24) can also be synthesized systematically with a correct-by-construction approach. We simply start with a hopeless attempt to prove $\langle ctrl\rangle(x=x^+ \wedge v=v^+)$ for the bouncing-ball controller *ctrl*:

$$\cfrac{\vdash (x = 0 \rightarrow x{=}x^+ \wedge -cv{=}v^+) \vee (x \neq 0 \rightarrow x{=}x^+ \wedge v{=}v^+)}{\cfrac{\vdash \langle ?x = 0 \rangle (x{=}x^+ \wedge -cv{=}v^+) \vee \langle ?x \neq 0 \rangle (x{=}x^+ \wedge v{=}v^+)}{\cfrac{\vdash \langle ?x = 0 \rangle \langle v := -cv \rangle (x{=}x^+ \wedge v{=}v^+) \vee \langle ?x \neq 0 \rangle (x{=}x^+ \wedge v{=}v^+)}{\cfrac{\vdash \langle ?x = 0; v := -cv \rangle (x{=}x^+ \wedge v{=}v^+) \vee \langle ?x \neq 0 \rangle (x{=}x^+ \wedge v{=}v^+)}{\vdash \langle ?x = 0; v := -cv \cup ?x \neq 0 \rangle (x{=}x^+ \wedge v{=}v^+)}{\scriptstyle (\cup)}}{\scriptstyle (;)}}{\scriptstyle (:=)}}{\scriptstyle (?)}$$

This formula cannot possibly be proved because not every value of new positions x^+ and velocities v^+ is reachable from every initial position x and velocity v in the bouncing-ball controller. Fortunately! But the purpose is not to prove the runtime monitor formula offline once and for all, but instead to check whether it evaluates to *true* at runtime and, thereby, to complete the above offline proof at runtime. The resulting monitor condition is the remaining premise at the top of the proof:

$$(x = 0 \rightarrow x{=}x^+ \wedge -cv{=}v^+) \vee (x \neq 0 \rightarrow x{=}x^+ \wedge v{=}v^+) \tag{19.7}$$

While this is syntactically different than the manually constructed controller monitor (19.2), the two are provably equivalent and can be obtained by minor simplification when taking into account that the evolution domain constraint guarantees $x \geq 0$. But monitor (19.7) has been systematically constructed and is already accompanied by a correctness proof, because it implies $\langle ctrl \rangle (x{=}x^+ \wedge v{=}v^+)$ by the above dL proof.

All dL-generated runtime monitors are correct-by-construction. How conservative they are can be read off from an inspection of their proofs. If only equivalence axioms and proof rules have been used, then the runtime monitors are exact. This is the case in the proof of Example 19.6. Otherwise, when implication axioms or proof rules are used, then the monitor may be conservative and might cause unnecessary false alarms, but at least its positive answers are perfectly reliable by proof.

The logical transformation generating correct-by-construction model monitors or other models involving differential equations and/or loops is slightly more involved but follows very similar principles [5]. The basic idea is to unroll loops once by axiom $\langle * \rangle$ and either skip or follow differential equations in a proof for one control cycle of the appropriate runtime to monitor reaction time.

19.6 Summary

Even if this chapter merely scratched the surface of the technical aspects of synthesizing provably correct runtime monitors [5], it, nevertheless, held particularly valuable lessons in store for the mindful CPS enthusiast. Meddling with models is an inevitable part of working out the design of a cyber-physical system. But, as the name suggests, such models have to include a sufficiently adequate model of the relevant part of the physical world, which is quite a nontrivial challenge in and of itself.

Fortunately, the logical foundations of safe model transfer, nevertheless, provide a way of exploiting differential dynamic logic to generate runtime monitors that are accompanied by correctness proofs implying that if they evaluate to *true*, then the actual system run fits a provably safe CPS model and is, thus, safe itself. Beyond the high-level ideas behind the ModelPlex approach making this idea reality [5], one of the most important take-home lessons is that the combination of offline verification with runtime monitoring concludes proofs about true CPS runs at runtime. This approach enables tradeoffs that analyze simpler models offline while safeguarding them for suitability with ModelPlex monitors at runtime. Except for the fact that the runtime monitors of overly simplistic models may cause more alarms for discrepancies, such a combination leads to better analysis results for simpler models without paying the full price that we pay when entirely ignoring important effects in pure offline models. Of course, even runtime monitors can only provide limited safety recovery when starting out with models with unbounded errors. When using a bouncing-ball model to describe the flight of an airplane, one should not be surprised to find a significant discrepancy when trying it out for real. At the very least, the bounce on the ground won't proceed as planned.

Exercises

19.1 (Correct bouncing-ball monitors). Prove correctness of the monitors for the bouncing ball from Sect. 19.4 in dL's sequent calculus. That is, prove that truth of the controller monitor (19.2) implies existence of a corresponding execution of the controller. Show that truth of the plant monitor (19.5) implies existence of a run of the plant.

19.2 (Ping-pong monitors). Create the controller monitor and model monitor for the verified ping-pong models of the event-triggered design in Chap. 8 and for the time-triggered design in Chap. 9. Convince yourself that they are correct, i.e., can lead to a corresponding dL proof. Discuss the pragmatic difference between the monitors resulting from the event-triggered models and those from the time-triggered designs. Is there a discrepancy that one monitor discovers that the other one does not?

19.3 (Model monitor correctness). Prove that truth of the model monitor (19.6) for the bouncing ball implies existence of a corresponding run of its model (4.24).

19.4 (Controller monitor generation). Extract the respective controller monitor and prove it correct for the models from Exercises 3.9, 4.22, 9.14, and 12.5.

19.5 (*Monitor synthesis). Describe a proof strategy that synthesizes a runtime monitor for an HP that it receives as input. Is the resulting monitor correct-by-construction? Is it possible to change the approach such that the proof strategy synthesizes both a runtime monitor together with a proof of the correctness of the monitor?

References

[1]   George E. P. Box. Science and statistics. *Journal of the American Statistical Association* **71**(356) (1976), 791–799. DOI: 10.1080/01621459.1976.10480949.

[2]   Albert Einstein. Die Feldgleichungen der Gravitation. *Sitzungsberichte der Preussischen Akademie der Wissenschaften zu Berlin* (1915), 844–847.

[3]   Sarah M. Loos. Differential Refinement Logic. PhD thesis. Computer Science Department, School of Computer Science, Carnegie Mellon University, 2016.

[4]   Sarah M. Loos and André Platzer. Differential refinement logic. In: *LICS*. Ed. by Martin Grohe, Eric Koskinen, and Natarajan Shankar. New York: ACM, 2016, 505–514. DOI: 10.1145/2933575.2934555.

[5]   Stefan Mitsch and André Platzer. ModelPlex: verified runtime validation of verified cyber-physical system models. *Form. Methods Syst. Des.* **49**(1-2) (2016). Special issue of selected papers from RV'14, 33–74. DOI: 10.1007/s10703-016-0241-z.

[6]   John von Neumann. The mathematician. In: *Works of the Mind*. Ed. by R. B Haywood. Vol. 1. 1. Chicago: University of Chicago Press, 1947, 186–196.

[7]   André Platzer. Differential dynamic logic for hybrid systems. *J. Autom. Reas.* **41**(2) (2008), 143–189. DOI: 10.1007/s10817-008-9103-8.

[8]   André Platzer. Logics of dynamical systems. In: *LICS*. Los Alamitos: IEEE, 2012, 13–24. DOI: 10.1109/LICS.2012.13.

[9]   André Platzer. A complete uniform substitution calculus for differential dynamic logic. *J. Autom. Reas.* **59**(2) (2017), 219–265. DOI: 10.1007/s108 17-016-9385-1.

[10] Erwin Schrödinger. An undulatory theory of the mechanics of atoms and molecules. *Phys. Rev.* **28** (1926), 1049–1070. DOI: 10.1103/PhysRev.2 8.1049.

Chapter 20
Virtual Substitution & Real Equations

Synopsis This chapter investigates decision procedures for real arithmetic, which serve as an important technology for proving the arithmetic questions that arise during cyber-physical systems analysis. The fact that first-order properties of real arithmetic are even decidable is one of the big miracles of logic on which CPS analysis depends. While a blackbox use of quantifier elimination often suffices, this chapter looks under the hood to understand why and how real arithmetic can be decided. This leads to a better appreciation of the working principles and complexity challenges in real arithmetic. The focus in this chapter will be on the case of linear and quadratic equations, which conceptually elegant virtual substitution techniques handle.

20.1 Introduction

Cyber-physical systems are important technical concepts for building better systems around us. Their safe design requires careful specification and verification, which this textbook provides using differential dynamic logic and its proof calculus [29–31, 33] discussed in Parts I and II. The proof calculus for differential dynamic logic has a number of powerful axioms and proof rules (especially in Chaps. 5, 6, 11, and 12). In theory, the *only* difficult problem in proving hybrid systems safety is finding their invariants or differential invariants [29, 32, 33] (also see Chap. 16). In practice, however, the handling of real arithmetic is another challenge that all CPS verification faces, even though the problem is easier in theory. How arithmetic interfaces with proofs by way of the proof rule ℝ for real arithmetic has already been discussed in Sect. 6.5. But how does the handling of real arithmetic by quantifier elimination really work?

This chapter discusses one technique for deciding interesting formulas of first-order real arithmetic. Understanding how such techniques for real arithmetic work is interesting for at least two reasons. First of all, it is important to understand why this miracle happens at all that something as complicated and expressive as first-

© Springer International Publishing AG, part of Springer Nature 2018 577
A. Platzer, *Logical Foundations of Cyber-Physical Systems*,
https://doi.org/10.1007/978-3-319-63588-0_20

order logic of real arithmetic ends up being decidable, so that a computer program can always tell us whether any real-arithmetic formula we dream up is *true* or *false*. But this chapter is also helpful to get an intuition about how real-arithmetic decision procedures work. With such an understanding, you are better prepared to identify the limitations of these techniques, learn when they are likely not to work out in due time, and get a sense of what you can do to help arithmetic prove more complicated properties. For complex proofs, it is often very important to use your insights and intuitions about the system to help a verification tool along to scale your verification results to more challenging systems in feasible amounts of time. An understanding of how arithmetic decision procedures work helps to focus such insights on the parts of the arithmetic analysis that has a big computational impact. Quite substantial impact has been observed for handling the challenges of real arithmetic [27, 30, 34].

There are a number of different approaches to understanding real arithmetic and its decision procedures beyond Tarski's original result from the 1930s [45], which was a major conceptual breakthrough but algorithmically impractical.[1] There is an algebraic approach using cylindrical algebraic decompositions [6, 7], which leads to practical procedures, but is highly nontrivial. Simple and elegant model-theoretic approaches use semantic properties of logic and algebra [22, 38], which are easy to understand, but do not lead to any particularly useful algorithms. There is a reasonably simple Cohen-Hörmander algorithm [5, 21] that, unfortunately, does not generalize well into a practical algorithm even if it works at small scale and has even been turned into a proof-producing algorithm [25]. Other simple but inefficient decision procedures are also described elsewhere [14, 24]. Finally, there is virtual substitution [48], a syntactical approach that fits well the understanding of logic that we have developed in this textbook and leads to highly efficient algorithms (although only for formulas with limited degrees). As a good compromise promoting accessibility and practicality, this chapter, thus, focuses on virtual substitution [48]. There are also approaches that focus on checking polynomial certificates for the validity of universal real arithmetic without existential quantifiers [19, 34]. These are simple and in principle capable of proving all valid formulas of the purely universal fragment of real arithmetic [43], but do not have the same generalizable insights that virtual substitution provides for how to eliminate quantifiers. See, e.g., [1, 2, 28, 34] for an overview of other techniques for real arithmetic.

The results in this chapter are from the literature [30, 48]. It adds substantial intuition and motivation that is helpful for following the technical development. The most important learning goals of this chapter are:

Modeling and Control: This chapter has an indirect impact on CPS models and controls by informing the reader about the consequences of the analytic complexity resulting from different arithmetical modeling tradeoffs. There is always more than one way of writing down a model. It becomes easier to find the right

---

[1] The significance of Tarski's result comes from his proof that real arithmetic is decidable at all and quantifier elimination even possible. The complexity of his procedure is entirely impractical compared to decision procedures that were invented later.

tradeoffs for expressing a CPS model with some knowledge of and intuition for the working principles of the workhorse of quantifier elimination that ultimately handles the resulting arithmetic.

Computational Thinking: The primary purpose of this chapter is to understand how arithmetical reasoning, which is crucial for CPS, can be done rigorously and automatically. Developing an intuition for the working principles of real-arithmetic decision procedures can be very helpful for developing strategies to verify CPS models at scale. The chapter also serves the purpose of learning to appreciate the miracle that quantifier elimination in real arithmetic provides by contrasting it with closely related problems of arithmetic that have fundamentally different challenges. We will also again see a conceptually very important device in the logical trinity: the flexibility of moving back and forth between syntax and semantics at will. We have seen this principle in action already in the case of differential invariants in Chap. 10, where we moved back and forth between analytic differentiation $\frac{d}{dt}$ and syntactic differentials $(\cdot)'$ by way of the differential lemma (Lemma 10.2) as we saw fit. This time, we leverage the same conceptual device for real arithmetic (rather than differential arithmetic) by working with virtual substitutions to bridge the gap between semantic operations that are inexpressible otherwise in the first-order logic of real arithmetic. Virtual substitutions will again allow us to move back and forth at will between syntax and semantics.

CPS Skills: This chapter has an indirect impact on CPS skills, because it gives some intuition and insights into useful pragmatics of CPS analysis for modeling and analysis tradeoffs that enable CPS verification at scale.

rigorous arithmetical reasoning
miracle of quantifier elimination
logical trinity for reals
switch between syntax & semantics at will
virtual substitution lemma
bridge gap between semantics and inexpressibles

analytic complexity
modeling tradeoffs

verifying CPS at scale

20.2 Framing the Miracle

First-order logic is an expressive logic in which many interesting properties and concepts can be expressed, analyzed, and proven. It is certainly significantly more expressive than propositional logic, which is decidable by NP-complete SAT solving [8], because propositional logic has no quantifiers and not even variables but only propositional connectives $\neg, \wedge, \vee$, etc. Propositional logic merely has arity 0 predicate symbols such as p, q, r that express tautologies like $p \wedge (q \vee r) \leftrightarrow (p \wedge q) \vee (p \wedge r)$.

In classical (uninterpreted) *first-order logic* (FOL), no symbol (except possibly equality) has a special meaning. There are only predicate symbols $p, q, r, \ldots$ and function symbols $f, g, h, \ldots$ whose meaning is subject to interpretation. And the domain that quantifiers range over is subject to interpretation, too. In particular, a formula of first-order logic is only valid if it holds true for all interpretations of all predicate and function symbols and all domains. Uninterpreted first-order logic corresponds to the fragment of dL that has propositional connectives and quantifiers (quantifying over any arbitrary domain, not necessarily the reals) as well as function and predicate symbols (Chap. 18) but no modalities or arithmetic.

In contrast, *first-order logic of real arithmetic* ($FOL_{\mathbb{R}}$ from Chap. 2) is interpreted, because all its symbols have a special fixed interpretation. The only predicate symbols are $=, \geq, >, \leq, <, \neq$ and they mean exactly equality, greater-or-equals, greater-than, etc., and the only function symbols are $+, -, \cdot$, which mean exactly addition, subtraction, and multiplication of real numbers. Furthermore, the universal and existential quantifiers quantify over the set $\mathbb{R}$ of all real numbers.[2]

The first special interpretation for symbols that comes to mind may not necessarily be addition and multiplication on real numbers but possibly the natural numbers $\mathbb{N}$ with $+$ for addition and $\cdot$ for multiplication on natural numbers where quantifiers range over the natural numbers. That gives the *first-order logic of natural numbers* ($FOL_{\mathbb{N}}$). Is $FOL_{\mathbb{N}}$ easier or harder than FOL? How does $FOL_{\mathbb{N}}$ compare to $FOL_{\mathbb{R}}$ where the only difference is that variables and quantifiers range over the reals instead of natural numbers? How do they both compare to $FOL_{\mathbb{Q}}$, the first-order logic of rational numbers? $FOL_{\mathbb{Q}}$ is like $FOL_{\mathbb{R}}$ and $FOL_{\mathbb{N}}$, except that all variables and quantifiers range over the rational numbers $\mathbb{Q}$ instead of over $\mathbb{R}$ and $\mathbb{N}$, respectively. How do those subtly different flavors of first-order logic compare? How difficult is it to prove validity of logical formulas in each case?

> Before you read on, see if you can find the answer for yourself.

Brief explanations of the meaning of decidability notions are summarized in Table 20.1. Uninterpreted first-order logic FOL is semidecidable, because there is a (sound and complete [16]) proof procedure that is able to prove all valid formulas of first-order logic [20]. If this proof procedure produces a proof, the output "yes" is justified by the soundness of the proof calculus. If it does not produce a proof, then the algorithm may or may not notice that it cannot ever find a proof, but nonter-

---

[2] Respectively over another real-closed field, but this does not change validity [45].

Table 20.1 Overview of decidability notions (e.g., for the validity problem)

Problem is	under the condition that
Decidable	There is an algorithm that always terminates and correctly says yes or no
Undecidable	There is no correct algorithm that always terminates
Semidecidable	There is a correct algorithm that terminates at least for all valid formulas
Cosemidecidable	There is a correct algorithm terminating at least for all invalid formulas

mination is acceptable for semidecidable problems if the correct answer would be "no." If an input formula is valid then the completeness of the proof procedure will guarantee that a proof will eventually be found for FOL, so this algorithm always terminates for input formulas that are valid and will, thus, ultimately say "yes." Of course, this is not actually helpful in practice unless the proof procedure is clever about its proof search.

The natural numbers are more difficult. Actually much more difficult! By Gödel's incompleteness theorem [17], first-order logic $\text{FOL}_\mathbb{N}$ of natural numbers does *not* have a sound and complete effective axiomatization. $\text{FOL}_\mathbb{N}$ is neither semidecidable nor cosemidecidable [4]. There is neither an algorithm that can prove all valid formulas of $\text{FOL}_\mathbb{N}$ nor one that can disprove all formulas of $\text{FOL}_\mathbb{N}$ that are not valid. Whatever algorithm we design for $\text{FOL}_\mathbb{N}$ it must fail to produce a correct answer for some valid formula as well as for some formula that is not valid. One way of understanding some of the inherent challenges with the logic of natural numbers in retrospect is to use the fact that not all questions about programs can be answered effectively (for example the halting problem of Turing machines is undecidable) [4, 46], in fact "none" can [36]. One can then encode questions about classical programs into the first-order logic of natural numbers. In such a reduction the natural number would, e.g., encode the state and tape of a Turing machine, while the $\text{FOL}_\mathbb{N}$ formula itself encodes the program of the Turing machine. We cannot prove all such formulas, because we cannot predict all behavior of all Turing machines.

Yet, a miracle happened! Alfred Tarski proved in 1930 that reals are much better behaved than natural numbers and that $\text{FOL}_\mathbb{R}$ is decidable, even though this seminal result remained unpublished for many years and only appeared in 1951 [44, 45]. We will follow a much more recent and simpler development of real arithmetic proving here than Tarski's original breakthrough, but will not achieve the same level of completeness, because virtual substitution only works for limited polynomial degrees.

The first-order logic $\text{FOL}_\mathbb{Q}$ of rational numbers was shown to be undecidable [39, 40], even though rational numbers may appear to be so close to real numbers. Rationals are lacking something important: completeness (in the topological sense). The square root $\sqrt{2}$ of 2 is a witness for $\exists x\,(x^2 = 2)$ but only a real number, not a rational one. So the formula $\exists x\,(x^2 = 2)$ is valid in $\text{FOL}_\mathbb{R}$ but not valid in $\text{FOL}_\mathbb{Q}$.

The first-order logic $\text{FOL}_\mathbb{C}$ of complex numbers, though, is again perfectly decidable [3, 45]. See Table 20.2 for a summary of how first-order logic behaves depending on the domain of quantification.

In between, there are a few additional fragments of logic that are better behaved and worth a short mention. *Linear real arithmetic* (i.e., no multiplication)

Table 20.2 The miracle of reals: overview of FOL validity problems

Logic	Domain	Validity
FOL	uninterpreted	semidecidable
$FOL_\mathbb{N}$	natural numbers	not semidecidable or cosemidecidable
$FOL_\mathbb{Q}$	rational numbers	not semidecidable or cosemidecidable
$FOL_\mathbb{R}$	real numbers	decidable
$FOL_\mathbb{C}$	complex numbers	decidable

with just equations, conjunctions, and existential quantifiers is decidable, because its generalization $FOL_\mathbb{R}$ is decidable. But the point is that $FOL_\mathbb{R}$ formulas that are only formed with $+, =, \wedge, \exists$ can already be solved by Gaussian elimination, because they only express the existence of solutions of linear equation systems. Linear real arithmetic with weak inequalities, conjunctions, and existential quantifiers is decidable by Fourier-Motzkin elimination [15], which Joseph Fourier invented in 1826 by generalizing Gaussian elimination with a way of flipping inequalities as needed when multiplying with negative quantities. The idea was subsequently reinvented by Dines and again by Motzkin [10, 26] and formed the basis for linear-programming optimization [13]. Linear real arithmetic is conceptually easier than nonlinear real arithmetic, because only nonlinear real arithmetic can tell the difference between the real and the rational numbers: $\exists x (x^2 = 2)$ is *true* over $\mathbb{R}$ but *false* over $\mathbb{Q}$, because $\pm\sqrt{2}$ are not rational numbers. It takes nonlinear arithmetic to notice such a difference, though, because linear real arithmetic (with rational coefficients) always has rational solutions if it has any solutions at all. The complexity of linear problems is hard regardless [47].

Presburger arithmetic, which is like $FOL_\mathbb{N}$ but without multiplication, has been shown to be decidable independently by Presburger in 1929 and by Skolem in 1931 [35, 42]. While multiplication can certainly be rephrased as repeated additions, there is no bound on the number of additions needed to represent the multiplication $n \cdot m$ and, thus, also no finite formula that expresses $n \cdot m$ with only addition. In fact, Presburger arithmetic also includes unary predicate symbols that check whether their argument is divisible by a given constant number, for example, whether a number is even, whether it is divisible by 3, etc., but that does not change its decidability.

That the validity problem of real-arithmetic $FOL_\mathbb{R}$ is decidable is a miracle. But it crucially depends on quantification ranging over real numbers (or other real-closed fields) and on addition and multiplication being the only arithmetic operations (besides comparison operators, propositional connectives, and quantifiers, or other definable operators such as subtraction). If we were to include the exponential function e^x then decidability is an open problem since Tarski, despite considerable progress [12]. That explains why we do not allow variable powers x^y for variables x and y but merely natural numbers as powers, because those are definable, for example x^3 for $x \cdot x \cdot x$. Several other extensions of $FOL_\mathbb{R}$ are undecidable [37], for example extensions with the trigonometric function $\sin x$, because its roots characterize an isomorphic copy of the natural numbers.

20.3 Quantifier Elimination

Alfred Tarski's seminal insight for deciding real arithmetic is based on quantifier elimination, i.e., the successive elimination of quantifiers from formulas so that the remaining formula is equivalent but structurally significantly easier, because it has fewer quantifiers. Why does eliminating quantifiers help? When evaluating whether a logical formula is true or false in a given state (i.e., an assignment of real numbers to all its free variables), then arithmetic comparisons and polynomial terms are easy, because all we need to do is plug the numbers in and compute according to their semantics from Sect. 2.7.2. For example, for a state ω with $\omega(x) = 2$, we can easily evaluate the logical formula

$$x^2 > 2 \wedge 2x < 3 \vee x^3 < x^2$$

to *false* by following the semantics, which ultimately plugs in 2 as the value for x:

$$\omega[\![x^2 > 2 \wedge 2x < 3 \vee x^3 < x^2]\!] = 2^2 > 2 \wedge 2 \cdot 2 < 3 \vee 2^3 < 2^2 = \textit{false}$$

Similarly, in a state v with $v(x) = -1$, the same formula evaluates to *true*:

$$v[\![x^2 > 2 \wedge 2x < 3 \vee x^3 < x^2]\!] = (-1)^2 > 2 \wedge 2 \cdot (-1) < 3 \vee (-1)^3 < (-1)^2 = \textit{true}$$

But quantifiers are a difficult matter, because they require us to check for *all* possible values of a variable (in the case of $\forall x F$) or to find exactly the right value for a variable that makes the formula true (in the case of $\exists x F$). The easiest formulas to evaluate are the ones that have no free variables (because then their value does not depend on the state ω by Sect. 5.6.5) and that also have no quantifiers (because then there are no choices for the values of the quantified variables during the evaluation). Quantifier elimination can take a logical formula that is closed, i.e., has no free variables, and equivalently remove its quantifiers, so that it becomes easy to evaluate the formula to *true* or *false*. Quantifier elimination even works when formulas still have free variables. Then it will eliminate all quantifiers in the formula but the original free variables will remain free in the resulting formula, unless it is simplified in the quantifier elimination process.

> **Definition 6.3 (Quantifier elimination).** A first-order logic theory (such as first-order logic $\text{FOL}_\mathbb{R}$ over the reals) admits *quantifier elimination* if, for each formula P, a quantifier-free formula $\text{QE}(P)$ can be effectively associated with P that is equivalent, i.e., $P \leftrightarrow \text{QE}(P)$ is valid.

That is, a first-order theory admits quantifier elimination iff there is a computer program that outputs a quantifier-free formula $\text{QE}(P)$ for any input formula P in that theory such that the input and output are equivalent (so $P \leftrightarrow \text{QE}(P)$ is valid) and such that the output $\text{QE}(P)$ is quantifier-free (and has no free variables that are not already free in the input formula P). Tarski's seminal result shows that quantifier elimination is computable and first-order real arithmetic is decidable [45]:

> **Theorem 6.2 (Tarski's quantifier elimination).** *The first-order logic of real arithmetic admits quantifier elimination and is, thus, decidable.*

The operation QE is further assumed to evaluate ground formulas (i.e., those without variables), yielding a decision procedure for closed formulas of $FOL_\mathbb{R}$ (i.e., formulas without free variables). For a closed formula P, all it takes is to compute its quantifier-free equivalent $QE(P)$ by quantifier elimination. The formula P is closed, so has no free variables or other uninterpreted symbols, and neither will $QE(P)$. Hence, P and its equivalent $QE(P)$ are either equivalent to *true* or to *false*. Yet, $QE(P)$ is quantifier-free, so which case holds can be found out by evaluating the (variable-free) concrete arithmetic in $QE(P)$ as in the above examples.

Example 20.1. Quantifier elimination uses the special structure of real arithmetic to express quantified arithmetic formulas equivalently without quantifiers and without using more free variables. For instance, QE yields the following equivalence:

$$QE(\exists x\,(2x^2 + c \le 5)) \equiv c \le 5$$

In particular, the formula $\exists x\,(2x^2 + c \le 5)$ is not valid, but only true if $c \le 5$ holds, as has been so aptly described by the outcome of the above quantifier elimination result.

Example 20.2. Quantifier elimination can be used to find out whether a first-order formula of real arithmetic is valid. Take $\exists x\,(2x^2 + c \le 5)$, for example. A formula is valid iff its universal closure is, i.e., the formula obtained by universally quantifying all free variables. After all, valid means that a formula is true for all interpretations. Hence, consider the universal closure $\forall c\,\exists x\,(2x^2 + c \le 5)$, which is a closed formula, because it has no free variables. Quantifier elimination can, for example, lead to

$$QE(\forall c\,\exists x\,(2x^2 + c \le 5)) \equiv QE(\forall c\,QE(\exists x\,(2x^2 + c \le 5))) \equiv QE(\forall c\,(c \le 5)) \equiv$$
$$-100 \le 5 \wedge 5 \le 5 \wedge 100 \le 5$$

The resulting formula still has no free variables but is now quantifier-free, so it can simply be evaluated arithmetically. Since the conjunct $100 \le 5$ evaluates to *false*, the universal closure $\forall c\,\exists x\,(2x^2 + c \le 5)$ is equivalent to *false* and, hence, the original formula $\exists x\,(2x^2 + c \le 5)$ is not valid (although still satisfiable for $c = 1$).

Geometrically, quantifier elimination corresponds to projection; see Fig. 20.1. Note that, when using QE, we usually assume it already evaluates ground arithmetic, so that the only two possible outcomes of applying QE to a closed formula without free variables are the formula *true* and the formula *false*.

Alfred Tarski's result that quantifier elimination over the reals is possible and that real arithmetic is decidable was groundbreaking. The only issue is that the complexity of Tarski's decision procedure is non-elementary, i.e., cannot be bounded by any tower of exponentials $2^{2^{\cdot^{\cdot^{\cdot^n}}}}$, which makes it completely impractical. Still, it was a seminal breakthrough because it showed reals to be decidable at all. It took further

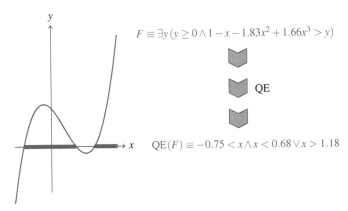

Fig. 20.1 The geometric counterpart of quantifier elimination for $\exists y$ is projection onto the x axis

advances [5, 14, 21, 24, 41] and a major breakthrough by George Collins in 1975 [6] before practical procedures were found [6, 7, 48]. The virtual substitution technique shown in this chapter has been implemented in Redlog [11]. Ideas from SMT solving are combined with nonlinear real arithmetic [23] in the SMT solver Z3.

20.3.1 Homomorphic Normalization for Quantifier Elimination

The first insight for defining quantifier elimination is to understand that the quantifier elimination operation commutes with almost all logical connectives, so that QE only needs to be defined for existential quantifiers. Consequently, as soon as we understand how to eliminate existential quantifiers, universal quantifiers can be eliminated as well just by double negation, because $\forall x A$ is equivalent to $\neg \exists x \neg A$:

$$QE(A \wedge B) \equiv QE(A) \wedge QE(B)$$
$$QE(A \vee B) \equiv QE(A) \vee QE(B)$$
$$QE(\neg A) \equiv \neg QE(A)$$
$$QE(\forall x A) \equiv QE(\neg \exists x \neg A)$$

These transformations isolate existential quantifiers for quantifier elimination. In particular, it is sufficient if quantifier elimination focuses on existentially quantified variables. When using the QE operation inside out, i.e., when using it repeatedly to eliminate the innermost quantifier to get a quantifier-free equivalent and then again eliminating the innermost quantifier, the quantifier elimination is solved if only we manage to solve it for $\exists x A$ with a quantifier-free formula A. If A is not quantifier-free yet, its quantifiers can be eliminated from inside out:

$$\text{QE}(\exists x A) \equiv \text{QE}(\exists x \, \text{QE}(A)) \qquad\qquad \text{if } A \text{ is not quantifier-free}$$

It is possible, although not necessary and not even necessarily helpful, to simplify the form of A as well. The following transformations transform the (quantifier-free) *kernel* after a quantifier into negation normal form using De Morgan's equivalences:

$$\text{QE}(\exists x\,(A \lor B)) \equiv \text{QE}(\exists x A) \lor \text{QE}(\exists x B)$$
$$\text{QE}(\exists x\, \neg(A \land B)) \equiv \text{QE}(\exists x\,(\neg A \lor \neg B))$$
$$\text{QE}(\exists x\, \neg(A \lor B)) \equiv \text{QE}(\exists x\,(\neg A \land \neg B))$$
$$\text{QE}(\exists x\, \neg\neg A) \equiv \text{QE}(\exists x A)$$

This transformation can make matters worse in practice, because conversions between disjunctive and conjunctive normal forms may have exponential results. We will make use of it just to reduce the number of cases that still need to be covered. Distributivity can be used to simplify the form of the quantifier-free kernel A to disjunctive normal form and split existential quantifiers over disjuncts:

$$\text{QE}(\exists x\,(A \land (B \lor C))) \equiv \text{QE}(\exists x\,((A \land B) \lor (A \land C)))$$
$$\text{QE}(\exists x\,((A \lor B) \land C)) \equiv \text{QE}(\exists x\,((A \land C) \lor (B \land C)))$$

The only remaining case to address is the case $\text{QE}(\exists x\,(A \land B))$ where $A \land B$ is a purely conjunctive formula (yet it can actually have any number of conjuncts, not just two). Finally, using the following normalizing equivalences,

$$p = q \equiv p - q = 0$$
$$p \leq q \equiv p - q \leq 0$$
$$p < q \equiv p - q < 0$$
$$p \neq q \equiv p - q \neq 0$$
$$p \geq q \equiv q \leq p$$
$$p > q \equiv q < p$$
$$\neg(p \leq q) \equiv p > q$$
$$\neg(p < q) \equiv p \geq q$$
$$\neg(p = q) \equiv p \neq q$$
$$\neg(p \neq q) \equiv p = q$$

it is possible to normalize all atomic formulas equivalently to one of the forms $p = 0, p < 0, p \leq 0, p \neq 0$ with right-hand side 0. Since $p \neq 0$ is equivalent to $p < 0 \lor -p < 0$, disequations $\neq$ are unnecessary *in theory* as well (although they are quite useful to retain in practice). Now, all that remains to be done is to focus on the core question of equivalently eliminating existential quantifiers from a conjunction of these normalized atomic formulas.

20.3.2 Substitution Base

Virtual substitution is a quantifier elimination technique that is based on substituting extended terms into formulas virtually, i.e., without the extended terms[3] actually occurring in the resulting constraints. Virtual substitution pretends that the language has additional constructs, but then replaces them by other equivalents in the substitution process.

> Virtual substitution in $\text{FOL}_{\mathbb{R}}$ essentially leads to an equivalence of the form
>
> $$\exists x F \leftrightarrow \bigvee_{t \in T} A_t \wedge F_x^t \qquad (20.1)$$
>
> for a suitable *finite* set T of extended terms that depends on the formula F and that gets substituted into F virtually, i.e., in a way that results in standard real-arithmetic terms, not extended terms. The additional formulas A_t are compatibility conditions that may be necessary to make sure the respective substitutions are meaningful.

Such an equivalence is how quantifier elimination can work. Certainly if the right-hand side of (20.1) is true, then t is a witness for $\exists x F$. The key to establishing an equivalence of the form (20.1) is to ensure that if F has a solution at all (in the sense of $\exists x F$ being true), then F must also already hold for one of the cases in the set T. That is, T must cover all representative cases. There might be many more solutions, but if there is one at all, one of the possibilities in T must be a solution as well. If we were to choose all real numbers $T \overset{\text{def}}{=} \mathbb{R}$, then (20.1) would be trivially valid, but then the right-hand side would not be a formula because it is uncountably infinitely long, which is even worse than the quantified form on the left-hand side. But if a finite set T is sufficient for the equivalence (20.1) and the extra formulas A_t are quantifier-free, then the right-hand side of (20.1) is structurally simpler than the left-hand side, even if it may be (sometimes significantly) less compact.

The various ways of virtually substituting various forms of extended reals e into logical formulas equivalently without having to mention the actual extended reals is the secret of virtual substitution. The first step is to see that it is enough to define substitutions only on atomic formulas of the form $p = 0, p < 0, p \leq 0$ (or, just as well, on $p = 0, p > 0, p \geq 0$). If σ denotes such a substitution of term θ for variable x, then σ lifts to arbitrary first-order formulas homomorphically:

$$\sigma(A \wedge B) \equiv \sigma A \wedge \sigma B$$
$$\sigma(A \vee B) \equiv \sigma A \vee \sigma B$$
$$\sigma(\neg A) \equiv \neg \sigma A$$
$$\sigma(\forall y A) \equiv \forall y \sigma A \qquad\qquad \text{if } x \neq y \text{ and } y \notin \theta$$

---

[3] Being an *extended real term* really means it is not a real term, but somehow closely related. We will see more concrete extended real terms and how to get rid of them again later.

$$\sigma(\exists y A) \equiv \exists y\, \sigma A \qquad\qquad\qquad \text{if } x \neq y \text{ and } y \notin \theta$$
$$\sigma(p = q) \equiv \sigma(p - q = 0)$$
$$\sigma(p < q) \equiv \sigma(p - q < 0)$$
$$\sigma(p \leq q) \equiv \sigma(p - q \leq 0)$$
$$\sigma(p > q) \equiv \sigma(q - p < 0)$$
$$\sigma(p \geq q) \equiv \sigma(q - p \leq 0)$$
$$\sigma(p \neq q) \equiv \sigma(\neg(p - q = 0))$$

This lifting applies the substitution σ to all subformulas (with minor twists on quantifiers for admissibility to avoid capture of variables) and with normalization of atomic formulas into the canonical forms $p = 0, p < 0, p \leq 0$ for which σ has been assumed to already have been defined.

From now on, all that remains to be done to define a substitution or virtual substitution is to define it on atomic formulas of the remaining forms $p = 0, p < 0, p \leq 0$ for terms p and the above construction will take care of substituting in any first-order formulas. Of course, the above construction is only helpful for normalizing atomic formulas that are not already of one of those forms, so the term q above can be assumed not to be the term 0, otherwise $\sigma(p < 0)$ would create a useless $\sigma(p - 0 < 0)$.

20.3.3 Term Substitutions for Linear Equations

This is as far as we can push quantifier elimination generically without looking more closely at the shape of the actual polynomials that are involved. Let's start with an easy case where one of the formulas in the conjunction in the scope of the existential quantifier is a linear equation. Consider a formula of the form

$$\exists x (bx + c = 0 \wedge F) \qquad (x \notin b,c) \qquad\qquad (20.2)$$

where x does not occur in the terms b,c (otherwise $bx + c$ would not be linear if b is, say, $5x$). Let's consider what a mathematical solution to this formula might look like. The only solution that the conjunct $bx + c = 0$ has is $x = -c/b$. Hence, the left conjunct in (20.2) only holds for $x = -c/b$, so formula (20.2) can only be true if F also holds for that single solution $-c/b$ in place of x. That is, formula (20.2) holds only if $F_x^{-c/b}$ does. Hence, (20.2) is equivalent to the formula $F_x^{-c/b}$, which is quantifier-free.

So, how can we eliminate the quantifier in (20.2) equivalently?

Before you read on, see if you can find the answer for yourself.

Most certainly, $F_x^{-c/b}$ is quantifier-free. But it is not exactly always equivalent to (20.2) and, thus, does not necessarily qualify as its quantifier-eliminated form. Oh no! What we wrote down is a good intuitive start, but does not make any sense at all if $b = 0$, for then $-c/b$ would be an ill-advised division by zero. Performing such divisions by zero sounds like a fairly shaky start for an equivalence transformation such as quantifier elimination, and it certainly sounds like a shaky start for anything that is supposed to ultimately turn into a proof.

Let's start over. The first conjunct in (20.2) has the solution $x = -c/b$ **if** $b \neq 0$. In that case, indeed, (20.2) is equivalent to $F_x^{-c/b}$, because the only way for (20.2) to be true then is exactly when the second conjunct F holds for the only solution of the first conjunct, i.e., when $F_x^{-c/b}$ holds. How do we know whether b is zero?

If b were a concrete number such as 5 or a term such as $2 + 4 - 6$ then it is easy to tell whether b is 0 or not. But if b is a term with other variables, such as $y^2 + y - 2z$, then it is really hard to say whether its value might be zero or not, because that depends on what values the variables y and z have. Certainly if b is the zero polynomial, we know for sure that is 0. Or b may be a polynomial that can never be zero, such as a sum of squares plus a positive constant. In general, we may have to retain a logical disjunction and have one formula that considers the case where $b \neq 0$ and another formula that considers the case where $b = 0$. After all, logic is quite good at keeping its options separate with disjunctions or other logical connectives.

If $b = 0$, then the first conjunct in (20.2) is independent of x and has all numbers for x as solutions if $c = 0$ and, otherwise, has no solution at all if $c \neq 0$. In the latter case, $b = 0, c \neq 0$, (20.2) is *false*, because its first conjunct is already *false*. In the former case, $b = c = 0$, however, the first conjunct $bx + c = 0$ is trivial and does not impose any constraints on x, nor does it help us to find a quantifier-free equivalent of (20.2). In that case $b = c = 0$, the trivial constraint will be dropped, and the remaining formula F will be considered recursively instead to see, e.g., whether it contains other linear equations that help identify its solution.

In the non-degenerate case $b \neq 0$ with $x \notin b, c$, the input formula (20.2) can be rephrased into a quantifier-free equivalent over $\mathbb{R}$ as follows.

Theorem 20.1 (Virtual substitution of linear equations). *If $x \notin FV(b)$ and $x \notin FV(c)$, then the following equivalence is valid over $\mathbb{R}$:*

$$b \neq 0 \rightarrow \left(\exists x (bx + c = 0 \wedge F) \leftrightarrow b \neq 0 \wedge F_x^{-c/b} \right) \qquad (20.3)$$

All it takes is, thus, the ability to substitute the term $-c/b$ for x in the formula F. The division $-c/b$ that will occur in $F_x^{-c/b}$ for ordinary term substitutions can cause technical annoyances but at least it is well-defined, because $b \neq 0$ holds in any context on which $-c/b$ is used. Instead of pursuing the looming question of how exactly this substitution of a fraction in $F_x^{-c/b}$ works, we already make the question more general by moving to the quadratic case right away, because that case will include an answer for the appropriate logical treatment of fractions as well.

Before proceeding to the quadratic case, first observe that the uniform substitutions from Chap. 18 provide a particularly elegant way of phrasing Theorem 20.1

axiomatically if divisions are in the term language (suitably guarded to only be used when the divisor is nonzero).

Lemma 20.1 (Uniform substitution of linear equations). *The linear equation axiom is sound, where b,c are arity 0 function symbols:*

$$\exists \text{lin } b \neq 0 \to \left(\exists x (b \cdot x + c = 0 \wedge q(x)) \leftrightarrow q(-c/b)\right)$$

Proof. If the assumption $b \neq 0$ is *true*, then, since the value of b is independent of x, the *only* value for variable x that satisfies the linear equation $b \cdot x + c = 0$ is its mathematical solution $-c/b$, which is well-defined since $b \neq 0$. Consequently, the conjunction $b \cdot x + c = 0 \wedge q(x)$ is true for some x iff $q(-c/b)$ is true, since $-c/b$ is the only solution of $b \cdot x + c = 0$. $\square$

Axiom $\exists$lin uses a unary predicate symbol q and arity 0 function symbols b,c, whose values, thus, cannot depend on the quantified variable x, so that $b \cdot x + c = 0$ is linear. Recall from Chap. 18, that uniform substitutions would clash if they were to replace the arity 0 function symbols b or c with terms that mention x in a context where x is bound by $\exists x$, which enforces linearity also after uniform substitution. But other variables can still be used in substitutes for b,c, just not the bound variable x.

Example 20.3. Since the linear cofactor $y^2 + 4$ is easily shown to be nonzero (it is a sum of squares with a strictly positive offset), the following formula

$$\exists x \left((y^2 + 4) \cdot x + (yz - 1) = 0 \wedge x^3 + x \geq 0\right)$$

is by axiom $\exists$lin equivalent to the quantifier-free formula:

$$\left(-\frac{yz - 1}{y^2 + 4}\right)^3 + \left(-\frac{yz - 1}{y^2 + 4}\right) \geq 0$$

While the chapter proceeds, can you already envision a way of restating this resulting quantifier-free formula equivalently without using fractions or quantifiers?

20.4 Square Root $\sqrt{\cdot}$ Virtual Substitutions for Quadratics

Next consider quadratic equations in a formula of the form

$$\exists x (ax^2 + bx + c = 0 \wedge F) \qquad (x \notin \text{FV}(a), \text{FV}(b), \text{FV}(c)) \qquad (20.4)$$

where x does not occur free in the terms a,b,c. Pursuing arguments analogously to the linear case, we identify the solutions of the quadratic equation and substitute them into F. The generic solution of its first conjunct is $x = (-b \pm \sqrt{b^2 - 4ac})/(2a)$, but that, of course, again depends on whether a can be evaluated to zero, in which case linear solutions may be possible and the division by $2a$ is most certainly not well-defined; see Fig. 20.2.

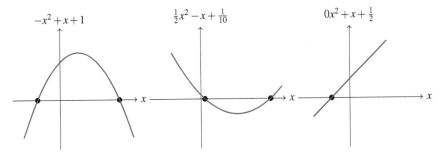

$-x^2+x+1$ $\frac{1}{2}x^2-x+\frac{1}{10}$ $0x^2+x+\frac{1}{2}$

Fig. 20.2 Roots of different quadratic functions p

Whether term a can be zero may again sometimes be hard to say when a is actually a polynomial term that has roots, but does not always evaluate to 0 either (which only the zero polynomial would). So let's be more careful right away this time to find an equivalent formulation for all possible cases of a, b, c. The cases to consider are where the first conjunct is either a constant equation (in which case the equation imposes no interesting constraint on x) or a linear equation (in which case $x = -c/b$ is the solution by Sect. 20.3.3) or a proper quadratic equation with $a \neq 0$ (in which case $x = (-b \pm \sqrt{b^2 - 4ac})/(2a)$ are the solutions). The trivial equation $0 = 0$ when $a = b = c = 0$ is again useless, so another part of F has to be considered in that case, and the equation $c = 0$ for $a = b = 0, c \neq 0$ is again *false*, so immediately refutes the existence of a solution of (20.4).

When $ax^2 + bx = 0$ is either a proper linear or a proper quadratic equation, its respective solutions single out the only points that can solve (20.4), so the only points in which it remains to be checked whether the second conjunct F also holds.

Theorem 20.2 (Virtual substitution of quadratic equations). *For quantifier-free formula F with $x \notin FV(a), FV(b), FV(c)$, the following equivalence is valid over $\mathbb{R}$:*

$$a \neq 0 \vee b \neq 0 \vee c \neq 0 \rightarrow$$
$$\left(\exists x (ax^2 + bx + c = 0 \wedge F) \leftrightarrow \right.$$
$$a = 0 \wedge b \neq 0 \wedge F_x^{-c/b}$$
$$\left. \vee\, a \neq 0 \wedge b^2 - 4ac \geq 0 \wedge \left(F_x^{(-b+\sqrt{b^2-4ac})/(2a)} \vee F_x^{(-b-\sqrt{b^2-4ac})/(2a)} \right) \right)$$

Hold on, we fortunately noticed just in time when writing down the formula in Theorem 20.2 that $(-b + \sqrt{b^2 - 4ac})/(2a)$ only ever makes actual sense in the reals if $b^2 - 4ac \geq 0$, because the square root is otherwise imaginary, which is really rather hard to find in FOL$_\mathbb{R}$. A quadratic equation only has a solution in the reals if its discriminant $b^2 - 4ac$ is nonnegative.

The resulting formula on the right-hand side of the bi-implication in Theorem 20.2 is quantifier-free and, thus, can be chosen for $\text{QE}(\exists x\,(ax^2+bx+c=0 \wedge F))$ as long as it is not the case that $a=b=c=0$.

> The important thing to notice is that $(-b \pm \sqrt{b^2-4ac})/(2a)$ is not a polynomial term, nor even a rational term, because it involves a square root $\sqrt{\cdot}$. Hence, the equivalence in Theorem 20.2 is not a formula of first-order real arithmetic unless we do something about its square roots and divisions!

If nonnegative square roots were allowed as expressions, then the same idea as in Lemma 20.1 would turn Theorem 20.2 into a uniform substitution axiom. Recall from Chap. 2 that the terms of $\text{FOL}_\mathbb{R}$ are polynomials with rational coefficients in $\mathbb{Q}$. So $4x^2 + \frac{1}{7}x - 1.41$ is a polynomial term of $\text{FOL}_\mathbb{R}$. But $4x^2 + \frac{1}{y}x - 1.41$ is not, because of the division by variable y, which should make us panic about y possibly being zero in any case. And $4x^2 + \frac{1}{7}x - \sqrt{2}$ is not a polynomial term with rational coefficients either, because of the square root $\sqrt{2}$. And $4x^2 + \sqrt{y}x - 2$ is totally off.

> **Note 82 (Semantic domains versus syntactic expressions)** The domains that the quantifiers $\forall$ and $\exists$ of first-order logic $\text{FOL}_\mathbb{R}$ of real arithmetic quantify over include reals like $\sqrt{2}$. But the terms and logical formulas themselves are syntactically restricted to be built from polynomials with rational coefficients. Square roots (and all higher roots) are already part of the semantic domain $\mathbb{R}$, but not allowed directly in the syntax of $\text{FOL}_\mathbb{R}$.

Of course, it is still easy to write down a formula such as $\exists x\,(x^2 = 5)$ that indirectly makes sure that x will have to assume the value $\sqrt{5}$, but that formula mentions a quantifier again, so requires extra effort during quantifier elimination.

20.4.1 Square Root Algebra

Square roots are really not part of real arithmetic. They can be defined by appropriate quadratures. For example, the positive root $x = \sqrt{y}$ can be defined by the formula $x^2 = y \wedge y \geq 0$. Let's find out how square roots such as $(-b \pm \sqrt{b^2-4ac})/(2a)$ can be substituted into first-order formulas systematically without the need to involve any square roots in the resulting formula. The first step in understanding how to virtually substitute expressions of the general shape $(a+b\sqrt{c})/d$ into a formula is to investigate how to substitute them into the polynomials that occur in the formula.

> **Definition 20.1 (Square root algebra).** A *square root expression* is an expression of the form
> $$(a+b\sqrt{c})/d$$

with polynomials $a,b,c,d \in \mathbb{Q}[x_1,\ldots,x_n]$ with rational coefficients in the variables $x_1,\ldots,x_n$ if, for well-definedness, $d \neq 0 \wedge c \geq 0$. Square root expressions with the same $\sqrt{c}$ can be added and multiplied symbolically by considering them as algebraic objects:[a]

$$((a+b\sqrt{c})/d) + ((a'+b'\sqrt{c})/d') = ((ad'+da') + (bd'+db')\sqrt{c})/(dd')$$
$$((a+b\sqrt{c})/d) \cdot ((a'+b'\sqrt{c})/d') = ((aa'+bb'c) + (ab'+ba')\sqrt{c})/(dd')$$
$$(20.5)$$

---

[a] Despite the poor notation, please don't mistake the primes for derivatives here. The name a' is not the derivative of a here but just meant as a name for a polynomial term that happens to go by the misleading name a'.

Another way of saying this is that square root expressions with the same $\sqrt{c}$ provide an addition and a multiplication operation that leads to square root expressions in the same $\sqrt{c}$. Substituting $(a+b\sqrt{c})/d$ for a variable x in a polynomial term p, thus, leads to a square root expression $p_x^{(a+b\sqrt{c})/d} = (\tilde{a}+\tilde{b}\sqrt{c})/\tilde{d}$ with the same $\sqrt{c}$, because the arithmetic resulting from evaluating the polynomial only requires addition and multiplication using (20.5).[4] After all, a polynomial is represented as a term involving only addition and multiplication (remembering that $a - b$ is $a + (-1) \cdot b$).

Symbolic addition and multiplication makes it possible to substitute a square root expression for a variable in a polynomial. Yet, the result $p_x^{(a+b\sqrt{c})/d}$ is still a square root expression, which cannot be written down directly in first-order real arithmetic. But, at least, substituting a square root expression $(a+b\sqrt{c})/d$ into a polynomial p for x leads to some square root expression $p_x^{(a+b\sqrt{c})/d} = (a'+b'\sqrt{c})/d'$ of the same $\sqrt{c}$.

Example 20.4 (Quadratic roots into quadratic polynomials). As a simple example, let us substitute the square root expression $(-b+\sqrt{b^2-4ac})/(2a)$ into the quadratic polynomial $ax^2 + bx + c$ by the symbolic computation (20.5):

$$(ax^2+bx+c)_x^{(-b+\sqrt{b^2-4ac})/(2a)}$$
$$= a((-b+\sqrt{b^2-4ac})/(2a))^2 + b((-b+\sqrt{b^2-4ac})/(2a)) + c$$
$$= a((b^2+b^2-4ac+(-b-b)\sqrt{b^2-4ac})/(4a^2)) + (-b^2+b\sqrt{b^2-4ac})/(2a) + c$$
$$= (ab^2+ab^2-4a^2c+(-ab-ab)\sqrt{b^2-4ac})/(4a^2) + (-b^2+2ac+b\sqrt{b^2-4ac})/(2a) + c$$
$$= ((ab^2+ab^2-4a^2c)2a + (-b^2+2ac)4a^2 + ((-ab-ab)2a+b4a^2)\sqrt{b^2-4ac})/(8a^3)$$
$$= (2a^2b^2 + 2a^2b^2 - 8a^3c + -4a^2b^2 + 8a^3c + (-2a^2b - 2a^2b + 4a^2b)\sqrt{b^2-4ac})/(8a^3)$$
$$= (0+0\sqrt{b^2-4ac})/(8a^3) = 0$$

---

[4] In practice, the polynomial addition and multiplication operations for a polynomial p are performed by Horner's scheme for dense polynomials p and by repeated squaring for sparse polynomials p. This avoids redundant cases when, e.g., considering x^3 and x^2.

The result is the zero expression! How did that happen? Come to think of it, we could have foreseen this, because the square root expression $(-b+\sqrt{b^2-4ac})/(2a)$ we just substituted into the polynomial ax^2+bx+c is its root and has to yield 0.

The polynomial evaluation resulting from these square root expression computations will substitute a square root expression into a polynomial. The next step is to handle the comparison to 0 of the resulting square root expression in atomic formulas $p\sim 0$ for some $\sim\,\in\{=,\leq,<\}$. That works by characterizing it using the square root expression $p_x^{(a+b\sqrt{c})/d}$:

$$(p\sim 0)_x^{(a+b\sqrt{c})/d} \equiv (p_x^{(a+b\sqrt{c})/d}\sim 0)$$

In order to save some notational effort, suppose the square root expression $p_x^{(a+b\sqrt{c})/d}$ is again $(a+b\sqrt{c})/d$, which is, of course, only accurate for the polynomial $p(x)=x$, but cuts down on the number of primes in the symbol names. All that remains to be done is to rewrite the square root expression comparison $(a+b\sqrt{c})/d\sim 0$ to an equivalent in FOL$_\mathbb{R}$ in a way that does not use square root expressions anymore.

Definition 20.2 (Square root comparisons). Assume $d\neq 0\wedge c\geq 0$ for well-definedness. For square-root-free expressions ($b=0$) with just divisions, i.e., those of the form $(a+0\sqrt{c})/d$ alias a/d, the following equivalences hold:

$$a/d=0\equiv a=0$$
$$a/d\leq 0\equiv ad\leq 0$$
$$a/d<0\equiv ad<0$$

For square root expressions $(a+b\sqrt{c})/d$ with arbitrary polynomial b, the following equivalences hold, assuming $d\neq 0\wedge c\geq 0$ for well-definedness:

$$(a+b\sqrt{c})/d=0\equiv ab\leq 0\wedge a^2-b^2c=0$$
$$(a+b\sqrt{c})/d\leq 0\equiv ad\leq 0\wedge a^2-b^2c\geq 0\vee bd\leq 0\wedge a^2-b^2c\leq 0$$
$$(a+b\sqrt{c})/d<0\equiv ad<0\wedge a^2-b^2c>0\vee bd\leq 0\wedge(ad<0\vee a^2-b^2c<0)$$

In the cases for $b=0$, the sign of ad determines the sign, except that $d\neq 0$ implies that $a=0$ is enough in the first case. The first line for arbitrary b characterizes that $(a+b\sqrt{c})/d=0$ holds iff a,b have different signs (possibly 0) and their squares cancel, because $a^2=b^2c$, which implies $a=-b\sqrt{c}$. The second line characterizes that ≤ 0 holds iff $a^2\geq b^2c$ so that a will dominate the overall sign, where a has a different sign than d by $ad\leq 0$, or if $a^2\leq b^2c$ so that $b\sqrt{c}$ will dominate the overall sign, where b has a different sign than d (possibly 0) by $bd\leq 0$. The square $a^2-b^2c=a^2-b^2\sqrt{c}^2$ is the square of the absolute value of the involved terms, which uniquely identifies the truth-values along with the accompanying sign conditions. The third line characterizes that <0 holds iff a strictly dominates, because $a^2>b^2c$ and the dominant a,d have different nonzero signs or if b,d have different signs and either a, d have different nonzero signs as well (so a, b have the same sign or 0 but strictly different than d) or $b\sqrt{c}$ strictly dominates the sign because $a^2 < b^2 c$. The last case involves extra care for the required sign conditions to avoid the $= 0$ case. Essentially, the condition holds when d has strictly opposing sign to a whose square dominates the square $b^2 c$ to $b\sqrt{c}$ or when d has opposing sign to b and either d has strictly opposing sign to a or $b\sqrt{c}$ dominates a.

20.4.2 Virtual Substitutions of Square Roots

The combination of polynomial evaluation according to Definition 20.1 and subsequent square root comparisons by Definition 20.2 defines the substitution of a square root $(a + b\sqrt{c})/d$ for x into atomic formulas and can be lifted to all first-order logic formulas as explained in Sect. 20.3.2. The important thing to note is that the result of this substitution does not introduce square root expressions or divisions even though the square root expression $(a + b\sqrt{c})/d$ has square root $\sqrt{c}$ and division $/d$. Substitution of a square root $(a + b\sqrt{c})/d$ for x into a (quantifier-free) first-order formula F then works by virtually substituting into all atomic formulas (Sect. 20.3.2). The result of such a *virtual* substitution is denoted by $F_{\bar{x}}^{(a+b\sqrt{c})/d}$.

It is crucial to note that the *virtual substitution* of the square root expression $(a + b\sqrt{c})/d$ for x in F giving $F_{\bar{x}}^{(a+b\sqrt{c})/d}$ is semantically equivalent to the result $F_x^{(a+b\sqrt{c})/d}$ of the literal substitution replacing x with $(a + b\sqrt{c})/d$, but operationally different, because the virtual substitution never introduces square roots or divisions. Because of their semantic equivalence, we use almost the same notation. The result $F_{\bar{x}}^{(a+b\sqrt{c})/d}$ of the *virtual substitution* is defined by square root comparisons (Definition 20.2) after polynomial evaluation (Definition 20.1). It is better behaved than the result of the literal substitution $F_x^{(a+b\sqrt{c})/d}$, because it stays within $\text{FOL}_\mathbb{R}$ proper instead of requiring an extension of the language with square root expressions.

> **Lemma 20.2 (Virtual substitution lemma for square roots).** *The result* $F_{\bar{x}}^{(a+b\sqrt{c})/d}$ *of the* virtual substitution *is semantically equivalent to the result* $F_x^{(a+b\sqrt{c})/d}$ *of the literal substitution. A language extension yields this validity:*
>
> $$F_x^{(a+b\sqrt{c})/d} \leftrightarrow F_{\bar{x}}^{(a+b\sqrt{c})/d}$$

Keep in mind, though, that the result $F_{\bar{x}}^{(a+b\sqrt{c})/d}$ of virtual substitution is a proper formula of $\text{FOL}_\mathbb{R}$, while the literal substitution $F_x^{(a+b\sqrt{c})/d}$ can only even be considered as a formula in an extended logic that allows for a syntactic representation of divisions and square root expressions within a context in which they are meaningful (no divisions by zero, no imaginary roots).

A more useful semantic rendition of the virtual substitution lemma shows

$\omega_x^r \in [\![F]\!]$ iff $\omega \in [\![F_{\bar{x}}^{(a+b\sqrt{c})/d}]\!]$ where $r = (\omega[\![a]\!] + \omega[\![b]\!]\sqrt{\omega[\![c]\!]})/\omega[\![d]\!] \in \mathbb{R}$

which is an equivalence of the value of the result of a virtual substitution in any state ω in which the value of F in the semantic modification of the state ω with the value of the variable x is changed to the (real) value that the expression $(a+b\sqrt{c})/d$ would have if only it were allowed in FOL$_\mathbb{R}$.

Using Lemma 20.2, Theorem 20.2 continues to hold when using the *square root virtual substitutions* $F_{\bar{x}}^{(-b\pm\sqrt{b^2-4ac})/(2a)}$ that change Theorem 20.2 to produce a valid formula of first-order real arithmetic, without scary square root expressions. In particular, since the fraction $-c/b$ also is a (somewhat impoverished) square root expression $(-c+0\sqrt{0})/b$, the FOL$_\mathbb{R}$ formula $F_{\bar{x}}^{-c/b}$ in Theorem 20.2 can be formed and rephrased equivalently using the square root virtual substitution as well. Hence, the quantifier-free right-hand side in Theorem 20.2 does not introduce square roots or divisions, but happily remains a proper formula in FOL$_\mathbb{R}$.

With this virtual substitution, the right-hand side of the bi-implication in Theorem 20.2 can be chosen as $QE(\exists x(ax^2+bx+c=0 \wedge F))$ *if it is not the case that* $a=b=c=0$. When using square root virtual substitutions, divisions can be avoided in the quantifier elimination (20.3) for the linear case. Thus, the right-hand side of (20.3) can be chosen as $QE(\exists x(bx+c=0 \wedge F))$ if it is not the case that $b=c=0$.

Example 20.5 (Quadratic curiosity). Using quantifier elimination to check under which circumstances the quadratic equality from (20.4) evaluates to *true* requires a nontrivial number of algebraic and logical computations to handle the virtual substitution of the respective roots of $ax^2+bx+c=0$ into F.

Just out of curiosity, what would happen if we tried to apply the same virtual substitution coming from this equation to $ax^2+bx+c=0$ itself instead of to F? Imagine, for example, that $ax^2+bx+c=0$ shows up a second time in F. Let's only consider the case of quadratic solutions, i.e., where $a \neq 0$. And let's only consider the root $(-b+\sqrt{b^2-4ac})/(2a)$. The other cases are left as an exercise. First virtually substitute $(-b+\sqrt{b^2-4ac})/(2a)$ into the polynomial ax^2+bx+c leading to the symbolic square root expression arithmetic from Example 20.4:

$$(ax^2+bx+c)_{\bar{x}}^{(-b+\sqrt{b^2-4ac})/(2a)} = (0+0\sqrt{b^2-4ac})/1 = 0$$

So $(ax^2+bx+c)_{\bar{x}}^{(-b+(8a^3)\sqrt{b^2-4ac})/(2a)}$ is the zero square root expression? That is actually exactly as expected by construction, because $(-b\pm\sqrt{b^2-4ac})/(2a)$ is supposed to be a root of ax^2+bx+c in the case where $a \neq 0 \wedge b^2-4ac \geq 0$. In particular, if ax^2+bx+c occurs again in F as either an equation or inequality, its virtual substitute in the various cases just ends up being:

$$(ax^2+bx+c=0)_{\tilde{x}}^{(-b+\sqrt{b^2-4ac})/(2a)} \equiv ((0+0\sqrt{b^2-4ac})/1=0) \equiv (0=0) \equiv true$$

$$(ax^2+bx+c\le 0)_{\tilde{x}}^{(-b+\sqrt{b^2-4ac})/(2a)} \equiv ((0+0\sqrt{b^2-4ac})/1\le 0) \equiv (0\cdot 1\le 0) \equiv true$$

$$(ax^2+bx+c< 0)_{\tilde{x}}^{(-b+\sqrt{b^2-4ac})/(2a)} \equiv ((0+0\sqrt{b^2-4ac})/1< 0) \equiv (0\cdot 1< 0) \equiv false$$

$$(ax^2+bx+c\ne 0)_{\tilde{x}}^{(-b+\sqrt{b^2-4ac})/(2a)} \equiv ((0+0\sqrt{b^2-4ac})/1\ne 0) \equiv (0\ne 0) \equiv false$$

And that makes sense as well. After all, the roots of $ax^2+bx+c=0$ satisfy the weak inequality $ax^2+bx+c\le 0$ but not the strict inequality $ax^2+bx+c< 0$. In particular, Theorem 20.2 could substitute the roots of $ax^2+bx+c=0$ also into the full formula $ax^2+bx+c=0\wedge F$ under the quantifier, but the formula resulting from the left conjunct $ax^2+bx+c=0$ always simplifies to *true* so that only the virtual substitution into F remains, where actual logic with real arithmetic happens.

The above computations are all that is needed for Theorem 20.2 to show the following quantifier elimination equivalences:

$$a\ne 0 \rightarrow (\exists x\,(ax^2+bx+c=0\wedge ax^2+bx+c=0) \leftrightarrow b^2-4ac\ge 0\wedge true)$$
$$a\ne 0 \rightarrow (\exists x\,(ax^2+bx+c=0\wedge ax^2+bx+c\le 0) \leftrightarrow b^2-4ac\ge 0\wedge true)$$

With similar computations for the case $(-b-\sqrt{b^2-4ac})/(2a)$, this also justifies

$$a\ne 0 \rightarrow (\exists x\,(ax^2+bx+c=0\wedge ax^2+bx+c< 0) \leftrightarrow b^2-4ac\ge 0\wedge false)$$
$$a\ne 0 \rightarrow (\exists x\,(ax^2+bx+c=0\wedge ax^2+bx+c\ne 0) \leftrightarrow b^2-4ac\ge 0\wedge false)$$

Consequently, in a context where $a\ne 0$ is known, for example because it is a term such as 5 or y^2+1, Theorem 20.2 and simplification yields the following quantifier elimination results:

$$QE(\exists x\,(ax^2+bx+c=0\wedge ax^2+bx+c=0)) \equiv b^2-4ac\ge 0$$
$$QE(\exists x\,(ax^2+bx+c=0\wedge ax^2+bx+c\le 0)) \equiv b^2-4ac\ge 0$$
$$QE(\exists x\,(ax^2+bx+c=0\wedge ax^2+bx+c< 0)) \equiv false$$
$$QE(\exists x\,(ax^2+bx+c=0\wedge ax^2+bx+c\ne 0)) \equiv false$$

In a context where $a\ne 0$ is not known, more cases become possible and the disjunctive structure in Theorem 20.2 remains, leading to a case distinction on whether $a=0$ or $a\ne 0$.

Example 20.6 (Nonnegative roots of quadratic polynomials). Consider the formula

$$\exists x\,(ax^2+bx+c=0\wedge x\ge 0) \tag{20.6}$$

for the purpose of eliminating quantifiers using Theorem 20.2. For simplicity, again assume $a \neq 0$ is known, e.g., because $a = 5$. Since $a \neq 0$, Theorem 20.2 will only consider the square root expression $(-b + \sqrt{b^2 - 4ac})/(2a)$ and the corresponding $(-b - \sqrt{b^2 - 4ac})/(2a)$ but no linear roots. The first thing that happens during the virtual substitution of those roots into the remaining formula $F \equiv (x \geq 0)$ is that the construction in Sect. 20.3.2 will flip $x \geq 0$ to a base case $-x \leq 0$. In that base case, the substitution of the square root expression $(-b + \sqrt{b^2 - 4ac})/(2a)$ into the polynomial $-x$ leads to the following square root computations by (20.5):

$$-(-b+\sqrt{b^2-4ac})/(2a) = ((-1 + 0\sqrt{b^2-4ac})/1) \cdot ((-b+\sqrt{b^2-4ac})/(2a))$$
$$= (b - \sqrt{b^2 - 4ac})/(2a)$$

Observe how the unary minus operator expands to multiplication by -1, whose representation as a square root expression is $(-1 + 0\sqrt{b^2 - 4ac})/1$ for square root $\sqrt{b^2 - 4ac}$. The virtual square root substitution of this square root expression yields

$$(-x \leq 0)_{\bar{x}}^{(b - \sqrt{b^2 - 4ac})/(2a)}$$
$$\equiv b2a \leq 0 \wedge b^2 - (-1)^2(b^2 - 4ac) \geq 0 \vee -1 \cdot 2a \leq 0 \wedge b^2 - (-1)^2(b^2 - 4ac) \leq 0$$
$$\equiv 2ba \leq 0 \wedge 4ac \geq 0 \vee -2a \leq 0 \wedge 4ac \leq 0$$

For the second square root expression $(-b - \sqrt{b^2 - 4ac})/(2a)$, the corresponding polynomial evaluation leads to

$$-(-b-\sqrt{b^2-4ac})/(2a) = ((-1 + 0\sqrt{b^2-4ac})/1) \cdot ((-b-\sqrt{b^2-4ac})/(2a))$$
$$= (b + \sqrt{b^2 - 4ac})/(2a)$$

The virtual square root substitution of this square root expression, thus, yields

$$(-x \leq 0)_{\bar{x}}^{(b + \sqrt{b^2 - 4ac})/(2a)}$$
$$\equiv b2a \leq 0 \wedge b^2 - 1^2(b^2 - 4ac) \geq 0 \vee 1 \cdot 2a \leq 0 \wedge b^2 - 1^2(b^2 - 4ac) \leq 0$$
$$\equiv 2ba \leq 0 \wedge 4ac \geq 0 \vee 2a \leq 0 \wedge 4ac \leq 0$$

Consequently, since $a \neq 0$, Theorem 20.2 implies the quantifier elimination equivalence:

$$a \neq 0 \rightarrow \left(\exists x (ax^2 + bx + c = 0 \wedge x \geq 0) \right.$$
$$\leftrightarrow b^2 - 4ac \geq 0 \wedge$$
$$\left. (2ba \leq 0 \wedge 4ac \geq 0 \vee -2a \leq 0 \wedge 4ac \leq 0 \vee 2ba \leq 0 \wedge 4ac \geq 0 \vee 2a \leq 0 \wedge 4ac \leq 0) \right)$$

Consequently, in a context where $a \neq 0$ is known, Theorem 20.2 yields the following quantifier elimination results:

$$\text{QE}(\exists x\,(ax^2+bx+c=0 \wedge x \geq 0))$$

$$\equiv b^2-4ac \geq 0 \wedge$$

$$(2ba \leq 0 \wedge 4ac \geq 0 \vee -2a \leq 0 \wedge 4ac \leq 0 \vee \underline{2ba{\leq}0 \wedge 4ac{\geq}0} \vee 2a \leq 0 \wedge 4ac \leq 0)$$

$$\equiv b^2-4ac \geq 0 \wedge (ba \leq 0 \wedge ac \geq 0 \vee a \geq 0 \wedge ac \leq 0 \vee a \leq 0 \wedge ac \leq 0)$$

The sign conditions that this formula expresses make sense when considering that the original quantified formula (20.6) expresses that the quadratic equation has a nonnegative root, which is only true under some conditions on its parameters.

20.5 Optimizations

Virtual substitutions admit a number of useful optimizations that make them more practical. When substituting a square root expression $(a+b\sqrt{c})/d$ for a variable x in a polynomial p, the resulting square root expression $p_x^{(a+b\sqrt{c})/d} = (\tilde{a}+\tilde{b}\sqrt{c})/\tilde{d}$ will end up occurring with a higher power of the form $\tilde{d}=d^k$ where k is the degree in p of variable x. This is easy to see by inspecting the definitions of addition and multiplication from (20.5). Such larger powers of d can be avoided using the equivalences $(pq^3 \sim 0) \equiv (pq \sim 0)$ and, if $q \neq 0$, using $(pq^2 \sim 0) \equiv (p \sim 0)$ for arithmetic relations $\sim\; \in \{=,>,\geq,\neq,<,\leq\}$. Since $d \neq 0$ needs to be assumed for well-definedness of a square root expression $(a+b\sqrt{c})/d$, the degree of d in the result $F_x^{(a+b\sqrt{c})/d}$ of the virtual substitution can, thus, be lowered to either 0 or 1 depending on whether it ultimately occurs as an even or odd power (Exercise 20.9). If d occurs as an odd power, its occurrence can be lowered to degree 1. If d occurs as an even power, its occurrence can be reduced to degree 0, which makes it disappear entirely.

A minor but important optimization to retain a low polynomial degree [48] for sign comparisons results from the fact that the odd power e^{2n+1} has the same sign as e and that an even power e^{2n} has the same sign as e^2. In particular if $e \neq 0$, then the even power e^{2n} has the same sign as 1.

The significance of lowering degrees does not just come from the conceptual and computational impact that large degrees have on the problem of quantifier elimination, but, for the case of virtual substitution, also from the fact that virtual substitution only works for certain bounded but common degrees.

20.6 Summary

This chapter showed part of the miracle of quantifier elimination and that quantifier elimination is possible in first-order real arithmetic. This technique works for formulas that normalize into an appropriate form as long as the technique can latch on to a linear or quadratic equation for all quantified variables. There can be higher-degree or inequality occurrences of the variables as well within the formula F of Theorem 20.2, but there has to be at least one linear or quadratic equation. Commuting the formula so that it has the required form is easily done if such an equation is anywhere at all. What is to be done if there is no quadratic equation but only other quadratic inequalities is the topic of the next chapter.

It is also foreseeable that the virtual substitution approach will ultimately run into difficulties for pure high-degree polynomials, because those generally have no radicals to solve the equations. That is where other more algebraic quantifier elimination techniques come into play that are beyond the scope of this textbook.

Virtual substitution of square root expressions uses symbolic computations:

$$((a+b\sqrt{c})/d) + ((a'+b'\sqrt{c})/d') = ((ad'+da') + (bd'+db')\sqrt{c})/(dd')$$
$$((a+b\sqrt{c})/d) \cdot ((a'+b'\sqrt{c})/d') = ((aa'+bb'c) + (ab'+ba')\sqrt{c})/(dd')$$

The following expansions were the core of eliminating square root expressions by virtual substitutions. For square root expressions $(a+b\sqrt{c})/d$ with $d \neq 0 \wedge c \geq 0$ for well-definedness, the following equivalences rewrite to eliminate square roots:

$$(a+b\sqrt{c})/d = 0 \equiv ab \leq 0 \wedge a^2 - b^2c = 0$$
$$(a+b\sqrt{c})/d \leq 0 \equiv ad \leq 0 \wedge a^2 - b^2c \geq 0 \vee bd \leq 0 \wedge a^2 - b^2c \leq 0$$
$$(a+b\sqrt{c})/d < 0 \equiv ad < 0 \wedge a^2 - b^2c > 0 \vee bd \leq 0 \wedge (ad < 0 \vee a^2 - b^2c < 0)$$

20.7 Appendix: Real Algebraic Geometry

This textbook follows a logical view of cyber-physical systems. It can be helpful to develop an intuition about what geometric objects the various logical concepts correspond to. What is most interesting in this context is real algebraic geometry [2] as it relates to real arithmetic [1]. General algebraic geometry is also very elegant and beautiful, especially over algebraically closed fields [9, 18].

The geometric counterpart of polynomial equations is real affine algebraic varieties. Every set F of polynomials defines a geometric object, its variety, i.e., the set of points on which all those polynomials are zero.

Definition 20.3 (Real affine algebraic variety). $V \subseteq \mathbb{R}^n$ is an *affine variety* iff, for some set $F \subseteq \mathbb{R}[X_1, \ldots, X_n]$ of polynomials over $\mathbb{R}$

$$V = V(F) := \{x \in \mathbb{R}^n \ : \ f(x) = 0 \text{ for all } f \in F\}$$

Affine varieties are subsets of $\mathbb{R}^n$ that are definable by a set of polynomial equations.

The converse construction is that of the vanishing ideal, which describes the set of all polynomials that are zero on a given set V.

Definition 20.4 (Vanishing ideal). $I \subseteq \mathbb{R}[X_1, \ldots, X_n]$ is the *vanishing ideal* of $V \subseteq \mathbb{R}^n$:

$$I(V) := \{f \in \mathbb{R}[X_1,\dots,X_n] \ : \ f(x) = 0 \text{ for all } f \in V\}$$

i.e., all polynomials that are zero on all of V.

Affine varieties and vanishing ideals are related by

$$S \subseteq V(I(S)) \qquad\qquad \text{for any set } S \subseteq \mathbb{R}^n$$
$$V = V(I(V)) \qquad\qquad \text{if } V \text{ is an affine variety}$$
$$F \subseteq G \Rightarrow V(F) \supseteq V(G)$$

Affine varieties and vanishing ideals are intimately related by Hilbert's Nullstellensatz over algebraically closed fields such as $\mathbb{C}$ and by Stengle's Nullstellensatz over real-closed fields such as $\mathbb{R}$.

The affine varieties corresponding to a number of interesting polynomials are illustrated in Fig. 20.3.

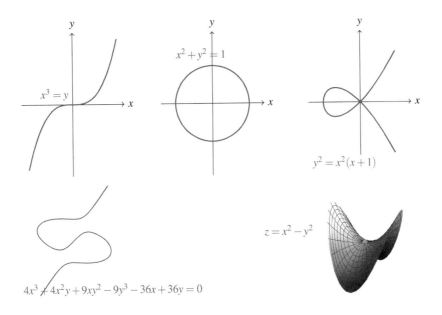

Fig. 20.3 Polynomial equations describe (real) affine (algebraic) varieties

Exercises

20.1. Definition 20.1 defined addition and multiplication of square root expressions. How can subtraction $((a+b\sqrt{c})/d) - ((a'+b'\sqrt{c})/d')$ and negation $-((a+b\sqrt{c})/d)$ be defined?

20.2. Assuming that $b \neq 0$, construct a quantifier-free equivalent for the existence of a nonnegative root of the general linear equation. That is, perform linear quantifier elimination on

$$\exists x\, (bx + c = 0 \wedge x \geq 0)$$

and state the result without using fractions. What is the result when, instead, assuming $b = 0 \wedge c \neq 0$?

20.3. Note 82 stated that the terms and formulas of $\mathrm{FOL}_\mathbb{R}$ can only be built from polynomials with rational coefficients. Show that it suffices to, indeed, only allow the numbers 0 and 1 but no other. First assume that a fraction operator $/$ can be used (at least on terms without variables).

20.4. Example 20.5 showed that $ax^2 + bx + c = 0$ simplifies to *true* for the virtual substitution of the root $(-b + \sqrt{b^2 - 4ac})/(2a)$. Show that the same thing happens for the root $(-b - \sqrt{b^2 - 4ac})/(2a)$ and, if $a = 0$, the root $(-c + 0\sqrt{0})/b$.

20.5. Example 20.5 argued that the simplification of $ax^2 + bx + c = 0$ to *true* for the virtual substitution of the root $(-b + \sqrt{b^2 - 4ac})/(2a)$ is to be expected, because the real number to which $(-b + \sqrt{b^2 - 4ac})/(2a)$ evaluates is a root of $ax^2 + bx + c = 0$ in the case where $a \neq 0 \wedge b^2 - 4ac \geq 0$. Yet, what happens in the case where the extra assumption $a \neq 0 \wedge b^2 - 4ac \geq 0$ does not hold? What is the value of the virtual substitution in that case? Is that a problem? Discuss carefully!

20.6. Use Theorem 20.2 to eliminate quantifiers in the following formulas, assuming $a \neq 0$ is known:

$$\exists x\, (ax^2 + bx + c = 0 \wedge x < 1)$$
$$\exists x\, (ax^2 + bx + c = 0 \wedge x^3 + x \leq 0)$$

20.7. How does Example 20.6 change when removing the assumption that $a \neq 0$?

20.8. Would first-order logic of real arithmetic miss the presence of π? That is, if we deleted π from the domain and made all quantifiers range only over $\mathbb{R} \setminus \{\pi\}$, would there be any formula that notices by having a different truth-value? If we deleted $\sqrt[3]{5}$ from the domain, would $\mathrm{FOL}_\mathbb{R}$ notice?

20.9. Consider the process of substituting a square root expression $(a + b\sqrt{c})/d$ for a variable x in a polynomial p. Let k be the degree in p of variable x, so that d occurs as d^k with power k in the result $p_x^{(a+b\sqrt{c})/d} = (\tilde{a} + \tilde{b}\sqrt{c})/\tilde{d}$. Let $\delta = 1$ when k is odd and $\delta = 0$ when k is even. Show that the following optimization can be used for the virtual substitution. Assume $d \neq 0 \wedge c \geq 0$ for well-definedness. For square-root-free expressions $(b = 0)$ with just divisions, i.e., those of the form $(a + 0\sqrt{c})/d$, the following equivalences hold:

$$(a + 0\sqrt{c})/d = 0 \equiv a = 0$$
$$(a + 0\sqrt{c})/d \leq 0 \equiv ad^\delta \leq 0$$

$$(a+0\sqrt{c})/d < 0 \equiv ad^{\delta} < 0$$
$$(a+0\sqrt{c})/d \neq 0 \equiv a \neq 0$$

Assume $d \neq 0$ and $c \geq 0$ for well-definedness. For any square root expression $(a+b\sqrt{c})/d$ with arbitrary b, the following equivalences hold:

$$(a+b\sqrt{c})/d = 0 \equiv ab \leq 0 \wedge a^2 - b^2 c = 0$$
$$(a+b\sqrt{c})/d \leq 0 \equiv ad^{\delta} \leq 0 \wedge a^2 - b^2 c \geq 0 \vee bd^{\delta} \leq 0 \wedge a^2 - b^2 c \leq 0$$
$$(a+b\sqrt{c})/d < 0 \equiv ad^{\delta} < 0 \wedge a^2 - b^2 c > 0 \vee bd^{\delta} \leq 0 \wedge (ad^{\delta} < 0 \vee a^2 - b^2 c < 0)$$
$$(a+b\sqrt{c})/d \neq 0 \equiv ab > 0 \vee a^2 - b^2 c \neq 0$$

References

[1] Saugata Basu, Richard Pollack, and Marie-Françoise Roy. *Algorithms in Real Algebraic Geometry*. 2nd. Berlin: Springer, 2006. DOI: 10.1007/3-540-33099-2.

[2] Jacek Bochnak, Michel Coste, and Marie-Francoise Roy. *Real Algebraic Geometry*. Vol. 36. Ergeb. Math. Grenzgeb. Berlin: Springer, 1998. DOI: 10.1007/978-3-662-03718-8.

[3] Claude Chevalley and Henri Cartan. Schémas normaux; morphismes; ensembles constructibles. In: *Séminaire Henri Cartan*. Vol. 8. 7. Numdam, 1955, 1–10.

[4] Alonzo Church. A note on the Entscheidungsproblem. *J. Symb. Log.* **1**(1) (1936), 40–41.

[5] Paul J. Cohen. Decision procedures for real and *p*-adic fields. *Communications in Pure and Applied Mathematics* **22** (1969), 131–151. DOI: 10.1002/cpa.3160220202.

[6] George E. Collins. Quantifier elimination for real closed fields by cylindrical algebraic decomposition. In: *Automata Theory and Formal Languages*. Ed. by H. Barkhage. Vol. 33. LNCS. Berlin: Springer, 1975, 134–183. DOI: 10.1007/3-540-07407-4_17.

[7] George E. Collins and Hoon Hong. Partial cylindrical algebraic decomposition for quantifier elimination. *J. Symb. Comput.* **12**(3) (1991), 299–328. DOI: 10.1016/S0747-7171(08)80152-6.

[8] Stephen A. Cook. The complexity of theorem-proving procedures. In: *STOC*. Ed. by Michael A. Harrison, Ranan B. Banerji, and Jeffrey D. Ullman. New York: ACM, 1971, 151–158. DOI: 10.1145/800157.805047.

[9] David A. Cox, John Little, and Donal O'Shea. *Ideals, Varieties and Algorithms: An Introduction to Computational Algebraic Geometry and Commu-*

tative Algebra. Undergraduate Texts in Mathematics. New York: Springer, 1992.

[10]   Lloyd Dines. Systems of linear inequalities. *Ann. Math.* **20**(3) (1919), 191–199.

[11]   Andreas Dolzmann and Thomas Sturm. Redlog: computer algebra meets computer logic. *ACM SIGSAM Bull.* **31**(2) (1997), 2–9. DOI: 10.1145/26 1320.261324.

[12]   Lou van den Dries and Chris Miller. On the real exponential field with restricted analytic functions. *Israel J. Math.* **85**(1-3) (1994), 19–56. DOI: 10.1 007/BF02758635.

[13]   Richard J. Duffin. On Fourier's analysis of linear inequality systems. In: *Pivoting and Extension: In honor of A.W. Tucker.* Ed. by M. L. Balinski. Berlin: Springer, 1974, 71–95. DOI: 10.1007/BFb0121242.

[14]   Erwin Engeler. *Foundations of Mathematics: Questions of Analysis, Geometry and Algorithmics.* Berlin: Springer, 1993. DOI: 10.1007/978-3-642-78052-3.

[15]   Jean-Baptiste Joseph Fourier. Solution d'une question particulière du calcul des inégalités. *Nouveau Bulletin des Sciences par la Société Philomatique de Paris* (1826), 99–100.

[16]   Kurt Gödel. Die Vollständigkeit der Axiome des logischen Funktionenkalküls. *Monatshefte Math. Phys.* **37** (1930), 349–360. DOI: 10.1007/BF016967 81.

[17]   Kurt Gödel. Über formal unentscheidbare Sätze der Principia Mathematica und verwandter Systeme I. *Monatshefte Math. Phys.* **38**(1) (1931), 173–198. DOI: 10.1007/BF01700692.

[18]   Joe Harris. *Algebraic Geometry: A First Course.* Graduate Texts in Mathematics. Berlin: Springer, 1995. DOI: 10.1007/978-1-4757-2189-8.

[19]   John Harrison. Verifying nonlinear real formulas via sums of squares. In: *TPHOLs.* Ed. by Klaus Schneider and Jens Brandt. Vol. 4732. LNCS. Berlin: Springer, 2007, 102–118. DOI: 10.1007/978-3-540-74591-4_9.

[20]   Jacques Herbrand. Recherches sur la théorie de la démonstration. *Travaux de la Société des Sciences et des Lettres de Varsovie, Class III, Sciences Mathématiques et Physiques* **33** (1930), 33–160.

[21]   Lars Hörmander. *The Analysis of Linear Partial Differential Operators II.* Vol. 257. Grundlehren der mathematischen Wissenschaften. Berlin: Springer, 1983.

[22]   Nathan Jacobson. *Basic Algebra I.* 2nd ed. San Francisco: Freeman, 1989.

[23]   Dejan Jovanović and Leonardo Mendonça de Moura. Solving non-linear arithmetic. In: *Automated Reasoning - 6th International Joint Conference, IJCAR 2012, Manchester, UK, June 26-29, 2012. Proceedings.* Ed. by Bernhard Gramlich, Dale Miller, and Ulrike Sattler. Vol. 7364. LNCS. Berlin: Springer, 2012, 339–354. DOI: 10.1007/978-3-642-31365-3_27.

[24]   Georg Kreisel and Jean-Louis Krivine. *Elements of mathematical logic: Model Theory.* 2nd ed. Amsterdam: North-Holland, 1971.

[25]   Sean McLaughlin and John Harrison. A proof-producing decision procedure for real arithmetic. In: *CADE*. Ed. by Robert Nieuwenhuis. Vol. 3632. LNCS. Springer, 2005, 295–314. DOI: 10.1007/11532231_22.

[26]   Theodore Samuel Motzkin. Beiträge zur Theorie der Linearen Ungleichungen. PhD thesis. Basel, Jerusalem, 1936.

[27]   Leonardo Mendonça de Moura and Grant Olney Passmore. The strategy challenge in SMT solving. In: *Automated Reasoning and Mathematics - Essays in Memory of William W. McCune*. Ed. by Maria Paola Bonacina and Mark E. Stickel. Vol. 7788. LNCS. Berlin: Springer, 2013, 15–44. DOI: 10.1007/978-3-642-36675-8_2.

[28]   Grant Olney Passmore. Combined Decision Procedures for Nonlinear Arithmetics, Real and Complex. PhD thesis. School of Informatics, University of Edinburgh, 2011.

[29]   André Platzer. Differential dynamic logic for hybrid systems. *J. Autom. Reas.* **41**(2) (2008), 143–189. DOI: 10.1007/s10817-008-9103-8.

[30]   André Platzer. *Logical Analysis of Hybrid Systems: Proving Theorems for Complex Dynamics*. Heidelberg: Springer, 2010. DOI: 10.1007/978-3-642-14509-4.

[31]   André Platzer. Logics of dynamical systems. In: *LICS*. Los Alamitos: IEEE, 2012, 13–24. DOI: 10.1109/LICS.2012.13.

[32]   André Platzer. The complete proof theory of hybrid systems. In: *LICS*. Los Alamitos: IEEE, 2012, 541–550. DOI: 10.1109/LICS.2012.64.

[33]   André Platzer. A complete uniform substitution calculus for differential dynamic logic. *J. Autom. Reas.* **59**(2) (2017), 219–265. DOI: 10.1007/s10817-016-9385-1.

[34]   André Platzer, Jan-David Quesel, and Philipp Rümmer. Real world verification. In: *CADE*. Ed. by Renate A. Schmidt. Vol. 5663. LNCS. Berlin: Springer, 2009, 485–501. DOI: 10.1007/978-3-642-02959-2_35.

[35]   Mojżesz Presburger. Über die Vollständigkeit eines gewissen Systems der Arithmetik ganzer Zahlen, in welchem die Addition als einzige Operation hervortritt. *Comptes Rendus du I Congrès de Mathématiciens des Pays Slaves* (1929), 92–101.

[36]   H. Gordon Rice. Classes of recursively enumerable sets and their decision problems. *Trans. AMS* **74**(2) (1953), 358–366. DOI: 10.2307/1990888.

[37]   Daniel Richardson. Some undecidable problems involving elementary functions of a real variable. *J. Symb. Log.* **33**(4) (1968), 514–520. DOI: 10.2307/2271358.

[38]   Abraham Robinson. *Complete Theories*. 2nd ed. Studies in logic and the foundations of mathematics. North-Holland, 1977, 129.

[39]   Julia Robinson. Definability and decision problems in arithmetic. *J. Symb. Log.* **14**(2) (1949), 98–114. DOI: 10.2307/2266510.

[40]   Julia Robinson. The undecidability of algebraic rings and fields. *Proc. AMS* **10**(6) (1959), 950–957. DOI: 10.2307/2033628.

[41]   Abraham Seidenberg. A new decision method for elementary algebra. *Annals of Mathematics* **60**(2) (1954), 365–374. DOI: 10.2307/1969640.

[42] Thoralf Skolem. Über einige Satzfunktionen in der Arithmetik. *Skrifter utgitt av Det Norske Videnskaps-Akademi i Oslo, I. Matematisk naturvidenskapelig klasse* **7** (1931), 1–28.

[43] Gilbert Stengle. A Nullstellensatz and a Positivstellensatz in semialgebraic geometry. *Math. Ann.* **207**(2) (1973), 87–97. DOI: `10.1007/BF0136214 9`.

[44] Alfred Tarski. Sur les ensembles définissables de nombres réels I. *Fundam. Math.* **17**(1) (1931), 210–239.

[45] Alfred Tarski. *A Decision Method for Elementary Algebra and Geometry.* 2nd. Berkeley: University of California Press, 1951.

[46] Alan M. Turing. Computability and λ-definability. *J. Symb. Log.* **2**(4) (1937), 153–163. DOI: `10.2307/2268280`.

[47] Volker Weispfenning. The complexity of linear problems in fields. *J. Symb. Comput.* **5**(1-2) (1988), 3–27. DOI: `10.1016/S0747-7171(88)80003 -8`.

[48] Volker Weispfenning. Quantifier elimination for real algebra — the quadratic case and beyond. *Appl. Algebra Eng. Commun. Comput.* **8**(2) (1997), 85–101. DOI: `10.1007/s002000050055`.

Chapter 21
Virtual Substitution & Real Arithmetic

Synopsis This chapter advances the understanding of real arithmetic by generalizing the ideas from the previous chapter to linear and quadratic *inequalities*. As in the previous chapter, the main workhorse will again be virtual substitutions that pretend to substitute a generalized expression into a logical formula by equivalently rephrasing each occurrence. The required virtual substitutions will, however, go beyond square root substitutions but cover infinities and infinitesimals, instead, in order to capture the fact that inequalities can also be satisfied without satisfying equality.

21.1 Introduction

Reasoning about cyber-physical systems and hybrid systems requires understanding and handling their real arithmetic, which can be challenging, because cyber-physical systems can have complex behavior. Differential dynamic logic and its proof calculus [6–8] reduce the verification of hybrid systems to real arithmetic. How arithmetic interfaces with proofs has already been discussed in Chap. 6. How real arithmetic with linear and quadratic equations can be handled by virtual substitution has been shown in Chap. 20. This chapter shows how virtual substitution for quantifier elimination in real arithmetic extends to the case of linear and quadratic inequalities.

The results in this chapter are based on the literature [13]. The chapter adds substantial intuition and motivation that is helpful for following the technical development. More information about virtual substitution can be found in the literature [13]. See, e.g., [1, 2, 5, 9] for an overview of other techniques for real arithmetic.

The most important learning goals of this chapter are:

Modeling and Control: This chapter refines the indirect impact that the previous chapter had on CPS models and controls by informing the reader about the consequences of the analytic complexity resulting from different arithmetical modeling tradeoffs. There are subtle analytic consequences from different arithmetic formulations of similar questions that can have an impact on finding the right tradeoffs for expressing a CPS model. In practical terms, a safe distance

© Springer International Publishing AG, part of Springer Nature 2018
A. Platzer, *Logical Foundations of Cyber-Physical Systems*,
https://doi.org/10.1007/978-3-319-63588-0_21

of car x to a stop light m could equally well be captured as $x \leq m$ or as $x < m$, for example, if only we knew the impact of this decision on the resulting real arithmetic.

Computational Thinking: The primary purpose of this chapter is to understand how arithmetical reasoning, which is crucial for CPS, can be done rigorously and automatically not just for the equations considered in Chap. 20 but also for inequalities. While formulas involving sufficiently many quadratic equations among other inequalities can be handled with the techniques from Chap. 20, such extensions are crucial for proving arithmetic formulas that involve only inequalities, which happens rather frequently in the world of CPS, where many questions concern inequality bounds on distances. Developing an intuition for the working principles of real-arithmetic decision procedures can be very helpful for developing strategies to verify CPS models at scale. We will again see the conceptually very important device of the logical trinity: the flexibility of moving back and forth between syntax and semantics at will. Virtual substitutions will again allow us to move back and forth at will between syntax and semantics. This time, however, square roots will not be all there is to it, but the logical trinity will lead us to ideas from nonstandard analysis to bridge the gap to semantic operations that are inexpressible otherwise in first-order logic of real arithmetic.

CPS Skills: This chapter has an indirect impact on CPS skills, because it discusses useful pragmatics of CPS analysis for modeling and analysis tradeoffs that enable CPS verification at scale.

rigorous arithmetical reasoning
miracle of quantifier elimination
logical trinity for reals
switch between syntax & semantics at will
virtual substitution lemma
bridge gap between semantics and inexpressibles
infinities & infinitesimals

analytic complexity   verifying CPS at scale
modeling tradeoffs

21.2 Recap: Square Root $\sqrt{\cdot}$ Virtual Substitutions for Quadratics

Recall the way to handle quantifier elimination for linear or quadratic equations from Chap. 20 by virtually substituting in its symbolic solutions $x = -c/b$ or $x = (-b \pm \sqrt{b^2 - 4ac})/(2a)$, respectively

> **Theorem 20.2 (Virtual substitution of quadratic equations).** *For quantifier-free formula F with $x \notin FV(a), FV(b), FV(c)$, the following equivalence is valid over* $\mathbb{R}$:
>
> $$a \neq 0 \vee b \neq 0 \vee c \neq 0 \rightarrow$$
>
> $$\left(\exists x \, (ax^2 + bx + c = 0 \wedge F) \leftrightarrow \right.$$
>
> $$a = 0 \wedge b \neq 0 \wedge F_x^{-c/b}$$
>
> $$\left. \vee \, a \neq 0 \wedge b^2 - 4ac \geq 0 \wedge \left(F_x^{(-b+\sqrt{b^2-4ac})/(2a)} \vee F_x^{(-b-\sqrt{b^2-4ac})/(2a)} \right) \right)$$

When using virtual substitutions of square roots from Chap. 20, the resulting formula on the right-hand side of the bi-implication is quantifier-free and can be chosen for $QE(\exists x \, (ax^2 + bx + c = 0 \wedge F))$ as long as it is not the case that $a = b = c = 0$. In case $a = b = c = 0$, another formula in F needs to be considered for directing quantifier elimination by commuting and reassociating $\wedge$, because the equation $ax^2 + bx + c = 0$ is noninformative if $a = b = c = 0$, e.g., when a, b, c are the zero polynomial or even if they just have a common root.

The equivalent formula on the right-hand side of the bi-implication in Theorem 20.2 is a formula in the first-order logic of real arithmetic when using the virtual substitution of square root expressions defined in Chap. 20.

21.3 Infinity ∞ Virtual Substitution

Theorem 20.2 addresses the case where the quantified variable occurs in a linear or quadratic equation, in which case it is efficient to use Theorem 20.2, because there are at most three symbolic points to consider corresponding to the respective solutions of the equation. But what do we do if the quantified variable only occurs in inequalities? Then Theorem 20.2 does not help the slightest bit. Consider a formula of the form

$$\exists x \, (ax^2 + bx + c \leq 0 \wedge F) \qquad (x \notin FV(a), FV(b), FV(c)) \qquad (21.1)$$

where x does not occur in a, b, c. Under the conditions from Theorem 20.2, the possible solutions $-c/b, (-b + \sqrt{d})/(2a), (-b - \sqrt{d})/(2a)$ from Theorem 20.2 continue to be options for solutions of (21.1), because one way of satisfying the weak inequality $ax^2 + bx + c \leq 0$ is by satisfying the equation $ax^2 + bx + c = 0$. So if F is *true* for any of those solutions of the quadratic equation (under the auspices of the additional constraints on a, b, c), then (21.1) holds as well.

Yet, even if those points do not work out, the weak inequality in (21.1) allows for more possible solutions than the equation does. For example, if $a = 0, b > 0$, then sufficiently small values of x would satisfy $0x^2 + bx + c \leq 0$. Also, if $a < 0$, then sufficiently small values of x would satisfy $ax^2 + bx + c \leq 0$, because x^2 grows faster than x and, thus the negative ax^2 ultimately overcomes any contribution of bx and c to the value of $ax^2 + bx + c$. But if we literally substituted each such smaller value of x into F, that would quickly diverge into the full substitution $\bigvee_{t \in T} F_x^t$ for the uninsightful case of all real numbers $T \stackrel{\text{def}}{=} \mathbb{R}$ from Chap. 20. So we have to be more clever than that.

Now, one possibile way of pursuing this line of thought may be to substitute smaller and smaller values for x into (21.1) and see if one of those happens to work. There is a much better way though. The only really small value that has to be substituted into (21.1) for x to see whether it happens to work is one that is so negative that it is smaller than all others: $-\infty$, which is the lower limit of all negative real numbers. Alternatively, $-\infty$ can be understood as being "always as negative as needed, i.e., more negative than anything else." Think of $-\infty$ as being built out of elastic rubber so that it always ends up being smaller when compared to any actual real number, because the elastic number $-\infty$ simply shrinks every time it is compared to any other number. Analogously, ∞ is the upper limit of all real numbers or "always as positive as needed, i.e., more positive than anything else." The elastic rubber version of understanding ∞ is such that ∞ always grows as needed every time it is compared to any other number.

Let $\infty, -\infty$ be *positive and negative infinities*, respectively, i.e., choose extra elements $\infty, -\infty \notin \mathbb{R}$ with $-\infty < r < \infty$ for all $r \in \mathbb{R}$. Formulas of real arithmetic can be substituted with $\pm\infty$ for a variable x in the compactified reals $\mathbb{R} \cup \{\infty, -\infty\}$. Yet, just like with square root expressions, $\pm\infty$ do not actually need to ever truly occur in the resulting formula, because substitution of infinities into formulas can be defined differently. For example, $(x + 5 > 0)_x^{-\infty}$ will be *false*, while $(x + 5 < 0)_x^{-\infty}$ is *true*.

Definition 21.1 (Infinite virtual substitution). *Substitution of the infinity* $-\infty$ for x into an atomic formula for a polynomial $p \stackrel{\text{def}}{=} \sum_{i=0}^{n} a_i x^i$ with polynomials a_i that do not contain x is defined by the following equivalences:

$$(p = 0)_{\bar{x}}^{-\infty} \equiv \bigwedge_{i=0}^{n} a_i = 0 \tag{21.2}$$

$$(p \leq 0)_{\bar{x}}^{-\infty} \equiv (p < 0)_{\bar{x}}^{-\infty} \vee (p = 0)_{\bar{x}}^{-\infty} \tag{21.3}$$

$$(p < 0)_{\bar{x}}^{-\infty} \equiv p(-\infty) < 0 \tag{21.4}$$

$$(p \neq 0)_{\bar{x}}^{-\infty} \equiv \bigvee_{i=0}^{n} a_i \neq 0 \tag{21.5}$$

Lines (21.2) and its dual (21.5) use that the only equation of real arithmetic that infinities $\pm\infty$ satisfy is the trivial equation $0 = 0$. Line (21.3) uses the equivalence $p \le 0 \equiv p < 0 \vee p = 0$ and is equal to $(p < 0 \vee p = 0)_{\bar{x}}^{-\infty}$ by the substitution base from Sect. 20.3.2. Line (21.4) uses a simple inductive definition based on the *degree*, $\deg(p)$, the highest power of the variable x in the polynomial p, to characterize whether p is ultimately negative at $-\infty$ (or sufficiently negative numbers):

Let $p \stackrel{\text{def}}{=} \sum_{i=0}^{n} a_i x^i$ with polynomials a_i that do not contain x. Whether p is ultimately negative at $-\infty$, suggestively written $p(-\infty) < 0$, is easy to characterize by induction on the degree of the polynomial:

$$p(-\infty) < 0 \stackrel{\text{def}}{\equiv} \begin{cases} p < 0 & \text{if } \deg(p) \le 0 \\ (-1)^n a_n < 0 \vee (a_n = 0 \wedge (\sum_{i=0}^{n-1} a_i x^i)(-\infty) < 0) & \text{if } \deg(p) > 0 \end{cases}$$

$p(-\infty) < 0$ is true in a state in which $\lim_{x \to -\infty} p(x) < 0$.

The first line captures that the sign of polynomials of degree 0 in the variable x does not depend on x, so $p(-\infty) < 0$ iff the polynomial of degree 0 in x is negative (which may still depend on the value of other variables in $p = a_0$ but not on x). The second line captures that the sign at $-\infty$ of a polynomial of degree $n = \deg(p) > 0$ is determined by the degree-modulated sign of its leading coefficient a_n, because for x of sufficiently big absolute value, the value of $a_n x^n$ will dominate all lower-degree values, whatever their coefficients are. For even $n > 0$, $x^n > 0$ while $x^n < 0$ for odd n at $-\infty$. In case the leading coefficient a_n evaluates to zero, the value of p at $-\infty$ depends on the value at $-\infty$ of the remaining polynomial $\sum_{i=0}^{n-1} a_i x^i$ of lower degree, which can be determined recursively as $(\sum_{i=0}^{n-1} a_i x^i)(-\infty) < 0$. Note that the degree of the 0 polynomial is sometimes considered to be $-\infty$, which explains why $\deg(p) \le 0$ is used in line 1 instead of $\deg(p) = 0$.

Substitution of ∞ for x into an atomic formula can be defined similarly, except that the sign factor $(-1)^n$ disappears, because $x^n > 0$ at ∞ whatever value $n > 0$ has. Substitution of ∞ or of $-\infty$ for x into other first-order formulas is then defined on this basis as in Sect. 20.3.2.

Example 21.1 (Sign of quadratic polynomials at $-\infty$). Using this principle to check systematically under which circumstances the quadratic inequality from (21.1) evaluates to *true* yields the answer from our earlier ad-hoc analysis of what happens for sufficiently small values of x:

$$(ax^2 + bx + c < 0)_{\bar{x}}^{-\infty} \equiv (-1)^2 a < 0 \vee a = 0 \wedge ((-1)b < 0 \vee b = 0 \wedge c < 0)$$
$$\equiv a < 0 \vee a = 0 \wedge (b > 0 \vee b = 0 \wedge c < 0)$$
$$(ax^2 + bx + c \le 0)_{\bar{x}}^{-\infty} \equiv (ax^2 + bx + c < 0)_{\bar{x}}^{-\infty} \vee a = b = c = 0$$
$$\equiv a < 0 \vee a = 0 \wedge (b > 0 \vee b = 0 \wedge c < 0) \vee a = b = c = 0$$

One representative example for each of those disjuncts is illustrated in Fig. 21.1. In the same way, the virtual substitution can be used to see under which circumstances

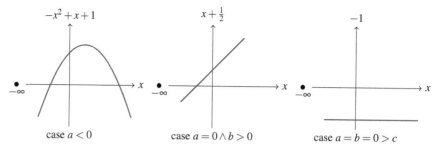

Fig. 21.1 Illustration of the value of different quadratic functions p where $p_{\bar{x}}^{-\infty} \equiv true$

the remainder formula F from (21.1) also evaluates to *true* for sufficiently small values of x, which is the case exactly when $F_{\bar{x}}^{-\infty}$ evaluates to *true*.

The crucial thing to note is again that the *virtual substitution* of infinities $\pm\infty$ for x in F giving $F_{\bar{x}}^{\pm\infty}$ from Definition 21.1 is semantically equivalent to the result $F_{x}^{\pm\infty}$ of the literal substitution replacing x with $\pm\infty$, but operationally different, because the virtual substitution never introduces actual infinities so remains in proper $\text{FOL}_{\mathbb{R}}$.

> **Lemma 21.1 (Virtual substitution lemma for infinities).** *The result $F_{\bar{x}}^{-\infty}$ of the virtual substitution is semantically equivalent to the result $F_{x}^{-\infty}$ of the literal substitution. A language extension yields this validity:*
>
> $$F_{x}^{-\infty} \leftrightarrow F_{\bar{x}}^{-\infty}$$

Keep in mind that the result $F_{\bar{x}}^{-\infty}$ of virtual substitution is a proper formula of $\text{FOL}_{\mathbb{R}}$, while the literal substitution $F_{x}^{-\infty}$ can only be considered a formula in an extended logic such as $\text{FOL}_{\mathbb{R}\cup\{-\infty,\infty\}}$ that allows for infinite quantities. The same property holds for $F_{\bar{x}}^{\infty}$.

Note that the situation is, in a sense, the converse of Lemma 20.2, where the square root expressions were already in the semantic domain $\mathbb{R}$, and just had to be made accessible in the syntactic formulas via virtual substitutions. In Lemma 21.1, instead, virtual substitutions already know more about infinities $\pm\infty$ than the semantic domain $\mathbb{R}$ does, which is why the semantic domain needs an extension to $\mathbb{R}\cup\{-\infty,\infty\}$ for the alignment in Lemma 21.1.

21.4 Infinitesimal ε Virtual Substitution

Theorem 20.2 addresses the case where the quantified variable occurs in a linear or quadratic equation and the virtual substitution in Sect. 21.3 adds the case of sufficiently small values of x to handle $ax^2 + bx + c \leq 0$. Consider a formula of the form

$$\exists x\,(ax^2 + bx + c < 0 \wedge F) \qquad (x \notin \text{FV}(a), \text{FV}(b), \text{FV}(c)) \qquad (21.7)$$

Expedition 21.1 (Infinite challenges with infinities in extended reals)

The set $\mathbb{R} \cup \{-\infty, \infty\}$ is seemingly easily written down as a semantic domain of extended reals. What exactly do we mean by it, though? We mean the set of reals to which we adjoin two new elements, denoted $-\infty$ and ∞, which are the minimum and maximum elements of the ordering $\leq$:

$$\forall x (-\infty \leq x \leq \infty) \tag{21.6}$$

This turns $\mathbb{R} \cup \{-\infty, \infty\}$ into a complete lattice, because every subset has a supremum and an infimum. The extended reals are a compactification of $\mathbb{R}$. But where does that leave the other arithmetic properties of $\mathbb{R}$? What is $\infty + 1$ or $\infty + x$ when ∞ is already infinitely big? The compatibility of $\leq$ with $+$ expects $\infty \leq \infty + x$ at least for all $x \geq 0$. By (21.6) also $\infty + x \leq \infty$. Because ∞ is so infinitely big, the same $\infty + x = \infty$ is expected even for all x, except $-\infty$. The compatibility of $\leq$ with $\cdot$ expects $\infty \leq \infty \cdot x$ at least for all $x \geq 1$. By (21.6) also $\infty \cdot x \leq \infty$. Since ∞ is infinitely big, the same $\infty \cdot x = \infty$ is expected even for all $x > 0$:

$$\begin{array}{ll}
\infty + x = \infty & \text{for all } x \neq -\infty \\
-\infty + x = -\infty & \text{for all } x \neq \infty \\
\infty \cdot x = \infty & \text{for all } x > 0 \\
\infty \cdot x = -\infty & \text{for all } x < 0 \\
-\infty \cdot x = -\infty & \text{for all } x > 0 \\
-\infty \cdot x = \infty & \text{for all } x < 0
\end{array}$$

This extension sounds reasonable. But the resulting set $\mathbb{R} \cup \{-\infty, \infty\}$ is not a field! Otherwise ∞ would have an additive inverse. But what x would satisfy $\infty + x = 0$? One might guess $x = -\infty$, but then one would also expect $0 = \infty + (-\infty) = \infty + (-\infty + 1) = (\infty + (-\infty)) + 1 = 0 + 1 = 1$, which is not a good idea to adopt for proving anything at all in a sound way. Instead, problematic terms remain explicitly undefined:

$$\begin{array}{l}
\infty - \infty = \text{undefined} \\
0 \cdot \infty = \text{undefined} \\
\pm\infty / \pm\infty = \text{undefined} \\
1/0 = \text{undefined}
\end{array}$$

Since these conventions make infinities somewhat subtle, we happily remember that the only thing we need them for is to make sense of inserting sufficiently negative (or sufficiently positive) numbers into inequalities to satisfy them. That is still mostly harmless.

In this case, the roots from Theorem 20.2 will not help, because they satisfy the equation $ax^2 + bx + c = 0$ but not the strict inequality $ax^2 + bx + c < 0$. The virtual substitution of $-\infty$ for x from Sect. 21.3 still makes sense to consider, because the arbitrarily small negative numbers that it corresponds to might indeed satisfy F and $ax^2 + bx + c < 0$. If $-\infty$ does not work, however, the solution of (21.7) might be *near* one of the roots of $ax^2 + bx + c = 0$, just *slightly off* so that $ax^2 + bx + c < 0$ is actually satisfied rather than the equation $ax^2 + bx + c = 0$. How far off? Well, saying that exactly is again difficult, because any particular real number might already be too large in absolute value, depending on the constraints in the remainder of F. Again, this calls for quantities that are always as small as we need them to be.

Sect. 21.3 used a negative quantity that is so small that it is smaller than all negative numbers and hence infinitely small (but infinitely large in absolute value). The negative infinity $-\infty$ is smaller no matter what other number we compare it with. Analyzing (21.7) needs *positive* quantities that are infinitely small and hence also infinitely small in absolute value. Infinitesimals are positive quantities that are always smaller than all positive real numbers, i.e., "always as small as needed." Think of them as built out of elastic rubber so that they always shrink as needed when compared with any actual positive real number so that the infinitesimals end up being smaller than positive reals. Of course, the infinitesimals are much bigger than negative numbers. Another way of looking at infinitesimals is that they are the multiplicative inverses of $\pm\infty$.

A *positive infinitesimal* ε is positive ($\infty > \varepsilon > 0$) and an extended real that is *infinitesimal*, i.e., positive but smaller than all positive real numbers ($\varepsilon < r$ for all $r \in \mathbb{R}$ with $r > 0$).

> **Note 83 (Infinitesimals in polynomials)** All nonzero univariate polynomials $p \in \mathbb{R}[x]$ with real coefficients satisfy the following cases infinitesimally near any real point $\zeta \in \mathbb{R}$:
>
> 1. $p(\zeta + \varepsilon) \neq 0$
>   That is, infinitesimals ε are always so small that they never yield roots of any equation, except the trivial zero polynomial. Whenever it looks like there might be a root, the infinitesimal just becomes a bit smaller to avoid satisfying the equation. Nonzero univariate polynomials $p(x)$ only have finitely many roots, so the infinitesimals will take care to avoid all of them by becoming just a little smaller.
> 2. If $p(\zeta) \neq 0$ then $p(\zeta)p(\zeta + \varepsilon) > 0$.
>   That is, p has constant sign on infinitesimal neighborhoods of nonroots ζ. If the neighborhood around ζ is small enough (and for an infinitesimal it will be), then the polynomial will not change sign on that interval, because the sign will only change after passing one of the roots.
> 3. $0 = p(\zeta) = p'(\zeta) = p''(\zeta) = \ldots = p^{(k-1)}(\zeta) \neq p^{(k)}(\zeta)$ then $p^{(k)}(\zeta)p(\zeta+\varepsilon) > 0$.
>   That is the first nonzero derivative of p at ζ determines the sign of p in small enough neighborhoods of ζ (infinitesimal neighborhoods will be small enough), because the sign only changes after passing a root.

Definition 21.2 (Infinitesimal virtual substitution). *Substitution of an infinitesimal expression $e + \varepsilon$ with a square root expression $e = (a + b\sqrt{c})/d$ and a positive infinitesimal ε for x into a polynomial $p = \sum_{i=0}^{n} a_i x^i$ with polynomials a_i that do not contain x is defined by the following equivalences:*

$$(p = 0)_x^{e+\varepsilon} \equiv \bigwedge_{i=0}^{n} a_i = 0 \tag{21.8}$$

$$(p \leq 0)_x^{e+\varepsilon} \equiv (p < 0)_x^{e+\varepsilon} \vee (p = 0)_x^{e+\varepsilon} \tag{21.9}$$

$$(p < 0)_x^{e+\varepsilon} \equiv (p^+ < 0)_x^{e} \tag{21.10}$$

$$(p \neq 0)_x^{e+\varepsilon} \equiv \bigvee_{i=0}^{n} a_i \neq 0 \tag{21.11}$$

Lines (21.8) and its dual (21.11) use that infinitesimal offsets satisfy no equation except the trivial equation $0=0$ (Case 1 of Note 83), which makes infinitesimals and infinities behave the same as far as equations go. Line (21.9) again uses the equivalence $p \leq 0 \equiv p < 0 \vee p = 0$. Line (21.10) checks whether the sign of p at the square root expression e is already negative (which will make p inherit the same negative sign after an infinitesimal offset at $e + \varepsilon$ by Case 2) or will immediately become negative using a recursive formulation of immediately becoming negative that uses higher derivatives (which determine the sign by Case 3). The lifting to arbitrary quantifier-free formulas of real arithmetic is again by substitution into all atomic subformulas and equivalences such as $(p > q) \equiv (p - q > 0)$ as defined in Chap. 20. Note that, for the case $(p < 0)_x^{e+\varepsilon}$, the (non-infinitesimal) square root expression e gets virtually substituted in for x into a formula $p^+ < 0$, which characterizes whether p becomes negative at or immediately after x (which will be virtually substituted by the intended square root expression e momentarily).

Whether p is immediately negative at x, i.e., negative itself or 0 and with a derivative p' that makes it negative on an infinitesimal interval $(x, x + \varepsilon]$, suggestively written $p^+ < 0$, can be characterized recursively:

$$p^+ < 0 \overset{\text{def}}{\equiv} \begin{cases} p < 0 & \text{if } \deg(p) \leq 0 \\ p < 0 \vee (p = 0 \wedge (p')^+ < 0) & \text{if } \deg(p) > 0 \end{cases}$$

$p^+ < 0$ is true in a state in which $\lim_{y \to x^+} p(x) = \lim_{y \searrow x} p(x) = \lim_{\substack{y > x \\ y \to x}} p(x) < 0$ holds for the limit of p at x from the right.

The first line captures that the sign of polynomials of degree 0 in the variable x does not depend on x, so they are negative at x iff the polynomial $p = a_0$ that has degree 0 in x is negative (which may still depend on the value of other variables in a_0). The second line captures that the sign at $x + \varepsilon$ of a non-constant polynomial is still negative if it is negative at x (because $x + \varepsilon$ is not far enough away from x for any sign change by Case 2) or if x is a root of p but its derivative p' at x is immediately negative, since the first nonzero derivative at x determines the sign near x by Case 3.

Example 21.2 (Sign of quadratic polynomials after root). Using this principle to check under which circumstances the quadratic strict inequality from (21.7) evaluates to *true* at the point $(-b+\sqrt{b^2-4ac})/(2a)+\varepsilon$, i.e., right after its quadratic root $(-b+\sqrt{b^2-4ac})/(2a)$, leads to the following computation:

$$(ax^2+bx+c)^+ < 0$$
$$\equiv\ ax^2+bx+c < 0 \vee ax^2+bx+c = 0 \wedge (2ax+b < 0 \vee 2ax+b = 0 \wedge 2a < 0)$$

with successive derivatives to break ties (i.e., 0 signs in previous derivatives). Hence,

$$(ax^2+bx+c < 0)_{\bar{x}}^{(-b+\sqrt{b^2-4ac})/(2a)+\varepsilon} \equiv ((ax^2+bx+c)^+ < 0)_{\bar{x}}^{(-b+\sqrt{b^2-4ac})/(2a)} \equiv$$

$$(ax^2+bx+c < 0 \vee ax^2+bx+c = 0 \wedge (2ax+b < 0 \vee 2ax+b = 0 \wedge 2a < 0))_{\bar{x}}^{(-b+\sqrt{b^2-4ac})/(2a)}$$

$$\equiv 0\cdot 1 < 0 \vee 0 = 0 \wedge (\underbrace{(0 < 0 \vee 4a^2 \leq 0 \wedge (0 < 0 \vee -4a^2(b^2-4ac) < 0))}_{(2ax+b<0)_{\bar{x}}^{(-b+\sqrt{b^2-4ac})/(2a)}} \vee \underbrace{0 = 0}_{(2ax+b=0)_{\bar{x}}^{\cdots}} \wedge \underbrace{2a1 < 0}_{(2a<0)_{\bar{x}}^{\cdots}})$$

$$\equiv 4a^2 \leq 0 \wedge -4a^2(b^2-4ac) < 0 \vee 2a < 0$$

because the square root virtual substitution of its own root $(-b+\sqrt{b^2-4ac})/(2a)$ into ax^2+bx+c gives $(ax^2+bx+c)_{\bar{x}}^{(-b+\sqrt{b^2-4ac})/(2a)} = 0$ by construction (compare Example 20.5). The virtual substitution into another polynomial $2ax+b$ gives

$$(2ax+b)_{\bar{x}}^{(-b\pm\sqrt{b^2-4ac})/(2a)} \equiv 2a\cdot(-b\pm\sqrt{b^2-4ac})/(2a)+b$$
$$= (-2ab+\pm 2a\sqrt{b^2-4ac})/(2a)+b$$
$$= (\cancel{-2ab}+\cancel{2ab}+\pm 2a\sqrt{b^2-4ac})/(2a)$$
$$= (0+\pm 2a\sqrt{b^2-4ac})/(2a)$$

The resulting formula can be further simplified internally to

$$(ax^2+bx+c < 0)_{\bar{x}}^{(-b+\sqrt{b^2-4ac})/(2a)+\varepsilon} \equiv 4a^2 \leq 0 \wedge -4a^2(b^2-4ac) < 0 \vee 2a < 0$$
$$\equiv 2a < 0$$

because the first conjunct $4a^2 \leq 0 \equiv a = 0$ and, with $a = 0$, the second conjunct simplifies to $-4a^2(b^2-4ac)_a^0 = -0(b^2) < 0$, which is impossible in the reals. This answer makes sense. Indeed, exactly if $2a < 0$ will a quadratic polynomial still evaluate to $ax^2+bx+c < 0$ right after its second root $(-b+\sqrt{b^2-4ac})/(2a)$. Fig. 21.2 illustrates how this relates to the parabola pointing downwards, because of $2a < 0$.

Formulas such as this one ($2a < 0$) are the result of a quantifier elimination procedure. If the formula after quantifier elimination is either *true* or *false*, then you know for sure that the formula is valid (*true*) or unsatisfiable (*false*), respectively.

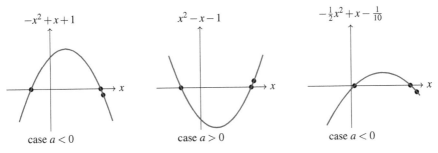

Fig. 21.2 Illustration of the sign after the second root for quadratic functions p

If the result of quantifier elimination is *true*, for example, KeYmaera X completes proof branches (marked by proof rule $\mathbb{R}$ in our sequent proofs). However, quantifier elimination can also return other formulas, such as $2a < 0$, which are equivalent to the formula to which quantifier elimination has been applied. In particular, they identify under exactly which circumstance that corresponding quantified formula is true. This can be very useful for identifying the missing assumptions to make a proof work and the corresponding statement true.

> **Note 84 (Quantifier elimination identifies requirements)** If the outcome of quantifier elimination is the formula *true*, the corresponding formula is valid. If it is the formula *false*, the corresponding formula is not valid (and even unsatisfiable). In between, i.e., when quantifier elimination results in a logical formula that is sometimes false and sometimes true, then this formula identifies exactly the missing requirements that are needed to make the desired formula true. This can be useful to synthesize missing requirements. Take care, however, not to work with universal closures, in which case *true* and *false* are the only possible outcomes.

The crucial thing to note about the process is that the *virtual substitution* of infinitesimal expressions $e + \varepsilon$ for x in F giving $F_{\bar{x}}^{e+\varepsilon}$ from Definition 21.2 is semantically equivalent to the result $F_x^{e+\varepsilon}$ of the literal substitution replacing x with $e + \varepsilon$, but operationally different, because it never introduces actual infinitesimals.

> **Lemma 21.2 (Virtual substitution lemma for infinitesimals).** *The result $F_{\bar{x}}^{e+\varepsilon}$ of the virtual substitution is semantically equivalent to the result $F_x^{e+\varepsilon}$ of the literal substitution. A language extension yields this validity:*
>
> $$F_x^{e+\varepsilon} \leftrightarrow F_{\bar{x}}^{e+\varepsilon}$$

Keep in mind that the result $F_{\bar{x}}^{e+\varepsilon}$ of virtual substitution is a proper formula of FOL$_\mathbb{R}$, while the literal substitution $F_x^{e+\varepsilon}$ could only be considered a formula in an extended logic such as FOL$_{\mathbb{R}[\varepsilon]}$ that allows for infinitesimal quantities from nonstandard analysis. Computationally more efficient substitutions of infinitesimals have been reported elsewhere [3].

Expedition 21.2 (Nonstandard analysis: infinite challenges with infinitesimal ε)

Infinite quantities in the extended reals $\mathbb{R} \cup \{-\infty, \infty\}$ already needed some attention to stay away from undefined expressions. Infinitesimals are infinitely more subtle than infinities. Real numbers are Archimedean, i.e., for every nonzero $x \in \mathbb{R}$, there is an $n \in \mathbb{N}$ such that

$$\underbrace{|x + x + \cdots + x|}_{n \text{ times}} > 1$$

Infinitesimals are non-Archimedean, because it does not matter how often you add ε, it still won't sum to one. There is a myriad of ways of making sense of infinitesimal quantities in nonstandard analysis, including surreal numbers, superreal numbers, and hyperreals. In a sense, infinitesimal quantities can be considered to be multiplicative inverses of infinities, but bring up many subtleties. For example, if an infinitesimal ε is added to $\mathbb{R}$, then the following terms need to denote values and satisfy ordering relations:

$$\varepsilon^2 \quad \varepsilon \quad x^2 + \varepsilon \quad (x+\varepsilon)^2 \quad x^2 + 2\varepsilon x + 5\varepsilon + \varepsilon^2$$

Fortunately, a rather tame version of infinitesimals is enough for the context of virtual substitution. The crucial properties of infinitesimals we need are [4]:

$$\varepsilon > 0$$
$$\forall x \in \mathbb{R}\, (x > 0 \rightarrow \varepsilon < x)$$

That is, the infinitesimal ε is positive and smaller than all positive reals.

21.5 Quantifier Elimination by Virtual Substitution for Quadratics

The following quantifier elimination technique due to Weispfenning [13] works for formulas with a quantified variable that occurs at most quadratically.

Theorem 21.1 (Virtual substitution of quadratic constraints). *Let F be a quantifier-free formula in which all atomic formulas are of quadratic form $ax^2 + bx + c \sim 0$ for polynomials a, b, c that do not mention variable x (that is, $x \notin FV(a), FV(b), FV(c)$) with some comparison operator $\sim \,\in \{=, \leq, <, \neq\}$ and corresponding discriminant $d = b^2 - 4ac$. Then $\exists x F$ is equivalent over $\mathbb{R}$ to the following quantifier-free formula:*

$$F_{\bar{x}}^{-\infty}$$

$$\vee \bigvee_{ax^2+bx+c\{\overset{=}{\leq}\}0\in F} \left(a{=}0 \wedge b{\neq}0 \wedge F_{\bar{x}}^{-c/b} \vee a{\neq}0 \wedge d{\geq}0 \wedge (F_{\bar{x}}^{(-b+\sqrt{d})/(2a)} \vee F_{\bar{x}}^{(-b-\sqrt{d})/(2a)})\right)$$

$$\vee \bigvee_{ax^2+bx+c\{\overset{\neq}{\leq}\}0\in F} \left(a{=}0 \wedge b{\neq}0 \wedge F_{\bar{x}}^{-c/b+\varepsilon} \vee a{\neq}0 \wedge d{\geq}0 \wedge (F_{\bar{x}}^{(-b+\sqrt{d})/(2a)+\varepsilon} \vee F_{\bar{x}}^{(-b-\sqrt{d})/(2a)+\varepsilon})\right)$$

Proof. The proof is an extended form of the proof reported in the literature [13]. The proof first considers the literal substitution of square root expressions, infinities, and infinitesimals and then, as a second step, uses that the virtual substitutions that avoid square root expressions, infinities, and infinitesimals are equivalent (Lemma 20.2, 21.1 and 21.2). Let G denote the quantifier-free right-hand side so that the validity of the following formula needs to be shown:

$$\exists x\, F \leftrightarrow G \qquad (21.12)$$

The implication from the quantifier-free formula G to $\exists x\, F$ in (21.12) is obvious, because each disjunct of the quantifier-free formula has a conjunct of the form F_x^t for some (extended) term t, even if it may be a square root expression or infinity or term involving infinitesimals. Whenever a formula of the form F_x^t is true, $\exists x\, F$ holds with that t as a witness, even when t is a square root expression, infinity, or infinitesimal.

The converse implication from $\exists x\, F$ to the quantifier-free formula G in (21.12) depends on showing that the quantifier-free formula G covers all possible representative cases and that the accompanying constraints on a, b, c, d are necessary so that they do not constrain solutions in unjustified ways.

One key insight is that it is enough to prove (21.12) for the case where all variables in F except x have concrete numeric real values, because the equivalence (21.12) is valid iff it is true in all states. So considering one concrete state at a time is enough. By a fundamental property of real arithmetic called *o-minimality*, the set

$$\mathscr{S}(F) = \{\omega(x) \in \mathbb{R} \,:\, \omega \in [\![F]\!]\}$$

of all real values for x that satisfy F forms a finite union of (pairwise disjoint) intervals, because the polynomials in F only change signs at their roots. There are only finitely many roots, now that the polynomials have become univariate, i.e., with the only variable x, since all free variables are evaluated to concrete real numbers in ω. Without loss of generality (by merging overlapping or adjacent intervals), all those intervals are assumed to be maximal, i.e., no bigger interval would satisfy F. So F actually changes its truth-value at most at the lower and upper endpoints of these intervals (unless the interval is unbounded). *Polynomials only change signs at their roots!*

The endpoints of these intervals are of the form $-c/b, (-b+\sqrt{d})/(2a), (-b-\sqrt{d})/(2a)$ or $\infty, -\infty$ for any of the polynomials $ax^2 + bx + c$ in F, because all polynomials in F are at most quadratic and all roots of those polynomials are of one of the above forms. In particular, if $-c/b$ is an endpoint of an interval of $\mathscr{S}(F)$

for a polynomial $ax^2 + bx + c$ in F, then $a = 0, b \neq 0$, because that is the only case where $-c/b$ satisfies F, which has only at most quadratic polynomials. Likewise, if $(-b + \sqrt{d})/(2a)$ and $(-b - \sqrt{d})/(2a)$ are endpoints of intervals of $\mathscr{S}(F)$ for a polynomial $ax^2 + bx + c$ in F, then both imply that $a \neq 0$ and discriminant $d \geq 0$, otherwise there is no such solution in the reals. Consequently, all the side conditions for the roots in the quantifier-free formula G are necessary.

Now consider one interval $I \subseteq \mathscr{S}(F)$ (if there is none, $\exists x F$ is *false* and so will G be). If I has no lower bound in $\mathbb{R}$, then $F_{\tilde{x}}^{-\infty}$ is true by construction (by Lemma 21.1, the virtual substitution $F_{\tilde{x}}^{-\infty}$ is equivalent to the literal substitution $F_x^{-\infty}$ in $\pm\infty$-extended real arithmetic). Otherwise, let $\alpha \in \mathbb{R}$ be the lower bound of I. If $\alpha \in I$ (i.e., I is closed at the lower bound), then α is of the form $-c/b, (-b + \sqrt{d})/(2a), (-b - \sqrt{d})/(2a)$ for some equation $(ax^2 + bx + c = 0) \in F$ or some weak inequality $(ax^2 + bx + c \leq 0) \in F$ from F. Since the respective extra conditions on a, b, c, d hold, the quantifier-free formula G evaluates to true. If, otherwise, $\alpha \notin I$ (i.e., I is open at the lower bound α), then α is of the form $-c/b, (-b + \sqrt{d})/(2a), (-b - \sqrt{d})/(2a)$ for some disequation $(ax^2 + bx + c \neq 0) \in F$ or some strict inequality $(ax^2 + bx + c < 0) \in F$. Hence, the interval I cannot be a single point. So, one of the infinitesimal increments $-c/b + \varepsilon, (-b + \sqrt{d})/(2a) + \varepsilon$, or $(-b - \sqrt{d})/(2a) + \varepsilon$ is in $I \subseteq \mathscr{S}(F)$, because infinitesimals are smaller than all positive real numbers, so smaller than the interval length. Since the respective conditions a, b, c, d hold, the quantifier-free formula G is again true. Hence, in either case, the quantifier-free formula is equivalent to $\exists x F$ in state ω. Since the state ω assigning concrete real numbers to all free variables of $\exists x F$ was arbitrary, the same equivalence holds for all states ω, which means that the quantifier-free formula G is equivalent to $\exists x F$. That is $G \leftrightarrow \exists x F$ is valid, i.e., $\vDash G \leftrightarrow \exists x F$.   □

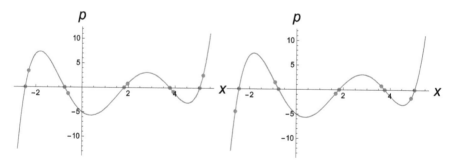

Fig. 21.3 Illustration of roots e and infinitesimal offsets $e + \varepsilon$ checked by virtual substitution along with $-\infty$ (**left**). Illustration of roots e and infinitesimal offsets $e - \varepsilon$ that could be checked along with $+\infty$ instead (**right**)

The order of the interval endpoints that the proof of Theorem 21.1 uses in addition to $-\infty$ is illustrated in Fig. 21.3(left). Observe that exactly one representative point is placed in each of the regions of interest, $-\infty$, each of the roots r, and just infinitesimally after the roots at $r + \varepsilon$. Alternatively, Theorem 21.1 could be rephrased to work with ∞, at each root r, and always before the roots at $r - \varepsilon$; see Fig. 21.3(right) and Exercise 21.4. The illustrations in Fig. 21.3 show the ordering situation for a higher-degree polynomial p even if Theorem 21.1 only makes use of the argument for $p = ax^2 + bx + c$ up to degree 2. Quantifier elimination procedures for higher degrees are still based on this fundamental principle, but require more subtle algebraic computations. The source of the trouble is Abel-Ruffini's impossibility theorem that there are, generally, no algebraic solutions to polynomial equations of degree ≥ 5. That is, the fact that we can characterize the roots of polynomials with roots was specific to degree ≤ 4 even when admitting nested roots.

Finally note that it is quite possible that the considered polynomial p does not single out the appropriate root e or off-root $e + \varepsilon$ that satisfies F to witness $\exists x F$. Then none of the points illustrated in Fig. 21.3 will satisfy F, because only a point other than $e + \varepsilon$ in the open interval between two roots will work.

> **Note 85 (No rejection without mention)** The key argument underlying all quantifier elimination procedures in some way or another is that all parts of F that are not satisfied for any of the points in Fig. 21.3 that p brings about would have to mention another polynomial q with different roots $\tilde{e}$ and different off-roots $\tilde{e} + \varepsilon$ that will then enter the big disjunction in Theorem 21.1.

Example 21.3. The example of nonnegative roots of quadratic polynomials from Example 20.6 in Chap. 20 used Theorem 20.2 to construct and justify the quantifier elimination equivalence

$$QE(\exists x\,(ax^2 + bx + c = 0 \wedge x \geq 0))$$
$$\equiv b^2 - 4ac \geq 0 \wedge (ba \leq 0 \wedge ac \geq 0 \vee a \geq 0 \wedge ac \leq 0 \vee a \leq 0 \wedge ac \leq 0)$$

under the assumption $a \neq 0$. Specializing to a case similar to Fig. 21.2 gives

$$QE(\exists x\,(x^2 - x + c = 0 \wedge x \geq 0)) \equiv (-1)^2 - 4c \geq 0 \wedge (c \geq 0 \vee c \leq 0) \equiv 1 - 4c \geq 0$$
$$\equiv c \leq \frac{1}{4}$$

By Theorem 21.1, the same square root expression substitution as in Example 20.6 in Chap. 20 will happen for the atomic formula $x^2 - x + c \leq 0$ except that the case of $-\infty$ will be added as well as the root 0 that results from considering the linear atomic formula $-x \geq 0$:

$$QE(\exists x\,(x^2 - x + c \leq 0 \wedge x \geq 0)) \equiv$$
$$\underbrace{(x^2 - x + c \leq 0 \wedge \ldots)_{\overset{-\infty}{x}}}_{false} \vee 1 - 4c \geq 0 \vee \underbrace{(x^2 - x + c \leq 0 \wedge x \geq 0)_{\overset{0}{x}}}_{c \leq 0 \wedge 0 \geq 0} \equiv 1 - 4c \geq 0$$

Note that the additional disjunction $c \leq 0$ coming from the root 0 of $-x$ is in this case subsumed by the previous disjunct $1 - 4c \geq 0$. Hence, adding the roots of $-x$ did not modify the answer in this case. When adding a third conjunct $-x + 2 = 0$, this handling of all roots becomes critical:

$$QE(\exists x\,(x^2-x+c\le 0 \wedge x\ge 0 \wedge -x+2=0))$$

Since the first two polynomials x^2-x+c and $-x$ are still the same, the same virtual substitutions will happen as before. Except that they now fail on the new conjunct $-x+2=0$, because the root 0 of the polynomial $-x$ from the second conjunct does not satisfy $-x+2=0$ and because the virtual substitution of the roots $(-1\pm\sqrt{1-4c})/2$ of the first polynomial x^2-x+c fails:

$$(-x+2=0)_{\bar{x}}^{(-1\pm\sqrt{1-4c})/2} \equiv ((1+\mp 1\sqrt{1-4c})/2+2=0) \equiv ((3+\mp 1\sqrt{1-4c})/2=0)$$
$$\equiv \mp 3 \le 0 \wedge 3^2-(\mp 1)^2(1-4c)=0 \equiv -3\le 0 \wedge 3^2-(-1)^2(1-4c)=0 \equiv 8-4c=0$$

The latter is only possible for $c=2$, which is ruled out by the discriminant condition $1-4c\ge 0$ that precedes it. And, indeed, neither the roots of the quadratic polynomial illustrated in Fig. 21.2 nor the roots of $-x$ nor $-\infty$ are the right points to consider to satisfy the last conjunct. Of course, the last conjunct expresses that constraint by saying $-x+2=0$ quite explicitly. Never mind that this is an equation for now. Either way, the atomic formula clearly reveals that $-x+2$ is the polynomial that it cares about. So its roots might be of interest and will, indeed, by considered in the big disjunction of Theorem 21.1 as well. Since $-x+2$ is a visibly linear polynomial, its solution is $x=-2/-1=2$ which is even kind enough to be a standard real number so that literal substitution is sufficient and no virtual substitution is needed. Consequently, the substitution of this root $x=2$ of the last conjunct into the full formula quickly yields

$$(x^2-x+c\le 0 \wedge x\ge 0 \wedge -x+2=0)_x^2 \equiv 2^2-2+c\le 0 \wedge 2\ge 0 \wedge 0=0 \equiv 2+c\le 0$$

This provides an answer that the quadratic polynomial x^2-x+c itself could not foresee because it depends on the polynomial $-x+2$ to even take this root into consideration. By Theorem 21.1, the overall result of quantifier elimination, thus, is the combination of the cases considered separately above:

$$QE(\exists x\,(x^2-x+c\le 0 \wedge x\ge 0 \wedge -x+2=0))$$
$$\equiv \underbrace{(x^2-x+c\le 0 \wedge \ldots)_{\bar{x}}^{-\infty}}_{\textit{false}}$$
$$\vee\, 1-4c\ge 0 \wedge \underbrace{(\cdots \wedge -x+2=0)_{\bar{x}}^{(-1\pm\sqrt{1-4c})/2}}_{8-4c=0}$$
$$\vee -1\ne 0 \wedge \underbrace{(x^2-x+c\le 0 \wedge x\ge 0)_x^0}_{c\le 0 \wedge 0\ge 0} \wedge \underbrace{(-x+2=0)_x^0}_{2=0}$$
$$\vee -1\ne 0 \wedge \underbrace{(x^2-x+c\le 0 \wedge x\ge 0 \wedge -x+2=0)_x^2}_{2+c\le 0} \qquad \equiv 2+c\le 0 \equiv c\le -2$$

In this particular case, observe that Theorem 20.2 using $-x+2=0$ as the key formula would have been most efficient, because that would have gotten the answer right away without fruitless disjunctions. This illustrates that it pays off to pay attention with real arithmetic and always choose the computationally most parsimonious approach. But the example also illustrates that the same computation would happen if the third conjunct had been $-x+2\le 0$, in which case Theorem 20.2 would not have helped.

21.6 Optimizations

Optimizations are possible for virtual substitutions [13] if there is only one quadratic occurrence of x, and that occurrence is not in an equation. If that occurrence is in an equation, Theorem 20.2 already showed what to do. If there is only one occurrence of a quadratic inequality, the following variation of Theorem 21.1 works, which uses exclusively linear fractions.

Note 86 ([13]) Let $\left(Ax^2+Bx+C\left\{\genfrac{}{}{0pt}{}{\le}{\ne}{<}\right\}0\right)\in F$ be the only quadratic occurrence of x. In that case, $\exists x\,F$ is equivalent over $\mathbb{R}$ to the following quantifier-free formula:

$$A=0\wedge B\ne 0\wedge F_{\tilde{x}}^{-C/B}\vee A\ne 0\wedge F_{\tilde{x}}^{-B/(2A)}$$
$$\vee F_{\tilde{x}}^{-\infty}\vee F_{\tilde{x}}^{\infty}$$
$$\vee\bigvee_{(0x^2+bx+c\{\genfrac{}{}{0pt}{}{=}{\le}\}0)\in F}(b\ne 0\wedge F_{\tilde{x}}^{-c/b})$$
$$\vee\bigvee_{(0x^2+bx+c\{\genfrac{}{}{0pt}{}{\ne}{<}\}0)\in F}(b\ne 0\wedge(F_{\tilde{x}}^{-c/b+\varepsilon}\vee F_{\tilde{x}}^{-c/b-\varepsilon}))$$

The clou in this case is that the extremal values of Ax^2+Bx+C are at the roots of the derivative

$$(Ax^2+Bx+C)'=2AX+B\overset{!}{=}0,\text{ i.e., }x=-\frac{B}{2A}$$

Since the only quadratic occurrence in Note 86 is not an equation, this extremal value is the only point of the quadratic polynomial that matters. In this case, $F_{\tilde{x}}^{-B/(2A)}$ will substitute $-B/(2A)$ for x in the only quadratic polynomial as follows:

$$\left(Ax^2+Bx+C\left\{\genfrac{}{}{0pt}{}{\le}{<}{\ne}\right\}0\right)_{\tilde{x}}^{-B/(2A)}\equiv\left(A\frac{(-B)^2}{4A^2}+\frac{-B^2}{2A}+C\left\{\genfrac{}{}{0pt}{}{\le}{<}{\ne}\right\}0\right)\equiv\left(\frac{-B^2}{4A}+C\left\{\genfrac{}{}{0pt}{}{\le}{<}{\ne}\right\}0\right)$$

The formula resulting from Note 86 might be bigger than that of Theorem 21.1 but *it does not increase the polynomial degree*, which can be crucial for nested quantifiers.

Further optimizations are possible if some signs of a, b are known, because several cases in the quantifier-free expansion then become impossible and can be simplified to *true* or *false* immediately. This helps simplify the formula in Theorem 21.1, because one of the cases $a = 0$ versus $a \neq 0$ might drop. But it also reduces the number of disjuncts in $F_{\bar{x}}^{-\infty}$, see Example 21.1, and in the virtual substitutions of square roots (Chap. 20) and of infinitesimals (Sect. 21.4), which can lead to significant simplifications.

Theorem 21.1 also applies to polynomials of higher degrees in x if they factor to polynomials of at most quadratic degree in x [13]. Degree reduction is also possible by renaming based on the greatest common divisor of all powers of x that occur in F. If a quantified variable x occurs only with exponents that are multiples of an odd number d then virtual substitution can use $\exists x F(x^d) \equiv \exists y F(y)$. If x only occurs with degrees that are multiples of even number d then $\exists x F(x^d) \equiv \exists y (y \geq 0 \wedge F(y))$. It helps reduce the number of cases in Theorem 21.1 that infinitesimals $+\varepsilon$ are only needed if x occurs in strict inequalities in F. The cases $F_{\bar{x}}^{(-b+\pm\sqrt{d})/(2a)}$ are only needed if x occurs in equations or weak inequalities.

21.7 Summary

Virtual substitution is one technique for eliminating quantifiers in real arithmetic. It works for linear and quadratic constraints and can be extended to some cubic cases [12]. Virtual substitution can be applied repeatedly from inside out to eliminate quantifiers. In each case, however, virtual substitution requires the eliminated variable to occur with small enough degree only. Even if that was the case initially, it may no longer be the case after eliminating the innermost quantifier, because the degrees of the formula resulting from virtual substitution may increase. In that case, degree optimizations and simplifications may sometimes work. If not, then other quantifier elimination techniques need to be used, which are based on semialgebraic geometry or model theory. Virtual substitution alone always works for mixed quadratic-linear formulas, i.e., those in which all quantified variables occur linearly except for one variable that occurs quadratically. In practice, however, many other cases turn out to work well with virtual substitution.

By inspecting Theorem 21.1 and its optimizations, we also observe that it is interesting to look at only closed sets or only open sets, corresponding to formulas with only $\leq$ and $=$ or formulas with only $<$ and $\neq$ conditions, respectively, because half of the cases then drop out of the expansion in Theorem 21.1. Furthermore, if the formula $\exists x F$ only mentions strict inequalities $<$ and disequations $\neq$, then all virtual substitutions will involve infinitesimals or infinities. While both are conceptually more demanding than virtual substitutions with mere square root expressions, the advantage is that both infinitesimals and infinities rarely satisfy any equations (except when they are trivial because all coefficients are zero). In that case, most formulas simplify tremendously. That is an indication in the virtual substitution method of a more general phenomenon: existential arithmetic with strict inequalities or, dually, validity of universal arithmetic with weak inequalities, is computationally easier.

21.8 Appendix: Semialgebraic Geometry

The geometric counterparts of polynomial equations or quantifier-free first-order formulas with polynomial equations are affine varieties. The geometric counterparts of first-order formulas of real arithmetic that may mention inequalities are called semialgebraic sets in real algebraic geometry [1, 2]. By quantifier elimination, the class of sets definable with quantifiers is the same as the class of sets definable without quantifiers. Hence, the formulas of first-order real arithmetic exactly define semialgebraic sets.

Definition 21.3 (Semialgebraic Set). $S \subseteq \mathbb{R}^n$ is an *semialgebraic set* iff it is defined by a finite intersection of polynomial equations and inequalities or any finite union of such sets:

$$S = \bigcup_{i=1}^{t} \bigcap_{j=1}^{s} \{x \in \mathbb{R}^n \; : \; p(x) \sim 0\} \quad \text{where } \sim \, \in \{=, \geq, >\}$$

The geometric counterpart of the quantifier elimination result is that semialgebraic sets are closed under projection (the other closure properties are obvious in logic), which is the Tarski-Seidenberg theorem [10, 11].

> **Theorem 21.2 (Tarski-Seidenberg).** *Semialgebraic sets are closed under finite unions, finite intersections, complements, and projection to linear subspaces.*

The semialgebraic sets corresponding to a number of interesting systems of polynomial inequalities are illustrated in Fig. 21.4.

Exercises

21.1. Consider the first-order real-arithmetic formula

$$\exists x \, (ax^2 + bx + c \leq 0 \wedge F) \tag{21.13}$$

The virtual substitution of the roots of $ax^2 + bx + c = 0$ according to Sect. 20.4 as well as of $-\infty$ according to Sect. 21.3 leads to

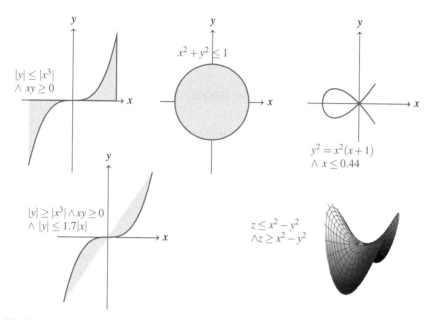

Fig. 21.4 Systems of polynomial inequalities describe semialgebraic sets

$$F_{\bar{x}}^{-\infty} \lor a{=}0 \land b{\neq}0 \land F_{\bar{x}}^{-c/b}$$

$$\lor a{\neq}0 \land b^2 - 4ac{\geq}0 \land \left(F_{\bar{x}}^{(-b+\sqrt{b^2-4ac})/(2a)} \lor F_{\bar{x}}^{(-b-\sqrt{b^2-4ac})/(2a)}\right)$$

But when F is $-ax^2 + bx + c < 0$, then none of those cases necessarily works. Does that mean the result of the virtual substitution is not equivalent to (21.13)? Where is the catch in this argument?

21.2. Perform quantifier elimination by virtual substitution to compute

$$QE(\exists x\,(x^2 - x + c \leq 0 \land x \geq 0 \land -x + 2 \leq 0))$$

21.3. Consider the first-order real-arithmetic formula

$$\exists x\,(ax^2 + bx + c \leq 0 \land ax^2 + bx + c = 0)$$

Compare the results of using Theorem 20.1 and Theorem 21.1 on this formula. Which theorem is more efficient? What happens in the case of

$$\exists x\,(ax^2 + bx + c \leq 0 \land ax^2 + bx + c = 0 \land x \geq 0)$$

21.4 (Virtual substitution on the right). Develop and prove a virtual substitution formula for quadratic polynomials analogous to Theorem 21.1 that uses the points illustrated in Fig. 21.3(right) instead of Fig. 21.3(left).

21.5 (Infinitesimals in polynomials). Use the Taylor series

$$p(\zeta + \varepsilon) = \sum_{n=0}^{\infty} \frac{p^{(n)}(\zeta)}{n!} (\zeta + \varepsilon - \zeta)^n = \sum_{n=0}^{\infty} \frac{p^{(n)}(\zeta)}{n!} \varepsilon^n = \sum_{n=0}^{\deg(p)} \frac{p^{(n)}(\zeta)}{n!} \varepsilon^n$$

of univariate polynomial $p \in \mathbb{R}[x]$ around $\zeta \in \mathbb{R}$ evaluated at $\zeta + \varepsilon$ (since ε is small enough to be in the domain of convergence of the Taylor series) to show Note 83.

References

[1]   Saugata Basu, Richard Pollack, and Marie-Françoise Roy. *Algorithms in Real Algebraic Geometry*. 2nd. Berlin: Springer, 2006. DOI: 10.1007/3-540-33099-2.

[2]   Jacek Bochnak, Michel Coste, and Marie-Francoise Roy. *Real Algebraic Geometry*. Vol. 36. Ergeb. Math. Grenzgeb. Berlin: Springer, 1998. DOI: 10.1007/978-3-662-03718-8.

[3]   Christopher W. Brown and James H. Davenport. The complexity of quantifier elimination and cylindrical algebraic decomposition. In: *ISSAC*. Ed. by Dongming Wang. New York: ACM, 2007, 54–60. DOI: 10.1145/1277548.1277557.

[4]   Leonardo Mendonça de Moura and Grant Olney Passmore. Computation in real closed infinitesimal and transcendental extensions of the rationals. In: *Automated Deduction - CADE-24 - 24th International Conference on Automated Deduction, Lake Placid, NY, USA, June 9-14, 2013. Proceedings*. Ed. by Maria Paola Bonacina. Vol. 7898. LNCS. Berlin: Springer, 2013, 178–192. DOI: 10.1007/978-3-642-38574-2_12.

[5]   Grant Olney Passmore. Combined Decision Procedures for Nonlinear Arithmetics, Real and Complex. PhD thesis. School of Informatics, University of Edinburgh, 2011.

[6]   André Platzer. Differential dynamic logic for hybrid systems. *J. Autom. Reas.* **41**(2) (2008), 143–189. DOI: 10.1007/s10817-008-9103-8.

[7]   André Platzer. *Logical Analysis of Hybrid Systems: Proving Theorems for Complex Dynamics*. Heidelberg: Springer, 2010. DOI: 10.1007/978-3-642-14509-4.

[8]   André Platzer. Logics of dynamical systems. In: *LICS*. Los Alamitos: IEEE, 2012, 13–24. DOI: 10.1109/LICS.2012.13.

[9]   André Platzer, Jan-David Quesel, and Philipp Rümmer. Real world verification. In: *CADE*. Ed. by Renate A. Schmidt. Vol. 5663. LNCS. Berlin: Springer, 2009, 485–501. DOI: 10.1007/978-3-642-02959-2_35.

[10] Abraham Seidenberg. A new decision method for elementary algebra. *Annals of Mathematics* **60**(2) (1954), 365–374. DOI: 10.2307/1969640.

[11] Alfred Tarski. *A Decision Method for Elementary Algebra and Geometry*. 2nd. Berkeley: University of California Press, 1951.

[12] Volker Weispfenning. Quantifier elimination for real algebra — the cubic
      case. In: *ISSAC*. New York: ACM, 1994, 258–263.
[13] Volker Weispfenning. Quantifier elimination for real algebra — the quadratic
      case and beyond. *Appl. Algebra Eng. Commun. Comput.* **8**(2) (1997), 85–101.
      DOI: 10.1007/s002000050055.

Index

© Springer International Publishing AG, part of Springer Nature 2018
A. Platzer, *Logical Foundations of Cyber-Physical Systems*,
https://doi.org/10.1007/978-3-319-63588-0

Operators of Differential Dynamic Logic (dL)

dL	Operator	Meaning
$e \geq \tilde{e}$	greater or equals	true if value of e greater-or-equal to $\tilde{e}$
$\neg P$	negation / not	true if P is false
$P \wedge Q$	conjunction / and	true if both P and Q are true
$P \vee Q$	disjunction / or	true if P is true or if Q is true
$P \to Q$	implication / implies	true if P is false or Q is true
$P \leftrightarrow Q$	bi-implication / equivalent	true if P and Q are both true or both false
$\forall x\, P$	universal quantifier / for all	true if P is true for all values of variable x
$\exists x\, P$	existential quantifier / exist	true if P is true for some value of variable x
$[\alpha]P$	$[\cdot]$ modality / box	true if P is true after all runs of HP α
$\langle\alpha\rangle P$	$\langle\cdot\rangle$ modality / diamond	true if P is true after some run of HP α

Statements and effects of Hybrid Programs (HPs)

HP Notation	Operation	Effect
$x := e$	discrete assignment	assigns current value of term e to variable x
$x := *$	nondet. assignment	assigns any real value to variable x
$x' = f(x)\,\&\,Q$	continuous evolution	follow differential equation $x' = f(x)$ within evolution domain Q for any duration
$?Q$	state test / check	test first-order formula Q at current state
$\alpha;\beta$	seq. composition	HP β starts after HP α finishes
$\alpha \cup \beta$	nondet. choice	choice between alternatives HP α or HP β
α^*	nondet. repetition	repeats HP α any $n \in \mathbb{N}$ times

Semantics of dL formula P is the set of states $[\![P]\!] \subseteq \mathscr{S}$ in which it is true

$$[\![e \geq \tilde{e}]\!] = \{\omega \in \mathscr{S} : \omega[\![e]\!] \geq \omega[\![\tilde{e}]\!]\}$$
$$[\![P \wedge Q]\!] = [\![P]\!] \cap [\![Q]\!]$$
$$[\![P \vee Q]\!] = [\![P]\!] \cup [\![Q]\!]$$
$$[\![\neg P]\!] = [\![P]\!]^{\complement} = \mathscr{S} \setminus [\![P]\!]$$
$$[\![\langle\alpha\rangle P]\!] = [\![\alpha]\!] \circ [\![P]\!] = \{\omega : v \in [\![P]\!] \text{ for some state } v \text{ such that } (\omega, v) \in [\![\alpha]\!]\}$$
$$[\![[\alpha]P]\!] = [\![\neg\langle\alpha\rangle\neg P]\!] = \{\omega : v \in [\![P]\!] \text{ for all states } v \text{ such that } (\omega, v) \in [\![\alpha]\!]\}$$
$$[\![\exists x\, P]\!] = \{\omega : v \in [\![P]\!] \text{ for some state } v \text{ that agrees with } \omega \text{ except on } x\}$$
$$[\![\forall x\, P]\!] = \{\omega : v \in [\![P]\!] \text{ for all states } v \text{ that agree with } \omega \text{ except on } x\}$$

Semantics of HP α is relation $[\![\alpha]\!] \subseteq \mathscr{S} \times \mathscr{S}$ between initial and final states

$$[\![x := e]\!] = \{(\omega, v) : v = \omega \text{ except that } v[\![x]\!] = \omega[\![e]\!]\}$$
$$[\![?Q]\!] = \{(\omega, \omega) : \omega \in [\![Q]\!]\}$$
$$[\![x' = f(x)\,\&\,Q]\!] = \{(\omega, v) : \varphi(0) = \omega \text{ except at } x' \text{ and } \varphi(r) = v \text{ for a solution } \varphi:[0,r] \to \mathscr{S} \text{ of any duration } r \text{ satisfying } \varphi \models x' = f(x) \wedge Q\}$$
$$[\![\alpha \cup \beta]\!] = [\![\alpha]\!] \cup [\![\beta]\!]$$
$$[\![\alpha;\beta]\!] = [\![\alpha]\!] \circ [\![\beta]\!] = \{(\omega, v) : (\omega, \mu) \in [\![\alpha]\!], (\mu, v) \in [\![\beta]\!]\}$$
$$[\![\alpha^*]\!] = [\![\alpha]\!]^* = \bigcup_{n\in\mathbb{N}} [\![\alpha^n]\!] \text{ with } \alpha^{n+1} \equiv \alpha^n;\alpha \text{ and } \alpha^0 \equiv\ ?true$$

© Springer International Publishing AG, part of Springer Nature 2018
A. Platzer, *Logical Foundations of Cyber-Physical Systems*,
https://doi.org/10.1007/978-3-319-63588-0

Axiomatization (dL)

$\langle \cdot \rangle \ \langle \alpha \rangle P \leftrightarrow \neg [\alpha] \neg P$

$[:=] \ [x:=e]p(x) \leftrightarrow p(e)$

$[?] \ [?Q]P \leftrightarrow (Q \to P)$

$['] \ [x' = f(x)]p(x) \leftrightarrow \forall t \geq 0 [x:=y(t)]p(x) \quad (y'(t) = f(y))$

$[\cup] \ [\alpha \cup \beta]P \leftrightarrow [\alpha]P \wedge [\beta]P$

$[;] \ [\alpha;\beta]P \leftrightarrow [\alpha][\beta]P$

$[*] \ [\alpha^*]P \leftrightarrow P \wedge [\alpha][\alpha^*]P$

$K \ [\alpha](P \to Q) \to ([\alpha]P \to [\alpha]Q)$

$I \ [\alpha^*]P \leftrightarrow P \wedge [\alpha^*](P \to [\alpha]P)$

$V \ p \to [\alpha]p \qquad\qquad\qquad (FV(p) \cap BV(\alpha) = \emptyset)$

$M[\cdot] \ \dfrac{P \to Q}{[\alpha]P \to [\alpha]Q}$

$G \ \dfrac{P}{[\alpha]P}$

Differential equation axioms

DW $[x' = f(x) \& Q]P \leftrightarrow [x' = f(x) \& Q](Q \to P)$

DI $([x' = f(x) \& Q]P \leftrightarrow [?Q]P) \leftarrow (Q \to [x' = f(x) \& Q](P)')$

DC $([x' = f(x) \& Q]P \leftrightarrow [x' = f(x) \& Q \wedge C]P) \leftarrow [x' = f(x) \& Q]C$

DE $[x' = f(x) \& Q]P \leftrightarrow [x' = f(x) \& Q][x' := f(x)]P$

DG $[x' = f(x) \& Q]P \leftrightarrow \exists y [x' = f(x), y' = a(x) \cdot y + b(x) \& Q]P$

$+' \ (e + k)' = (e)' + (k)'$

$\cdot' \ (e \cdot k)' = (e)' \cdot k + e \cdot (k)'$

$c' \ (c())' = 0 \qquad\qquad\qquad\qquad \text{(for numbers or constants } c())$

$x' \ (x)' = x' \qquad\qquad\qquad\qquad\quad \text{(for variable } x \in \mathcal{V})$

Differential equation proof rules

dW $\dfrac{Q \vdash P}{\Gamma \vdash [x' = f(x) \& Q]P, \Delta}$

dI $\dfrac{Q \vdash [x' := f(x)](F)'}{F \vdash [x' = f(x) \& Q]F}$

dC $\dfrac{\Gamma \vdash [x' = f(x) \& Q]C, \Delta \quad \Gamma \vdash [x' = f(x) \& (Q \wedge C)]P, \Delta}{\Gamma \vdash [x' = f(x) \& Q]P, \Delta}$

Sequent calculus proof rules

$\neg$R $\dfrac{\Gamma,P \vdash \Delta}{\Gamma \vdash \neg P,\Delta}$ $\wedge$R $\dfrac{\Gamma \vdash P,\Delta \quad \Gamma \vdash Q,\Delta}{\Gamma \vdash P \wedge Q,\Delta}$ $\vee$R $\dfrac{\Gamma \vdash P,Q,\Delta}{\Gamma \vdash P \vee Q,\Delta}$

$\neg$L $\dfrac{\Gamma \vdash P,\Delta}{\Gamma,\neg P \vdash \Delta}$ $\wedge$L $\dfrac{\Gamma,P,Q \vdash \Delta}{\Gamma,P \wedge Q \vdash \Delta}$ $\vee$L $\dfrac{\Gamma,P \vdash \Delta \quad \Gamma,Q \vdash \Delta}{\Gamma,P \vee Q \vdash \Delta}$

$\rightarrow$R $\dfrac{\Gamma,P \vdash Q,\Delta}{\Gamma \vdash P \rightarrow Q,\Delta}$ id $\dfrac{}{\Gamma,P \vdash P,\Delta}$ WR $\dfrac{\Gamma \vdash \Delta}{\Gamma \vdash P,\Delta}$

$\rightarrow$L $\dfrac{\Gamma \vdash P,\Delta \quad \Gamma,Q \vdash \Delta}{\Gamma,P \rightarrow Q \vdash \Delta}$ cut $\dfrac{\Gamma \vdash C,\Delta \quad \Gamma,C \vdash \Delta}{\Gamma \vdash \Delta}$ WL $\dfrac{\Gamma \vdash \Delta}{\Gamma,P \vdash \Delta}$

$\forall$R $\dfrac{\Gamma \vdash p(y),\Delta}{\Gamma \vdash \forall x\, p(x),\Delta}$ $(y \notin \Gamma,\Delta,\forall x\, p(x))$ $\exists$R $\dfrac{\Gamma \vdash p(e),\Delta}{\Gamma \vdash \exists x\, p(x),\Delta}$ (arbitrary term e)

$\forall$L $\dfrac{\Gamma,p(e) \vdash \Delta}{\Gamma,\forall x\, p(x) \vdash \Delta}$ (arbitrary term e) $\exists$L $\dfrac{\Gamma,p(y) \vdash \Delta}{\Gamma,\exists x\, p(x) \vdash \Delta}$ $(y \notin \Gamma,\Delta,\exists x\, p(x))$

CER $\dfrac{\Gamma \vdash C(Q),\Delta \quad \vdash P \leftrightarrow Q}{\Gamma \vdash C(P),\Delta}$ =R $\dfrac{\Gamma,x = e \vdash p(e),\Delta}{\Gamma,x = e \vdash p(x),\Delta}$

CEL $\dfrac{\Gamma,C(Q) \vdash \Delta \quad \vdash P \leftrightarrow Q}{\Gamma,C(P) \vdash \Delta}$ =L $\dfrac{\Gamma,x = e,p(e) \vdash \Delta}{\Gamma,x = e,p(x) \vdash \Delta}$

Derived axioms and derived rules

$\wedge'$ $(P \wedge Q)' \leftrightarrow (P)' \wedge (Q)'$

$\vee'$ $(P \vee Q)' \leftrightarrow (P)' \wedge (Q)'$

$[]\wedge$ $[\alpha](P \wedge Q) \leftrightarrow [\alpha]P \wedge [\alpha]Q$

$[*]$ $[\alpha^*]P \leftrightarrow P \wedge [\alpha^*][\alpha]P$

$[**]$ $[\alpha^*;\alpha^*]P \leftrightarrow [\alpha^*]P$

$[:=]_=$ $\dfrac{\Gamma,y = e \vdash p(y),\Delta}{\Gamma \vdash [x := e]p(x),\Delta}$ (y new)

iG $\dfrac{\Gamma \vdash [y := e]p,\Delta}{\Gamma \vdash p,\Delta}$ (y new)

dG $\dfrac{\Gamma \vdash \exists y\, [x' = f(x),y' = a(x) \cdot y + b(x) \,\&\, Q]P,\Delta}{\Gamma \vdash [x' = f(x) \,\&\, Q]P,\Delta}$

dA $\dfrac{\vdash J \leftrightarrow \exists y\, G \quad G \vdash [x' = f(x),y' = a(x) \cdot y + b(x) \,\&\, Q]G}{J \vdash [x' = f(x) \,\&\, Q]J}$

CPSIA information can be obtained
at www.ICGtesting.com
Printed in the USA
LVHW082143181121
703804LV00007B/280